Methods in Microbiology
Volume 36

Recent titles in the series

Volume 24 *Techniques for the Study of Mycorrhiza*
JR Norris, DJ Reed and AK Varma

Volume 25 *Immunology of Infection*
SHE Kaufmann and D Kabelitz

Volume 26 *Yeast Gene Analysis*
AJP Brown and MF Tuite

Volume 27 *Bacterial Pathogenesis*
P Williams, J Ketley and GPC Salmond

Volume 28 *Automation*
AG Craig and JD Hoheisel

Volume 29 *Genetic Methods for Diverse Prokaryotes*
MCM Smith and RE Sockett

Volume 30 *Marine Microbiology*
JH Paul

Volume 31 *Molecular Cellular Microbiology*
P Sansonetti and A Zychlinsky

Volume 32 *Immunology of Infection, 2nd edition*
SHE Kaufmann and D Kabelitz

Volume 33 *Functional Microbial Genomics*
B Wren and N Dorrell

Volume 34 *Microbial Imaging*
T Savidge and C Pothoulakis

Volume 35 *Extremophiles*
FA Rainey and A Oren

Methods in Microbiology

Volume 36
Yeast Gene Analysis

Second Edition

Edited by

Ian Stansfield

School of Medical Sciences
University of Aberdeen
Foresterhill, Aberdeen
Scotland, UK

and

Michael JR Stark

Division of Gene Regulation & Expression
College of Life Sciences
University of Dundee, Dundee
Scotland, UK

Amsterdam • Boston • Heidelberg • London • New York • Oxford
Paris • San Diego • San Francisco • Singapore • Sydney • Tokyo
Academic Press is an imprint of Elsevier

Academic Press is an imprint of Elsevier
84 Theobald's Road, London WC1X 8RR, UK
Radarweg 29, PO Box 211, 1000 AE Amsterdam, The Netherlands
Linacre House, Jordan Hill, Oxford OX2 8DP, UK
30 Corporate Drive, Suite 400, Burlington, MA 01803, USA
525 B Street, Suite 1900, San Diego, CA 92101-4495, USA

Second edition 2007

ISBN: 978-0-12-369478-2 (Hardbound – this volume)
ISBN: 978-0-12-369479-9 (Paperback – this volume)
ISSN: 0580-9517 (Series)

For information on all Academic Press publications
visit our website at books.elsevier.com

Printed and bound in The Netherlands

07 08 09 10 11 10 9 8 7 6 5 4 3 2 1

***Cover image:* Yeast colony sectoring; a *Saccharomyces cerevisiae (ade2ade3)* strain transformed with a plasmid carrying a wild type copy of the *ADE3* gene forms colonies on agar plates that exhibit red/white sectoring due to random loss of the plasmid (see Chapter 15 'Smart Genetic Screens') I. Stansfield, University of Aberdeen**

Contents

Series Advisors viii

Contributors ix

Preface xiii
I Stansfield and MJR Stark

1. Introduction to Functional Analysis in Yeast 1
   SG Oliver

2. Yeast Genetics and Strain Construction 23
   I Stansfield and MJR Stark

3. Yeast Transformation 45
   D Truong and RD Gietz

4. A Guided Tour to PCR-based Genomic Manipulations of *S. cerevisiae* (PCR-targeting) 55
   CI Maeder, P Maier and M Knop

5. Studying Essential Genes: Generating and Using Promoter Fusions and Conditional Alleles 79
   MJR Stark

6. Yeast Hybrid Approaches 103
   E Izumchenko, M Wolfson, EA Golemis and IG Serebriiskii

7. Array-Based Yeast Two-Hybrid Screening for Protein–Protein Interactions 139
   SV Rajagopala, B Titz and P Uetz

8. Reporter Genes and Their Uses in Studying Yeast Gene Expression 165
   T von der Haar, LJ Jossé and LJ Byrne

9. Transcript Analysis: A Microarray Approach 189
   A Hayes, JI Castrillo, SG Oliver, A Brass and LAH Zeef

10. GFP-based Microscopic Approaches for Whole Chromosome Analysis in Yeasts 221
    Q Gao, TU Tanaka and X He

11. Immunological Methods 241
EH Hettema and KR Ayscough

12. Measuring the Proximity of Proteins in Living Cells by Fluorescence Resonance Energy Transfer between CFP and YFP 269
TN Davis and EGD Muller

13. Identification, Characterization, and Phenotypic Analysis of Covalently Linked Cell Wall Proteins 281
FM Klis, P De Groot and S Brul

14. Yeast Protein Microarrays 303
J Ptacek and M Snyder

15. Smart Genetic Screens 331
M Breitenbach, JR Dickinson and P Laun

16. High-Throughput Strain Construction and Systematic Synthetic Lethal Screening in *Saccharomyces cerevisiae* 369
AHY Tong and C Boone

17. Chemical Genomic Tools for Understanding Gene Function and Drug Action 387
C Nislow and G Giaever

18. RNA Gene Analysis 415
C Saveanu, M Fromont-Racine and A Jacquier

19. Analysis of Gene Function of Mitochondria. 445
S Duvezin-Caubet, AS Reichert and W Neupert

20. Yeast Prions and Their Analysis *In Vivo*. 491
MF Tuite, LJ Byrne, L Jossé, F Ness, N Koloteva-Levine and B Cox

21. Metabolic Control in the Eukaryotic Cell, a Systems Biology Perspective 527
JI Castrillo and SG Oliver

22. Phylogenetic Footprinting 551
PF Cliften

23. *Saccharomyces cerevisiae* as a Tool for Human Gene Function Discovery 577
HR Waterham and RJA Wanders

24. Bioinformatic Prediction of Yeast Gene Function 597
I Lee, R Narayanaswamy and EM Marcotte

25. Yeast Genetic Strain and Plasmid Collections 629
K-D Entian and P Kötter

26. Yeast Gene Analysis: The Remaining Challenges 667
MJR Stark and I Stansfield

Index 685

Colour Plate Section can be found at the back of this book

Series Advisors

Gordon Dougan The Wellcome Trust Sanger Institute, Wellcome Trust Genome Campus, Hinxton, Cambridge CBIO ISA, UK

Graham J Boulnois Schroder Ventures Life Science Advisers (UK) Limited, 71 Kingsway, London WC2B 6ST, UK

Jim Prosser School of Medical Sciences, University of Aberdeen, Cruickshank Building, St Machar Drive Aberdeen, AB24 3UU, UK

Ian R Booth School of Medical Sciences, University of Aberdeen, Institute of Medical Sciences, Foresterhill, Aberdeen AB25 2ZD, UK

David A Hodgson Department of Biological Sciences, University of Warwick, Conventry CV4 7AL, UK

David H Boxer University of Dundee, Dundee DD1 4HN, UK

Contributors

Kathryn R Ayscough Department of Molecular Biology and Biotechnology, University of Sheffield, Firth Court, Western Bank, Sheffield S10 2TN, UK

Charles Boone Banting and Best Department of Medical Research and Department of Medical Genetics and Microbiology, Terrance Donnelly Centre for Cellular and Biomolecular Research, University of Toronto, 160 College Street, Toronto, Canada M5S 3E1

Andy Brass Faculty of Life Sciences and School of Computer Science, The University of Manchester, Oxford Road, Manchester M13 9PT, UK

Michael Breitenbach Department of Cell Biology, Division of Genetics, University of Salzburg, Hellbrunnerstrasse 34, 5020 Salzburg, Austria

Stanley Brul Swammerdam Institute for Life Sciences, University of Amsterdam, BioCentrum Amsterdam, Nieuwe Achtergracht 166, 1018 WV Amsterdam, The Netherlands

Lee J Byrne Department of Biosciences, University of Kent, Canterbury, Kent CT2 7NJ, UK

Juan I Castrillo Centre for the Analysis of Biological Complexity (CABC), Faculty of Life Sciences, Michael Smith Building, The University of Manchester, Oxford Road, Manchester M13 9PT, UK

Paul F Cliften Department of Biology, Utah State University, 5305 Old Main Hill, Logan, UT 84322, USA

Brian Cox Department of Biosciences, University of Kent, Canterbury, Kent CT2 7NJ, UK

Trisha N Davis Department of Biochemistry, University of Washington, Seattle, WA 98195, USA

Piet De Groot Swammerdam Institute for Life Sciences, University of Amsterdam, BioCentrum Amsterdam, Nieuwe Achtergracht 166, 1018 WV Amsterdam, The Netherlands

J Richard Dickinson Cardiff School of Biosciences, Main Building, Museum Avenue, Cardiff CF10 3TL, UK

Stéphane Duvezin-Caubet Adolf-Butenandt-Institut für Physiologische Chemie, Ludwig-Maximilians-Universität München, Butenandtstraße 5, 81377 München, Germany

Karl-Dieter Entian Center of Exellence: Macromolecular Complexes and Institute for Molecular Biosciences, Johann Wolfgang Goethe University, Max-von-Laue Str. 9, 60438 Frankfurt/Main, Germany

Micheline Fromont-Racine Unité de Génétique des Interactions Macromoléculaires, Institut Pasteur (CNRS-URA2171), 25-28 rue du Dr. Roux 75724, Paris cedex 15, France

Qi Gao Department of Human and Molecular Genetics, Baylor College of Medicine, Houston, TX 77030, USA

Guri Giaever Department of Pharmaceutical Sciences, Donnelley CCBR, University of Toronto, Ontario M5S3E1, Canada

R Daniel Gietz Department of Biochemistry and Medical Genetics, University of Manitoba, T250-770 Bannatyne Avenue, Winnipeg, Manitoba, Canada R3E 0W3

Erica A Golemis Division of Basic Sciences, Fox Chase Cancer Center, 333 Cottman Ave., Philadelphia, PA 19111, USA

Andrew Hayes Faculty of Life Sciences, The University of Manchester, Michael Smith Building, Oxford Road, Manchester M13 9PT, UK

Xiangwei He Department of Human and Molecular Genetics, Baylor College of Medicine, Houston, TX 77030, USA

Ewald H Hettema Department of Molecular Biology and Biotechnology, University of Sheffield, Firth Court, Western Bank, Sheffield S10 2TN, UK

Eugene Izumchenko Department of Microbiology and Immunology, Ben Gurion University, Beer Sheva 84105, Israel

Alain Jacquier Unité de Génétique des Interactions Macromoléculaires, Institut Pasteur (CNRS-URA2171), 25-28 rue du Dr. Roux 75724, Paris cedex 15, France

Lyne J Jossé Department of Biosciences, University of Kent, Canterbury, Kent CT2 7NJ, UK

Frans M Klis Swammerdam Institute for Life Sciences, University of Amsterdam, BioCentrum Amsterdam, Nieuwe Achtergracht 166, 1018 WV Amsterdam, The Netherlands

Michael Knop Cell Biology and Biophysics Unit, EMBL, Meyerhofstraße 1, 69117 Heidelberg, Germany

Nadejda Koloteva-Levine Department of Biosciences, University of Kent, Canterbury, Kent CT2 7NJ, UK

Peter Kötter Institute for Molecular Biosciences, Johann Wolfgang Goethe University, Max-von-Laue Str. 9, 60438 Frankfurt/Main, Germany

Peter Laun Department of Cell Biology, Division of Genetics, University of Salzburg, Hellbrunnerstrasse 34, 5020 Salzburg, Austria

Insuk Lee Center for Systems and Synthetic Biology, Institute for Cellular & Molecular Biology, 2500 Speedway, University of Texas at Austin, Austin, TX 78712, USA

Celine I. Maeder Cell Biology and Biophysics Unit, EMBL, Meyerhofstraße 1, 69117 Heidelberg, Germany

Peter Maier Cell Biology and Biophysics Unit, EMBL, Meyerhofstraße 1, 69117 Heidelberg, Germany

Edward M Marcotte Center for Systems and Synthetic Biology, Institute for Cellular & Molecular Biology, 2500 Speedway, University of Texas at Austin, Austin, TX 78712, USA

Eric GD Muller Department of Biochemistry, University of Washington, Seattle, WA 98195, USA

Rammohan Narayanaswamy Center for Systems and Synthetic Biology, Institute for Cellular & Molecular Biology, 2500 Speedway, University of Texas at Austin, Austin, TX 78712, USA

Frederique Ness I.B.G.C 1, Rue Camille Saint Saens, 33077 Bordeaux, France

Walter Neupert Adolf-Butenandt-Institut für Physiologische Chemie, Ludwig-Maximilians-Universität München, Butenandtstraße 5, 81377 München, Germany

Corey Nislow Banting and Best Department of Medical Research, Donnelley CCBR, University of Toronto, Ontario M5S3E1, Canada

Stephen G Oliver Centre for the Analysis of Biological Complexity (CABC), Faculty of Life Sciences, Michael Smith Building, The University of Manchester, Oxford Road, Manchester M13 9PT, UK

Jason Ptacek Department of Molecular Biophysics & Biochemistry, Yale University, New Haven, CT 06520-8103, USA

Seesandra V Rajagopala Institut für Toxikologie und Genetik, Forschungszentrum Karlsruhe, Hermann-von-Helmholtz-Platz 1, D-76344 Eggenstein-Leopoldshafen, Germany and The Institute of Genomic Research (TIGR), Rockville, MD 20850, USA

Andreas S Reichert Adolf-Butenandt-Institut für Physiologische Chemie, Ludwig-Maximilians-Universität München, Butenandtstraße 5, 81377 München, Germany

Cosmin Saveanu Unité de Génétique des Interactions Macromoléculaires, Institut Pasteur (CNRS-URA2171), 25-28 rue du Dr. Roux 75724, Paris cedex 15, France

Ilya G Serebriiskii Division of Basic Sciences, Fox Chase Cancer Center, 333 Cottman Avenue, Philadelphia, PA 19111, USA

Michael Snyder Department of Molecular, Cellular & Developmental Biology, Yale University, New Haven, CT 06520-8103, USA

Ian Stansfield School of Medical Sciences, Institute of Medical Sciences, University of Aberdeen, Foresterhill, Aberdeen AB25 2ZD, UK

Michael JR Stark Division of Gene Regulation and Expression, College of Life Sciences, University of Dundee, MSI/WTB Complex, Dow Street, Dundee DD1 5EH, UK

Tomoyuki U Tanaka Division of Gene Regulation and Expression, College of Life Sciences, University of Dundee, MSI/WTB Complex, Dow Street, Dundee DD1 5EH, UK

Björn Titz Institut für Toxikologie und Genetik, Forschungszentrum Karlsruhe, Hermann-von-Helmholtz-Platz 1, D-76344 Eggenstein-Leopoldshafen, Germany

Amy Hin Yan Tong Banting and Best Department of Medical Research, and Department of Medical Genetics and Microbiology, University of Toronto, 112 College Street, Toronto, ON, Canada M5G 1L6

Duy Truong Department of Biochemistry and Medical Genetics, University of Manitoba, T250-770 Bannatyne Avenue, Winnipeg, Manitoba, Canada

Mick F Tuite Department of Biosciences, University of Kent, Canterbury, Kent CT2 7NJ, UK

Peter Uetz Institut für Toxikologie und Genetik, Forschungszentrum Karlsruhe, Hermann-von-Helmholtz-Platz 1, D-76344 Eggenstein-Leopoldshafen, Germany

Tobias von der Haar Department of Biosciences, University of Kent, Canterbury, Kent CT2 7NJ, UK

Ronald JA Wanders Laboratory Genetic Metabolic Diseases, Departments of Pediatrics and Clinical Chemistry, Academic Medical Centre, Meibergdreef 9, 1105 AZ Amsterdam, The Netherlands

Hans R Waterham Laboratory Genetic Metabolic Diseases, Departments of Pediatrics and Clinical Chemistry, Academic Medical Centre, Meibergdreef 9, 1105 AZ Amsterdam, The Netherlands

Marina Wolfson Department of Microbiology and Immunology, Ben Gurion University, Beer Sheva 84105, Israel

Leo AH Zeef Faculty of Life Sciences, The University of Manchester, Michael Smith Building, Oxford Road, Manchester M13 9PT, UK

Preface

The first edition of this volume was published in 1998, just over 2 years following completion of the *Saccharomyces cerevisiae* genome sequence and at the start of a period of concerted effort towards genome-wide functional analysis. With the genome sequencing complete, an integrated series of functional genomics approaches was needed, with the ultimate aim of assigning a role and biochemical function to each of the approximately 6000 proteins encoded by the genome.

The first edition of *Yeast Gene Analysis* (Brown and Tuite, 1998) was a very successful compendium of the current functional analysis methodologies at that time. However, since then a number of factors have combined to necessitate the update represented by this second edition. Firstly, great strides have been made through the introduction of many new, high-throughput and genome-wide approaches to yeast gene analysis that were either in their infancy or not even in existence when the first edition was published, such as microarray-based transcriptional profiling, high-throughput two-hybrid techniques, synthetic genetic array analysis and the development of protein microarrays. Secondly, a plethora of new molecular genetic and cell biology methodologies have been developed and applied to yeast functional genomics, including the use of TAP (tandem affinity purification)-tagging and improved mass-spectrometry analysis to study protein complexes and fluorescence resonance energy transfer (FRET), to name just two. Finally, many of the more established techniques applied to yeast functional analysis have been extensively developed, improved or re-evaluated over the past decade. The publication of this second edition therefore represents an ideal opportunity to assemble critical appraisals of these advances, together with experimental details as appropriate.

How has our understanding of the genome changed, and how much progress in functional analysis has been made, since publication of the first edition? The inventory of yeast genes has fluctuated a little, both through the elimination of falsely identified ORFs using comparative genomics (Cliften *et al.*, 2003; Kellis *et al.*, 2003; – see Chapter 22, this volume), and through new additions following the analysis of small ORFs (Kastenmayer *et al.*, 2006) and tiling array approaches (David *et al.*, 2006). However, excluding those ORFs still annotated as "dubious" the current size of the ORF set has not significantly changed. That over 800 potential genes remain within the "dubious" category underlines that there is still work to be done before we have a definitive gene set. The editors of the first edition of this book (Mick Tuite and Alistair Brown) included a table taken from Goffeau *et al.* (1996) summarising our genome-wide knowledge of yeast gene function at that time. This indicated 4343 ORFs that could be 'verified' because either they encoded known proteins (2611 ORFs) or showed similarity to other known or unknown proteins (1732 ORFs). Since at the time of writing the *Saccharomyces* Genome Database (SGD; http://www.yeastgenome.org/) lists 4425 verified ORFs, how much progress have we really made since then? Although these statistics highlight a significant group of genes about which there is still little or no information available, these simplistic comparisons fortunately do not reveal the extent to which functional information on the *S. cerevisiae* gene set has been tremendously expanded by both high-throughput and smaller scale, focused studies over the past decade. For many of the characterised ORFs, we now know the localisation of the gene product, the likely interaction partners, loss of function phenotype, genetic interactions and how transcription of the gene varies under a wide range of conditions, or at least a useful subset of such information. Using SGD data (August 2006), there are 4160 genes assigned gene ontology (GO) term for molecular function. Even more ORFs have been assigned a biological process GO term (4824). However, despite this progress there are still 2112 ORFs with an unknown molecular function. Moreover, for some ORFs, even when a GO molecular function term has been assigned our knowledge regarding the precise function is sketchy to say the least. Functional analysis in yeast has made significant advances, but there is clearly still plenty to do.

Perhaps the best way to summarise the current state of play is to say that we now know a little about a lot of genes and a lot about a few genes. The experience of the last 8 years' efforts in yeast functional genomics now allows us to appraise the scale of the task of reversing this situation so that we can know a great deal about many (and ultimately, all) yeast genes. What has become clear is the value of high-throughput, genome-wide methodologies, both to provide an overall framework in which to carry out functional analysis and to ask specific questions on a genome-wide scale. When investigators identify a new gene in a focused study, the information on its localisation, transcriptional regulation and protein–protein interactions is frequently now already known, allowing rapid progress to be made in working outwards from a core process, gene set or protein complex towards identifying new gene products interacting with their system. Such approaches represent efficient ways to define an initial list of all the cellular components associated with a given gene, protein, process or environmental response. The new edition of *Yeast Gene Analysis* therefore brings together contributions covering both the high-throughput methodologies and those more suited to a gene-by-gene approach, representing an excellent primer for those adopting either style of research.

While the techniques covered by this new edition have obviously been expanded and revised, the new version is now even more focused on the budding yeast *S. cerevisiae*. Although since the first edition the genomes of both *Schizosaccharomyces pombe* and *Candida albicans* have been completed (Wood *et al.*, 2002; Jones *et al.*, 2004), each of these organisms is now associated with its own burgeoning suite of species-specific molecular genetic methods for the analysis of gene function and as a result, *C. albicans* and *S. pombe* functional analysis could now each easily merit the entire coverage of a volume in their own right. We therefore made a conscious decision to restrict coverage in the second edition almost entirely to *S. cerevisiae*, especially given the expanded range of budding yeast functional genomics approaches. However, we have retained the format of the original volume, each chapter providing a critical review of techniques and protocols placed in context by authors with first-hand knowledge, and in many cases the authors have been instrumental in developing the approaches themselves. Over half the book consists of completely new chapters covering novel techniques and methodologies developed since the first edition was published, while all of the remaining chapters have been extensively revised and updated. The aim has been to present a comprehensive analysis of the pros and cons of each approach and its applicability, with experimental protocols provided where methodologies are not fully or straightforwardly described elsewhere. Each chapter also includes hints and advice from experienced practitioners and developers, handy tips that should make adoption of new methodology by a lab that much easier. The spirit of the first edition, which attempted to make yeast molecular genetics approachable for the newcomer, has again been retained by including chapters describing platform methodologies of strain construction and classical yeast genetics (Chapter 2), yeast transformation (Chapter 3), targeted gene manipulation (Chapter 4) and immunological approaches (Chapter 11). We hope that this will inspire sufficient confidence among non-yeast experts to encourage them to make forays into yeast territory when required, and perhaps, having read this volume, consider more extensive use of yeast in which to explore their areas of interest in what is still arguably one of the most powerful model eukaryotic systems.

The assembly of this volume has only been made possible by those yeast researchers, all experts in their fields, who gave generously of their time and patience. The broad geographic spread of contributors is a testament to the community-wide involvement with this book. We would like to thank all the authors of chapters in this volume for contributing with efficiency, speed and good nature. Finally, we thank those publishing editors at Elsevier who have been centrally involved with this project, Claire Minto and Lisa Tickner, for keeping the book as close as possible to schedule.

Ian Stansfield and Michael J. R. Stark
August 2006

References

Brown, A. J. and Tuite, M. F. (1998). *Yeast Gene Analysis: Methods in Microbiology*. Vol. 26 Elsevier, London.

Cliften, P., Sudarsanam, P., Desikan, A., Fulton, L., Fulton, B., Majors, J., Waterston, R., Cohen, B. A. and Johnston, M. (2003). Finding functional features in *Saccharomyces genomes* by phylogenetic footprinting. *Science* **301**, 71–76.

David, L., Huber, W., Granovskaia, M., Toedling, J., Palm, C. J., Bofkin, L., Jones, T., Davis, R. W. and Steinmetz, L. M. (2006). A high-resolution map of transcription in the yeast genome. *Proc. Natl. Acad. Sci. USA* **103**, 5320–5325.

Goffeau, A., Barrel, B. G., Bussey, H., Davis, R. W., Dujon, B., Feldmann, H., Galibert, F., Hoheisel, J. D., Jacq, C., Johnston, M., *et al.* (1996). Life with 6000 genes. *Science* **274**, 563–567.

Jones, T., Federspiel, N. A., Chibana, H., Dungan, J., Kalman, S., Magee, B. B., Newport, G., Thorstenson, Y. R., Agabian, N., Magee, P. T. *et al.* (2004). The diploid genome sequence of *Candida albicans*. *Proc. Natl. Acad. Sci. USA* **101**, 7329–7334.

Kastenmayer, J. P., Ni, L., Chu, A., Kitchen, L. E., Au, W. C., Yang, H., Carter, C. D., Wheeler, D., Davis, R. W., Boeke, J. D., Snyder, M. A. and Basrai, M. A. (2006). Functional genomics of genes with small open reading frames (sORFs) in *S. cerevisiae*. *Genome Res.* **16**, 365–373.

Kellis, M., Patterson, N., Endrizzi, M., Birren, B. and Lander, E. S. (2003). Sequencing and comparison of yeast species to identify genes and regulatory elements. *Nature* **423**, 241–254.

Wood, V., Gwilliam, R., Rajandream, M. A., Lyne, M., Lyne, R., Stewart, A., Sgouros, J., Peat, N., Hayles, J., Baker, S., *et al.* (2002). The genome sequence of *Schizosaccharomyces pombe*. *Nature* **415**, 871–880.

1 Introduction to Functional Analysis in Yeast

Stephen G Oliver
Centre for the Analysis of Biological Complexity (CABC), Faculty of Life Sciences, Michael Smith Building, The University of Manchester, Manchester, UK

♦♦♦

CONTENTS

The genome sequence and how it changed the world of yeast
Yeast genetics – new perspectives
Transcriptomics and an emerging RNA world
Proteomics
Metabolomics
The minimal genome and the e-yeast

ABBREVIATIONS

CGH	comparative genome hybridisation
ITS	internal transcribed spacer
SAGE	serial analysis of gene expression

♦♦♦♦♦♦ I. THE GENOME SEQUENCE AND HOW IT CHANGED THE WORLD OF YEAST

A. Where We were at the Beginning, and Where We are Now

Ten years ago, the world of yeast research changed forever with the publication of the complete DNA sequence of *Saccharomyces cerevisiae* (Goffeau *et al.*, 1996). However, it was already possible to discern just how it would change some four or five years earlier when the DNA sequence of the first chromosome (chromosome III) was completed (Oliver *et al.*, 1992). When we embarked on the sequencing of the baker's yeast genome, there were just 1000 genes on the *S. cerevisiae* genetic map (Mortimer *et al.*, 1989); the complete

METHODS IN MICROBIOLOGY, VOLUME 36
0580-9517 DOI:10.1016/S0580-9517(06)36001-1

genome sequence revealed some 5885 protein-encoding genes (Goffeau *et al.*, 1996; Mewes *et al.*, 1997), whose predicted products were at least 100 amino acids in length. In 1996, some 43% (2531) of these genes were either of known function (or their function could be inferred by a high level of amino acid sequence similarity of their protein products with gene products of known function in other organisms); 20% (1177) encoded products which showed some similarity with those of known genes in other organisms; while 37% (2177) were of completely unknown function. Today, there are less than 1000 yeast genes for which there is not some functional annotation. That is a measure of the genome revolution. While classical molecular genetics rather than high-throughput functional genomics has revealed the function of many of these genes, their very existence was unknown at the start of the sequencing project, and functional genomics has provided abundant tools that have complemented and accelerated more classical analyses.

B. Genome-Based Phylogenetics

The sequencing of, first, the *S. cerevisiae* genome and then the genomes of other yeasts and filamentous fungi (Fitzpatrick *et al.*, 2006) has given us a completely new view of the position of *Saccharomyces cerevisiae* in the fungal world – how its genome evolved and the nature of its relationship with other hemiascomycetes. Molecular phylogeny has revolutionised our view of evolution – particularly for microorganisms, where the fossil record is not very useful. Phylogenetic trees are usually based on the comparison of the sequence of a single gene, or part of a gene, between species. For the yeasts, the ribosomal ITS sequence has been employed frequently for phylogenetic analyses. However, the problem is that if the sequences of different genes are compared, then different trees may be constructed (Sicheritz-Ponten and Andersson, 2001; Kurtzman and Robnett, 2003) and it is not clear which one is the best representation of the evolutionary history of the group of species being studied. The availability of complete genome sequences presents the opportunity to employ all of the genes (or, at least, a large number of them) to construct the phylogeny (Kurtzman and Robnett, 2003; Rokas *et al.*, 2003) with the expectation that this will produce robust phylogenies that do not change as additional information is incorporated.

C. Comparative Genomic Hybridisation

One of the first new high-throughput approaches that were developed in response to the availability of complete genome sequences was that of hybridisation-array technology (see Section III). While this has been extensively used to analyse gene transcription in a comprehensive manner (transcriptomics; see Chapter 9, this volume), it may also be used to analyse genome composition by hybridising labelled preparations of genomic DNA to the (c)DNA or oligonucleotide arrays. This approach was pioneered by Elizabeth Winzeler to look at strain-to-strain variation in *Saccharomyces cerevisiae* using Affymetrix arrays that consisted of *ca.* 300 000 25-mer oligonucleotides covering all of the open reading frames (ORFs) defined by the complete genome sequence (Winzeler *et al.*, 2003). The set of laboratory strains investigated demonstrated that most variation occurred in sub-telomeric regions and frequently involved genes concerned with fermentation and transport. The recent availability of tiling arrays for *Saccharomyces cerevisiae* means that such inter-strain comparisons may now be carried out at the level of single-nucleotide polymorphisms (Gresham *et al.*, 2006).

Comparative Genomic Hybridisation (CGH) is a method whereby whole-genome-based phylogenies may be constructed by using hybridisation arrays to compare the genomic DNA of a number of species to that of a reference organism. A major advantage of CGH is that it '... circumvents the need for sequencing multiple closely related genomes' (Murray *et al.*, 2001) and may be used to type new specimens from field surveys. Edwards-Ingram *et al.* (2004) used a microarray-containing PCR products corresponding to each of the ORFs of *S. cerevisiae* to build a phylogeny of *Saccharomyces 'sensu stricto'* species, *S. castellii* (a member of the *Saccharomyces 'sensu lato'*), and the highly contested 'species' *S. boulardii.* The phylogeny (constructed on the basis of the presence and absence calls for each ORF) demonstrated that *S. boulardii* was not a separate species, but a strain of *S. cerevisiae* that had lost all intact Ty1/Ty2 retrotransposons. With genome sequences now available for at least 19 members of the hemiascomycetes, and cheaper methods of generating oligonucleotide-based microarrays, we can expect the CGH method to be extended to encompass all yeast species.

D. Comparative Genomics

Because of the small size of their genomes and their scientific, industrial, medical, and agricultural importance, there are more complete genome sequences available for the fungi (including the yeasts) than any other group of eukaryotes – some 42 whole-genome sequences, in total, according to a recent count (Fitzpatrick *et al.*, 2006). Moreover, the yeasts and filamentous fungi cover a huge evolutionary range. The filamentous ascomycetes and the budding yeasts diverged from one another some 900–1000 million years ago (Hedges *et al.*, 2004), and the Saccharomycotina alone are more evolutionarily diverged than the Chordate phylum of the animal kingdom (Goffeau, 2004). Thus students of *Saccharomyces cerevisiae* are fortunate in having whole-genome sequences for quite closely (the *Saccharomyces 'sensu stricto'*, a small group of species that inter-breed with *S. cerevisiae*) and distantly related (the hemiascomycetes, or 'true') yeasts available for comparative studies.

Comparative genomics has, in recent years, had an enormous impact on the functional analysis of *Saccharomyces cerevisiae* at both the conceptual and practical levels. At the conceptual level, the most important consequence of the availability of whole genomes from multiple yeast species has been the confirmation, from the sequences of *Kluyveromyces waltii* (Kellis *et al.*, 2004), *Ashbya gossypii* (Dietrich *et al.*, 2004), *Kluyveromyces lactis* (Dujon *et al.*, 2004), of Ken Wolfe's original contention (Wolfe and Shields, 1997) that there was a whole-genome duplication in the evolutionary history of the *Saccharomyces* clade. The existence of this duplication event needs to be considered when carrying out functional analyses, at the same time realising that genes that have been retained in duplicate may have diverged in either their regulation or function, or both (Delneri *et al.*, 1999; Harrison *et al.*, 2006).

It is in the area of gene regulation that comparative genomics has had its greatest practical impact on functional analysis. Given the extreme difficulty of inferring transcription factor target sites by purely informatic analyses, the ability to compare orthologous upstream regions of genes across a number of closely related species, such as the *Saccharomyces 'sensu stricto'* group, is of enormous utility. This approach has been elegantly exploited by teams at the sequencing centres at both Washington University (Cliften *et al.*, 2003) and Whitehead/MIT (Kellis *et al.*, 2003), and is further elucidated by Paul Cliften in this volume (see Chapter 22).

♦♦♦♦♦♦ II. YEAST GENETICS – NEW PERSPECTIVES

A. Why Yeast is such a Good Genetic System?

Yeast is an excellent genetic system for two reasons. First, it has an almost perfect life cycle for classical genetic studies; it is able to grow vegetatively in both the haploid and diploid phases. The crossing of haploids of opposite mating type is easily accomplished following the isolation of strains (notably S288c; Mortimer and Johnston, 1986) that were stably heterothallic due to a mutation in the *HO* gene that prevents mating-type switching. Meiosis may be induced easily in diploid strains by nitrogen starvation, and the products of meiosis can be examined directly by dissecting the haploid spores from the ascus and germinating them to produce haploid clones. The only thing that yeast lacks is the ordered tetrads of *Neurospora crassa*, and some other filamentous Ascomycetes, and that disadvantage counted for little once a full set of centromere-linked genetic markers was available. The second major advantage of yeast as a genetic system, at least as far as molecular genetics is concerned, is the facility and fidelity with which it carries out mitotic recombination.

B. The Full Dimensions of the Life Cycle and Yeast's Mating Strategies

We are used to thinking of yeast as having a simple sexual system with just two mating types (**a** and α) and a pairing between individuals of opposite mating type being required to generate a diploid. However, *Saccharomyces cerevisiae* is more sexually versatile than that (Knop, 2006). The classic mating event between two cells having different alleles at the *MAT* locus is properly termed amphimixis. In addition, yeast has two other routes to diploid formation: (i) haplo-selfing, which occurs in homothallic strains as a result of mating-type switching – a daughter cell mating with its mother that has switched to the opposite mating type; (ii) automixis (or intra-tetrad mating), in which sibling spores within the ascus mate before the ascus wall has been broken down. These three different mating routes have profoundly different genetic consequences: amphimixis promotes heterozygosity; haplo-selfing immediately renders all genes in the diploid homozygous, apart from *MAT* itself; while intra-tetrad mating results in all genes, except for those linked to the *MAT* locus, having a 1/3 chance of becoming homozygous following any given mating event. We have no idea as to the relative frequency with which these three mating routes are used by yeast in the wild, but knowing those frequencies would be a great help in the interpretation of laboratory haploinsufficiency analyses (see Section II.C). The fact that we are largely ignorant of the natural history of *Saccharomyces cerevisiae* used to be an embarrassment – now, it is definite impediment to further progress of research on this remarkable experimental organism.

C. The Deletion Collection and What it Means

Perhaps the major resource generated since the completion of the *S. cerevisiae* genome sequence, and absolutely dependent on that sequence, is the collection of deletion mutants (Winzeler *et al.*, 1999; Giaever *et al.*, 2002), one for every protein-encoding gene predicted by the complete genome sequence, which was constructed as a result of a collaboration of laboratories belonging to the EUROFAN Consortium (Oliver, 1996a) in Europe and others in the USA and Canada. An early product of this exercise was a figure for the number of essential (at least, for growth on a YPD plate) genes in the yeast genome – some 19% of the total (Giaever *et al.*, 2002).

It was the inspiration of Ron Davis (Shoemaker *et al.*, 1996) to tag each mutant with two unique 20 bp sequences, one each side of the replacement marker (Wach *et al.*, 1994; Wach, 1996), which act as 'molecular bar-codes' to identify each mutant unambiguously. This not only means that curators can perform effective quality control of the collection, and individual laboratories easily assure themselves that they are working with the correct mutant(s), but it also allows the identification of the presence and, indeed, the proportions of individual mutants in a mixed population of such mutants. This has allowed the determination of quantitative phenotypes as a result of competition analyses (e.g. Giaever *et al.*, 2002; Delneri *et al.*, 2006; Chapter 17, this volume).

The deletion mutants were generated in diploid cells so that essential genes could be deleted. The initial product of the exercise, then, was a collection of *ca.* 5800 heterozygous (or hemizygous) deletants. Normally, one would not expect to detect a phenotype in such mutants. While this may be true in qualitative terms, it is not necessarily true quantitatively. It would be expected, to a first approximation, that the result of reducing the copy number of an individual gene from two to one would be to halve the concentration, in the diploid cell, of the protein that it encodes. In these circumstances, there may be simply insufficient of that protein available to sustain growth at wild-type rates. In the terms of classical genetics, such mutants are said to display a 'haploinsufficiency' phenotype. However, the possibility that the hemizygote will grow faster than the wild type should not be excluded, and we have termed such an improved phenotype 'haploproficient' (Delneri *et al.*, 2006; Oliver, 2006a,b). Competition experiments between hemizygous mutants have been used to identify the target sites of drugs (Giaever *et al.*, 1999, 2004; Baetz *et al.*, 2004) and identify genes with a high degree of control over yeast's growth rate (Delneri *et al.*, 2006; Oliver, 2006a,b) – in the parlance of Metabolic Control Analysis (Teusink *et al.*, 1998; Chapter 21, this volume), *h*igh *f*lux *c*ontrol (HFC) genes.

D. Other Collections

There is now a wealth of resources available for functional analyses with *S. cerevisiae* (Delneri, 2004). Before the deletion collection was constructed, Mike Snyder and his colleagues (Ross-Macdonald *et al.*, 1999) had constructed a comprehensive library of yeast mutants using transposon-based mutagenesis. However, the elegant use of such mutant collections for functional analysis by 'genetic footprinting' (Smith *et al.*, 1996) has largely been superseded by the use of the bar-coded deletion collection. Ross-Macdonald *et al.* (1997) extended the transposon-tagging technique in order to generate GFP fusions to monitor protein production and location, and a comprehensive GFP-fusion library has now been constructed in a directed manner by Erin O'Shea and her colleagues (Huh *et al.*, 2003). The latter has allowed a comprehensive assessment of protein levels (Ghaemmaghami *et al.*, 2003) and localisation, and has permitted investigations into such problems as the relationship between the levels of expression of individual proteins and the cell-to-cell variance of such levels (Bar-Even *et al.*, 2006).

Essential genes are a problem for the deletion collection since they may only be studied in the heterozygous (sometimes referred to as hemizygous) state. As we have seen, this has its uses; nevertheless, it has proved valuable to place the expression of essential genes under the control of a regulatable promoter such as *Tet*O (Belli *et al.*, 1998a,b) or *GAL1* (Liu *et al.*, 1992). Expression may then be switched off and the effect of depletion of the essential protein product observed (see Chapter 5 of this volume for a more extensive discussion). Collections of strains with essential genes under *Tet*O control (Mnaimneh *et al.*, 2004; Wishart *et al.*, 2006) have been constructed and a strain collection with all ORFs under *GAL1* control has been employed to uncover over-expression phenotypes (Sopko *et al.*, 2006). Moreover, any yeast protein can now be purified (in principle, at least) by using a strain collection in which each yeast ORF has been tagged, at its N-terminus, with GST-His_6 to facilitate purification (Zhu *et al.*, 2001; Gelperin *et al.*, 2005). The use of this tagged-protein library in the production of protein arrays is discussed in Chapter 17 of this volume.

◆◆◆◆◆◆ III. TRANSCRIPTOMICS AND AN EMERGING RNA WORLD

Probably the most popular of the novel kinds of analysis developed in the post-genomics era has been that of transcriptomics (see Chapter 9, this volume). The availability of the complete genome sequence led to two main methods whereby, in a single experiment, both the presence and level of transcripts from *all* of yeast's genes could be analysed. There are two main ways of achieving this – *s*erial *a*nalysis of *g*ene *e*xpression (SAGE: Velculescu *et al.*, 1997; Varela *et al.*, 2005) and hybridisation-array analysis (see below). SAGE is much the more difficult of the two techniques – it requires a considerable degree of molecular biological skill to perform and results are generated from sequencing large numbers of catenated cDNA tags. However, SAGE has a powerful advantage over arrays – it involves no assumptions about what constitutes a gene. Thus SAGE analysis is truly comprehensive and *unprejudiced*. Early studies (Velculescu *et al.*, 1997) gave glimpses of an RNA world within the yeast cell of which we were completely ignorant. This world included, for instance, antisense transcripts of known genes; nobody knew quite what to make of it or whether, even, to believe in it at all. We should have had more faith.

Hybridisation-array analysis, whether using PCR products (Schena *et al.*, 1995), or short (Lockhart *et al.*, 1996) or long (Hauser *et al.*, 1999) oligonucleotides as probes, has the distinct disadvantage (compared with SAGE) that it requires that we decide what a gene is. Moreover, most studies have ignored untranslated transcripts since oligo-dT primers are used to generate the first cDNA strands from the RNAs in the cell extracts. Thus array analysis has missed the RNA world, although there is now hope that the use of tiling arrays will make good this deficiency. The great advantage of hybridisation-array technology is that it is easy to use, and many hundreds of experiments have been performed; the data from which are (increasingly) made publicly available in computable formats.

A salutary effect of this plethora of data is that the molecular biology community has been forced to rediscover statistics. Many early array studies are unreliable (at least, in quantitative terms) since essentially no statistical analysis was performed on the data. It was common to count as significant any change in transcript level equal to (or greater than) two-fold, without taking any account of the variance in the data. Indeed, often insufficient biological repeats were performed, or none at all. Fortunately, things have changed and Chapter 9 describes current approaches to the normalisation and analysis of hybridisation-array data. Problems now tend to be biological, rather than technological, and stem from poor experimental design. An experiment to analyse the expression of all of the genes (by hybridisation-array analysis), rather one or a few genes (by, say, Northern analysis), has to be designed with far greater rigour and attention to detail than has usually been the case in the past. The reader is referred to Hayes *et al.* (2002) and Lim *et al.* (2003) for an example of the profound effect that removal of confounding variables can have on the outcome of transcriptome analyses.

♦♦♦♦♦♦ IV. PROTEOMICS

A. Identification and Quantification

Proteomics should be the most useful level of functional genomic analysis since proteins (like metabolites, but unlike mRNA molecules) are functional entities within the cell, and (like mRNAs, but unlike metabolites) there is a direct link between proteins and genes. Two-dimensional gel electrophoresis has been used to analyse the yeast proteome for more than 30 years (O'Farrell, 1975; Elliott and McLaughlin, 1979; Garrels, 1979) but, despite sterling efforts (e.g. Perrot *et al.*, 1999), it is clear that the technology is simply not up to the job of providing the sort of comprehensive coverage that modern functional analysis demands. Fortunately, alternative methods, which combine liquid chromatography (LC) or capillary zone electrophoresis with mass spectrometry (e.g. Washburn *et al.*, 2001), are capable of accessing a much higher proportion of the total yeast proteome. Identification of the proteins corresponding to the peptides released from the LC column into the mass spectrometer is still not straightforward. However, the construction of tools for mass analysis that have been customised for the *Saccharomyces cerevisiae* proteome is a great help (McLaughlin *et al.*, 2006). There remain, at least, two major problems – quantitation and post-translational events.

A number of novel methods for the relative and absolute quantitation of proteins in cell extracts, using MS analysis, have been developed in recent years. Two are worthy of mention here. First, the iTRAQ technique (Ross *et al.*, 2004) developed by Applied Biosystems is an elegant and robust, if expensive, solution to the problem. The development of sound statistical techniques for data normalisation and analysis (Shadforth *et al.*, 2005) has been of central importance to its exploitation in real experiments. Second, QCat (Beynon *et al.*, 2005) enables the absolute quantitation of the levels of individual proteins. It does this by using an internal standard that consists of a polypeptide, encoded by a synthetic gene, which contains the catenated sequences of a number of tryptic peptides – one from each of the natural proteins that the experimenter wishes to quantify. Each of these Q peptide sequences is diagnostic for a particular protein and has been detected previously in the mass spectra of natural proteins. This QCat polypeptide is then synthesised in *E. coli* grown in a medium containing ^{15}N label. A known amount of the labelled QCat artificial protein is added to the cell extract, and the whole mixture is digested with trypsin before analysis by mass spectrometry. The concentrations of individual proteins in the cell extract may then be determined by reference of the diagnostic peptide in the natural protein to its density-labelled equivalent released from the QCat protein. It is to be expected that either commercial interests, or the yeast research community acting in concert, will cause QCat proteins for *S. cerevisiae* to be designed and made widely available.

B. The Interactome – the Data Avalanche and the Data Shortage

The majority of processes in the yeast cell are mediated by protein–protein interactions, including signal transduction pathways and the regulation of gene expression. Therefore, considerable research effort has been expended in elucidating the protein interaction networks of *S. cerevisiae*, in the hope that knowledge of their structure and topology will help us to understand their functions and evolutionary history (Wuchty, 2004). A number of large protein interaction datasets are now available – some (Uetz *et al.*, 2000; Ito *et al.*, 2001) based on the yeast two-hybrid system (Fields and Song, 1989; see Chapter 7 in this volume) and others based on various combinations of affinity chromatography and mass spectrometry (Gavin *et al.*, 2002, 2006; Ho *et al.*, 2002; Krogan *et al.*, 2006). These datasets all contain a fair amount of noise, usually in the form of false-positive interactions and it is prudent to combine different types of evidence in order to validate any particular interaction of interest (von Mering *et al.*, 2002). A number of yeast interaction databases have been established that should facilitate such validation exercises. Unfortunately, most do not explicitly define what is meant by 'an interaction' and some even conflate protein–protein interactions with genetic interactions, such as synthetic lethality (see Chapter 16, this volume). This is particularly unfortunate since it is rarely the case that there is any physical interaction between the protein products of two genes that interact genetically. Fortunately a new, curated dataset of yeast protein–protein interaction has become available recently (Reguly *et al.*, 2006).

In order to derive general rules from these large protein interaction datasets, researchers have used a graph theory approach (Barabasi and Oltvai, 2004), in which the proteins within the network are represented as nodes, with interacting proteins connected by undirected links (edges). Such an analysis has demonstrated that protein interaction networks have 'scale-free' properties – that is the network is a mixture of highly connected 'hub' proteins and more sparsely connected 'peripheral' proteins. Most of these analyses tend to be overly simplistic and, moreover, are very sensitive to both the definition of 'interaction' and the dataset selected for the analysis. For instance, it has been claimed that there is a negative correlation between the connectivity of a yeast protein and the average connectivity of its binding partners, and this has been represented as an adaptation that prevents the propagation of deleterious perturbations through the network (Maslov and Sneppen, 2002). A number of workers have shown that such a conclusion is sensitive to both the dataset detected and the general model used to represent protein interactions (Hakes *et al.*, 2005; Pereira-Leal *et al.*, 2005; Batada *et al.*, 2006). While it is clear that graph theory will be of great use in analysing protein interaction and other biological networks, its results should be interpreted with caution. The field is sorely in need of more hard data; in the case of protein–protein interactions, this should be in the form of X-ray structures for more protein complexes.

◆◆◆◆◆◆ V. METABOLOMICS

The metabolome is, in many ways, the closest level of 'omic analysis to function (Oliver *et al.*, 1998). Moreover, methods of physical analysis, such as NMR spectrometry and mass spectrometry, especially when combined with prior chromatographic or electrophoretic separations, hold out a realistic hope of analysing the great majority of yeast's metabolites (see Chapter 21, this volume). There are two kinds of metabolome in yeast that may currently be analysed (Kell *et al.*, 2005; Nielsen and Oliver, 2005): the endometabolome – all the low-molecular-weight intermediates inside the cell, and the exometabolome – all such compounds that are excreted from the cell into the growth medium. The importance of the exometabolome should not be overlooked – for a start, it's what makes beer and wine taste good! From an experimental viewpoint, the exometabolome is much less complex than the endometabolome (and therefore is easier to analyse) and, since extraction is unnecessary, it is very amenable to automation for high-throughput studies.

Both endometabolome (Raamsdonk *et al.*, 2001; Bundy *et al.*, 2006) and exometabolome (Allen *et al.*, 2003; Bundy *et al.*, 2006) have been used to classify deletion mutants in order to associate genes of related function. The exometabolome has also been used to classify drugs according to their site of action. Indeed, it is tempting to speculate that the target site of any drug or inhibitor that affects the yeast cell should be identifiable by a combination of exometabolomic and haploinsufficiency (see Section II.C) analyses.

For all the considerable advantages of metabolomics, there remain three major problems that will hinder its exploitation, probably for some time to come. First, there is the lack of a direct connection with the genome. This may be solved experimentally by the use of mutants, sometimes in combination with inhibitors. It may also be solved bioinformatically by integrating metabolomic data with other kinds of 'omic data. So far there have been attempts to do this with the transcriptome (Patil and Nielsen, 2005; Pir *et al.*, 2006) and with both transcriptome and proteome (Castrillo *et al.*, 2007), but much remains to be done; facile and robust data integration pipelines have yet to be developed. The second problem is one of quantitation – there is a very real need to develop community-wide protocols for metabolite quantification (Neilsen and Oliver, 2005). This will probably involve the preparation and distribution of standard mixtures of pure metabolic compounds. However, given the chemical heterogeneity of metabolites, the achievement of effective standard approaches to their quantitation is a significant technological challenge. It is also a sociological one – community-wide standards for analyses, compound libraries, and the recording of metadata (Jenkins *et al.*, 2004) need to be established and widely adopted.

The final problem is likely to take the longest to solve. At the moment, we measure the endometabolome as a homogeneous whole. But, of course, it is nothing of the sort; the metabolites are distributed between the cytosol and the membrane-bound organelles (vacuole, mitochondria, peroxisomes, nucleus). Eventually, we will need to measure the distribution of metabolites between these compartments and, indeed, some metabolic models (Duarte *et al.*, 2004) already partition the endometabolome. What is easy to do *in silico* is very hard to do in the laboratory. Yeast, largely because of its tough cell wall, is not a very favourable organism with which to perform sub-cellular fractionation. Carrying out such fractionations without the loss of metabolites from, or cross-contamination between, compartments represents a huge technical challenge.

♦♦♦♦♦♦ VI. THE MINIMAL GENOME AND THE E-YEAST

The comprehensive identification of synthetic gene interactions by Charlie Boone and his colleagues (see Chapter 16, this volume; Tong *et al.*, 2001, 2004), as well as the exploitation of the stoichiometric model to predict of gene–gene and higher order synthetic interactions in genes encoding components of the yeast metabolic network (Harrison *et al.*, 2006), mean that it should be possible to predict the minimal set of genes necessary for yeast's metabolism when it grows, say, on a YPD plate, and to construct a strain with that minimal gene set. That the construction of a minimal metabolism is, at least, theoretically possible has been demonstrated in an *in silico* evolution exercise in which, starting with the gene set encoding the enzymes of the *E. coli* metabolic network, minimal metabolic networks were evolved that satisfied the nutritional environments of the obligately endosymbiotic bacteria, *Buchnera*, and *Wigglesworthia* (Pál *et al.*, 2006). There were two main lessons from that theoretical study, and both were foreshadowed by an earlier assessment of the nature of a minimal genome (Oliver, 1996b).

First, there is no single minimal metabolic network. There are many solutions to the problem of the minimal set and Pál *et al.* (2006) generated 500 different solutions from their simulations. However, all solutions shared a common core of some 88% of their genes. There are many different solutions because, in biology (unlike in engineering), history counts – the *order* in which the genes are deleted, or lost, determines which genes may be removed later in the process and still retain viability. In classical evolutionary theory, such a phenomenon is described as 'contingency' – i.e. the ability to lose gene *x* is contingent upon whether gene *y* has already been lost.

Second, the number of genes in the minimal set required to produce a functioning metabolic network was *ca.* 260, which is twice as many as the number of genes in the *E. coli* metabolic network that have been shown to be essential by single-gene deletion analyses. In other words, there are more essential functions than there are essential genes.

However, it seems clear that we should be able to use the stoichiometric model of the yeast metabolic network (Forster *et al.*, 2003; Kuepfer *et al.*, 2005), and the facile techniques of gene replacement using re-cyclable markers (Güldener *et al.*, 1996; Toh-e, 1995; Storici *et al.*, 1999; Delneri *et al.*, 2000), to construct a yeast strain with a minimal metabolism. Why should one bother to do such a thing – apart from sheer technical bravado? First, it is an excellent way to test the stoichiometric model. Even the prediction of synthetic interactions has revealed deficiencies in the metabolic model and in the annotation of the yeast genome (Harrison *et al.*, 2006), and it is certainly important to refine both. Second, a yeast strain with a minimal metabolism should be a useful research tool. For instance, many of the early achievements of systems biology will be in the modelling of metabolism using classical ordinary or partial differential equations. The major impediment to this modelling exercise is that, far from having too much data (as is often claimed), we have far too little. Most of the parameters required to construct such models (things like binding coefficients, and kinetic constants for enzyme reactions) are completely unknown and, at present, have to be estimated. If such models are to be genuinely predictive, then these parameter values will have to be determined empirically, and the acquisition of such data will be very hard work. While serious efforts are underway (including at the Manchester Centre for Integrative Systems Biology) to acquire such data for yeast, it will be even more difficult to garner it for less malleable organisms and, especially, for humans. One use of a yeast strain with a minimal genome, then, will be to replace specific components of the metabolic network with their human equivalents. If we know the parameter values for the original yeast component, then we should be able to estimate those values for its human equivalent with accuracy and confidence. Thus, much time, effort, and expense will be saved in constructing predictive models of the human metabolic network.

Need all of this stop at metabolic networks? There is already much talk of the e-yeast – a comprehensive computer model of the yeast cell that makes accurate predictions and has genuine explanatory power. While we do not know if such an enterprise will be successful, it seems certain that, even if it eventually fails, it will do so in a way that is extremely informative. Currently, in addition to models of individual metabolic pathways (notably glycolysis, Teusink *et al.*, 2000; Pritchard and Kell, 2002; and sphingolipid biosynthesis, Alvarez-Vasquez *et al.*, 2005), there are good mathematical models for the yeast cell cycle (e.g. Ciliberto *et al.*, 2003), and for some signal transduction pathways (e.g. Kofahl and Klipp, 2004; Klipp *et al.*, 2005). What has not been done, as yet, is to integrate, say, signal transduction and metabolism in a single model. I have suggested earlier (Oliver, 2006) that what is needed now is a coarse-grained model of the yeast cell, which can act as a framework for the integration of the sub-system models (of glycolysis or the cell cycle) that are the components of the larger endeavour.

In my 2006 article, I suggested an approach, based on Metabolic Control Analysis, whereby the components of this coarse-grained model could be identified. But what form should this crude, but comprehensive, model take? It clearly cannot be based on ordinary differential equations, since the parameterisation problem will defeat us. Nor, despite its appealing simplicity and remarkable predictive facility, can the (steady-state) stoichiometric modelling approach be used – that can really only be applied to metabolic networks. The answer may well lie in a *logical* model, of the kind that we used to model metabolism in the Robot Scientist project (King *et al.*, 2004). This approach need not be confined to metabolism, and the form of the model is not so very far removed from the boxes-and-arrows type of models beloved of molecular biologists, and that (alone) should recommend it. The real message is that, if it is achievable, there will be many paths to the e-yeast, and we shall learn a lot exploring each of them. Thus, there is plenty for the young yeast molecular geneticist to do, and old stagers who claim that everything important has been done and only the details remain to be tidied up should be firmly, if politely, ignored.

Acknowledgements

This chapter was written during a period as a CNRS Visiting Director of Research at Institut de Génétique et Microbiologie, Université Paris XI, Orsay, France. I am grateful to Monique Bolotin-Fukuhara for her guidance and support during that period.

Research in my own laboratory on functional analysis and systems biology in yeast has been supported by funds from BBSRC, EC, EPSRC, UK Department of Trade & Industry (through a Beacon Award), the Wellcome Trust, and AstraZeneca plc.

References

Allen, J., Davey, H. M., Broadhurst, D., Heald, J. K., Rowland, J. J., Oliver, S. G. and Kell, D. B. (2003). High-throughput classification of yeast mutants for functional genomics using metabolic footprinting. *Nat. Biotechnol.* **21**, 692–696.

Alvarez-Vasquez, F., Sims, K. J., Cowart, L. A., Okamoto, Y., Voit, E. O. and Hannun, Y. A. (2005). Simulation and validation of modelled sphingolipid metabolism in *Saccharomyces cerevisiae*. *Nature* **433**, 425–430.

Baetz, K., McHardy, L., Gable, K., Tarling, T., Reberioux, D., Bryan, J., Andersen, R. J., Dunn, T., Hieter, P. and Roberge, M. (2004). Yeast genome-wide drug-induced haploinsufficiency screen to determine drug mode of action. *Proc. Natl. Acad. Sci. USA* **101**, 4525–4530.

Bar-Even, A., Paulsson, J., Maheshri, N., Carmi, M., O'Shea, E., Pilpel, Y. and Barkai, N. (2006). Noise in protein expression scales with natural protein abundance. *Nat. Genet.* **38**, 636–643.

Barabasi, A. L. and Oltvai, Z. N. (2004). Network biology: understanding the cell's functional organization. *Nat. Rev. Genet.* **5**, 101–113.

Batada, N. N., Regally, T., Breitkreutz, A., Boucher, L., Breitkreutz, B. J., Hurst, L. D. and Tyers, M. (2006). Stratus not altocumulus: a new view of the yeast protein interaction network. *PLOS Biol.* **4**, e317 DOI: 10.1371/journal.pbio.0040317.

Belli, G., Aldea, M. and Herrero, E. (1998a). Functional analysis of yeast essential genes using a promoter substitution cassette and the tetracycline-regulatable dual expression system. *Yeast* **14**, 1127–1138.
Belli, G., Gari, E., Piedrafita, L., Aldea, M. and Herrero, E. (1998b). An activator/repressor dual system allows tight tetracycline-regulated gene expression in budding yeast. *Nucleic Acids Res.* **26**, 942–947.
Beynon, R. J., Doherty, M. K., Pratt, J. M. and Gaskell, S. J. (2005). Multiplexed absolute quantification in proteomics using artificial QCAT proteins of concatenated signature peptides. *Nat. Methods* **2**, 587–589.
Bundy, J. G., Papp, B., Harmston, R., Browne, R. A., Clayson, E.M., Burton, N., Reece, R. J., Oliver, S. G. and Brindle, K. M. (2006). Identification of metabolic modules in *Saccharomyces cerevisiae* using NMR-based metabolite profiling. *Genome Res.* (In Press.)
Castrillo, J. I., Zeef, L. A., Hoyle, D. C., Zhang, N., Hayes, A., Gardner, D. C. J., Cornell, M. J., Petty, J., Hakes, L., Wardleworth, L. *et al.* (2007). Growth control of the eukaryote cell: A systems biology study in yeast. *J. Biol.* (In Press.)
Ciliberto, A., Novak, B. and Tyson, J. J. (2003). Mathematical model of the morphogenesis checkpoint in budding yeast. *J. Cell Biol.* **163**, 1243–1254.
Cliften, P., Sudarsanam, P., Desikan, A., Fulton, L., Fulton, B., Majors, J., Waterston, R., Cohen, B. A. and Johnston, M. (2003). Finding functional features in *Saccharomyces* genomes by phylogenetic footprinting. *Science* **301**, 71–76.
Delneri, D. (2004). The use of yeast mutant collections in genome profiling and large-scale functional analysis. *Curr. Genom.* **5**, 59–65.
Delneri, D., Gardner, D. C. J. and Oliver, S. G. (1999). Analysis of the seven-member *AAD* gene set demonstrates that genetic redundancy in yeast may be more apparent than real. *Genetics* **153**, 1591–1600.
Delneri, D., Hoyle D. C., Gkargkas, K., Cross, E. J. M., Rash, B., Zeef, L., Leong, H.-S., Hayes, A., Kell, D. B., Griffith, G. W. and Oliver, S. G. (2006). Identification and characterisation of high flux control (HFC) genes of *Saccharomyces cerevisiae* through competition analysis in continuous cultures. (Submitted for publication.)
Delneri, D., Tomlin, G. C., Wixon, J. L., Hutter, A., Sefton, M., Louis, E. J. and Oliver, S. G. (2000). Exploring redundancy in the yeast genome: an improved strategy for use of the *cre-lox*P system. *Gene* **252**, 127–135.
Dietrich, F. S., Voegeli, S., Brachat, S., Lerch, A., Gates, K., Steiner, S., Mohr, C., Pohlmann, R., Luedi, P., Choi, S. *et al.* (2004). The *Ashbya gossypii* genome as a tool for mapping the ancient *Saccharomyces cerevisiae* genome. *Science* **304**, 304–307.
Duarte, N. C., Herrgard, M. J. and Palsson, B. O. (2004). Reconstruction and validation of *Saccharomyces cerevisiae* iND750, a fully compartmentalized genome-scale metabolic model. *Genome Res.* **14**, 1298–1309.
Dujon, B., Sherman, D., Fischer, G., Durrens, P., Casaregola, S., Lafontaine, I., de Montigny, J., Marck, C., Neuveglise, C., Talla, E. *et al.* (2004). Genome evolution in yeasts. *Nature* **430**, 35–44.
Edwards-Ingram, L. C., Gent, M. E., Hoyle, D. C., Hayes, A., Stateva, L. I. and Oliver, S. G. (2004). Comparative genomic hybridisation provides new insights into the molecular taxonomy of the *Saccharomyces 'sensu stricto'* complex. *Genome Res.* **14**, 1043–1051.
Elliott, S. G. and McLaughlin, C. S. (1979). Synthesis and modification of proteins during the cell cycle of the yeast *Saccharomyces cerevisiae*. *J. Bacteriol.* **137**, 1185–1190.

Fields, S. and Song, O. (1989). A novel genetic system to detect protein–protein interactions. *Nature* **340**, 245–246.

Fitzpatrick, D. A., Logue1, M. E., Stajich, J. E. and Butler, G. (2006). A fungal phylogeny based on 42 complete genomes derived from supertree and combined gene analysis. *BMC Evol. Biol.* **6**, 99 doi:10.1186/1471-2148-6-99.

Forster, J., Famili, I., Palsson, B. O. and Nielsen, J. (2003). Genome-scale reconstruction of the *Saccharomyces cerevisiae* metabolic network. *Omics* **7**, 193–202.

Garrels, J. I. (1979). 2-dimensional gel-electrophoresis and computer-analysis of proteins synthesized by clonal cell-lines. *J. Biol. Chem.* **254**, 7961–7977.

Gavin, A. C., Aloy, P., Grandi, P., Krause, R., Boesche, M., Marzioch, M., Rau, C., Jensen, L. J., Bastuck, S., Dumpelfeld, B. *et al.* (2006). Proteome survey reveals modularity of the yeast cell machinery. *Nature* **440**, 631–636.

Gavin, A. C., Bosche, M., Krause, R., Grandi, P., Marzioch, M., Bauer, A., Schultz, J., Rick, J. M., Michon, A. M., Cruciat, C. M. *et al.* (2002). Functional organization of the yeast proteome by systematic analysis of protein complexes. *Nature* **415**, 141–147.

Gelperin, D. M., White, M. A., Wilkinson, M. L., Kon, Y., Kung, L. A., Wise, K. J., Lopez-Hoyo, N., Jiang, L., Piccirillo, S., Yu, H. *et al.* (2005). Biochemical and genetic analysis of the yeast proteome with a movable ORF collection. *Genes Dev.* **19**, 2816–2826.

Ghaemmaghami, S., Huh, W. K., Bower, K., Howson, R. W., Belle, A., Dephoure, N., O'Shea, E. K. and Weissman, J. S. (2003). Global analysis of protein expression in yeast. *Nature* **425**, 737–741.

Giaever, G., Chu, A. M., Ni, L., Connelly, C., Riles, L., Veronneau, S., Dow, S., Lucau Danila, A., Anderson, K., Andre, B., Arkin, A. P. *et al.* (1999). Genomic profiling of drug sensitivities via induced haploinsufficiency. *Nat. Genet.* **21**, 278–283.

Giaever, G., Chu, A. M., Ni, L., Connelly, C., Riles, L., Veronneau, S., Dow, S., Lucau-Danila, A., Anderson, K., Andre, B. *et al.* (2002). Functional profiling of the *Saccharomyces cerevisiae* genome. *Nature* **418**, 387–391.

Giaever, G., Flaherty, P., Kumm, J., Proctor, M., Nislow, C., Jaramillo, D. F., Chu, A. M., Jordan, M. I., Arkin, A. P. and Davis, R. W. (2004). Chemogenomic profiling: identifying the functional interactions of small molecules in yeast. *Proc. Natl. Acad. Sci. USA* **101**, 793–798.

Goffeau, A. (2004). Evolutionary genomics: seeing double. *Nature* **430**, 25–26.

Goffeau, A., Barrell, B. G., Bussey, H., Davis, R. W., Dujon, B., Feldmann, H., Galibert, F., Hoheisel, J. D., Jacq, C., Johnston, M. *et al.* (1996). Life with 6000 genes. *Science* **274**, 546–563.

Gresham, D., Ruderfer, D. M., Pratt, S. C., Schacherer, J., Dunham, M. J., Botstein, D. and Kruglyak, L. (2006). Genome-wide detection of polymorphisms at nucleotide resolution with a single DNA microarray. *Science* **311**, 1932–1936.

Güldener, U., Heck, S., Fiedler, T., Beinhauer, J. and Hegemann, J. H. (1996). A new efficient gene disruption cassette for repeated use in budding yeast. *Nucleic Acids Res.* **24**, 2519–2524.

Hakes, L., Robertson, D. L. and Oliver, S. G. (2005). Effect of dataset selection on the topological interpretation of protein interaction networks. *BMC Genomics* **6**, 131.

Harrison, R., Papp, B., Pál, C., Oliver, S. G. and Delneri, D. (2006). Plasticity of genetic interactions in metabolic networks of yeast. *Proc. Natl. Acad. Sci. USA* doi: 10.1073/pnas.0607153104.

Hauser, N. C., Scheideler, M., Matysiak, S., Vingron, M. and Hoheisel, J. D. (1999). DNA arrays for transcriptional profiling. In: *Automation: Genomic and Functional Analyses. Methods in Microbiology*, vol. 28 (A. G. Craig and J. D. Hoheisel, eds), pp. 193–204. Academic Press, London.

Hayes, A., Zhang, N., Wu, J., Butler, P. R., Hauser, N. C., Hoheisel, J. D., Lim, F., Sharrocks, A. D. and Oliver, S. G. (2002). Hybridization array technology coupled with chemostat culture: tools to interrogate gene expression in *S. cerevisiae*. *Methods* **26**, 281–290.

Hedges, S. B., Blair, J. E., Venturi, M. L. and Shoe, J. L. (2004). A molecular timescale of eukaryote evolution and the rise of complex multicellular life. *BMC Evol. Biol.* **4**, 2.

Ho, Y., Gruhler, A., Heilbut, A., Bader, G. D., Moore, L., Adams, S. L., Millar, A., Taylor, P., Bennett, K., Boutilier, K. *et al.* (2002). Systematic identification of protein complexes in *Saccharomyces cerevisiae* by mass spectrometry. *Nature* **415**, 180–183.

Huh, W.-K., Falvo, J. V., Gerke, L. C., Carroll, A. S., Howson, R. W., Weissman, J. S. and O'Shea, E. K. (2003). Global analysis of protein localization in budding yeast. *Nature* **425**, 686–691.

Ito, T., Chiba, T., Ozawa, R., Yoshida, M., Hattori, M. and Sakaki, Y. (2001). A comprehensive two-hybrid analysis to explore the yeast protein interactome. *Proc. Natl. Acad. Sci. USA* **98**, 4569–4574.

Jenkins, H., Hardy, N., Beckmann, M., Draper, J., Smith, A. R., Taylor, J., Fiehn, O., Goodacre, R., Bino, R., Hall, R. *et al.* (2004). A proposed framework for the description of plant metabolomics experiments and their results. *Nat. Biotechnol.* **22**, 1601–1606.

Kell, D. B., Brown, M., Davey, H. M., Dunn, W. B., Spasic, I. and Oliver, S. G. (2005). Metabolic footprinting and systems biology: the medium is the message. *Nat. Rev. Microbiol.* **3**, 557–565.

Kellis, M., Birren, B. W. and Lander, E. S. (2004). Proof and evolutionary analysis of ancient genome duplication in the yeast *Saccharomyces cerevisiae*. *Nature* **428**, 617–624.

Kellis, M., Patterson, N., Endrizzi, M., Birren, B. and Lander, E. S. (2003). Sequencing and comparison of yeast species to identify genes and regulatory elements. *Nature* **423**, 241–254.

King, R. D., Whelan, K. E., Jones, F. M., Reiser, P. G. K., Bryant, C. H., Muggleton, S. H., Kell, D. B. and Oliver, S. G. (2004). Functional genomic hypothesis generation and experimentation by a robot scientist. *Nature* **427**, 247–255.

Klipp, E., Nordlander, B., Kruger, R., Gennemark, P. and Hohmann, S. (2005). Integrative model of the response of yeast to osmotic shock. *Nat. Biotechnol.* **23**, 975–982.

Knop, M. (2006). Evolution of the hemiascomycete yeasts: on life styles and the importance of inbreeding. *Bioassays* **28**, 696–708.

Kofahl, B. and Klipp, E. (2004). Modelling the dynamics of the yeast pheromone pathway. *Yeast* **21**, 831–850.

Krogan, N. J., Cagney, G., Yu, H., Zhong, G., Guo, X., Ignatchenko, A., Li, J., Pu, S., Datta, N., Tikuisis, A. P. *et al.* (2006). Global landscape of protein complexes in the yeast *Saccharomyces cerevisiae*. *Nature* **440**, 637–643.

Kuepfer, L., Sauer, U. and Blank, L. M. (2005). Metabolic functions of duplicate genes in *Saccharomyces cerevisiae*. *Genome Res.* **15**, 1421–1430.

Kurtzman, C. P. and Robnett, C. J. (2003). Phylogenetic relationships among yeasts of the '*Saccharomyces* complex' determined from multigene sequence analyses. *FEMS Yeast Res.* **3**, 417–432.

Lim, F. L., Hayes, A., West, A. G., Pic-Taylor, A., Darieva, Z., Morgan, B. A., Oliver, S. G. and Sharrocks, A. D. (2003). Mcm1p-induced DNA bending regulates the formation of ternary transcription factor complexes. *Mol. Cell. Biol.* **23**, 450–461.

Liu, H. P., Krizek, J. and Bretscher, A. (1992). Construction of a GAL1-regulated yeast cDNA expression library and its application to the identification of genes whose overexpression causes lethality in yeast. *Genetics* **132**, 665–673.

Lockhart, D. J., Dong, H., Byrne, M. C., Follettie, M. T., Gallo, M. V., Chee, M. S., Mittmann, M., Wang, C., Kobayashi, M., Horton, H. *et al.* (1996). Expression monitoring by hybridization to high-density oligonucleotide arrays. *Nat. Biotechnol.* **14**, 1675–1680.

Maslov, S. and Sneppen, K. (2002). Specificity and stability in topology of protein networks. *Science* **296**, 910–913.

McLaughlin, T., Siepen, J. A., Selley, J., Lynch, J. A., Lau, K. W., Yin, H. J., Gaskell, S. J. and Hubbard, S. J. (2006). PepSeeker: a database of proteome peptide identifications for investigating fragmentation patterns. *Nucleic Acids Res.* **34**, D649–D654.

Mewes, H. W., Albermann, K., Bähr, M., Frishman, D., Gleissner, D., Hani, J., Heumann, K., Kleine, K., Maierl, A., Oliver, S. G. *et al.* (1997). Overview of the yeast genome. *Nature* **387**, 7–65.

Mnaimneh, A., Davierwala, A. P., Haynes, J., Moffat, J., Peng, W., Zhang, W., Yang, X., Pootoolal, J., Chua, G., Lopez, A. *et al.* (2004). Exploration of essential gene functions via titratable promoter alleles. *Cell* **118**, 31–44.

Mortimer, R. K. and Johnston, J. R. (1986). Genealogy of principal strains of the yeast. Genetic stock center. *Genetics* **113**, 35–43.

Mortimer, R. K., Schild, D., Contopoulou, C. R. and Kans, J. A. (1989). Genetic map of *Saccharomyces cerevisiae*, edition 10. *Yeast* **5**, 321–403.

Murray, A. E., Lies, D., Li, G., Nealson, K., Zhou, J. and Tiedje, J. M. (2001). DNA/DNA hybridization to microarrays reveals gene-specific differences between closely related microbial genomes. *Proc. Natl. Acad. Sci. USA* **98**, 9853–9858.

Nielsen, J. and Oliver, S. (2005). The next wave in metabolome analysis. *Trends Biotechnol.* **23**, 544–546.

O'Farrell, P. H. (1975). High-resolution 2-dimensional electrophoresis of proteins. *J. Biol. Chem.* **250**, 4007–4021.

Oliver, S. G. (1996a). A network approach to the systematic analysis of gene function. *Trends Genet.* **12**, 241–242.

Oliver, S. G. (2006). From genomes to systems: the path with yeast. *Philos. Trans. R. Soc. Lond. B Biol. Sci.* **361**, 477–482.

Oliver, S. G. (2006). From DNA sequence to biological function. *Nature* **379**, 597–600.

Oliver, S. G., van der Aart, Q. J., Agostoni-Carbone, M. L., Aigle, M., Alberghina, L., Alexandraki, D., Antoine, G., Anwar, R., Ballesta, J. P., Benit, P. *et al.* (1992). The complete DNA sequence of yeast chromosome III. *Nature* **357**, 38–46.

Oliver, S. G., Winson, M. K., Kell, D. B. and Baganz, F. (1998). Systematic functional analysis of the yeast genome. *Trends Biotechnol.* **16**, 373–378.

Pál, C., Papp, B., Lercher, M. J., Csermely, P., Oliver, S. G. and Hurst, L. D. (2006). Chance and necessity in the evolution of minimal metabolic networks. *Nature* **440**, 667–670.

Patil, K. R. and Nielsen, J. (2005). Uncovering transcriptional regulation of metabolism by using metabolic network topology. *Proc. Natl. Acad. Sci. USA* **102**, 2685–2689.

Pereira-Leal, J. B., Audit, B., Peregrin-Alvarez, J. M. and Ouzounis, C. A. (2005). An exponential core in the heart of the yeast protein interaction network. *Mol. Biol. Evol.* **22**, 421–425.

Perrot, M., Sagliocco, F., Mini, T., Monribot, C., Schneider, U., Shevchenko, A., Mann, M., Jeno, P. and Boucherie, H. (1999). Two-dimensional gel protein database of *Saccharomyces cerevisiae* (update 1999). *Electrophoresis* **20**, 2280–2298.

Pir, P., Kırdar, B., Hayes, A., Önsan, Z.İ., Ülgen, K.Ö. and Oliver, S. G. (2006). Integrative investigation of metabolic and transcriptomic data. *BMC Bioinformatics* **7**, 203 doi:10.1186/1471-2105-7-203.

Pritchard, L. and Kell, D. B. (2002). Schemes of flux control in a model of *Saccharomyces cerevisiae* glycolysis. *Eur. J. Biochem.* **269**, 3894–3904.

Raamsdonk, L. M., Teusink, B., Broadhurst, D., Zhang, N., Hayes, A., Walsh, M. C., Berden, J. A., Brindle, K. M., Kell, D. B., Rowland, J. J. *et al.* (2001). A functional genomics strategy that uses metabolome data to reveal the phenotype of silent mutations. *Nat. Biotechnol.* **19**, 45–50.

Reguly, T., Breitkreutz, A., Boucher, L., Breitkreutz, B. J., Hon, G. C., Myers, C. L., Parsons, A., Friesen, H., Oughtred, R., Tong, A., Stark, C. *et al.* (2006). Comprehensive curation and analysis of global interaction networks in *Saccharomyces cerevisiae. J. Biol.* **5**, 11.

Rokas, A., Williams, B. L., King, N. and Carroll, S. B. (2003). Genome-scale approaches to resolving incongruence in molecular phylogenies. *Nature* **425**, 798–804.

Ross, P. L., Huang, Y. N., Marchese, J. N., Williamson, B., Parker, K., Hattan, S., Khainovski, N., Pillai, S., Dey, S., Daniels, S. *et al.* (2004). Multiplexed protein quantitation in *Saccharomyces cerevisiae* using amine-reactive isobaric tagging reagents. *Mol. Cell. Proteomics* **3**, 1154–1169.

Ross-Macdonald, P., Coelho, P. S., Roemer, T., Agarwal, S., Kumar, A., Jansen, R., Cheung, K. H., Sheehan, A., Symoniatis, D., Umansky, L. *et al.* (1999). Large-scale analysis of the yeast genome by transposon tagging and gene disruption. *Nature* **402**, 413–418.

Ross-Macdonald, P., Sheehan, A., Roeder, G. S. and Snyder, M. (1997). A multipurpose transposon system for analyzing protein production, localization, and function in *Saccharomyces cerevisiae. Proc. Natl. Acad. Sci. USA.* **94**, 190–195.

Schena, M., Shalon, D., Davis, R. W. and Brown, P. O. (1995). Quantitative monitoring of gene expression patterns with a complementary DNA microarray. *Science* **270**, 467–470.

Shadforth, I. P., Dunkley, T. P. J., Lilley, K. S. and Bessant, C. (2005). i-Tracker: for quantitative proteomics using iTRAQ™. *BMC Genomics* **6**, 145.

Shoemaker, D. D., Lashkari, D. A., Morris, D., Mittmann, M. and Davis, R. W. (1996). Quantitative phenotypic analysis of yeast deletion mutants using a highly parallel molecular bar-coding strategy. *Nat. Genet.* **14**, 450–456.

Sicheritz-Ponten, T. and Andersson, S. G. (2001). A phylogenomic approach to microbial evolution. *Nucleic Acids Res.* **29**, 545–552.

Smith, V., Chou, K. N., Lashkari, D., Botstein, D. and Brown, P. O. (1996). Functional analysis of the genes of yeast chromosome V by genetic footprinting. *Science* **274**, 2069–2074.

Sopko, R., Huang, D., Preston, N., Chua, G., Papp, B., Kafadar, K., Snyder, M., Oliver, S. G., Cyert, M., Hughes, T. R. *et al.* (2006). Mapping pathways and phenotypes by systematic gene overexpression. *Mol. Cell* **21**, 319–330.

Storici, F., Coglievina, M. and Bruschi, C. V. (1999). A 2-micron DNA-based marker recycling system for multiple gene disruption in the yeast *Saccharomyces cerevisiae*. *Yeast* **15**, 271–283.

Teusink, B., Baganz, F., Westerhoff, H. V. and Oliver, S. G. (1998). Metabolic control analysis as a tool in the elucidation of the function of novel genes. In: *Yeast Gene Analysis. Methods in Microbiology*, vol. 26 (A. J. P. Brown and M. F. Tuite, eds), pp. 297–336. Academic Press, London.

Teusink, B., Passarge, J., Reijenga, C. A., Esgalhado, E., van der Weijden, C. C., Schepper, M., Walsh, M. C., Bakker, B. M., van Dam, K., Westerhoff, H. V. *et al.* (2000). Can yeast glycolysis be understood in terms of *in vitro* kinetics of the constituent enzymes? Testing biochemistry. *Eur. J. Biochem.* **267**, 5313–5329.

Toh-e, A. (1995). Construction of a marker gene cassette which is repeatedly usable for gene disruption in yeast. *Curr. Genet.* **27**, 293–297.

Tong, A. H., Evangelista, M., Parsons, A. B., Xu, H., Bader, G. D., Page, N., Robinson, M., Raghibizadeh, S., Hogue, C. W., Bussey, H. *et al.* (2001). Systematic genetic analysis with ordered arrays of yeast deletion mutants. *Science* **294**, 2364–2368.

Tong, A. H., Lesage, G., Bader, G. D., Ding, H., Xu, H., Xin, X., Young, J., Berriz, G. F., Brost, R. L., Chang, M *et al.* (2004). Global mapping of the yeast genetic interaction network. *Science* **303**, 808–813.

Uetz, P., Giot, L., Cagney, G., Mansfield, T. A., Judson, R. S., Knight, J. R., Lockshon, D., Narayan, V., Srinivasan, M., Pochart, P. *et al.* (2000). A comprehensive analysis of protein–protein interactions in *Saccharomyces cerevisiae*. *Nature* **403**, 623–627.

Varela, C., Cardenas, J., Melo, F. and Agosin, E. (2005). Quantitative analysis of wine yeast gene expression profiles under winemaking conditions. *Yeast* **22**, 369–383.

Velculescu, V. E., Zhang, L., Zhou W., Vogelstein, J., Basrai, M. A., Bassett, D. E., Hieter, P., Vogelstein, B. and Kinzler, K. W. (1997). Characterization of the yeast transcriptome. *Cell* **88**, 243–251.

von Mering, C., Krause, R., Snel, B., Cornell, M., Oliver, S. G., Fields, S. and Bork, P. (2002). Comparative assessment of large-scale data sets of protein–protein interactions. *Nature* **417**, 399–403.

Wach, A. (1996). PCR-synthesis of marker cassettes with long flanking homology regions for gene disruptions in *S. cerevisiae*. *Yeast* **12**, 259–265.

Wach, A., Brachat, A., Pohlmann, R. and Philippsen, P. (1994). New heterologous modules for classical or PCR-based gene disruptions in *Saccharomyces cerevisiae*. *Yeast* **10**, 1793–1808.

Washburn, M. P., Wolters, D. and Yates, J. R., 3rd. (2001). Large-scale analysis of the yeast proteome by multidimensional protein identification technology. *Nat. Biotechnol.* **19**, 242–247.

Winzeler, E. A., Castillo-Davis, C. I., Oshiro, G., Liang, D., Richards, D. R., Zhou, Y. and Hartl, D. L. (2003). Genetic diversity in yeast assessed with whole-genome oligonucleotide arrays. *Genetics* **163**, 79–89.

Winzeler, E. A., Shoemaker, D. D., Astromoff, A., Liang, H., Anderson, K., Andre, B., Bangham, R., Benito, R., Boeke, J. D., Bussey, H. *et al.* (1999). Functional characterization of the *S. cerevisiae* genome by gene deletion and parallel analysis. *Science* **285**, 901–906.

Wishart, J. A., Osborn, M., Gent, M. E., Yen, K., Vujovic, Z., Gitsham, P., Zhang, N., Miller, J. R. and Oliver, S. G. (2006). The relative merits of the *tet*O2 and *tet*O7 promoter systems for the functional analysis of heterologous genes in yeast and a compilation of essential yeast genes with *tet*O2 promoter substitutions. *Yeast* **23**, 325–331.

Wolfe, K. H. and Shields, D. C. (1997). Molecular evidence for an ancient duplication of the entire yeast genome. *Nature* **387**, 708–713.

Wuchty, S. (2004). Evolution and topology in the yeast protein interaction network. *Genome Res.* **14**, 1310–1314.

Zhu, H., Bilgin, M., Bangham, R., Hall, D., Casamayor, A., Bertone, P., Lan, N., Jansen, R., Bidlingmaier, S., Houfek, T. *et al.* (2001). Global analysis of protein activities using proteome chips. *Science* **293**, 2101–2105.

2 Yeast Genetics and Strain Construction

Ian Stansfield[1] and Michael JR Stark[2]

[1] School of Medical Sciences, Institute of Medical Sciences, University of Aberdeen, Foresterhill, Aberdeen, UK; [2] Division of Gene Regulation and Expression, College of Life Sciences, University of Dundee, Dundee, UK

◆◆

CONTENTS

Introduction
Growing Yeast
Planning Strain Construction
Yeast Mating
Sporulating Yeast
Tetrad Dissection and Phenotype Analysis
Modifying Strains by Plasmid Integration
Concluding Remarks

◆◆◆◆◆◆ I. INTRODUCTION

The yeast *Saccharomyces cerevisiae* is an excellent model system for the study of many aspects of eukaryote cell biology, genetics and biochemistry. It has a compact, 16-chromosome genome, of which genes represent approximately 70%. This in combination with a sophisticated armoury of molecular genetic techniques for its genetic manipulation, makes yeast a very approachable model system. Nevertheless, despite the ease with which genes can be knocked out, integrated, overexpressed and tagged in yeast, classical yeast genetic methods are still used routinely for constructing new yeast strains designed for particular experiments. This chapter discusses basic yeast strain construction and is particularly aimed at those researchers who may not be familiar with the methods for yeast mating, sporulation and tetrad analysis, which are needed to generate strains containing new combinations of alleles.

The standard approach to constructing new *S. cerevisiae* strains involves carrying out genetic crosses between haploid strains so

METHODS IN MICROBIOLOGY, VOLUME 36
0580-9517 DOI:10.1016/S0580-9517(06)36002-3

that new combinations of genetic markers can be generated. In *S. cerevisiae*, the haploid form can take one of the two mating types, denoted **a** and α. An **a**-type yeast can mate with an α-type yeast to form a diploid (Lindegren and Lindegren, 1943). Budding yeast are stable as diploids, but will undergo sporulation through a reductive meiotic division if starved for nitrogen. From each diploid cell, four haploid ascospores result from this process, two of each mating type, packaged into a single tetrahedral structure called an ascus. Yeast can replicate asexually for an indefinite period in both haploid and diploid forms.

Wild haploid yeast, that is to say not normal laboratory strains, can freely alternate between the two mating types and are described as heterothallic (sexually self-fertile). For this reason, yeast isolated from the wild would naturally tend to be diploid. The process of mating-type switching has been intensively studied and is mediated by a gene conversion event that copies mating type genetic information from one of the two silent chromosomal loci on chromosome III to the *MAT* locus, where it is actively expressed (see Haber, 1998 for a review). Thus, **a** mating type yeast are genetically *MAT***a**, while α mating type are *MAT*α. This gene conversion event is triggered by a double-stranded chromosomal break made by the yeast *HO* gene product, an endonuclease with a 24 nucleotide recognition site (Shibata *et al.*, 1984; Nickoloff *et al.*, 1986). However, almost all laboratory-adapted strains carry a mutant *ho* allele, preventing expression of an active endonuclease, and thus ensuring that they exist as stable **a** or α strains in the haploid state. In the following sections, the process of exploiting the yeast life cycle to create new strains and genetic backgrounds will be described.

♦♦♦♦♦♦ II. GROWING YEAST

Yeast are typically grown on either complete or defined media. Complete medium (YPD; 2% w/v peptone, 1% w/v yeast extract, 2% w/v glucose [dextrose] and for plates, 2 % w/v agar) contains all nutritional requirements including a rich source of nitrogen, amino acids and cofactors in the form of peptone and yeast extract. The carbon source can be varied at will, for instance from glucose to galactose (YPG).

Two types of defined medium are commonly used for growing yeast. Synthetic Dextrose (SD: 0.67% Difco yeast nitrogen base without amino acids, 2% w/v glucose and for plates, 2% w/v agar) and Synthetic Complete Dextrose or 'dropout' medium (SCD: SD medium containing 'dropout' mix). The yeast nitrogen base contains all vitamins and cofactors required, and a nitrogen source in the form of ammonium sulphate. Variants of yeast nitrogen base are also available without added ammonium sulphate, to allow researchers to define their own nitrogen source for particular experiments. The carbon source can again be varied according to experimental requirements and need not be glucose. When using SD medium, the amino acids or nucleotides required by a particular yeast strain to complement any auxotrophies have to be added separately. The concentrations at which these are added can vary a good deal from lab to lab, but the following concentrations (Sherman, 2002) bear some resemblance to intracellular amino acid pool sizes in yeast ($20\,\text{mg}\,\text{l}^{-1}$ W, H, R, M; $30\,\text{mg}\,\text{l}^{-1}$ Y, I, K; $50\,\text{mg}\,\text{l}^{-1}$ L; $60\,\text{mg}\,\text{l}^{-1}$ F; $100\,\text{mg}\,\text{l}^{-1}$ D, E; $150\,\text{mg}\,\text{l}^{-1}$ V; $200\,\text{mg}\,\text{l}^{-1}$ T; $400\,\text{mg}\,\text{l}^{-1}$ S; and $20\,\text{mg}\,\text{l}^{-1}$ adenine and uracil; where amino acids are defined by their single-letter code). Several commercial sources exist for the 'dropout mix' used in SC medium (e.g. Formedium: http://www.formedium.com/), which in its complete form contains all of the above nutritional supplements. Alternatively, dropout mix can be prepared from the individual components using a standard recipe (Amberg *et al.*, 2005). The point about defined media such as SD and SC in the context of strain construction is that they can be used to select for a diploid strain generated in a cross and to monitor the auxotrophic markers contributed to the progeny by the two parental strains – frequently, the particular alleles that it is wished to combine in a new strain will have been marked by one of the commonly used auxotrophic marker genes. In the case of SCD medium, dropout mixes in which a single supplement (e.g. tryptophan or uracil) or combinations of supplements have been omitted can be made or purchased, such that the presence of the corresponding marker gene (*TRP1* or *URA3*, for example) can be followed by the ability of strains to grow on the appropriate dropout plates.

◆◆◆◆◆◆ III. PLANNING STRAIN CONSTRUCTION

At its simplest, designing a new strain involves selecting the two haploid yeast strains that, between them contain the range of markers required in the final haploid. Once these haploids have been mated and the ensuing diploid sporulated spores containing the desired combination of markers can be selected from the population. The frequency with which a particular combination of alleles will arise in the population of spores from such a genetic cross is defined simply by Mendelian genetic principles. For a cross requiring the assortment of three different markers, one eighth of spores will carry the desired combination of alleles, but if combining six different loci is required then only 1 in 64 (i.e. 2^6) spores will have the correct genotype. However, this assumes independent assortment of the different markers. It is therefore crucial when planning the required final strain that the issues of genetic linkage are considered, and this can easily be achieved using the *Saccharomyces* Genome Database (http://www.yeastgenome.org/) to identify the location of all the genetic markers involved in the cross. If two of the genes do lie on the same chromosome, then provided the genetic map distance between them is at least 50 cM they should show essentially random assortment. If an experimental value for the genetic distance is not available, the likelihood of seeing free assortment in a genetic cross can be estimated by assuming that 3 kb of physical-map distance that corresponds to about 1 cM of genetic distance (Mortimer *et al.*, 1989). It should be borne in mind that this relationship is not preserved throughout the genome, and that there are variations in recombination frequency in some regions (hot-spots). Where there is genetic linkage between some of the markers involved in the cross then more (possibly many more) tetrads will need to be dissected to find the desired recombinants.

If a particular locus is required in the final strain, but it is either not marked or carries a phenotype that is hard to score, it may be advantageous to mark that locus using an easily detectable positive selectable marker such as resistance to the aminoglycoside G418 (geneticin), encoded by the KanMX cassette (Wach *et al.*, 1994), or resistance to nourseothricin encoded by the NatR cassette (Goldstein and McCusker, 1999). These markers can be integrated in a targeted manner using homologous recombination to mark a particular locus (see Chapter 4, this volume). When doing so, care should be taken not to interfere with the transcription terminator or promoter sequences of the gene or its neighbours.

An important aspect of strain construction is the consideration of the yeast genetic background of the parental strains that are to be crossed. Unfortunately there is no such thing as a standard wild-type *S. cerevisiae* strain, although much work has been carried out using a small number of different parental strains such as W303, S288C, SK1 and Σ1278B (see Sherman, 2002). However, these 'standard' genetic backgrounds differ in a number of important ways and it is therefore important to work in a single genetic background for any particular series of experiments – indeed, many labs work exclusively with strains in a single background (e.g. W303). A good example of such differences between different genetic backgrounds is provided by the *SSD1* gene, first discovered because some alleles support viability (*SSD1-v* alleles), while others (*ssd1-d* alleles) confer lethality to *sit4* knockouts (Sutton *et al.*, 1991). Subsequently, it has been found that the *SSD1* status in fact affects the phenotype of a wide range of other mutations (see Stark, 2004). There are also plenty of other examples of differences between strain backgrounds that may affect the outcome of an experiment (Sherman, 2002). When introducing a desired allele from one genetic background to another, it is therefore important that only the desired allele and not other genetic differences are transferred. To ensure this as much as possible, a parent strain that is not in the desired background should first be backcrossed to the desired genetic background several times (e.g. six). This involves crossing the new strain with the desired background, sporulating the resulting diploid and identifying the haploid segregants with the required mutation, and then repeating the procedure several times. With each backcross, the genetic contribution from the 'wrong' background is approximately halved and ideally, with sufficient backcrosses, only the relevant section of the chromosome carrying the mutation of interest would remain from the new strain. Crossing different backgrounds often results in a high level of spore inviability and this may well be seen in the initial backcrosses.

♦♦♦♦♦♦ IV. YEAST MATING

A. Establishing Yeast Mating Type

Before yeast can be mated, it may be necessary to confirm the mating type of the strains concerned. Yeast mating type can of course be established often by simply attempting the mating reaction. However, some strains mate with low frequency, and it can be useful to have an alternative way of determining mating type. One approach that can be taken is to use a yeast strain such as RC629 (Chan *et al.*, 1983) as a mating type indicator strain. RC629 is a *MAT***a** strain that carries a mutation in the *SST1* (*BAR1*) gene, and is therefore hypersensitive to α-mating pheromone, the 13-amino acid peptide secreted by *MAT*α yeast. The encoded Sst1 protein is a secreted aspartyl protease that is specific for the *MAT*α peptide and in *sst1* mutant strains, the α-mating pheromone encountered by the RC629 strain is not degraded, causing prolonged G1 arrest and growth inhibition. This phenomenon can be used as the basis of a simple test for mating type (see Protocol 1). Patching yeast of unknown mating type on a lawn of RC629 that has been spread immediately beforehand allows the patched strains to secrete mating pheromone into the agar as the surrounding lawn of tester strain grows up. Patches of α-type yeast cause a zone of growth inhibition of the RC629 lawn (Figure 1A).

An alternative and more commonly used way to check mating type of any haploid strain containing at least one auxotrophic marker involves the use of haploid mating type tester strains (Protocol 2). This has the advantage over the first method in that the mating type of *MAT***a** strains can be independently confirmed rather than just deduced by failure to score positive as *MAT*α. Tester strains typically contain a single, unusual auxotrophic marker (e.g. *his1*) that must not be present in the strains to be tested, such that when they mate with any auxotrophically marked lab strain that is not also *his1* they will generate a prototrophic diploid (i.e. heterozygous for *his1* and for any other mutations contributed by the strains to be tested). Strains DC14 (*MAT***a** *his1*) and DC17 (*MAT*α *his1*) are commonly used as tester strains (see e.g. Clark *et al.*, 1993), but if these strains are not available then similar strains can be obtained from the American Type Culture Collection (ATCC strains 208012

Protocol 1. Determination of yeast mating type using *sst1* (*bar1*) strains.

1. Using a sterile loop, take a small amount of strain RC629 colony material, and resuspend in 1 ml of sterile water. The actual amount of RC629 material is not critical, but a suspension with an optical density (600 nm) of between 0.5 and 1.0 seems to work well.
2. Take 100 µl of this suspension, and spread onto a YPD agar plate using a sterile spreader.
3. Onto this spread lawn of RC629, immediately patch 0.5 cm^2 areas of the strains whose mating type is to be tested. These should have been freshly grown overnight prior to the patching. Use generous amounts (1–2 whole colonies) of the strains to be tested in order that their growth gets a head start over that of the RC629 lawn.
4. Results should be clear after 24–36 h. *MATα* strain patches will inhibit growth of RC629, and be surrounded by a cleared zone in the lawn (see Figure 1A). The lawn will grow right up to the perimeter of the *MAT***a** patches.
5. *Note*: The method works best using material from an RC629 stock plate that is not freshly grown. This allows the patched strains to be tested to grow and secrete mating pheromone into the agar before the RC629 lawn becomes established.

and 208013). The test involves resuspending a good-sized loopful of a fresh colony from each tester strain separately in 1 ml water and then spreading 200 µl of each suspension on a separate SD agar plate. The strains whose mating type is to be tested are patched onto the surface of these plates, which are then grown at 26°C for 2–3 days. *MAT***a** strains patched onto the *MATα* tester plate will mate with the tester strain to generate prototrophic diploid cells, which will then grow to form a patch, and *MATα* strains patched onto the *MAT***a** tester plate will similarly mate and grow. Since both the tester strain and the patched strains carry auxotrophic mutations, neither can grow on SD agar in the absence of mating, thus *MAT***a** cells patched on the *MAT***a** tester will not grow to form a patch. Figure 1B shows an example of such a test in which a large number of strains have been replica-plated from a YPD agar plate onto a *MAT***a** tester plate.

B. Crossing Yeast Strains

Yeast can be mated quite simply by taking two cultures of opposite mating type that have been freshly grown overnight on solid medium, and mixing them together on fresh medium for between 5 and 8 h. If so desired, progress of the mating reaction can be checked by microscope observation from 5 h onwards (see Protocol 3),

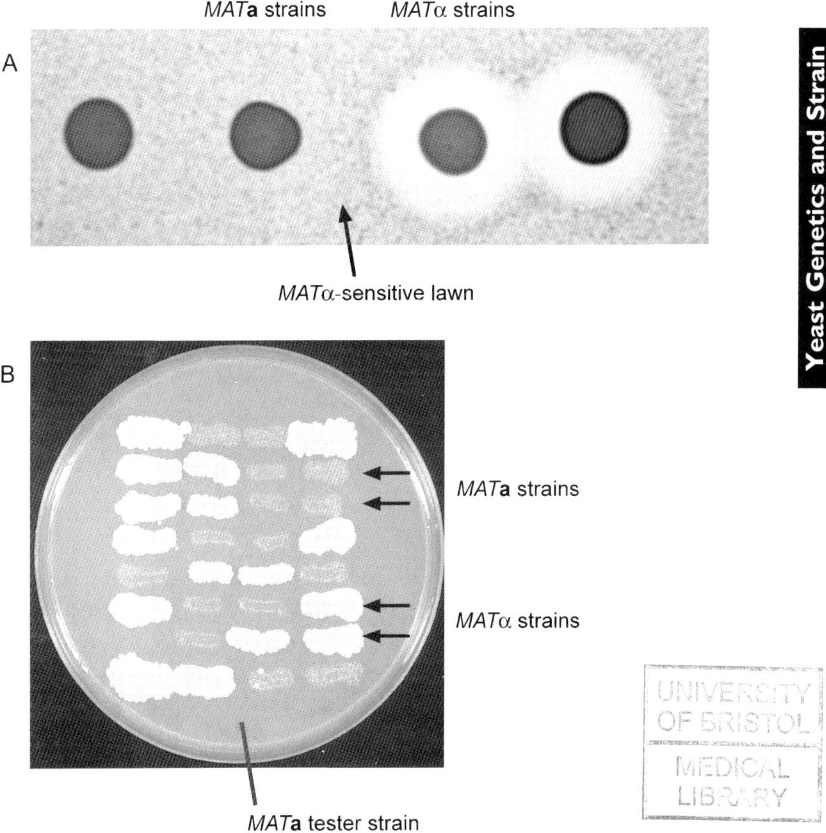

Figure 1. Determination of yeast mating type using an α-sensitive yeast. (A) Determination of yeast mating type using an α-sensitive yeast. Yeast cells of unknown mating type are patched onto a lawn of strain RC629, a *MAT***a** *sst1-2* mutant that is supersensitive to α-mating pheromone. An α-mating type strain will inhibit growth of the α-pheromone sensitive RC629 lawn, generating a halo around the patched strain. The lawn will grow right up to the perimeter of an **a**-mating type strain. (B) Mating type determination by crossing with a tester strain. An SD agar plate was seeded with a lawn of DC14 cells (*MAT***a** *his1*), and then patches of 32 haploid yeast strains (carrying several auxotrophic markers NOT including *his1*) replica-plated onto this lawn. The plate was incubated at 26°C for 3 days. Since, neither the tested strains nor the DC14 tester strain can grow on unsupplemented SD agar, growth only occurs where one of the strains has mated with the tester to form a prototrophic diploid. The reciprocal pattern would be obtained using the DC17 *MATα* tester.

looking for the fusion of the 'shmooed' (elongated) haploid cells (see examples in Figure 2A and B). However, where each of the two parent strains contain unique auxotrophies, it is usually sufficient just to allow cells to mate for 10 h (or overnight for convenience), and then streak them onto suitable dropout plates that will enable the diploid mating products to grow, but prevent the growth of either haploid parent. This takes advantage of a prototrophy in the *MAT***a** strain complementing an auxotrophy in the mating *MATα*

Protocol 2. Determination of yeast mating type by mating with tester strains.

1. Resuspend freshly grown cells of DC14 (*MAT***a** *his1*) and DC17 (*MATα his1*) or similar tester strains separately as described in Protocol 1 and spread 200 μl each onto a separate, unsupplemented SD agar plate.
2. When the two tester plates are dry, patch or replica-plate the strains to be tested onto each of them and incubate 2–4 days at 26°C.
3. Patched strains should, if haploid, robustly form a thick patch of cells on just one of the two tester strains and will have the opposite mating type to the tester strain on which they grew.
4. *Note*: Sometimes a few small colonies may form on the tester plate of the same mating type if it is incubated for too long.

partner, and *vice versa*. For instance, the strains BY4741 (*MAT***a** *his3Δ1 leuΔ0 met15Δ0 ura3Δ0*) and BY4742 (*MATα his3Δ1 leu2Δ0 lys2Δ0 ura3Δ0*), in which the complete yeast viable deletion collection is available, can be mated on complete medium as described. The resulting diploids, heterozygous for both the *met15Δ0* and *lys2Δ0* alleles, and thus phenotypically *Met*$^+$ *Lys*$^+$, can then be selected by streaking for single colonies on SCD medium lacking both methionine and lysine; neither parent can grow on such medium. A similar approach can be taken using any recessive mutation in place of an auxotrophic marker, for example exploiting cross-complementation of a temperature-sensitive mutation contributed by one parent and selecting the diploids at the restrictive temperature. Plasmids carrying selectable markers that complement auxotrophies can also be employed to engineer a situation such that each strain caries auxotrophies that will be cross-complemented in the resulting diploid. Thus, if it was desired to cross a *trp1 his3* strain with one that is just *trp1*, the *trp1 his3* strain could be transformed to *Trp*$^+$ with a *TRP1* plasmid such as YCplac22 (Gietz and Sugino, 1988) to drive selection of the diploid on SCD lacking tryptophan and histidine. Loss of the plasmid can subsequently be achieved by non-selective growth of the diploid followed by testing up to 100 independent colonies for reacquisition of, in this case, a Trp$^-$ phenotype due to plasmid loss.

Alternatively, or when selection for the diploid is not possible, zygotes can be selected using a micromanipulator after 5–8 h mating. A small amount of mating yeast is taken using the edge of a wire loop and a dilute suspension of the mating mix made in sterile distilled water (typically 400 μl). An aliquot of this can be spotted onto the edge of a fresh agar plate, from where the zygotes can be picked up using a micromanipulator and put down on a new, sterile area of the same plate to form a diploid colony. Figure 2 shows some

Protocol 3. Yeast mating.

1. Grow the *MAT***a** and *MAT*α haploid strains separately overnight, on complete medium (YPD agar medium: 2% w/v glucose, 2% w/v peptone, 1% w/v yeast extract, 2% w/v agar). Transformed strains can be grown on selective defined medium to maintain plasmid selection, although plasmid loss during overnight growth on YPD is in our experience, limited.
2. The following day, mix equal small amounts (1/5 of a typical sized colony is sufficient) of each strain together on an area of fresh YPD agar. Take care to ensure mixing is thorough, to enhance the frequency of mating. Patch the two parent strains separately on a different area of the plate.
3. Incubate at 26–30°C for 5 h – overnight and then re-streak each patch of cells onto an SCD plate lacking at least one component required uniquely by each parent strain (e.g. SCD minus tryptophan and uracil for *MAT***a** *his3 trp1 URA3* × *MAT*α *his3 TRP1 ura3*).
4. If it is impossible to use auxotrophic selection to obtain the diploid (e.g. because the markers used are dominant such as KanMX or because both parents carry the same auxotrophic markers), diploid cells can be obtained by selecting zygotes using a tetrad dissection microscope. After 5–8 h mating check under the microscope for the formation of shmoos (elongated, pear-shaped cells), mating figures (dumb-bell figures) and zygotes (3-pointed dumb-bells formed by mating of an **a** and an α cell followed by emergence of a bud as the diploid cell begins to proliferate). Zygotes typically develop by 8–10 h after mixing. Spread the mating mixture along one side of a YPD agar plate, identify zygotes by their characteristic 3-point dumb-bell morphology and micromanipulate them to a clear area of the plate. After germination, verify that they are diploid by absence of expression of a mating type (Protocol 2) and/or ability to sporulate (Protocol 4).

typical zygotes, characterised by a dumb-bell shaped cell formed by fusion of the two mating partners (A, B), often with a new bud emerging close to the point of fusion (C, D). Note that the 'neck' is very broad in contrast to a large-budded haploid cell.

If the strains being mated are transformed with a plasmid, pre-growth of the haploids and the mating reaction itself can be performed on selective medium (SD or SCD) to maintain plasmid selection, but mating will be more efficient on complete YPD medium. Our experience is that frequencies of plasmid loss when transformed cells are grown for limited periods on complete medium are low.

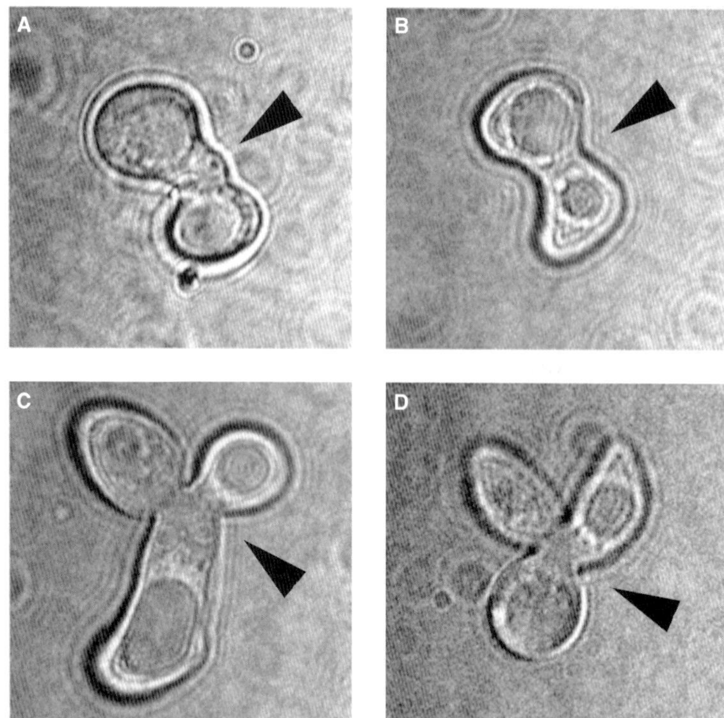

Figure 2. Zygotes generated after 8 h of mating between DC14 and DC17. Cells from freshly grown colonies of DC14 and DC17 were mixed on YPD agar and photographed at 1000 × magnification after 8 h. A, B: Early zygotes formed by fusion of two mating cells. C, D: Later zygotes in which a new bud is emerging. Note the wide neck between the two fused haploid parent cells (arrowheads).

♦♦♦♦♦♦ V. SPORULATING YEAST

Yeast will initiate sporulation in response to nitrogen starvation while growing on a non-fermentative carbon source, typically acetate. Several different methods have been described, but here we will limit ourselves to describing two, one using liquid cultures (Protocol 4) and the second using agar plates (Protocol 5). In the first method, yeast are normally first grown on acetate as a sole carbon and energy source, but with peptone and yeast extract included to provide a rich nitrogen source. Once the yeast culture is adapted to growth on acetate, and while in the logarithmic phase of growth, the actively growing culture is harvested, and resuspended in a solution of potassium acetate, effectively mimicking a sudden nitrogen starvation (see Protocol 4). During these culturing periods, yeast should be well aerated to allow growth on acetate, a non-fermentative carbon source requiring mitochondrial oxidative metabolism. Typically, sporulation is complete by about 3 days after growth on acetate is initiated. Microscope observation should be used to track the appearance of tetrads, appearing as tetrahedral

Protocol 4. Sporulation of yeast diploids (liquid culture).

1. Grow diploid yeast strain overnight at 30°C in a 250 ml flask containing 10 ml of presporulation medium (1% w/v yeast extract, 2 % w/v peptone, 1% w/v potassium acetate), until an OD_{600} of approximately 0.6–1.0 is reached.
2. Harvest the culture by centrifugation (3000 × g, 5 min), wash with 10 ml sterile distilled water.
3. Repeat this wash procedure 2 more times. This part of the procedure must be thorough to ensure complete removal of the pre-sporulation medium, which contains a nitrogen source.
4. Resuspend the yeast in 10 ml of sporulation medium (1% w/v potassium acetate), decant into a 250 ml flask, incubate at 30°C in an orbital incubator for 3–4 days.
5. *Note*: High levels of aeration are important for sporulation, hence the reason why small volumes of culture are placed in large flasks. Higher cell densities also improve sporulation efficiency – it is therefore important to allow the pre-sporulation culture to reach the correct cell density.

Protocol 5. Sporulation of yeast diploids on agar plates.

1. Streak diploid cells on GNA plates and grow for 1–2 days at 29°C. Cells must be actively growing to get good sporulation. GNA plates contain (per l) 50 g D-Glucose, 30 g Difco Nutrient Broth, 10 g Difco Yeast Extract and 20 g agar.
2. Patch cells onto VB Sporulation agar and incubate at 26°C for 3–5 days. VB Sporulation agar contains (per l) 8.2 g sodium acetate (100 mM final), 1.9 g KCl, 1.2 g NaCl, 15 g agar and 0.35 g $MgSO_4$ (or 1.4 ml 1 M stock) added after autoclaving the other components).
3. *Note*: Both types of plate should be freshly made for it to have the best chance of success. It seems to work even without adding the *his*, *leu* and *ura* supplements that the diploid strains in this background will need.

bundles of spores. Different strains will sporulate with differing efficiencies. For example, diploids generated from crosses between BY4741 and BY4742, the strains in which the systematic yeast knockout collection has been made (see http://web.uni-frankfurt.de/fb15/mikro/euroscarf/index.html), are relatively poor sporulators compared with W303 strains. To sporulate cells on solid medium (Protocol 5), they are first grown in patches (e.g. ~5–10 mm square) on a very rich medium (GNA agar) and then these cells re-patched onto acetate plates (VB sporulation agar). Again sporulation should occur in 3–4 days, although some strains are slower.

At this stage, spores can be dissected using a micromanipulator. To begin this process, ascospores need to be released enzymatically from the ascus, a process described in Protocol 6. Various enzymes are available for digestion of the ascus to release the spores, including lyticase (Sigma) and Zymolyase (Zymo Research or MP Biomedicals), both containing a β-1,3-glucanase from *Arthrobacter luteus*. It is important to choose a preparation with sufficient specific activity to be effective, such as a partially pure lyticase preparation (with an activity of $\geq$2,000 U mg^{-1} protein; Sigma) or Zymolyase 100 T (MP Biomedicals). Once the spore preparation has been digested, individual spores can be dissected using a micromanipulator. This involves mechanically picking up individual spores from the spore preparation, and placing them on a new area of sterile agar to grow into a colony. It is crucial for proper genetic analysis to know that a particular group of four spores derives from a single original tetrad. For this reason, normally all four spores from a tetrad that is well separated from its neighbours on the agar are picked up at one time, and then individually put down, each on a separate area of sterile agar to form a set of four colonies. Over-digestion of the asci can result in individual spores, which have a hydrophobic spore coat and are therefore rather sticky, aggregating

Protocol 6. Spore dissection.

1. Take 50 μl of a sporulated culture, and add 15 μl of lyticase solution (10 mg ml^{-1} in sterile water – use partially purified lyticase powder for this solution, not the crude preparations). Digest the spore coat for between 10 and 15 min at room temperature. We routinely take samples every 2 min from 8 min to 14 min, to ensure that at least one sample is obtained where digestion is optimal.
2. Alternatively, resuspend a loopful of sporulated cells in 95 μl sterile water and add 5 μl 10 mg/ml Zymolyase 100 T solution and incubate on the bench for 5–10 at room temperature.
3. As the staggered digestion period progresses, spot 3 μl spots onto YPD agar. Alternatively, if you are confident about the digestion time, spot 10 μl of the digest on one side of the plate and spread across the full width of the plate with a sterile loop or toothpick.
4. Dissect using a micromanipulator. Usually at least 10 tetrads would be dissected, with many more as required depending on the number of markers involved and whether any of them show genetic linkage (see main text).
5. Spore germination and growth into a small colony typically takes 2–3 days at 26–30°C.

into groups that may look like genuine tetrads, so it is important to use the minimum digestion time needed to allow relatively easy rupture of the ascus coat during tetrad dissection.

The exact method for picking the four spores in a tetrad, disaggregating them and placing each one onto a defined location to develop into a new colony, will vary depending upon the type of micromanipulator being used; readers are referred to the manufacturer's instructions for their particular instrument (e.g. see Saunders-Singer, 1996). One version, produced by the Singer Instrument Company Ltd., works especially well and is shown as an example in Figure 3. It comprises a microscope with motorised stage, to which is mounted the micromanipulator itself. The agar Petri dish on which the digested spore preparation has been spotted is placed inverted on the stage, and groups of four spores are

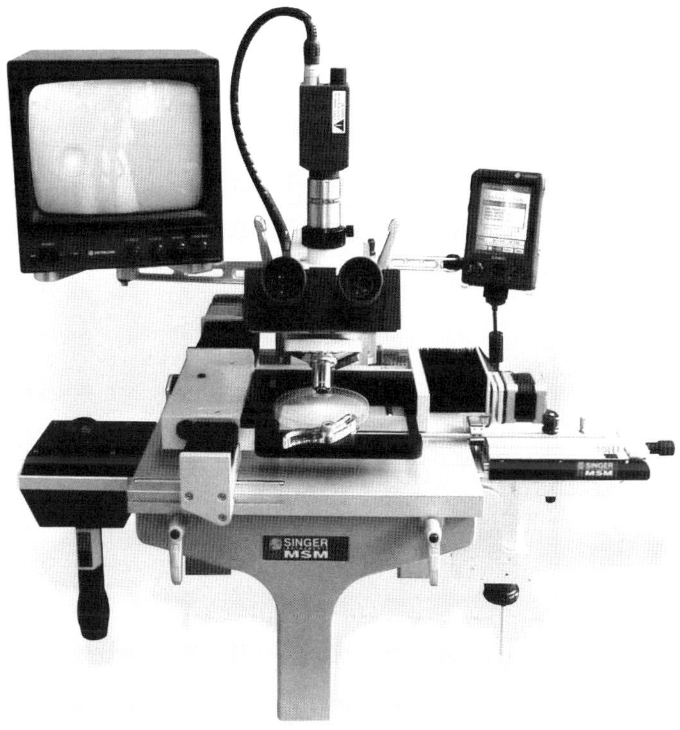

Figure 3. Micromanipulators for yeast tetrad dissection. Yeast genetic analysis requires micromanipulation equipment. In tetrad analysis, the four ascospores have to be released from the ascus by enzymatic digestion of the spore coat. Each of the four spores is individually dissected, and placed on the points of an orthogonal array on the surface of an agar plate. An example of a micromanipulator suitable for yeast tetrad dissection is the Singer MSM System 300. It is equipped with a microscope whose motor-driven stage holds the Petri dish. Control of stage movement is by a joystick via a PC, enabling orthogonal arrays of dissected spores to be generated. A special micromanipulator attached to the stage enables dissection of the tetrad, with individual placement of the spores onto a sterile area of the agar where they form a colony.

picked using a vertically mounted fine glass needle mounted on the micromanipulator. The glass needle has a tip with a flat end suitable for picking spores. This equipment allows for spores to be easily picked, and then placed in ordered arrays by a computer-controlled motorised stage, after which they germinate and grow, forming neat arrays of colonies in groups of four, each group representing the spores from one individual tetrad (Figure 4).

In addition to micromanipulation of spores, it is also possible to carry out random spore analysis, which as the name suggests involves surveying a population of randomised spores for one with the required phenotype(s). In this instance, the spore preparation is digested to completion using lyticase so that all spores are released from their asci. This process can be helped by vortexing and then ultrasonicating the spore mix after the digestion. A suitable dilution of the spore preparation is then spread onto an agar plate for outgrowth and colony formation. In order to eliminate non-sporulated diploids from this process, various procedures have been developed to enrich for spores. These include ether killing, in which the sporulated culture is emulsified with ether, killing the vegetative diploid cells while leaving the more resistant spore population intact

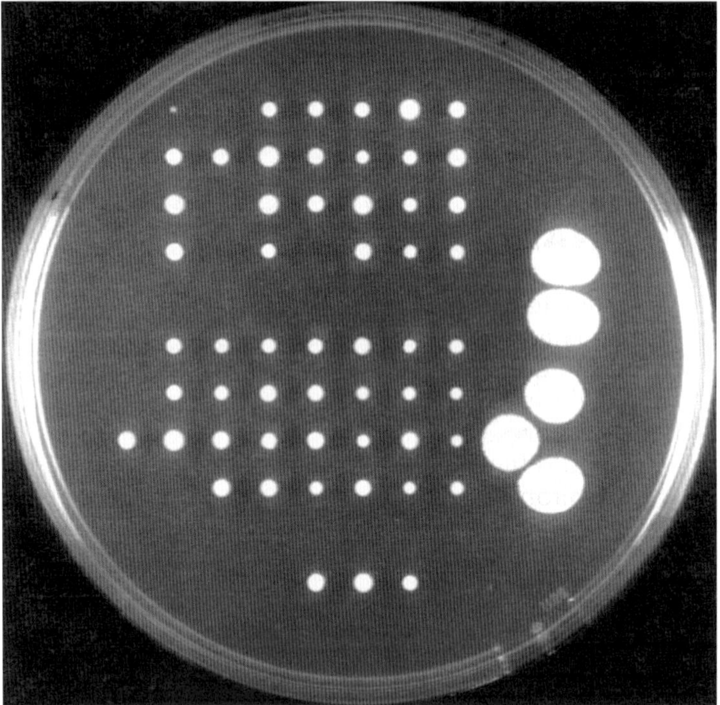

Figure 4. Agar plate with dissected tetrads. Tetrads were selected from the 'pools' of lyticase-digested sporulated culture placed on the right of the plate, and the four spores in each tetrad placed in a grid arrangement on the left of the plate. Tetrad members are arranged vertically. In this experiment, most tetrads contained four viable spores.

(Rockmill *et al.*, 1991). Alternatively, the toxic arginine analogue canavanine can be used to select for haploid spores in the case that the diploid is heterozygous for the *can1* allele (see Amberg *et al.*, 2005 for a protocol). The *CAN1* gene encodes the yeast arginine permease, and strains carrying the wild-type allele are sensitive to canavanine. Outgrowth of spores, and any non-sporulated diploids, on medium containing canavanine (40 μg ml^{-1}) kills all but the *can1* haploids through incorporation of canavanine into cellular proteins (Sherman and Roman, 1963).

As a general rule, while random spore methods can be used successfully to isolate novel recombinant haploid yeast strains, tetrad dissection is a much more reliable method. Using tetrad dissection, the segregation pattern of all the markers in the cross can be easily followed and the level of viability of different combinations of alleles can be monitored. For example, if the desired combination of alleles leads to a slow growth of phenotype or is even synthetically lethal, this will be immediately apparent by tetrad analysis. In comparison, using random spore methods it is possible to isolate clones where secondary events may have occurred to enable the particular combination of alleles to be more easily selected, for example a second site suppressor mutation, a chromosome gain event or, if the genetic background of the diploid is not completely homogeneous, fortuitous combinations of alleles of completely unknown genes may be selected.

♦♦♦♦♦♦ VI. TETRAD DISSECTION AND PHENOTYPE ANALYSIS

Once tetrad dissection has been performed, the dissected and germinated spores can be subjected to phenotypic analysis to identify the required strain carrying the right combination of alleles. This might involve screening for the phenotype of marker genes such as *TRP1* or KanMX associated with gene knockouts or genes modified by addition of an epitope tag (see e.g. Chapter 4, this volume), or monitoring the temperature-sensitivity associated with a conditional Ts^{-} allele of a gene. This is best achieved by picking the germinated spores using sterile toothpicks and re-streaking them into a grid pattern (see Figure 5) on pairs of replicate YPD plates. One of these will serve as a stock plate for temporary storage of the strains, while the other can be used to replica-plate the strains onto other plates for phenotypic testing using a replicator block and sterile velvets.

In a cross involving two strains of gentotype *aB* and *Ab*, there are three potential arrangements of alleles in the tetrad; the parental ditype (PD) in which no reassortment has occurred (spore genotypes *aB*, *aB*, *Ab*, *Ab*), the non-parental ditype (NPD) in which both

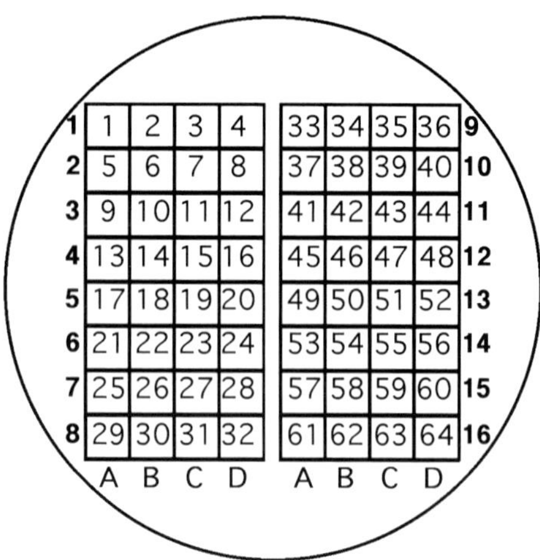

Figure 5. Grid pattern used for re-streaking germinated spores following tetrad dissection.

alleles have reassorted (spore genotypes *ab*, *ab*, *AB*, *AB*) and the tetratype (TT), in which the both parental genotypes and both reassorted genotypes are each represented (spore genotypes *aB*, *ab*, *Ab*, *AB*). If the genes *A* and *B* are unlinked, and thus subject to random assortment, a diagnostic ratio of PD:NPD:TT tetrads of 1:1:4 is obtained. If *A* and *B* are linked, then parental ditypes will be found with increased frequency relative to the NPD and TT types, with a more dramatic reduction in the NPD class. The extent to which PD types dominate therefore reflects the degree of linkage, which is defined as follows (Perkins, 1949):

$$\text{Linkage (centiMorgans)} = \frac{100}{2} \times \frac{[\text{TT}] + [6 \times \text{NPD}]}{[\text{PD} + \text{NPD} + \text{TT}]}$$

In practice, this means that where it is desired to recombine alleles of two unlinked genes in a genetic cross, most tetrads should contain one desired recombinant.

The segregation pattern described above can be exploited when two loci in a cross are marked with the same marker gene; for example, two of the systematic deletion strains each carrying a different KanMX-marked gene knockout. If both knockouts are required in the new strain, then identifying an NPD tetrad (2 $G418^R$:2 $G418^S$ spores) should ensure that each resistant spore carries both knockouts. Similarly, if only one is needed, then choosing a PD tetrad (4 $G418^R$:0 $G418^S$ spores) ensures a 50% chance that any one spore has the desired knockout. Such knockouts must, however, be verified by PCR using standard procedures (see for example http://www-sequence.stanford.edu/group/yeast_deletion_project/protocols.html).

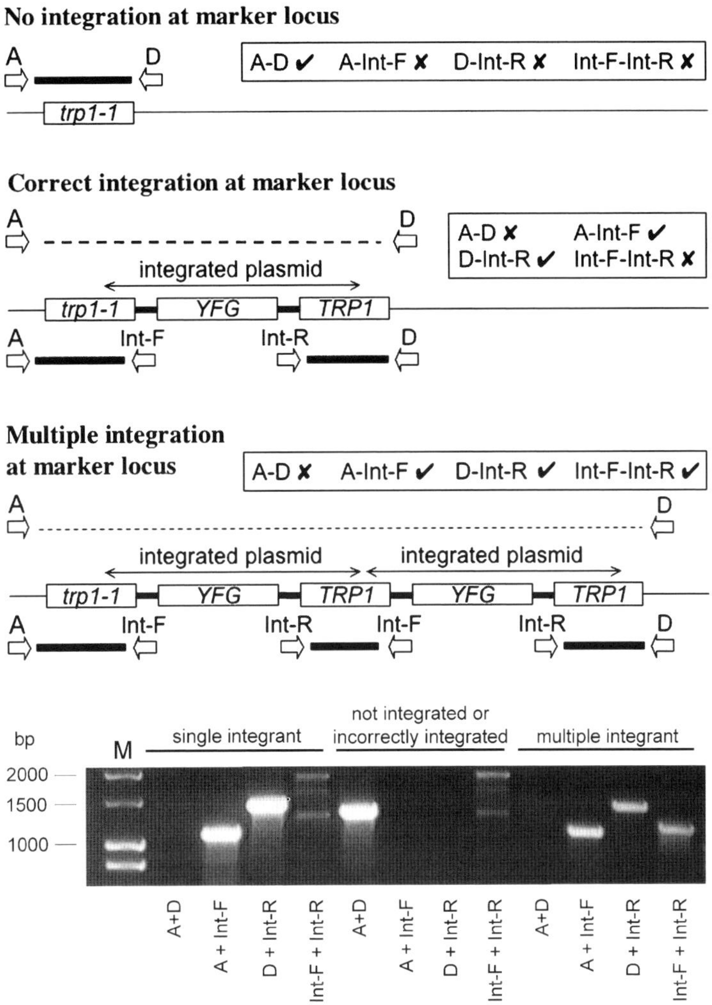

Figure 6. Monitoring correct integration of a YIp plasmid. A PCR-based strategy is shown using two plasmid-specific primers (Int-F and Int-R) and two marker gene-specific primers, indicating the expected PCR products obtained for correct integration, multiple integration and no integration using genomic DNA as template (Protocol 7). The gel shows the results of such PCR analysis for integration of a construct at the *trp1* locus (see Table 1 for Primer sequences). Some faint, non-specific products can be seen in the In-F+Int-R reactions when multiple integration has not occurred.

Sometimes it may be necessary to generate a strain that carries several different mutations or marked genetic loci. For example, in a recent study of chromosome behaviour by fluorescence time-lapse microscopy (Tanaka *et al.*, 2005), strains were generated that contained P_{MET3}-*CDC20*, P_{GAL}-*CEN3-tetO*, *tetR*-GFP, YFP-*TUB1*, YFP-*NIC96*, *KIP2*-4GFP and one of a variety of other mutant alleles. Such strain construction clearly requires careful planning and screening of many dissected tetrads. If generating the correct combination proves difficult, the process can be broken down into simpler steps to increase the number of desired loci that are present in both parents in the final cross. Furthermore, where two linked loci are involved, it may be useful first to carry out a cross that recombines these together before trying to introduce additional loci, since once the two linked loci have been recombined they will tend to co-segregate in further genetic crosses.

◆◆◆◆◆◆ VII. MODIFYING STRAINS BY PLASMID INTEGRATION

In addition to using genetic crosses to generate new yeast strains, it may also be necessary to introduce new features such as epitope or fluorescent protein tags by transformation. Chapter 4 in this volume provides details of this methodology and a summary of the wide variety of tags that are available to yeast researchers. However, it should be remembered that since yeast transformation is inherently mutagenic, genetic crosses are the method of choice for introducing such a tagged gene into a new strain once a suitably tagged version is already available. A gene can thus be tagged once, and then the same construct introduced into other strains by genetic crosses, minimising the chance that any extraneous mutations may be unwittingly introduced into some isolates. In some cases, it may also be necessary to integrate a copy of a gene using a yeast integrative plasmid (YIp); for example, to integrate a copy of a heterologous gene into a diploid yeast strains heterozygous for a specific gene knockout, so that complementation of the knockout can be tested following tetrad analysis. Unlike the gene conversion event following introduction of a fragment designed to delete or add a tag to a gene (see Chapter 4), transformation with a YIp construct, whether integration is targeted by linearising in the marker gene or other

Table 1. Primers for verifying integration of YIplac plasmids at their marker loci

Primer[a]	Sequence
Int-F	CATGTGTCAGAGGTTTTCACCGTC
Int-R	CAGCATCTTTTACTTTCACCAGCG
TRP1-A	AGAGACCAATCAGTAAAAATCAACG
TRP1-D	GCGAAAAGACGATAAATACAAGAAA
URA3-A	AATGTGGCTGTGGTTTCAGGG
URA3-D	CTGTTACTTGGTTCTGGCGAGG
LEU2-A	TTGTCCTGTACTTCCTTGTTCATGTG
LEU2-D	CCTAACTTTTTGTGTGGTGCCCTC

[a]'A' and 'D' primers are those designed by the yeast genome deletion consortium (see http://www-sequence.stanford.edu/group/yeast_deletion_project/protocols.html). If integration is targeted to different genetic locus, the gene-specific 'A' and 'D' primers for that locus can be substituted. The optimal annealing temperature for these primers is 54–55°C.

yeast sequences that it carries, frequently leads to multiple integration events at the targeted locus. In our experience, this problem can occur in as many as 30–50% of transformants selected solely on the basis of prototropy for the YIp marker, leading to undesirable differences in the copy number of the gene that is being introduced that may influence the outcome of an experiment. Some transformants may also not reflect integration events at the desired locus, for example, resulting from a gene conversion event between the YIp marker gene and the chromosomal marker locus. Thus, correct

Protocol 7. Colony PCR analysis.

1. Aliquot 10 µl samples of SPZ Buffer containing 2.5 mg/ml Zymolyase 100 T into 0.2 ml PCR tubes. SPZ (50 ml) is made by mixing 30 ml 2 M sorbitol, 4.05 ml 1 M Na_2HPO_4, 0.95 ml NaH_2PO_4 and 15 ml water (final pH 7.5). Samples (5 ml) can be supplemented with Zymolyase and stored in 500 µl aliquots at –20°C ready for use.
2. Using a yellow pipette tip, pick a small amount of cells from a colony and resuspend in 10 µl sample of SPZ Buffer/Zymolyase by pipetting up and down. The correct amount of cells gives a noticeably turbid suspension but is not too thick. *Note*: DNA isolated from untransformed cells present even at a low level in a colony can confound colony PCR analysis, so transformants should be replated at least once under selection to give single colonies before extracting template DNA.
3. Place in a PCR machine and run on a 3-step programme: 30 min at 37°C, 5 min at 95°C and 15°C 'forever'.
4. Mix each sample with 95 µl water to generate a template preparation that can be frozen and re-used.
5. PCR reaction: Use 2 µl template in a 20 µl reaction. Conditions will depend on the particular primers, but a typical reaction might be as follows.
6. For 20 reactions (2 µl template + 18 µl of the following mix for each):
   288 µl water
   40 µl 10 × buffer
   12 µl 50 mM $MgCl_2$
   8 µl dNTPs (each at 10 mM)
   4 µl each primer at 100 pmol/µl
   4 µl *Taq* polymerase (5 U/µl)
   94°C 2 min
   30 cycles of 94°C, 45 s; 54°C, 45 s; 72°C, 2 min
   72°C 5 min
   15°C 'forever'
7. Run the entire reaction on an agarose gel for a strong band of product where present.

integration of YIps must always be verified. Figure 6 shows a simple PCR-based strategy for distinguishing single and multiple integration events at some of the commonly used marker loci from strains where integration has not occurred. Suitable primers for use with the YIplac series of plasmids (Gietz and Sugino, 1988) are shown in Table 1 and Protocol 7 gives details for how to perform yeast colony PCR.

♦♦♦♦♦♦ VIII. CONCLUDING REMARKS

While the facility with which the yeast genome can be engineered using homologous recombination to insert or delete genes or specific alleles has made the construction of yeast strains very easy, it can often be simpler, faster and more reliable to resort to the classical strain construction techniques described here. The simplicity of the yeast life cycle, and the ease with which it can be manipulated, makes the application of the suite of methods here achievable in most labs, including for those researchers without experience of working with *S. cerevisiae*.

References

Amberg, D. C., Burke, D. J. and Strathern, J. N. (2005). *Methods in Yeast Genetics: A Cold Spring Harbor Laboratory Course Manual*. Cold Spring Harbor Laboratory Press, Cold Spring Harbor, NY.

Chan, R. K., Melnick, L. M., Blair, L. C. and Thorner, J. (1983). Extracellular suppression allows mating by pheromone-deficient sterile mutants of *Saccharomyces cerevisiae*. *J. Bacteriol.* **155**, 903–906.

Clark, K. L., Dignard, D., Thomas, D. Y. and Whiteway, M. (1993). Interactions among the subunits of the G protein involved in *Saccharomyces cerevisiae* mating. *Mol. Cell Biol.* **13**, 1–8.

Gietz, R. D. and Sugino, A. (1988). New yeast–*Escherichia coli* shuttle vectors constructed with in vitro mutagenized yeast genes lacking six-base pair restriction sites. *Gene* **74**, 527–534.

Goldstein, A. L. and McCusker, J. H. (1999). Three new dominant drug resistance cassettes for gene disruption in *Saccharomyces cerevisiae*. *Yeast* **15**, 1541–1553.

Haber, J. E. (1998). Mating-type gene switching in *Saccharomyces cerevisiae*. *Annu. Rev. Genet.* **32**, 561–599.

Lindegren, C. C. and Lindegren, G. (1943). A new method for hybridizing yeast. *Proc. Natl. Acad. Sci. USA* **29**, 306–308.

Mortimer, R. K., Schild, D., Contopoulou, C. R. and Kans, J. A. (1989). Genetic map of *Saccharomyces cerevisiae*, edition 10. *Yeast* **5**, 321–403.

Nickoloff, J. A., Chen, E. Y. and Heffron, F. (1986). A 24-base-pair DNA sequence from the MAT locus stimulates intergenic recombination in yeast. *Proc. Natl. Acad. Sci. USA* **83**, 7831–7835.

Perkins, D. D. (1949). Biochemical mutants in the smut fungus *Ustilago maydis*. *Genetics* **34**, 607–626.

Rockmill, B., Lambie, E. J. and Roeder, G. S. (1991). Spore enrichment. *Methods Enzymol.* **194**, 146–149.

Saunders-Singer, C. (1996). Ascus dissection. In: *Yeast Protocols*, Vol. 53 (I. H. Evans, ed.), pp. 146–149. Humana Press, Totowa, NJ.

Sherman, F. (2002). Getting started with yeast. *Methods Enzymol.* **350**, 3–41.

Sherman, F. and Roman, H. (1963). Evidence for two types of allelic recombination in yeast. *Genetics* **48**, 255–261.

Shibata, T., Watabe, H., Kaneko, T., Iino, T. and Ando, T. (1984). On the nucleotide sequence recognized by a eukaryotic site-specific endonuclease, Endo.SceI from yeast. *J. Biol. Chem.* **259**, 10499–10506.

Stark, M. J. R. (2004). Protein phosphorylation and dephosphorylation. In: *The Metabolism and Molecular Physiology of Saccharomyces cerevisiae* (J. R. Dickinson and M. Schweizer, eds), pp. 284–375. CRC Press, London.

Sutton, A., Immanuel, D. and Arndt, K. T. (1991). The *SIT4* protein phosphatase functions in late G1 for progression into S phase. *Mol. Cell Biol.* **11**, 2133–2148.

Tanaka, K., Mukae, N., Dewar, H., van Breugel, M., James, E. K., Prescott, A. R., Antony, C. and Tanaka, T. U. (2005). Molecular mechanisms of kinetochore capture by spindle microtubules. *Nature* **434**, 987–994.

Wach, A., Brachat, A., Pohlmann, R. and Philippsen, P. (1994). New heterologous modules for classical or PCR-based gene disruptions in *Saccharomyces cerevisiae*. *Yeast* **10**, 1793–1808.

3 Yeast Transformation

Duy Truong and R Daniel Gietz
Department of Biochemistry and Medical Genetics, University of Manitoba, Winnipeg, Manitoba, Canada

♦♦♦

CONTENTS

Introduction
Reagents and Solutions
Rapid Transformation
High Efficiency Transformation
Microtitre Plate Transformation
Preparation of Transformation-Competent Frozen Yeast Cells
Summary

LIST OF ABBREVIATIONS

LiOAc	Lithium acetate
PEG	Polyethylene glycol
ss	Single-stranded

♦♦♦♦♦♦ I. INTRODUCTION

The term "transformation" was first coined by Griffith (1928), describing a change in the phenotype of a pathogenic bacterium. Some 50 years later, yeast transformation was first accomplished by Hinnen *et al.* (1978) followed shortly thereafter by Beggs (1978), using a method requiring the production of spheroplasts. This method removed the cell walls from the yeast and used polyethylene glycol (PEG) and $CaCl_2$ to produce transformants. In 1983, Ito *et al.* (1983) reported the use of alkaline cations and PEG to transform intact yeast cells. In 1993, Schiestl *et al.* (1993) showed that modifications to the LiOAc/Carrier ssDNA/PEG method could generate millions of transformants per microgram of input plasmid DNA. Intact yeast cells are usually transformed by two procedures: the LiOAc/Carrier

METHODS IN MICROBIOLOGY, VOLUME 36
0580-9517 DOI:10.1016/S0580-9517(06)36003-5

ssDNA/PEG method (Gietz and Woods, 2002), and electroporation (Thompson *et al.*, 1998). For a recent review of yeast transformation see Gietz and Woods (2001). In this chapter, we outline a number of applications of this technique including (1) rapid transformation; (2) high efficiency transformation; (3) microtitre plate transformation; and (4) a protocol for producing frozen competent yeast cells for quick efficient transformation. This method of transformation can generate yields up to 5×10^6 transformants/μg plasmid DNA/10^8 cells with many commonly used laboratory strains of yeast.

The protocols described here have been standardized and the steps reduced from previous published versions. Growth and transformation efficiency are improved if liquid cultures are grown in $2 \times$ strength YPD plus adenine medium (YPAD). One of the most critical factors affecting transformation efficiency is duration of heat shock at 42°C. Times given in these protocols are averages. However, if your strain does not perform as desired, test for the optimal heat shock time.

♦♦♦♦♦♦ II. REAGENTS AND SOLUTIONS

The following reagents and solutions are required for all four LiOAc/ssDNA/PEG protocols.

A. Lithium Acetate (1.0 M)

Dissolve 10.2 g of lithium acetate dihydrate (Sigma Chemical Co. Ltd., St Louis, MO, USA; Catalogue #L-6883) in 100 ml of water, sterilize by autoclaving for 15 min and store at room temperature.

B. PEG MW 3350 (50% w/v)

Dissolve 100 g of PEG 3350 (Sigma Chemical Co. Ltd., St Louis, MO, USA; Catalogue #P-3640) in 60 ml of distilled/deionised water in a 500 ml beaker with stirring. When the solution has completely dissolved make the volume up to 200 ml in a graduated cylinder and mix thoroughly by inversion. Transfer the solution to a suitable glass bottle and autoclave for 15 min to sterilize. Store, securely capped, at room temperature. Evaporation of water from the solution will increase the concentration of PEG and severely reduce the yield of transformants.

C. Single-stranded Carrier DNA (2.0 mg/ml)

Dissolve 200 mg of salmon sperm DNA (Sigma Chemical Co. Ltd., St Louis, MO, USA; Catalogue #D-1626) in 100 ml of TE buffer

(10 mM Tris–HCl, 1 mM Na_2EDTA, pH 8.0) by drawing up and down in a 25 ml pipette and then mixing at 4°C on a magnetic stirring plate for 1–2 h or until dissolved. Store in 1.0 ml samples at –20°C. Denature the carrier DNA in a boiling water bath for 5 min and chill in ice/water before use. Be sure to pierce the top of the tube prior to boiling to avoid any problems with exploding caps. Boiled samples, stored at –20°C, can be re-used or re-boiled three times without significant loss of activity.

D. Yeast Growth Media

Yeast strains are grown on plates of YPAD agar (YPD supplemented with 100 mg adenine hemisulphate per liter; Sherman, 2002). The yeast cells to be transformed are usually re-grown in liquid 2 × YPAD medium (2% Bacto yeast extract, 4% Bacto peptone, 4% glucose and adenine hemisulphate, 100 mg/liter). SC selection medium is adjusted to pH 5.6 with 1.0 M NaOH and autoclaved (Rose, 1987).

♦♦♦♦♦♦ III. RAPID TRANSFORMATION

This method (Protocol 1) can be used when large numbers of transformants are not important. It can be set up quickly with no pre-culture using plates from the bench top or refrigerator. The transformation efficiencies can range from a few hundred to a few thousand per microgram input plasmid DNA. For those that wish to get the higher transformation efficiencies a pre-culture can be used. Inoculate the yeast strain onto a 2 cm^2 patch of YPAD agar and incubate overnight at 30°C. Alternatively, the yeast strain can be inoculated into 5 ml of liquid medium (2 × YPAD or SC selection medium) and incubated on a shaker at 30°C and 200 rpm. The protocol is useful for putting plasmids into a specific yeast strain.

♦♦♦♦♦♦ IV. HIGH EFFICIENCY TRANSFORMATION

This method (Protocol 2) can be used to generate millions of transformants in a single reaction to screen yeast genome equivalents or cDNA libraries. It has applications for the transformation of integrating plasmids, DNA fragments and oligonucleotides used to manipulate yeast genome. In addition, it can be used to optimize the transformation conditions for any specific yeast strain. In this protocol, the yeast strain is pre-cultured to allow for fresh inoculum for the growth culture. The transformation efficiency (transformants/μg plasmid/10^8 cells) can be calculated after determining the number of transformants in the transformation reaction. We have noticed that

Protocol 1. Rapid transformation.

1. Denature a tube of carrier DNA in a boiling water bath for 5 min and chill in ice water.
2. Scrape a 50 μl blob of yeast from any plate and suspend the cells in 1 ml of sterile water in a 1.5 ml microfuge tube. The type of culture used will determine the efficiency. The older the culture the lower the transformation efficiency. The suspension should contain about 5×10^8 cells. Cells grown overnight in 2 × YPAD broth will reach a titre between 1 and 2×10^8/ml; the titre in SC medium will be about 5×10^7/ml. Harvest 2 ml of YPAD culture and 5 ml of a SC culture.
3. Pellet the cells at top speed in a microcentrifuge for 30 s and discard the supernatant.
4. Add the following components of the Transformation Mix (T Mix) to the cell pellet in the order listed.

Component	Volume (μl)
PEG 3350 50% (w/v)	240
LiOAc 1.0 M	36
Boiled carrier ssDNA (2 mg/ml)	50
Plasmid DNA (0.1–1 μg) plus water	34
Total volume	360

5. Vortex the samples until the cells have been fully resuspended.
6. Incubate the tube in a water bath at 42°C for 40–60 min. Extending the heat shock to 180 min will increase the transformation yield for some strains.
7. Pellet cells at top speed in a microcentrifuge for 30 s and remove the liquid with a micropipette.
8. Resuspend the cells in 1 ml of sterile water by stirring with a micropipette tip and then vortex mixing vigorously.
9. Pipette 20 and 200 μl samples onto plates of appropriate SC selection medium, incubate at 30°C for 3–4 days and isolate transformants. The 20 μl samples should be pipetted into 200 μl puddles of sterile water.

the transformation efficiency declines as plasmid concentration is increased, although the actual yield of transformants increases but not in a linear fashion (Gietz *et al.*, 1995). It is often best to scale up transformation reactions to generate the required number of transformants instead of using higher plasmid concentrations in a single reaction. For complex library screens, such as yeast two-hybrid screens (Gietz, 2006), one can generate large numbers of

Protocol 2. High-efficiency yeast transformation.

Day 1

Pre-culture the yeast strain in 5 ml of liquid medium (2 × YPAD or SC selection medium) and incubate overnight on a rotary shaker at 200 rpm and 30°C. It is also helpful to place a bottle of 2 × YPAD and a 250 ml Erlenmeyer flask in the 30°C incubator in preparation for the next day's transformation.

Day 2

1. Determine the titre of the yeast pre-culture using a haemocytometer or cell counter. Dilute the culture 1 in 10 in sterile water and pipette a small amount into the haemocytometer and count. Use the appropriate factor to calculate the titre. Alternatively, pipette 10 μl of cells into 1.0 ml of water in a spectrophotometer cuvette and measure the OD at 600 nm. An OD_{600} of 0.1 is approximately 1×10^6 cells/ml.
2. Inoculate 50 ml of the pre-warmed 2 × YPAD in a sterile 250 ml Erlenmeyer flask with 2.5×10^8 cells to give a starting cell density of 5×10^6 cells/ml. Incubate the flask on a rotary or reciprocating shaker at 30°C and 200 rpm for about 4 h, or until at least 2 cell divisions have taken place (i.e. 2×10^7 cells/ml).
3. Harvest the cells by centrifugation at 3000*g* for 5 min. Wash the cells in 25 ml of sterile water and re-suspend in 1 ml of sterile water.
4. Prepare a sample of carrier ssDNA by incubating in a boiling water bath for 5 min and then moving to an ice/water bath.
5. Centrifuge the cell suspension in a 1.5 ml microfuge tube for 30 s and discard the supernatant. Resuspend the cell pellet to a final volume of 1.0 ml in sterile water by mixing vigorously.
6. For each transformation, pipette 100 μl samples (approximately 10^8 cells) into a 1.5 ml microfuge tube and pellet the cells by centrifugation at top speed for 30 s. Remove the supernatant.
7. While collecting and washing the yeast cells for transformation, prepare sufficient T mix plus one extra. Add the ingredients below together and mix using a vortex mixer. The plasmid DNA can be added directly to each tube if desired, in which case adjust the volumes accordingly.

Transformation mix reagents	Volume (μl)
1. PEG 3350 50% w/v	240
2. LiOAc 1.0 M	36

3. Boiled SS-carrier DNA (2 mg/ml)	50
4. Plasmid DNA plus water	34
Total volume	360

8. Add 360 µl of T mix to each transformation tube and resuspend the cells by vortex mixing vigorously. If adding plasmid DNA at this stage, add the adjusted volume of T mix first and then the plasmid DNA.
9. Incubate the tubes in a 42°C water bath from 20 to 40 min. Each yeast strain has its own heat shock optimum, so consider testing your strain to determine the optimum.
10. Pellet cells at top speed in a microcentrifuge for 30 s and remove the T mix with a micropipette.
11. Pipette 1.0 ml of sterile water into each tube; stir the pellet with a micropipette tip and vortex vigorously.
12. Plate appropriate amount of the cell suspension onto SC selection medium. We generally plate 2, 20 and 200 µl onto the appropriate plates. The 2 and 20 µl samples should be pipetted into 200 µl puddles of sterile water on the plates.
13. Incubate the plates at 30°C for 3–4 days to allow the transformants to form colonies.

transformants by scaling up the transformation reaction up to 120-fold. The heat shock time of a scaled-up transformation reaction should be extended from 40 to 60 min depending on the yeast strain.

♦♦♦♦♦♦ V. MICROTITRE PLATE TRANSFORMATION

The two methods listed below (Protocols 3 and 4) can be used to accomplish transformation in 96-well microtitre plate format. The Plate Growth Protocol (Protocol 3) can be used to transform a plasmid into many different yeast strains and the Liquid Growth Protocol (Protocol 4) can be used to introduce many different plasmids or constructs into a single strain. Both these protocols require a microtitre plate centrifuge, a 96-prong replicator (VWR Cat number 62409–606), 150 mm Petri plates, an eight-channel micropipette (Eppendorf™ or TiterTek™) and sterile reagent troughs.

♦♦♦♦♦♦ VI. PREPARATION OF TRANSFORMATION-COMPETENT FROZEN YEAST CELLS

This method can be used to produce frozen competent yeast cells when a single yeast strain is used repeatedly. Yeast cultures

Protocol 3. Plate Growth Protocol.

1. To inoculate and grow the yeast colonies in the 96-well plate grid, sterilize the 96-well replicator by dipping the prongs in 95% ethanol and carefully passing through a flame. Carefully place onto a large 150 mm YPAD plate to imprint the positions of the prongs. The colonies can now be patched onto the corresponding positions. The plates are incubated overnight at 30°C. These plates can now be used as a cell source for transformation.
2. Dispense 150 μl samples of sterile water into the wells of a sterile microtitre plate.
3. Sterilize the 96-well replicator as described above, cool the prongs by dipping into the wells containing the sterile water. Shake off the excess water.
4. Carefully align and place the tips of the sterilized replicator onto the patches of yeast inoculum. Move the replicator gently in small circles to transfer cells to the replicator, being careful not to damage the agar surface or cross-contaminate the inoculum patches. Remove the replicator and inspect the prongs to ensure each contains sufficient inoculum. Carefully place the replicator into the microtitre plate wells containing sterile water and agitate to suspend the cells. The average number of cells/well with our replicator is $\sim 1 \times 10^7$, a second transfer usually doubles the number. Mark the orientation of the microtitre plate.
5. Centrifuge at 3000 rpm for 10 min using a microtitre plate rotor with an appropriate balance plate (if necessary).
6. Remove the water by aspiration with a sterile micropipette tip attached to an aspirator. Be careful not to touch the cell pellet with the tip. Alternatively, shake the water out of the wells into a sink. This takes practice but is much faster than aspiration!
7. Boil Carrier ssDNA (2 mg/μl) for 5 min and chill in ice/water.
8. Prepare T Mix minus PEG. The volumes below are for a single well and 100 wells (96 + 4 extra for pipette error).

Component	1 Well (μl)	100 Wells (ml)
LiOAc 1.0 M	15.0	1.5
Carrier DNA (2 mg/ml)	20.0	2.0
Plasmid DNA+Water	15.0	1.5
Total volume	50.0	5.0

Note: we use 20 ng plasmid DNA per well; however, more can be added.

9. Pipette 50 μl T Mix minus PEG to each well. Clamp the plate and lid on a rotary shaker and agitate at 350–400 rpm for 2 min to re-suspend the cell pellets.

10. Pipette 100 µl PEG 3350 (50% w/v) into each well. Clamp the plate on the rotary shaker at 350–400 rpm for 5 min to ensure that the cell suspension is homogeneous.
11. Place the microtitre plate in a plastic bag or seal it with Parafilm™ and incubate in 42°C incubator for 1–4 h, depending on the yeast strain. You are encouraged to test this variable for your yeast strain.
12. Centrifuge the microtitre plate as before and remove the T Mix by aspiration or a combination of the shaking method and aspiration.
13. The transformation reactions can be sampled as follows:
    a. *Quantitative samples.* Pipette 100 µl of sterile water into each well. Clamp the plate on the rotary shaker at 400 rpm for 5 min to re-suspend the cells. Pipette 5 µl samples into 100 µl puddles on regular plates of SC selection medium.
    b. *Qualitative samples.* Pipette 50 µl of water into the wells. Resuspend the cells and use the sterile replicator to print onto plates of SC selection medium. The transfer volume is approximately 10 µl. Additional samples can be overlaid with care if required.
14. Incubate the plates at 30°C for 2–4 days and recover transformants. We have obtained up to 8000 transformants/well with 1×10^7 cells/well, 20 ng plasmid and an incubation of 4 h at 42°C.

Protocol 4. Liquid Growth Protocol.

This protocol is used when transforming a single strain with multiple plasmids or DNA constructs. The yeast culture is grown overnight and re-grown for two divisions as in the High Efficiency Transformation Protocol. To reproduce the conditions of the high efficiency transformation protocol each well should receive 4×10^7 cells. A complete microtitre plate (96 wells) will then require 200 ml of re-grown culture and 8 µg plasmid. The cells of the re-grown culture should be harvested, washed and re-suspended in sterile water at concentrations of 4×10^8 cells/ml.

1. Dispense 100 µl samples (4×10^7 cells) of the suspension of re-grown cells into each well of the microtitre plate. Centrifuge and remove the supernatant.
2. Continue from Step 7 of Protocol 3 with the following changes:
    a. Increase the amount of plasmid in the T Mix minus PEG accordingly.
    b. Incubate the plates at 42°C for 60 min.
3. Sample the wells using the quantitative or qualitative method in Protocol 3.

Protocol 5. Transformation-competent frozen yeast cells.

Preparation

1. Grow the yeast strain overnight and then re-grow in 2 × YPAD to a titre of 2×10^7 cells/ml as described in Protocol 2. One hundred samples of 1×10^8 frozen competent cells will require 500 ml of re-grown culture (1×10^{10}cells).
2. Harvest the cells by centrifugation at 3000*g* for 5 min, wash the cells in 0.5 volumes of sterile water, resuspend in 0.01 volumes of sterile water, transfer to a suitable sterile centrifuge tube and pellet the cells at 3000*g* for 5 min.
3. Re-suspend the cell pellet in 0.01 volumes of Frozen Competent Cell (FCC) solution (5% v/v glycerol, 10% v/v DMSO). Use good quality DMSO.
4. Dispense 50 µl samples into an appropriate number of 1.5 ml microfuge tubes.
5. Place the microfuge tube samples into a 100-tube styrofoam rack with lid (Sarstedt # 95.064.249). It is best to place this container upright in a larger box (Styrofoam or cardboard) with additional insulation such as Styrofoam chips or newspaper to reduce the air space around the sample box. This will result in the samples freezing slowly, which is essential for high survival rates.
6. Put the large Styrofoam container in a −80°C freezer overnight. The Styrofoam rack containing the frozen yeast cells can now be removed from the freezing container and stored at −80°C. These cells can be stored for up to one year at −80°C with little loss of transformation efficiency.

Use of frozen competent cells

Frozen competent cells are transformed using the High Efficiency Transformation protocol (Protocol 2) but with the following differences:

1. Thaw cell samples in a 42°C water bath for 15 s.
2. Pellet cells at top speed in a microcentrifuge for 2 min and remove the supernatant.
3. Add 360 µl of FCC T Mix (260 µl 50% PEG 3350, 36 µl 1.0 M LiOAc, 50 µl carrier ssDNA, 14 µl of plasmid DNA and water) and vortex mix vigorously to re-suspend the cell pellet. Note the difference in PEG concentration.
4. Incubate in a 42°C water bath for 20–60 min depending on the strain. Centrifuge, remove the supernatant and re-suspend the cell pellet in 1 ml of sterile water.
5. Plate appropriate dilutions onto SC selection medium (See Protocol 2, Step 12).

re-grown for a least two division can be used to produce transformation-competent cells that can be frozen and stored for later use (Protocol 5).

♦♦♦♦♦♦ VII. SUMMARY

The efficient transformation of yeast has allowed much progress in the field of yeast molecular biology. For techniques such as two-hybrid analysis and yeast genome modification, transformation is an essential step. We have presented a protocol for rapid transformation that is useful for putting plasmids into yeast strains. An efficient transformation protocol is also presented, which can be used for library screening and genome modification. In addition, we have developed two microtitre plate protocols, which can be used and adapted with current high throughput methodology. Finally, we have developed a protocol for the production of frozen competent yeast cells. This protocol allows large quantities of yeast cells that are competent for transformation to be prepared and frozen. Highly efficient transformation can then be performed at a moment's notice without the required pre-culture and re-growth steps.

References

Beggs, J. D. (1978). Transformation of yeast by a replicating hybrid plasmid. *Nature* **275**, 104–109.

Gietz, R. D. (2006). Yeast two-hybrid system screening. *Methods Mol Biol.* **313**, 345–372.

Gietz, R. D., Schiestl, R. H., Willems, A. and Woods, R. A. (1995). Studies on the mechanism of high efficiency transformation of intact yeast cells. *Yeast* **11**, 355–360.

Gietz, R. D. and Woods, R. A. (2001). Genetic transformation of yeast. *Biotechniques* **30**, 816–820.

Gietz, R. D. and Woods, R. A. (2002). Transformation of yeast by lithium acetate/single-stranded carrier DNA/polyethylene glycol method. *Methods Enzymol.* **350**, 87–96.

Griffith, F. (1928). The significance of pneumococcal types. *J. Hyg. (London)* **27**, 113–159.

Hinnen, A., Hicks, J. B. and Fink, G. R. (1978). Transformation of yeast. *Proc. Natl. Acad. Sci. USA* **75**, 1929–1933.

Ito, H., Fukuda, Y., Murata, K. and Kimura, A. (1983). Transformation of intact yeast cells treated with alkali cations. *J. Bacteriol.* **153**, 163–168.

Rose, M. D. (1987). Isolation of genes by complementation in yeast. *Methods Enzymol.* **152**, 481–504.

Schiestl, R. H., Manivasakam, P., Woods, R. A. and Gietz, R. D. (1993). Introducing DNA into yeast by transformation. *Methods* **5**, 79–85.

Sherman, F. (2002). Getting started with yeast. *Methods Enzymol.* **350**, 3–41.

Thompson, J. R., Register, E., Curotto, J., Kurtz, M. and Kelly, R. (1998). An improved protocol for the preparation of yeast cells for transformation by electroporation. *Yeast* **14**, 565–571.

4 A Guided Tour to PCR-based Genomic Manipulations of *S. cerevisiae* (PCR-targeting)

Celine I Maeder, Peter Maier and Michael Knop
Cell Biology and Biophysics Unit, EMBL, Meyerhofstraße 1, 69117 Heidelberg, Germany

A Guided Tour to PCR-based Genomic Manipulations of *S. cerevisiae*

♦♦♦

CONTENTS

Introduction
Application of PCR targeting for functional studies
Available cassettes for PCR targeting
Primers for use with PCR targeting
Application of PCR targeting for genome-wide studies
Critical methods
Concluding remarks

List of abbreviations

PCR Polymerase chain reaction

♦♦♦♦♦♦ I. INTRODUCTION

PCR-based targeting is a method to introduce foreign DNA or specific alterations into the yeast genome using homologous recombination between the targeted locus and homologous sequence information provided by PCR primers. It enables various genomic manipulations: gene deletions, gene truncations, C- and N-terminal gene tagging, promoter substitutions and *in vivo* site-directed mutagenesis. Carrying out a particular manipulation relies on the availability of specific templates for PCR, called modules or cassettes, which provide characteristic features. PCR-based targeting has greatly accelerated directed gene studies and has been a key technique for genome-wide functional analysis. The intention of this

METHODS IN MICROBIOLOGY, VOLUME 36
0580-9517 DOI:10.1016/S0580-9517(06)36004-7

chapter is to provide an overview of the different applications of this method and the available cassettes, and to list critical methods in a user-friendly manner. In addition, we will elucidate a number of considerations that have to be made while planning and conducting functional studies that rely on PCR-based targeting.

When analyzing a new gene, one of the first things a researcher usually needs is a rapid and reliable method to detect the expression of the gene. The researcher also needs to detect the corresponding protein within the cell, in cell extracts obtained under different growth conditions, or in association with other proteins upon protein complex purification. Although a specific antibody is an excellent tool in the long run, it is not the fastest way to validate a working hypothesis. One fast and reliable method to detect a protein (or several proteins in parallel) is the fusion of the target gene at its chromosomal location with easily detectable tags using PCR-based targeting. The method relies on efficient homologous recombination between chromosomal DNA and a piece of foreign DNA, which requires a minimum of 36 bp of sequence identity in order to occur with a reasonable frequency. The foreign piece of DNA is generated through a PCR using primers with 5′ ends that provide sequence homology to the target locus in the genome. The primers further contain 3′ sequences that allow PCR amplification of specific templates, which provides the desired functionality, for example, a tag and a selectable marker. The markers to be used for selection of faithful recombination events between the correct genomic locus and the short homologous sequences that derive from the primers should have no sequence homology to other parts of the genome in order to minimize unwanted integration events. Baudin *et al.* (1993) were probably the first to propose a method where an auxotrophy complementing gene from another yeast species is used as the selection marker. Subsequently, dominant markers encoding antibiotic resistance were developed for use in *S. cerevisiae* using PCR targeting for gene deletion and gene fusion (Wach *et al.*, 1994). Most of these markers rely on a transcriptional regulation of a resistance-conferring open reading frame through a promoter and a terminator taken from the *TEF1* gene of *Ashbya gossypii*. The method has been developed further by the construction of template plasmids that combine new selectable markers with tags, such as GFP, that can be PCR targeted in order to create C-terminal gene fusions (Wach *et al.*, 1994; Schneider *et al.*, 1995; Longtine *et al.*, 1998; Goldstein and McCusker, 1999; Goldstein *et al.*, 1999; Knop *et al.*, 1999). The bandwidth of application of this method has been expanded by the construction of cassettes that provide a whole variety of tags, new resistance markers, or promoters in combination with different tags for N-terminal fusions (Longtine *et al.*, 1998; Ito-Harashima and McCusker, 2004; Janke *et al.*, 2004; Sheff and Thorn, 2004; Vorachek-Warren and McCusker, 2004; Sato *et al.*, 2005). The inclusion of flanking *loxP* sites enables the removal of the selectable marker using transient expression of the *cre* recombinase. The *loxP* sites have also permitted N-terminal tagging as well as internal tagging using the endogenous promoter (Gauss *et al.*, 2005).

Using these cassettes for functional analysis in *S. cerevisiae* and other yeasts has now become routine. This is reflected by the fact that most "yeast papers" cite one or more of the papers mentioned above. Furthermore, a few reviews on the issue of PCR targeting have been written that summarize most aspects from a point of view that discusses the advantages and applications of this method (Wendland, 2003; Davis, 2004). Here we aim to help newcomers and more experienced "yeast researchers" alike to make successful use of this strategy, by providing shortcuts to different applications and an overview of the available cassettes and critical methods. We have also put emphasis on discussing in detail important concerns that one has to consider when using this technique. These concerns mainly relate to clean microbiological techniques and genetic issues associated with the genomic manipulation of cells. Furthermore, we discuss problems that often emerge when using PCR targeting and ways to solve them.

♦♦♦♦♦♦ II. APPLICATION OF PCR TARGETING FOR FUNCTIONAL STUDIES

A. Gene Deletion

The complete deletion of a gene is one of the most frequent applications of PCR targeting (Figure 1). It enables one to target the deletion in a precise manner, such that the whole coding sequence of a gene is deleted. Available selection markers, as well as many of their features, are listed in Tables 1 and 2. When planning a deletion, several considerations have to be taken. One has always to consider the genomic context of a particular gene. For example, in the case of overlapping ORFs, a deletion of one gene might also lead to the deletion of the other gene. It is not always the case that predicted overlapping ORFs encode for transcribed genes; however, phenotypic analysis of such a deletion will always bear a certain level of uncertainty about the real cause of any observed phenotypes. To prevent this one could test the consistency of the results using different deletions, either of the full gene or of only a part of the gene that is predicted not to affect the expression of the overlapping ORF. The ORF maps provided by the yeast genome database (www.yeastgenome.org) are of great help in planning the deletions. Another important consideration is the phenotypic consequences of a gene deletion. Deleting an essential gene in a haploid strain is simply not a good idea. However, it might well also be that a specific gene that has been found not to be essential in genome-wide approaches is in fact essential in another genetic strain background. There are many genes for which this is indeed the case. In addition,

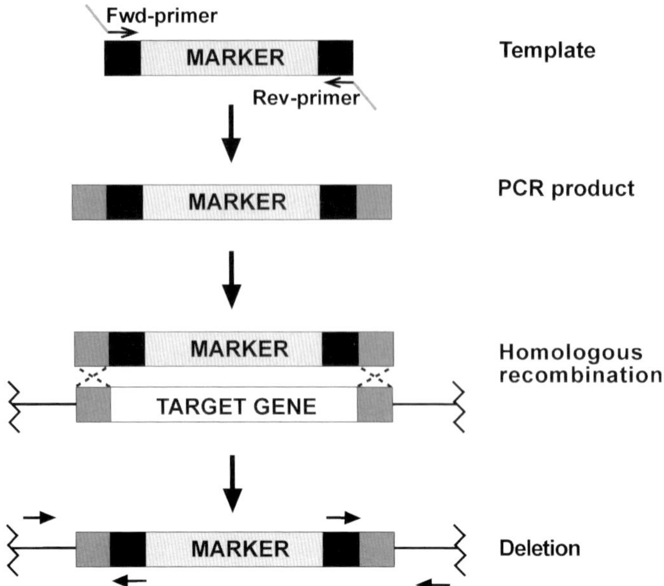

Figure 1. Gene deletion. Cassettes to be used for gene deletion are described in Tables 1 and 2. Primers for PCR amplification of the cassette are designed such that the 5′end of the forward primer contains ~45–55 nucleotides upstream of the ATG (including the ATG) of the target gene while the 3′ end contains ~20 nucleotides of homology to the plasmid template. The 5′ end of the reverse primer contains ~50 nucleotides of the reverse complement after the stop codon (including the stop codon) while the 3′ end contains ~20 nucleotides of homology to the plasmid template. After PCR amplification of the cassette, it is transformed into yeast. Cells that integrated the cassette by homologous recombination into their genome are selected on medium selective for presence of the marker gene. Correct recombination into the target gene locus is verified by two chromosomal PCRs. The 5′ junction is checked using a target gene specific primer of ~21 nucleotides, which anneals upstream of the cassette integration site, together with a primer specific for the 5′end of the PCR cassette. The 3′ junction is checked using a target gene specific primer of ~21 nucleotides, which anneals downstream of the cassette integration site, and a primer specific for the 3′end of the PCR cassette.

pre-existing genomic manipulations of the strain of interest may cause synthetic phenotypes with an otherwise non-essential gene deletion. There are also many situations in which this is the case. To account for this, it may be preferred to perform the deletion in a diploid background, and to analyze the segregation pattern upon sporulation and tetrad dissection for phenotypic consequences that are linked to the deletion.

B. C-terminal Tagging

Another popular use of PCR-targeting is the construction of genomic gene fusions with DNA sequences that encode for a polypeptide (tag) with specific properties (Figure 2). Popular tags are fluorescent proteins, such as GFP, and short peptide sequences that

Table 1. Dominant markers for PCR targeting

Cassette	ORF	Promotor/ terminator	Selective medium	Reference	Notes
kanMX1	kan^R, Tn903	*TEF1* (*A. gossypii*)	YPD+G418	Wach *et al*. (1994)	Hybrid kan^R cassette flanked by 460 bp long direct repeats (derived from 3′ end of *AgLEU2* gene), facilitates recycling of marker, ORF encodes aminoglycoside phosphotransferase
kanMX2	kan^R, Tn903	*TEF1* (*A. gossypii*)	YPD+G418	Wach *et al*. (1994)	Hybrid kan^R allele, deletion of Tn903-derived kan^R 3′-UTR
kanMX3	kan^R, Tn903	*TEF1* (*A. gossypii*)	YPD+G418	Wach *et al*. (1994)	Based on *kanMX2*, additionally flanked by 460 bp long direct repeats (derived from 3′ end of *AgLEU2* gene), facilitates recycling of marker
kanMX4	kan^R, Tn903	*TEF1* (*A. gossypii*)	YPD+G418	Wach *et al*. (1994)	Based on *kanMX2*, additionally flanked by more restriction sites
kanMX6	kan^R, Tn903	*TEF1* (*A. gossypii*)	YPD+G418	Wach (1996)	Similar to *kanMX4*, but single restriction site is exchanged
loxP–kanMX–loxP	kan^R, Tn903	*TEF1* (*A. gossypii*)	YPD+G418	Güldener *et al*. (1996)	Based on *kanMX4*, additionally flanked by 34 bp long *loxP* sites (derived from bacteriophage P1), facilitates recycling of marker

(*Continued*)

Table 1. *(Continued)*

Cassette	ORF	Promotor/ terminator	Selective medium	Reference	Notes
loxP–ble^R–loxP	ble^R, Tn5	*TEF1* (*A. gossypii*)	YPD+phleo	Güldener *et al.* (2002)	ORF encodes a gene mediating bleomycin resistance
hphMX3/MX4	*hph* (*K. pneumoniae*)	*TEF1* (*A. gossypii*)	YPD+hygro	Goldstein and McCusker (1999)	ORF encodes hygromycin B phosphotransferase
hphNT1	*hph* (*K. pneumoniae*)	*TEF1* (*A. gossypii*)/ *CYC1* (*S. cerevisiae*)	YPD+hygro	Janke *et al.* (2004)	*TEF*-terminator of MX cassettes exchanged to that of *CYC1*
natMX3/MX4	nat1 (*S. noursei*)	*TEF1* (*A. gossypii*)	YPD+nat	Goldstein and McCusker (1999)	ORF encodes nourseothricin-*N*-acetyltransferase, ORF is GC rich
natNT2	*nat1* (*S. noursei*)	*TEF1* (*A. gossypii*)/ *ADH1* (*S. cerevisiae*)	YPD+nat	Janke *et al.* (2004)	*TEF*-terminator of MX cassettes exchanged to that of *ADH1*
dsdA-MX4	dsdA (*E. coli*)	*TEF1* (*A. gossypii*)	SD+D-Ser	Vorachek-Warren and McCusker (2004)	ORF encodes D-serine deaminase
loxP-dsdAMX4-loxP	dsdA (*E. coli*)	*TEF1* (*A. gossypii*)	SD+D-Ser	Vorachek-Warren and McCusker (2004)	Recyclable marker
patMX3/MX4	pat (*S. viridochromogenes*)	*TEF1* (*A. gossypii*)	SDP+bialaphos/ SDP+glufosinate	Goldstein and McCusker (1999)	ORF encodes phosphinothricine-*N*-acetyltransferase

Selective media: YPD (Amberg *et al.*, 2005)+200 mg/l geneticin (G418); YPD+7.5 mg/l phleomycin (phleo); YPD+300 mg/l hygromycin (hygro); YPD+100 mg/l nourseothricin (nat, ClonNAT); SD+D-Ser = 1.7 g/l Difco YNB without $(NH_4)_2SO_4$ and without amino acids, 5 g/l L-proline, 0.5 g/l D-serine (Sigma), 20 g/l dextrose; SDP+bialaphos/ SDP+glufosinate = SDP (1.7 g/l Difco YNB without $(NH_4)_2SO_4$ and without amino acids, 1 g/l L-proline, 20 g/l dextrose)+200 mg/l bialaphos or SDP+600–800 mg/l glufosinate. Suppliers: geneticin (G418): Invitrogen; hygromycin B: Cayla, Toulouse, FR (www.cayla.com); Roche, Calbiochem–Novabiochem; nourseothricin: (ClonNAT), Werner Bioagents, Jena-Cospeda, GER (www.webioage.com); bialaphos: Shinyo Sangyo Co., Tokyo, JP; glufosinate: Sigma-Aldrich; phleomycin: Cayla, FR, Invitrogen.

Table 2. Auxotrophy markers for PCR targeting

Cassette	Origin of ORF	Promotor/ terminator	Selective medium	Counterselective medium	Reference	Notes
AgLEU2	*LEU2* (*A. gossypii*)	*LEU2* (*A. gossypii*)	SC-Leu	–	Wach *et al.* (1994)	Marker poorly expressed, therefore slow growth of transformants, ORF codes for β-isopropyl malate dehydrogenase
AgLEU2 MX1	*LEU2* (*A. gossypii*)	*LEU2* (*A. gossypii*)	SC-Leu	–	Wach *et al.* (1994)	Marker poorly expressed, therefore slow growth of transformants; cassette flanked by 460 bp long direct repeats (derived from 3′ end of *AgLEU2* gene), facilitates marker recycling
loxP–LEU2–loxP	*LEU2* (*K. lactis*)	*LEU2* (*K. lactis*)	SC-Leu	–	Güldener *et al.* (2002)	Recyclable marker
CaURA3MX3/MX4	*URA3* (*C. albicans*)	*TEF1* (*A. gossypii*)	SC-Ura	SD+5-FOA	Goldstein *et al.* (1999)	ORF encodes orotidine-5′-phosphate decarboxylase, MX3 recyclable marker
loxP–KlURA3–loxP	*URA3* (*C. albicans*)	*TEF1* (*A. gossypii*)	SC-Ura	SD+5-FOA	Goldstein *et al* (1999)	Recyclable marker
LYS5MX3/MX4	*LYS5* (*S. cerevisiae*)	*TEF1* (*A. gossypii*)	SC-Lys	SDP+AAA	Ito-Harashima and McCusker (2004)	ORF encodes phosphopantetheinyl transferase, which activates *LYS2* (α-aminoadipate reductase), MX3 is recyclable

(*Continued*)

Table 2. *(Continued)*

Cassette	Origin of ORF	Promotor/ terminator	Selective medium	Counterselective medium	Reference	Notes
CaLYS5MX3	*LYS5* (*C. albicans*)	*LYS5* (*C. albicans*)	SC-Lys	SDP+AAA	Ito-Harashima and McCusker (2004)	
loxP–LYS5MX–loxP	*LYS5* (*S. cerevisiae*)	*TEF1* (*A. gossypii*)	SC-Lys	SDP+AAA	Ito-Harashima and McCusker (2004)	Recyclable marker
loxP–CaLYS5MX–loxP	*LYS5* (*C. albicans*)	*TEF1* (*A. gossypii*)	SC-Lys	SDP+AAA	Ito-Harashima and McCusker (2004)	Recyclable marker
loxP–LYS2–loxP	*LYS2* (*S. cerevisiae*)	*LYS2* (*S. cerevisiae*)	SC-Lys	SDP+AAA	Delneri *et al.* (2000)	Recyclable marker
TRP1	*TRP1* (*S. cerevisiae*)	*TRP1* (*S. cerevisiae*)	SC-Trp	SDP+AAA	Longtine *et al.* (1998)	ORF codes for a phosphoribosylanthranilate isomerase
KlTRP1	*TRP1* (*K. lactis*)	*TRP1* (*K. lactis*)	SC-Trp	SD+FAA	Knop *et al.* (1999)	
HIS3MX6[1]	*HIS5* (*S. pombe*)	*TEF1* (*A. gossypii*)	SC-His	–	Wach *et al.* (1997)	ORF encodes imidazoleglycerol-phosphate dehydratase (encoded by *HIS3* in *S. cerevisiae* and *HIS5* in *S. pombe*)
loxP–HIS3MX–loxP	*HIS5* (*S. pombe*)	*TEF1* (*A. gossypii*)	SC-His	–	Delneri *et al.* (2000)	Recyclable marker

Selective media: SC (synthetic complete medium) minus indicated amino acid. Counterselective media: SD+5-FOA = SD+1 mg/ml 5-fluoroorotic acid (5-FOA); SD+FAA = SD+0.5 mg/ml 5-fluoroanthranilic acid (5-FAA); SDP+AAA = SDP+2 mg/ml α-aminoadipate (AAA).

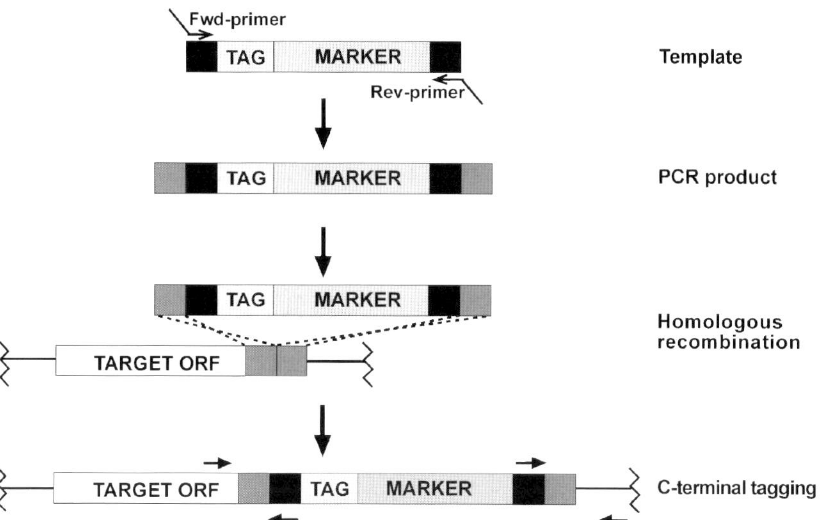

Figure 2. C-terminal tagging. Primers for PCR amplification of the cassette are designed such that the 5′end of the forward primer contains ~45–55 nucleotides of homology to the coding sequence upstream of the stop codon (excluding the stop codon) of the target gene while the 3′ end contains ~20 nucleotides of homology to the plasmid template designed to maintain the reading frame between the target gene and the tag. The 5′ end of the reverse primer contains ~50 nucleotides of homology to the reverse complement after the stop codon (including the stop codon) of the target gene while the 3′ end contains ~20 nucleotides of homology to the plasmid template. After PCR amplification of the cassette, it is transformed into yeast. Cells that integrated the cassette by homologous recombination into their genome are selected on medium selective for the presence of the marker gene. Correct recombination into the target gene locus is verified by two chromosomal PCRs as described in the legend to Figure 1. A wealth of cassettes that fulfill many needs with respect to different tags are described in a variety of publications including Knop *et al.* (1999), Longtine *et al.* (1998), Janke *et al.* (2004) and Sheff and Thorn (2004).

are recognized by commercially available and well-established monoclonal antibodies. A whole variety of such tags have been introduced in the field of yeast research (see literature cited in the legend to Figure 2). The application of this strategy to perform genome-wide localization of proteins using GFP or the measurement of protein abundance using Protein A demonstrated that in many cases the tag does not obviously interfere with the function of a protein (Ghaemmaghami *et al.*, 2003; Huh *et al.*, 2003). However, the simple detection of the expression product and the absence of a strong growth defect does not prove anything about the real functionality of the protein fusion on a quantitative level for essential genes and, of course, not at all for non-essential genes. This has to be kept in mind when working with strains that express C-terminally tagged proteins. For individual cases, it is necessary to validate the functionality of the protein fusion in order to draw firm conclusions from results obtained with such a construct. This might not be that easy, especially in situations where nothing is known about the function of a protein. A more generally applicable way to validate the functionality is to test the ability of a gene fusion to complement the wild-type gene in a strain background where the gene is essential. The availability of genome-wide datasets about synthetic lethal interactions (Tong *et al.*, 2001; data available through www.yeastgenome.org) helps to address this issue in an efficient way. It must also be considered that different tags may have different effects on the functionality of a protein. Furthermore, tags may influence the localization of a protein, or affect one of many functions of a protein. Because of this, any result obtained with a tagged protein needs to be verified using an independent method, wherever possible.

C. Promoter Substitution and N-terminal Tagging

Expression of a gene from a foreign promoter is another powerful method to analyze the function of genes. Classically, the regulatable *GAL1* promoter has been used to study the consequences of gene depletions, e.g. for the analysis of cell cycle dependent regulation of protein stability, but also as a means of constructing a conditional allele. Other regulatable promoters have been used for analogous investigations, including the *MET3*, *MET25*, or *CUP1* promoters. Furthermore, the use of constitutive promoters with different strengths offers additional possibilities for functional studies. In order to combine this strategy with PCR targeting, a number of cassettes have been generated that enable the genomic substitution of the endogenous promoter of a gene with a promoter of choice. In addition, several such promoters are also available in combination with different tags, which leads to the expression of N-terminally tagged proteins under the control of a specific promoter (Figure 3). As always, when using PCR targeting, the consequences of changing the level of expression of genes have to be considered. Artifacts may arise that are difficult to interpret, leading to results that are of limited value. In particular, the overexpression of a gene using a strong promoter in order to localize the protein within the cell is not a good idea (although frequently done). In many cases this has been proven to lead to mislocalization of a protein, due either to untimely expression or saturation of binding sites. Aggregation and even mistargeting of a protein to other sites inside the cell have also been observed, and have led to misleading results in publications. However, endogenous protein expression is sometimes so low that the sensitivity of detection offered by antibodies or GFP is not high enough to enable the localization of a protein. In this case, it is recommended first to try tags that allow more sensitive detection, such as tandem HA or *myc* tags or tandem GFP constructs, before attempting to overexpress a protein.

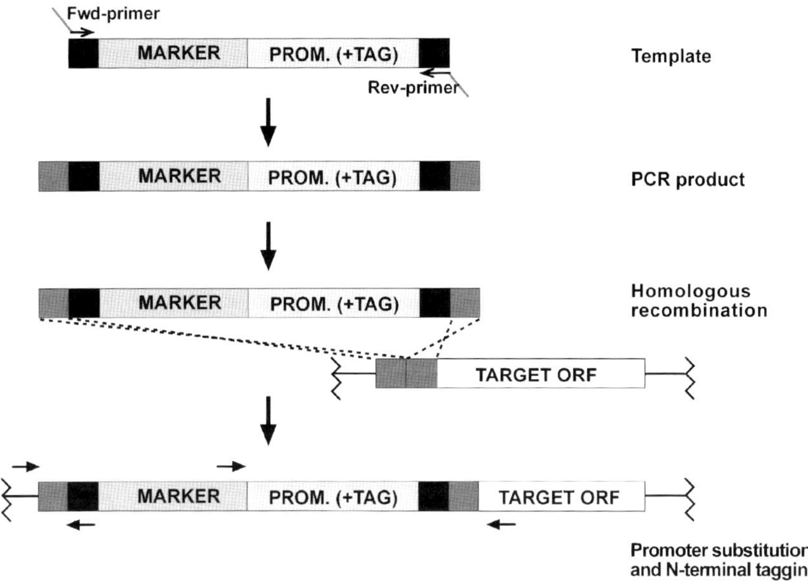

Figure 3. Promoter substitution and N-terminal tagging. The forward primer is designed in analogy to the forward primer used for gene deletion (Figure 1). The 5′ end of the reverse primer contains ~45–55 nucleotides of homology to the reverse complement after the start codon (no start codon for cassettes from Janke *et al.*, 2004) of the target gene while the 3′ end contains ~20 nucleotides of homology to the plasmid template. After PCR amplification of the cassette, it is transformed into yeast. Cells that integrated the cassette by homologous recombination into their genome are selected on medium selective for the presence of the marker gene. Correct recombination into the target gene locus is verified by two chromosomal PCRs. The 5′ junction is checked using a target gene-specific primer of ~21 nucleotides, which anneals upstream of the cassette integration site, together with a primer specific for the 5′ end of the PCR cassette. The 3′ junction is checked using a target gene-specific primer of ~21 nucleotides, which anneals downstream of the cassette integration site, and a primer specific for the 3′ end of the PCR cassette. A series of cassettes that fulfill many needs with respect to different promoters and tags are described in publications including Longtine *et al.* (1998) and Janke *et al.* (2004).

D. Gene Truncations

PCR targeting is also a rapid method for mapping the domain architecture of a protein. C-terminal truncations can be performed by deletion of only a fraction of the gene in such a way that a stop codon followed by a transcription terminator and a selectable marker is introduced into the strains (Figure 4). Using PCR targeting, deletion studies are greatly accelerated and allow studying the effects of truncations on chromosomally expressed protein with unaltered expression levels. Although the construction of N-terminal truncations can be pursued in the same way (Figure 4), they require either the application of the *loxP* system or the introduction of a promoter. This can be either one of the available promoters present in pre-existing constructs, or, alternatively, a specific cassette constructed for this purpose, containing the endogenous promoter

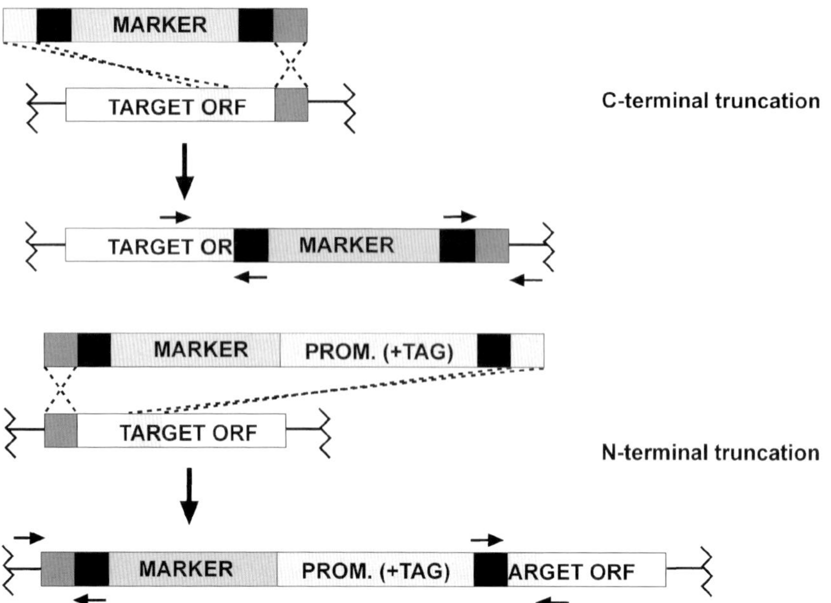

Figure 4. Gene truncation. For C-terminal gene truncation studies primers for PCR amplification of the cassette are designed such that the 5′ end of the forward primer contains ~45–55 nucleotides of homology to the coding sequence just upstream of the first codon of the part of the gene that is to be deleted. The primer can provide a new stop codon where no tag is to be added at the C-terminus, but the stop codon is omitted in cases where a cassette with a tag is being used to enable detection of the truncated protein. The reverse primer is designed as described for gene deletion or C-terminal tagging.

For N-terminal truncation studies, the forward primer is constructed as described for promoter substitution and N-terminal tagging (Figure 3). The 5′ end of the reverse primer contains ~45–55 nucleotides of homology to the coding sequence immediately after the last codon of the part of the ORF that is to be deleted, while the 3′ end contains ~20 nucleotides of homology to the plasmid template. For N-terminal tagging, cassettes have to provide promoters that ideally match more or less the expression strength of the gene of interest. The cassettes may or may not contain a tag in order to detect the truncated protein. A selection of cassettes with different promoters of different strength are described in Janke *et al.* (2004). For C-terminal deletions, the cassettes used should also provide a terminator, because disruption of a terminator may lead to mRNA instability and thus to lower expression of a gene. All cassettes containing the promoter of *TEF1* from *A. gossypii* (all pFA6-MX or NT plasmids: Wach *et al.*, 1994; Goldstein and McCusker, 1999; Goldstein *et al.*, 1999; Janke *et al.*, 2004) also provide a terminator from the gene upstream of the *TEF1* gene (ADL369c), as well as ~60 nucleotides intervening sequence. This fragment may help to terminate the transcript, but care has to be taken with regard to altered expression levels, because termination is not perfect.

of the gene under study. Although the use of PCR targeting for truncation studies is straightforward, it is less flexible for further manipulations to the gene, such as making internal deletions or the use of site-directed mutagenesis to scan for important residues. In these cases, a plasmid-borne strategy is usually recommended.

E. Marker Rescue Using the *cre/loxP* System: N-terminal and Internal Tags

The insertion of *loxP* sites by homologous recombination to the flanks of a piece of DNA, e.g. a selectable marker, allows the induced and selective excision of the intervening stretch of DNA upon expression of the Cre recombinase in the cell. After excision, one *loxP* site is left behind in the genome. This strategy has been successfully applied for repeated use of the same marker in a strain. In addition, the strategy can be used to insert DNA into the genome at sites where a marker would disturb the expression and function of a gene, e.g. at the N-terminus or inside an ORF (Prein *et al.*, 2000; Puig *et al.*, 2001; Gauss *et al.*, 2005). It can also be applied for N-terminal truncation of a gene under its endogenous promoter. The strategy requires two consecutive steps and is straightforward (Figure 5). However, the resulting strains have to be carefully analyzed for unwanted genomic rearrangement events that might have been induced by the Cre recombinase. Ideally, a successfully manipulated strain is crossed once with a wild type followed by tetrad analysis. Any larger genomic rearrangements will lead to lowered spore viability due to problems associated with meiotic recombination and thus spore genomic abnormalities. This analysis is pivotal to situations where the strains would be used for meiosis analysis. Another issue regarding this strategy is that it involves an intermediate step where the targeted gene is inactivated. If the manipulation is to be applied to an essential gene, it therefore has to be performed in a diploid strain background (Gauss *et al.*, 2005).

F. *In Vivo* Site-directed Mutagenesis

PCR targeting enables the alteration of genomic sequence in a base-specific manner at virtually any genomic locus (Storici *et al.*, 2001). Single base pair alterations, insertions and deletions can be constructed without leaving any foreign helper DNA (such as a selectable marker or *loxP* sites) in the genome. This time, a two-step strategy (Figure 6) is applied. First, a double selection cassette containing at least one counter-selectable marker in addition to a second marker is introduced near the site where the alteration is planned. Then the strain is transformed with a piece of DNA either generated from annealed primers or excised from a cloned fragment and carrying the desired alteration. Providing that the transformed DNA contains enough homology to both sides of the inserted marker, it will recombine and lead to the loss of the marker. This event can be assayed by counter-selection of the first marker and validated for a concomitant loss of the second marker, demonstrating that resistance is not caused by a mutation in the background of the strain. This strategy, as compared to plasmid-borne strategies, is considerably more time-consuming, but enables the generation of stable strains that differ only by the introduced mutation. Such

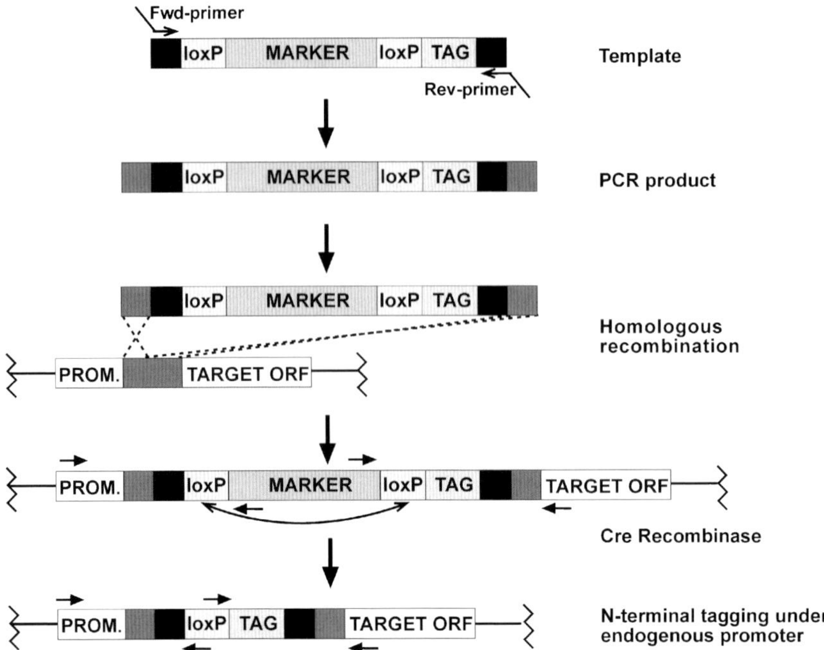

Figure 5. *cre/loxP* system: N-terminal and internal tagging, marker rescue. For N-terminal tagging primers for PCR amplification of the cassette are designed such that the 5′ end of the forward primer contains ~45–55 nucleotides upstream of the ATG of the target gene while the 3′ end contains ~20 nucleotides of homology to the plasmid template. The target site is chosen to allow homologous recombination in frame with the ORF. The 5′ end of the reverse primer contains ~45–55 nucleotides of the reverse complement downstream of the start codon of the target gene while the 3′ end contains ~20 nucleotides of homology to the plasmid template. For internal tagging, the extended 5′ ends of the primers are homologous to a region within the gene. For further details for primer design please refer to the relevant publications (listed at the end of this legend). After PCR amplification of the cassette, it is transformed into yeast. Cells that integrated the cassette by homologous recombination into their genome are selected on medium selective for the presence of the marker gene. Correct recombination into the target gene locus is verified by two chromosomal PCRs. The 5′ junction is checked using a target gene-specific primer of ~21 nucleotides, which anneals upstream of the cassette integration site, together with a primer which anneals downstream of the first *loxP* site in the PCR cassette. The 3′ junction is checked using a target gene-specific primer of ~21 nucleotides, which anneals downstream of the cassette integration site, and a primer which anneals upstream of the second *loxP* site in the PCR cassette.

In order to remove the transient gene disruption via the integrated marker cassette, positive clones are transformed with a plasmid expressing the Cre recombinase under the *GAL1* promoter. After induction of the recombinase with galactose, the marker cassette is excised via recombination, thereby restoring the expression of the target gene. Correct marker rescue is again checked by two chromosomal PCRs. The 5′ junction is checked using a target gene specific primer of ~21 nucleotides, which anneals upstream of the cassette integration site, together with a primer which anneals in the *loxP* site. The 3′ junction is checked using a target gene specific primer of ~21 nucleotides, which anneals downstream of the cassette integration site, together with a primer which anneals in the *loxP* site. The following publications report on cassettes to be used with the *cre/loxP* system: De Antoni and Gallwitz (2000); Delneri *et al.* (2000); Prein *et al.* (2000); Güldener *et al.* (2002) and Gauss *et al.* (2005).

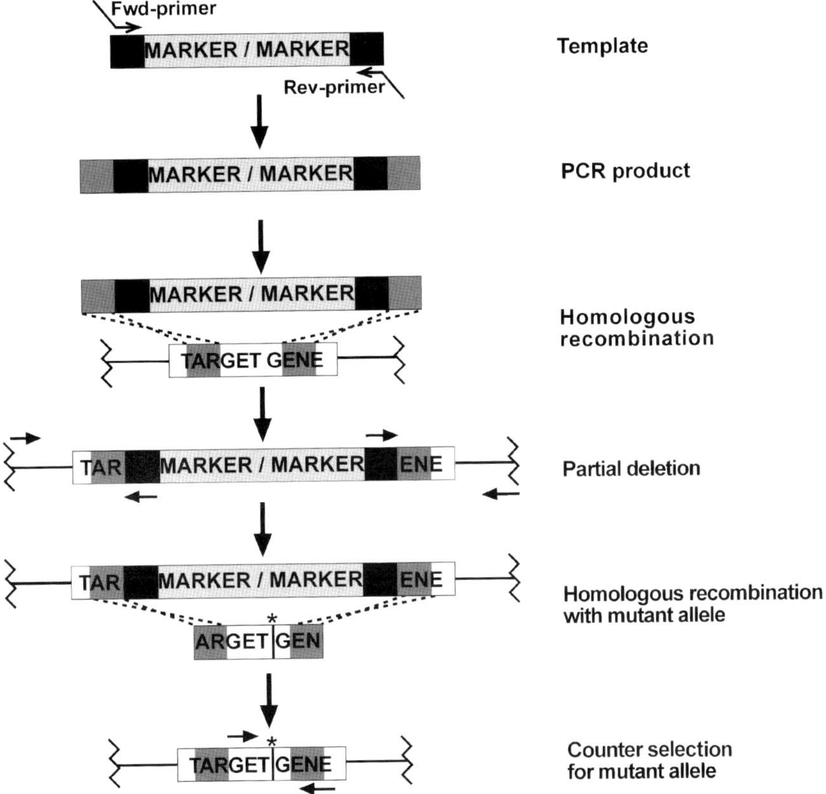

Figure 6. *In vivo* site-directed mutagenesis. Primers for PCR amplification of the cassette are designed such that the 3′ end of the forward primer contains ~20 nucleotides of homology to the plasmid template while the 5′ end contains ~45–55 nucleotides upstream to the region of the target gene where the point mutation will be introduced. The 3′ end of the reverse primer contains ~20 nucleotides of homology to the plasmid template while the 5′ end contains ~45–55 nucleotides of the reverse complement downstream to region of the target gene where the point mutation will be introduced. After PCR amplification, the double marker cassette (e.g. KanMX/*URA3*: Storici *et al.*, 2001) is transformed into yeast. Cells that integrated the cassette by homologous recombination into their genome are selected on medium selective for one of the two marker genes. Correct recombination into the target gene locus is verified by two chromosomal PCRs. The 5′ junction is checked using a target gene-specific primer of ~21 nucleotides, which anneals upstream of the cassette integration site, together with a primer specific for the PCR cassette. The 3′ junction is checked using a target gene-specific primer of ~21 nucleotides, which anneals downstream of the cassette integration site, and a primer specific for the PCR cassette.

In a second yeast transformation step, the mutated allele (containing homology to sequences of the target gene adjacent to the marker cassette; either a piece of DNA excised from a plasmid, or a pair of annealed primers) is co-transformed with an empty plasmid. Cells are first selected for the uptake of plasmid. Single colonies that derive from cells that have concomitantly recombined the co-transformed DNA with the mutant allele are then selected for loss of markers by counter-selection on 5′-FOA and subsequent replica plating on plates that indicate the absence of the second marker. Only cells that can grow on 5′-FOA, but not on the plate that assays the presence of the second marker, have correctly integrated the mutant DNA fragment. The region encompassing the point mutation can be amplified from genomic DNA and sequenced for verification.

strains are key for functional studies that require quantitative investigations (e.g. for "systems biology").

♦♦♦♦♦♦ III. AVAILABLE CASSETTES FOR PCR TARGETING

There are numerous cassettes available that together govern a wide range of properties. Tables 1 and 2 provide an overview about the properties and features of different marker cassettes that contain little or no homology to the genomic sequence of *S. cerevisiae*. Furthermore, these tables refer to literature and other information relevant for their use. Cassettes to be used for other applications of PCR-targeting can be found in the references cited in the legends to Figures 2–5. Given the large variety of cassettes, it is a challenge merely to select the right cassette. There are a number of criteria that have to be considered when attempting to use PCR targeting. An important issue is the availability of the cassettes. Most of the cassettes are freely floating within the yeast community, and knowing someone who has a certain cassette is usually the fastest way to get hold of it. Some cassettes can be obtained for a small handling fee through Euroscarf (http://web.uni-frankfurt.de/fb15/mikro/euroscarf/). An important point that needs careful consideration is the iterative usage of primers for use with different cassettes. In fact, most of the cassettes from one publication are not compatible with cassettes from other publications with respect to the required primer annealing sites. Sometimes, even the cassettes from one publication are not compatible with each other. Since primer prices are an important factor to be considered, careful evaluation regarding the future use of certain primers is required.

♦♦♦♦♦♦ IV. PRIMERS FOR USE WITH PCR TARGETING

To design primers for the amplification of particular cassettes it is advisable to refer to the instructions provided by the authors of the publication concerning the cassette. One important thing to consider is the length of the primer sequence that is homologous to the targeted locus. For many applications 45 bp of homology is sufficient. However, this applies only to cases where standard laboratory yeast strains are used that have been shown to allow efficient transformation to occur, e.g. derivatives of S288c or W303. Other yeast strains can show considerably lower transformation efficiency (e.g. strain SK1). In this case, 55 bp of homology is preferred. Higher amounts of PCR products used in the transformation further improve the yield. Also, if the homologous chromosomal region contains high amounts of repetitive and AT-rich sequences, such as in terminator regions, it is often helpful to expand the region of homology up to 55 bp or to the next GC pair. *S. cerevisiae* strains other than the sequenced S288c strain may exhibit polymorphism (Gu *et al.*, 2005) or even genomic rearrangements that can prevent the faithful recombination of a PCR product with its target locus. This issue will probably be solved in the future when more and more full genome sequences of other yeast strains become available.

The last, and probably one of the most important points to mention, concerns the quality of the oligonucleotides. From experience, it can be said that the quality of oligonucleotides is subject to dramatic variations from company to company, but also over time from the same company. This is due to the fact that high competition in the field of commercial oligonucleotides synthesis has lowered prices for oligonucleotides and forced companies to reduce production costs, sometimes affecting quality. This alone can give rise to a failure rate of up to 20% already for the amplification of the cassette. Faulty primers can be spotted by testing for their inability to form a PCR product when combined with other well-established primers. Replacement primers can usually be obtained from the manufacturer and (as judged from our own experience) these will normally function properly. The situation becomes more difficult when no transformants are obtained. Faulty primers, genomic polymorphisms or phenotypic consequences of the manipulation, as well as technical issues related to transformation can all be causes. A control transformation of an S288c strain should help to eliminate possibilities.

♦♦♦♦♦♦ V. APPLICATION OF PCR TARGETING FOR GENOME-WIDE STUDIES

PCR targeting has been key for functional genome-wide studies, such as protein localization or gene deletion (Davis, 2004). In principle, any application of PCR targeting can be used for such studies. However, it is highly recommended to perform proof-of-principle studies in order to verify that the chosen cassette and the methods used are able to yield the expected results in a reproducible manner. For such genome-wide studies it may also be useful to construct a cassette *a priori* that provides exactly the desired functionality, rather than using a previously described cassette.

A. Multiple Genomic Manipulations

The ease with which single genomic manipulations can be conducted enables strain constructions that result in several genomic manipulations being present in one strain. Manipulations can be performed in iterative rounds. Since several dominant marker cassettes contain the same promoter and terminator to control expression of different resistance genes, a second cassette will have a high likelihood of integrating at the site of a pre-existing cassette. To prevent this, one has to either select for the presence of both cassettes simultaneously, or alternatively, use cassettes that do not contain homologous sequences. For example, auxotrophic markers such as the *TRP1* gene from *K. lactis* (*klTRP1*) can easily be combined with a *kanMX6* marker, or alternatively, the *natNT2* and *hph1NT1* cassettes (Janke *et al.*, 2004) can be combined with the *kanMX6* cassette.

♦♦♦♦♦♦ VI. CRITICAL METHODS

The following section provides a short list of critical methods that help to get started with PCR targeting. For all of the protocols provided here, many alternatives are available, many of which are probably equivalent. The protocols listed here have a long reputation as being reliable and reproducible. Only steps that are essential for obtaining yeast strains containing validated genomic manipulations using PCR targeting are discussed.

A. PCR Amplification

Protocol 1 gives a standard PCR protocol that reliably yields DNA product. For more details consult previous publications (Knop *et al.*, 1999; Janke *et al.*, 2004). An excellent alternative protocol is provided by Goldstein and McCusker (1999). For markers based on the GC-rich *S. noursei nat1* gene a PCR method that includes DMSO and Triton X-100 is usually required (see Janke *et al.*, 2004, for details). After PCR amplification, 3 µl of the reaction are analyzed on an agarose gel and should give a bright band. For yeast transformation, the DNA is usually ethanol precipitated and resuspended in a 1/5th volume of water, using 5 µl for each transformation.

B. Frozen Competent Yeast Cells and Transformation

The transformation procedure shown in Protocol 2 is based on the method of Schiestl and Gietz (1989) as described in Chapter 3. A modification of the protocol allows freezing of competent cells for further use. For basics in manipulation and growth of yeast cells, please refer to Amberg *et al.* (2005). Regarding the selection of transformants, the following points should be borne in mind:

- If auxotrophic markers are used for selection of homologous recombination, cells can be directly plated onto synthetic complete medium lacking the corresponding amino acid.
- If the PCR product contains an antibiotic resistance marker, cells are resuspended in approximately 3 ml of YPAD and allowed to recover at

Protocol 1. PCR amplification of cassettes.

1. Reaction mix (100 µl)

10 µl	10 × Buffer 1 (500 mM Tris–HCl (pH 9.2), 22.5 mM $MgCl_2$, 160 mM NH_4SO_4)
6.4 µl	10 µM Primer 1
6.4 µl	10 µM Primer 2
3.5 µl	10 mM dNTPs
1.0 µl	Template (~100 µg/ml)
71.5 µl	Water

Hot start with 2 U *Taq* polymerase and 0.4 U Vent polymerase

2. Amplification program

i.	5 min	97°C	
ii.	1 min	97°C	10 cycles
iii.	30 sec	54°C	
iv.	x min	68°C	
v.	1 min	97°C	20 cycles, + 20 sec / cycle
vi	30 sec	54°C	
vii.	x min	68°C	

where x ≅ 1 min/1000 bp

30°C while shaking for at least 4–6 h. Cells are then harvested by centrifugation and plated onto selective medium.

- Colonies usually appear on the plates after two days. Using antibiotics selection, a high background of presumably transiently transformed cells is often visible. This often leads to depletion of nutrients from the medium and thus prevents growth of the real transformants. Replica plating and another overnight incubation solves this problem.

C. Identification of Positive Clones

1. Chromosomal DNA isolation and verification PCR

Protocol 3 is adapted from Finley and Brent (1995) and allows the reliable validation of chromosomal alterations by PCR. Alternative (and shorter protocols) that do not involve a DNA extraction step are often unreliable, especially in cases where a PCR product above approx 700 bp is expected.

2. Protein extraction and immunoblotting

Validation of clones where the protein of interest should be tagged can be done by protein extraction and immunoblotting with an

Protocol 2. Yeast transformation.

Buffers

- SORB: 100 mM LiOAc, 10 mM Tris–HCl (pH 8.0), 1 mM EDTA (pH 8.0), 1 M sorbitol (special grade for molecular biology from Merck), pH adjusted to pH 8.0 with diluted acetic acid, sterile filtered.
- PEG: 100 mM LiOAc, 10 mM Tris–HCl (pH 8.0), 1 mM EDTA (pH 8.0), 40% PEG 3350 (Sigma), sterile filtered.
- Carrier DNA: Salmon sperm DNA (Gibco) is denatured at 95°C for 10 min and cooled on ice. DNA can be stored –20°C and re-used.

Competent yeast cells

1. Grow an overnight culture in YPAD medium or selective medium (SD) if the cells contain a plasmid.
2. In either instance, dilute the overnight culture 50-fold in YDAP and re-grow to an OD_{600} of 0.8–2.0 ($1–2 \times 10^7$ cells per ml) at 30°C in YPAD (~6 h).
3. Harvest yeast cells by centrifugation at 500G for 5 min at room temperature (RT).
4. Wash cells once in 1 volume of sterile water and collect by centrifugation at 500G for 5 min at RT.
5. Wash cells once in 0.2 volume of SORB and collect by centrifugation at 500G for 5 min at RT.
6. Fully aspirate the SORB and resuspend cells in a total volume of 360 µl SORB per 50 ml culture.
7. Add 40 µl of carrier DNA.
8. Aliquot cells into appropriate volumes (e.g. 50 µl, at RT). Cells can either be directly used for transformation or stored at −80°C (do not shock freeze).

Transformation

1. Mix 1 to 5 µl of DNA with 50 µl competent yeast cells.
2. Add a six-fold volume of PEG (e.g. 300 µl) and mix the samples by inversion.
3. Incubate cells at RT for 30–60 min.
4. Add DMSO (1/9th volume, e.g. 40 µl) and mix the samples by inversion.
5. Heat shock the cells at 42°C for 15 min.
6. Collect the cells by centrifugation (2 min, 500 g) and resuspend for plating (but see text for note concerning antibiotic resistance markers).

Protocol 3. DNA extraction for PCR verification.

Buffers

- S-buffer: 10 mM K_2HPO_4 (pH 7.2), 10 mM EDTA, 50 mM β-mercaptoethanol.
- 50 × stock of Zymolyase 100 T (10 mg/ml).
- Lysis solution: 25 mM Tris–HCl (pH7.5), 25 mM EDTA, 2.5% SDS.
- 3 M potassium acetate.

DNA extraction

1. Resuspend a small amount of cells (from plate) in 100 μl of S-buffer with Zymolyase (use a sterile micropipette tip for the cells).
2. Incubate at 37°C for 30 min.
3. Add 55 μl of Lysis solution.
4. Incubate at 65°C for 30 min.
5. Add 86 μl of 3 M potassium acetate, chill on ice for 15 min.
6. Spin at 4°C for 10 min at maximum speed in a microfuge.
7. Transfer the supernatant into a new tube and add 400 μl of 100% ethanol.
8. Incubate on ice for 10 min.
9. Spin at 4°C for 10 min at maximum speed in a microfuge.
10. Wash pellet with 500 μl of 70% ethanol (centrifuge for 5 min at maximum speed in a microfuge).
11. Dry pellet (e.g. Speedvac vacuum centrifuge) and resuspend in 30 μl water.
12. Use 1 μl of DNA solution for consecutive PCR.
13. *For long verification PCRs:* Protocol 1 yields reliable results. *Taq* polymerase is sufficient; no addition of a proofreading polymerase is necessary.
14. *For short verification PCRs:* standard *Taq*-PCR protocols using 2.5 mM $MgCl_2$ work well. Primers that have annealing temperature above 55°C are more reliable.

antibody that recognizes the tag. Protocol 4 describes a protein extraction procedure using a NaOH/β-mercaptoethanol/TCA protocol (Knop *et al.*, 1999).

3. Fluorescence microscopy

Proteins modified with a fluorophore can be verified for proper construction by fluorescence microscopy. For this reason strains are grown in sterile filtered synthetic complete medium, which gives low background fluorescence as compared with YPD. Cells are analyzed under a fluorescence microscope using the appropriate filter sets (for further details, see Amberg *et al.* (2005) and Chapter 11, this volume).

Protocol 4. Protein extraction for Western blot analysis.

Buffers

- 1.85 M NaOH, 7.5% (v/v) β-mercaptoethanol (added freshly each time).
- 55% (w/v) TCA (trichloroacetic acid, stored in the dark).
- HU-buffer: 8 M urea, 5% SDS, 200 mM Tris–HCl (pH 6.8), 1 mM EDTA bromophenol blue as coloring and pH indicator, 1.5% (w/v) DTT).

Protocol

1. Collect samples corresponding to 0.5–3 OD_{600} of cells.
2. Harvest cells by centrifugation and resuspend pellet in 1 ml of cold water.
3. Add 150 μl of 1.85 M NaOH/7.5% β-mercaptoethanol and incubate on ice for 15 min.
4. Add 150 μl of 55% TCA and incubate for 15 min on ice.
5. Spin samples at 13 000 rpm for 10 min at 4°C.
6. Remove supernatant.
7. Centrifuge again for 10 sec at 13 000 rpm at 4°C to completely remove residual TCA.
8. Resuspend pellet in 30–100 μl HU-buffer per OD_{600} of cells.
9. Denature samples at 65°C for 10 min (if HU-buffer turns yellow, add 1–3 μl of 2 M Tris). Make sure that the pellets are completely resuspended.
10. Spin samples for 5 min at 13 000 rpm at room temperature.
11. Analyze samples corresponding to 0.2–0.5 OD_{600} of cells by SDS-PAGE followed by immunoblotting.

♦♦♦♦♦♦ VII. CONCLUDING REMARKS

PCR targeting has proven to be a powerful method for functional studies with yeast. It certainly was one of the most important keys to the continuous success of this model organism for cell biological and functional genomics studies. In the future, we will see an even greater number of new cassettes constructed by different laboratories, but less and less will probably become published as separate papers.

References

Amberg, D., Burke, D. and Strathern, J. (2005). *Methods in Yeast Genetics. A Cold Spring Harbor Laboratory Course Manual.* Cold Spring Harbor Laboratory Press, New York.

Baudin, A., Ozier-Kalogeropoulos, O., Denouel, A., Lacroute, F. and Cullin, C. (1993). A simple and efficient method for direct gene deletion in *Saccharomyces cerevisiae*. *Nucleic Acids Res.* **21**, 3329–3330.

Davis, T. N. (2004). Protein localization in proteomics. *Curr. Opin. Chem. Biol.* **8**, 49–53.

De Antoni, A. and Gallwitz, D. (2000). A novel multi-purpose cassette for repeated integrative epitope tagging of genes in *Saccharomyces cerevisiae*. *Gene* **246**, 179–185.

Delneri, D., Tomlin, G. C., Wixon, J. L., Hutter, A., Sefton, M., Louis, E. J. and Oliver, S. G. (2000). Exploring redundancy in the yeast genome: an improved strategy for use of the *cre-loxP* system. *Gene* **252**, 127–135.

Finley, R. and Brent, R. (1995). Interaction trap cloning with yeast. In: *DNA Cloning – a Practical Approach. Vol. 2: Expression Systems*(D. Glover and B. Hames, eds), pp. 169–203. Oxford University Press, Oxford.

Gauss, R., Trautwein, M., Sommer, T. and Spang, A. (2005). New modules for the repeated internal and N-terminal epitope tagging of genes in *Saccharomyces cerevisiae*. *Yeast* **22**, 1–12.

Ghaemmaghami, S., Huh, W. K., Bower, K., Howson, R. W., Belle, A., Dephoure, N., O'Shea, E. K. and Weissman, J. S. (2003). Global analysis of protein expression in yeast. *Nature* **425**, 737–741.

Goldstein, A. L. and McCusker, J. H. (1999). Three new dominant drug resistance cassettes for gene disruption in *Saccharomyces cerevisiae*. *Yeast* **15**, 1541–1553.

Goldstein, A. L., Pan, X. and McCusker, J. H. (1999). Heterologous *URA3MX* cassettes for gene replacement in *Saccharomyces cerevisiae*. *Yeast* **15**, 507–511.

Gu, Z., David, L., Petrov, D., Jones, T., Davis, R. W. and Steinmetz, L. M. (2005). Elevated evolutionary rates in the laboratory strain of *Saccharomyces cerevisiae*. *Proc. Natl. Acad. Sci. USA* **102**, 1092–1097.

Güldener, U., Heck, S., Fielder, T., Beinhauer, J. and Hegemann, J. H. (1996). A new efficient gene disruption cassette for repeated use in budding yeast. *Nucleic Acids Res.* **24**, 2519–2524.

Güldener, U., Heinisch, J., Koehler, G. J., Voss, D. and Hegemann, J. H. (2002). A second set of *loxP* marker cassettes for Cre-mediated multiple gene knockouts in budding yeast. *Nucleic Acids Res.* **30**, e23.

Huh, W. K., Falvo, J. V., Gerke, L. C., Carroll, A. S., Howson, R. W., Weissman, J. S. and O'Shea, E. K. (2003). Global analysis of protein localization in budding yeast. *Nature* **425**, 686–691.

Ito-Harashima, S. and McCusker, J. H. (2004). Positive and negative selection *LYS5MX* gene replacement cassettes for use in *Saccharomyces cerevisiae*. *Yeast* **21**, 53–61.

Janke, C., Magiera, M. M., Rathfelder, N., Taxis, C., Reber, S., Maekawa, H., Moreno-Borchart, A., Doenges, G., Schwob, E., Schiebel, E. and Knop, M. (2004). A versatile toolbox for PCR-based tagging of yeast genes: new fluorescent proteins, more markers and promoter substitution cassettes. *Yeast* **21**, 947–962.

Knop, M., Siegers, K., Pereira, G., Zachariae, W., Winsor, B., Nasmyth, K. and Schiebel, E. (1999). Epitope tagging of yeast genes using a PCR-based strategy: more tags and improved practical routines. *Yeast* **15**, 963–972.

Longtine, M. S., McKenzie, A., 3rd, Demarini, D. J., Shah, N. G., Wach, A., Brachat, A., Philippsen, P. and Pringle, J. R. (1998). Additional modules for versatile and economical PCR-based gene deletion and modification in *Saccharomyces cerevisiae*. *Yeast* **14**, 953–961.

Prein, B., Natter, K. and Kohlwein, S. D. (2000). A novel strategy for constructing N-terminal chromosomal fusions to green fluorescent protein in the yeast *Saccharomyces cerevisiae*. *FEBS Lett.* **485**, 29–34.
Puig, O., Caspary, F., Rigaut, G., Rutz, B., Bouveret, E., Bragado-Nilsson, E., Wilm, M. and Seraphin, B. (2001). The tandem affinity purification (TAP) method: a general procedure of protein complex purification. *Methods* **24**, 218–229.
Sato, M., Dhut, S. and Toda, T. (2005). New drug-resistant cassettes for gene disruption and epitope tagging in *Schizosaccharomyces pombe*. *Yeast* **22**, 583–591.
Schiestl, R. H. and Gietz, R. D. (1989). High efficiency transformation of intact yeast cells using single stranded nucleic acids as a carrier. *Curr. Genet.* **16**, 339–346.
Schneider, B. L., Seufert, W., Steiner, B., Yang, Q. H. and Futcher, A. B. (1995). Use of polymerase chain reaction epitope tagging for protein tagging in *Saccharomyces cerevisiae*. *Yeast* **11**, 1265–1274.
Sheff, M. A. and Thorn, K. S. (2004). Optimized cassettes for fluorescent protein tagging in *Saccharomyces cerevisiae*. *Yeast* **21**, 661–670.
Storici, F., Lewis, L. K. and Resnick, M. A. (2001). *In vivo* site-directed mutagenesis using oligonucleotides. *Nat. Biotechnol.* **19**, 773–776.
Tong, A. H., Evangelista, M., Parsons, A. B., Xu, H., Bader, G. D., Page, N., Robinson, M., Raghibizadeh, S., Hogue, C. W., Bussey, H., Andrews, B., Tyers, M. and Boone, C. (2001). Systematic genetic analysis with ordered arrays of yeast deletion mutants. *Science* **294**, 2364–2368.
Vorachek-Warren, M. K. and McCusker, J. H. (2004). *DsdA* (D-serine deaminase): a new heterologous MX cassette for gene disruption and selection in *Saccharomyces cerevisiae*. *Yeast* **21**, 163–171.
Wach, A. (1996). PCR-synthesis of marker cassettes with long flanking homology regions for gene disruptions in *S. cerevisiae*. *Yeast* **12**, 259–265.
Wach, A., Brachat, A., Alberti-Segui, C., Rebischung, C. and Philippsen, P. (1997). Heterologous *HIS3* marker and GFP reporter modules for PCR-targeting in *Saccharomyces cerevisiae*. *Yeast* **13**, 1065–1075.
Wach, A., Brachat, A., Pohlmann, R. and Philippsen, P. (1994). New heterologous modules for classical or PCR-based gene disruptions in *Saccharomyces cerevisiae*. *Yeast* **10**, 1793–1808.
Wendland, J. (2003). PCR-based methods facilitate targeted gene manipulations and cloning procedures. *Curr. Genet.* **44**, 115–123.

5 Studying Essential Genes: Generating and Using Promoter Fusions and Conditional Alleles

Michael JR Stark
Division of Gene Regulation & Expression, College of Life Sciences, University of Dundee, Dundee. DD1 5EH. UK

♦♦♦

CONTENTS

Introduction
Using Regulated Promoters to Study Essential Genes
Heat-Inducible Degron Fusions
Generating Conditional Alleles

List of Abbreviations

Cs^-	Cold sensitive
5-FOA	5-fluoroorotic acid
PCR	Polymerase chain reaction
Ts^-	Temperature sensitive
YCp	Yeast centromeric plasmid
YIp	Yeast integrative plasmid
YFG	"Your favourite gene"

♦♦♦♦♦♦ I. INTRODUCTION

When studying an essential gene, complete loss of function is by definition lethal to the yeast cell and so the effect of loss of function cannot be studied using a gene knockout. Alternative approaches are therefore needed for functional studies that allow for conditional loss of function, enabling cells to be grown under conditions where gene function is preserved and then switching to conditions where gene function has been inactivated for functional analysis. This

METHODS IN MICROBIOLOGY, VOLUME 36
0580-9517 DOI:10.1016/S0580-9517(06)36005-9

Chapter will address three commonly-used strategies for inactivating gene function: depletion of a gene product by promoter shut-off, 'N-degron' fusions and generation of conditional mutant alleles of the gene in question. The use of conditional alleles in particular constitutes a powerful approach for functional analysis of an essential yeast gene since it not only allows for loss of function to be studied but also provides the basis for further genetic approaches. In what follows, '*YFG*' ('your favourite gene') will be used to refer to the essential yeast gene to be studied following the 'nomenclature' of Sikorski and Boeke (1991).

A. Fusions to Regulated Promoters and 'N-degron' Fusions

In promoter shut-off experiments, the endogenous *YFG* is replaced by a copy that has been fused to a regulatable promoter. The galactose-inducible *GAL1-GAL10* promoter (Johnston and Davis, 1984) has been used in many studies although, as discussed below, the natural *GAL1-GAL10* promoter is not necessarily the best choice and several other alternatives are available. By switching growth conditions to ones in which the promoter is repressed (or no longer induced), the effect of depleting the gene product from the cell can be studied. While this may be helpful in understanding the function of the gene, if the gene product is a particularly stable protein it may persist at levels sufficient to carry out its function for several generations. A simple promoter shut-off experiment will not therefore necessarily enable the investigator to observe rapid loss of function. However, the N-degron approach (Labib *et al.*, 2000; Kanemaki *et al.*, 2003) combines the use of a regulated promoter (*CUP1*) with a heat-inducible protein-destabilising genetic element (Dohmen *et al.*, 1994) such that rapid loss of function may nonetheless be achieved by a combination of transcriptional repression and rapid protein degradation. Where the function of an essential gene and its product is already well understood, promoter shut-off strategies can also be useful for achieving a particular experimental design. For example, P_{GAL}-*CDC20* and P_{GAL}-*CDC6* gene fusions can be used for cell synchronisation in mitotic metaphase and blocking DNA replication respectively (see Tanaka *et al.*, 2002).

B. Conditional Alleles

Conditional alleles typically confer either a temperature-sensitive (Ts^-) or cold-sensitive (Cs^-) phenotype such that sufficient gene function remains to support growth at normal temperatures (24–30°C), but gene function is lost at higher or lower growth temperatures respectively. Thus mutant cells can be grown under permissive conditions and then shifted to the non-permissive ('restrictive') temperature to examine what happens as the gene function is lost. Conditional mutants can also provide the basis for genetic analysis, for example synthetic lethal, extragenic and multicopy suppressor screens that can be used to identify other genes whose products function in the same process as the protein of interest. This Chapter will describe approaches based on the localised mutagenesis method of Muhlrad *et al.* (1992) for selecting conditional alleles starting with virtually any essential yeast gene. Random mutations can be generated using polymerase chain reaction (PCR) under conditions promoting misincorporation of nucleotides during amplification, followed by generation of a library of mutant strains by *in vivo* gapped plasmid repair and plasmid shuffling. This library can then be screened for clones conferring a conditional growth phenotype. While the emphasis of this Chapter is on conditional mutations, the method is of course of general applicability for the production of any type of mutations given a suitable selectable phenotype (e.g. resistance to an inhibitor, osmosensitivity etc.). Finally, the Chapter will consider analysis of conditional mutants once they have been generated.

♦♦♦♦♦♦ II. USING REGULATED PROMOTERS TO STUDY ESSENTIAL GENES

The *GAL1-GAL10* promoter, used in many studies of gene function, is repressed during growth on glucose and induced on galactose, enabling modulation of the expression level by around 1000-fold. To examine what happens when expression of *YFG* is switched off, the coding sequence of *YFG* can be fused to the *GAL* promoter and a strain generated in which this is the sole source of *YFG* function. The easiest way to achieve this is to modify the endogenous gene by direct transformation of a yeast strain with a suitable PCR fragment as described in Chapter 4. A variety of templates have been described (Longtine *et al.*, 1998; Janke *et al.*, 2004) that enable the *GAL* promoter to be amplified so that it can be targeted, in tandem with an upstream selectable marker, to replace the native promoter of *YFG*. Frequently, such templates also include an N-terminal triple-HA epitope tag, allowing detection of the protein product of *YFG* by Western blotting and therefore permitting verification that it has been depleted following promoter shut-off. While in principal such a P_{GAL1}-*YFG* strain can be generated by transformation of a haploid strain followed by selection on galactose-containing medium, it is much safer to introduce the construct into a diploid strain first and then generate the required haploid by tetrad analysis on galactose-containing plates. In this way, the required construct can be introduced without any risks of secondary genetic events driven by selective pressure for immediate functionality or by problems of the high expression level from the *GAL1* promoter; any such problems should become immediately apparent following tetrad dissection.

Furthermore, the long primers used to amplify the promoter swap cassette can sometimes contain mistakes (see comments in Chapter 4, this volume) that could lead to lethality following integration at the desired locus due to changing the reading frame of the fusion. Performing the promoter swap in a diploid strain allows transformants to be obtained and provides a source of DNA that can be amplified and sequenced to check for faulty primer sequences should the promoter fusion turn out to be lethal following tetrad dissection. Alternatively, a P_{GAL}-*YFG* fusion could be made on a yeast integrative or centromeric plasmid, transformed into a diploid strain heterozygous for a *yfgΔ* deletion and then tetrad analysis performed to identify segregants in which the *yfgΔ* gene knockout and the plasmid-borne construct co-segregate. pBM150 (Johnston and Davis, 1984) is a suitable YCp that carries the *GAL1-GAL10* promoter as a *Bam*HI-*Eco*RI restriction fragment while pRB1438 (generated in the Botstein lab) has the *GAL1-10* promoter inserted into the polylinker region of pRS316 (see e.g. Kozminski *et al.*, 2000). Whichever method is used to generate the P_{GAL1}-*YFG* gene fusion, cells can be grown in medium containing 2% galactose and 2% raffinose (the latter being added to support better growth) and then transferred to medium containing 2% glucose to achieve the shut-off.

The native *GAL1* promoter is, however, not necessarily ideal for performing this type of experiment since it is very strong when induced and will in almost all cases lead to overexpression of the *YFG* product, which may itself affect the behaviour of the cells. Overexpression may be detrimental or even lethal to the cell, or may simply lead to such a high level of protein that depletion after addition of glucose to perform the shut-off takes too long. In some cases, even very low levels of expression may suffice to provide *YFG* function and cells reliant on a P_{GAL}-*YFG* fusion may remain capable of essentially normal growth even on glucose medium. To address this problem, derivatives of the *GAL1* promoter (termed *GAL1*S and *GAL1*L) have been developed that lead to properly regulated expression but at a level at least an order of magnitude lower than that achieved using the native *GAL1* promoter (Mumberg *et al.*, 1994), and these promoters have been built into PCR templates suitable for use in genomic promoter substitution approaches (Janke *et al.*, 2004). Alternatively, weaker, regulated promoters from genes such as *MET3* (Mountain *et al.*, 1991) and *MET25* (Mountain *et al.*, 1991; Mumberg *et al.*, 1994) can be used for modulating the expression levels of a variety of gene products in yeast cells (see e.g. Amon *et al.*, 1994; Black *et al.*, 1995; Spang *et al.*, 1995). These promoters are much weaker than P_{GAL} and yet are tightly repressed by addition of 2 mM methionine to the growth medium: for example, a P_{MET3}-*URA3* construct transformed into a *ura3* mutant strain failed to confer 5-FOA sensitivity on methionine-supplemented medium (Mountain *et al.*, 1990). Use of the *MET3* promoter therefore enables more rapid depletion of the *YFG* product following promoter shutoff and avoids the possible complications of overexpression when the promoter is on. Again, the basis of the method is to generate a strain in which, for example, a P_{MET3}-*YFG* fusion, either integrated or on a YCp vector, is the sole source of *YFG*. This is achieved using methionine-free medium such that the P_{MET3}-*YFG* fusion is expressed in the desired strain, and the above comments regarding the use of a diploid strain for the initial construct are also relevant here. The standard collections of templates for PCR-mediated gene targeting do not contain P_{MET3} or P_{MET25} promoter substitution templates, but a suitable equivalent could easily be made from an existing template: a 491-bp *Xba*I-*Eco*RV fragment (-561 to -70 relative to the start ATG codon) of the *MET3* promoter contains the relevant sequences (Figure 1). Alternatively, this segment can be cloned into a suitable construct in which the start of the *YFG* open reading frame and the *YFG* upstream region are separated by a unique restriction site to enable targeting of the promoter to the genomic *YFG* locus (see Figure 1). Once a P_{MET3}-*YFG* strain has been generated, cells can then be supplemented with methionine to observe the effect of Yfg protein depletion. Even using P_{MET3}, however, it may still take much longer to deplete cells of a stable protein than to inactivate a conditionally mutated gene product. If a suitable antibody is available or *YFG* has been epitope-tagged, then loss of the gene product following promoter shut-off can be readily followed by Western blot analysis as with the P_{GAL} fusions discussed above.

One potential problem with both the *MET3* and *GAL* promoters is that promoter shutoff requires changing the growth medium – adding methionine or glucose respectively, and this may lead to pleiotropic changes in cellular metabolism that could obscure the effects that are being monitored, particularly where these involve examining global gene expression patterns using microarrays. To avoid this problem a tetracycline-regulated promoter has been developed (Belli *et al.*, 1998b) that is available both in the form of yeast plasmids (Gari *et al.*, 1997-available through EUROSCARF) and templates for PCR-mediated promoter replacement as described above (Belli *et al.*, 1998a; Yen *et al.*, 2003). The system is based on expression in yeast of a tetracycline-inactivatable *tet* repressor (*tetR*)-VP16 transactivator fusion (tTA) coupled with replacement of the *YFG* promoter with *tetO* sequences to which tTA binds in the absence of doxycycline (a tetracycline analogue). When doxycycline is added, it binds to tTA and releases it from the *tetO* sequences in the substituted promoter, leading to loss of gene expression. Doxycycline is neutral in terms of its effect on global gene expression and so is ideal for use for controlling the expression of a specific gene in a microarray studies (Wishart *et al.*, 2005). One set of templates for promoter substitution includes both the KanMX4 selectable marker and a tTA expression cassette upstream of the *tetO* sequences that will replace the *YFG* promoter following targeted integration into the genome (Belli *et al.*, 1998a). Versions are available with either 2 or 7 repeats of the *tetO* sequence in the substituted promoter

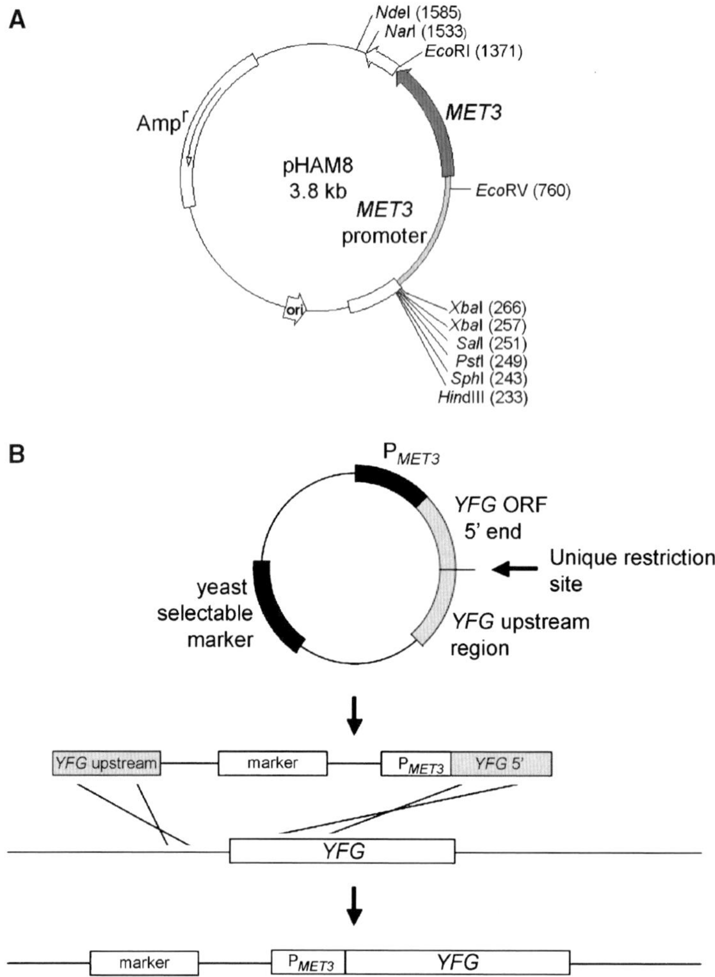

Figure 1. Tools for protein depletion experiments. (A) pHAM8 is a convenient source of the *MET3* promoter for subcloning (Mountain *et al.*, 1991). All the signals required for tightly-regulated expression of downstream genes are contained within the 495-bp *Xba*I-*Eco*RV fragment of the *MET3* promoter. (B) A strategy for generating a construct that can be used to integrate the *MET3* promoter upstream of a yeast gene. A P_{MET3}-*YFG* fusion involving the 5′ region of the open reading frame (*YFG*) is constructed on a yeast integrative plasmid in tandem with a region upstream of the *YFG* ORF such that these two sequences are separated by a unique restriction site. When linearised at this site and transformed into yeast, the construct directs replacement of the *YFG* promoter with P_{MET3}.

(termed $tetO_2$ and $tetO_7$ respectively). While this can work well it does require amplification of a 3.9 kb PCR product. An alternative system, in which the tTA expression cassette is integrated into the yeast genome first (at the *TRP1* locus) and the promoter swap cassette contains just the KanMX marker and the *tetO* sequences, is therefore easier to use (Yen *et al.*, 2003). A range of strains in which individual genes are doxycycline-regulated from a $tetO_7$ promoter is already available (Mnaimneh *et al.*, 2004), as is a second series in which a *tetO*$_2$ promoter has been used (Wishart *et al.*, 2006).

♦♦♦♦♦♦ III. HEAT-INDUCIBLE DEGRON FUSIONS

A very useful strategy for studying essential genes that combines the use of a copper-inducible promoter (*CUP*1) with a heat-inducible protein destabilising element (Dohmen *et al.*, 1994) is the heat-inducible degron method (Sanchez-Diaz *et al.*, 2004 – see Figure 2). The degron itself consists of ubiquitin fused to a mutated mouse dihydrofolate reductase (DHFR) sequence and followed by a single *myc* epitope. The degron is integrated at the start of the *YFG* coding region so that it is expressed as a fusion protein with the *YFG* product. The chimeric protein is rapidly processed in yeast by a deubiquitylating enzyme to reveal an amino-terminal arginine residue, which destabilises the remainder of the fusion polypeptide

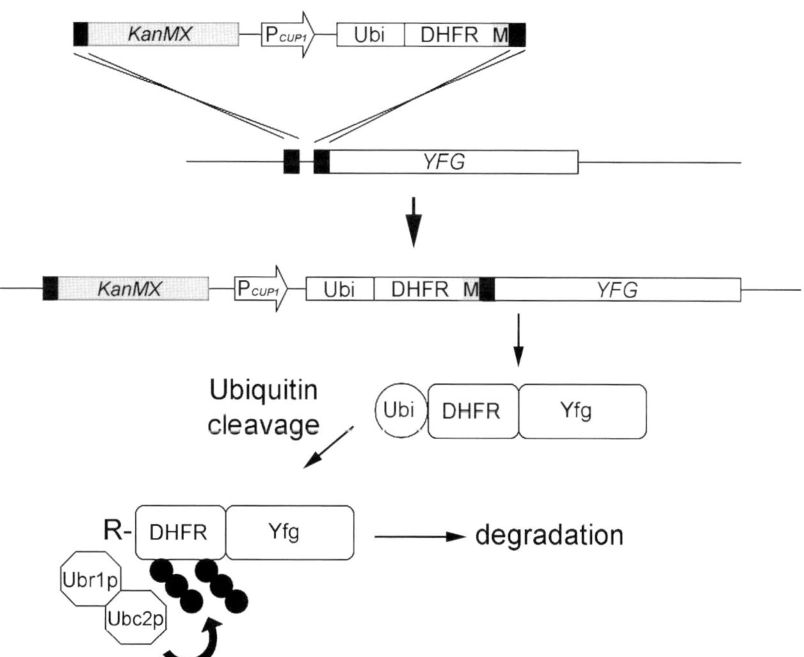

Figure 2. Heat-inducible degron tagging. Insertion of a PCR fragment generated using primers containing 5′ identity (■) to the gene of interest (*YFG*) so as to fuse the degron to *YFG* is depicted. The selectable marker (KanMX), *CUP1* promoter (P_{CUP1}), N-terminal Ubiquitin (Ubi), Ts^- mouse dihydrofolate reductase sequence (DHFR) and *myc* tag (M) are indicated. Cleavage of the ubiquitin moiety following translation reveals a destabilising N-terminal arginine (R) that is recognised by Ubr1p and modified by Ubc2p on certain lysine residues by addition of polyubiquitin chains (●) that target the fusion protein for proteasomal degradation.

(Bachmair and Varshavsky, 1989). Destabilisation is brought about by multiple ubiquitin conjugation by Ubc2p onto the DHFR-encoded region in response to the amino-terminal arginine, which is recognised by Ubr1p. Ubiquitylation is then followed by rapid, ubiquitin-dependent proteolysis. The DHFR sequence is thermosensitive, such that the ubiquitylation and degradation is strongly stimulated when the cells expressing the fusion are shifted from 24°C to 37°C. The yeast strains used for degron tagging can be induced to overexpress *UBR1*, since this improves the turnover of the protein of interest following the temperature shift. A cassette in which the KanMX selectable marker is followed by the P_{CUP1}-degron sequence is readily available through EUROSCARF (http://web.uni-frankfurt.de/fb15/mikro/euroscarf/index.html) and provides a suitable template for PCR-mediated targeting of the degron to the gene of interest using the procedures described in Chapter 4. Strains containing a P_{GAL1}-*UBR1* fusion integrated at the *HIS3* locus are also available, as is a plasmid for introducing the construct into other yeast genetic backgrounds (Sanchez-Diaz *et al.*, 2004).

The methods involved in generating a degron-tagged construct have been fully described by Sanchez-Diaz *et al.* (2004). Briefly, the KanMX-P_{CUP1}-degron cassette is amplified and integrated into a suitable P_{GAL1}-*UBR1*-containing yeast strain so as to target it to the gene being studied, selecting for transformants on YPD agar plates (see Chapter 2) containing 0.1 mM $CuSO_4$ to induce the *CUP1* promoter. This is best performed in a diploid strain such as YKM165 (Sanchez-Diaz *et al.*, 2004) so that transformants can be obtained in the absence of any selective pressure due to problems of construct functionality as already mentioned above. Haploid strains containing both the P_{GAL1}-*UBR1* construct (His^+) and the degron-tagged gene of interest ($G418^R$) can then be identified following tetrad dissection (Chapter 2, this volume). PCR performed on genomic DNA extracted from the tagged strains is then performed to verify correct insertion of the degron cassette (Sanchez-Diaz *et al.*, 2004). To test the degron construct, growth at 24°C and 37°C is compared on YPD plates containing 0.1 mM $CuSO_4$ (degron fusion expressed and *UBR1* overexpression off) or YPGalactose plates without $CuSO_4$ (degron fusion not induced and Ubr1p overexpressed). Under the latter conditions, in excess of 60% of strains generated during a large study of essential genes were found to give a robust, temperature-sensitive phenotype due to loss of function of the relevant protein (Kanemaki *et al.*, 2003), and around one hundred existing strains are available through EUROSCARF. The inbuilt *myc* tag can be used to follow protein depletion following activation of the degron, but if gives insufficient sensitivity then an antibody against the protein of interest itself might prove advantageous. For using the strains experimentally, they can be grown at 24°C in raffinose-containing liquid medium to facilitate induction of P_{GAL1}-*UBR1*, which requires only 30–40 minutes following addition of galactose when cells have been grown in the absence of glucose repression. Shifting to 37°C should then once again give a robust phenotype due to rapid depletion of the tagged protein.

♦♦♦♦♦♦ IV. GENERATING CONDITIONAL ALLELES

A. Conditional Alleles: Ts⁻ Versus Cs⁻

Will any essential genes necessarily yield Ts⁻ or Cs⁻ alleles? In one study where a yeast strain was mutagenised and Ts⁻ mutations on chromosome I were systematically identified, Harris and Pringle (1991) showed that despite using a range of mutagens with different specificities, many essential genes failed to yield Ts⁻ alleles. In some cases this may be due to the nature of the gene product; for some proteins it may be particularly difficult to generate changes that affect function without causing complete inactivation. However, by focusing on specific essential genes in isolation and using polymerase chain reaction (PCR) to generate a large, unbiased spectrum of mutations, it is quite likely that many essential genes will ultimately yield the desired mutant allele. The method described in this Chapter has been successfully used in this and many other laboratories to obtain Ts⁻ alleles in genes encoding a range of proteins including enzymes, structural proteins and components of multiprotein complexes (Stirling *et al.*, 1994; MacKelvie *et al.*, 1995; Connelly and Hieter, 1996; Geissler *et al.*, 1996; Lewis and Pelham, 1996; Evans and Stark, 1997). In this laboratory, we have to date a 100% record of obtaining Ts⁻ alleles in a variety of yeast genes.

Regarding the analysis of conditional alleles, it is probably easier to use Ts⁻ strains than Cs⁻ mutants. Wild-type strains grow rather slowly below 20°C and it can sometimes be more difficult to obtain a clear distinction between wild-type and mutant than with a Ts⁻ strain. Ultimately, it is the difference between the restrictive temperature of the mutant and the maximum (for Ts⁻ alleles) or minimum (for Cs⁻ alleles) growth temperature of the wild-type strain which counts and the greater this difference, the more clear-cut the results will be. Thus most laboratory strains of yeast fail to grow above 38°C and while 37°C can be used successfully as the restrictive temperature for a Ts⁻ mutant, a more severe allele that fails to grow above 35°C might be more useful. Similarly, a Cs⁻ strain that can't grow below 20–23°C would be easier to handle than one with a restrictive temperature of 16°C. Folklore has it that Cs⁻ alleles are harder to find: they may require a more restricted type of alteration to the protein, for example an alteration specifically at the interface of a protein-protein interaction. However, they are potentially just as useful as Ts⁻ mutations and just as different Ts⁻ alleles can be of great use (see below), the ability to obtain Cs⁻ as well as Ts⁻ alleles could be particularly valuable. Note that strains lacking *TRP1* function, a common feature of many lab strains, show some cold sensitivity already (Singh and Manney, 1974) and so this may be a complicating factor.

B. Generating a Suitable Yeast Strain for Mutant Screening

Figure 3 shows the overall scheme devised by Muhlrad *et al.* (1992) for generating a library of mutant alleles of an essential gene. The

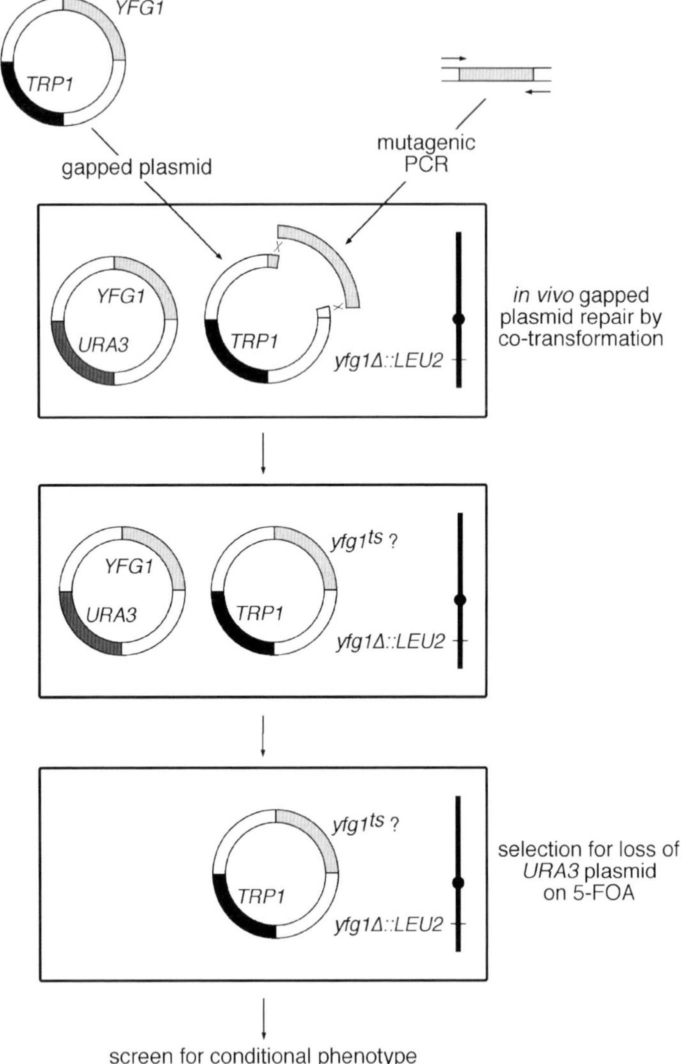

Figure 3. Generating conditional alleles of an essential gene. Scheme for generating conditional alleles of an essential gene (*YFG*) by mutagenic PCR and gapped plasmid repair *in vivo*. For the purpose of this example, the chromosomal locus has been deleted with *LEU2* and the gapped plasmid carries *TRP1*, but other markers may be used as appropriate. The only requirement is that the resident plasmid in the starting yeast strain should carry the *URA3* marker to facilitate selection for its loss in the final step using 5-FOA.

starting point for the scheme is the production of a haploid yeast strain in which the gene in question has been deleted from the genome, but which is kept alive by a wild-type copy of the gene on a yeast centromeric plasmid (YCp) carrying the *URA3* gene. Vectors such as YCplac33 (Gietz and Sugino, 1988) and pRS316 (Sikorski and Hieter, 1989) are suitable for constructing the required plasmid carrying *YFG*. The yeast strain is best generated by tetrad dissection of a diploid strain that has been made heterozygous for a complete *yfgΔ* knockout by one-step gene disruption (Chapter 4, this volume) and then transformed with the *YFG URA3* YCp, screening for haploid progeny in which the null allele and the plasmid co-segregate. Such progeny should fail to papillate when patched onto 5-FOA medium (Sikorski and Boeke, 1991), indicating that the strain is fully dependent on the plasmid-borne wild-type *YFG* allele. The marker used to generate the knockout allele is not important, although clearly *URA3* must be reserved to enable 5-FOA counterselection in the later step (Figure 3). If for some reason *URA3* cannot be used, *LYS2* and the $Lys2^+$ counterselective agent α-aminoadipic acid could be employed instead (Sikorski and Boeke, 1991).

C. Primer Design and Plasmid Gapping Strategy

Figure 4 shows the strategy that underlies the *in vivo* gapped plasmid repair step. The region of the gene to be mutated is defined by two restriction sites (A and B in Figure 4), which ideally will enable removal of the bulk of the coding region of the gene, or just the domain that it is desired to mutate. While it is possible to obtain

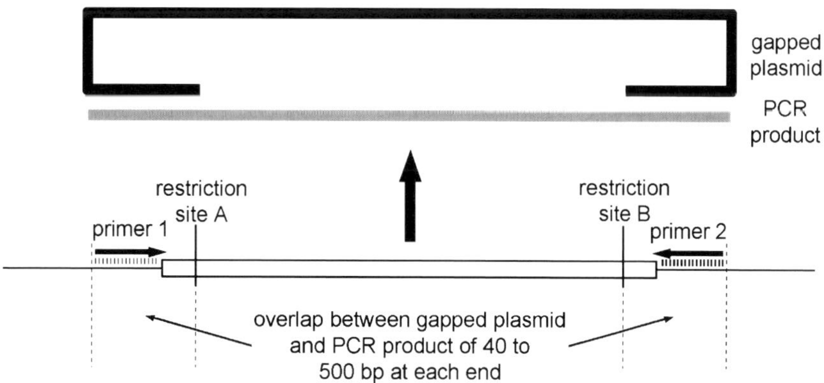

Figure 4. Primer design and plasmid gapping strategy. Primers should be designed that flank the gene to be mutated such that the entire coding region (or one particular domain of interest) is amplified. For *in vivo* gapped plasmid repair, the regions of overlap between the gapped plasmid and the PCR product should be at least 40 bp at each end and preferably greater. The availability of the restriction sites (A, B) used to generate the gapped plasmid will therefore play some role in determining the selection of PCR primers.

mutations in the regions of overlap between the PCR product and the gapped plasmid, we have never found mutations in these regions in our own work. If nothing is known about which regions of the gene are critical for function, then the larger the section of the coding region excised the better (since this will maximise the region available for mutagenesis). Conversely, if restriction sites A and B lie outside the coding region, then it may be possible to identify mutations which do not change the protein but that affect the promoter activity or mRNA stability; while these might be of interest, they may not be as useful as mutations which generate an altered protein. Mutagenesis is therefore best limited to the coding region by suitable choice of sites and primers. Restriction sites A and B clearly must not cut the vector backbone or remaining regions of the gene and if no suitable sites are available, they must be introduced (or perhaps a vector site removed) by site-directed mutagenesis, a procedure for which many rapid and easy-to-use kits are now available. Another option would be to replace the coding region with a short linker containing a unique restriction site when constructing the YCp.

The design of the PCR primers follows standard procedures that aim at optimizing the melting temperature (preferably at least 65°C) and avoiding primer dimers or weak priming from other sites in the template. Several of the commonly-used DNA analysis packages contain facilities for doing this, or alternatively a freeware or shareware programme such as Amplify 3 (written by Bill Engels, University of Wisconsin at Madison: Engels, 2005) can be used. Primers of around 17 bases work well. For the method to work there must clearly be sufficient overlap between the gapped plasmid and PCR fragment for efficient homologous recombination to occur. As with any yeast genetic manipulations involving gene targeting through flanking homology, this overlap should preferably be at least 40 bp at each end and if one overlap is short (e.g. under 100 bp) it will help if the other is much longer. We have not rigorously tested the minimum requirements, but in one case (MacKelvie *et al.*, 1995) the overlaps were 46 and 374 bp while in another instance (Evans and Stark, 1997) they were 148 and 187 bp; both instances yielded large libraries of potentially mutant genes. If the template plasmid and the gapped plasmid are identical, then vector sequences can be used for PCR priming if so desired. The largest PCR fragment we have used was 2.5 kb, but there is no reason why longer genes could not be mutated using one set of primers. For very large genes, it may be better to split them into two or more regions if problems are encountered with the PCR step. This might also reduce the change of obtaining multiple mutations in each conditional allele.

D. Mutagenic PCR

A library of *YFG* mutant alleles can be rapidly and easily generated by PCR amplification of *YFG* under conditions that favour misincorporation of dNTPs into the amplimer. This is achieved by using biased dNTP ratios and by including $MnCl_2$ in the amplification reaction (see Protocol 1). Although other methods of mutagenesis have been described previously in procedures similar to that presented here (e.g. Sikorski and Boeke, 1991), PCR has the advantage that it is easy, safe and can generate an unbiased range

Protocol 1. PCR mutagenesis.

1. Set up mutagenic and non-mutagenic (control) PCR reactions in parallel. Standard mutagenic conditions are as follows:
   (i) Ratio of (dTTP+dCTP) : (dATP+dGTP) = 5 : 1 (or can try 1 : 5).
   (ii) Include 0.1–0.5 mM $MnCl_2$ in the reaction .
   (iii) Use double the normal level of *Taq* polymerase in reaction (i.e. 0. 05 U/μl).
2. For each template-primer combination, optimise the following:
   (i) The ratio of $MgCl_2$: $MnCl_2$ in the reaction. This can greatly affect the frequency of mutations obtained and the PCR product yield.
   (ii) The PCR cycle conditions for denaturation, annealing and extension.
3. Typical conditions (for a 100 μl reaction) are as follows:

	Stock Solution	Final Conditions	Volume (μl)
primer 1	200 ng/μl	4 ng/μl	2
primer 2	200 ng/μl	4 ng/μl	2
template[a]	10 ng/μl	0.1–0.2 ng/μl	1–2
Reaction Buffer[b]	10 ×	1 ×	10
$MgCl_2$[b]	25 mM	2 mM	8
50 × dNTP mix[c]	see above	see above	2
Water	-	-	68–72
$MnCl_2$	10 mM	0.3–0.5 mM	3–5[d]
Taq polymerase	5 U/μl	0.05 U/μl	1

[a]template is a circular plasmid carrying the relevant gene.
[b]Some PCR buffers may already include $MgCl_2$.
[c]diluted from 100 mM stocks e.g., 50 × stock "mutagenic" dNTP mix: 25 mM dTTP, 25 mM dCTP, 5 mM dATP, 5 mM dGTP; 50 × stock "normal" dNTP mix: 25 mM each dNTP.
[d]mutagenic reaction only.

4. Mix, heat at 94°C for 5 min and then carry out PCR using the pre-optimised settings, for example 30 cycles of 45–60 s at 94°C, 2 min at 55°C and 72°C for 2–2.5 min (N.B. allow at least 1 min per kb), holding at 72°C for 20 min for the last cycle.

of mutations. Furthermore, mutagenesis is localised to the gene of interest in a way not possible with chemical mutagenesis of the gene present on a plasmid. Finally, in some cases (e.g. where a specific domain is evident in a protein, mutagenesis by PCR can be localised just to the region encoding this domain rather than the whole gene by suitable design of primers. Protocol 1 suggests some typical conditions both for the composition of the PCR reaction and the cycling conditions, but ultimately the best conditions need to be determined empirically. The type of enzyme used for PCR is clearly of importance: since the aim is to generate mutations, a non-proof-reading enzyme such as *Taq* polymerase should be used. We have found it best to use a range of perhaps three different $MnCl_2$ concentrations, optimising the cycling conditions for a good yield of product at each $MnCl_2$ concentration. The dNTP and $MnCl_2$ concentrations can greatly affect the yield of mutants and also the average number of mutational changes per allele. Since it is normally better to work with alleles containing single point alterations, conditions that minimise misincorporation while still yielding a few conditional mutants (e.g. at a level of 1–2 %) are generally preferable. Ultimately, it may be necessary to carry out a few attempts to determine conditions that generate conditional alleles at a reasonable frequency, sequence a few of these to determine the overall level of mutational change and then refine the process as required.

E. Gapped Plasmid Repair *in vivo*

The next step is to use the PCR product to generate a library of *YFG* alleles that can be screened for conditionality. In principle, this could be done by cloning the PCR product into a YCp, introducing the ensuing library of plasmids into the yeast strain generated as above and following the latter part of the procedure shown in Figure 3. However, it is far easier to generate the required plasmids by *in vivo* recombination in yeast and thereby circumvent an extra step using *E. coli*. This is achieved by co-transformation of the PCR fragment and a gapped plasmid whose ends are homologous to the amplified product. When both DNA molecules enter a competent yeast cell, high-efficiency homologous recombination between the PCR fragment and the gapped plasmid lead to reconstruction of a circular plasmid carrying the entire gene. Protocol 2 shows suitable conditions for achieving this and clearly requires yeast cells that are highly competent for transformation (see Chapter 3, this volume). Transformants should be selected at the normal growth temperature just for the gap-repaired plasmid (i.e. include uracil in the plates), thereby allowing for loss of the *URA3 YFG* YCp from some fraction of cells in each colony. An important control in this process is to transform the competent cells with the gapped vector alone, which will always generate a background level of transformants by recircularisation or by gap-repair using the null allele in the genome or

Protocol 2. Gapped plasmid repair, plasmid shuffling and selection.

1. The host yeast strain carries a wild-type copy of the gene in question on a *URA3-CEN* plasmid (YCp) and a gene deletion in the chromosome (see Figure 3).
2. Prepare gapped plasmid (a YCp carrying the gene but with a different selectable marker e.g. *TRP1*) by digestion and gel purification using a product such as Qiagen QIAquick gel extraction kit.
3. Check the PCR product (Protocol 1) for purity by gel electrophoresis. If a clean band of the expected size is obtained then it can be used following removal of the unused primers, otherwise it should be gel purified as above.
4. For yeast transformation, use the high efficiency protocol (Chapter 2, Protocol 2) with 10 µl PCR product (roughly 1–5 µg) and 50–100 ng gapped plasmid. Use non-mutated product as a control to show that any Ts^-/Cs^- clones resulted from the mutagenic PCR.
5. Pellet cells and resuspend in 200 µl sterile water for plating.
6. Plate out on medium selective for the gapped plasmid. Include alongside controls with gapped plasmid alone and a zero DNA sample. Presence of the PCR fragment should stimulate the transformation frequency 5–10-fold.
7. Patch out 500–1000 transformants onto selective plates containing 5-FOA (Sikorski and Boeke, 1991) and then when these have grown, patch out again on 5-FOA medium. Screen these strains on selective plates at 23–26°C and either 35–37°C (to screen for Ts^-) or 14–20°C (to screen for Cs^-).
8. *Very important*: check any potential mutant alleles by isolating the putative mutant plasmid from each Ts^- or Cs^- strain, followed by retransformation into the starting 'shuffle' strain and reselection on 5-FOA as above. Genuine mutant alleles should generate essentially 100% conditional strains after retesting in this way.
9. Sequence the insert of the mutant plasmid to identify the mutation(s).

the copy of *YFG* on the *URA3* YCp. Ideally, presence of the PCR fragment should stimulate transformation frequency at least 5–10-fold, indicating that efficient gap-repair using the incoming fragment has occurred. The number of transformants required for a reasonable probability of obtaining mutants is between 500 and 5000.

F. Plasmid Shuffling and Selection for Mutant Alleles

By this stage, a library of yeast transformants will have been generated but since each contains the *URA3 YFG* YCp as well as the gap-repaired plasmid and since most conditional alleles will be recessive, any effect of a mutant *YFG* will not be evident. To uncover any mutant phenotypes, the *URA3 YFG* YCp must first be lost using 5-FOA counterselection or 'plasmid shuffling' (Sikorski and Boeke, 1991), as described in Protocol 2. A minimum of 500 transformants should be patched out onto 5-FOA medium (Sikorski and Boeke, 1991) selective for the gap repaired plasmid to enable growth of cells in which the gap-repaired plasmid has become the sole source of *YFG* function in the cell. Since it is likely that a significant fraction of gap-repaired plasmids will encode unconditionally dysfunctional *YFG* alleles, some of the transformants will not generate 5-FOA-resistant progeny at this stage. The temperature chosen for this step is important, since it will define the permissive conditions for any conditional mutations and alleles not supporting growth at this temperature will be lost at this stage. While 26°C might be typically the best temperature to use, it could be advantageous to try slightly lower temperatures for Ts^- screens or slightly higher ones for Cs^- screens. A second round of growth on 5-FOA medium ensures complete loss of the *URA3 YFG* YCp.

All 5-FOA resistant colonies should next be replica plated or restreaked onto several plates selective both for the gap-repaired plasmid and the genomic knockout; one of these is grown under the permissive conditions and the others at a range of potentially restrictive temperatures for both Ts^- and Cs^- alleles. In this way, colonies that fail to grow specifically on any of the latter plates can be identified as containing candidate conditional *YFG* alleles.

G. Verification of Mutants

For each candidate mutant strain, the plasmid should be recovered from a small liquid culture of cells grown at the permissive temperature and recovered by transformation of *E. coli*. Substantial quantities of DNA can then be prepared and the alleles verified by re-transforming the shuffle strain with each recovered plasmid, selecting cells that have lost the *URA3 YFG* YCp on 5-FOA medium and then re-testing for conditionality as above. Shuffling should be done using at least 4 or 5 primary yeast transformants and testing for conditionality carried out using several Ura^- derivatives of each, so that a clear and representative result can be obtained. Verification of the plasmids is critical since not every strain that appeared Ts^- or Cs^- initially will consistently continue to do so. Once conditional alleles have been verified in this manner, the site of mutational alteration can be determined by DNA sequencing. Sequencing confers two benefits: firstly it confirms that the coding region has been mutated and, secondly, it demonstrates how many mutations are present. As discussed above, sometimes the PCR conditions may be sufficiently mutagenic to yield multiple changes and so sequencing one or more alleles at the first opportunity enables the procedure to be repeated using less error-prone conditions for the PCR step. Where multiple missense mutations have been obtained, the mutation responsible for the conditional phenotype may be identified by subcloning (to produce alleles containing single point changes) or by site-directed mutagenesis of a wild-type copy. However, it is not always the case that a single mutation will alone be responsible for the conditional phenotype (e.g. Evans and Stark, 1997) and so repeating the mutagenesis may be preferable as the means of obtaining conditional phenotypes resulting from single point changes.

A final aspect to verification of the mutant alleles concerns their phenotype. The screen described above is of necessity a growth/no growth test performed on plates, but the conditional mutants obtained may show considerable differences in their behaviour. In particular, they may differ in how rapidly the mutant protein is inactivated at the restrictive temperature. This may be critical, especially where the possible effects of mutations on the cell division cycle is being examined and where arrest in the first cell cycle following temperature shift is usually needed to obtain clear-cut results. Furthermore, mutations that are conditional on plates may not have such a clear-cut effect in liquid medium, which will be used for much of the future work with them. Thus candidate mutant alleles should also be checked in small-scale liquid culture and those that cease proliferation most rapidly are likely to be of greatest use. In some instances, prior knowledge of the gene function may suggest what sort of phenotype is to be expected following rapid loss of function. In such a case, cells can be examined at the restrictive conditions and alleles identified that seem to confer the expected effect. This could be especially important where a protein has several functions, only one of which is currently of interest to the investigators.

H. Strategies for Integrating Mutant Alleles at Their Genomic Locus

The above procedure should reliably yield conditional yeast strains in which the sole source of the mutant protein is encoded by a YCp. While such strains may be very useful, ultimately it is probably best to generate strains in which the mutant allele is integrated, either ectopically in a strain deleted for the normal copy or preferably at its normal genomic locus. The copy number of a YCp can vary somewhat and so the only way of ensuring a stable, single copy of the mutant allele is by integration into the genome. Several possible strategies are available for achieving this, all of which require that the correct integration is properly verified, usually by PCR. It may also be beneficial to verify the integrated mutant allele by DNA sequencing of a suitable PCR product, thereby directly demonstrating presence of the mutation in the genome. Finally, it is worth remembering that a mutation that was Ts on a YCp at 1–4 copies per cell may be lethal when integrated in single-copy.

1. Ectopic integration

The mutant allele can be subcloned onto a yeast integrative plasmid (YIp) and integrated at the site of the YIp marker gene in a diploid strain that is heterozygous for a *YFG* knockout. Haploid segregants in which the integrated copy of the gene is the sole source of *YFG* function can then be identified following tetrad analysis. Alternatively, if a suitable marker is available the integration could be done in the haploid plasmid-dependent mutant strain obtained following the original mutagenesis. Following transformation, transformants that have lost the plasmid and thus rely solely on an integrated copy of the *YFG* conditional allele can be identified following nonselective growth and replica-plating. This latter approach avoids possible gene conversion of the integrated allele, which can occur at an unexpectedly high frequency (see Schneider *et al.*, 1995), since the cell contains no wild-type copy of *YFG* at any stage in the process. Both strategies may result in multiple integrations of the YIp and this should be checked using a PCR-based strategy such as that described in Chapter 2.

2. Pop-in, pop out

This is the procedure of choice for moving a conditional *YFG* allele to the *YFG* genomic locus and requires the generation of a YIp carrying *URA3* and the conditional *YFG* allele. The procedure is outlined in Figure 5. Integration of this construct in a haploid strain at the *YFG* locus is then targeted by cutting the YIp at a suitable unique restriction site within *YFG* and results in tandem wild-type and mutant copies of *YFG* separated by the vector sequences. 5-FOA selection can then be used to screen for spontaneous excision of the YIp by homologous recombination. This should occur at different sites throughout *YFG* in different isolates, so as to leave the site of the conditional mutation in the remaining genomic copy in a proportion of the cells that have looped out *URA3* (see Struhl, 1983). Candidate strains will thus appear as Ts^- (or Cs^-) 5-FOA-resistant derivatives. Note, however, that if the conditional *YFG* allele contains multiple point mutations and that more than one of these is needed for the conditional phenotype, targeting integration by cutting at a restriction site that lies between the key sites may severely restrict the range of events that can loop out *URA3* to generate the required combination of mutations in the remaining genomic copy. Thus with pop-in pop-out it is advantageous to have identified the sites of the mutations first, or at the very least to attempt the procedure by cutting at a number of possible sites to target the initial integration event.

3. Integration of a marked *YFG* fragment

While ectopic integration of a *YFG* allele by integration of a YIp clone generates a marked allele, the pop in–pop out procedure

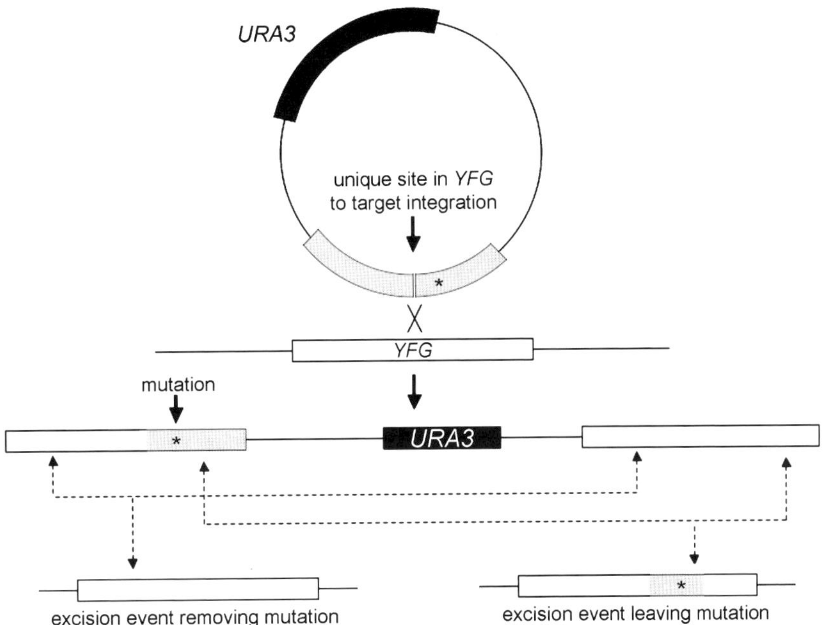

Figure 5. Pop in-pop-out method for integration of a conditional allele at its genomic locus. A yeast integrative plasmid (YIp) carrying the *URA3* marker and the mutant *yfg* allele is linearised within the latter to target integration to the *YFG* locus, leading to tandem wild-type and mutant copies. Counterselection against the *URA3* marker using 5-FOA selects for eviction of the YIp sequences by homologous recombination between the tandem *YFG* copies. Depending on the precise locus of the recombination event leading to excision of the YIp, a proportion of the Ura$^-$ cells will retain the conditional mutation (*) in the genome.

for integration at the *YFG* locus result in strains where the conditional allele is not marked and so it can only be followed in subsequent genetic crosses by its conditional phenotype. Where it is desirable to mark the mutant allele at its genomic locus, this can be achieved by insertion of a suitable marker gene 3′ to the gene of interest and transforming a haploid strain with a linear fragment encompassing the whole *YFG* mutant sequence, the adjacent marker and more distal sequences homologous to the *YFG* locus (Figure 6). A caveat here is that clearly the site of integration of the marker must not affect the function either of *YFG* or any downstream gene. Although integration of the marker could occur via a recombination event between the site of the *YFG* mutation and the marker gene, in practice a sufficiently high proportion of events that incorporate both the mutant *yfg* and the marker can be found among the transformants, typically 30% or more. Such events can be identified by screening transformants for the conditional phenotype conferred by the *YFG* allele and presence of the mutations verified by DNA sequencing of a suitable PCR product amplified from the integrated copy (see Figure 6).

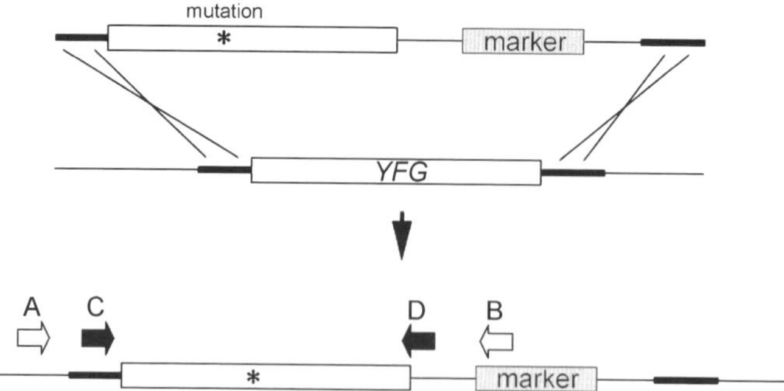

Figure 6. Integration of a marked conditional allele at its genomic locus. A DNA fragment is generated by cloning or by fusion PCR in which a suitable marker (e.g. NatR: Goldstein and McCusker, 1999) is located downstream of the conditional *yfg* allele, such that the terminal sequences of the fragment (corresponding to regions upstream and downstream of *YFG*) can direct recombination of the *yfg*-marker fragment with its genomic locus. In some of the transformants, the recombination event will occur as shown and carry the mutation (*) into the genome. Such transformants should express the conditional phenotype associated with the *yfg* allele. Presence of the mutation can be checked by generating a PCR fragment from genomic DNA (see Chapter 2, Protocol 7) using a primer upstream of the transforming fragment (A) and one within the marker (B), then using internal primers (e.g. C, D) to generate sequence data.

I. Analysis of Conditional Mutants

Having generated one or more conditional mutants, the way is paved for analysing the effect of the mutations and then using them in further experimental strategies. Frequently, the investigator will have at least some idea of what effect loss of function may have and will have experimental assays set up to monitor some facet of cell biology in which the *YFG* product is thought to play a role. Otherwise, examination of the morphology of cells under the microscope following shift to the restrictive temperature is a good starting point. The mutant allele may block cell cycle progression or lead to defective morphological development, in which case it may be useful to examine the effect of the mutation in synchronous cultures. Two methods are commonly used for this, α-factor block-release and centrifugal elutriation. The former has the advantage that it is straightforward and requires no specialist equipment, although the latter method simply selects small cells at the beginning of the cell cycle and does not require perturbation with mating pheromone. Immunofluorescence microscopy using probes for DNA, actin, tubulin and other intracellular components or GFP-tagged marker proteins (e.g. *GAR1*-GFP: Verheggen *et al.*, 2001 as a marker for the nucleolus) may also be of great use in uncovering the cellular defect under restrictive conditions (Pringle *et al.*, 1991; see Chapter 11). If an antibody is available that recognises the *YFG* product (or if the gene has been epitope-tagged) then its abundance or intracellular localisation can be examined following temperature shift. Where the polypeptide is known to engage in protein-protein interactions, the effect of the mutation on these can also be tested (e.g. by co-immune precipitation or by building the mutant allele into the two-hybrid assay). It is not uncommon to find that conditional alleles of genes encode mutant proteins that are quite significantly defective in either their protein-protein interactions or their biochemical activity, even when isolated from extracts made from cells grown at a permissive temperature where cell growth is relatively normal.

It is worth pointing out that the phenotype of a conditional mutation may depend on how far beyond the minimum restrictive temperature the strain is shifted and so it may be important to try more than one set of conditions. Equally, different alleles may have different phenotypes; intragenic complementation between different alleles would suggest that the protein product has different functional domains and that each complementing mutation has affected a different function of the protein. The availability of different conditional alleles is also of great use when they are used for further genetic screens. Thus dosage or extragenic suppressors that simply bypass a mutation may non-specifically suppress any conditional allele (or a deletion) of *YFG*. In comparison, the ability to identify suppressors mutations of either type that show specificity for one (or a subset) of mutations implies that the suppression is more specific, making it more likely that the components identified in the screen have a direct involvement with the *YFG* product. Likewise, allele specificity of synthetic lethal interactions gives increased confidence that the genetic interaction represents a direct and meaningful biochemical one.

References

Amon, A., Irniger, S. and Nasmyth, K. (1994). Closing the cell cycle circle in yeast: G2 cyclin proteolysis initiated at mitosis persists until the activation of G1 cyclins in the next cycle. *Cell* **77**, 1037–1050.

Bachmair, A. and Varshavsky, A. (1989). The degradation signal in a short-lived protein. *Cell* **56**, 1019–1032.

Belli, G., Gari, E., Aldea, M. and Herrero, E. (1998a). Functional analysis of yeast essential genes using a promoter-substitution cassette and the tetracycline-regulatable dual expression system. *Yeast* **14**, 1127–1138.

Belli, G., Gari, E., Piedrafita, L., Aldea, M. and Herrero, E. (1998b). An activator/repressor dual system allows tight tetracycline-regulated gene expression in budding yeast. *Nucleic Acids Res* **26**, 942–947.

Black, S., Andrews, P. D., Sneddon, A. A. and Stark, M. J. R. (1995). A regulated *MET3-GLC7* gene fusion provides evidence of a mitotic role for *Saccharomyces cerevisiae* protein phosphatase 1. *Yeast* **11**, 747–759.

Connelly, C. and Hieter, P. (1996). Budding yeast *SKP1* encodes an evolutionarily conserved kinetochore protein required for cell cycle progression. *Cell* **86**, 275–285.

Dohmen, R. J., Wu, P. and Varshavsky, A. (1994). Heat-inducible degron: a method for constructing temperature-sensitive mutants. *Science* **263**, 1273–1276.

Engels, W. R. (2005). Amplify 3. Retrieved November 29, 2006 from the University of Wisconsin website: http://engels.genetics.wise.edu/amplify/.

Evans, D. R. H. and Stark, M. J. R. (1997). Mutations in the *Saccharomyces cerevisiae* Type 2A Protein Phosphatase Catalytic Subunit reveal roles in cell wall integrity, actin cytoskeleton organisation and mitosis. *Genetics* **145**, 227–241.

Gari, E., Piedrafita, L., Aldea, M. and Herrero, E. (1997). A set of vectors with a tetracycline-regulatable promoter system for modulated gene expression in *Saccharomyces cerevisiae. Yeast* **13**, 837–848.

Geissler, S., Pereira, G., Spang, A., Knop, M., Soues, S., Kilmartin, J. and Schiebel, E. (1996). The spindle pole body component Spc98p interacts with the gamma-tubulin-like Tub4p of *Saccharomyces cerevisiae* at the sites of microtubule attachment. *EMBO J* **15**, 3899–3911.

Gietz, R. D. and Sugino, A. (1988). New yeast-*Escherichia coli* shuttle vectors constructed with *in vitro* mutagenised yeast genes lacking six-base pair restriction sites. *Gene* **74**, 527–534.

Goldstein, A. L. and McCusker, J. H. (1999). Three new dominant drug resistance cassettes for gene disruption in *Saccharomyces cerevisiae. Yeast* **15**, 1541–1553.

Harris, S. D. and Pringle, J. R. (1991). Genetic analysis of *Saccharomyces cerevisiae* chromosome I: on the role of mutagen specificity in delimiting the set of genes identifiable using temperature-sensitive-lethal mutations. *Genetics* **127**, 279–285.

Janke, C., Magiera, M. M., Rathfelder, N., Taxis, C., Reber, S., Maekawa, H., Moreno-Borchart, A., Doenges, G., Schwob, E., Schiebel, E. *et al.* (2004). A versatile toolbox for PCR-based tagging of yeast genes: new fluorescent proteins, more markers and promoter substitution cassettes. *Yeast* **21**, 947–962.

Johnston, M. and Davis, R. W. (1984). Sequences that regulate the divergent *GAL1-GAL10* promoter in *Saccharomyces cerevisiae. Mol Cell Biol* **4**, 1440–1448.

Kanemaki, M., Sanchez-Diaz, A., Gambus, A. and Labib, K. (2003). Functional proteomic identification of DNA replication proteins by induced proteolysis *in vivo. Nature* **423**, 720–724.

Kozminski, K. G., Chen, A. J., Rodal, A. A. and Drubin, D. G. (2000). Functions and functional domains of the GTPase Cdc42p. *Mol Biol Cell* **11**, 339–354.

Labib, K., Tercero, J. A. and Diffley, J. F. (2000). Uninterrupted *MCM2-7* function required for DNA replication fork progression. *Science* **288**, 1643–1647.

Lewis, M. J. and Pelham, H. R. B. (1996). SNARE-mediated retrograde traffic from the Golgi complex to the endoplasmic reticulum. *Cell* **85**, 205–215.

Longtine, M. S., McKenzie, A., Demarini, D. J., Shah, N. G., Wach, A., Brachat, A., Philippsen, P. and Pringle, J. R. (1998). Additional modules for versatile and economical PCR-based gene deletion and modification in *Saccharomyces cerevisiae. Yeast* **14**, 953–961.

MacKelvie, S. H., Andrews, P. D. and Stark, M. J. R. (1995). The *Saccharomyces cerevisiae* gene *SDS22* encodes a potential regulator of the mitotic function of yeast type 1 protein phosphatase. *Mol Cell Biol* **15**, 3777–3785.

Mnaimneh, S., Davierwala, A. P., Haynes, J., Moffat, J., Peng, W. T., Zhang, W., Yang, X., Pootoolal, J., Chua, G., Lopez, A. *et al.* (2004). Exploration of essential gene functions via titratable promoter alleles. *Cell* **118**, 31–44.

Mountain, H. A., Bystrom, A. S., Larsen, J. T. and Korch, C. (1991). Four major transcriptional responses in the methionine/threonine biosynthetic pathway of *Saccharomyces cerevisiae*. *Yeast* **7**, 781–803.

Mountain, H. A., Heiber, M., Korch, C. and Byström, A. S. (1990). A *URA3* gene fusion for isolating methionine-specific regulatory genes. *Yeast* **6**, S275.

Muhlrad, D., Hunter, R. and Parker, R. (1992). A rapid method for localized mutagenesis of yeast genes. *Yeast* **8**, 79–82.

Mumberg, D., Muller, R. and Funk, M. (1994). Regulatable promoters of *Saccharomyces cerevisiae*: comparison of transcriptional activity and their use for heterologous expression. *Nucleic Acids Res* **22**, 5767–5768.

Pringle, J. R., Adams, A. E. M., Drubin, D. G. and Haarer, B. K. (1991). Immunofluorescence Methods for Yeast. *Methods in Enzymology* **194**, 565–602.

Sanchez-Diaz, A., Kanemaki, M., Marchesi, V. and Labib, K. (2004). Rapid depletion of budding yeast proteins by fusion to a heat-inducible degron. *Sci STKE* **2004**, PL8.

Schneider, B. L., Seufert, W., Steiner, B., Yang, Q. H. and Futcher, A. B. (1995). Use of polymerase chain reaction epitope tagging for protein tagging in *Saccharomyces cerevisiae*. *Yeast* **11**, 1265–1274.

Sikorski, R. S. and Boeke, J. D. (1991). *In Vitro* mutagenesis and plasmid shuffling-from cloned gene to mutant yeast. *Methods Enzymol* **194**, 302–318.

Sikorski, R. S. and Hieter, P. (1989). A system of shuttle vectors and yeast host strains designed for efficient manipulation of DNA in *Saccharomyces cerevisiae*. *Genetics* **122**, 19–27.

Singh, A. and Manney, T. R. (1974). Genetic analysis of mutations affecting growth of *Saccharomyces cerevisiae* at low temperature. *Genetics* **77**, 651–659.

Spang, A., Courtney, I., Grein, K., Matzner, M. and Schiebel, E. (1995). The Cdc31p-binding protein Kar1p is a component of the half bridge of the yeast spindle pole body. *J Cell Biol* **128**, 863–877.

Stirling, D. A., Welch, K. A. and Stark, M. J. R. (1994). Interaction with calmodulin is required for the function of Spc110p, an essential component of the yeast spindle pole body. *EMBO J* **13**, 4329–4342.

Struhl, K. (1983). The new yeast genetics. *Nature* **305**, 391–397.

Tanaka, T. U., Rachidi, N., Janke, C., Pereira, G., Galova, M., Schiebel, E., Stark, M. J. and Nasmyth, K. (2002). Evidence that the Ipl1-Sli15 (Aurora kinase-INCENP) complex promotes chromosome bi-orientation by altering kinetochore-spindle pole connections. *Cell* **108**, 317–329.

Verheggen, C., Mouaikel, J., Thiry, M., Blanchard, J. M., Tollervey, D., Bordonne, R., Lafontaine, D. L. and Bertrand, E. (2001). Box C100/D small nucleolar RNA trafficking involves small nucleolar RNP proteins, nucleolar factors and a novel nuclear domain. *EMBO J* **20**, 5480–5490.

Wishart, J. A., Hayes, A., Wardleworth, L., Zhang, N. and Oliver, S. G. (2005). Doxycycline, the drug used to control the tet-regulatable promoter system, has no effect on global gene expression in *Saccharomyces cerevisiae*. *Yeast* **22**, 565–569.

Wishart, J. A., Osborn, M., Gent, M. E., Yen, K., Vujovic, Z., Gitsham, P., Zhang, N., Ross Miller, J. and Oliver, S. G. (2006). The relative merits of the tetO2 and tetO7 promoter systems for the functional analysis of heterologous genes in yeast and a compilation of essential yeast genes with tetO2 promoter substitutions. *Yeast* **23**, 325–331.

Yen, K., Gitsham, P., Wishart, J., Oliver, S. G. and Zhang, N. (2003). An improved tetO promoter replacement system for regulating the expression of yeast genes. *Yeast* **20**, 1255–1262.

6 Yeast Hybrid Approaches

Eugene Izumchenko[1,2], Marina Wolfson[2], Erica A Golemis[1] and Ilya G Serebriiskii[1]

[1] *Division of Basic Sciences, Fox Chase Cancer Center, 333 Cottman Ave., Philadelphia, PA 19111, USA;*
[2] *Department of Microbiology and Immunology, Ben Gurion University, Beer Sheva 84105, Israel*

Yeast Hybrid Approaches

◆◆

CONTENTS

Introduction: the origins and composition of the yeast two-hybrid (Y2H) system
Selecting a system for Y2H assays
How to use the "classic" Y2H
Specialized approaches
Conclusion

List of abbreviations

AD	activation domain
ORF	open reading frame
DBD	DNA-binding domain
ER	endoplasmic reticulum
Y2H	yeast two-hybrid
Y1H	yeast one-hybrid
5FOA	5-fluoroотic acid
α-AA	α-aminoadipic acid
ONPG	2-nitrophenyl-β-D-galactopyranoside
RTA	repressed transactivator
GFP	green fluorescent protein
DIS	double interaction screen
SRS	Sos-recruitment system
RRS	Ras-recruitment system
SUS	split-ubiquitin system
Nub	N-terminal ubiquitin
Cub	C-terminal ubiquitin
UPR	unfolded protein response
MYTHS	membrane yeast two-hybrid screen
SCINEX-P	screening for interactions between extracellular proteins
TAP	tandem affinity purification
GR	glucocorticoid receptor
Dex	dexamethasone

METHODS IN MICROBIOLOGY, VOLUME 36
0580-9517 DOI:10.1016/S0580-9517(06)36006-0

DHFR	dihydrofolate reductase
MTX	methotrexate
CDK	cyclin-dependent kinase

♦♦♦♦♦♦ I. INTRODUCTION: THE ORIGINS AND COMPOSITION OF THE YEAST TWO-HYBRID (Y2H) SYSTEM

Most of the chapters in this book describe the use of genetic techniques to gain insights into the natural processes of yeast. By contrast, this chapter focuses on the exploitation of yeast genetics to create artificial selection/reporter systems suitable for yielding insights into the signaling function of any organism.

The Y2H system was designed in the late 1980s as a means of detecting direct interactions between binary pairs of proteins. There were two compelling reasons to develop such a technology. One was to facilitate analysis of the physical requirements for interaction between two defined proteins already known or suspected to interact. The second was to allow the identification of novel physical interaction partners for proteins of interest, thereby obtaining clues as to the biological function of these proteins. Among the advantages of using yeast for such purposes were the low cost and speed of assays involving this organism, and the availability of a large number of well-characterized and effective expression and reporter systems. The spark of inspiration for the Y2H lay in the selection and assembly of components that would allow robust and accurate reflection of protein interactions.

Work in the early 1980s had established that it was possible to generate novel transcription factors of predetermined DNA binding specificity by creating chimeric or hybrid domains from two different proteins. This was demonstrated using DNA-binding domains (DBDs) provided by the bacterial repressor protein LexA, or the yeast transcriptional activator Gal4p, that were fused to transcriptional activation domains provided either from Gal4p, or other heterologous activating sequences (Brent and Ptashne, 1984, 1985; Silver *et al.*, 1986). As a second important observation, it was shown that a chimeric transcriptional activating domain (Gal80p fused to acidic activating sequences) lacking a DBD was able to activate transcription when it was brought into the proximity of a promotor based on an association with the Gal4p DBD containing Gal80p interaction sequences (Ma and Ptashne, 1988). Adapting these physical components, in 1989 Fields and Song demonstrated that it was in principle possible to replace the interacting components with

Durfee, T., Becherer, K., Chen, P. L., Yeh, S. H., Yang, Y., Kilburn, A. E., Lee, W. H. and Elledge, S. J. (1993). The retinoblastoma protein associates with the protein phosphatase type 1 catalytic subunit. *Genes Dev.* **7**, 555–569.

Ehrhard, K. N., Jacoby, J. J., Fu, X. Y., Jahn, R. and Dohlman, H. G. (2000). Use of G-protein fusions to monitor integral membrane protein–protein interactions in yeast. *Nat. Biotechnol.* **18**, 1075–1079.

Endo, T., Sasaki, A., Minoguchi, M., Joo, A. and Yoshimura, A. (2003). CIS1 interacts with the Y532 of the prolactin receptor and suppresses prolactin-dependent STAT5 activation. *J. Biochem. (Tokyo)* **133**, 109–113.

Estojak, J., Brent, R. and Golemis, E. A. (1995). Correlation of two-hybrid affinity data with *in vitro* measurements. *Mol. Cell. Biol.* **15**, 5820–5829.

Fan, Q., Li, J., Kariuki, M. and Cui, L. (2004). Characterization of PfPuf2, member of the Puf family RNA-binding proteins from the malaria parasite Plasmodium falciparum. *DNA Cell Biol.* **23**, 753–760.

Fearon, E. R., Finkel, T., Gillison, M. L., Kennedy, S. P., Casella, J. F., Tomaselli, G. F., Morrow, J. S. and Van Dang, C. (1992). Karyoplasmic interaction selection strategy: a general strategy to detect protein–protein interactions in mammalian cells. *Proc. Natl. Acad. Sci. USA* **89**, 7958–7962.

Feng, S. Y., Ota, K., Yamada, Y., Sawabu, N. and Ito, T. (2004). A yeast one-hybrid system to detect methylation-dependent DNA–protein interactions. *Biochem. Biophys. Res. Commun.* **313**, 922–925.

Fessart, D., Simaan, M. and Laporte, S. A. (2005). c-Src regulates clathrin adapter protein 2 interaction with beta-arrestin and the angiotensin II type 1 receptor during clathrin-mediated internalization. *Mol. Endocrinol.* **19**, 491–503.

Fetchko, M. and Stagljar, I. (2004). Application of the split-ubiquitin membrane yeast two-hybrid system to investigate membrane protein interactions. *Methods* **32**, 349–362.

Fields, S. and Song, O. (1989). A novel genetic system to detect protein–protein interactions. *Nature* **340**, 245–246.

Fujiwara, K., Poikonen, K., Aleman, L., Valtavaara, M., Saksela, K. and Mayer, B. J. (2002). A single-chain antibody/epitope system for functional analysis of protein–protein interactions. *Biochemistry* **41**, 12729–12738.

Gardiner, L., Coyle, B. J., Chan, W. C. and Soultanas, P. (2005). Discovery of antagonist peptides against bacterial helicase–primase interaction in *B. stearothermophilus* by reverse yeast three-hybrid. *Chem. Biol.* **12**, 595–604.

Gavin, A. C., Bosche, M., Krause, R., Grandi, P., Marzioch, M., Bauer, A., Schultz, J., Rick, J. M., Michon, A. M., Cruciat, C. M., Remor, M., Hofert, C., Schelder, M., Brajenovic, M., Ruffner, H., Merino, A., Klein, K., Hudak, M., Dickson, D., Rudi, T., Gnau, V., Bauch, A., Bastuck, S., Huhse, B., Leutwein, C., Heurtier, M. A., Copley, R. R., Edelmann, A., Querfurth, E., Rybin, V., Drewes, G., Raida, M., Bouwmeester, T., Bork, P., Seraphin, B., Kuster, B., Neubauer, G. and Superti-Furga, G. (2002). Functional organization of the yeast proteome by systematic analysis of protein complexes. *Nature* **415**, 141–147.

Gingras, A. C., Aebersold, R. and Raught, B. (2005). Advances in protein complex analysis using mass spectrometry. *J. Physiol.* **563**, 11–21.

Giot, L., Bader, J. S., Brouwer, C., Chaudhuri, A., Kuang, B., Li, Y., Hao, Y. L., Ooi, C. E., Godwin, B., Vitols, E., Vijayadamodar, G., Pochart, P., Machineni, H., Welsh, M., Kong, Y., Zerhusen, B., Malcolm, R., Varrone, Z., Collis, A., Minto, M., Burgess, S., McDaniel, L., Stimpson, E., Spriggs, F., Williams, J., Neurath, K., Ioime, N., Agee, M., Voss, E., Furtak, K.,

Renzulli, R., Aanensen, N., Carrolla, S., Bickelhaupt, E., Lazovatsky, Y., DaSilva, A., Zhong, J., Stanyon, C. A., Finley, R. L., Jr., White, K. P., Braverman, M., Jarvie, T., Gold, S., Leach, M., Knight, J., Shimkets, R. A., McKenna, M. P., Chant, J. and Rothberg, J. M. (2003). A protein interaction map of *Drosophila melanogaster*. *Science* **302**, 1727–1736.

Golemis, E. A. and Adams, P. D. (2005). *Protein–Protein Interactions: A Molecular Cloning Manual*. Cold Spring Harbor Laboratory Press, Cold Spring Harbor, New York.

Gordon, S. M., Alon, N. and Buchwald, M. (2005). FANCC, FANCE and FANCD2 form a ternary complex essential to the integrity of the Fanconi anemia DNA damage response pathway. *J. Biol. Chem.* **280**, 36118–36125.

Grossel, M. J., Wang, H., Gadea, B., Yeung, W. and Hinds, P. W. (1999). A yeast two-hybrid system for discerning differential interactions using multiple baits. *Nat. Biotechnol.* **17**, 1232–1233.

Gunde, T. and Barberis, A. (2005). Yeast growth selection system for detecting activity and inhibition of dimerization-dependent receptor tyrosine kinase. *Biotechniques* **39**, 541–549.

Gunde, T., Tanner, S., Auf der Maur, A., Petrascheck, M. and Barberis, A. (2004). Quenching accumulation of toxic galactose-1-phosphate as a system to select disruption of protein–protein interactions *in vivo*. *Biotechniques* **37**, 844–852.

Guo, D., Hazbun, T. R., Xu, X. J., Ng, S. L., Fields, S. and Kuo, M. H. (2004). A tethered catalysis, two-hybrid system to identify protein–protein interactions requiring post-translational modifications. *Nat. Biotechnol.* **22**, 888–892.

Gyuris, J., Golemis, E., Chertkov, H. and Brent, R. (1993). Cdi1, a human G1 and S phase protein phosphatase that associates with Cdk2. *Cell* **75**, 791–803.

Hanamoto, T., Ozaki, T., Furuya, K., Hosoda, M., Hayashi, S., Nakanishi, M., Yamamoto, H., Kikuchi, H., Todo, S. and Nakagawara, A. (2005). Identification of protein kinase A catalytic subunit beta as a novel binding partner of p73 and regulation of p73 function. *J. Biol. Chem.* **280**, 16665–16675.

Heinrich, J. N., Kwak, S. P., Howland, D. S., Chen, J., Sturner, S., Sullivan, K., Lipinski, K., Cheng, K. Y., She, Y., Lo, F. and Ghavami, A. (2005). Disruption of ShcA signaling halts cell proliferation-characterization of ShcC residues that influence signaling pathways using yeast. *Cell Signal* **18**, 795–806.

Heiska, L. and Carpen, O. (2005). Src phosphorylates ezrin at tyrosine 477 and induces a phosphospecific association between ezrin and a kelch-repeat protein family member. *J. Biol. Chem.* **280**, 10244–10252.

Henthorn, D. C., Jaxa-Chamiec, A. A. and Meldrum, E. (2002). A GAL4-based yeast three-hybrid system for the identification of small molecule-target protein interactions. *Biochem. Pharmacol.* **63**, 1619–1628.

Hirst, M., Ho, C., Sabourin, L., Rudnicki, M., Penn, L. and Sadowski, I. (2001). A two-hybrid system for transactivator bait proteins. *Proc. Natl. Acad. Sci. USA* **98**, 8726–8731.

Hittelman, A. B., Burakov, D., Iniguez-Lluhi, J. A., Freedman, L. P. and Garabedian, M. J. (1999). Differential regulation of glucocorticoid receptor transcriptional activation via AF-1-associated proteins. *EMBO J.* **18**, 5380–5388.

Honma, T. and Goto, K. (2001). Complexes of MADS-box proteins are sufficient to convert leaves into floral organs. *Nature* **409**, 525–529.

Hook, B., Bernstein, D., Zhang, B. and Wickens, M. (2005). RNA–protein interactions in the yeast three-hybrid system: affinity, sensitivity, and enhanced library screening. *RNA* **11**, 227–233.

Huang, A., Ho, C. S., Ponzielli, R., Barsyte-Lovejoy, D., Bouffet, E., Picard, D., Hawkins, C. E. and Penn, L. Z. (2005). Identification of a novel c-Myc protein interactor, JPO2, with transforming activity in medulloblastoma cells. *Cancer Res.* **65**, 5607–5619.

Huang, J. and Schreiber, S. L. (1997). A yeast genetic system for selecting small molecule inhibitors of protein–protein interactions in nanodroplets. *Proc. Natl. Acad. Sci. USA* **94**, 13396–13401.

Hubsman, M., Yudkovsky, G. and Aronheim, A. (2001). A novel approach for the identification of protein–protein interaction with integral membrane proteins. *Nucleic Acids Res.* **29**, E18.

Hussey, S. L., Muddana, S. S. and Peterson, B. R. (2003). Synthesis of a beta-estradiol-biotin chimera that potently heterodimerizes estrogen receptor and streptavidin proteins in a yeast three-hybrid system. *J. Am. Chem. Soc.* **125**, 3692–3693.

Inouye, C., Dhillon, N., Durfee, T., Zambryski, P. C. and Thorner, J. (1997). Mutational analysis of STE5 in the yeast *Saccharomyces cerevisiae*: application of a differential interaction trap assay for examining protein–protein interactions. *Genetics* **147**, 479–492.

Inouye, C., Remondelli, P., Karin, M. and Elledge, S. (1994). Isolation of a cDNA encoding a metal response element binding protein using a novel expression cloning procedure: the one hybrid system. *DNA Cell Biol.* **13**, 731–742.

Ismaili, N., Blind, R. and Garabedian, M. J. (2005). Stabilization of the unliganded glucocorticoid receptor by TSG101. *J. Biol. Chem.* **280**, 11120–11126.

Ito, T., Chiba, T., Ozawa, R., Yoshida, M., Hattori, M. and Sakaki, Y. (2001). A comprehensive two-hybrid analysis to explore the yeast protein interactome. *Proc. Natl. Acad. Sci. USA* **98**, 4569–4574.

Iyer, K., Burkle, L., Auerbach, D., Thaminy, S., Dinkel, M., Engels, K. and Stagljar, I. (2005). Utilizing the split-ubiquitin membrane yeast two-hybrid system to identify protein–protein interactions of integral membrane proteins. *Sci. STKE* **2005**, pl3.

Jaaro, H., Levy, Z. and Fainzilber, M. (2005). A genome wide screening approach for membrane-targeted proteins. *Mol. Cell. Proteomics* **4**, 328–333.

Jaeger, S., Eriani, G. and Martin, F. (2004). Critical residues for RNA discrimination of the histone hairpin binding protein (HBP) investigated by the yeast three-hybrid system. *FEBS Lett.* **556**, 265–270.

Jiang, R. and Carlson, M. (1996). Glucose regulates protein interactions within the yeast SNF1 protein kinase complex. *Genes Dev.* **10**, 3105–3115.

Johnsson, N. and Varshavsky, A. (1994). Split ubiquitin as a sensor of protein interactions *in vivo*. *Proc. Natl. Acad. Sci. USA* **91**, 10340–10344.

Joung, J. K., Ramm, E. I. and Pabo, C. O. (2000). A bacterial two-hybrid selection system for studying protein–DNA and protein–protein interactions. *Proc. Natl. Acad. Sci. USA* **97**, 7382–7387.

Kato-Stankiewicz, J., Hakimi, I., Zhi, G., Zhang, J., Serebriiskii, I., Guo, L., Edamatsu, H., Koide, H., Menon, S., Eckl, R., Sakamuri, S., Lu, Y., Chen, Q. Z., Agarwal, S., Baumbach, W. R., Golemis, E. A., Tamanoi, F. and Khazak, V. (2002). Inhibitors of Ras/Raf-1 interaction identified by two-hybrid screening revert Ras-dependent transformation phenotypes in human cancer cells. *Proc. Natl. Acad. Sci. USA* **99**, 14398–14403.

Kaufer, N. F., Fried, H. M., Schwindinger, W. F., Jasin, M. and Warner, J. R. (1983). Cycloheximide resistance in yeast: the gene and its protein. *Nucleic Acids Res.* **11**, 3123–3135.

Kawachi, H., Fujikawa, A., Maeda, N. and Noda, M. (2001). Identification of GIT1/Cat-1 as a substrate molecule of protein tyrosine phosphatase zeta /beta by the yeast substrate-trapping system. *Proc. Natl. Acad. Sci. USA* **98**, 6593–6598.

Keegan, K. and Cooper, J. A. (1996). Use of the two hybrid system to detect the association of the protein-tyrosine-phosphatase, SHPTP2, with another SH2-containing protein, Grb7. *Oncogene* **12**, 1537–1544.

Kohler, F. and Muller, K. M. (2003). Adaptation of the Ras-recruitment system to the analysis of interactions between membrane-associated proteins. *Nucleic Acids Res.* **31**, e28.

Kornbluth, S., Jove, R. and Hanafusa, H. (1987). Characterization of avian and viral p60src proteins expressed in yeast. *Proc. Natl. Acad. Sci. USA* **84**, 4455–4459.

Kumar, A., Agarwal, S., Heyman, J. A., Matson, S., Heidtman, M., Piccirillo, S., Umansky, L., Drawid, A., Jansen, R., Liu, Y., Cheung, K. H., Miller, P., Gerstein, M., Roeder, G. S. and Snyder, M. (2002). Subcellular localization of the yeast proteome. *Genes Dev.* **16**, 707–719.

Le, S., Sternglanz, R. and Greider, C. W. (2000). Identification of two RNA-binding proteins associated with human telomerase RNA. *Mol. Biol. Cell* **11**, 999–1010.

Leanna, C. A. and Hannink, M. (1996). The reverse two-hybrid system: a genetic scheme for selection against specific protein/protein interactions. *Nucleic Acids Res.* **24**, 3341–3347.

Lee, J. W., Choi, H. S., Gyuris, J., Brent, R. and Moore, D. D. (1995). Two classes of proteins dependent on either the presence or absence of thyroid hormone for interaction with the thyroid hormone receptor. *Mol. Endocrinol.* **9**, 243–254.

Li, J. J. and Herskowitz, I. (1993). Isolation of ORC6, a component of the yeast origin recognition complex by a one-hybrid system. *Science* **262**, 1870–1874.

Li, S., Armstrong, C. M., Bertin, N., Ge, H., Milstein, S., Boxem, M., Vidalain, P. O., Han, J. D., Chesneau, A., Hao, T., Goldberg, D. S., Li, N., Martinez, M., Rual, J. F., Lamesch, P., Xu, L., Tewari, M., Wong, S. L., Zhang, L. V., Berriz, G. F., Jacotot, L., Vaglio, P., Reboul, J., Hirozane-Kishikawa, T., Li, Q., Gabel, H. W., Elewa, A., Baumgartner, B., Rose, D. J., Yu, H., Bosak, S., Sequerra, R., Fraser, A., Mango, S. E., Saxton, W. M., Strome, S., Van Den Heuvel, S., Piano, F., Vandenhaute, J., Sardet, C., Gerstein, M., Doucette-Stamm, L., Gunsalus, K. C., Harper, J. W., Cusick, M. E., Roth, F. P., Hill, D. E. and Vidal, M. (2004). A map of the interactome network of the metazoan *C. elegans*. *Science* **303**, 540–543.

Licitra, E. J. and Liu, J. O. (1996). A three-hybrid system for detecting small ligand–protein receptor interactions. *Proc. Natl. Acad. Sci. USA* **93**, 12817–12821.

Lin, H., Abida, W. M., Sauer, R. T. and Cornish, V. W. (2000). Dexamethasone–methotrexate: an efficient chemical inducer of protein dimerization *in vivo*. *J. Am. Chem. Soc.* **122**, 4247–4248.

Lundbladt, V. (2004). An Est Fest: using baker's yeast for cancer gene discovery. American Association for Cancer Research Annual Meeting, Orange County Convention Center, Orlando, FL, USA.

Ma, J. and Ptashne, M. (1988). Converting a eukaryotic transcriptional inhibitor into an activator. *Cell* **55**, 443–446.

Mark-Danieli, M., Laham, N., Kenan-Eichler, M., Castiel, A., Melamed, D., Landau, M., Bouvier, N. M., Evans, M. J. and Bacharach, E. (2005). Single point mutations in the zinc finger motifs of the human immunodeficiency virus type 1 nucleocapsid alter RNA binding specificities of the gag protein and enhance packaging and infectivity. *J. Virol.* **79**, 7756–7767.

Marsolier, M. C., Prioleau, M. N. and Sentenac, A. (1997). A RNA polymerase III-based two-hybrid system to study RNA polymerase II transcriptional regulators. *J. Mol. Biol.* **268**, 243–249.

Marsolier, M. C. and Sentenac, A. (1999). RNA polymerase III-based two-hybrid system. *Methods Enzymol.* **303**, 411–422.

Martin, F., Michel, F., Zenklusen, D., Muller, B. and Schumperli, D. (2000). Positive and negative mutant selection in the human histone hairpin-binding protein using the yeast three-hybrid system. *Nucleic Acids Res.* **28**, 1594–1603.

Matsumoto, A., Comatas, K. E., Liu, L. and Stamler, J. S. (2003). Screening for nitric oxide-dependent protein–protein interactions. *Science* **301**, 657–661.

Monshausen, M., Putz, U., Rehbein, M., Schweizer, M., DesGroseillers, L., Kuhl, D., Richter, D. and Kindler, S. (2001). Two rat brain staufen isoforms differentially bind RNA. *J. Neurochem.* **76**, 155–165.

Morra, M., Lu, J., Poy, F., Martin, M., Sayos, J., Calpe, S., Gullo, C., Howie, D., Rietdijk, S., Thompson, A., Coyle, A. J., Denny, C., Yaffe, M. B., Engel, P., Eck, M. J. and Terhorst, C. (2001). Structural basis for the interaction of the free SH2 domain EAT-2 with SLAM receptors in hematopoietic cells. *EMBO J.* **20**, 5840–5852.

Morris, J. R., Keep, N. H. and Solomon, E. (2002). Identification of residues required for the interaction of BARD1 with BRCA1. *J. Biol. Chem.* **277**, 9382–9386.

Nakamura, T., Komiya, M., Sone, K., Hirose, E., Gotoh, N., Morii, H., Ohta, Y. and Mori, N. (2002). Grit, a GTPase-activating protein for the Rho family, regulates neurite extension through association with the TrkA receptor and N-Shc and CrkL/Crk adapter molecules. *Mol. Cell. Biol.* **22**, 8721–8734.

Naya, F. J., Stellrecht, C. M. and Tsai, M. J. (1995). Tissue-specific regulation of the insulin gene by a novel basic helix-loop-helix transcription factor. *Genes Dev.* **9**, 1009–1019.

Obrdlik, P., El-Bakkoury, M., Hamacher, T., Cappellaro, C., Vilarino, C., Fleischer, C., Ellerbrok, H., Kamuzinzi, R., Ledent, V., Blaudez, D., Sanders, D., Revuelta, J. L., Boles, E., Andre, B. and Frommer, W. B. (2004). K+ channel interactions detected by a genetic system optimized for systematic studies of membrane protein interactions. *Proc. Natl. Acad. Sci. USA* **101**, 12242–12247.

Oldenburg, K. R., Vo, K. T., Michaelis, S. and Paddon, C. (1997). Recombination-mediated PCR-directed plasmid construction *in vivo* in yeast. *Nucleic Acids Res.* **25**, 451–452.

Osborne, M. A., Dalton, S. and Kochan, J. P. (1995). The yeast tribrid system – genetic detection of trans-phosphorylated ITAM-SH2-interactions. *Biotechnology (NY)* **13**, 1474–1478.

Ozenberger, B. A. and Young, K. H. (1995). Functional interaction of ligands and receptors of the hematopoietic superfamily in yeast. *Mol. Endocrinol.* **9**, 1321–1329.

Park, Y. W., Tan, S. L. and Katze, M. G. (1999). Differential sensitivity to 5-fluorootic acid as a screen for bait RNA-independent false positives in a yeast three-hybrid system. *Biotechniques* **26**, 1102–1106.

Pause, A., Peterson, B., Schaffar, G., Stearman, R. and Klausner, R. D. (1999). Studying interactions of four proteins in the yeast two-hybrid system: structural resemblance of the pVHL/elongin BC/hCUL-2 complex with the ubiquitin ligase complex SKP1/cullin/F-box protein. *Proc. Natl. Acad. Sci. USA* **96**, 9533–9538.

Paziewska, A., Wyrwicz, L. S. and Ostrowski, J. (2005). The binding activity of yeast RNAs to yeast Hek2p and mammalian hnRNP K proteins, determined using the three-hybrid system. *Cell. Mol. Biol. Lett.* **10**, 227–235.

Petrascheck, M., Castagna, F. and Barberis, A. (2001). Two-hybrid selection assay to identify proteins interacting with polymerase II transcription factors and regulators. *Biotechniques* **30**, 296–298 300, 302.

Plaza, S., Prince, F., Jaeger, J., Kloter, U., Flister, S., Benassayag, C., Cribbs, D. and Gehring, W. J. (2001). Molecular basis for the inhibition of Drosophila eye development by Antennapedia. *EMBO J.* **20**, 802–811.

Pollock, S., Kozlov, G., Pelletier, M. F., Trempe, J. F., Jansen, G., Sitnikov, D., Bergeron, J. J., Gehring, K., Ekiel, I. and Thomas, D. Y. (2004). Specific interaction of ERp57 and calnexin determined by NMR spectroscopy and an ER two-hybrid system. *EMBO J.* **23**, 1020–1029.

Puthalakath, H., Strasser, A. and Huang, D. C. (2001). Rapid selection against truncation mutants in yeast reverse two-hybrid screens. *Biotechniques* **30**, 984–988.

Putz, U., Skehel, P. and Kuhl, D. (1996). A tri-hybrid system for the analysis and detection of RNA–protein interactions. *Nucleic Acids Res.* **24**, 4838–4840.

Rain, J. C., Selig, L., De Reuse, H., Battaglia, V., Reverdy, C., Simon, S., Lenzen, G., Petel, F., Wojcik, J., Schachter, V., Chemama, Y., Labigne, A. and Legrain, P. (2001). The protein–protein interaction map of *Helicobacter pylori*. *Nature* **409**, 211–215.

Ray, M. R., Wafa, L. A., Cheng, H., Snoek, R., Fazli, L., Gleave, M. and Rennie, P. S. (2005). Cyclin G-associated kinase: a novel androgen receptor-interacting transcriptional coactivator that is overexpressed in hormone refractory prostate cancer. *Int. J. Cancer* **118**, 1108–1119.

Reeder, M. K., Serebriiskii, I. G., Golemis, E. A. and Chernoff, J. (2001). Analysis of small GTPase signaling pathways using p21-activated kinase mutants that selectively couple to Cdc42. *J. Biol. Chem.* **276**, 40606–40613.

Rual, J. F., Venkatesan, K., Hao, T., Hirozane-Kishikawa, T., Dricot, A., Li, N., Berriz, G. F., Gibbons, F. D., Dreze, M., Ayivi-Guedehoussou, N., Klitgord, N., Simon, C., Boxem, M., Milstein, S., Rosenberg, J., Goldberg, D. S., Zhang, L. V., Wong, S. L., Franklin, G., Li, S., Albala, J. S., Lim, J., Fraughton, C., Llamosas, E., Cevik, S., Bex, C., Lamesch, P., Sikorski, R. S., Vandenhaute, J., Zoghbi, H. Y., Smolyar, A., Bosak, S., Sequerra, R., Doucette-Stamm, L., Cusick, M. E., Hill, D. E., Roth, F. P. and Vidal, M. (2005). Towards a proteome-scale map of the human protein–protein interaction network. *Nature* **437**, 1173–1178.

Sandrock, B. and Egly, J. M. (2001). A yeast four-hybrid system identifies Cdk-activating kinase as a regulator of the XPD helicase, a subunit of transcription factor IIH. *J. Biol. Chem.* **276**, 35328–35333.

Sayos, J., Martin, M., Chen, A., Simarro, M., Howie, D., Morra, M., Engel, P. and Terhorst, C. (2001). Cell surface receptors Ly-9 and CD84 recruit the

X-linked lymphoproliferative disease gene product SAP. *Blood* **97**, 3867–3874.

Sengupta, D. J., Wickens, M. and Fields, S. (1999). Identification of RNAs that bind to a specific protein using the yeast three-hybrid system. *RNA* **5**, 596–601.

SenGupta, D. J., Zhang, B., Kraemer, B., Pochart, P., Fields, S. and Wickens, M. (1996). A three-hybrid system to detect RNA–protein interactions *in vivo*. *Proc. Natl. Acad. Sci. USA* **93**, 8496–8501.

Serebriiskii, I. G. (2005). Yeast two-hybrid system for studying protein–protein interactions. In: *Protein–Protein Interactions: A Molecular Cloning Manual* (E. A. Golemis and P. D. Adams, eds)938. Cold Spring Harbor Laboratory Press, Adams.

Serebriiskii, I. G., Fang, R., Latypova, E., Hopkins, R., Vinson, C., Joung, J. K. and Golemis, E. A. (2005). A combined yeast/bacteria two-hybrid system: development and evaluation. *Mol. Cell. Proteomics* **4**, 819–826.

Serebriiskii, I. G. and Golemis, E. A. (2000). Uses of lacZ to study gene function: evaluation of beta-galactosidase assays employed in the yeast two-hybrid system. *Anal. Biochem.* **285**, 1–15.

Serebriiskii, I. G. and Golemis, E. A. (2001a). Two-hybrid system and false positives. Approaches to detection and elimination. *Methods Mol. Biol.* **177**, 123–134.

Serebriiskii, I. G. and Golemis, E.A. (2001b). Two-hybrid system false positives and approaches to their detection and elimination. In: *Methods in Molecular Biology: Two-Hybrid Systems, Methods and Protocols* (P. N. MacDonald, ed.), Ch.9, pp.123–134. Humana Press, Totowa, NJ.

Serebriiskii, I. G., Mitina, O., Pugacheva, E. N., Benevolenskaya, E., Kotova, E., Toby, G. G., Khazak, V., Kaelin, W. G., Chernoff, J. and Golemis, E. A. (2002). Detection of peptides, proteins, and drugs that selectively interact with protein targets. *Genome Res.* **12**, 1785–1791.

Serebriiskii, I., Khazak, V. and Golemis, E. A. (1999). A two-hybrid dual bait system to discriminate specificity of protein interactions. *J. Biol. Chem.* **274**, 17080–17087.

Shih, H. M., Goldman, P. S., DeMaggio, A. J., Hollenberg, S. M., Goodman, R. H. and Hoekstra, M. F. (1996). A positive genetic selection for disrupting protein–protein interactions: identification of CREB mutations that prevent association with the coactivator CBP. *Proc. Natl. Acad. Sci. USA* **93**, 13896–13901.

Silver, P. A., Brent, R. and Ptashne, M. (1986). DNA binding is not sufficient for nuclear localization of regulatory proteins in *Saccharomyces cerevisiae*. *Mol. Cell. Biol.* **6**, 4763–4766.

Sonoda, J. and Wharton, R. P. (2001). *Drosophila* brain tumor is a translational repressor. *Genes Dev.* **15**, 762–773.

Stagljar, I., Korostensky, C., Johnsson, N. and te Heesen, S. (1998). A genetic system based on split-ubiquitin for the analysis of interactions between membrane proteins *in vivo*. *Proc. Natl. Acad. Sci. USA* **95**, 5187–5192.

Stanyon, C. A., Liu, G., Mangiola, B. A., Patel, N., Giot, L., Kuang, B., Zhang, H., Zhong, J. and Finley, R. L., Jr. (2004). A *Drosophila* protein-interaction map centered on cell-cycle regulators. *Genome Biol.* **5**, R96.

Steyn, A. J., Joseph, J. and Bloom, B. R. (2003). Interaction of the sensor module of *Mycobacterium tuberculosis* H37Rv KdpD with members of the Lpr family. *Mol. Microbiol.* **47**, 1075–1089.

Stocklein, W. and Piepersberg, W. (1980). Binding of cycloheximide to ribosomes from wild-type and mutant strains of *Saccharomyces cerevisiae*. *Antimicrob. Agents Chemother.* **18**, 863–867.

Suzuki, H., Fukunishi, Y., Kagawa, I., Saito, R., Oda, H., Endo, T., Kondo, S., Bono, H., Okazaki, Y. and Hayashizaki, Y. (2001). Protein–protein interaction panel using mouse full-length cDNAs. *Genome Res.* **11**, 1758–1765.

Thaminy, S., Miller, J. and Stagljar, I. (2004). The split-ubiquitin membrane-based yeast two-hybrid system. *Methods Mol. Biol.* **261**, 297–312.

Tirode, F., Malaguti, C., Romero, F., Attar, R., Camonis, J. and Egly, J. M. (1997). A conditionally expressed third partner stabilizes or prevents the formation of a transcriptional activator in a three-hybrid system. *J. Biol. Chem.* **272**, 22995–22999.

Uetz, P., Giot, L., Cagney, G., Mansfield, T. A., Judson, R. S., Knight, J. R., Lockshon, D., Narayan, V., Srinivasan, M., Pochart, P., Qureshi-Emili, A., Li, Y., Godwin, B., Conover, D., Kalbfleisch, T., Vijayadamodar, G., Yang, M., Johnston, M., Fields, S. and Rothberg, J. M. (2000). A comprehensive analysis of protein–protein interactions in *Saccharomyces cerevisiae*. *Nature* **403**, 623–627.

Urech, D. M., Lichtlen, P. and Barberis, A. (2003). Cell growth selection system to detect extracellular and transmembrane protein interactions. *Biochim. Biophys. Acta* **1622**, 117–127.

Van Criekinge, W., van Gurp, M., Decoster, E., Schotte, P., Van de Craen, M., Fiers, W., Vandenabeele, P. and Beyaert, R. (1998). Use of the yeast three-hybrid system as a tool to study caspases. *Anal. Biochem.* **263**, 62–66.

Vidal, M., Brachmann, R. K., Fattaey, A., Harlow, E. and Boeke, J. D. (1996). Reverse two-hybrid and one-hybrid systems to detect dissociation of protein–protein and DNA–protein interactions. *Proc. Natl. Acad. Sci. USA* **93**, 10315–10320.

Vojtek, A. B., Hollenberg, S. M. and Cooper, J. A. (1993). Mammalian Ras interacts directly with the serine/threonine kinase Raf. *Cell* **74**, 205–214.

Wafa, L. A., Cheng, H., Rao, M. A., Nelson, C. C., Cox, M., Hirst, M., Sadowski, I. and Rennie, P. S. (2003). Isolation and identification of l-dopa decarboxylase as a protein that binds to and enhances transcriptional activity of the androgen receptor using the repressed transactivator yeast two-hybrid system. *Biochem. J.* **375**, 373–383.

Watt, P. M., Thomas, W. and Hopkins, R. M. (2003). Methods of constructing and screening diverse expression libraries. United States Patent Application, USA.

West, M., Flanery, D., Woytek, K., Rangasamy, D. and Wilson, V. G. (2001). Functional mapping of the DNA binding domain of bovine papillomavirus E1 protein. *J. Virol.* **75**, 11948–11960.

Wittke, S., Lewke, N., Muller, S. and Johnsson, N. (1999). Probing the molecular environment of membrane proteins *in vivo*. *Mol. Biol. Cell* **10**, 2519–2530.

Wolf, S. S., Roder, K. and Schweizer, M. (1996). Construction of a reporter plasmid that allows expression libraries to be exploited for the one-hybrid system. *Biotechniques* **20**, 568–574.

Xu, C. W., Mendelsohn, A. R. and Brent, R. (1997). Cells that register logical relationships among proteins. *Proc. Natl. Acad. Sci. USA* **94**, 12473–12478.

Xu, P., Jacobs, A. R. and Taylor, S. I. (1999). Interaction of insulin receptor substrate 3 with insulin receptor, insulin receptor-related receptor, insulin-like growth factor-1 receptor, and downstream signaling proteins. *J. Biol. Chem.* **274**, 15262–15270.

Yamada, M., Suzuki, K., Mizutani, M., Asada, A., Matozaki, T., Ikeuchi, T., Koizumi, S. and Hatanaka, H. (2001). Analysis of tyrosine phosphorylation-dependent protein–protein interactions in TrkB-mediated intracellular signaling using modified yeast two-hybrid system. *J. Biochem. (Tokyo)* **130**, 157–165.

Young, K., Lin, S., Sun, L., Lee, E., Modi, M., Hellings, S., Husbands, M., Ozenberger, B. and Franco, R. (1998). Identification of a calcium channel modulator using a high throughput yeast two-hybrid screen. *Nat. Biotechnol.* **16**, 946–950.

Yu, Y., Li, W., Su, K., Yussa, M., Han, W., Perrimon, N. and Pick, L. (1997). The nuclear hormone receptor Ftz-F1 is a cofactor for the *Drosophila* homeodomain protein Ftz. *Nature* **385**, 552–555.

Yu, Z., Fotouhi-Ardakani, N., Wu, L., Maoui, M., Wang, S., Banville, D. and Shen, S. H. (2002). PTEN associates with the vault particles in HeLa cells. *J. Biol. Chem.* **277**, 40247–40252.

Yu, Z., Lai, C. M., Maoui, M., Banville, D. and Shen, S. H. (2001). Identification and characterization of S2V, a novel putative siglec that contains two V set Ig-like domains and recruits protein-tyrosine phosphatases SHPs. *J. Biol. Chem.* **276**, 23816–23824.

Yussa, M., Lohr, U., Su, K. and Pick, L. (2001). The nuclear receptor Ftz-F1 and homeodomain protein Ftz interact through evolutionarily conserved protein domains. *Mech. Dev.* **107**, 39–53.

Zhang, J. and Lautar, S. (1996). A yeast three-hybrid method to clone ternary protein complex components. *Anal. Biochem.* **242**, 68–72.

7 Array-Based Yeast Two-Hybrid Screening for Protein–Protein Interactions

Seesandra V Rajagopala, Björn Titz and Peter Uetz
Institut für Genetik, Forschungszentrum Karlsruhe, Karlsruhe, Germany

♦♦

CONTENTS

Introduction
Steps involved in an array-based yeast two-hybrid screen
Screening for protein interactions using a yeast colony array
Conclusion

List of abbreviations

AD	Activation domain
ORF	Open reading frame
DBD	DNA-binding domain

♦♦♦♦♦♦ I. INTRODUCTION

The yeast two-hybrid system is probably the most widely used system to detect direct protein–protein interactions. Up to now, screening of a random (genomic or cDNA) library was the method of choice to identify novel interactions for a protein of interest (see Chapter 6). However, more recently, an array-based variation of this original principle has been increasingly used. While classical two-hybrid screens used random libraries, array-based screens simply use defined arrays of prey clones that can be screened with a bait of interest (Figure 1). This approach can not only be applied to a few proteins but also to whole genomes.

METHODS IN MICROBIOLOGY, VOLUME 36
0580-9517 DOI:10.1016/S0580-9517(06)36007-2

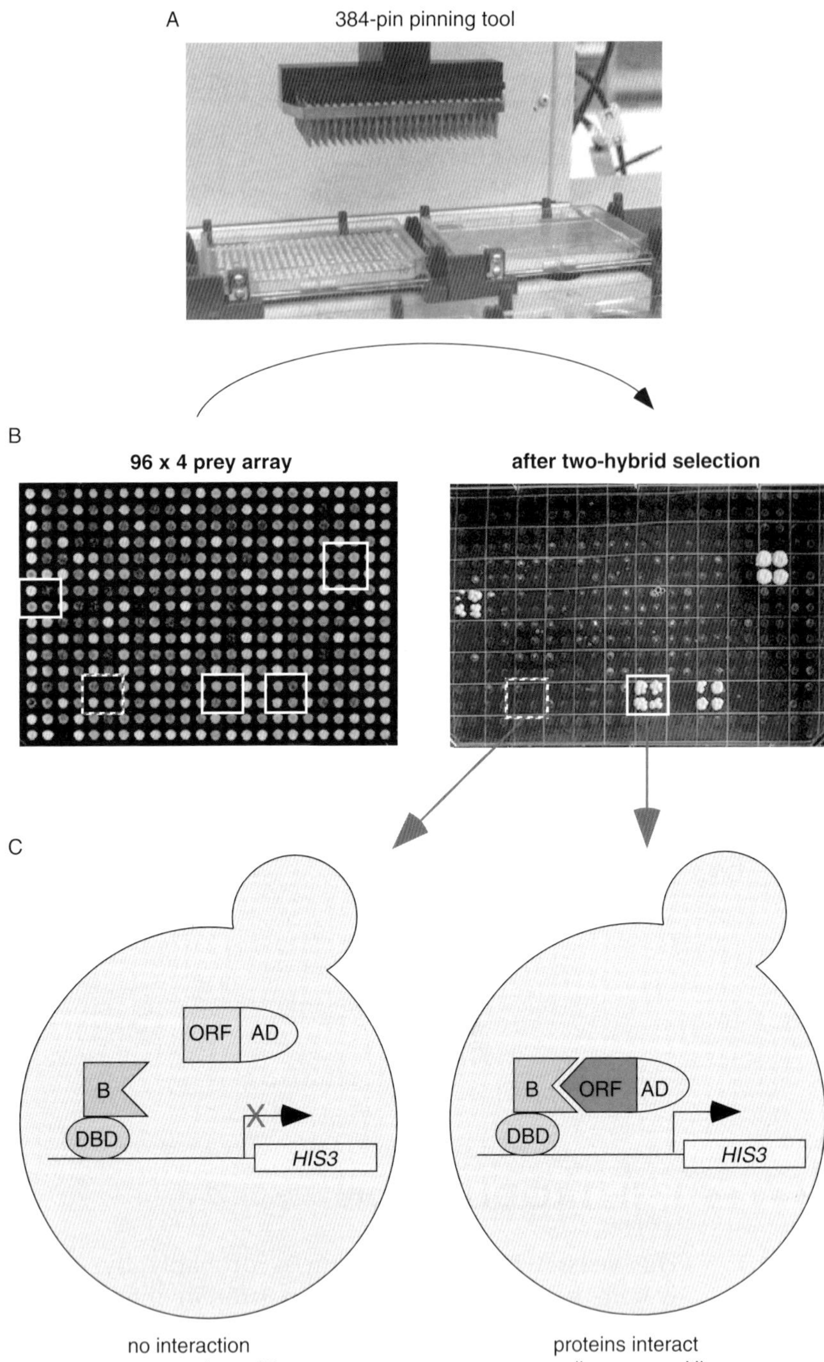
A
384-pin pinning tool
B
96 x 4 prey array
after two-hybrid selection
C
ORF
AD
B
DBD
HIS3
B
ORF
AD
DBD
HIS3
no interaction
no growth on -His
proteins interact
cells grow on -His

A. What is an Array-based Screen?

In an array, a number of defined prey proteins are tested for interactions with a bait protein (Figure 1). Usually, the bait protein is expressed in one yeast strain and the prey is expressed in another yeast strain of different mating type. The two strains are then mated so that the two proteins are expressed in the resulting diploid cell (Figure 1). The assays are done side-by-side, so they can be well controlled, i.e., compared. As the identity of the preys is usually known, no sequencing is required after positives have been identified. However, the prey clones need to be obtained or made up-front. This can be done for a few genes or for a whole genome, e.g., an ORFeome (i.e. all ORFs of a genome).

B. Why Array-based Screens?

Since many two-hybrid tests are done in parallel under the same condition, their results can be directly compared. In an array, each element has a known identity and thus it is immediately clear which two proteins are interacting when positives are selected. In addition, it is often immediately clear if an interaction is stronger than another one (but see below). Most importantly, since all of these assays are done in an ordered array, background signals can be easily distinguished from true signals (Figure 1B). Until recently, it was much easier to construct a random library and screen it than to construct many individual clones and screen them individually. However, now whole genomes become increasingly available as ordered clone sets in a variety of vectors. Modern cloning systems also allow direct transfer of entry clones into many specialized vectors (see below). For most model organisms such genome-scale clone collections are already available or will be soon. One of the first applications of such clone collections is often a protein interaction screen so it is likely that a prey library is already available for your favorite organism!

◀

Figure 1. Scheme of an array-based two-hybrid screen. The Y2H system tests for a protein interaction of a bait protein (DNA-binding domain [= DBD] fusion) with a prey protein (activation-domain [= AD] fusion). Only if bait and prey protein interact, an active transcription factor is reconstituted and transcription of a reporter gene is activated (C. lower right). In the array-based Y2H system a bait protein is tested against the whole systematic prey library. This library consists of individual yeast colonies at specific positions of an array (e.g. in 384 format, B. middle left); each colony carries a specific prey construct of a specific ORF. Systematic testing is done by robotic transfer of yeast cells starting with the mating of bait and prey strains; diploid cells are selected on specific plates and, finally, transferred to plates selecting for the activation of a reporter gene; here the activation of the *HIS3* gene is detected by scoring the growth on histidine deficient plates (B. middle right). The white rectangle on the selective plate marks a positive interaction between the bait and the prey at this specific position of the array (test is done in quadruplicates). (Adapted after Uetz *et al.*, 2000.)

In fact, in some cases only an array screen may do the job. For example, if you have a bait protein that activates transcription on its own, a carefully controlled array may be the only way to distinguish between signal and background (see Figure 1B). Similarly, weak interactors may only be detectable when compared with a comparably weak background.

C. Limitations of Array Screens

1. False negatives

Two-hybrid screens are not perfect. It is quite unlikely that you will detect all interactors of your bait protein in an array screen. In fact, it appears that array screens actually overlook more interactions than library screens (in other words: array screens have more false negatives). The main reason for that is probably the use of full-length open reading frames (ORFs) in most array screens. When several fragments (or domains) of a protein are used, many more surfaces can be exposed for interactions. These surfaces may be revealed *in vivo* under certain circumstances, e.g., when the protein is phosphorylated.

In other cases, false negatives may arise from steric hindrance of the two fusion proteins, so that physical interaction or subsequent transcriptional activation is prevented. Other explanations for false negatives include instability of proteins or failure of nuclear localization; absence of a prey protein from a library; and inappropriate post-translational modification of a bait or prey, prohibiting an interaction.

2. False positives

As with many assay systems, the two-hybrid system has the potential to produce false positives (i.e., reporter gene activity where no specific protein–protein interaction is involved). Frequently, such false positives are associated with bait proteins that act as transcriptional activators. False positives may also be caused by proteins that have the propensity to take part in non-specific interactions (for largely unknown reasons). Some bait or prey proteins may affect general colony viability, and hence enhance the ability of a cell to grow under selective conditions and activate the reporter gene. Mutations or other random events of unknown nature may be invoked as potential explanations as well. Overall, extremely few cases of false positives can be explained mechanistically (although many may simply interact non-physiologically!). A number of procedures have been developed to identify or avoid false positives, including the utilization of multiple reporters, independent methods of specificity testing or simply repeating assays to make sure a result is reproducible (see below).

D. Requirements for a Screen

Although the protocols in this chapter are based on the DNA-binding and activation domain (AD) of the yeast Gal4 protein, other DNA-binding domains can be used.

In the LexA two-hybrid system, the DNA-binding domain (DBD) is provided by the entire prokaryotic LexA protein, which normally functions as a repressor in *Escherichia coli* when it binds to LexA operators. In the yeast two-hybrid system, the LexA protein usually does not act as a repressor as the promoter with its binding sites is not constitutively active. The AD in the LexA two-hybrid system is a heterologous 88-residue acidic peptide that activates transcription in yeast. An interaction between the target protein (fused to the DNA-BD) and a library-encoded protein (fused to an AD) creates a novel transcriptional activator with binding affinity for LexA operators (see Chapter 6).

In general, every component of the "classic" two-hybrid system can be replaced by different components: For example, the reporter gene does not need to be *HIS3*. *LEU2*, an enzyme involved in leucine biosynthesis, can also be used. The reporter does not have to be a biosynthetic enzyme at all; green fluorescent protein (GFP) has been successfully used as a reporter gene, β-galactosidase (*lacZ*) is common, and many others are under investigation. Finally, the two-hybrid system does not need to be based on transcription. Johnsson and Varshavsky (1994) developed a related system that is based on reconstituting artificially split ubiquitin, a protein that tags other proteins for degradation. As long as the function of a protein can be used as a selective marker, it is theoretically possible to divide it into fragments, and drive the reassociation of the two fragments by exogenous "B and P" proteins that are attached to each half. Several other variations have been developed and are described elsewhere (Drees, 1999; Frederickson, 1998; Bartel and Fields, 1997; Table 1; see also Chapter 6, this volume).

E. Applications

Originally, the two-hybrid system was invented to demonstrate the association of two proteins (Fields and Song, 1989). Later, it was demonstrated that completely new protein interactions can be identified with this system, even when there are no candidates for an interaction with a given bait. Over time, it has become clear that the

Table 1. What you need for an array-based screen

1. Bait plasmid(s)
2. Prey plasmid(s)
3. Bait strain
4. Prey strain
5. Rich media and 3 selective media; plates
6. Pinning device (optional but necessary when large numbers are tested)

ability to conveniently perform unbiased library screens is the most powerful application of the system. With whole-genome arrays, such unbiased screens can be expanded to defined, non-redundant sets of proteins. Arrays, like traditional two-hybrid screens, can also be adapted to a variety of related questions, such as the identification of mutants that prevent or allow interactions (Schwartz *et al.*, 1998), the screening for drugs that affect protein interactions (Vidal and Legrain, 1999; Vidal and Endoh, 1999), the identification of RNA-binding proteins (Sengupta *et al.*, 1996), or the semi-quantitative determination of binding affinities (Estojak *et al.*, 1995). The system can also be exploited to map binding domains (Rain *et al.*, 2001), to study protein folding (Raquet *et al.*, 2001) or map interactions within a protein complex (Cagney *et al.*, 2001). Finally, recent large-scale projects have been successful in systematically mapping interactions within whole proteomes (yeast: Uetz *et al.*, 2000; Ito *et al.*, 2001; Giot *et al.*, 2003). These studies have shown for the first time that most proteins in a cell are actually connected to each other.

In combination with structural genomics, gene expression data, and metabolic profiling, the enormous amount of data in these interaction networks should allow us eventually to model complex biological phenomena in molecular detail. An ultimate goal of this work is to understand the interplay of DNA, RNA, and proteins, together with small molecules, in a dynamic and realistic way.

F. Genome-wide Yeast Two-hybrid Screening

The construction of an entire proteome array of an organism that can be screened *in vivo* under uniform conditions is a challenge. When proteins are screened at a genome scale, automated robotic procedures are necessary (see below).

The procedure can be modified for manual use or for use with alternative screenings strategies such as synthetic lethal screens (see Chapter 16). With minor modifications, the array can be used to screen for protein interactions with DNA, RNA, or even small-molecule inhibitors of the yeast two-hybrid interactions.

The protocols described here were established for yeast proteins, but they can be applied to any other genome or subset thereof; for example, viral and bacterial genomes have been screened for interactions in our lab. Different high-throughput cloning methods used to generate two-hybrid clones, i.e., proteins with AD fusions (preys) and the DBD fusions (baits), are presented. The steps of the process involve the construction of the array (Protocol 1) and screening of the array by either manual or robotic manipulation (Protocols 2–4), including the selection of positives and scoring of results.

High-throughput screening projects deal with a large number of proteins, therefore hands-on time and amount of resources become an important issue. Options to reduce the screening effort are discussed. A prerequisite for array-based genome-wide screen is the existence of a cloned ORFeome; we will briefly mention strategies to

Protocol 1. Yeast Transformation for Bait and Prey Construction.

This protocol is suitable for 100 yeast transformations, and may be scaled up or down as needed. Selection of the transformed yeast cells requires leucine or tryptophan-free media ("-Leu" or "-Trp", depending on the selective marker on the plasmid). Moreover, at least one of the haploid strains must contain a two-hybrid reporter gene under Gal4p control.

Materials required

- Salmon sperm DNA;
- DMSO;
- Competent host yeast strains, e.g. AH109 (for baits), and Y187 (for preys);
- Lithium Acetate (0.1 M);
- Selective plates (depending on the selective markers); and
- 96PEG solution.

Carrier DNA (salmon sperm DNA). Dissolve 7.75 mg/ml salmon sperm DNA (e.g. Sigma D1626) in water and store at –20°C following a 15 min 121°C autoclave cycle.

Preparation of 96 PEG solution (100 ml). Mix 45.6 g PEG (Sigma P3640), 6.1 ml of 2 M LiOAc (Lithium acetate), 1.14 ml of 1 M *Tris* pH 7.5 and 232 μl 0.5 M EDTA; make up to 100 ml with sterile water and autoclave.

Preparation of Competent Yeast Cells

1. Inoculate 50 ml YEPD liquid medium with 200 μl liquid stock of yeast strains (e.g. AH109, Y187 or any other appropriate yeast strain; we use Y187 strains for preys and AH109 for baits) in a 250 ml flask and grow overnight with shaking at 30°C (minimum 15 h, maximum 24 h)
2. Spin out cells in 50 ml conical tube (3500 rpm, 5 min at room temperature); pour off supernatant and dissolve the pellet by adding 2 ml LiOAc (0.1 M); and transfer resuspended yeast to two 1.5 ml microfuge tubes. Spin out yeast and resuspend in a total volume of 1.8 ml LiOAc (0.1 M).

Preparation of "CT110" for yeast transformation

Materials required

- 20.73 ml 96PEG;
- 0.58 ml boiled salmon sperm DNA (boil frozen salmon sperm DNA at 95 °C for 5 min and cool on ice before use);
- 2.62 ml DMSO; and

- 200 ng of linearized vector DNA (for co-transformation and homologous recombination in yeast).

1. Mix the above listed solutions in a 50 ml Falcon tube; add DMSO last and mix quickly after adding by shaking hard and vortex for 30 s.
2. Add all the competent yeast cells prepared above and mix hard by hand or by vortexing for 1 min. Immediately pipet 245 µl into each of 96 wells of a 96-well dish (e.g. Costar 3596).
3. Now add 50–100 ng of plasmid or 5 µl of PCR products (in case of co-transformation and homologous recombination in yeast) and positive control (empty vector) and negative control (only CT110). Seal the 96-well plate with plastic or aluminum tape and vortex for 4 min.
4. Incubate at 42°C for 30 min.
5. Spin the 96-well plate for 10 min at 2000 rpm; discard the supernatant and aspirate with eight-channel wand or by tapping on cotton napkin for a couple of times. Add 150 µl of sterile water to all 96 wells, resuspend and plate them on selective plates (35 mm) with -Leu (or -Trp) agar. Incubate the plates at 30°C for 2–3 days. After one day the colonies start to appear; pick the colonies after 2–3 days and make glycerol (20%) stocks (−80°C).

create such ORFeomes. Many ORFeome projects are currently being done. We expect readily available complete ORFeomes for all major model organisms in the near future.

♦♦♦♦♦♦ II. STEPS INVOLVED IN AN ARRAY-BASED YEAST TWO-HYBRID SCREEN

A. Strategic Planning

Before starting an array-based screen, the size and character of the array must be designed and the ultimate aims of the experiment need to be considered. Factors that may be varied include the form of protein array (e.g. full length protein or single domain, choice of epitope tags, etc.). Similarly, the arrayed proteins may be related (e.g. a family or pathway of related proteins, orthologs of a protein from different species, the entire protein compliment of a model organism). In our experience, certain protein families work extremely well (e.g. splicing proteins), while others do not appear to work at all (e.g. many metabolic enzymes). We recommend carrying out a small-scale pilot study, incorporating positive and negative controls, before committing to a full-scale project.

Although high-throughput screening projects can be performed manually, automation is strongly recommended. Highly repetitive tasks are not only boring and straining but also error prone when done manually. If you do not have local access to robotics you may have to collaborate with a laboratory that does.

B. Generation of a Protein Array Suitable for High-Throughput Screening

Once the set of proteins to be included in the array is defined the coding genes need to be PCR-amplified and cloned into Y2H bait and prey vectors. In order to facilitate the cloning of a large number to proteins, site-specific recombination-based systems are commonly used (e.g. Gateway or Univector cloning, Walhout *et al.*, 2000; Liu *et al.*, 1998; see Figure 2, see Colour Plate section). Some of these systems require expensive enzymes and vectors although both may be produced in the lab.

1. Cloning by homologous recombination in yeast

An alternative to site-directed systems is the cloning by homologous recombination directly in yeast (Cagney *et al.*, 2000). A two-step PCR protocol is used to make DNA with sufficient homology to vector DNA at the terminal ends to allow homologous recombination in the yeast cell (Figure 2). In the 'first round' PCR reaction the ORF is amplified with primers that contain ~20 nucleotide tails which are homologues to sequences in the two-hybrid vectors. In the second-round PCR ~50-nucleotide tails are attached to the first-round PCR product that are homologous to the destination vector-cloning site (Figure 2A). The PCR product is then transformed into the yeast cells together with the linearized vector and the recombination event between them takes place inside the yeast cell. The advantage of this strategy is its much-reduced cost. The disadvantage is that plasmids are not available as purified DNA but have to be recovered from yeast which can be time-consuming and inefficient.

2. Univector-plasmid-fusion system (UPS)

Similar to the Gateway system, the UPS system requires an entry vector containing the ORF. The UPS uses Cre–*lox*P-based site-specific recombination to catalyze plasmid fusion between the entry "univector" and destination vectors containing, e.g., specific promoters, fusion proteins, and selection markers. Cre is a site-specific recombinase, which catalyzes the recombination between two 34 bp *loxP* sequences (Figure 2B).

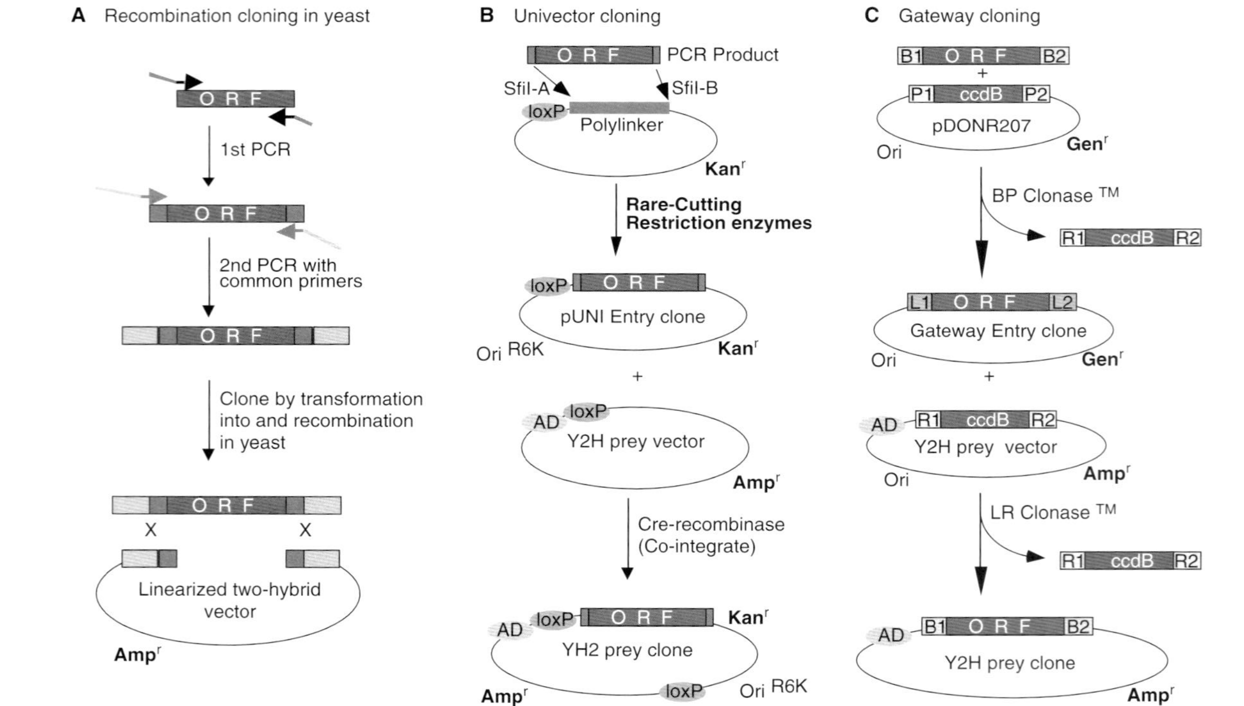

Figure 2. Cloning strategies for creating baits and preys. (A) *Homologous recombination*. ORFs are amplified (first PCR) with specific primers that generate a product with common 5′ and 3′ 20-nucleotide tails. A second PCR generate a product with common 5′ and 3′ 70-nucleotide tails. The common 70-nucleotide ends allow cloning into linearized two-hybrid expression vectors by co-transformation into yeast. The endogenous yeast recombination machinery performs the recombination reaction and results in a circular plasmid. (B) *Univector plasmid fusion system*. ORFs are amplified with specific primers that generate a product with common 5′ and 3′ Rare-Cutting restriction sites. The PCR product is cloned into a pUNI entry vector by DNA ligation. Cre–*loxP*-mediated site-specific recombination fuses the pUNI entry clone and yeast two-hybrid expression plasmids (bait/prey) at the *loxP* site. As a result, the gene of interest is placed under the control of the yeast two-hybrid expression vector promoter. (C) *Gateway cloning*. The ORFs are amplified with specific primers that generate a product with common 5′ and 3′ recombination sites (*attB*). The entry clones are made by recombining the ORFs of interest with the flanking *att*B sites into the *att*P sites of a suitable Gateway entry vector (such as pDONR201 or pDONR207) mediated by the Gateway BP Clonase (Invitrogen) Enzyme Mix. Subsequently, the fragment in the entry clone can be transferred to any yeast two-hybrid expression vector that contains the *att*R sites by mixing both plasmids and by using the Gateway LR Clonase Enzyme Mix. (See color plate section).

The pUNI plasmid is the entry vector of this system, the vector into which the gene of interest is inserted. The pHOST plasmid is the recipient vector containing the appropriate transcriptional regulatory sequences that will eventually control the expression of the gene of interest in the designated host cells. A recombinant expression construct is made through Cre–*loxP*-mediated site-specific recombination that fuses pUNI and pHOST into a dimeric fusion plasmid. A crucial feature of the pUNI plasmid is its conditional origin of replication derived from the plasmid R6Kgamma that allows its propagation only in bacterial hosts expressing the *pir* gene (encoding the essential replication protein p), and thereby the selection for and propagation of dimeric pUNI–pHOST vectors (Liu *et al.*, 1998, 2000) (Figure 2b).

3. *Gateway*® cloning

Gateway® (Invitrogen) cloning provides another fast and efficient way of cloning the ORFs (Walhout *et al.*, 2000). It is based on the site-specific recombination properties of bacteriophage lambda (Landy, 1989); recombination is mediated between so-called attachment sites (*att*) of DNA molecules: between *attB* and *attP* sites or between *attL* and *attR* sites. The first step to Gateway® cloning is inserting your gene of interest into a specific entry vector. This entry clone is a plasmid containing your gene of interest flanked by *attL* recombination sites. These *attL* sites can be recombined with *attR* sites on a destination vector resulting in a plasmid for functional protein expression in a specific host. One way of obtaining the initial entry clones is by recombining a PCR product of the ORF flanked by *attB* sites with the *attP* sites of a pDONR vector (Figure 2C).

4. The ORFeome

The starting point of an array-based Y2H screening is the construction of an ORFeome. An ORFeome represents all ORFs of a genome – in our case: the selected gene set – individually cloned into entry vectors of this recombination-based cloning system. More and more ORFeomes are available and can be directly used for generating the Y2H bait and prey constructs. These ORFeome range from small viral genomes, e.g., KSHV and VZV (Uetz et al. 2006) to bacterial genomes, e.g., *Treponema pallidum* (McKevitt *et al.*, 2003), and to genomes of multicellular eukaryotes, e.g., *Caenorhabditis elegans* (Lamesch *et al.*, 2004), Human (Rual *et al.*, 2004), and plant (Gong *et al.*, 2004). However, not all genes of interest are already available in entry vectors. However, they can be cloned into the entry vector by multiple strategies such as classical ligation or recombination. Both entry vector construction and the subsequent destination vector cloning can be done for multiple ORFs in parallel.

The whole procedure can be automated using 96-well plates so that whole ORFeomes can be processed in parallel.

5. The prey array

The Y2H array is set up from an ordered set of AD-containing strains (preys), rather than BD-containing strains (baits), because the former do not generally result in self-activation of transcription. The prey constructs are assembled by transfer of the ORFs from entry vectors into specific prey vectors by the used recombination system. Several prey vectors for the UPS and the Gateway® system are available. In our lab, we use the pLP-GADT7 vector (Clontech) for the UPS and a Gateway® adapted pGADT7 vector (Clontech), pGADT7 g, for the Gateway® system (Figure 3). An alternative is the direct cloning of prey constructs by homologous recombination

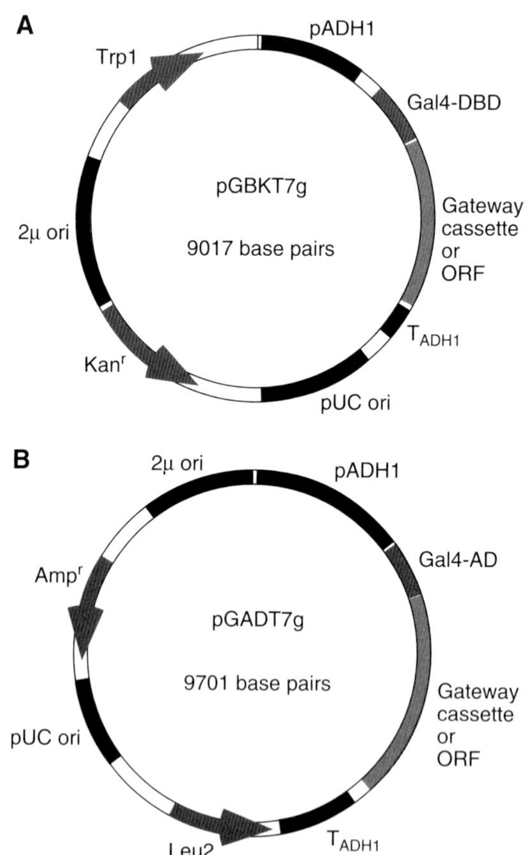

Figure 3. Maps of commonly used bait and prey vectors. The two-hybrid vectors used in the protocols described here, pGBKT7g and pGADT7g. (Please see text for further details.)

Table 2. Yeast strains and their genotypes

Y187 (*MATα, ura3- 52, his3- 200, ade2- 101, trp1- 901, leu2- 3, 112, gal4Δ, Met–, gal80Δ,URA3::GAL1*$_{UAS}$ *-GAL1*$_{TATA}$ *-lacZ*) (after Harper *et al.*, 1993).
AH109 (*MATa, trp1-901, leu2-3, 112, ura3-52, his3-200, gal4Δ, gal80 ΔLYS2::GAL1*$_{UAS}$*-GAL1*$_{TATA}$*-HIS3, GAL2*$_{UAS}$*-GAL2*$_{TATA}$*-ADE2, URA3::MEL1*$_{UAS}$*-MEL1*$_{TATA}$*-lacZ*) (after James *et al.*, 1996).

in yeast (see above). These prey constructs are transformed into haploid yeast cells (Protocol 1); we use the Y187 strain (mating type alpha) (Table 2). Finally, individual yeast colonies, each carrying one specific prey construct, are arrayed on agar plates in a 96- or 384-format (Figure 1).

6. Bait construction

Baits are also constructed by recombination-based transfer of the ORFs into specific bait vectors or, alternatively, directly by homologous recombination in yeast. Bait vectors used in our lab are the pLP-GBKT7 (Clontech) vector with an ampicillin resistance, pLP-GBKT7Amp, for UPS and an adapted pGBKT7 (Clontech), pGBKT7g, for Gateway® cloning. The bait constructs are also transformed into haploid yeast cells (Protocol 1); we use the AH109 strain (mating type **a**) (Table 2). After self-activation testing, the baits can be tested for interactions against the Y2H prey array.

Note: Bait and prey must be transformed into yeast strains of opposite mating type, to combine bait and prey plasmids by mating.

C. Self Activation Test

Prior to the two-hybrid analyses, the bait yeast strains should be examined for self-activation. Self-activation is defined as a detectable bait-dependent reporter gene activation in the absence of any prey interaction partner. Weak-to-intermediate strength self-activator baits can be used in two-hybrid array screens because the corresponding bait–prey interactions confer stronger signals than the self-activation background. In case of the *HIS3* reporter gene the self-activation background can be titrated by adding different concentrations of 3-AT, a competitive inhibitor of His3p.

Self-activation of all the baits is examined on plates containing different concentrations of 3-AT (Protocol 2). The lowest concentration of 3-AT that suppresses growth in this test is used for the interaction screen (see below) because it avoids background growth whereas true interactions are still detected.

Protocol 2. Bait self activation test.

The aim of this test is to measure the background reporter activity (here: *HIS3*) of bait proteins in absence of an interacting prey protein. This measurement is used for choosing the selection conditions used in Protocol 3.

Materials required

- Full medium and selective media agar in single-well microtiter plates (Omnitray plates, Nunc);
- YEPD plates;
- –LT plates (i.e. plates without leucine and without tryptophan)
- Selective plates without Trp, Leu, and His, but with different concentrations of 3-AT, e.g., 0 mM, 1 mM, 3 mM, 10 mM, 50 mM, and 100 mM (-LTH/3-AT plates)
- Prey strain carrying the empty prey plasmid, e.g., Y187 strain with pLP-GADT7 plasmid (Clontech)

Procedure

1. Bait strains are arrayed onto a single-well Omnitray agar plate; either the standard 96-spot format or the 384-spot format is used.

   Baits are first inoculated at the different positions of a 96-well plate as liquid culture, then cells are transferred (manually or with robot) to solid agar single-well plates (Omnitray plates). In this step, the 96-well format can also be converted into the 384-colony format, this will position each bait in quadruplicates on the 384-colony formatted plate. Full media agar (YEPD agar) can be used; however, for long term storage of the array selective agar (-Trp) is suggested to prevent loss of plasmids.

2. The arrayed bait strains are mated with a prey strain carrying the empty prey plasmid, e.g., Y187 strain with pLP-GADT7 plasmid (Clontech). Mating is conducted according to the standard screening protocol (Protocol 3).

   Note: Compared to Protocol 3 bait and prey strains are exchanged during mating.

3. After selecting for diploid yeast cells (on –LT agar) the cells are transferred to media selecting for the His3p reporter gene activity (Protocol 3). The –LTH transfer is done to several

selective plates with increasing concentrations of the competitive inhibitor of His3p, 3-Aminotriazole (3-AT).

Suggested are 3-AT concentrations of 0, 1, 3, 10, 25, 50 and 100 mM.

4. These –LTH/3-AT plates are incubated for 1 week at 30°C. The self-activation level of each bait is assessed: the lowest 3-AT concentration that completely prevents colony growth is noted. As this concentration of 3-AT suppresses reporter activation in absence of an interacting prey this 3-AT concentration is added to –LTH plates in the actual interaction screen (Protocol 3).

♦♦♦♦♦♦ III. SCREENING FOR PROTEIN INTERACTIONS USING A YEAST COLONY ARRAY

The Y2H prey array can be screened for protein interactions by a mating procedure that can be carried out manually or using robotics (Protocol 3). A yeast strain expressing a single candidate protein as a DBD fusion is mated to all the colonies in the prey array. After mating, the colonies are transferred to diploid-specific medium, and then to two-hybrid interaction selective medium. To manually screen with more than one bait, replicate copies of the array are used. For large number of baits, robotic screening is recommended.

In many cases, a hand-held 96 or 384-pin replicating tool can be used for routine transfer of colonies for screening. For large projects, however, a robotic workstation (e.g. Biomek 2000, Beckman Coulter) may be used to speed up the screening procedures and to maximize reproducibility. A 384- or 768-pin stainless steel replicating tool (e.g. high-density replication tool; Beckman Coulter) can be used to transfer the colonies from one plate to another. Between the transfer steps, the pinning tool must be sterilized by sequential immersion into a 10–20% (v/v) bleach solution (20 s), sterile water (1 s), 95% (v/v) ethanol (20 s), and sterile water (1 s). Immersion of the pins into these solutions must be sufficient to ensure complete sterilization. When automatic pinning devices are used, solutions need to be checked and refilled occasionally (especially ethanol which evaporates faster than the others).

Note that not all plasticware is compatible with robotic devices although most modern robots can be reprogrammed to accept different consumables. In the procedure described here, the prey array is gridded on 86 × 128 mm single-well microtiter plates (e.g. OmniTray, Nalge Nunc International) in a 384-colony format (Protocol 4).

Protocol 3. Two-hybrid screening protocol.

Materials required

- 20% (v/v) bleach (1 % Sodium hypochlorite);
- 95% (v/v) ethanol;
- Single-well microtiter plate (e.g., OmniTray; Nalge Nunc) containing solid (a) YEPD + Adenine medium, (b) -Leu –Trp, (c) –His –Leu –Trp and (d) –His –Leu –Trp+different concentrations of 3AT;
- 384-Pin Replicator for manual screening (Nalge Nunc International) or Robot;
- Bait liquid culture (DBD fusion-expressiong yeast strain); and
- Yeast prey array on solid YEPD plates.

Screening Procedure

Sterilization

Sterilize a 384-Pin Replicator by dipping the pins into 20% bleach for 20 s, sterile water for 1 s, 95% ethanol for 20 s, and sterile water again for 1 s. Repeat this sterilization after each transfer.

Preparing prey array for screening

Use the sterile replicator to transfer the yeast prey array from selective plates to single-well microtiter plates containing solid YEPD medium and grow the array overnight in 30 °C incubator.

If duplicate or quadruplicate colonies were not used to construct the array (see array construction), the entire experiment should be done using duplicate arrays. Ideally, the template prey array should be kept on selective plates.

Preparing bait liquid culture (DBD fusion-expression yeast strain)

Inoculate 20 ml of liquid YEPD medium in a 250 ml conical flask with a bait strain and grow overnight in 30°C shaker.

If the Bait strain is frozen, it is streaked or pinned on a rich media or selective plate (-Trp) solid medium and grown 1–2 days at 30°C. Baits from this plate are then used to inoculate the liquid YEPD medium.

Mating procedure

Pour the overnight liquid bait culture into sterile Omnitray plate. Dip the sterilized pins of the pin-replicator (thick pins should be used to pin baits) into the bait liquid culture and place directly onto a fresh single-well microtiter plate containing solid YEPD media. Repeat with the required number of plates and allow the yeast to dry onto the plates for 10–20 min

Pick up the prey array (i.e., AD) yeast colonies with sterilized pins (thin pins should be used to pin the preys) and transfer them directly onto the baits pinned onto the YEPD plate, so that each of the 384 bait spots per plate receives different prey yeast cells (i.e. a different AD fusion protein). Incubate 1–2 days at 30°C to allow mating.

Mating will take place in <15 h, but a longer period is recommended, because some bait strains show poor mating efficiency. Adding adenine into the bait culture before mating increases mating efficiency of some baits.

Seletion of Diploids
For the selection of diploids, transfer the colonies from YEPD-mating plates to single-well microtiter plates containing -Leu-Trp medium using the sterilized pinning tool (thin pins should be used in this step). Grow for 2–3 days at 30°C until the colonies are >1 mm in diameter.

This step is an essential control step because only diploid cells containing the LEU2 and TRP1 markers on the prey and bait vectors, respectively, will grow on this medium. This step also helps recovery of the colonies and increases the efficiency of the next selection step.

Interaction selection

Transfer the colonies from –Leu–Trp plates to a single-well microtiter plate containing solid –His –Leu –Trp agar, using the sterilized-pinning tool. If the baits are self-activating, they have to be transferred to –His –Leu –Trp + a specific concentration of 3-AT (Protocol 4). Incubate at 30°C for 6–10 days.
Score the interactions by looking for growing colonies that are significantly above background by size and that are present as duplicate or quadruplicate colonies (see Figure 1).

The plates should be examined every day. Most two-hybrid positive colonies appear within 3–5 days, but occasionally positive interactions can be observed later. Very small colonies are usually designated as background; however, there is no absolute measure to distinguish between the background and real positives. When there are many (i.e.,>30) large colonies per array of 6000 positions, we consider these baits as "random" activators. In this case, the screen should be repeated

Scoring can be done manually or using automated image analysis procedures. When using image analysis, care must be taken not to score contaminated colonies as positives.

Protocol 4. Yeast media and selective plates.

YEPD liquid medium. (10 g Yeast extract, 20 g Peptone, 20 g Glucose). Make up to 1 l with sterile water, and autoclave.

YEPD solid medium. (10 g Yeast extract, 20 g Peptone, 20 g Glucose, and 16 g Agar). Make up to 1 l with sterile water and autoclave. After autoclaving add 4 ml of 1% Adenine solution (1% in 0.1 M NaOH), and pour 40 ml into each sterile Omnitray plate under sterile hood and let them solidify.

Minimal media for selective plates

Medium concentrate. (8.5 g Yeast Nitrogen base, 25 g Ammonium sulfate, 100 g Glucose, 7 g Dropout mix [see below]). Make up to 1 l with sterile water, and sterile filter (e.g., Millipore sterile filter).

Making selective plates

For 1 l of medium autoclave 16 g agar in 800 ml water, cool the medium to 60–70° C then add 200 ml medium concentrate. Depending on the required selective plates you have to add the missing amino acids or 3-AT (3-amino-1,2,4-triazole):

- Trp plates: 8.3 ml Leucine and 8.3 ml Histidine from the stock solution (see below);
- Leu plates: 8.3 ml Tryptophan 8.3 ml Histidine from the stock solution;
- Leu–Trp plates: 8.3 ml Histidine from the stock solution;
- Leu–Trp–His plates: nothing needs to be added; and
- Leu–Trp–His+3mM 3-AT plates: 6 ml of 3-AT (3-amino-1,2,4-triazole, 0.5 M) to a final concentration of 3 mM.

Dropout mix (–His, –Leu, –Trp)

1 g Methionine, 1 g Arginine, 2.5 g Phenylalanine, 3 g Lysine, 3 g Tyrosine, 4 g Isoleucine, 5 g Glutamic acid, 5 g Aspartic acid, 7.5 g Valine, 10 g Threonine, 20 g Serine, 1 g Adenine, 1 g Uracil. Mix all components and store under dry, sterile conditions.

Amino acid stock solution

Histidine (His): Dissolve 4 g of Histidine in 1 l sterile water and sterile filter;

Leucine (Leu): dissolve 7.2 g of Leucine in 1 l sterile water and sterile filter; and

Tryptophan (Trp): dissolve 4.8 g of Tryptophan in 1 l sterile water and sterile filter.

A. Time Considerations

In the simplest case a set of baits is tested individually against a set of preys. For 10 baits and 10 preys this results in $10 \times 10 = 100$ individual tests (e.g. when all components of a protein complex are tested against each other). For a viral genome of 100 genes already 10,000 tests are required. Thus, the number of tests grows exponentially with the number of baits and preys.

As a consequence, automation will be required for larger projects. For example, in our laboratory a single Biomek 2000 robot is sufficient for testing per week about 50 baits against a bacterial genome of 1000 ORFs or all 100 proteins of a viral genome against each other. Note that each interaction also should be tested at least twice, just to make sure that the result is reproducible. This doubles the number of tests to be done. In fact, for a smaller project, we recommend to do each test four times, e.g., by spotting quadruplicates of each prey (Figure 1).

In larger projects, all tests can be done once but then each positive protein pair needs to be retested later, ideally in a coordinated effort to verify all positives. This time, quadruplicates can be used.

In theory, the colony density of the array can be increased as well, e.g., from 384 to 768 or even 1536 colonies per plate. However, this approach requires a higher precision of the robot, smaller colony sizes, and thus can reduce the number of detected interactions, e.g., due to a smaller number of transferred cells. While we have used 768-spot arrays on microtiter-sized plates, 1536 spots turned out to be too error-prone with our equipment.

B. Pooling Strategies

A different screening strategy, the pooling strategy, has the potential to accelerate screening a lot, but might also have the disadvantage of increasing the number of false negatives. In the first step, sets of proteins (pools, rather than single proteins) are tested for interactions against each other. In the second step, positive pool are taken and the proteins defining this pool are individually tested for their binary interactions (as in the classical array based Y2H strategy). Depending on the pool size the first level of screening is very fast and, as only a few interactions are expected for each protein, only a few pools need to be tested for binary interactions in the second step. Such a pooling strategy was established by Zhong *et al.* (Zhong *et al.*, 2003). Pools of prey proteins were tested against single bait proteins and it was shown that the pools can contain 96 or even more proteins. The authors calculate that the Y2H array screening of the whole yeast genome (~6000 proteins) requires only 1/24 of time and effort when using the proposed pooling strategy.

C. Retesting

A major consideration when using the Y2H system is the number of false positives. The major source for false positives are non-reproducible signals, which arise through little-understood mechanisms. In array screens and probably in random library screen, more than 90% of all signals can be non-reproducible background (Uetz, 2002). Thus, simple retesting by repeated mating can identify most false positives. We routinely use at least duplicate retesting, although quadruplicates should be used if possible (see above). Retesting is done by manually mating the interaction pair to be tested and by comparing the activation strength of this pair with the activation strength of a control, usually the bait mated with the strain that contains the empty prey vector (Protocol 5).

D. Alternative Reporter Genes

Y2H interactions can be reproduced using other reporter genes in addition to the one used in the actual screen. Examples include β-galactosidase or *ADE2* (for selection on adenine deficient medium). Due to the use of different promoters these reporter genes may have different activation requirements and Y2H interactions reproduced with different reporter genes are assumed to be more reliable. The β-galactosidase reporter has the advantage of giving a semi-quantitative output of the activation strength (Protocol 6). Other reporters might be advantageous and can be transformed into yeast as additional plasmids, or by using alternative strains, which contain the reporter as integrated construct. For example, the strain AH109 carries an α-galactosidase reporter gene that produces an enzyme, which is secreted into the medium. Therefore, these cells do not require cell lysis for detection.

E. Evaluation of Raw Results

Filtering of raw results significantly improves the data quality of the protein interaction set. For filtering at least three parameters should be considered. First, protein interactions that cannot be reproduced should be discarded. Second, for each prey the number of different interacting baits is calculated. Preys interacting with a large number of baits are assumed to be non-specific and thus can be discarded. However, the concrete cut-off number depends also on the nature of baits that are screened: if a large family of proteins is screened it is not surprising that many of them find the same bait. As a rough guideline the number of baits interacting with a certain prey should not be larger than 5% of the bait number. The third parameter is the background activation activity of the tested bait. The activation strength of interaction pairs must be significantly higher than with all other (background) pairs (see Figure 1). In principle, at least with

Protocol 5. Retest of protein interactions.

Testing for reproducibility of interactions greatly increases the reliability of the interaction data. This protocol is used for specifically retesting interaction pairs detected in an array screen.

Materials required

- 96-well microtiter plates (U- or V- shaped);
- YEPD medium and YEPD agar in Omnitrays (Nunc);
- Selective agar plates (–LT, –LTH with 3-AT);
- Prey yeast strain carrying empty prey plasmid, e.g. pLP-GADT7 in Y187 strain;
- Bait and prey strains to be retested.

Procedure

1. Re-array bait and prey strains of each interaction pair to be tested into 96-well microtiter plates. Use individual 96-well plates for the baits as well as for the preys. For each retested interaction fill one well of the bait plate and one corresponding well of the prey plate with 150 μl YEPD.
2. For each retested interaction inoculate the bait strain into a well of the 96-well bait plate and the prey strain at the corresponding position of the 96-well prey plate. For example, bait at position B2 of bait plate and prey at position B2 of prey plate.
3. Incubate the plates o/n at 30°C. In addition, inoculate the prey strain with the empty prey vector, e.g., strain Y187 with plasmid pLP-GADT7, into 50 ml YEPD; and incubate this strain o/n in a shaker at 30°C.
4. Mate the baits grown in the bait plate with their corresponding preys in the prey plate. In addition, mate each bait with the prey strain carrying an empty prey vector as a background activation control. The mating is done according to Protocol 3 using the bait and prey 96-well plates directly as the source plates.

   First the baits are transferred from their 96-well plate to two YEPD plates (interaction test and control plate) using a 96-well replication tool. Let plate dry for 10–20 min. Then transfer preys from their 96-well plate onto the first YEPD plate and the empty prey vector control strain onto the second YEPD plate

   The transfers to selective plates and incubations are done according to Protocol 3. As before, test different baits with different activation strengths on a single plate and pin the diploid cells onto –LTH plates with different concentrations of 3-AT. For choosing the 3-AT range the activation strengths (Protocol 2) serve as a guideline.

5. After incubating for ~1 week at 30°C on –LTH/3-AT plates the interactions are scored.

   Positive interactions show a clear colony growth at a certain level of 3-AT, whereas no growth should be seen in the control (bait mated with empty vector strain).

Protocol 6. β-Galactosidase filter lift assay This protocol was adapted from the Breeden lab (http://www.fhcrc.org/science/labs/breeden/Methods/b-GAL_filterassay.html).

Materials required

- Selective plate (–LT) with diploid yeast colonies (*Protocol 3*);
- *The diploid cells carry the bait and prey combinations to be tested for activation of the β-galactosidase reporter;*
- Omnitray plate (Nunc);
- Nitrocellulose membrane and Whatman paper;
- Z-buffer: 60 mM Na_2HPO_4 (anhyd.); 60 mM NaH_2PO_4; 10 mM KCl; 1 mM $MgSO_4$;
- X-GAL solution: 40 mg/ml in DMF (dimethylformamide).

Procedure

1. Use the same diploid plate as transfer to *HIS3* selective plate in Protocol 3. As a control the same bait strains mated with a prey strain containing an empty vector (following mating steps of Protocol 3) should be used.
2. Cut a nitrocellulose membrane to the dimensions of an Omnitray plate (Nunc). Place nitrocellulose membrane on top of diploid yeast colonies and leave for 10 s.
3. Use tweezers to lift filter and slowly submerge in liquid nitrogen for 1 min.
4. Place membrane into empty Omnitray plate (Nunc) to thaw.
5. Cut Whatman paper to same size as nitrocellulose membrane. Soak Whatman paper with 2 ml Z-buffer, to which 35 μl X-Gal-solution was added.
6. Overlay nitrocellulose filter with Whatman paper and remove air bubbles.
7. Incubate at 30°C for 10–60 min.
8. Evaluate: A blue staining indicates the activation of the β-galactosidase reporter and, therefore a positive interaction.

the *HIS3* reporter, no activation (no colony growth) should be observed in non-interacting pairs.

In addition to these parameters, more sophisticated statistical evaluations of the raw results have been suggested (Bader *et al.*, 2004).

♦♦♦♦♦♦ IV. CONCLUSION

The array-based Y2H system is well suited for *de novo* detection of protein interactions. The testing of protein subsets, e.g., functional subgroups of proteins, up to whole genomes is possible. The prerequisite is the existence of an ORFeome or subsets thereof, preferentially in a recombination-based cloning system. A common objection against Y2H is its supposed high rate of false positives. However, the array-based Y2H approach reduces this rate by simple retesting and by evaluation of the background activation strength. Hands-on time and amount of resources used grow exponentially with the number of tested proteins. This is a disadvantage for large genome sizes. However, the usage of a pooling strategy compensates for this. Taken together, the array-based Y2H system is the method of choice for genome-wide testing of binary protein interactions.

Acknowledgments

We thank Jürgen Haas for providing the vectors pGBKT7g and pGADT7g. This work was supported by a grant (Ue50/2) from the *Deutsche Forschungsgemeinschaft*. Björn Titz is a fellow in the *Studienstiftung des Deutschen Volkes*.

References

Bader, J. S., Chaudhuri, A., Rothberg, J. M. and Chant, J. (2004). Gaining confidence in high-throughput protein interaction networks. *Nat. Biotechnol.* **22**, 78–85.

Bartel, P. and Fields, S. (1997). *Advances in Molecular Biology.* The yeast two-hybrid system, pp. 344, Oxford University Press, Oxford, New York.

Cagney, G., Uetz, P. and Fields, S. (2000). High-throughput screening for protein–protein interactions using two-hybrid assay. *Methods Enzymol.* **328**, 3–14.

Cagney, G., Uetz, P. and Fields, S. (2001). Two-hybrid analysis of the *Saccharomyces cerevisiae* 26S proteasome. *Physiol. Genomics* **7**, 27–34.

Drees, B. L. (1999). Progress and variations in two-hybrid and three-hybrid technologies. *Curr. Opin. Chem. Biol.* **3**, 64–70.

Estojak, J., Brent, R. and Golemis, E. A. (1995). Correlation of two-hybrid affinity data with *in vitro* measurements. *Mol. Cell. Biol.* **15**, 5820–5829.

Fields, S. and Song, O. (1989). A novel genetic system to detect protein–protein interactions. *Nature* **340**, 245–246.

Frederickson, R. M. (1998). Macromolecular matchmaking: advances in two-hybrid and related technologies. *Curr. Opin. Biotechnol.* **9**, 90–96.

Giot, L., Bader, J. S., Brouwer, C., Chaudhuri, A., Kuang, B., Li, Y., Hao, Y. L., Ooi, C. E., Godwin, B., Vitols, E., Vijayadamodar, G., Pochart, P., Machineni, H., Welsh, M., Kong, Y., Zerhusen, B., Malcolm, R., Varrone, Z., Collis, A., Minto, M., Burgess, S., McDaniel, L., Stimpson, E., Spriggs, F., Williams, J., Neurath, K., Ioime, N., Agee, M., Voss, E., Furtak, K., Renzulli, R., Aanensen, N., Carrolla, S., Bickelhaupt, E., Lazovatsky, Y., DaSilva, A., Zhong, J., Stanyon, C. A., Finley, R. L., Jr., White, K. P., Braverman, M., Jarvie, T., Gold, S., Leach, M., Knight, J., Shimkets, R. A., McKenna, M. P., Chant, J. and Rothberg, J. M. (2003). A protein interaction map of *Drosophila melanogaster*. *Science* **302**, 1727–1736.

Gong, W., Shen, Y. P., Ma, L. G., Pan, Y., Du, Y. L., Wang, D. H., Yang, J. Y., Hu, L. D., Liu, X. F., Dong, C. X., Ma, L., Chen, Y. H., Yang, X. Y., Gao, Y., Zhu, D., Tan, X., Mu, J. Y., Zhang, D. B., Liu, Y. L., Dinesh-Kumar, S. P., Li, Y., Wang, X. P., Gu, H. Y., Qu, L. J., Bai, S. N., Lu, Y. T., Li, J. Y., Zhao, J. D., Zuo, J., Huang, H., Deng, X. W. and Zhu, Y. X. (2004). Genome-wide ORFeome cloning and analysis of *Arabidopsis* transcription factor genes. *Plant Physiol.* **135**, 773–782.

Harper, J. W., Adami, G. R., Wei, N., Keyomarsi, K. and Elledge, S. J. (1993). The p21 Cdk-interacting protein Cip1 is a potent inhibitor of G1 cyclin-dependent kinases. *Cell* **75**, 805–816.

Ito, T., Chiba, T., Ozawa, R., Yoshida, M., Hattori, M. and Sakaki, Y. (2001). A comprehensive two-hybrid analysis to explore the yeast protein interactome. *Proc. Natl. Acad. Sci. USA* **98**, 4569–4574.

James, P., Halladay, J. and Craig, E. A. (1996). Genomic libraries and a host strain designed for highly efficient two-hybrid selection in yeast. *Genetics* **144**, 1425–1436.

Johnsson, N. and Varshavsky, A. (1994). Ubiquitin-assisted dissection of protein transport across membranes. *EMBO J.* **13**, 2686–2698.

Lamesch, P., Milstein, S., Hao, T., Rosenberg, J., Li, N., Sequerra, R., Bosak, S., Doucette-Stamm, L., Vandenhaute, J., Hill, D. E. and Vidal, M. (2004). *Caenorhabditis elegans* ORFeome version 3.1: increasing the coverage of ORFeome resources with improved gene predictions. *Genome Res.* **14**, 2064–2069.

Landy, A. (1989). Dynamic, structural, and regulatory aspects of lambda site-specific recombination. *Annu. Rev. Biochem.* **58**, 913–949.

Liu, Q., Li, M. Z., Leibham, D., Cortez, D. and Elledge, S. J. (1998). The univector plasmid-fusion system, a method for rapid construction of recombinant DNA without restriction enzymes. *Curr. Biol.* **8**, 1300–1309.

Liu, Q., Li, M. Z., Liu, D. and Elledge, S. J. (2000). Rapid construction of recombinant DNA by the univector plasmid-fusion system. *Methods Enzymol.* **328**, 530–549.

McKevitt, M., Patel, K., Smajs, D., Marsh, M., McLoughlin, M., Norris, S. J., Weinstock, G. M. and Palzkill, T. (2003). Systematic cloning of *Treponema pallidum* open reading frames for protein expression and antigen discovery. *Genome Res.* **13**, 1665–1674.

Rain, J. C., Selig, L., De Reuse, H., Battaglia, V., Reverdy, C., Simon, S., Lenzen, G., Petel, F., Wojcik, J., Schachter, V., Chemama, Y., Labigne, A. and Legrain, P. (2001). The protein–protein interaction map of *Helicobacter pylori*. *Nature* **409**, 211–215.

Raquet, X., Eckert, J. H., Muller, S. and Johnsson, N. (2001). Detection of altered protein conformations in living cells. *J. Mol. Biol.* **305**, 927–938.
Rual, J. F., Hirozane-Kishikawa, T., Hao, T., Bertin, N., Li, S., Dricot, A., Li, N., Rosenberg, J., Lamesch, P., Vidalain, P. O., Clingingsmith, T. R., Hartley, J. L., Esposito, D., Cheo, D., Moore, T., Simmons, B., Sequerra, R., Bosak, S., Doucette-Stamm, L., Le Peuch, C., Vandenhaute, J., Cusick, M. E., Albala, J. S., Hill, D. E. and Vidal, M. (2004). Human ORFeome version 1.1: a platform for reverse proteomics. *Genome Res.* **14**, 2128–2135.
Schwartz, H., Alvares, C. P., White, M. B. and Fields, S. (1998). Mutation detection by a two-hybrid assay. *Hum. Mol. Genet.* **7**, 1029–1032.
SenGupta, D. J., Zhang, B., Kraemer, B., Pochart, P., Fields, S. and Wickens, M. (1996). A three-hybrid system to detect RNA-protein interactions *in vivo*. *Proc. Natl. Acad. Sci. USA* **93**, 8496–8501.
Uetz, P., Dong, Y. A., Zeretzke, C., Atzler, C., Baiker, A., Berger, B., Rajagopala, S. V., Roupelieva, M., Rose, D., Fossum, E. and Haas, J. (2006). Herpesviral protein networks and their interaction with the human proteome. *Science* **311**, 239–242.
Uetz, P. (2002). Two-hybrid arrays. *Curr. Opin. Chem. Biol.* **6**, 57–62.
Uetz, P., Giot, L., Cagney, G., Mansfield, T. A., Judson, R. S., Knight, J. R., Lockshon, D., Narayan, V., Srinivasan, M., Pochart, P., Qureshi-Emili, A., Li, Y., Godwin, B., Conover, D., Kalbfleisch, T., Vijayadamodar, G., Yang, M., Johnston, M., Fields, S. and Rothberg, J. M. (2000). A comprehensive analysis of protein–protein interactions in *Saccharomyces cerevisiae*. *Nature* **403**, 623–627.
Vidal, M. and Endoh, H. (1999). Prospects for drug screening using the reverse two-hybrid system. *Trends Biotechnol.* **17**, 374–381.
Vidal, M. and Legrain, P. (1999). Yeast forward and reverse 'n'-hybrid systems. *Nucleic Acids Res.* **27**, 919–929.
Walhout, A. J., Temple, G. F., Brasch, M. A., Hartley, J. L., Lorson, M. A., van den Heuvel, S. and Vidal, M. (2000). GATEWAY recombinational cloning: application to the cloning of large numbers of open reading frames or ORFeomes. *Methods Enzymol.* **328**, 575–592.
Zhong, J., Zhang, H., Stanyon, C. A., Tromp, G. and Finley, R. L., Jr. (2003). A strategy for constructing large protein interaction maps using the yeast two-hybrid system: regulated expression arrays and two-phase mating. *Genome Res.* **13**, 2691–2699.

8 Reporter Genes and Their Uses in Studying Yeast Gene Expression

Tobias von der Haar, Lyne J Jossé and Lee J Byrne
Department of Biosciences, University of Kent, Canterbury CT2 7NJ, UK

♦♦♦

Reporter Genes and Their Uses in Studying Yeast Gene Expression

CONTENTS

Introduction to reporter genes
Generating reporter gene constructs
Reporter genes available for work in yeast
Problems and pitfalls
Conclusions

Abbreviations

UTR Untranslated region
ORF Open reading frame
CAT Chloramphenicol acetyltransferase
PCR Polymerase chain reaction

♦♦♦♦♦♦ I. INTRODUCTION TO REPORTER GENES

A. Principles of Using Reporter Genes

To anyone studying gene expression and its regulation, it will sooner or later become necessary to characterise the action of specific regulatory elements in a quantitative way. A direct way to achieve this is by simply probing for natural expression products that are controlled by a genetic element of interest, like proteins or mRNAs. For example, the expression level of a protein controlled by a certain genetic element may be detected by Western blotting. If the amount of protein that is detected changes between any two conditions, this may indicate changes in the activity of the genetic element.

METHODS IN MICROBIOLOGY, VOLUME 36
0580-9517 DOI:10.1016/S0580-9517(06)36008-4

Rather than using this direct approach, however, it is often more desirable to use the activity of the genetic element under study to control expression of a non-natural gene product, which is then termed a reporter gene. In general this is done for two reasons: first, reporter gene expression can usually be quantified via simple and easy-to-use assays. Second, placing of a particular genetic element in the artificial context of a reporter gene construct allows the isolation of its effect on gene expression from the effects of other, coupled genetic elements. Such an isolation of effects is often required because activity levels of natural genes are regulated at multiple levels, and by multiple regulatory elements (promoters, sequences that regulate mRNA stability, etc.).

B. Examples of the Use of Reporter Genes in Yeast

In the following, we will use a few examples of studies on transcriptional, translational and post-translational regulation of gene expression to illustrate some general principles of the use of reporter genes.

Many of the reporters developed for use in yeast were originally designed to investigate promoter activity. A good example for this is the original description of the chloramphenicol acetyltransferase (CAT) assay in yeast (Mannhaupt *et al.*, 1988). The authors of this study demonstrated the validity of CAT as a yeast reporter gene by constructing vectors that contained the coding region of the bacterial *cat* gene, but no promoter. The promoter of *PHO5*, a gene that is induced when yeast is grown in phosphate-free medium, was then inserted into the plasmids 5′ of the CAT ORF. In phosphate-containing medium, neither the promoterless nor the *PHO5* promoter-containing plasmids conferred detectable CAT activity. In contrast, upon transfer of the yeast to phosphate-free medium, the *PHO5*-containing but not the promoterless plasmids generated measurable amounts of CAT activity. This principle has since been used in numerous studies to identify and delineate promoter sequences, and to study their regulation under specific conditions.

A second example, a study of the 5′ leader of the *GCN4* mRNA, demonstrates that the same approach can also be used to study post-transcriptional regulation. *GCN4* encodes a transcription factor that is produced when amino acid levels become too low to support optimal levels of protein synthesis in growing yeast cells, and it induces a number of genes required for the synthesis of new amino acids (reviewed in Hinnebusch, 2005). It is now known that this factor is highly regulated at the translational level: when amino acid starvation occurs, synthesis of this protein increases although levels of the corresponding mRNA remain constant. The main regulatory element under these conditions is the 5′ untranslated region (5′-UTR) of the gene, which allows efficient translation of the *GCN4* ORF only under conditions of reduced eIF2 activity, when translation of most other mRNAs becomes repressed. Much of the work that led to the elucidation of the regulatory mechanism relied on fusions of the promoter and 5′-UTR of *GCN4* to *E. coli* β-galactosidase (*lacZ*). Using this reporter construct, the conclusion that regulation of *GCN4* occurs post-transcriptionally was drawn because the reporter gene activity, but not the levels of the corresponding transcript, changed under conditions of amino acid starvation (Hinnebusch, 1984). More detailed insights into the regulatory mechanism were then generated by combining the reporter assay with mutagenesis of the 5′-UTR as well as knockouts or mutants of transacting factors that act on this region.

A last example demonstrates the use of a reporter gene for studying post-translational regulation. In order to discover new sequence elements that act as target sites for the proteasomal protein degradation system, Gilon *et al.* (1998) generated fusions of β-galactosidase with random C-terminal extensions. Since all fusions were expressed from identical and active promoter elements, transformation of the respective fusions into yeast cells resulted in colonies showing a blue colour when grown on X-gal plates, due to high expression levels of the reporter enzyme. However, colonies showed a white colour if the fused element limited the expression of or destabilised the expressed protein. Reporter fusions isolated from the initially recovered white colonies were then transferred to a yeast strain in which proteasome function was partially destroyed. True proteasomal target sequences (as opposed to sequences that reduced expression levels through other mechanisms) could be identified in this second strain by showing that loss of proteasome function restored expression of the β-galactosidase, and therefore led to a blue colony colour.

Besides the examples discussed above, many other types of regulatory elements have been studied with reporter genes. Indeed, published examples can now be found for most of the levels at which gene expression is regulated (Table 1).

In general terms, the establishment of a reporter assay requires the selection of a suitable reporter gene, placing the reporter under the control of a genetic element of interest (usually on a plasmid), introduction of the relevant construct(s) into yeast cells, growth of the cells and determination of reporter expression levels via a suitable assay. More in-depth information on these steps is given in the following sections.

♦♦♦♦♦♦ II. GENERATING REPORTER GENE CONSTRUCTS

At the DNA level, a basic reporter gene construct for use in yeast consists of the minimal set of genetic elements that can drive protein

Table 1. Examples of gene expression levels analysed via reporter genes

Gene expression level analysed	Genetic element combined with reporter	Published examples
Transcription	Promoter	Walmsley *et al.* (1997)
mRNA end processing	3′-UTR	Duvel *et al.* (1999)
mRNA splicing	Intron	Luukkonen and Seraphin (1999)
Translation initiation	Altered translation start codon	Donahue *et al.* (1988)
	5′-UTR	Hinnebusch (1984)
	3′-UTR	Rajkowitsch *et al.* (2004)
Translation termination	Nonsense mutation within reading frame	Firoozan *et al.* (1991)
mRNA stability	3′-UTR	Muhlrad and Parker (1999)
Protein	Entire protein ORF	Huh *et al.* (2003)
Localisation	Putative localisation sequences	
Protein stability	Protein fusion to protease target sequences	Gilon *et al.* (1998)

expression in a eukaryotic system: a promoter, a 5′-UTR, the coding sequence of the reporter gene product itself, a 3′-UTR and a transcriptional terminator (Figure 1). On top of this basic set of features, additional elements such as splice sites and introns may also be present in the construct.

The procedure by which a genetic element of interest is introduced into a basic reporter gene construct depends to some extent on what kind of element is studied. For many kinds of frequently studied genetic elements like promoters, 5′-UTRs, etc., modular sets of plasmid-borne reporter constructs are available that allow for the easy introduction of foreign DNA sequences by standard restriction-enzyme-based procedures. This approach gives the researcher some control over the placement of a foreign sequence in relation to the reporter gene open reading frame, although this is limited by the nature and position of restriction sites contained in the original construct. Where exact placement of the genetic element in relation to the reporter gene is important (as for example in the case of introns), or where a new combination of promoters/reporters, etc. needs to be designed from scratch, an alternative procedure that we have found very useful is the so-called splicing by overlap extension (SOE-PCR; Horton *et al.*, 1989). In principle, this procedure allows the fusion of multiple, unrelated DNA sequences with single nucleotide accuracy. A detailed description of this procedure is given in Figure 2 and Protocol 1.

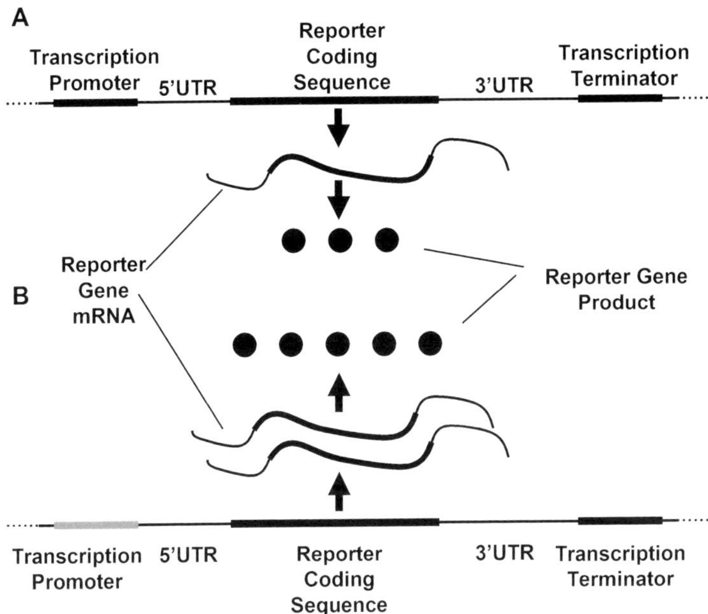

Figure 1. Principle approach of quantifying gene expression via reporter constructs. A general reporter construct comprises a promoter region, 5′-UTR, coding sequence, 3′-UTR and transcription terminator **(A)**. When introduced into a yeast cell, a construct of this type will give a certain basic expression level of the reporter gene product. If one of the elements of the basic construct is exchanged for another element (panel **B,** in this case the promoter), expression levels of the reporter gene product will change. After quantification of the gene product, the activity of promoter B can be set in relation to the activity of promoter A.

◆◆◆◆◆◆ III. REPORTER GENES AVAILABLE FOR WORK IN YEAST

A. Growth Rate Reporters

In principle, any essential gene can be used as a reporter in the corresponding knockout yeast strains because in these strains the ability to grow depends on expression of the respective gene. The growth rate thus provides an easy readout for expression of the gene. In practice, many of the auxotrophic markers that are commonly used with yeast plasmids (in *S. cerevisiae*, e.g. *URA3*, *HIS3*, *LEU2*, etc.) have been used for this purpose. These genes are particularly useful when the reporter activity must be genetically selectable, for example when large numbers of candidate sequences will be screened for a certain genetic activity. Because growth rates can be assessed very easily and cheaply, growth rate reporters are also sometimes used for the semi-quantitative analysis of gene expression, for example to rank different genetic elements in order of activity. However, the exact relationship between expression

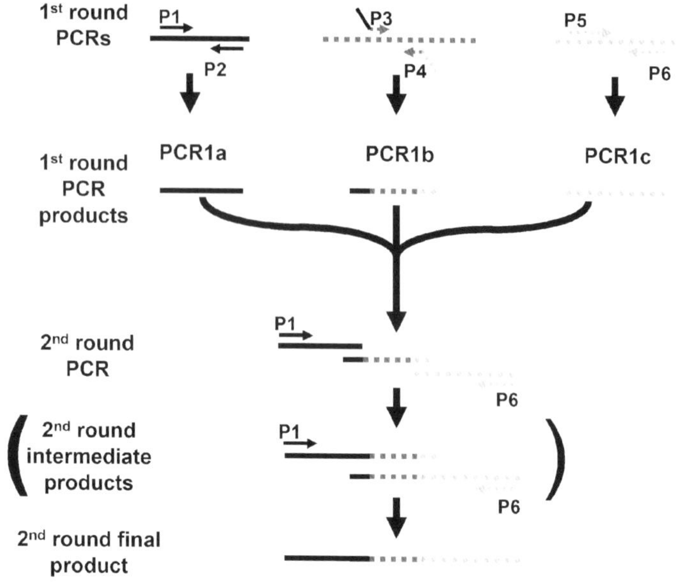

Figure 2. Generating reporter gene constructs via SOE-PCR. This figure depicts the steps described in Protocol 1. In a first round of PCR reactions, the individual DNA sequences that will be fused together are amplified in individual reactions. The DNA oligomers amplifying the central portion of the fusion (P3 and P4) are designed to introduce ends into the corresponding PCR product 1b that overlap with the 5′ and 3′ ends, respectively, of PCR products 1a and 1c. Following purification of the individual PCR products from these reactions, the latter are then mixed and a single PCR reaction is conducted with DNA oligomers P1 and P6, which anneal to the very 5′ and 3′ ends of the final fusion product. Owing to the overlapping ends introduced by oligos P3 and P4, the first few PCR cycles will generate fusions of the respective DNA fragments. Once a complete fusion product is formed, this will be exponentially amplified by oligos P1 and P6, so that the vast majority of final PCR product consists of the fusion construct.

levels and growth rate is often unknown, so that any quantitative conclusions from growth rate measurements of this kind should be treated with caution.

B. Enzyme-Based Reporters

All enzyme-based reporter genes are analysed and quantified via their ability to catalyse a reaction involving a particular chemical substrate. The latter may be a metabolite produced by the analysed organism itself, but is usually a chemical compound that needs to be somehow added to the cells. The enzyme reporter transforms this substrate into a product that is detectable by photometric, luminometric or other assays. The higher the expression level of the enzyme in the cell, the higher the reaction rate and the more detectable product is generated within a certain amount of time.

Some of the easiest-to-use reporter genes are the *S. cerevisiae ADE1* and *ADE2* genes. In terms of enzyme-based reporter assays, they are

Protocol 1. Generating Reporter Gene constructs by SOE-PCR.

This protocol assumes that three sequences A, B and C will be fused as depicted in Figure 2. Fusions of fewer or more sequences can be achieved using the same general principle.

1. Design your DNA oligomers for the first round PCRs. The oligomers should be designed so that sequences A and C are amplified exactly to the fusion boundaries. Oligomers for the amplification of sequence B should amplify this sequence exactly to the fusion boundaries, and they must also incorporate nucleotides corresponding to the end of sequence A at the 5′ end and nucleotides corresponding to the beginning for sequence C at the 3′ end of sequence B. The length of additional sequence should ideally be chosen so that the calculated annealing temperature of this part alone is 54°C or higher.
2. Using a standard PCR protocol, perform three separate PCR reactions that amplify the individual component sequences. Proofreading polymerases that do not append additional adenine residues at the end of the amplified sequence are preferable for this amplification step.
3. Separate the PCR products on a suitable agarose gel. Excise the PCR products and isolate them from the gel matrix.
4. For the second round PCR, mix together 20–50 ng of each of the first round PCR products. Amplify this mixture using only the two oligomers that anneal to the 5′ and 3′ ends of the final, fused sequence (P1 and P6 in Figure 1).
5. Load the PCR reaction onto a fresh agarose gel. Isolate the correct product from the gel matrix and clone into a suitable vector system.

a special case because they do not rely on the addition of exogenous substrate. The *ADE* genes of *S. cerevisiae* catalyse the multi-step conversion of 5-phosphoribosylpyrophosphate to adenosine-5′-monophosphate. One of the pathway intermediates that are produced during this reaction is a brightly red-coloured compound. Wild-type yeast cells are white because the red compound is normally rapidly converted into subsequent pathway intermediates by the action of the *ADE2* and *ADE1* genes, and does not accumulate. In the case of *ade1* or *ade2* mutants, any of the red compound that is produced can no longer be converted into the subsequent intermediate. It therefore accumulates, leading to brightly red-coloured colonies.

In order to produce the red intermediate in the first place, the adenine synthesis pathway of the respective cells needs to be activated. Conditions where sufficient adenine is in the medium for *ade1* or *ade2* mutants to survive, but where adenine levels are low enough to de-repress the adenine synthesis pathway, can usually be achieved by reducing the yeast extract concentration in normal YPD medium to 0.25%, or by supplementing minimal medium with one-quarter of the normal amount of adenine (i.e. to 5 rather than 20 mg/l). Colonies grown on plates made with such media will develop red colour after 2–3 days of growth, and the colour can be intensified by subsequent incubation of the plates at 4°C.

If the expression of *ADE1* or *ADE2* in the respective knockout strains is restored to low levels, this will lead to intermediate colony colours, and the change from bright red to white colonies with increasing levels of expression can be used as a semi-quantitative assay (see e.g. Parham *et al.*, 2001). To date, this assay is mainly confined to use in *Saccharomyces cerevisiae*, although published studies have shown that in both *Candida albicans* and *Schizosaccharomyces pombe* certain adenine minus mutants produce similar colour changes (Kurtz *et al.*, 1986; Leupold, 1958).

Two further reporter genes that have proven useful in several different yeast species and which, like *ADE1* or *ADE2*, can be detected without disrupting cell integrity are glucoamylases and exo-glucanases. The latter comprise a class of enzymes endogenous to many yeasts, where they can be used as a reporter if the endogenous activities are destroyed (e.g. in *S. cerevisiae exg1 exg2* strains; Cid *et al.*, 1994). The particular usefulness of the *EXG1* ORF as a reporter gene derives from the fact that the Exg1 protein is secreted into the periplasmic space and cell walls following its production. Upon addition of the fluorescent substrate FDGP (fluorescein di-b-D-galactopyranoside) to the culture medium, the cell wall-trapped enzyme generates a fluorescent signal (Cid *et al.*, 1994). Because this signal is cell-associated, Exg1 is an ideal reporter for flow-cytometric applications. Apart from *S. cerevisiae*, the usefulness of exo-glucanases as reporter has also been demonstrated in *S. pombe* (Molero *et al.*, 1999).

Unlike the exo-glucanases, which remain associated with the secreting cell, glucoamylases are enzymes that are produced in certain fungi and that are secreted into the growth medium. A glucoamylase-based reporter assay for use in *S. cerevisiae* was developed using the corresponding gene from the fungus *Aspergillus awamori* (Scorpione *et al.*, 1993). Glucoamylases hydrolyse starch to glucose, and this activity allows for easy detection of their expression on soluble starch plates. Upon incubation with iodine vapours, the starch and iodine react to give a deep purple colour. Glucoamylase-secreting yeast colonies can be easily identified on this medium because, following iodine staining, they show a clear halo in the area surrounding the colony where the starch has been degraded. A more quantitative assay for glucoamylase secreted into liquid culture supernatant can be performed based on commercially available glucose oxidase assays (Scorpione *et al.*, 1993). Apart from *S. cerevisiae*, this reporter has also been successfully used in *Pichia* (Mannazzu *et al.*, 1995).

In contrast to the examples mentioned above, other reporter systems require lysis or permeabilisation of the cells in order to quantitatively determine reporter gene expression. This group comprises three of the most popular reporters, namely the β-galactosidase, luciferase and CAT-based systems, which will be described in more detail below.

The CAT system utilises the enzyme chloramphenicol-3-*O*-acetyltransferase as reporter. The corresponding gene was isolated from a bacterial transposon (Tn*9*), which in its natural host confers resistance to the antibiotic chloramphenicol. The original assay for determination of this reporter relies on the acetylation of chloramphenicol in the presence of acetyl-coenzyme A (Gorman *et al.*, 1982). If radioactive chloramphenicol and cold acetyl-coenzyme A are added to yeast extracts containing CAT enzyme, the acetyl group is transferred to the antibiotic. Acetylated chloramphenicol can be separated from the non-acetylated form by thin-layer chromatography (TLC), and the amount of acetylated product formed can be quantified after autoradiography of the TLC plates. This assay is in principle very sensitive, but recent work has shown that it suffers from poorly understood non-linear effects when used with *S. cerevisiae* extracts (Alipour *et al.*, 1999), and the quantitative interpretation of results obtained with this procedure may thus be unreliable. As alternatives that avoid the use of radioactivity, protocols based on fluorescent substrates have been developed (see Young *et al.*, 1991, and several commercially available products) as well as an ELISA-based assay that directly detects the CAT protein rather than its enzymatic activity (Alipour *et al.*, 1999).

β-Galactosidase-based assays rely on a class of enzymes that catalyse the hydrolysis of lactose to galactose and glucose. By far the most commonly used gene is the *lacZ* gene from *E. coli*, although β-galactosidases from other organisms have also been employed (e.g. the *K. lactis LAC4* gene in *C. albicans*; Leuker *et al.*, 1992). The original assays for β-galactosidase activity rely on the conversion of chromogenic substrates to a visible or photometrically quantifiable dye. One of the simplest protocols for chromogenic detection of *lacZ*-expressing *S. cerevisiae* colonies involves growing the cells on medium containing X-Gal (5-bromo-4-chloro-3-indolyl-β-D-galactopyranoside). Colonies that express the reporter process the X-Gal to a blue dye, and can thus easily be distinguished from colonies that do not express the gene. However, while this assay is very simple it is also relatively unreliable, since the uptake of the substrate appears to vary greatly between colonies depending on parameters like the age and size of the colony (Serebriiskii and Golemis, 2000). A more quantitative assay relies on the preparation of cell extracts, and subsequent addition of ONPG (*o*-nitrophenyl-β-D-galactopyranoside). This compound is converted by β-galactosidase activities to a bright yellow dye, which can be easily quantified using a spectrophotometer. As the combination of a β-galactosidase reporter and the ONPG assay is one of the most popular ways of quantitatively assessing reporter gene expression, we have provided a detailed description of this procedure in Protocol 2. A means of detecting β-galactosidase

Protocol 2. Measuring β-galactosidase activity in *S. cerevisiae* cells with *o*-nitrophenyl-β-D-galactopyranoside (ONPG).

Reagents

- 1 M Na_2CO_3.
- 4 mg/ml ONPG. Note that ONPG can take 1–2 h to dissolve at room temperature. Pre-warm this solution to 37°C before use.
- Z-buffer – 60 mM Na_2HPO_4, 40 mM NaH_2PO_4, 10 mM KCl, 1 mM $MgSO_4$, 50 mM β-mercaptoethanol. Store at 4°C until required.
- 0.1% SDS.

Method

1. Grow a culture of yeast cells that have been transformed with your reporter construct in liquid medium under the desired conditions.
2. Remove 1 ml of the culture and determine the OD_{600}, using medium as the blank. If the OD is above 0.5, the culture needs to be diluted 1:10 in order to arrive at accurate values.
3. Transfer another 0.8 ml of the culture to a 1.5 ml tube and pellet the cells (1 min at room temperature and at maximum speed in a microcentrifuge). Discard the supernatant without disturbing the cell pellet.
4. Add 0.8 ml of Z-buffer to the cell pellet, cap the tube and resuspend the cells by vortexing vigorously.
5. Add 50 µl chloroform and 20 µl of 0.1% SDS. Cap the tube tightly and vortex the mix for 15 s.
6. Place the capped tube in a 37°C water bath in the fume hood for 5 min to pre-incubate the sample.
7. Start the reaction by adding 0.16 ml of the pre-warmed 4 mg/ml ONPG solution and continue incubation at 28°C for 15 min.
8. Stop the reaction by adding 0.4 ml of the 1 M Na_2CO_3 stock solution.
9. Remove the cell debris by centrifugation for 1 min in a microcentrifuge and then carefully transfer 1 ml of the supernatant to a fresh 1.5 ml tube.
10. Centrifuge for a further 5 min in a microfuge and then measure the optical density of the supernatant at 420 nm using Z-buffer as the blank.
11. Express the activity as β-galactosidase units:

$$\frac{OD_{420}(\text{Reaction})}{OD_{600}(\text{Culture})} \times \text{Volume assayed (ml)} \times \text{ time (min)}$$

12. *Note*: OD_{420} is the optical density of the product, *o*-nitrophenol. OD_{600} is the optical density of the culture at the time of assay. Volume is the amount of the culture used in the assay in ml. Time is in minutes.

activity with increased sensitivity is via the generation of fluorescent products, for example from the substrate fluorescein di-β-D-galactopyranoside. This principle is the basis for the commercial fluorescent β-galactosidase assays provided by a number of companies.

The last of the reporter genes described here are the luciferases, a group of enzymes of widespread origin that catalyse the oxidation of various substrates in a light-emitting reaction. Several different luciferases have been used as reporter genes in yeast, including those from *E. coli* (Boylan *et al.*, 1989), *Vibrio harveyi* (Kirchner *et al.*, 1989), the firefly *Photinus pyralis* (Vieites *et al.*, 1994) and from the sea pansy *Renilla reniformis* (Srikantha *et al.*, 1996). The bacterial luciferases are generally composed of two subunits, and although fusions of the two genes into one open reading frame have been shown to be active (Boylan *et al.*, 1989; Kirchner *et al.*, 1989), their use as reporter genes is not very common. In contrast, the *Photinus* and *Renilla* luciferases have become widely adapted as reporter genes in mammalian systems, and their use has recently also become more widespread in yeast. The great advantages of these two reporters are their extreme sensitivity, broad linearity and generally very low background, but they do require the use of specialised equipment to measure the light emission during the detection reaction. Dedicated luminometers provide the highest sensitivity and linearity for determining luciferase levels, but standard scintillation counters can also be used for this purpose (for details on the adaptation of scintillation counters for measuring luciferase activity see Fulton and Van Ness, 1993).

The detection reagent for firefly luciferase is relatively complex compared to the other enzyme reporters, as the reaction requires the presence of the substrate luciferin, excess ATP levels and Mg^{2+} ions. Upon addition of these reagents, luciferin emits a flash of light that decays after 1–2 s, and the original assay procedure therefore required that the sample be measured immediately after mixing. A number of modifications to the original assay have since been described in the literature and are sold as proprietary solutions from several companies that produce a longer lasting, steady glow, yielding more reproducible results and an easier assay procedure (reviewed in Bronstein *et al.*, 1994). Although *Renilla* luciferase requires less cofactors than its firefly counterpart, the majority of published studies now appear to use commercially available assay kits for both luciferases rather than home-made reagents.

An advantage of using luciferase as a reporter is the possibility of using both *Renilla* and firefly luciferases in combination in the same cell. This is the basis of the so-called dual-luciferase assay: typically, one of the two luciferases is employed as an endogenous control used to normalise the activity levels of the second luciferase, which is under the control of the genetic element of interest. The dual-luciferase system is very popular in mammalian cells, but has also been adapted for use in yeast (Grentzmann *et al.*, 1998).

C. Fluorescent Reporters

The discovery that the *Aequorea victoria* green fluorescent protein (GFP) could function in heterologous cell types (Chalfie *et al.*, 1994; Inouye and Tsuji, 1994) has added a powerful tool to the repertoire of reporter systems. GFP is detected by irradiating a sample with light at a certain excitation wavelength (395 nm). GFP present in the sample will be excited by the incident light, and the higher-energy excited state of the protein will return to the basic state by emitting light at a wavelength that is different from the exciting light (508 nm). GFP can thus be quantified in a fluorimeter via the intensity of the emitted light.

As with the other reporter systems described in this chapter, expression of GFP can be used to quantitatively monitor gene expression. Typically, levels of GFP expression are determined in living yeast cells by directly measuring the fluorescence of a yeast culture, as the fluorescence signal will increase with the amount of GFP expressed by the cells in the culture. However, there are several problems associated with this approach that stem from the so-called autofluorescence, i.e. a fluorescent signal produced by components of the medium and by cellular metabolites (reviewed by Billinton and Knight, 2001). Several remedies have been suggested to reduce the interference of the autofluorescence signal with the GFP-related signal, including the use of low-fluorescence culture media (Sheff and Thorn, 2004; Walmsley *et al.*, 1983) and fluorescence polarisation (Knight *et al.*, 2002). Modern monochromator-based fluorimeters help to additionally reduce the problem by allowing excitation/emission light bandwidths to be set within relatively narrow limits, thus better separating GFP fluorescence from autofluorescence. With combinations of these techniques, sensitive fluorimeters have been reported to detect the presence of less than 0.5% GFP-containing cells in a background of 99.5% GFP-free cells (Tecan Group Technical Notes). Because little sample preparation is required to measure the fluorescence, GFP is an excellent reporter for high-throughput applications: for example, it is possible to grow individual yeast clones in 96- or 384-well plates, and directly measure the fluorescence/absorbance ratios in the individual wells as a measure of GFP expression in the respective clones.

Although GFP is useful as a general reporter of gene expression levels, the great strength of this protein lies in additional applications that become possible in conjunction with microscopic techniques. Thus, fusions of GFP to other proteins are frequently used to study the localisation of a protein in living cells, by simply observing the fluorescent signal generated from a GFP-expressing cell under a light microscope. This particular application has recently been greatly extended by the development of variants of the original GFP, and of entirely new fluorescent protein moieties.

Since the original GFP protein was discovered, there has been a development of a range of variants to improve or optimise the protein for use in different biological systems, and at the time of writing, there are 31 GFP variants listed on the protein databank website (URL: http://www.rcsb.org/pdb/molecules/pdb42_report.html). Today, it is possible to use GFPs with increased fluorescence intensity and altered wavelength of emission, resulting in a different observed colour of fluorescence. Some examples of these variants along with the excitation and emission wavelengths are shown in Table 2.

However, the recent development of much smaller biarsenical ligands that bind to very short motifs circumvents the problem of putting a large GFP tag onto a target protein and provides an exciting alternative to bulky GFP moieties (Adams *et al.*, 2002). In this system, membrane-permeable fluorescein is modified so that As(III) groups are substituted at the 4′ and 5′ positions to make "FlAsH" (fluorescein arsenical helix binder). The interaction of a single arsenic with a pair of thiol groups is well known and, as such, FlAsH binds specifically with high affinity to the amino acid motif Cys–Cys–Xaa–Xaa–Cys–Cys (where Xaa is any amino acid except cysteine), through the rigid spacing of the two arsenics. Binding to endogenous cysteine pairs or lipoamide cofactors is minimised by the addition of micromolar levels of 1,2-dithiol antidotes, which outcompete FlAsH, whereas higher concentrations in the millimolar range will outcompete the CCXXCC motif and thereby strip FlAsH off the target protein if desired. Proteins visualised by this tagging system, have an additional advantage over GFP as their fluorescence is almost instantaneous, allowing real-time observation of specific biochemical events (Cavagnero and Jungbauer, 2005). GFP undergoes post-translational maturation, involving an

Table 2. Frequently used fluorescent reporters

GFP variant	Excitation wavelength (nm)	Emission wavelength (nm)	Reference
wtGFP	396	504	Chalfie *et al.* (1994)
GFP-S65T	489	509	Heim *et al.* (1995)
EGFP	488	509	Cormack *et al.* (1996)
yEGFP	490	510	Cormack *et al.* (1997)
EBFP	382	445	Heim and Tsien (1996)
CFP	434	474	Kohlwein (2000)
ECFP	430–437	475–478	Heim and Tsien (1996)
YFP	514	527	Miyawaki *et al.* (1997)
Topaz	514	527	Cubitt *et al.* (1999)
DsRed	558	583	Shaner *et al.* (2004)

intramolecular oxidative cyclisation that may result in a significant delay in observed fluorescence, although enhanced, faster maturing versions of yellow fluorescent protein (YFP) and GFP are under development (Bevis and Glick, 2002).

Analogues of FlAsH have also been developed that differ in their excitation properties, emission wavelengths and membrane permeability, thereby increasing the applicability of these compounds. Indeed, ReAsH is both fluorescent and suitable for detection by electron microscopy, providing a powerful tool for comparing optical and electron microscopy images (Adams *et al.*, 2002). These biarsenical ligand variants are available from Invitrogen (http://www.invitrogen.com), where they are marketed under the *Lumio*™ trade name. In addition, the reversible binding of FlAsH for tetracysteine sites allows a convenient method for affinity purification of recombinant proteins (Thorn *et al.*, 2000), and has also proved successful in direct in-gel detection of tagged protein (Adams *et al.*, 2002).

The critical consideration when fusing a protein to GFP, or inserting a CCXXCC biarsenical tag, is that the fusion protein retains its activity at comparable levels to wild-type protein and also its proper location within the cell. It is therefore recommended that the position of fusion be optimised for each individual protein. Typically, fusions are carried out at either the N- or C-terminus of a protein, but if a protein has well-defined domain architecture, it may be possible to place the GFP/biarsenical tag between domains and for it still to remain active. In any case, to ensure correct localisation and proper function, allowances must be made for signal sequences for ER translocation, organelle targeting and retention sequences, lipid or prenyl group attachment or other membrane-anchoring sequences. It is also possible to separate the GFP domain from a target protein functional domain by using a short peptide linker. However, this linker may affect the stability of the protein, increasing its susceptibility to proteolysis (Prescott *et al.*, 1999).

Observing GFP-fusion/CCXXCC-tagged proteins under the microscope can be very informative. Static images of an asynchronous population can provide some information of the behaviour and localisation of the tagged protein. Using time-lapse microscopy, however, one can observe the dynamic changes that may occur to a fusion protein over a period of time. A procedure for observing yeast cells by time-lapse microscopy is given in Protocol 3 and shown in Figure 3.

It is possible to observe images in both single and multiple planes and, using the appropriate software, to build up three-dimensional images of the distribution of fusion proteins within a cell. All of these applications depend upon several factors such as the location of the fusion protein, abundance of signal and the time span of the event being followed. To test the correct localisation of the fusion protein within the cell, it is advisable to use an additional independent second marker that co-localises with the tagged protein.

Protocol 3. A procedure for observing fluorescent-tagged proteins in live yeast cells by time-lapse microscopy.

1. Grow a culture of yeast cells that have been transformed with your fluorescent protein fusion construct in liquid medium with the appropriate selection and induce fusion protein expression.
2. Place a sterile glass slide in the bottom of a standard 90 mm × 90 mm × 15 mm Petri dish and cover with 10 ml molten selective agar medium. The molten agar medium should contain any necessary inducer (i.e. galactose, copper, etc.). This volume of agar will form a thin film of agar over the slide. Allow to set.
3. Take 100 µl of the culture, vortex briefly to separate any clumped cells and plate onto the surface of the agar. Allow to dry for 15–20 min.
4. Excise the slide from the agar and place a cover slip on top. Trim off the excess agar from around the cover slip.
5. Seal the cover slip using molten VALAP (1:1:1 mixture of Vaseline, lanolin and paraffin wax), extruded through the tip of a wide bore needle. Allow to cool and harden. The VALAP mixture gives a semi-gas permeable seal; therefore the cells under the cover slip are not growing in a completely anaerobic environment.
6. Observe cells using a confocal microscope, preferably fitted with a constant humidity/30°C temperature chamber. Excite at the appropriate wavelength for the fluorophore that is attached to the target protein and adjust the microscope emission filters to the correct wavelength for observing the fluorescence.
7. Use the minimum exposure time necessary to record a good fluorescent signal, as prolonged exposure to the confocal laser results in a decrease in cell viability.
8. If the slide is well sealed, it is possible to take images every 3 min for up to 16 h before drying out of the thin agar strip occurs.

This is frequently performed by immunofluorescence, where antibodies specific to a known protein within the suspected area of localisation are used to independently confirm localisation.

Fluorescent fusion proteins have also been successfully employed to test for interactions between proteins using fluorescence resonance energy transfer or FRET (Selvin, 2000). In this system, which is covered in detail in Chapter 12, this volume, there are two fluorescent fusion proteins, each with different emission spectra (e.g. CFP and YFP), one of which is used as a donor fluorophore and the other as an acceptor fluorophore. The donor fluorophore is excited by incident light, and if an acceptor is in close proximity, the

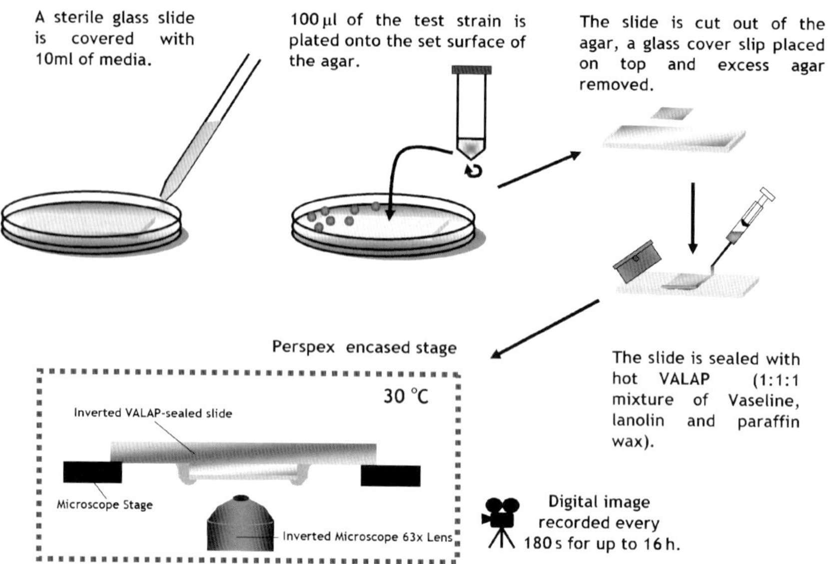

Figure 3. Diagram demonstrating a procedure for observing live yeast cells by time-lapse microscopy. See text and Protocol 3 for details.

excited state energy from the donor can be transferred. This leads to a reduction in the donor's fluorescence intensity and excited state lifetime, and an increase in the acceptor's emission intensity. A frequent donor/acceptor pair currently in use are the two GFP mutants: CFP and YFP (Miyawaki *et al.*, 1997). However, while they offer reasonable spectral separation and brightness (whilst not requiring harmful ultraviolet excitation), they have considerable overlap between emission spectra, resulting in substantial "cross-talk" of CFP emission in the YFP detection channel. The development of a growing class of sea coral fluorescent proteins (exemplified by DsRed) with genuine red emission, circumvents this problem, and acceptor/donor pairs of CFP and DsRed exhibit excellent separation of donor and acceptor emission spectra (Matz *et al.*, 1999; Rodrigues *et al.*, 2001).

In conclusion, the continuing expansion in the range and colour of fluorescent proteins, together with the development of innovative technologies (such as FlAsH and ReAsH), has only increased the power of fluorescence for use in reporter systems.

D. Adapting Reporter Genes to New Yeast Species

Unsurprisingly, a survey of the literature shows that the number of reporter systems available for a particular yeast species increases significantly with the popularity of that species as a research object. Thus, most of the available reporter assays were originally introduced to the yeast community for use in *S. cerevisiae*. However, many of the assays can be relatively easily adapted to new species:

the basic requirements for a reporter gene to work are that it can be efficiently expressed, that its activity is not represented in the organism's natural enzyme repertoire and that the signal produced by the reporter assay has no overlapping signals that arise, for example from endogenous metabolite pools. Most of these problems will have been taken into account in the original design of the reporter system, and the process of adapting it for use in relatively closely related species will thus mostly consist in placing the reporter cassette onto vector systems that are suitable for the new species.

In contrast, adapting a reporter system may be a significant problem if the new species differs significantly from the one for which it was originally developed. One such example is the group of *Candida* species that decode CUG as serine rather than the standard leucine, which includes a number of important human pathogens like *C. albicans*. As most reporter genes contain CUG codons in their open reading frames, they cannot necessarily be translated into active gene products in these yeasts. For *Candida*, this problem was addressed by various strategies such as codon optimisation (in the case of GFP; Cormack *et al.*, 1997), by using variants of a reporter that are insensitive to CUG-recoding (use of the *K. lactis LAC4* gene rather than the *E. coli lacZ* gene; Leuker *et al.*, 1992, use of the CUG-less *Renilla* luciferase; Srikantha *et al.*, 1996), or by use of homologous reporters in combination with knockouts of the corresponding genes (e.g. *URA3*; Myers *et al.*, 1995). Despite the initial difficulties, the combination of these strategies has resulted in a wide variety of reporter assays being available for work with *C. albicans* and related species.

♦♦♦♦♦♦ IV. PROBLEMS AND PITFALLS

A. Linearity

An important aspect in the use of reporter genes is to ensure that the amount of reporter protein produced by the cell is linearly related to the signal observed in the quantification assay. In the case of enzyme-based reporters, this is of particular importance since enzyme-catalysed reactions in general are not linear. The lower panels in Figure 4 show typical time-dependencies for the amount of product produced (Figure 4A) and for reaction rates over time (Figure 4B). For assays like the β-galactosidase/ONPG assay described above, where the reaction is stopped and the product detected after a certain reaction time, the signal is only proportional to the amount of reporter enzyme in the cell in the first part of the graph. In the extreme, when all the substrate has been converted, the signal will become constant irrespective of the reporter enzyme concentration. Similarly, luminescence-based assays that directly observe reaction rates rather than products only show a signal that is proportional to

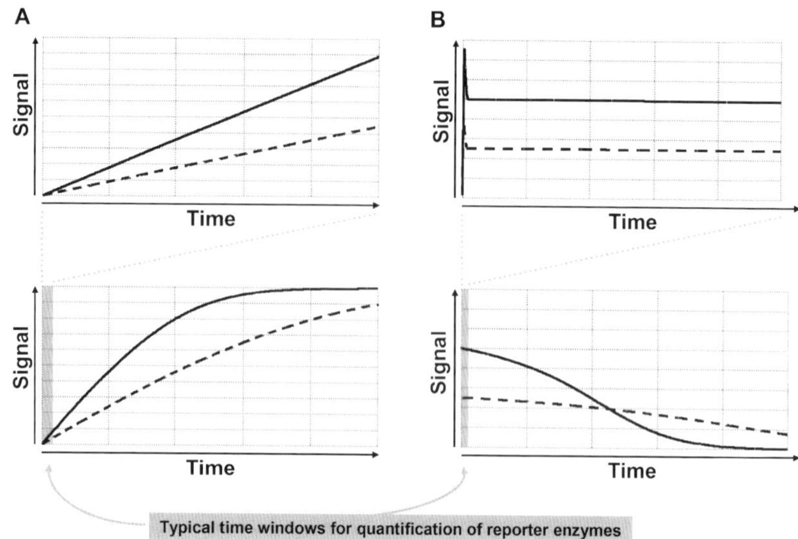

Figure 4. The development of enzymatic reactions over time and their use for the quantification of enzyme-based reporters. (A) Development of a signal where the reaction *product* is observed (e.g. the β-galactosidase/ONPG assay), and (B) where the reaction *rate* is observed (luminescence-based assays). The lower panels show the development of signals over long time periods: The amount of product eventually converges towards a maximum value when all the available substrate has been transformed into product, whereas the reaction rate converges towards zero (non-reversible reactions are assumed in both cases). Time windows that are commonly used for quantifying reporter gene expression are shaded grey in the lower panels and shown in detail in the upper panels. Within these time windows, the signal intensity is linearly related to the amount of reporter present in the cells.

enzyme concentration during the initial time points. In this case, if a single point of measurement is examined towards the right-hand side of the graph (Figure 4B), the relative enzyme levels in two cell extracts may even be inversely related to the actual signal observed.

The standard protocols employed for detecting reporter enzyme activity circumvent these problems by employing specific combinations of substrate concentrations, enzyme levels and reaction times. In general, substrate is used far in excess over enzyme levels: as an example, the conditions given in Protocol 2 result in ONPG/β-galactosidase ratios of roughly 10^5–10^6 at average expression levels. This ensures that enzyme activity, and not substrate availability, limits the rate of the reaction for significant lengths of time. Also, assay times are kept in regions that correspond to the far left of the graphs in Figure 4, and where the ratios of product or reaction rate to enzyme are proportional. Most standard protocols are fairly robust in this respect and result in linear signal:reporter relationships over a wide range of expression levels. However, unusual conditions such as very high expression levels may result in non-linear relationships. If the linearity of the assay is in doubt, the dependency of the signal on enzyme levels can easily be tested by generating a standard curve, for example by creating serial dilutions of cell extracts containing expressed reporter, or by spiking cell extracts with known amounts of purified reporter enzyme. Additional information on linearity can be generated by examining signal levels at several time points. Issues relating to linearity are discussed in more detail elsewhere (Serebriiskii and Golemis, 2000).

Particular problems may also arise from unforeseen and less controllable effects. In a recent study (Alipour *et al.*, 1999), the authors found that a widely used procedure for determining CAT reporter activity gave strongly non-linear results with *S. cerevisiae* cells because cell extracts from this organism contain a low-molecular-weight compound that appears to stimulate CAT enzyme activity. The facts that this effect could be demonstrated with *S. cerevisiae* but not with *S. pombe* extracts, and that it was only published after the relevant assay had been used with yeast for more than a decade, demonstrate the importance of examining the actual relationship between reporter levels and assay results whenever a reporter is used under untested conditions.

B. Clonal Variation

For a number of different reporter systems, significant variation in activity has been reported both between individual colonies derived from a single transformation, and between individual cells derived from a single colony (reviewed e.g. in Serebriiskii and Golemis, 2000). At least one of the causes of this variability is based on the fact that most reporter genes are introduced into yeast cells via plasmids, which themselves show some variation in copy numbers. The ratio of reporter signals generated in two different conditions therefore arises through a combination of the changes in genetic activity of the reporter construct, and changes in plasmid copy number. The latter may vary randomly, or may itself be a function of growth conditions.

In order to assess the influence of clonal variation on the results, it is good practice to analyse cultures derived from several different transformants for each condition or construct analysed. The usual practice of performing triplicate assays (i.e. assays with three independent transformants) per condition gives an initial idea of the variability, but larger numbers of assays may be necessary to decide whether differences in reporter activity between two conditions are statistically significant or not (Jacobs and Dinman, 2004).

Besides variations in plasmid copy number, the growth status of the initial yeast colonies used to start an experiment may have a significant influence on the final result. Yeast retains viability on plates for several weeks to months, although the growth status, as well as transcriptional and translational activity, vary widely between freshly grown and older colonies. In fact these parameters were shown to vary significantly even between the centre and edge of a yeast colony (Meunier and Choder, 1999).

In the case of β-galactosidase, it was suggested that due to the long half-life of this protein initial expression levels may significantly influence the intracellular levels of the reporter for several generations after inoculation into the fresh medium. The same considerations are likely to hold true for other reporters, which are generally stable proteins. It is therefore important to grow yeast cultures for as many generations as possible prior to reporter measurements, in order to allow potentially high initial levels of the reporter to be diluted out through cell divisions, and to ensure that reporter enzyme levels are solely dependent on ongoing synthesis. It should be stressed that the use of starter cultures that are grown overnight and from which the experimental culture is then inoculated can potentially contribute to correcting this problem, but only if the growth status of the different starter cultures at the time of inoculation of the final cultures is similar. In contrast, if at the time of inoculation of the final cultures one starter culture is in stationary phase but the other still in logarithmic growth, and the final cultures are then grown for only a few generations prior to quantification of the reporter, the results may again reflect differences in starting conditions rather than differences in genetic activities during growth in the final culture.

♦♦♦♦♦♦ V. CONCLUSIONS

Reporter genes are tools of ever-increasing importance in the study of yeast gene expression. Their versatility and ease of use have made them widely available, and ongoing developments will undoubtedly continue to increase their usefulness as well as their potential applications. The present overview over the available systems and their use is necessarily incomplete due to the vast amount of relevant literature. However, we have endeavoured to describe what we subjectively regard as the most important and most widely used assays in a way that allows interested researchers to refer back to the original publications for more in-depth information.

Acknowledgements

TvdH is supported by a Wellcome Trust Research Career Development Fellowship (075438/Z/04/Z). LJB is supported by a University of Kent Fellowship in Mathematical Biology.

References

Adams, S. R., Campbell, R. E., Gross, L. A., Martin, B. R., Walkup, G. K., Yao, Y., Llopis, J. and Tsien, R. Y. (2002). New biarsenical ligands and tetracysteine motifs for protein labeling *in vitro* and *in vivo*: synthesis and biological applications. *J. Am. Chem. Soc.* **124**, 6063–6076.

Alipour, H., Eriksson, P., Norbeck, J. and Blomberg, A. (1999). Quantitative aspects of the use of bacterial chloramphenicol acetyltransferase as a reporter system in the yeast *Saccharomyces cerevisiae*. *Anal. Biochem.* **270**, 153–158.

Bevis, B. J. and Glick, B. S. (2002). Rapidly maturing variants of the *Discosoma* red fluorescent protein (DsRed). *Nat. Biotechnol.* **20**, 83–87.

Billinton, N. and Knight, A. W. (2001). Seeing the wood through the trees: a review of techniques for distinguishing green fluorescent protein from endogenous autofluorescence. *Anal. Biochem.* **291**, 175–197.

Boylan, M., Pelletier, J. and Meighen, E. A. (1989). Fused bacterial luciferase subunits catalyze light emission in eukaryotes and prokaryotes. *J. Biol. Chem.* **264**, 1915–1918.

Bronstein, I., Fortin, J., Stanley, P. E., Stewart, G. S. and Kricka, L. J. (1994). Chemiluminescent and bioluminescent reporter gene assays. *Anal. Biochem.* **219**, 169–181.

Cavagnero, S. and Jungbauer, L. M. (2005). Painting protein misfolding in the cell in real time with an atomic-scale brush. *Trends Biotechnol.* **23**, 157–162.

Chalfie, M., Tu, Y., Euskirchen, G., Ward, W. W. and Prasher, D. C. (1994). Green fluorescent protein as a marker for gene expression. *Science* **263**, 802–805.

Cid, V. J., Alvarez, A. M., Santos, A. I., Nombela, C. and Sanchez, M. (1994). Yeast exo-β-glucanases can be used as efficient and readily detectable reporter genes in *Saccharomyces cerevisiae*. *Yeast* **10**, 747–756.

Cormack, B. P., Bertram, G., Egerton, M., Gow, N. A., Falkow, S. and Brown, A. J. (1997). Yeast-enhanced green fluorescent protein (yEGFP): a reporter of gene expression in *Candida albicans*. *Microbiology* **143**, 303–311.

Cormack, B. P., Valdivia, R. H. and Falkow, S. (1996). FACS-optimized mutants of the green fluorescent protein (GFP). *Gene* **173**, 33–38.

Cubitt, A. B., Woollenweber, L. A. and Heim, R. (1999). Understanding structure–function relationships in the *Aequorea victoria* green fluorescent protein. *Methods Cell Biol.* **58**, 19–30.

Donahue, T. F., Cigan, A. M., Pabich, E. K. and Valavicius, B. C. (1988). Mutations at a Zn(II) finger motif in the yeast eIF 2β gene alter ribosomal start-site selection during the scanning process. *Cell* **54**, 621–632.

Duvel, K., Egli, C. M. and Braus, G. H. (1999). A single point mutation in the yeast *TRP4* gene affects efficiency of mRNA 3′ end processing and alters selection of the poly(A) site. *Nucleic Acids Res.* **27**, 1289–1295.

Firoozan, M., Grant, C. M., Duarte, J. A. and Tuite, M. F. (1991). Quantitation of readthrough of termination codons in yeast using a novel gene fusion assay. *Yeast* **7**, 173–183.

Fulton, R. and Van Ness, B. (1993). Luminescent reporter gene assays for luciferase and beta-galactosidase using a liquid scintillation counter. *Biotechniques* **14**, 762–763.

Gilon, T., Chomsky, O. and Kulka, R. G. (1998). Degradation signals for ubiquitin system proteolysis in *Saccharomyces cerevisiae*. *EMBO J.* **17**, 2759–2766.

Gorman, C. M., Moffat, L. F. and Howard, B. H. (1982). Recombinant genomes which express chloramphenicol acetyltransferase in mammalian cells. *Mol. Cell. Biol.* **2**, 1044–1051.

Grentzmann, G., Ingram, J. A., Kelly, P. J., Gesteland, R. F. and Atkins, J. F. (1998). A dual-luciferase reporter system for studying recoding signals. *RNA* **4**, 479–486.

Heim, R., Cubitt, A. B. and Tsien, R. Y. (1995). Improved green fluorescence. *Nature* **373**, 663–664.

Heim, R. and Tsien, R. Y. (1996). Engineering green fluorescent protein for improved brightness, longer wavelengths and fluorescence resonance energy transfer. *Curr. Biol.* **6**, 178–182.

Hinnebusch, A. G. (1984). Evidence for translational regulation of the activator of general amino acid control in yeast. *Proc. Natl. Acad. Sci. USA* **81**, 6442–6446.

Hinnebusch, A. G. (2005). Translational regulation of *GCN4* and the general amino acid control of yeast. *Annu. Rev. Microbiol.* **59**, 407–450.

Horton, R. M., Hunt, H. D., Ho, S. N., Pullen, J. K. and Pease, L. R. (1989). Engineering hybrid genes without the use of restriction enzymes: gene splicing by overlap extension. *Gene* **77**, 61–68.

Huh, W. K., Falvo, J. V., Gerke, L. C., Carroll, A. S., Howson, R. W., Weissman, J. S. and O'Shea, E. K. (2003). Global analysis of protein localization in budding yeast. *Nature* **425**, 686–691.

Inouye, S. and Tsuji, F. I. (1994). *Aequorea* green fluorescent protein. Expression of the gene and fluorescence characteristics of the recombinant protein. *FEBS Lett.* **341**, 277–280.

Jacobs, J. L. and Dinman, J. D. (2004). Systematic analysis of bicistronic reporter assay data. *Nucleic Acids Res.* **32**, e160.

Kirchner, G., Roberts, J. L., Gustafson, G. D. and Ingolia, T. D. (1989). Active bacterial luciferase from a fused gene: expression of a *Vibrio harveyi luxAB* translational fusion in bacteria, yeast and plant cells. *Gene* **81**, 349–354.

Knight, A. W., Goddard, N. J., Billinton, N., Cahill, P. A. and Walmsley, R. M. (2002). Fluorescence polarization discriminates green fluorescent protein from interfering autofluorescence in a microplate assay for genotoxicity. *J. Biochem. Biophys. Methods* **51**, 165–177.

Kohlwein, S. D. (2000). The beauty of the yeast: live cell microscopy at the limits of optical resolution. *Microsc. Res. Tech.* **51**, 511–529.

Kurtz, M. B., Cortelyou, M. W. and Kirsch, D. R. (1986). Integrative transformation of *Candida albicans*, using a cloned *Candida ADE2* gene. *Mol. Cell. Biol.* **6**, 142–149.

Leuker, C. E., Hahn, A. M. and Ernst, J. F. (1992). β-Galactosidase of *Kluyveromyces lactis* (Lac4p) as reporter of gene expression in *Candida albicans* and *C. tropicalis*. *Mol. Gen. Genet.* **235**, 235–241.

Leupold, U. (1958). Studies on recombination in *Schizosaccharomyces pombe*. *Cold Spring Harb. Symp. Quant. Biol.* **23**, 161–170.

Luukkonen, B. G. and Seraphin, B. (1999). A conditional U5 snRNA mutation affecting pre-mRNA splicing and nuclear pre-mRNA retention identifies *SSD1/SRK1* as a general splicing mutant suppressor. *Nucleic Acids Res.* **27**, 3455–3465.

Mannazzu, I., Sudbery, P. E., Berardi, E. and Fatichenti, F. (1995). Promoter isolation in *Hansenula polymorpha*. *Annali Di Microbiol. Ed Enzimol.* **45**, 209–218.

Mannhaupt, G., Pilz, U. and Feldmann, H. (1988). A series of shuttle vectors using chloramphenicol acetyltransferase as a reporter enzyme in yeast. *Gene* **67**, 287–294.

Matz, M. V., Fradkov, A. F., Labas, Y. A., Savitsky, A. P., Zaraisky, A. G., Markelov, M. L. and Lukyanov, S. A. (1999). Fluorescent proteins from non-bioluminescent *Anthozoa* species. *Nat. Biotechnol.* **17**, 969–973.

Meunier, J. R. and Choder, M. (1999). *Saccharomyces cerevisiae* colony growth and ageing: biphasic growth accompanied by changes in gene expression. *Yeast* **15**, 1159–1169.

Miyawaki, A., Llopis, J., Heim, R., McCaffery, J. M., Adams, J. A., Ikura, M. and Tsien, R. Y. (1997). Fluorescent indicators for Ca2+ based on green fluorescent proteins and calmodulin. *Nature* **388**, 882–887.

Molero, G., Cid, V. J., Vivar, C., Nombela, C. and Sanchez-Perez, M. (1999). *Candida albicans* exoglucanase as a reporter gene in *Schizosaccharomyces pombe*. *FEMS Microbiol. Lett.* **175**, 143–148.

Muhlrad, D. and Parker, R. (1999). Aberrant mRNAs with extended 3′ UTRs are substrates for rapid degradation by mRNA surveillance. *RNA* **5**, 1299–1307.

Myers, K. K., Sypherd, P. S. and Fonzi, W. A. (1995). Use of *URA3* as a reporter of gene expression in *C. albicans*. *Curr. Genet.* **27**, 243–248.

Parham, S. N., Resende, C. G. and Tuite, M. F. (2001). Oligopeptide repeats in the yeast protein Sup35p stabilize intermolecular prion interactions. *EMBO J.* **20**, 2111–2119.

Prescott, M., Nowakowski, S., Nagley, P. and Devenish, R. J. (1999). The length of polypeptide linker affects the stability of green fluorescent protein fusion proteins. *Anal. Biochem.* **273**, 305–307.

Rajkowitsch, L., Vilela, C., Berthelot, K., Ramirez, C. V. and McCarthy, J. E. (2004). Reinitiation and recycling are distinct processes occurring downstream of translation termination in yeast. *J. Mol. Biol.* **335**, 71–85.

Rodrigues, F., van Hemert, M., Steensma, H. Y., Corte-Real, M. and Leao, C. (2001). Red fluorescent protein (DsRed) as a reporter in *Saccharomyces cerevisiae*. *J. Bacteriol.* **183**, 3791–3794.

Scorpione, R. C., De Camargo, S. S., Schenberg, A. C. and Astolfi-Filho, S. (1993). A new promoter–probe vector for *Saccharomyces cerevisiae* using fungal glucoamylase cDNA as the reporter gene. *Yeast* **9**, 599–605.

Selvin, P. R. (2000). The renaissance of fluorescence resonance energy transfer. *Nat. Struct. Biol.* **7**, 730–734.

Serebriiskii, I. G. and Golemis, E. A. (2000). Uses of *lacZ* to study gene function: evaluation of β-galactosidase assays employed in the yeast two-hybrid system. *Anal. Biochem.* **285**, 1–15.

Shaner, N. C., Campbell, R. E., Steinbach, P. A., Giepmans, B. N., Palmer, A. E. and Tsien, R. Y. (2004). Improved monomeric red, orange and yellow fluorescent proteins derived from *Discosoma* sp. red fluorescent protein. *Nat. Biotechnol.* **22**, 1567–1572.

Sheff, M. A. and Thorn, K. S. (2004). Optimized cassettes for fluorescent protein tagging in *Saccharomyces cerevisiae*. *Yeast* **21**, 661–670.

Srikantha, T., Klapach, A., Lorenz, W. W., Tsai, L. K., Laughlin, L. A., Gorman, J. A. and Soll, D. R. (1996). The sea pansy *Renilla reniformis* luciferase serves as a sensitive bioluminescent reporter for differential gene expression in *Candida albicans*. *J. Bacteriol.* **178**, 121–129.

Thorn, K. S., Naber, N., Matuska, M., Vale, R. D. and Cooke, R. (2000). A novel method of affinity-purifying proteins using a bis-arsenical fluorescein. *Protein Sci.* **9**, 213–217.

Vieites, J. M., Navarro-Garcia, F., Perez-Diaz, R., Pla, J. and Nombela, C. (1994). Expression and *in vivo* determination of firefly luciferase as gene reporter in *Saccharomyces cerevisiae*. *Yeast* **10**, 1321–1327.

Walmsley, R. M., Billinton, N. and Heyer, W. D. (1997). Green fluorescent protein as a reporter for the DNA damage-induced gene *RAD54* in *Saccharomyces cerevisiae*. *Yeast* **13**, 1535–1545.

Walmsley, R. M., Gardner, D. C. and Oliver, S. G. (1983). Stability of a cloned gene in yeast grown in chemostat culture. *Mol. Gen. Genet.* **192**, 361–365.

Young, S. L., Barbera, L., Kaynard, A. H., Haugland, R. P., Kang, H. C., Brinkley, M. and Melner, M. H. (1991). A nonradioactive assay for transfected chloramphenicol acetyltransferase activity using fluorescent substrates. *Anal. Biochem.* **197**, 401–407.

9 Transcript Analysis: A Microarray Approach

Andrew Hayes[1], Juan I Castrillo[1], Stephen G Oliver[1], Andy Brass[1,2] and Leo AH Zeef[1]

[1] Faculty of Life Sciences, The University of Manchester, Michael Smith Building, Oxford Road, Manchester M13 9PT, UK; [2] School of Computer Science, The University of Manchester, Kilburn Building, Oxford Road, Manchester M13 9PL, UK

CONTENTS

Introduction
Performing the microarray experiment
Experimental design
Computational analysis
Summary and future perspectives

I. INTRODUCTION

Microarray transcript analysis sets itself a very ambitious goal in attempting to measure transcription changes in more than 5000 yeast genes simultaneously. This is particularly challenging because these 5000 genes have an expression range of 10^{-3}–10^{2} transcripts per cell, a dynamic range of five orders of magnitude (Hereford and Rosbash, 1977). The potential for error is obviously large. In this chapter, we describe procedures that allow these errors to be minimized and also focus on computational analysis and its role both in the extraction of knowledge from vast tables of data, and in providing an assessment of the quality and reliability of that knowledge. Thus, a pipeline of analysis will be described that has been designed to minimize errors and assess data quality. Clearly, experimental design needs to be considered from the outset and so this will be discussed prior to data analysis. Good experimental design is the key to an effective microarray study and, since it is so important, it is imperative to be well aware of all experimental procedures, the challenges of data analysis, and commonly encountered problems before designing a microarray experiment.

METHODS IN MICROBIOLOGY, VOLUME 36
0580-9517 DOI:10.1016/S0580-9517(06)36009-6

The initial fundamental concept and limitation to be grasped is that, with current microarray technology, quantification is relative not absolute. The goal is always to measure relative transcript abundances between two or more samples. Because of this, the technique is extremely sensitive to errors arising from differences in sample handling. This places high demands on sample preparation and technical performance of the microarray experiment.

A second point to be made is that the current microarray technologies show poor congruence in results (Tan *et al.*, 2003). When 5000 genes are measured simultaneously, compromises in the accuracy and sensitivity will inevitably occur for certain genes when specific labelling and hybridization protocols are chosen. This means verification of results by alternative methods such as RT-qPCR should be seen as integral to a microarray experiment.

Many different experimental platforms are available for the interrogation of yeast gene expression using microarrays. The purpose of this chapter is neither to review all the different formats nor to recommend definitively any one over another. Rather, we aim to describe an analysis pipeline whose fundamental concepts are as far as possible generic for the analysis of the yeast transcriptome, regardless of the platform used. The examples we have used as illustrations in the following sections are from experiments performed using the commercially available Affymetrix GeneChip® system. For the sake of consistency, we have therefore concentrated on this platform throughout the chapter but stress the generic nature of the concepts described.

◆◆◆◆◆◆ II. PERFORMING THE MICROARRAY EXPERIMENT

A. Yeast Arrays

Affymetrix GeneChip® arrays comprise a set of oligonucleotide probes synthesized onto an array. The probes are complementary to portions of each open reading frame (ORF) in the genome. The array is mounted in a special plastic cartridge containing a square glass substrate. The array of oligonucleotides is on the inner glass surface and a chamber in the housing directly under the glass acts as a reservoir where the hybridization and washing steps occur. Details of the technology and the arrays available can be found on the company's website (http://www.affymetrix.com). Briefly, there are currently two types of GeneChip® yeast expression arrays available:

- The Yeast Genome S98 Array (YG-S98) is a single array containing probe sequences to monitor mRNA from over 6000 ORFs and other sequences from the yeast genome. Affymetrix state that the sequences used to design the array are mainly from *Saccharomyces*

Genome Database (SGD) with a download date of 12/98. Additional sequences were obtained from the Munich Information Center for Protein Sequence (MIPS). Also included are probes for potentially expressed sequences identified by Serial Analysis of Gene Expression (SAGE; Velculescu *et al.*, 1997). This YG-S98 array contains approximately 16 pairs of probes for each gene and these probe pairs are arranged in an adjacent manner.

- The Yeast Genome 2.0 Array is a later version of the yeast array manufactured in a higher density format and covers the genomes of two yeasts. According to the manufacturers, it can be used to study gene expression for 5841 *Saccharomyces cerevisiae* transcripts and 5031 *Schizosaccharomyces pombe* transcripts. The sequence information for this array was selected from public data sources GenBank® (May 2004) and Sanger Institute (June 2004) for the *S. cerevisiae* and *S. pombe* genomes, respectively. This array contains 11 pairs of probes for each gene. On this array, the probe pairs that comprise each probe set are randomized across the area of the array thus dramatically reducing the impact of any spatial biases or artefacts on the array.

There are, of course, limitations to the amount of genomic information that these arrays can interrogate. For example, non-coding RNA makes up 29.5% of the genome of *S. cerevisiae* (Goffeau *et al.*, 1996) and 42.5% of *S. pombe* (Wood *et al.*, 2002) and only a very small proportion of these are represented on the current arrays. It is anticipated that these non-coding RNAs will be shown to have a major impact in gene and network regulation (Mattick, 2004; Mattick and Gagen, 2005). As the technology advances, new tools such as "tiling arrays" are becoming available. At the time of writing, Affymetrix have just released *S. cerevisiae* tiling arrays (part no. 900646) which contain oligonucleotide probes covering the whole of the yeast genome at a resolution of 5 bp. These tiling arrays are certain to become increasingly important in the transcriptional analysis of yeast.

B. Extraction of RNA

For a microarray experiment to be successful the quality of the starting material is crucially important. There are many methods available for the extraction of total RNA from yeast cultures and these are generally based on chemical and/or physical disruption techniques. The method we favour is adapted from Hauser *et al.*, (1998) and Hayes *et al.*, (2002). The cells are mechanically disrupted at very low temperature and then the RNA is extracted using a monophasic chaotrope. This method has proven to be rapid, very efficient and, most importantly, yielded RNA that was reverse transcribed with consistently high efficiency. The biotinylated cRNA targets synthesized from starting material prepared in this way produce consistently reproducible results in subsequent GeneChip hybridizations. Protocol 1 gives details of the RNA extraction

Protocol 1. Extraction of RNA.

1. Pellet cells by centrifugation for 5 min at 5000 r.p.m. Remove the supernatant and re-suspend the cell pellet in the residual medium to form a slurry. Add this in a drop-wise manner directly into a 5 ml teflon flask (B. Braun Biotech, Germany) containing liquid nitrogen and a 7 mm diameter tungsten carbide ball.
2. Following evaporation of the liquid nitrogen re-assemble the flask and disrupt the cells by agitation at 1500 r.p.m. for 2 min in a micro-dismembrator U (B. Braun Biotech, Germany).
3. Dissolve the frozen powder in 1 ml of TriZol reagent (Sigma-Aldrich, UK), vortex for 1 min and then keep at room temperature for a further 5 min.
4. Add 0.2 ml chloroform, shake vigorously for 15 s, then incubate for 5 min at room temperature.
5. Following centrifugation at 12 000 r.p.m. for 5 min, precipitate the RNA (contained in the aqueous phase) with 0.5 volumes of propan-2-ol at room temperature for 15 min.
6. After a further centrifugation (12 000 r.p.m. for 10 min at 4°C), wash the RNA pellet twice with 70% v/v ethanol, briefly air-dry, and then re-dissolve in 0.5 ml diethyl pyrocarbonate (DEPC) treated water.
7. Precipitate the single-stranded RNA once more by addition of 0.5 ml of LiCl buffer (4 M LiCl, 20 mM Tris-HCl pH 7.5, 10 mM EDTA), thus removing tRNA and DNA from the sample.
8. After precipitation (−20°C for 1 h), and centrifugation (12 000 r.p.m., 30 min, 4°C), wash the RNA twice in 70% v/v ethanol prior to dissolving in a minimal volume of DEPC-treated water.
9. Measure the quantity and quality of the total RNA using a BioAnalyzer 2100 (Agilent Technologies Ltd., UK)-see Figure 1 for representative traces. Use 15 μg total array per reaction for the synthesis of labelled target.

method. We find a BioAnalyzer 2100 (Agilent Technologies Ltd., UK) particularly useful for quality control purposes. This is a capillary electrophoresis instrument that uses fluorescence to characterize size distribution of (in this case) nucleic acids. The results from this instrument characterize RNA not only in terms of amount, but more importantly in terms of integrity. The traces in Figure 1 illustrate the qualitative differences between intact and degraded RNA. At the time of writing there is no "gold standard" against which to define RNA quality prior to gene expression analysis; however, tools are currently being developed to address this (Imbeaud *et al.*, 2005).

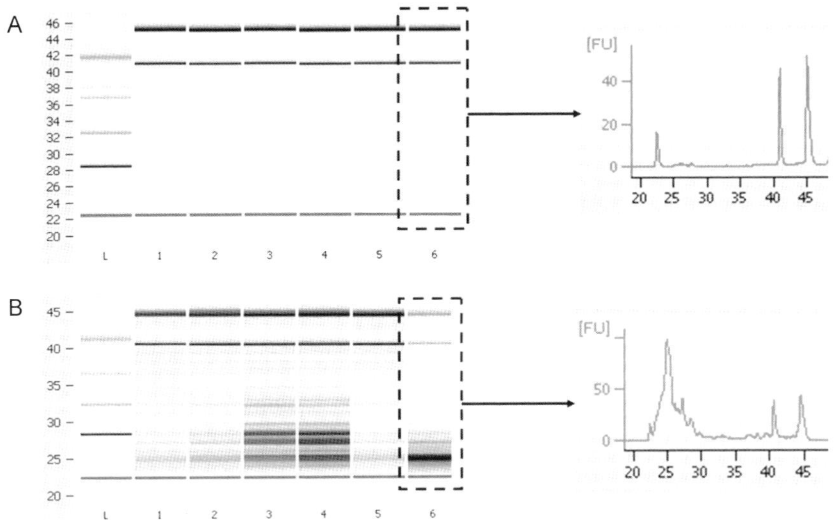

Figure 1. Gel and electropherogram profiles of *S. cerevisiae* total RNA samples acquired using a Bioanalyzer 2100. (A) Intact, good quality RNA. (B) RNA at various stages of degradation. FU = Fluorescence Units.

C. Labelling and Hybridization

Once RNA of sufficient quality has been prepared, this is used to synthesize biotinylated cRNA targets for hybridization to the microarrays. The labelled targets are prepared and fragmented exactly according to the manufacturer's instructions. A series of controls are then added to form the hybridization cocktail, which is hybridized to the arrays for 16 h at 45°C in the rotisserie oven at 60 r.p.m. The washing and staining steps are performed using the Affymetrix Fluidics Station 450 where a streptavidin–phycoerythrin conjugate is used to stain the arrays. The GeneChip Operating System (GCOS) software uses downloadable protocols to define the washing and staining procedures for each chip. The YG-S98 and Yeast 2.0 arrays require protocols "EukGE-WS2" and "Mini_euk2v3" respectively.

D. Scanning

Once the probe array has been processed in the fluidics station, the array is scanned using the Affymetrix GeneChip scanner 3000. There are several stages to the scanning process, which is again driven by the GCOS software. First, the probe array is scanned and the image data saved. This creates a .dat file, an example of which is shown in Figure 2A. Next the software computes the cell intensity data from the image data and creates a .cel file. This expression cell intensity data is then analysed and saved as a .chp file. The .chp file contains data analysis information for each probe set on the array as well as controls. The software can then generate an expression

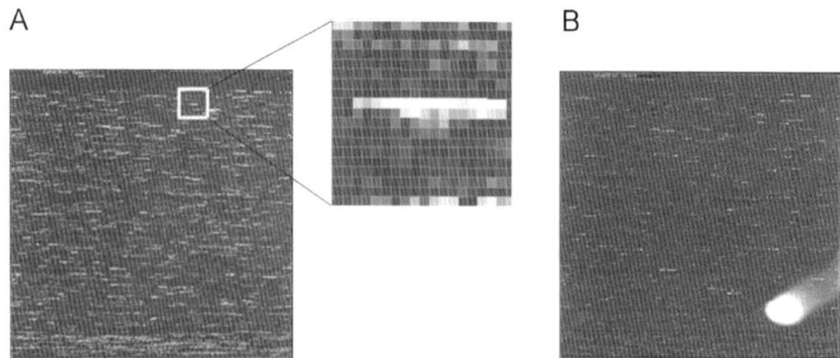

Figure 2. Images from .dat files of hybridized Affymetrix YGs_98 GeneChip arrays. (A) Image of a successful hybridization including a "zoomed-in" portion to illustrate the individual probes. (B) Image of a hybridized array which failed the initial QC procedure. The artifact responsible for this can be clearly seen in the lower right-hand corner of the array.

report file (.rpt). This will be discussed in more detail in the computational analysis section.

Occasionally, things will go wrong with the hybridizations. Again an example of this is given in Figure 2B, where an artefact can clearly be seen obscuring a large area of the array. In such cases, tools are available within the GCOS software to mask out this area. Thankfully, however, such problems are very rare.

♦♦♦♦♦♦ III. EXPERIMENTAL DESIGN

Experimental design needs to be considered from the outset. Experimental design entails far more than just "how many replicates should I run?" In the following section, the significance of replicates is considered within the broader context of errors. Good experimental design follows from knowing the sources of error and goes a long way towards minimising errors and increasing confidence in the data. It is important to understand errors as originating in two distinct ways: errors due to systematic bias and random errors.

A. Errors due to Systematic Bias

This type of error occurs when a factor other than the biological variable under study has influenced transcript measurements. For example, the day on which the experiment was run may explain the clustering of samples in a Principal Components Analysis (PCA) as mentioned later in Section 5 of Computational Analysis. For this reason, it is important to predict the origin of this kind of bias and build it into the experimental design. If all samples cannot be run on the same day, then half the controls and half the treated samples should be run on day one and the rest on day two, rather than all the controls on day one and all the treated samples on day two.

Managing systematic bias in this way does not eliminate the error, but its detrimental influence is reduced and analysis of variance (ANOVA) methods exist to reduce this influence even further. The following is a checklist of potential sources of systematic bias:

- If there are different experimenters, try to randomize the samples between them. For example, do not let John do all control replicates and Jane do all treatment replicates.
- If the experiment is done on separate days, do not do all control replicates on day 1 and all treatment replicates on day 2. Rather, do half of the control and treatment replicates on day 1 and the second half on day 2.
- If possible, keep some RNA back for verification with alternative methods such as quantitative RT-PCR.
- If you are performing the experiment in a laboratory (rather than, for instance, taking samples from the environment or an industrial fermenter), make full use of your ability to control the experimental conditions. Make sure the physical environment, volume of medium, flask/dish size and shape, temperature and location in the incubator are the same for all samples.
- Be aware of the effect of stress on cell morphology or growth rates.
- If using batch cultures, try to understand the influence of growth phase on gene expression. Consider using steady-state chemostat cultures.
- How fast does gene expression respond to environmental changes? Is the time taken to arrest growth and harvest cells going to influence transcript levels?
- Any disturbance to the environment will have consequences on gene expression. So make sure controls and treatment samples are disturbed equally. For example, if adding a chemical solution to the treatment samples, add an equal volume of the solvent to the control samples at exactly the same time intervals.
- How genetically stable are your strains? If unsure, consider doing quantitative RT-PCR on the samples before running them on expensive arrays. Use biological replicates to make sure you are not measuring a biological outlier, revertant or mutant strain.
- Make sure there are no contaminants.

B. Random Error and Replicates

Unlike systematic error, random error cannot be removed with good experimental design. Random error is generated by variations in every step of the experiment: biology, experimental procedures and equipment inconsistencies. In practice, it is usually the biological variation between individuals that is the major source of random error. The way to manage random error is by good experimental design and performing sufficient biological replicates (Figure 3). "How many replicates to do?" is not a simple question to answer. From a statistics point of view, it is possible to calculate the

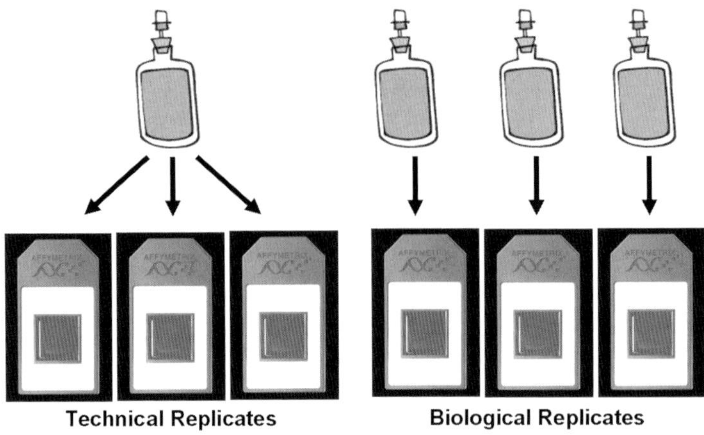

Figure 3. The difference between technical and biological replicates.

number of replicates required by doing a power analysis. However, one needs an estimate of variance for a power analysis. This requires prior knowledge and is not a simple matter if a microarray measures 5000 genes, all of which might have a different variance. Ideally there should be eight or more replicates but in practice, given the cost of arrays, three or four replicates are enough. As mentioned in the Computational Analysis section, statistical procedures have been developed taking low replication into account and making use of the high number measurements in a microarray to strengthen the variance estimate. It should be stressed that the number of replicates needed depends on the variance and that comes back to the design of the experiment and careful control of the conditions. Although replication is often considered a stick to beat the researcher with, it is worth considering the consequences of no replicates at all. No replication leads to a high number of false positive and false negatives in lists of differentially expressed genes generated by an analysis pipeline, and this means much time, effort and money wasted in pointless follow-on experiments.

C. False Positives, False Negatives and Replicates

Whatever the source of error, be it systematic or random, a list of differentially expressed genes may contain two types of misinformation (Figure 4). First, genes appear on the list when they should not. These are false-positive genes that, in reality, do not show differential expression at all. Fortunately there is a solution to the false positive error. Verification of differential expression with an independent technique removes the false positives. False positives are costly in terms of futile verification procedures targeted at genes that show no differential expression. Second, genes do not appear on the list of differentially expressed genes when they should. These are false negatives that, in reality, do show differential expression.

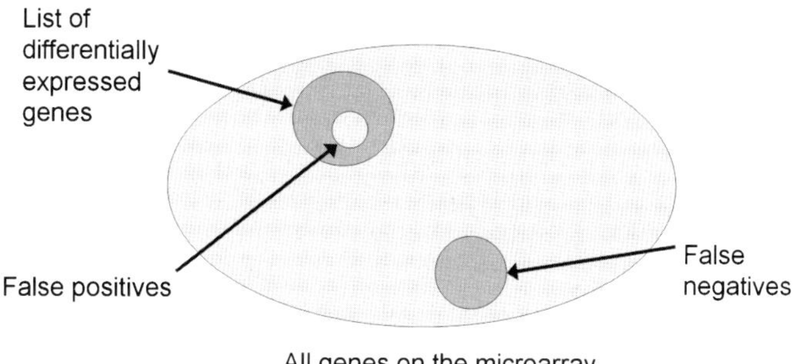

Figure 4. False positives and false negatives in lists of statistically significant genes.

False negatives are costly in terms of missing a valuable novel discovery. In general, the more replicates the stronger the statistical power and also the larger the set of genes that may be called significantly changed by the analysis pipeline. However, it is also important to realize that the gene may be a false negative because its expression level is below that measurable in a microarray analysis and no amount of replication can correct this. Even if the gene has high expression levels, differential expression may be masked by poor probe selection in array design leading to cross reaction with other RNA species or no hybridization at all to the probe set due to secondary structure or low melting temperatures. There is no real solution to false negatives due to poor probe set design, apart from redesigning the array. As a result, one should expect the list of false negatives to be large from a microarray experiment.

D. Pooling and Replicates

One strategy to reduce the cost of running many replicate arrays while still obtaining a confident measure for the average gene expression levels, is to pool separate yeast cultures or colonies prior to analysis. In many situations where yeast is cultured in a laboratory, it is possible to control conditions and genotype to such an extent that pooling is of little value. However, there are situations where the benefits of pooling are clear. By using many pooling batches, any biological variance between individual batches is averaged out. The measurement from a single array is therefore a better representation of a population of batches than a measurement of a single batch. By pooling, we have a good estimate of the average expression but we have no measure of how variable expression is between individual batches. This information is important in assessing statistical uncertainty and deciding how reliable and trustworthy the data are. Finally, pooling samples effectively destroys them in terms of being able to go back and measure or validate each individual batch. This can be circumvented by extracting RNA from each batch separately and then pooling a portion of each for the pool applied to the array.

E. Classification and Machine Learning

Any discussion of microarray experimental design would be incomplete without mention of using microarrays to classify samples. This experimental approach is a paradigm shift from an experiment where one is trying to identify genes changing between known samples. Instead, the expression profile as a whole is used as a tool to identify the original sample. An example of the use of this approach would be diagnostic identification of a dangerous pathogenic strain of yeast. The requirement for this procedure would be a large set of microarray data from pathogenic and non-pathogenic yeast samples. By employing machine-learning methods, this dataset is used to generate a classification model that will assign a new microarray measurement, from an unknown yeast sample, to either a pathogenic or non-pathogenic class. Many computational methods exist for classification, with support vector machines usually performing well (Pochet *et al.*, 2005). Performance usually improves by filtering the number of genes prior to running the machine learning methods (using procedures described in the Computational Analysis section below). An interesting additional analysis procedure related to this is "feature selection", identifying a small set of genes that are key to classifying our pathogenic and non-pathogenic strains (Li *et al.*, 2004; Liu *et al.*, 2005).

♦♦♦♦♦♦ IV. COMPUTATIONAL ANALYSIS

A data analysis pipeline will be described with the help of a dataset of Affymetrix GeneChip® arrays. Although the software algorithms described are specific for Affymetrix, the analysis pipeline and quality control steps are generic and applicable to all microarray methods. In this example, the sample dataset has 12 arrays; four of these are outliers that failed quality control tests. The aim of the experiment was to study differential yeast gene expression profiles between nitrogen- and carbon-limited growth conditions. The experimental design, using steady-state chemostat cultures in defined minimal media, was a simple two-condition control *vs.* treatment type. The nitrogen-limited (N) treatment samples had a growth-rate-determining supply of ammonium compared with glucose in excess, while the carbon-limited (C) control samples had a growth-rate-determining supply of glucose with ammonium in excess. The data analysis pipeline has the following steps and is graphically represented in Figure 5:

(1) Experimental Design
(2) Sample Quality Control
(3) Quality Control of Microarrays
(4) Low-Level Analysis: Expression Analysis and Normalization
(5) Valuation of the Whole Experiment

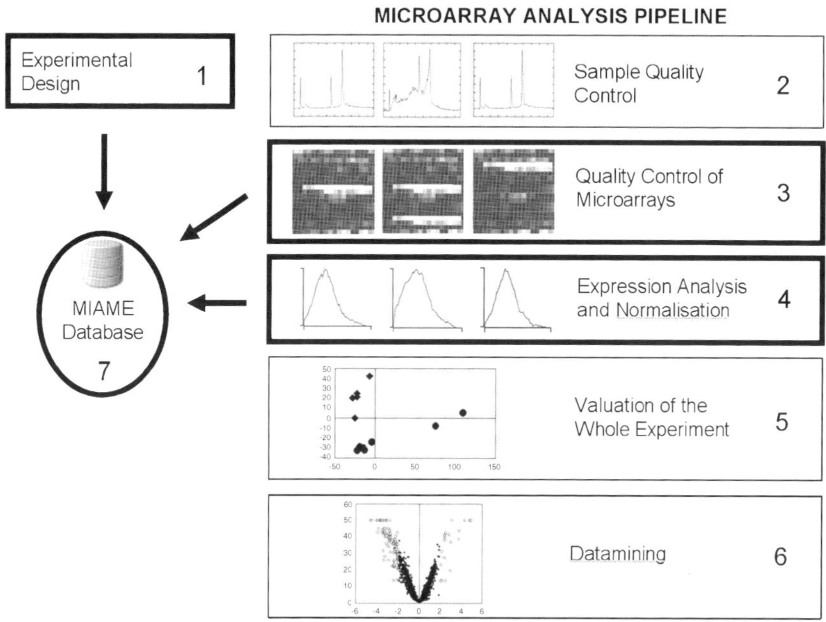

Figure 5. A pipeline of microarray analysis as described in this chapter. Steps outlined in bold represent information required for a MIAME-compliant submission to public repositories.

(6) High-Level Analysis: Data Mining
(7) Recording the experiment-MIAME

Following on from a description of each of these steps is a worked example. Steps (1) and (2) have both been described in the preceding sections and so will not be considered in further detail. In this next section we concentrate on step (3) onwards.

A. Quality Control of Microarrays

This section describes some of the tools and metrics available for assessing technical performance of the arrays in a dataset. Similar performance of arrays may be more important than meeting ideal target values for these metrics. This is because quantification is relative, as explained in the introduction.

1. dChip array summary file

An extremely valuable indicator of outliers in a dataset has been incorporated in the software dChip (Li & Wong, 2001) which is available free of charge to academic users (http://www.dchip.org/). Following the "Normalize" and "Model-based Expression" steps, dChip generates an array summary file which indicates array outliers in a dataset. A multi-array analysis is performed by comparing each probe set of each array. This is done by generating a profile of probe intensities for each probe set and then comparing this to the consensus probe response profile seen in most arrays of the dataset. An array is considered to be an outlier if more than 5% of the probe sets for that array are judged as outliers by the dChip outlier detection algorithm. In the example dataset shown in Table 1, four arrays have been flagged as outliers (*). This method is very effective in detecting aberrations in a variety of array parameters. Outlier detection by dChip correlates well with PCA (see Section 5 below).

2. Affymetrix report files

Affymetrix platforms include GCOS or MAS software that can generate report files for each array. Information contained in the report file for an array (.RPT file extension) gives several measures of quality as described in the Affymetrix MAS 5.0 User's Guide. Commonly used indicators are the 3′/5′ ratio of housekeeping controls, background level, and probe sets called present (P).

- 3′/5′ ratio of housekeeping controls: Affymetrix arrays, including the yeast YG_S98 GeneChip® used in this sample data, are designed as probe sets composed of 12–16 probes depending on array type (16 for YG_S98). In addition to the conventional probe sets designed to be within the most 3′ 600 base pairs of a transcript, additional probe sets in the 5′ region and middle portion of the transcript have also been selected for the yeast "housekeeping genes" *SRB4*, *SPT15* and *ACT1*. Signal intensity ratio of the 3′ probe set over the 5′ probe set is referred to as the 3′/5′ ratio. This ratio gives an indication of the integrity of your starting RNA, efficiency of amplification reactions, first-strand cDNA synthesis, and/or *in vitro* transcription of cRNA. A ratio of more than 3 may indicate problems. However, this may be condition specific, so it is important to compare the 3′/5′ ratios of all the arrays in a given dataset. Data from the report files have been included in Table 1. From the Table 1 data, it is clear that the report file quality metrics do not always match outliers as judged by dChip. High 3′/5′ ratios are an indicator of poor performance. However, a moderately high ratio of 5 for example may or may not indicate an outlier array.
- Background level: To calculate background the array is divided into 16 quadrants and background calculated for each. In this way an average and standard deviation of background can be calculated for an array.
- Probe sets are called present (P), marginal (M) or absent (A) by a statistical analysis of all the probes making up a probe set. A probe set will be called absent if it has low or erratic probe intensities.

Some computational tools exist for collecting and summarising these QC values from data files such as ArrayAssist Lite (Stratagene) and Simpleaffy (Wilson and Miller, 2005).

Table 1. Quality control data for Affymetrix YG_S98 GeneChips®. Data from an array summary file generated by dChip and Affymetrix GCOS report files (shaded regions)

Array	P(%)	Avg Background	Std Background	3′/5′ Actin	3′/5′ SPT15	3′/5′ SRB4	Median Intensity (un-normalized)	P(%)	% Array outlier	% Single outlier	Warning
C1	81.1	65.47	0.66	1.87	4.43	7.39	109	88.8	0.503	0.042	
C2	80.9	81.40	0.76	1.41	3.35	3.66	179	87.4	0.728	0.030	
C3	64.7	85.14	3.91	1.88	4.15	2.35	176	71.8	9.898	0.341	*
C4	84.7	62.10	0.39	1.45	3.29	3.05	114	91.2	0.589	0.399	
C5	58.4	76.49	2.28	2.94	2.84	2.98	130	67.2	15.779	0.429	*
C6	81.9	75.30	1.18	1.61	2.67	3.63	168	88.4	0.546	0.023	
N7	75.3	74.42	1.31	1.81	4.06	3.25	139	83.6	0.343	0.009	
N8	90.8	76.79	1.10	9.30	4.94	2.51	139	95.1	14.826	0.719	*
N9	78.8	86.52	3.07	1.40	5.14	3.99	167	85.9	0.043	0.004	
N10	69.6	197.81	9.15	1.66	5.30	3.65	521	70.6	1.125	0.245	
N11	74.6	77.64	1.22	1.72	3.72	4.07	147	83.0	0.075	0.005	
N12	67.9	73.56	0.49	1.55	3.08	2.78	123	76.8	0.161	0.021	
N13	87.7	78.83	1.17	54.59	19.52	8.49	190	92.3	10.455	0.233	*

B. Low-Level Analysis-Expression Analysis and Normalization

As shown in Figure 6, expression analysis is the summarization process determining the value for a probe set based on the range of signal intensities of the probes that make up that probe set. Normalization is the process of removing systematic bias between arrays (Quackenbush, 2002), correcting for total intensity differences between arrays that occur due to differences in labelling efficiency etc. (Figure 7A). Expression analysis and normalization have a significant impact on values obtained from arrays as shown in the scatter plots in Figure 7B for four commonly used methods. Expression analysis and normalization of Affymetrix GeneChip® arrays are actively researched areas and new algorithms to perform these tasks are often implemented in the open source R software environment (http://www.r-project.org/) and incorporated into the Bioconductor project (http://www.bioconductor.org/). The following are examples of Bioconductor scripts for doing expression analysis and normalization:

Start by choosing the working directory containing CEL files (**File, Change dir...**)

library(affy) # Loads affy package.

Data <- ReadAffy() # Reads all CEL data in working directory and stores them in list object "data".

eset <- rma(Data) # Creates expression values using RMA method. The generated data are stored in object eset in standard exprSet format.

write. Exprs(eset, file = "RMA. Txt") # Writes expression values to text file in working directory.

eset <- mas5(Data) # Creates expression values using MAS5 method.

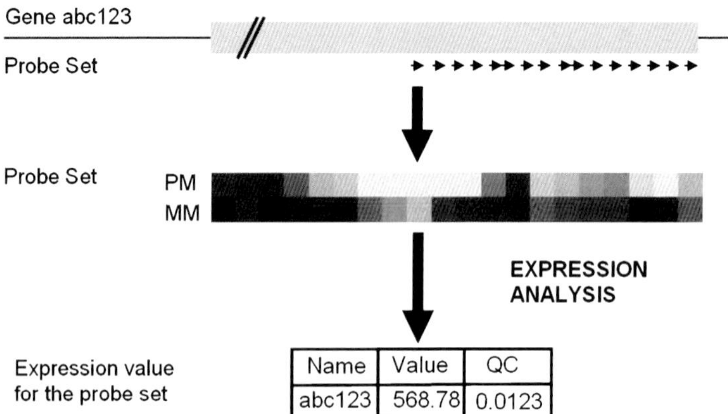

Figure 6. Schematic view of expression analysis of Affymetrix GeneChip® data showing the summarization of 16 probes to a single expression value. The perfect match probes (PM) are 25-mers complementary to the transcript whereas the mismatch probes (MM) are identical to the PM but with a homomeric mismatch at the central position 13.

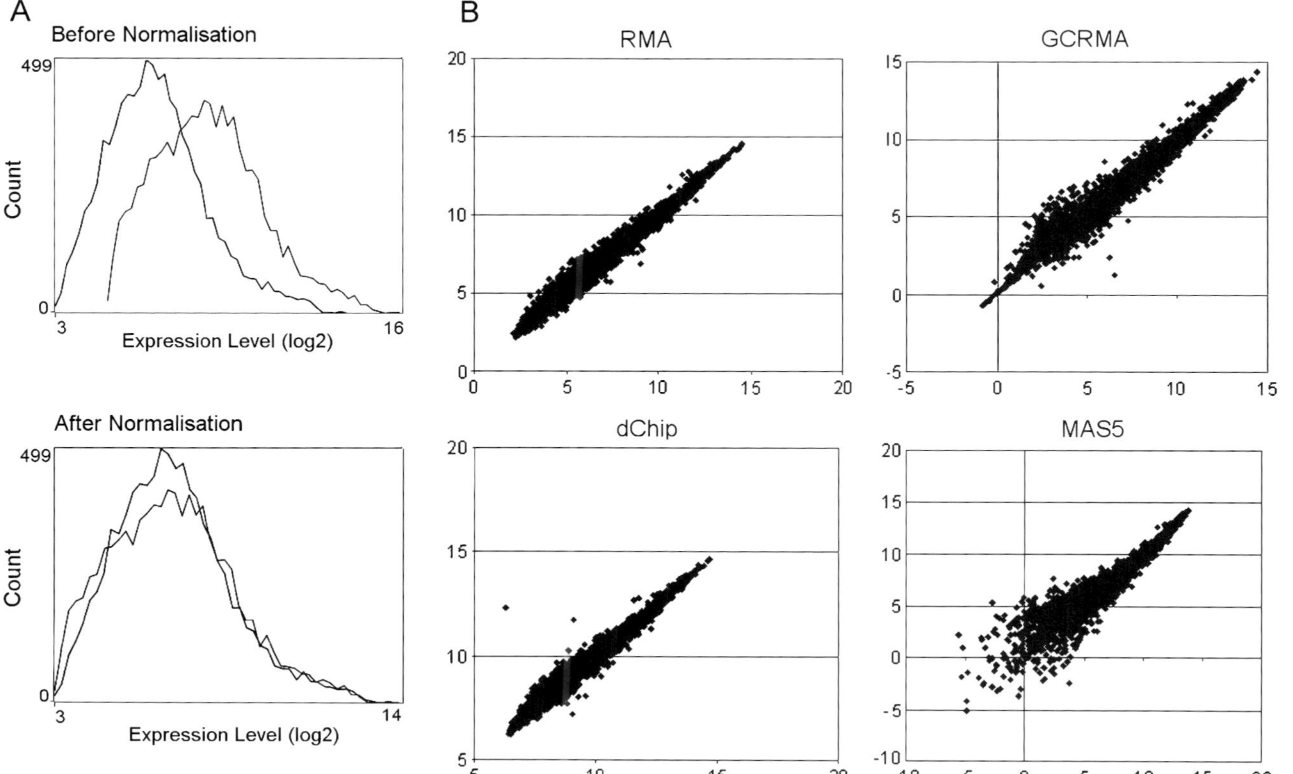

Figure 7. (A) Normalization of two arrays. The plots show the distribution of the data and correction for total intensity differences between arrays by making the mean expression level the same. (B) Scatter plots of probeset intensities (log_2) of replicate arrays of YG_S98 Affymetrix array. The arrays are C2 and C4 from Table 1 and are technical replicates (the same biological sample). Normalisztion was performed in the R software package using four commonly used summarization methods: RMA (Bolstad *et al.*, 2003), GCRMA (Wu *et al.*, 2004), dChip (Li and Wong, 2001) and MAS5 (Hubbell *et al.*, 2002).

write. Exprs(eset, file = "mas5. Txt")

eset <- expresso(Data, normalize. Method = "invariantset", bg. Correct = FALSE, pmcorrect. Method = "pmonly", summary. Method = "liwong") # Generates expression calls similar to dChip (MBEI) method from Li and Wong.

write. Exprs(eset, file = "dchip. Txt")

library(gcrma) # Loads GCRMA package.

eset <- rma(Data) # Creates expression values using GCRMA method.

write. Exprs(eset, file = "GCRMA. Txt")

Although RMA (Bolstad *et al.*, 2003) is currently popular as a means of normalization, it is impossible to state categorically that one method is better than another since performance is dataset- and criteria-dependent (Bolstad *et al.*, 2003; Cope *et al.*, 2004; Seo *et al.*, 2004; Shedden *et al.*, 2005; Liu *et al.*, 2005). In general, all procedures take the .CEL files from the Affymetrix platform, perform expression analysis and normalization, and then generate a spreadsheet list of values for each probe set.

C. Valuation of the Whole Experiment

After performing technical quality control and normalization, it is important to ascertain if the experiment is able to produce valuable data. This involves taking the actual biology of the experiment into account. Is there a measurable difference between groups of treated samples vs. control samples? If there is no true difference, it is still likely that genes will be found that are "significantly changed" between these two conditions simply due to random error and the fact that we are performing 5000 measurements at once. Which genes are known to change under the experimental conditions studied? Looking at the performance of a few key genes can satisfy you that the experiment will produce valuable data instead of random noise. A more powerful approach is to use computational techniques which identify patterns in the data. PCA is particularly well suited to this task, although various other clustering methods can be used as well. PCA is a way of reducing the high dimensionality of microarray data into a smaller set of components representing the main similarities and differences in a dataset. The reduction in dimensionality makes it easier to understand the relationships between the samples. For example, consider the arrays in Table 1 that are plotted in Figure 8A and B. In Figure 8A, all the arrays are used and we see that the arrays flagged as outliers by dChip are also outliers in a plot of principal components 2 vs. 1. When we remove the outliers from the dataset and perform a PCA again we obtain Figure 8B, in which the biology (control vs. treatment) is the major source of variance in the dataset (since control and treated samples cluster separately on component 1). A result such as that in Figure 9B provides reassurance and it allows

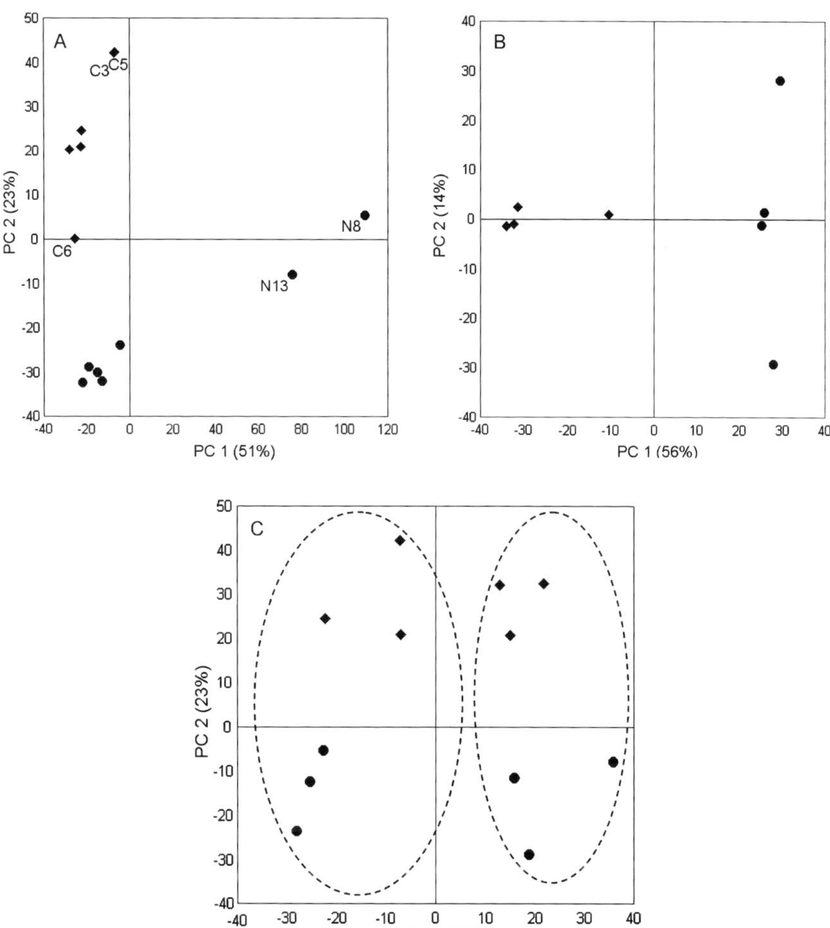

Figure 8. Principal component analysis (PCA) applied to microarray data. (A) PCA plots of the sample data performed with maxdView (free software from http://bioinf.man.ac.uk/microarray/maxd/). Data were normalized with RMA, imported into maxdView in log base 2 format (removing Affymetrix control probesets). PCA was performed with the SVD plug-in (note: PCA is available in R and most microarray software packages). (A): principal component (PC) 2 vs. PC1 with outlier arrays labelled. (B): PC2 vs. PC1 after removal of 4 outlier arrays indicated in Figure 9A. ●: C-limited samples; ♦: N-limited samples. (B) PCA analysis of an experiment where there is significant systematic bias. Arrays circled on the left were performed three months prior to the samples circled on the right.

continuation to the higher-level analysis stage with confidence that genes called significantly changed have, in fact, changed due to the biological conditions selected.

Performing a PCA can be very revealing if systematic bias has occurred in the execution of the microarray experiment. For example, instead of the experimental treatment appearing as the factor determining principal component 1, it is quite common to find the day the experiment was run or the person who extracted the RNA to be the factor determining the principal component, as shown in Figure 8C.

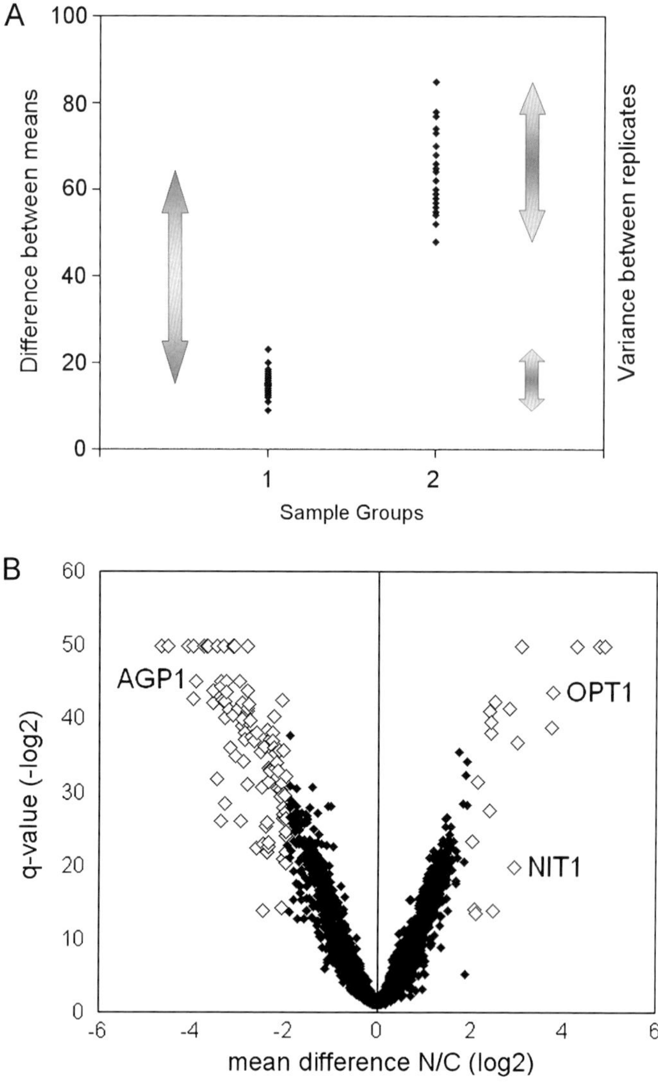

Figure 9. (A) Factors influencing statistical significance. The difference between means of each group is represented by the left arrow. The variances for replicate measurements within each group are represented by the arrows on the right. (B) Volcano plot with some selected relevant genes labelled. The plot was made with the sample data following *t*-test analysis and FDR correction (to get the final *q*-values) with the *q*-value software (Storey and Tibshirani, 2003). The x-axis shows the difference between the means in log base 2; the *y*-axis shows the *q*-value in -log base 2.

D. High-level Analysis: Data Mining

This last stage in the analysis pipeline is generally the first thing people imagine when first considering "microarray analysis". It is the process of extracting a list of genes whose expression has changed significantly under the conditions studied. It is currently an anarchic field with a huge selection of methods and software available. In this introduction, aimed at the average microarray user, we will attempt to point out general concepts, give specific recommendations and indicate further methods for extracting knowledge from microarray data. Despite the confusing proliferation of analysis methods, most high-level analyses can be summarized as follows:-

(1) Form a subset of genes that show statistically significant change
(2) Make clusters of this subset of genes based on expression patterns
(3) Find the biological significance of these genes or patterns

1. Statistical significance

To obtain the best results from this step in the analysis pipeline, some understanding of key statistical concepts is valuable. Statistical significance is usually judged by looking at the difference in means between the groups and the variance for replicate measurements within each group (Figure 9A). For microarrays, a suitable linear statistical test (such as the *t*-test or ANOVA) can be used. The simple microarray case study we have been using, a two-group study, could be analysed using the *t*-test. This test is called linear because it, and ANOVA, is based on the General Linear Model. Microarray data satisfies this model when it is in logarithmic space so the data should be log transformed before performing the test. The *t*-test analyzes genes independently of one another and generates a *p*-value to predict the statistical significance of change between the two sample groups (Figure 9A). To estimate variance, a reasonable number of replicates are required (eight or more). Given the cost of each microarray, this number of replicates is not within reach of the average small research laboratory. For this reason, statistical tests using other methods to estimate variance for a gene have been developed and out-perform the standard *t*-test e.g. cyberT (Baldi & Long, 2001), Limma (Smyth *et al.*, 2005), LPE (Jain *et al.*, 2003), SAM (Tusher *et al.*, 2001). Once the *t*-test has been performed and a False Discovery Rate (FDR) correction applied (see Section 3 below), it can be visualized with a volcano plot (see Figure 9B) to show the statistical distribution of the data. Ranking the data by statistical significance (*p*-value), as shown in Figure 10, is informative in showing how consistent the replicates are and what fraction of the results is significantly changed between conditions.

2. Other statistical tests

For more complex experimental designs an ANOVA test can be performed. For example, for a time or dose series (Figure 11A), a one-dimensional ANOVA test can be used. For a study with more than

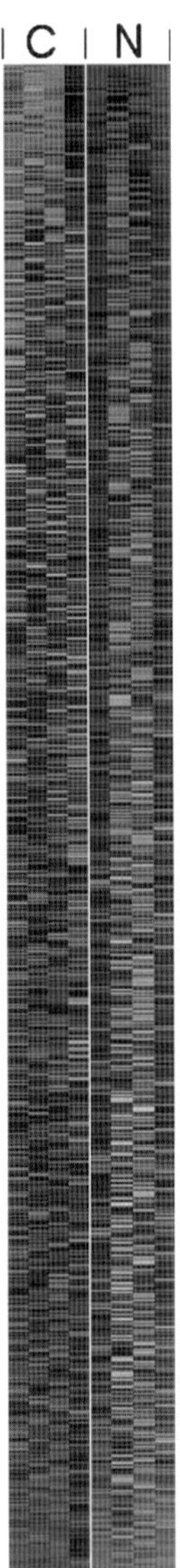

Figure 10. Ranking the sample data by statistical significance (*p*-value). For visualization purposes the data has been normalized by *z*-transformation (set the mean expression to zero and standard deviation to 1 for each gene in the dataset) and separated in terms of N- vs. C-limitation into up- (above) and down-regulated genes (below). These are colour coded in Red and Green respectively. (See color plate section).

one variable, for example, treatment and yeast strain (Figure 11B), a two-dimensional ANOVA could be used. The ANOVA test is preferred to several separate *t*-tests between conditions because all the data are combined in the estimate of variance when using an ANOVA. In addition, other tests (such as Wilcoxon's rank sums test) are preferred to the *t*-test by some analysts for their alternative method of assessing difference between groups of replicates (Figure 9A).

A

Time	0	10	20	30
Arrays	A B C	D E F	G H I	J K L

B

	Strain X	Strain Y
Control	A B C	D E F
Treatment	G H I	J K L

Figure 11. Complex experimental designs: (A) a time series (B) a study with more than one variable (treatment and yeast strain).

3. False discovery rate

As mentioned above, the *t*-test estimates statistical significance by treating each gene independently. However, once this is done we will want to form a subset of genes from our array that pass a threshold *p*-value of, say, 0.05. It is this step which leads to a multiple testing error (for a *p*-value of 0.05, a gene has an expected false prediction rate of 5% so 250 falsely predicted genes can be expected for that *p*-value threshold and an array measuring 5000 genes). This type of error, introduced by combining multiple test results, is dealt with by the fields of multiple testing error, FDR or family-wise error. The FDR of a set of predictions is the expected percent of false predictions in the set of predictions. Methods to estimate FDR do exist that correct for this type of error but they are often found to be too stringent for microarray data. Currently, bioinformatics journals regularly publish new methods for performing FDR corrections for microarray, and other high-throughput data-indicating that this is a rapidly evolving field. However, despite this lack of a standard method, it is important to be aware of this problem and use an FDR method when generating a list of significantly changed genes. In general, the FDR is very different from a *p*-value and, as such, a much higher FDR can be tolerated than with a *p*-value (an FDR of 0.5 might even be tolerable). FDR methods do not change the rank order in *p*-values but, rather, adjust the *p*-values depending on the full range of *p*-values found for the microarray dataset (Figure 12).

4. Removing unreliable probe set measurements

Once a suitable statistical test has been applied, there is a further step needed to remove genes with unreliably low signal intensities. This addresses the problem of gene expression at levels too low for the microarray to measure properly. The signal intensity is then close to background noise. Genes of this type sometimes pass a statistical test as "significantly changed" due to good consistency in the background noise rather than their own expression level. In the MAS5.0 approach to expression analysis, present (P) and absent (A) calls are generated based on a statistical analysis of detection of the

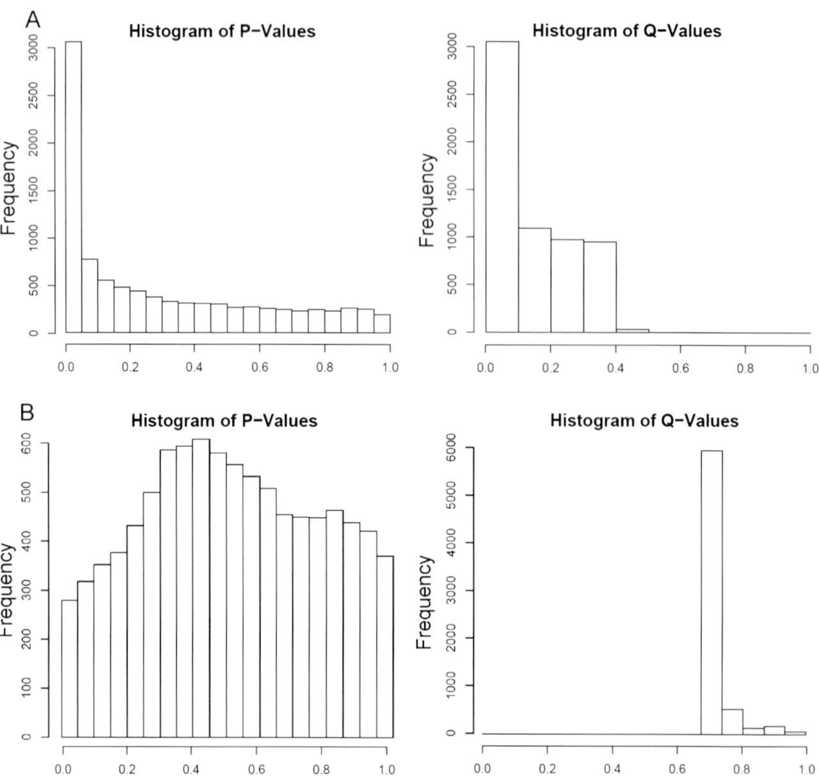

Figure 12. Histograms showing the effect of FDR correction of statistical significance by the *q*-value software (Storey and Tibshirani, 2003). (A) A dataset with many genes showing true differential expression. (B) A dataset with no genes showing true differential expression. Some genes are assigned a statistically significant *p*-value simply due to random noise over thousands of measurements. FDR correction detects and corrects this by assessing the range of *p*-values in the dataset.

probes making up a probe set. However, in practice, simply excluding all "Absent" probe sets in a dataset becomes very messy when there is a mixture of P and A calls between replicates for a particular probe set. Model-based normalization methods such as RMA handle the data in a different way and (as shown in Figure 7B) these methods give more reliability in assessing measurements at the low end of the expression scale, as shown by the better linearity of the scatter plots. However, there should still be little confidence in a measurement close to background noise. Again we are faced with an aspect of microarray analysis with no standard approach. Approaches using the MAS5 detection *p*-value have been proposed (Seo *et al.*, 2004). A simpler approach is to remove genes from a list with expression levels similar to background. "Similar" could be defined as less than twice the standard deviation of the background. From Table 1, we can see that the background and standard deviation of background is array-dependent, so an average or approximation could be taken.

At this stage, we reach a turning point in the analysis. We have done several quality control checks and now have a list of genes that we can say with confidence are changed under the conditions we are studying (Table 2). If we have a broad knowledge of the genes and their function, simply looking through this list will result in valuable conclusions. In practice, understanding the biological significance of microarray data is a large part of the analysis effort. Fortunately, additional computational methods exist to assist this process and are described below.

5. Clustering

Once a set of significantly changed genes has been assembled the next step is clustering. For our simple worked example, with two sample groups, there are really only two classes, up- or down-regulated by treatment. However, for time course experiments or multi-factorial designs, genes can be grouped into clusters based on their expression profiles such as those shown in Figure 13 (Eisen *et al.*, 1998; Tibshirani *et al.*, 2002; de Hoon *et al.*, 2004). Before clustering, it is advisable to reduce the dataset to significantly changed genes using linear statistical methods as explained above, rather than using all genes in the array (Quackenbush, 2002). Reducing the size of the dataset reduces the time it takes to run clustering algorithms but, more importantly, it improves accuracy by removing genes that do not show significant differential expression. It is advisable to normalize by z-transformation (set the mean expression to zero and standard deviation to 1 for each gene in the dataset) prior to clustering in order reduce intensity difference effects. There are a large number of clustering methods and software available. However, unless one has expert assistance, it is worth staying with tried and tested approaches such as k-means, hierarchical clustering or self organising maps. Some of the criticisms of clustering are (a) that the user usually has to input (guess) the number of clusters the algorithm must find prior to a run, and (b) given the random seeding of clusters at the start of each run, it is almost impossible to achieve exactly the same result twice.

Although not a clustering method, a simple mathematical technique exists for finding genes with similar expression profiles to a specific target gene. This can be implemented using the "Profile Filter" within maxdView (free software available from http://bioinf.man.ac.uk/microarray/maxd/). This method is particularly useful for identifying functionally related genes when one has a large dataset from diverse experimental conditions (Brown *et al.*, 2005).

6. Gene ontology and other annotation tools

All the hard work above leads up to the point in the analysis pipeline where the biological significance of the results can be studied.

Table 2. Abbreviated list of significantly changed genes in the sample data

Probe set	ORF	Description	Name	q-val	fc (N/C)[a]
10854_at	YJR152W	allantoate_permease	DAL5	0.000	13.53
7546_at	YPR194C	peptide_transporter	*OPT2*	0.000	8.02
11260_at	YJL212C	peptide_transporter | glutathione_transporter	*OPT1*	0.000	5.46
11211_at	YJL172W	carboxypeptidase_yscS	*CPS1*	0.000	4.48
5257_at	YGL256W	alcohol_dehydrogenase_isoenzyme_IV	*ADH4*	0.000	4.27
9283_at	YMR318C	medium_chain_alcohol_dehydrogenase	*ADH6*	0.000	4.12
4237_at	YIL165C	nitrilase	*NIT1*	0.000	2.81
4238_at	YIL164C	nitrilase	*NIT1*	0.000	2.66
4064_at	YIR028W	allantoin_permease	*DAL4*	0.002	2.62
10934_at	YJR095W	succinate-fumarate_transport_protein	*SFC1*	0.000	−5.12
4568_at	YHL040C	Transporter,_member_of_the_ARN_family	*ARN1*	0.000	−5.2
10117_at	YLR225C	Hypothetical ORF	—	0.000	−5.78
6128_f_at	YDR342C	hexose_transporter	*HXT6*	0.000	−7.03
9444_at	YMR174C	inhibitor_of_proteinase_Pep4p	*PAI3*	0.000	−7.25
6907_at	YCL025C	amino_acid_permease	*AGP1*	0.000	−9.08
10128_at	YLR193C	Hypothetical ORF	—	0.000	−10.33
7295_at	YBR072W	heat_shock_protein_26	*HSP26*	0.000	−11.06
7253_g_at	YBR116C	transketolase,_similar_to_TKL1	*TKL2*	0.000	−15.1
8599_at	YOL053C-A	Multistress_response_protein	*DDR2*	0.000	−15.75
9445_at	YMR175W	Salt-Induced_Protein	*SIP18*	0.000	−25.33

[a]fc (N/C); fold change in expression level under nitrogen- *vs.* carbon-limitation conditions.

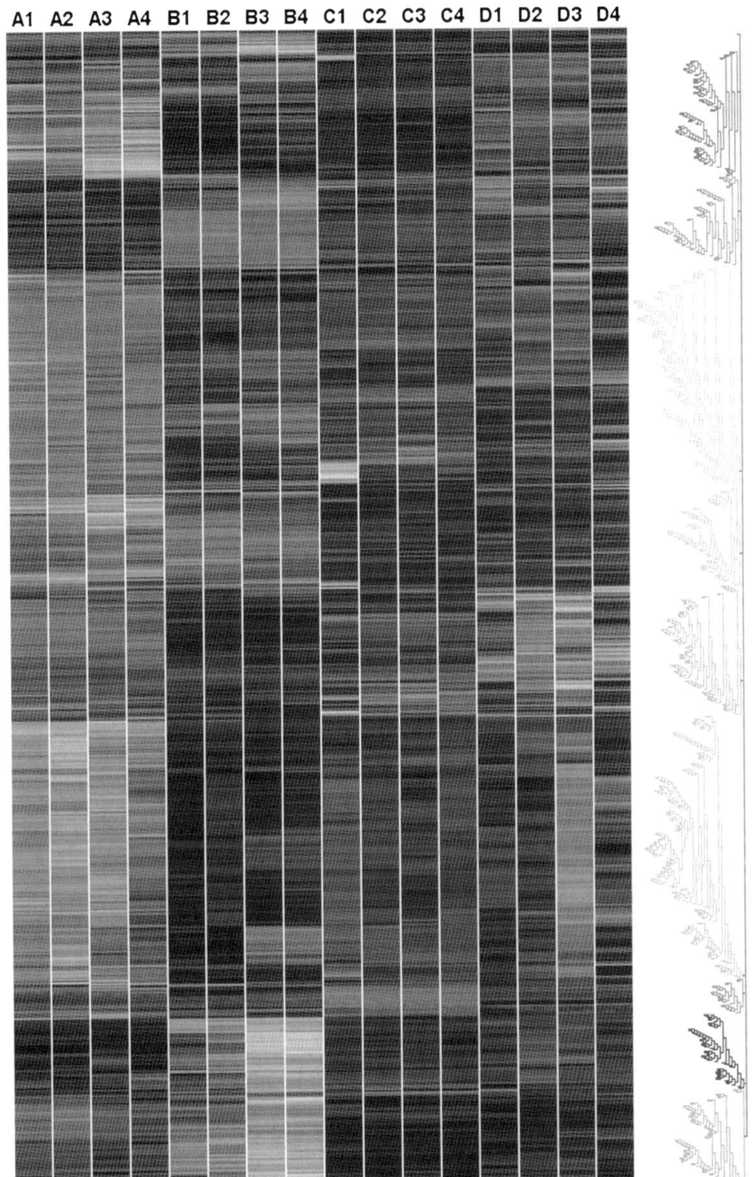

Figure 13. Clustering analysis of data from an experiment similar to that shown in Figure 14A. Following a one-way ANOVA analysis, the data were filtered down to 800 genes by statistical significance (p-value), fold change (fold change of 2 or more in any 2-way comparison of conditions) and expression level (mean expression level above 100). The $\log_2$ data values were normalized by z-transformation (set the mean expression to zero and standard deviation to 1 for each gene in the dataset) prior to clustering by the k-means method (performed in maxdView: http://bioinf.man.ac.uk/microarray/maxd/ using the XCluster plug-in http://genetics.stanford.edu/~sherlock/cluster.html). (See color plate section).

The Gene-Ontology database (GO: http://www.geneontology.org) provides a useful tool to annotate and analyse the function of large numbers of genes using a method GO term enrichment or over-representation. Briefly, the GO project provides a controlled vocabulary to describe gene and gene product attributes in any organism. By comparing the prevalence of GO categories in a subset of genes to their prevalence in the array as a whole, a list of biological attributes highly represented in that subset of genes can be found. This is a computationally intensive task involving in excess of 17 000 GO terms and 5000 genes. However, statistical significance can be calculated and graphical tools have been developed to help visualize results (Figure 14). Several free tools can be recommended that will analyse yeast data. The online tool GOstat is fast and includes FDR correction (Beissbarth and Speed, 2004), the visually attractive GenMAPP (Doniger *et al.*, 2003) is extremely useful for browsing GO categories and the software must be installed on a PC, as does GoMiner which also has a high-throughput online version available (Zeeberg *et al.*, 2005). The SGD (http://www.yeastgenome.org/) is responsible for most of the yeast GO annotation and has an online tool for GO analysis as well as a wealth of other essential annotation information such as gene homologues and pathways. The online NetAffx™ Analysis Centre from Affymetrix is another useful resource for correlation of GeneChip® array results with array design and annotation information (http://www.affymetrix.com/).

7. Worked example

Taking the worked example above, the following sequence of steps could be followed to satisfy the analysis factors we have explained:

(1) Perform RMA expression analysis using R or RMAExpress (*http://stat-www.berkeley.edu/users/bolstad/RMAExpress/RMAExpress.html*) on the eight arrays judged to be good in steps 1-3 above.
(2) Remove Affymetrix control probesets and antilog the data using Microsoft Excel or maxdView (the next step, cyberT, assumes the data is natural scale). Save the data in a tab-delimited text file format.
(3) Perform cyberT using the online tool (http://visitor.ics.uci.edu/genex/cybert/).
(4) FDR correction can be performed using the q-value software (Storey and Tibshirani, 2003). Create a subset of significantly changed genes using a q-value threshold of 0.1.
(5) Filter out low intensity genes by removing genes where the group mean is less than 100 (6.64 $\log_2$) in both control and treated samples using maxdView.
(6) Create two clusters of genes from this subset by separating up- and down-regulated genes (Table 2).

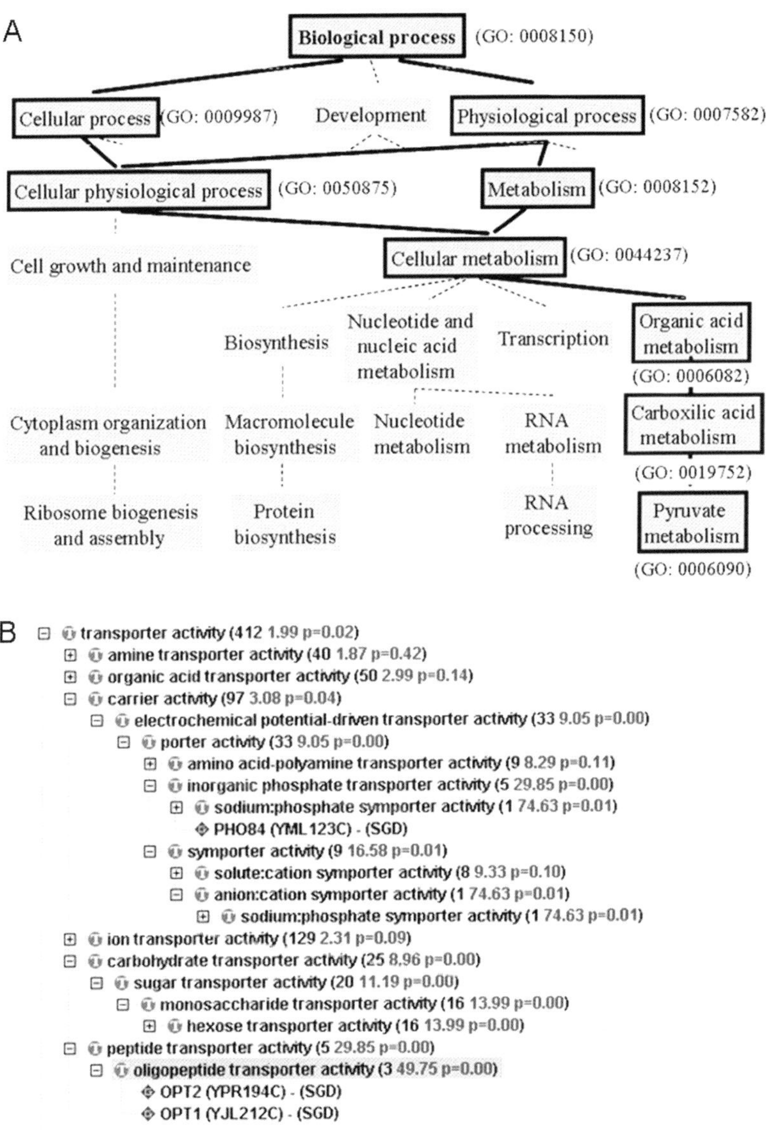

Figure 14. (A) Gene ontologies of *S. cerevisiae PDC1* gene (bold lines), integrated into a schematic gene ontology (GO) diagram of the biological process category (http://www.geneontology.org). (B) Example output of significantly over-represented GO terms using GoMiner.

8. Recording the experiment – MIAME

Given the exquisite sensitivity of yeast and all organisms to their environment, any transcript analysis experiment can only be repeated if recorded in detail. In addition, as we have seen, all the sample handling and analysis procedures can influence the results. MIAME describes the Minimum Information About a Microarray Experiment that is needed to enable the interpretation of the results of the experiment unambiguously and potentially to reproduce the experiment (Brazma *et al.*, 2001). It is currently a standard to which a microarray experiment must be recorded prior to submission to many highly-rated scientific journals (Ball *et al.*, 2002a, 2002b, 2002c). The microarray experiment needs to be submitted in a MIAME-compliant format to a public repository such as ArrayExpress (Parkinson *et al.*, 2005; *http://www.ebi.ac.uk/arrayexpress*) or Gene Expression Omnibus (*http://www.ncbi.nlm.nih.gov/projects/geo/*). The steps up to, and including, the expression analysis are all that need be recorded for a MIAME-compliant submission. The steps in the analysis pipeline following on from this do not make up part of the information required. It is anticipated that, in future, computational tools will be developed to interrogate microarray data stored in repositories. For this reason, and in an attempt to facilitate automatic submission of MIAME-compliant data to repositories, a markup language has been designed called MAGE-ML (Spellman *et al.*, 2002; Spellman, 2005). As with so many areas in microarray analysis, this field of recording the data and metadata (i.e. the data about the data) of a microarray experiment is in an early stage of development. Several tools are in development to facilitate storage of microarray data: MIAMExpress (*http://www.ebi.ac.uk/miamexpress/*); MADAM (*http://www.tm4.org/madam.html*); BioArray Software Environment (BASE: Saal *et al.*, 2002); The Longhorn Array Database (LAD: Killion *et al.*, 2003); MARS: Microarray analysis, retrieval, and storage system (Maurer *et al.*, 2005); maxdLoad2 and maxdBrowse (Hancock *et al.*, 2005). Of these, MIAMExpress is arguably the easiest place to start for a one-off submission to a public repository.

♦♦♦♦♦♦ V. SUMMARY AND FUTURE PERSPECTIVES

We have described a pipeline for the analysis of yeast gene expression using microarrays. This field is still rapidly developing. With the advent of newer and more powerful tools, the potential for this technology remains great and certainly the last word has yet to be said on many aspects of microarray analysis. In this chapter, a commercial microarray format was used to provide examples with which to illustrate our approach, but the concepts we have described are not limited to that microarray platform. Moreover, wherever possible, the software tools recommended are free (to academic groups) and open-source. The aim of this chapter has been to provide a generic approach to the microarray analysis of gene expression in yeast.

Acknowledgements

Leanne Wardleworth is gratefully thanked for technical assistance. Work described in this chapter was supported by grants from the Wellcome Trust, the Environmental Genomics theme of the Natural

Environment Research Council and the Investigating Gene Function initiative of the Biotechnology and Biological Sciences Research Council.

References

Baldi, P. and Long, A. D. (2001). A Bayesian framework for the analysis of microarray expression data: Regularized t-test and statistical inferences of gene changes. *Bioinformatics* **17**, 509–519.

Ball, C. A., Sherlock, G., Parkinson, H., Rocca-Sera, P., Brooksbank, C., Causton, H. C., Cavalieri, D., Gaasterland, T., Hingamp, P., Holstege, F. *et al.* (2002a). A guide to microarray experiments – an open letter to the scientific journals. *Lancet* **360**, 1019.

Ball, C. A., Sherlock, G., Parkinson, H., Rocca-Sera, P., Brooksbank, C., Causton, H. C., Cavalieri, D., Gaasterland, T., Hingamp, P., Holstege, F. *et al.* (2002b). An open letter to the scientific journals. *Bioinformatics* **18**, 1409.

Ball, C. A., Sherlock, G., Parkinson, H., Rocca-Serra, P., Brooksbank, C., Causton, H. C., Cavalieri, D., Gaasterland, T., Hingamp, P., Holstege, F. *et al.* (2002c). Standards for Microarray data. *Science* **298**, 539.

Beissbarth, T. and Speed, T. P. (2004). GOstat: Find statistically overrepresented Gene Ontologies within a group of genes. *Bioinformatics* **20**, 1464–1465.

Bolstad, B. M., Irizarry, R. A., Astrand, M. and Speed, T. P. (2003). A comparison of normalization methods for high density oligonucleotide array data based on variance and bias. *Bioinformatics* **19**, 185–193.

Brazma, A., Hingamp, P., Quackenbush, J., Sherlock, G., Spellman, P., Stoeckert, C., Aach, J., Ansorge, W., Ball, C. A., Causton, H. C. *et al.* (2001). Minimum information about a microarray experiment (MIAME)-toward standards for microarray data. *Nat. Genet.* **29**, 365–371.

Brown, D. M., Zeef, L. A. H., Ellis, J., Goodacre, R. and Turner, S. R. (2005). Identification of novel genes in *Arabidopsis* involved in secondary cell wall formation using expression profiling and reverse genetics. *Plant Cell* **17**, 2281–2295.

Cope, L. M., Irizarry, R. A., Jaffee, H. A., Wu, Z. J. and Speed, T. P. (2004). A benchmark for Affymetrix GeneChip expression measures. *Bioinformatics* **20**, 323–331.

de Hoon, M. J. L., Imoto, S., Nolan, J. and Miyano, S. (2004). Open source clustering software. *Bioinformatics* **20**, 1453–1454.

Doniger, S. W., Salomonis, N., Dahlquist, K. D., Vranizan, K., Lawlor, S. C. and Conklin, B. R. (2003). MAPPFinder: Using Gene Ontology and GenMAPP to create a global gene-expression profile from microarray data. *Genome Biol.* **4**.

Eisen, M. B., Spellman, P. T., Brown, P. O. and Botstein, D. (1998). Cluster analysis and display of genome-wide expression patterns. *Proc. Natl. Acad. Sci. USA* **95**, 14863–14868.

Goffeau, A., Barrell, B. G., Bussey, H., Davis, R. W., Dujon, B., Feldmann, H., Galibert, F., Hoheisel, J. D., Jacq, C., Johnston, M. *et al.* (1996). Life with 6000 genes. *Science* **274**, 546–567.

Hancock, D., Wilson, M., Velarde, G., Morrison, N., Hayes, A., Hulme, H., Wood, A. J., Nashar, K., Kell, D. B. and Brass, A. (2005). maxdLoad2 and maxdBrowse: Standards-compliant tools for microarray experimental annotation, data management and dissemination. *BMC Bioinformatics* **6**, 264.

Hauser, N. C., Vingron, M., Scheideler, M., Krems, B., Hellmuth, K., Entian, K. D. and Hoheisel, J. D. (1998). Transcriptional profiling on all open reading frames of *Saccharomyces cerevisiae*. *Yeast* **14**, 1209–1221.

Hayes, A., Zhang, N. S., Wu, J., Butler, P. R., Hauser, N. C., Hoheisel, J. D., Lim, F. L., Sharrocks, A. D. and Oliver, S. G. (2002). Hybridization array technology coupled with chemostat culture: Tools to interrogate gene expression in *Saccharomyces cerevisiae*. *Methods* **26**, 281–290.

Hereford, L. M. and Rosbash, M. (1977). Number and distribution of polyadenylated RNA sequences in yeast. *Cell* **10**, 453–462.

Hubbell, E., Liu, W. M. and Mei, R. (2002). Robust estimators for expression analysis. *Bioinformatics* **18**, 1585–1592.

Imbeaud, S., Graudens, E., Boulanger, V., Barlet, X., Zaborski, P., Eveno, E., Mueller, O., Schroeder, A. and Auffray, C. (2005). Towards standardization of RNA quality assessment using user-independent classifiers of microcapillary electrophoresis traces. *Nucleic Acids Res* **33**, e56.

Jain, N., Thatte, J., Braciale, T., Ley, K., O'Connell, M. and Lee, J. K. (2003). Local-pooled-error test for identifying differentially expressed genes with a small number of replicated microarrays. *Bioinformatics* **19**, 1945–1951.

Killion, P. J., Sherlock, G. and Iyer, V. R. (2003). The Longhorn array database (LAD): An open-source, MIAME compliant implementation of the Stanford microarray database (SMD). *BMC Bioinformatics* **4**, 32.

Li, C. and Wong, W. H. (2001). Model-based analysis of oligonucleotide arrays: Expression index computation and outlier detection. *Proc. Natl. Acad. Sci. USA* **98**, 31–36.

Li, T., Zhang, C. L. and Ogihara, M. (2004). A comparative study of feature selection and multiclass classification methods for tissue classification based on gene expression. *Bioinformatics* **20**, 2429–2437.

Liu, J. J., Cutler, G., Li, W. X., Pan, Z., Peng, S. H., Hoey, T., Chen, L. B. and Ling, X. F. B. (2005). Multiclass cancer classification and biomarker discovery using GA-based algorithms. *Bioinformatics* **21**, 2691–2697.

Mattick, J. S. (2004). RNA regulation: A new genetics?. *Nat. Rev. Genet.* **5**, 316–323.

Mattick, J. S. and Gagen, M. J. (2005). Accelerating networks. *Science* **307**, 856–858.

Maurer, M., Molidor, R., Sturn, A., Hartler, J., Hackl, H., Stocker, G., Prokesch, A., Scheideler, M. and Trajanoski, Z. (2005). MARS: Microarray analysis, retrieval, and storage system. *BMC Bioinformatics* **6**, 101.

Parkinson, H., Sarkans, U., Shojatalab, M., Abeygunawardena, N., Contrino, S., Coulson, R., Farne, A., Lara, G. G., Holloway, E., Kapushesky, M. *et al.* (2005). ArrayExpress – a public repository for microarray gene expression data at the EBI. *Nucleic Acids Res.* **33**, 553–555.

Pochet, N. L. M. M., Janssens, F. A. L., De Smet, F., Marchal, K., Suykens, J. A. K. and De Moor, B. L. R. (2005). MACBETH: A microarray classification benchmarking tool. *Bioinformatics* **21**, 3185–3186.

Quackenbush, J. (2002). Microarray data normalization and transformation. *Nat. Genet.* **32**, 496–501.

Saal, L., Troein, C., Vallon-Christersson, J., Gruvberger, S., Borg, A. and Peterson, C. (2002). BioArray Software Environment (BASE): A platform for comprehensive management and analysis of microarray data. *Genome Biol.* **3,** software0003.

Seo, J., Bakay, M., Chen, Y. W., Hilmer, S., Shneiderman, B. and Hoffman, E. P. (2004). Interactively optimizing signal-to-noise ratios in expression profiling: project-specific algorithm selection and detection p-value weighting in Affymetrix microarrays. *Bioinformatics* **20**, 2534–2544.

Shedden, K., Chen, W., Kuick, R., Ghosh, D., Macdonald, J., Cho, K. R., Giordano, T. J., Gruber, S. B., Fearon, E. R., Taylor, J. M. G. and Hanash, S. (2005). Comparison of seven methods for producing Affymetrix expression scores based on false discovery rates in disease profiling data. *BMC Bioinformatics* **6**, 26.

Smyth, G. K., Michaud, J. and Scott, H. S. (2005). Use of within-array replicate spots for assessing differential expression in microarray experiments. *Bioinformatics* **21**, 2067–2075.

Spellman, P. (2005). A status report on MAGE. *Bioinformatics* **21**, 3459–3460.

Spellman, P., Miller, M., Stewart, J., Troup, C., Sarkans, U., Chervitz, S., Bernhart, D., Sherlock, G., Ball, C., Lepage, M. *et al.* (2002). Design and implementation of microarray gene expression markup language (MAGE-ML). *Genome Biol.* **3,** research0046.

Storey, J. D. and Tibshirani, R. (2003). Statistical significance for genome-wide studies. *Proc. Natl. Acad. Sci. USA* **100**, 9440–9445.

Tan, P. K., Downey, T. J., Spitznagel, E. L., Xu, P., Fu, D., Dimitrov, D. S., Lempicki, R. A., Raaka, B. M. and Cam, M. C. (2003). Evaluation of gene expression measurements from commercial microarray platforms. *Nucleic Acids Res.* **31**, 5676–5684.

Tibshirani, R., Hastie, T., Narasimhan, B. and Chu, G. (2002). Diagnosis of multiple cancer types by shrunken centroids of gene expression. *Proc. Natl. Acad. Sci. USA* **99**, 6567–6572.

Tusher, V. G., Tibshirani, R. and Chu, G. (2001). Significance analysis of microarrays applied to the ionizing radiation response. *Proc. Natl. Acad. Sci. USA* **98**, 5116–5121.

Velculescu, V. E., Zhang, L., Zhou, W., Vogelstein, J., Basrai, M. A., Bassett, D. E., Hieter, P., Vogelstein, B. and Kinzler, K. W. (1997). Characterization of the yeast transcriptome. *Cell* **88**, 243–251.

Wilson, C. L. and Miller, C. J. (2005). Simpleaffy: A BioConductor package for Affymetrix Quality Control and data analysis. *Bioinformatics* **21**, 3683–3685.

Wood, V., Gwilliam, R., Rajandream, M. A., Lyne, M., Lyne, R., Stewart, A., Sgouros, J., Peat, N., Hayles, J., Baker, S. *et al.* (2002). The genome sequence of Schizosaccharomyces pombe. *Nature* **415**, 871–880.

Wu, Z. J., Irizarry, R. A., Gentleman, R., Martinez-Murillo, F. and Spencer, F. (2004). A model-based background adjustment for oligonucleotide expression arrays. *J. Am. Stat. Assoc.* **99**, 909–917.

Zeeberg, B. R., Qin, H. Y., Narasimhan, S., Sunshine, M., Cao, H., Kane, D. W., Reimers, M., Stephens, R. M., Bryant, D., Burt, S. K. *et al.* (2005). High-throughput GoMiner, an 'industrial-strength' integrative gene ontology tool for interpretation of multiple-microarray experiments, with application to studies of Common Variable Immune Deficiency (CVID). *BMC Bioinformatics* **6**, 168.

10 GFP-based Microscopic Approaches for Whole Chromosome Analysis in Yeasts

Qi Gao[1], Tomoyuki U Tanaka[2] and Xiangwei He[1]
[1] *Department of Human and Molecular Genetics, Baylor College of Medicine, Houston, TX, USA;*
[2] *School of Life Sciences, University of Dundee, Dundee, UK*

◆◆◆

GFP-based Microscopic Approaches

CONTENTS

Introduction
Microscopic techniques
GFP-tagging schemes
Representative GFP-based assays

List of Abbreviations

CFP	Cyan Fluorescent Protein
FRET	Fluorescence Resonance Energy Transfer
GFP	Green Fluorescent Protein
MT	Microtubule
ORF	Open Reading Frame
RFP	Red Fluorescent Protein
SPB	Spindle Pole Body
YFP	Yellow Fluorescent Protein

◆◆◆◆◆◆ I. INTRODUCTION

Microscopy has been a fittingly powerful approach for studying mitosis, which consists of a series of visually spectacular events. Yeasts (here we limit our discussion only to the budding yeast *Saccharomyces cerevisiae* and the fission yeast *Saccharomyces pombe*) have long been utilised as the model organisms for the study of mitosis, especially by genetic and biochemical approaches. The

METHODS IN MICROBIOLOGY, VOLUME 36
0580-9517 DOI:10.1016/S0580-9517(06)36010-2

microscopic approach, which has also contributed remarkably to our understanding of yeast mitosis, was nonetheless limited by the size of the yeast nucleus. However, recent developments in green fluorescent protein (GFP) tagging techniques as well as in microscope technology and image processing software have made it feasible to follow the whole process of chromosome segregation play-by-play, non-invasively, in live yeast cells. With suitable equipment and setup, we can now monitor individual major events, even several events simultaneously (or almost simultaneously). Quantitative measurement of the kinetic parameters is possible for events such as the motion of a single chromosome and spindle elongation. The combination of these new microscopic tools with the existing genetic and biochemical approaches has already led to exciting new insights into yeast mitosis and will no doubt make further significant contributions to our understanding of the molecular mechanisms of mitosis. This chapter aims to discuss some general practical concerns for GFP-based microscopy, as well as introducing some of the techniques that have been successfully used to monitor and quantify major mitotic events, in an order that roughly matches with the progression of mitosis.

A. Major Cellular Events During Mitosis in Budding and Fission Yeast

Both budding yeast and fission yeast undergo a closed mitosis, i.e., the nuclear envelope remains intact throughout the cell cycle. During cell division, sister chromatids are physically separated by the spindle. Interaction between the chromosomes and the spindle microtubules (MTs) is mediated through a specialised protein mega-complex called the kinetochore, which is assembled at the centromeric region of the chromosome. A series of events occur, that result in spindle assembly, chromosome segregation and subsequent nuclear division. As expected, these events are controlled by the cell cycle regulatory mechanisms, which further provide feedback to the central regulatory machinery via pathways such as the spindle assembly checkpoint. The precise and timely execution of these events ensures the high fidelity of equal chromosome segregation. Many of these mitotic events can now be directly monitored in live cells with proper GFP-based techniques. These techniques provide the means of functional assays for mutations in specific proteins that are involved in certain steps of mitosis. They also provide convenient tools to determine accurately the mitotic stage of an individual live cell. We here outline the major mitotic events in budding and fission yeasts. In the latter part of this chapter, we describe the corresponding assays for some of the mitotic events and the application of these assays.

1. Budding yeast

During the G1 phase of the cell cycle, a single spindle pole body (SPB) is anchored in the nuclear membrane. Short MTs are nucleated by SPB on both the nuclear and cytoplasmic sides. On the nuclear side, kinetochores are associated with the short MTs so that centromeres are localised in the vicinity of the SPB (see Knop *et al.*, 1999 for a review). Upon centromeric DNA replication in early S-phase, kinetochores disassemble and centromeres detach from MTs (Tanaka *et al.*, 2005). A unique feature of budding yeast mitosis is that the spindle begins to form during early S-phase, during which SPB duplication completes, resulting in side-by-side SPBs connected by a bridge (Knop *et al.*, 1999). Before the new SPB becomes operational, kinetochores are reassembled on both replicated sister centromeres and recaptured laterally by MTs (a single microtubule in many cases) extended mostly from the old SPB. Once captured, kinetochores are transported along MTs, predominantly towards spindle poles (Tanaka *et al.*, 2005). Later in S-phase, the new SPB becomes functional and starts nucleating spindle MTs. The kinetochores change their association with some of the MTs from the old SPB to those from the new one, and switch between them. At the end of S-phase, SPBs separate, which allows formation of a bipolar spindle. At the same stage, sister kinetochores establish bi-orientation on the spindle, attaching to MTs extended from opposing SPBs (Tanaka, 2005). After the establishment of bi-orientation and prior to anaphase, sister kinetochores oscillate along the spindle axis. Furthermore, a short region on the sister chromatids flanking the centromeres undergoes dynamic separation and re-association, termed ''transient separation'' or ''breathing''. At the onset of anaphase, triggered by the activation of the anaphase promotion complex (APC), which biochemically functions as a key component of the ubiquitination proteolysis machinery, sister kinetochores are irreversibly separated and move to the vicinity of the poles in anaphase A (Humphrey and Pearce, 2005). Rapid elongation of the spindle in anaphase B contributes most to the separation of sister chromatids (Winey and O'Toole, 2001). The new SPBs move into their respective daughter cells, and the mitotic exit network (MEN) is activated, which is necessary for cells to exit mitosis (McCollum and Gould, 2001).

2. Fission yeast

Compared to mitosis in budding yeast, several distinct aspects of mitosis in fission yeast are worth highlighting. In the fission yeast cell cycle, the G1- and S-phases are relatively short. G1 is usually cryptic in logarithmically dividing cultures of *S. pombe*, and S-phase directly follows completion of nuclear division, resulting in cells that are already in G2 when cytokinesis completes. Most of the fission yeast interphase is comprised of G2 (MacNeill and Nurse, 1997) whereas in the budding yeast cell cycle, there is a clear distinction between G1-phase and S-phase, but not between G2 and pro-metaphase (Pringle and Hartwell, 1981).

During interphase, fission yeast kinetochores are almost always tethered to the SPB anchored on the nuclear envelope, except perhaps for a short period during S-phase in which the centromeric DNA is replicated. The tethering to the SPB seems direct, without any MTs mediating the connection (Kniola *et al.*, 2001). SPB duplication is completed at the end of G2. Upon entry into mitosis, kinetochores are likely to be released from the SPBs (Petersen *et al.*, 2001; our unpublished observations). Nascent nuclear MTs radiate from the SPBs and quickly form a short-bipolar spindle, which then undergoes a short-initial phase of elongation (Nabeshima *et al.*, 1998). During spindle formation and initial elongation, kinetochores are re-captured by the spindle MTs and establish bi-orientation on the spindle. Oscillation and breathing of the kinetochores are also evident, although rather brief due to the short duration of metaphase. In anaphase, *S. pombe* experiences anaphase A and anaphase B in a similar manner to *S. cerevisiae*, resulting in chromosome segregation (Nabeshima *et al.*, 1998). Upon exit of mitosis, the spindle is disassembled, and cytoplasmic MTs re-form around the septum region.

♦♦♦♦♦♦ II. MICROSCOPIC TECHNIQUES

A. Some General Considerations for Choosing Suitable Microscopy Equipment

Chromosome segregation in yeast is a rapid, dynamic process: at certain stages, the chromosomes move at a velocity up to $\sim 0.3\,\mu m\,min^{-1}$ (Pearson *et al.*, 2001). Furthermore, mitosis in yeasts occurs in an enclosed nucleus that is spherical in shape with a diameter of about 2 μm prior to anaphase. These features of yeast mitosis impose high demand on the spatial resolution as well as the temporal resolution for the microscope and image acquisition hardware. A microscopic image is only two dimensional (2-D), but from a stack of such images acquired at different focal planes along the *z*-axis, a 3-D image can be reconstructed. The spatial resolution is almost always limited in the *z*-axis. It is thus essential to have a motorised stage control on the microscope that is capable of automated *z*-axis control with sufficient precision.

During time-lapse imaging, multiple stacks of images will be taken on the same living cell. Adverse side effects of repeated illumination become a serious concern at the later time points. Two issues require particular attention: photo-toxicity, which disturbs the normal cell physiology, and photobleaching, which damages the fluorophore (GFP), and thus diminishes the desired signal. To minimise photo damaging, a highly sensitive CCD camera should be used so that the light exposure time of the specimen is minimised.

The kinetics of mitosis is affected by temperature (Nabeshima *et al.*, 1998). Often, a temperature-sensitive mutant is the subject of study. It is thus desirable to have a temperature control device for the microscope stage or ideally, for the whole microscope unit. Readers are referred to a recent review for an in-detail discussion of the microscope hardware (Rines *et al.*, 2002). Finally, it is crucial that the microscope is properly aligned and well maintained on a regular basis. It occurs all too often that a microscope cannot perform even close to its designed capability simply due to mis-handling and poor maintenance.

B. Cell Growth

To obtain a reliable image of good quality, the physiological state of the cell is critical. A healthy culture assures that the phenotypes observed under the microscope are due to genetic variation rather than environmental complications. Also, a healthy yeast culture can reduce to an acceptable level the background auto-fluorescence emitted due to the accumulation of metabolic by-products. This phenomenon is exceptionally significant in yeasts, particularly in strains that are Ade$^-$, since an intermediate in adenine biosynthesis, phosphoribosylamino-imidazole, accumulates in Ade$^-$ cells and is highly fluorescent in the GFP channel (Ishiguro, 1989; Stotz and Linder, 1990). To reduce the auto-fluorescence, an Ade$^-$ strain must be grown in medium containing 20 $\mu g\, ml^{-1}$ adenine.

A good culture for microscopy is grown in rich medium or minimum medium with essential supplements. In some cases where gene expression is under the regulation of a certain promoter (e.g. the *GAL* promoter in *S. cerevisiae* or the *nmt* promoter in *S. pombe*), special attention should be paid to the concentration of the regulator. Both over-expression and under-expression might result in unhealthy yeast cultures. Usually, it is better to prepare samples for microscopy in liquid culture rather than on agar plates so that all cells in the population are exposed to identical nutrient conditions. A benchmark guide for growing a yeast culture for microscopic observation is to dilute a fresh pre-culture, to allow the diluted culture to grow for at least four generation times and to harvest the cells at a concentration within the range of $0.1{-}1 \times 10^7$ cells ml^{-1}.

C. Mounting the Cells

Depending on the type of experiment, two methods are commonly used to mount yeast cells on slides. For static imaging or short time-lapse (usually several minutes) on living cells, the cells are completely resuspended in a small amount of fresh medium to a desirable concentration. A small aliquot of the resuspended cells (about 1.8–2.2 μl for a one square inch size cover slip) is then applied directly to a dust-free glass slide. After laying the cover glass in place, the surface tension is sufficient to hold the cover glass and immobilize the cells. Sealing the cover slip by nail polish is not necessary and not recommended because of concerns regarding the potential cytotoxicity caused by the organic solvent in nail polish. The most frequently encountered problem is cell floating after mounting. This is usually caused by minute dust grains on the slide or the cover slip and aggregation of cells in the suspension. A new box of slides and cover glass and more vortexing of the cell suspension usually solve the problem. It is noteworthy that cells mounted on blank glass seem to be under a fair amount of pressure, which may affect certain cellular structures or physiological processes. We have noticed that in *S. pombe*, cytoplasmic MTs tend to depolymerise under pressure (our unpublished observations). Thus, cells that look "flattened" should be avoided. This method is good for both live cell and fixed cell with a GFP tag. To fix cells, add 0.6 vol 100% methanol to the yeast culture drop by drop while vortexing. After washing cells in PBS buffer, the cells are mounted onto glass slides as described above.

For longer time-lapse movies, an agarose pad is required to provide a nutrient supply and a cushion for maintaining cell growth. There are several successful protocols developed by different laboratories (e.g. see http://www.bio.unc.edu/faculty/salmon/lab/salmonprotocols.html). Protocol 1 summarises a simple procedure used in our laboratory.

A special technique for making slides that enables the medium to be changed during live cell analysis (Browning *et al.*, 2003) is particularly useful for experiments that utilise small molecule drugs. Basically, the cover slip is coated with lectin or Concanavalin A to make the surface sticky enough for cells to adhere, followed by inversion of the cover slip onto two parallel pieces of double-sided sticky tape mounted on a microscope slide. By this means, a chamber is formed between the cover slip and the slide through which medium and small molecule solutions can be changed by pipetting from one side and soaking out from the other side using tissue paper.

D. Image Acquisition Protocol

Often, the GFP signals are not easily visible with naked eye, especially when the GFP signal is localized as minute dots. Thus, as a general precaution to avoid photobleaching, eye observation should be restrained or totally bypassed in the initial step of locating the cell and focusing. "Snapshots" with the camera should be used instead. The imaging protocol should be set according to the goal of the individual experiment. The rule of thumb is to avoid aiming for unnecessarily high signal intensity and spatial resolution at the initial time points so as to minimise photo-toxicity and photobleaching. The extent of photo damaging may vary between different

Protocol 1. Agarose pads for time-lapse microscopy of yeast cells.

1. Weigh out 0.1 g agarose, add to 5 ml of fresh medium and mix in a 50 ml Falcon tube. Heat in a microwave oven at low-power level to avoid overboiling. After pad preparation, the remaining agarose mixture can be reused. However, repeated re-heating, which causes moisture loss, should be avoided.
2. Immediately spot 150 μl of melted agarose onto a clean slide. Put another slide on top of the agarose drop, perpendicular to the bottom slide. Gently press the top slide if needed, to spread the agarose droplet as much as possible.
3. After the agarose has solidified, slip off the top slide by gently pushing (no lifting). Trim the agarose pad with a razor blade to the size of about a 6 mm × 6 mm^2. Avoid areas that are visibly uneven or have air bubbles.
4. Using a 3 ml syringe with a 200 μl pipette tip as the needle, load some Vaseline (available in regular pharmacy stores), and gently paste a thin line of Vaseline around the agarose pad.
5. Spot 1 μl of cell suspension on the agarose pad. Put on a cover glass. Press very gently, if needed, so that the cover glass is in contact with the Vaseline lines and the cell suspension is spread out. Vaseline prevents loss of the moisture, but more importantly prevents drifting in the Z-dimension during image acquisition. Seal the cover glass with nail polish so that no Vaseline will accidentally contaminate the lens. Applying nail polish is safe in this instance because Vaseline seems to provide a sufficient barrier to prevent the nail polish from poisoning the cells.
6. For characterization of temperature-sensitive mutant cells at their restrictive temperature, slides with such an agarose pad can be prepared at room temperature and incubated at the restrictive temperature for developing the mutant phenotypes before imaging.

microscopes and should be determined empirically. For the DeltaVision microscope in our laboratory with an Osram 100 W mercury lamp as the illumination source, the "benchmark" imaging protocol is 0.3 μm space per section along the *z*-axis with 10 sections (to cover the depth of the nucleus) and 0.2 sec exposure per section. Usually, a cell stays healthy and divides normally when the total exposure time is within 2 min under our specific conditions. It also seems that with the same total amount of illumination time, cells are better off with short exposures (our unpublished observation). Image size and camera speed are of particular concern when performing time-lapse experiments. Depending on the camera speed and the image size, it commonly takes several or up to 20 sec to finish acquiring a stack of images at each designated time point (Rines *et al.*, 2002). The results should therefore be interpreted within the context of this "temporal resolution".

◆◆◆◆◆◆ III. GFP-TAGGING SCHEMES

A. *lacO*/GFP-*lacI* and *tetO*/*tetR*-GFP

Visualisation of individual chromosome is often difficult to achieve by DNA dye staining in yeasts because of the nuclear size limitation and the lack of sufficient chromosome condensation. The problem is partially solved by the usage of *in situ* hybridisation technique, but it depends on fixation and thus is not suitable for studies of chromosome dynamics and live cell analysis (Funabiki *et al.*, 1993; Guacci *et al.*, 1994). With the development of GFP-tagging techniques, a method has been developed to tag a defined chromosomal locus with GFP using a tandem repeat of prokaryotic operator DNA and a GFP fusion to the repressor that specifically binds to the operator. The chromosome locus with the operator insertion is then visible as green fluorescent dots within the nuclei of live or fixed cells.

The *lacO*/GFP-*lacI* system is one of two similar systems used to tag chromosomes. It contains two essential elements: *lac* repressor (*lacI*) fused with GFP and tandem repeats of *lac* operator (*lacO*) integrated into the chromosome (Belmont and Straight, 1998). *lacI* was modified by deletion of the amino-terminal region responsible for tetramerisation, addition of a nuclear localisation signal (NLS), and fusion with GFP at the amino-terminal of the *lacI* product. The expression of GFP-*lacI* is driven by a regulatable promoter such as HIS3 (Straight *et al.*, 1996) or a strong, constitutive promoter such as *URA3* (Belmont, 2001).

To ensure an adequate signal, a 10 kb segment of DNA that contains 256 tandem repeats of *lacO* is used. The *lacO* array can theoretically be integrated into any site in the genome by homologous recombination (e.g. Straight *et al.*, 1996; He *et al.* 2000). In fission yeast, a *lacO* array/GFP-*lacI* system is also employed. The *lacO* array is inserted at the *lys1*$^+$ locus that is tightly linked to the *cen1* locus and a GFP-*lacI* expression cassette is integrated into the genome at another locus (*his7*$^+$), thus making the *lys1*$^+$ locus visible (Nabeshima *et al.*, 1998).

A variation for tagging chromosomes is the *tetO* array/*tetR*-GFP system. Similar to the *lacO*/GFP-*lacI* system, 336 tandem repeats of tetracycline operator sequence are integrated into the genome, which are recognised by TetR fused to GFP at its carboxyl-terminal and with an NLS added to its amino-terminal (Michaelis *et al.*, 1997). Although very similar in principle, the GFP signal derived from the *lacO*/GFP-*lacI* system seems unstable at high temperature (36°C), at least in fission yeast, perhaps due to unstable binding between *lacO* and GFP-*lacI* (Nabeshima *et al.*, 1998). This limits its usage for characterising chromosome dynamics in temperature-sensitive mutants, whereas *tetO*/*tetR*-GFP does not suffer the same limitation, at least in budding yeast (e.g. He *et al.*, 2001).

One concern of tagging chromosomes using the operator/repressor system is the association between two loci both tagged with tandem repeats of exogenous *tetO* sequences (Fuchs *et al.*, 2002). This artificial association was detected only in interphase, but not in mitosis.

B. Conditional Centromere under Regulation of the *GAL* Promoter

In yeasts, a functional kinetochore is present on the chromosome almost throughout the entire cell cycle (see above). The initial phase of kinetochore assembly and its attachment to the spindle MTs is relatively brief and hard to observe in details in a normal cell cycle. A conditional centromere is thus desirable for the study of certain aspects of mitosis, such as the initial phase of kinetochore and spindle-microtubule attachment. In budding yeast, such a conditional centromere was constructed by placing the centromere of a chromosome immediately downstream of the inducible *GALI* promoter (Dewar *et al.*, 2004; Tanaka *et al.*, 2005). By utilising different growth conditions, either transcriptional induction with galactose as carbon source or transcriptional repression with glucose as carbon source can be used to switch centromere function off or on, respectively (Hill and Bloom, 1987). Combining a conditional centromere with the GFP-chromatin tagging technique allows for direct visualisation of the initial binding between the newly assembled kinetochore and a spindle microtubule (see below).

C. GFP Tagging of Kinetochore and Spindle Proteins

GFP is usually fused to either the carboxyl or amino terminal of the target protein. C-terminal tagging and replacing the endogenous gene with the fusion by homologous recombination is the favourite strategy (Bahler *et al.*, 1998; Wach *et al.*, 1997; see Chapter 4). Sixty-seven percent of all open reading frames in the budding yeast genome have been successfully tagged with GFP at their C-termini (Huh *et al.*, 2003). Alternative strategies include N-terminal tagging (see Chapter 4), insertion of a spacer between the target protein and the tag (Doyle and Botstein, 1996) or expressing the fusion protein ectopically. In the last case, since variation in the plasmid copy number from cell to cell will contribute to heterogeneity in expression level within a population, an integration strategy is recommended whenever possible (Bloom *et al.*, 1999).

The GFP-fusion technique is a powerful tool to tag proteins that can be used to study the localisation and dynamics of an unknown protein in live cells, resulting in informative implications regarding its functionality. On the other hand, certain proteins with a well characterised cellular location, after tagging with GFP, can be used as a microscopic marker for certain subcellular compartments or structures. For example, *atb2*$^{+}$ (one of the two α-tubulin genes in fission yeast) was tagged with GFP and localised onto MTs, allowing us to deduce spindle microtubule dynamics. A variety of constitutive kinetochore proteins, such as Mis6 and Mis12, are regarded as kinetochore markers, and when tagged with GFP reveal the behaviour of the kinetochores *in vivo* (Tatebe *et al.*, 2001). We will discuss the application of GFP-fusion protein in detail later.

D. GFP Variants, Double Tagging Schemes and FRET

Since the GFP was first cloned from the jellyfish *Aequorea victoria* in 1992, it has been introduced into a large variety of heterologous organisms including yeasts. In order to enhance its fluorescent properties and broaden its usage range, wild-type GFP has been modified in various ways. Specific amino acids have been changed to ensure efficient translation in fungi and efficient folding at a relatively higher temperature of 37°C (for a recent review, see Czymmek *et al.*, 2004). The GFP molecule was also successfully altered to shift the excitation and emission spectra, thus giving rise to different colour variants (Haseloff, 1999; Tsien, 1998). Among these are commercially available "enhanced fluorescent proteins", including ECFP (cyan), EGFP (green) and EYFP (yellow). New additions to the collection of fluorescent proteins include the red fluorescent protein (RFP), which also originated from a marine coral species and was further modified genetically to improve its signal intensity and to increase its maturation rate after translation in yeast (Janke *et al.*, 2004). In addition to genetic modification, sequentially repeated fluorescent proteins are also used (e.g. 4 × GFP) to give a much more intensified signal as a way to observe low-expression proteins or to distinguish target proteins from other single copy tagged proteins (Maekawa *et al.*, 2003).

The emergence of colour variants also enables multi-spectral imaging applications as well as fluorescence resonance energy transfer (FRET) experiments. Sufficient difference between the spectra of YFP and CFP or GFP and RFP allows these pairs of markers to be used simultaneously for careful determination of the relative positions of two proteins within the same cell at the same time. Equipped with the matching optical filters (e.g. from Chroma Technology Corp.), it is possible to record one colour signal through a specific channel without leak-through of the other colour. FRET, a technique based on non-radiative energy transfer between two fluorophores, can be applied to determine physical contact (or at least very close proximity) between two target molecules. An in-depth discussion of FRET can be found in Chapter 12, this volume.

◆◆◆◆◆◆ IV. REPRESENTATIVE GFP-BASED ASSAYS

A. Chromosome Loci Distribution and Dynamics in an Interphase Nucleus

Using the *lacO*/GFP-*lacI* chromosome tagging technique, almost any particular locus in the genome can be directly visualised. By combining the chromosome tag at various loci with a GFP-tagged copy of a nuclear pore protein (Nup49p) that labels the nuclear envelope, movement of the tagged locus relative to the nuclear periphery or the calculated centre of the nuclear plane can be recorded by time-lapse movie. The range of movement can be used to determine the possible spatial constraints for the locus of interest in the interphase nucleus (Heun *et al.*, 2001). As a result, the telomere regions have been shown to be limited to a region close to the nucleus periphery, and centromeric regions are confined to a sub-nuclear zone in both G1- and S-phase, consistent with previous reports (Marshall *et al.*, 1997; Tham *et al.*, 2001). On the other hand, replication origins are more mobile in G1-phase in an energy-dependent fashion, but become restrained in S-phase, dependent on DNA replication (Heun *et al.*, 2001).

It is noteworthy that, the same colour GFP tag can sometimes be used for both the chromosome tag and the reference tag, such as Nup49p for labelling the nuclear envelope (above) and Spc42p for the spindle pole (see below). The signal intensity and its distribution pattern are usually sufficient to distinguish the chromosome tag and the reference tag. Using the same colour facilitates fast image acquisition, especially for time-lapse experiments.

GFP-based Microscopic Approaches

B. Spindle Assembly and Spindle Elongation

Direct observation of spindle dynamics has been made possible by time-lapse microscopy of fluorescent protein fused to microtubule proteins. For instance, the non-essential α-tubulin (*atb2*$^+$) was chosen to construct the GFP-tubulin fusion as a fluorescent marker of MTs under regulation of the *nmt* promoter in *S. pombe*. In the presence of 2 μM thiamine, the fusion proteins were expressed at a level lower than the authentic *atb2*$^+$ without affecting the mitotic growth of the cells (Ding *et al.*, 1998). Similarly, the major α-tubulin (Tub1p) was fused with GFP at its amino terminus under control of the *HIS3* promoter in budding yeast (Straight *et al.*, 1997). Alternatively, spindle length can specifically be measured as the distance between fluorescent proteins localized at the two SPB, from which the spindle MTs are nucleated (Nabeshima *et al.*, 1998). This is based on the assumption that the spindle is straight, which is true in most cases. The spindle dynamics deduced from GFP-fusion proteins at SPB and from the GFP-tubulin fusion protein are consistent with one another.

In fission yeast, during interphase before the SPB is duplicated, the MTs are imaged as fibres extending along the long axis of the cell in the cytoplasm (Hagan, 1998). Fission yeast mitosis features the presence of the spindle in the nucleus, whose development is divided into three phases based on quantitative measurements of spindle elongation (Nabeshima *et al.*, 1998). Phase 1 refers to the period of spindle formation, corresponding to pro-metaphase. Upon commitment to mitosis, an intense dot of tubulin-GFP appears and rapidly elongates ($\sim$1 µm min^{-1} at 25°C) into a short bar approximately 1.5 µm in length, between the old SPB and the new one (Nabeshima *et al.*, 1998). This occurs as most cytoplasmic MTs quickly disappear. On the cytoplasmic face of the SPB, there are still bundles of multiple MTs extending tangentially at the ends of the metaphase or early anaphase spindle, termed astral MTs, which are responsible for spindle positioning and orientation (Hagan, 1998).

Phase 2 refers to the period during which the spindle length elongates very slowly ($\sim$0.09 µm min^{-1} at 25°C) from $\sim$1.5 µm to $\sim$3 µm, corresponding to metaphase and anaphase A (Nabeshima *et al.*, 1998). While the spindle length remains relatively constant, the kinetochore-attached MTs shorten, accompanied with the separation and poleward movement of sister chromatids at end of Phase 2. Photobleaching at the midpoint of the spindle in Phase 2 followed by fluorescence recovery in a short time indicates that the free plus ends of non-kinetochore MTs (which account for the majority of spindle MTs) overlapping in the middle of spindle are undergoing rapid polymerisation and depolymerisation (Mallavarapu *et al.*, 1999).

Phase 3 represents the period of rapid spindle extension, from 3 µm to 12–15 µm at a speed similar to Phase 1 (0.8 µm min^{-1} at 25°C), corresponding to anaphase B (Nabeshima *et al.*, 1998). Characterisation of spindle MTs dynamics by photobleaching indicates that the plus ends of the non-kinetochore MTs polymerise, which in turn increase the length of the spindle (Mallavarapu *et al.*, 1999).

Once the nuclei have been separated to the ends of the cell, the spindle breaks down and cytoplasmic MTs re-appear in the middle of the cell, termed the post-anaphase array (PAA; Hagan, 1998). The rate of spindle elongation increases with increased temperature, but the distinction of the three phases persists (Nabeshima *et al.*, 1998).

Budding yeast shows similar spindle kinetics when compared with fission yeast, except the assembly of a short bipolar spindle starts as early as G2-phase (reviewed in Winey and Byers, 1993). The mitotic spindle nucleates from the SPB, quickly reaches a length of 1–1.5 µm and is positioned parallel to the mother-bud axis by the interaction between astral MTs and the cortex (Segal *et al.*, 2000;

Straight *et al.*, 1997; Yeh *et al.*, 1995). The spindle slowly elongates from about 1.5 to about 3.0 μm until the initiation of anaphase, when the rate of spindle elongation significantly increases. This rapid phase of elongation slows down until the spindle reaches a final extent of 10–11 μm, at which point the spindle starts to depolymerise (Straight *et al.*, 1997; Yeh *et al.*, 1995).

C. Kinetochore Attachment to the Spindle Microtubules

In *S. cerevisiae*, centromeres are each attached to a single microtubule via kinetochores during most of the time in cell cycle. However, during S-phase, when the centromeres are being replicated, the kinetochores are believed to disassemble and the chromosomes are temporarily released from the MTs nucleated from the SPB (Tanaka *et al.*, 2005). Kinetochores are reassembled on both sister chromatids quickly after centromeric DNA replication and are recaptured by spindle MTs, eventually resulting in the establishment of bipolar attachment.

The recapturing process in wild-type cells is relatively brief and hard to monitor in details. However, by using an elaborate genetic design, the molecular mechanisms of the recapture event have been explored in budding yeast (Tanaka *et al.*, 2005). The particular yeast strain construction and growth condition manipulation allows a conditional *CEN3* to be inactivated when entering S phase so that an intact kinetochore is not assembled on chromosome III. The cells are then arrested at the metaphase stage and the inactivated *CEN3*, tagged using the *tetO*/*tetR*-GFP system, is seen floating within the nucleus without being captured by any MTs while the other normal chromosomes are attached to the spindle MTs. The conditional *CEN3* is then activated, and it can be seen attaching laterally to a single microtubule radiating from one of the SPBs before being transported poleward. Although much easier to observe in the engineered yeast strain background just described, the process of kinetochore capture by MTs has been verified in wild-type cells.

Also, this rather brief process seems rich in significant events. For example, during transportation, *CEN3* might sometimes transiently fall off the microtubule or change its association to another microtubule, but eventually it moves to the spindle pole. Shortly after *CEN3* reaches the spindle pole body, the GFP signal splits into two dots as the sister chromatids separate from each other when the chromosome bi-orients on the spindle. More molecular mechanisms await elucidation.

These sequential events can be monitored and measured quantitatively in a variety of selected mutant background, which would indicate the potential roles of the molecules of interest, when compared with the wild-type background. This work has led to the conclusion that before kinetochore capture, factors such as the Ran GTPase nucleotide exchange factor (Ran GEF) Prp20p and the plus-end tracking proteins (+TIPs: Stu2p, Bim1p and Bik1p) facilitate the non-preferential extension of nuclear MTs. The subsequent step of kinetochore capture laterally by a single nuclear microtubule requires the involvement of kinetochore components (the CBF3 complex, Ndc80 complex and Ctf19 complex) and Ran GTPase-activating protein (Ran GAP; Rna1p). A minus-end oriented motor Kar3p, probably in cooperation with other proteins, then transports the kinetochore along the microtubule. After arrival at the spindle, the *CEN3* establishes bi-orientation through a mechanism requiring the DASH complex and Ipl1p (Tanaka *et al.*, 2005).

As for fission yeast, in an undisturbed mitosis, the initial process of kinetochores attaching to the spindle is also too brief to be carefully characterised (Petersen *et al.*, 2001). Due to the much larger size of the centromeric DNA, a similar conditional centromere system as described above would be hard to implement. Nonetheless, it might be possible to utilise a unique tubulin mutation allele, *nda3-311*^{cs}, which completely depolymerises MTs at the restrictive temperature (20°C), and instantaneously forms a short spindle when switched back to the permissive temperature (36°C). We have observed that kinetochores scattered within the nuclei in *nda3-311*^{cs} cells at 20°C were still separated by the newly formed spindle when switched to 36°C (our unpublished observation). If the kinetochore capturing process can be directly visualised, this would provide an opportunity for characterising the initial binding between a kinetochore and the MTs in fission yeast.

D. Chromosome Bi-orientation and Transient Separation around the Centromere

It was found in budding yeast that the sister centromeric DNAs are able to separate transiently and reversibly to a distance nearly 1 μm when the short spindle is formed (Goshima and Yanagida, 2000; He *et al.*, 2000; Pearson *et al.*, 2001; Tanaka *et al.*, 2000). Using *tetO* array/*tetR*-GFP or *lacO* array/GFP-*lacI* systems to label different loci, together with a spindle pole marker such as Spc42-GFP, motion dynamics of each GFP-labelled chromosome locus relative to the SPB were recorded by time-lapse microscopy. Tags within 15–20 kb surrounding the centromere displayed frequent and reversible sister separation prior to anaphase. The transient separations on different chromosomes were uncorrelated (He *et al.*, 2000). During transient separation, significant stretching was detected in the chromosomal region near the centromere, indicating that chromosome elasticity may influence kinetochore motion (He *et al.*, 2000; Pearson *et al.*, 2001).

Importantly, the transient sister separation is reduced or even eliminated in mutants in which the functional kinetochores become inactive or in nocodazole-treated cells, which destabilises the MTs. Taken together, these results demonstrate that quantification of the transient sister separation in the *CEN*-proximal region can serve as a functional assay for the establishment of bi-orientation as well as the amount of the pulling force applied on sister kinetochores (He *et al.*, 2000). This approach has been applied to the characterisation of a number of kinetochore mutants, which revealed three classes of defect: a complete failure of chromosome–microtubule attachment (in *ndc80, nuf2* and *ndc10* mutant cells), in which the chromosome tag appears to detach completely from the spindle; a mono-polar attachment (in *dam1* and *ipl1* cells), in which the chromosome tag undergoes subtle movement within a limited range near one pole; and bi-polar microtubule attachment with a reduced tension across sister centromeres (in *stu2* cells), in which the centromere-proximal tag oscillates between the SPBs, thus indicating proper bi-orientation, but with reduced velocity and no transient sister separation suggesting diminished pulling force by the kinetochore MTs (He *et al.*, 2001).

Cohesins between the sister centromeres are required to ensure by providing physical connection between sister kinetochores necessary to generate tension when they bi-orient (Dewar *et al.*, 2004; Tanaka *et al.*, 2000). Kinetochores with the wrong attachment configurations lack tension. Ipl1p kinase promotes the turnover of kinetochore–spindle pole connections in a tension-dependent manner, which facilitates correction of erroneous attachments (Dewar *et al.*, 2004).

E. Measurement of the Velocity of Single Chromosome Movement

Kinetochores undergo constant motion throughout mitosis. They are transported towards the spindle pole after initial attachment to the spindle microtubule laterally (Tanaka *et al.*, 2005). In the following stage before sister chromatids separation, sister kinetochores undergo transient separation (see above) driven at least partially by microtubule dynamic instability (Pearson *et al.*, 2003). The nucleus or the spindle frequently rocks or drifts, which interferes the measurement of the chromosome movement relative to the spindle. It is thus necessary to image the motion of the labelled chromosome locus together with a reference marker, such as a SPB marker or labelled spindle MTs. Furthermore, spindle length provides a reliable and convenient index of the cell cycle stages. Detailed kinetics analysis suggested that the minus-end directed motor Kar3p is involved in the poleward transport of kinetochores along the MTs: in the dominant-negative *KAR3-1* mutant, the kinetochore showed longer and more frequent standstill, while in *KAR3* over-expression cells, the movement was more rapid and with shorter pauses (Tanaka *et al.*, 2005). Stu2p, on the other hand, modulates spindle microtubule dynamics, which further regulates the pre-anaphase oscillation of sister kinetochores (He *et al.*, 2001; Pearson *et al.*, 2003).

F. Anaphase Progression Rate and Lagging Chromosomes

Upon the entry of anaphase, the sister chromatids separate completely. The dynamics of chromosome motion can again be characterised by imaging centromeres tagged with *lacO*/GFP-*lacI* and a SPB fusion protein and measuring the rate of their relative movement. In budding yeast, careful quantification shows that anaphase A, defined by sister kinetochore segregation towards the spindle poles, coincides or happens immediately after the initiation of anaphase B, defined by prominent spindle elongation (Pearson *et al.*, 2001). There is a brief delay in the separation of chromosome arms in comparison to the kinetochores, perhaps due to the need to resolve chromosome cohesion. GFP-tagged chromosome arm exhibit a brief rapid poleward movement, even faster than the kinetochores in anaphase A, possibly due to the recoiling of the stretched chromosome (Pearson *et al.*, 2001).

Analysis of chromosome movement in anaphase has been performed in fission yeast as well. With a chromosome tag that is 30 kb away from *cen1*, sister separation was seen roughly to coincide with the rapid spindle elongation at the transition of Phase 2 and 3 (Nabeshima *et al.*, 1998; see above). Chromosomes are separated equally into daughter cells in anaphase in normal mitosis. Using specific kinetochore proteins as the markers (e.g. Nuf2-GFP), the two sets of kinetochores are clustered and seen as two bright fluorescent dots moving synchronously towards opposite spindle poles (Pearson *et al.*, 2001). However, in some mutant backgrounds in which the kinetochore–microtubule attachment is defective, one or more kinetochore dots are seen lagging behind the majority of the other chromosomes, a phenomenon known as lagging chromosomes. Lagging chromosomes can be readily detected using specific kinetochore protein-GFP fusions, which label all the kinetochores (Ekwall *et al.*, 1999; Liu *et al.*, 2005; Sanchez-Perez *et al.*, 2005).

G. Chromosome Mis-Segregation as the Final Readout of the Defects in Mitosis

The ultimate consequence of various types of mitotic defect is the unequal distribution of DNA mass in two daughter cells. There are several ways to detect and quantify this defect. DNA staining by specific dyes, such as DAPI (4′,6-diamidino-2-phenylindole) or Hoechst 33342, is easy and quick, and is suitable for detecting gross size differences between two daughter nuclei. However, this method is not sensitive to less dramatic defects and sometimes could be misleading, for example by failing to detect the defect in which sister chromatids do not separate but pairs of sister chromatids are randomly separated into two sets of DNA masses (as seen in *dam1* and *ipl1* mutants in *S. cerevisiae*). A good substitution for DNA dye staining is the single chromosome tag method. Because only a pair of sister chromatids are tagged in mitosis, the chromosome mis-segregation can be easily detected by the observation of two chromosome dots in one daughter nucleus and no dots in the other. Due to the clarity of the phenotype (one dot on each side *versus* both dots on the same side and no dot on the opposite side), this method provides unequivocal and consistent detection of chromosome mis-segregation. Noticeably, since only one pair of chromosomes among the 16 in total in budding yeast (and three in fission yeast) is visualised, this approach tends to under-estimate the defect in a mutant strains. In addition, nuclear division, revealed by additional DNA dye staining, or simply by the diffused nuclear signal of unbound GFP-*lacI*, will help to determine whether the non-disjunction is caused by defects in sister chromosome segregation or the failure of division of the whole nucleus (Biggins *et al.*, 1999; Liu *et al.*, 2005; Sanchez-Perez *et al.*, 2005).

Acknowledgement

The authors thank Eric Jiang for proofreading the manuscript.

References

Bahler, J., Wu, J. Q., Longtine, M. S., Shah, N. G., McKenzie 3rd, A., Steever, A. B., Wach, A., Philippsen, P. and Pringle, J. R. (1998). Heterologous modules for efficient and versatile PCR-based gene targeting in *Schizosaccharomyces pombe*. *Yeast* **14**, 943–951.

Belmont, A. S. (2001). Visualizing chromosome dynamics with GFP. *Trends Cell Biol.* **11**, 250–257.

Belmont, A. S. and Straight, A. F. (1998). *In vivo* visualization of chromosomes using *lac* operator-repressor binding. *Trends Cell Biol.* **8**, 121–124.

Biggins, S., Severin, F. F., Bhalla, N., Sassoon, I., Hyman, A. A. and Murray, A. W. (1999). The conserved protein kinase Ipl1 regulates microtubule binding to kinetochores in budding yeast. *Genes Dev.* **13**, 532–544.

Bloom, K. S., Beach, D. L., Maddox, P., Shaw, S. L., Yeh, E. and Salmon, E. D. (1999). Using green fluorescent protein fusion proteins to quantitate microtubule and spindle dynamics in budding yeast. *Methods Cell Biol.* **61**, 369–383.

Browning, H., Hackney, D. D. and Nurse, P. (2003). Targeted movement of cell end factors in fission yeast. *Nat. Cell Biol.* **5**, 812–818.

Czymmek, K. J., Bourett, T. M. and Howard, R. J. (2004). Fluorescent protein probes in fungi. In: *Microbiol Imaging Methods in Microbiology*, Vol. 34 (T. Savidge and C. Pothulakis, eds), pp. 27–62. Elsevier, London.

Dewar, H., Tanaka, K., Nasmyth, K. and Tanaka, T. U. (2004). Tension between two kinetochores suffices for their bi-orientation on the mitotic spindle. *Nature* **428**, 93–97.

Ding, D. Q., Chikashige, Y., Haraguchi, T. and Hiraoka, Y. (1998). Oscillatory nuclear movement in fission yeast meiotic prophase is driven by astral microtubules, as revealed by continuous observation of chromosomes and microtubules in living cells. *J. Cell Sci.* **111**, 701–712.

Doyle, T. and Botstein, D. (1996). Movement of yeast cortical actin cytoskeleton visualized *in vivo*. *Proc. Natl. Acad. Sci. USA* **93**, 3886–3891.

Ekwall, K., Cranston, G. and Allshire, R. C. (1999). Fission yeast mutants that alleviate transcriptional silencing in centromeric flanking repeats and disrupt chromosome segregation. *Genetics* **153**, 1153–1169.

Fuchs, J., Lorenz, A. and Loidl, J. (2002). Chromosome associations in budding yeast caused by integrated tandemly repeated transgenes. *J. Cell Sci.* **115**, 1213–1220.

Funabiki, H., Hagan, I., Uzawa, S. and Yanagida, M. (1993). Cell cycle-dependent specific positioning and clustering of centromeres and telomeres in fission yeast. *J. Cell Biol.* **121**, 961–976.

Goshima, G. and Yanagida, M. (2000). Establishing biorientation occurs with precocious separation of the sister kinetochores, but not the arms, in the early spindle of budding yeast. *Cell* **100**, 619–633.

Guacci, V., Hogan, E. and Koshland, D. (1994). Chromosome condensation and sister chromatid pairing in budding yeast. *J. Cell Biol.* **125**, 517–530.

Hagan, I. M. (1998). The fission yeast microtubule cytoskeleton. *J. Cell Sci.* **111**, 1603–1612.

Haseloff, J. (1999). GFP variants for multispectral imaging of living cells. *Methods Cell Biol.* **58**, 139–151.

He, X., Asthana, S. and Sorger, P. K. (2000). Transient sister chromatid separation and elastic deformation of chromosomes during mitosis in budding yeast. *Cell* **101**, 763–775.

He, X., Rines, D. R., Espelin, C. W. and Sorger, P. K. (2001). Molecular analysis of kinetochore-microtubule attachment in budding yeast. *Cell* **106**, 195–206.

Heun, P., Laroche, T., Shimada, K., Furrer, P. and Gasser, S. M. (2001). Chromosome dynamics in the yeast interphase nucleus. *Science* **294**, 2181–2186.

Hill, A. and Bloom, K. (1987). Genetic manipulation of centromere function. *Mol. Cell Biol.* **7**, 2397–2405.

Huh, W. K., Falvo, J. V., Gerke, L. C., Carroll, A. S., Howson, R. W., Weissman, J. S. and O'Shea, E. K. (2003). Global analysis of protein localization in budding yeast. *Nature* **425**, 686–691.

Humphrey, T. and Pearce, A. (2005). Cell cycle molecules and mechanisms of the budding and fission yeasts. *Methods Mol. Biol.* **296**, 3–29.

Ishiguro, J. (1989). An abnormal cell division cycle in an AIR carboxylase-deficient mutant of the fission yeast *Schizosaccharomyces pombe*. *Curr. Genet.* **15**, 71–74.

Janke, C., Magiera, M. M., Rathfelder, N., Taxis, C., Reber, S., Maekawa, H., Moreno-Borchart, A., Doenges, G., Schwob, E., Schiebel, E. and Knop, M. (2004). A versatile toolbox for PCR-based tagging of yeast genes: new fluorescent proteins, more markers and promoter substitution cassettes. *Yeast* **21**, 947–962.

Kniola, B., O'Toole, E., McIntosh, J. R., Mellone, B., Allshire, R., Mengarelli, S., Hultenby, K. and Ekwall, K. (2001). The domain structure of centromeres is conserved from fission yeast to humans. *Mol. Biol. Cell* **12**, 2767–2775.

Knop, M., Pereira, G. and Schiebel, E. (1999). Microtubule organization by the budding yeast spindle pole body. *Biol. Cell* **91**, 291–304.

Liu, X., McLeod, I., Anderson, S., Yates 3rd, J. R. and He, X. (2005). Molecular analysis of kinetochore architecture in fission yeast. *EMBO J.* **24**, 2919–2930.

MacNeill, S. A. and Nurse, P. (1997). Cell cycle control in fission yeast. In: *The Molecular and Cellular Biology of the Yeast Saccharomyces* (J. Pringle, J. R. Broach and E. W. Jones, eds), pp. 697–763. Cold Spring Harbor Press, New York.

Maekawa, H., Usui, T., Knop, M. and Schiebel, E. (2003). Yeast Cdk1 translocates to the plus end of cytoplasmic microtubules to regulate bud cortex interactions. *EMBO J.* **22**, 438–449.

Mallavarapu, A., Sawin, K. and Mitchison, T. (1999). A switch in microtubule dynamics at the onset of anaphase B in the mitotic spindle of *Schizosaccharomyces pombe*. *Curr. Biol.* **9**, 1423–1426.

Marshall, W. F., Straight, A., Marko, J. F., Swedlow, J., Dernburg, A., Belmont, A., Murray, A. W., Agard, D. A. and Sedat, J. W. (1997). Interphase chromosomes undergo constrained diffusional motion in living cells. *Curr. Biol.* **7**, 930–939.

McCollum, D. and Gould, K. L. (2001). Timing is everything: regulation of mitotic exit and cytokinesis by the MEN and SIN. *Trends Cell Biol.* **11**, 89–95.

Michaelis, C., Ciosk, R. and Nasmyth, K. (1997). Cohesins: chromosomal proteins that prevent premature separation of sister chromatids. *Cell* **91**, 35–45.

Nabeshima, K., Nakagawa, T., Straight, A. F., Murray, A., Chikashige, Y., Yamashita, Y. M., Hiraoka, Y. and Yanagida, M. (1998). Dynamics of centromeres during metaphase-anaphase transition in fission yeast: Dis1 is implicated in force balance in metaphase bipolar spindle. *Mol. Biol. Cell* **9**, 3211–3225.

Pearson, C. G., Maddox, P. S., Salmon, E. D. and Bloom, K. (2001). Budding yeast chromosome structure and dynamics during mitosis. *J. Cell Biol.* **152**, 1255–1266.

Pearson, C. G., Maddox, P. S., Zarzar, T. R., Salmon, E. D. and Bloom, K. (2003). Yeast kinetochores do not stabilize Stu2p-dependent spindle microtubule dynamics. *Mol. Biol. Cell* **14**, 4181–4195.

Petersen, J., Paris, J., Willer, M., Philippe, M. and Hagan, I. M. (2001). The *S. pombe* aurora-related kinase Ark1 associates with mitotic structures in a stage dependent manner and is required for chromosome segregation. *J. Cell Sci.* **114**, 4371–4384.

Pringle, J. R. and Hartwell, L. H. (1981). The *Saccharomyces cerevisiae* cell cycle. In: *The Molecular Biology of the Yeast Saccharomyces: Life Cycle and Inheritance* (J. N. Strathern, E. W. Jones and J. R. Broach, eds), pp. 97–143. Cold Spring Harbor Laboratory Press, New York.

Rines, D. R., He, X. and Sorger, P. K. (2002). Quantitative microscopy of green fluorescent protein-labeled yeast. *Methods Enzymol.* **351**, 16–34.

Sanchez-Perez, I., Renwick, S. J., Crawley, K., Karig, I., Buck, V., Meadows, J. C., Franco-Sanchez, A., Fleig, U., Toda, T. and Millar, J. B. (2005). The DASH complex and Klp5/Klp6 kinesin coordinate bipolar chromosome attachment in fission yeast. *EMBO J.* **24**, 2931–2943.

Segal, M., Clarke, D. J., Maddox, P., Salmon, E. D., Bloom, K. and Reed, S. I. (2000). Coordinated spindle assembly and orientation requires Clb5p-dependent kinase in budding yeast. *J. Cell Biol.* **148**, 441–452.

Stotz, A. and Linder, P. (1990). The *ADE2* gene from *Saccharomyces cerevisiae*: sequence and new vectors. *Gene* **95**, 91–98.

Straight, A. F., Belmont, A. S., Robinett, C. C. and Murray, A. W. (1996). GFP tagging of budding yeast chromosomes reveals that protein-protein interactions can mediate sister chromatid cohesion. *Curr. Biol.* **6**(12), 1599–1608.

Straight, A. F., Marshall, W. F., Sedat, J. W. and Murray, A. W. (1997). Mitosis in living budding yeast: anaphase A but no metaphase plate. *Science* **277**, 574–578.

Tanaka, K., Mukae, N., Dewar, H., van Breugel, M., James, E. K., Prescott, A. R., Antony, C. and Tanaka, T. U. (2005). Molecular mechanisms of kinetochore capture by spindle microtubules. *Nature* **434**, 987–994.

Tanaka, T., Fuchs, J., Loidl, J. and Nasmyth, K. (2000). Cohesin ensures bipolar attachment of microtubules to sister centromeres and resists their precocious separation. *Nat. Cell Biol.* **2**, 492–499.

Tanaka, T. U. (2005). Chromosome bi-orientation on the mitotic spindle. *Philos. Trans. R. Soc. Lond. B. Biol. Sci.* **360**, 581–589.

Tatebe, H., Goshima, G., Takeda, K., Nakagawa, T., Kinoshita, K. and Yanagida, M. (2001). Fission yeast living mitosis visualized by GFP-tagged gene products. *Micron.* **32**, 67–74.

Tham, W. H., Wyithe, J. S., Ko Ferrigno, P., Silver, P. A. and Zakian, V. A. (2001). Localization of yeast telomeres to the nuclear periphery is separable from transcriptional repression and telomere stability functions. *Mol. Cell* **8**, 189–199.

Tsien, R. Y. (1998). The green fluorescent protein. *Annu. Rev. Biochem.* **67**, 509–544.

Wach, A., Brachat, A., Alberti-Segui, C., Rebischung, C. and Philippsen, P. (1997). Heterologous *HIS3* marker and GFP reporter modules for PCR-targeting in *Saccharomyces cerevisiae*. *Yeast* **13**, 1065–1075.

Winey, M. and Byers, B. (1993). Assembly and functions of the spindle pole body in budding yeast. *Trends Genet.* **9**, 300–304.

Winey, M. and O'Toole, E. T. (2001). The spindle cycle in budding yeast. *Nat. Cell Biol.* **3**, E23–E27.

Yeh, E., Skibbens, R. V., Cheng, J. W., Salmon, E. D. and Bloom, K. (1995). Spindle dynamics and cell cycle regulation of dynein in the budding yeast, *Saccharomyces cerevisiae*. *J. Cell Biol.* **130**, 687–700.

11 Immunological Methods

Ewald H Hettema and Kathryn R Ayscough
Department of Molecular Biology and Biotechnology, University of Sheffield, Firth Court, Western Bank, Sheffield, S10 2TN, UK

♦♦♦

CONTENTS

Introduction
Western blotting
Immunoprecipitation
Immunofluorescence methods
Summary

Immunological Methods

List of Abbreviations

CHAPS	3-[(3-cholamidopropyl)dimethylammonio]-1-propanesulfonate
DAPI	4′,6′-diamidino 2-phenylindole dihydrochloride
GFP	green fluorescent protein
IP	immunoprecipitation
PBS	phosphate buffered saline
PVDF	polyvinylidene difluoride
SDS-PAGE	sodium dodecyl sulphate-polyacrylamide gel electrophoresis
TBS	Tris buffered saline
UV	ultraviolet

♦♦♦♦♦♦ I. INTRODUCTION

Antibodies have played a key role in generating much of the data on which we now base many of our hypotheses concerning functioning of specific proteins and protein complexes in cells. In this chapter we discuss the major methods that use antibodies – western blotting, immunoprecipitation and immunofluorescence. In particular, we aim to include details of factors that are of significance to studies in yeast but that are perhaps less relevant in other cell types.

The major methods will be discussed in turn, and full protocols and analysis of important stages are given. However, prior to this it

METHODS IN MICROBIOLOGY, VOLUME 36
0580-9517 DOI:10.1016/S0580-9517(06)36011-4

is appropriate to think about more general aspects of the use of antibodies that are of relevance to all methods included. Two major considerations at the outset are whether to raise the antibody to your own purified protein, or to obtain commercial antibodies. Often coupled to this is the question of whether or not to tag your protein, as commercial antibodies are more readily available to the commonly used epitope tags than to the majority of proteins found in cells.

A. Raising Antibodies to your Favourite Protein

There are good reasons why you might wish to raise your own antibodies. First, in general it is most likely that antibodies to the protein of interest have not been raised commercially. Second, the main alternative to raising your own antibodies is epitope tagging and this procedure may be problematic for your particular protein and affect its levels, interactions or localisation.

There are many excellent books and methods devoted to the purification of proteins for the production of antibodies, including the Harlow and Lane manuals (Harlow and Lane, 1988, 1998) that have become standard texts in many labs. There are also an increasing number of companies who offer the facility of raising high-quality polyclonal sera at competitive prices. Production and purification of antibodies is not a trivial task and the cost of commercial production should be balanced against the effort and time involved in producing them yourself. Whether or not you raise antibodies yourself, there are still some important considerations. First, depending on the preparation methods your protein antigen could be in a folded, native form or denatured on a gel or blot. An antibody raised against a denatured protein is usually most appropriate for western blotting, but for immunoprecipitation or immunofluorescence, epitopes are often more likely to be in a native conformation. Second, antibodies can be raised to peptides, domains or full-length proteins. With a larger region there is more chance of getting a good antigenic site, though peptide antisera might be a good choice when there are other closely related proteins in the cell. A decision as to whether to raise monoclonal or polyclonal antisera will probably depend on the expected use of the antiserum. A polyclonal antiserum is likely to contain many antibodies, some of which might react with other antigens in yeast cells, but on the positive side the serum will contain a range of antibodies to different sites. Monoclonal antibodies will be more specific but potentially more restricted in their use. Often these have been used to detect phosphorylated forms of specific peptides. The host species in which the antiserum is raised can have implications for cost and the amount of antiserum generated. Rabbits are most used for raising polyclonal antisera, but commonly available alternatives include mice and sheep. The choice of host might be important if one is considering double labelling if, for example, an existing antibody was raised in rabbit.

A final stage in generating antibodies suitable for high quality research purposes is that of purification. To ensure specificity, antibodies should be affinity purified. Micro-affinity purification of antibodies using antigen bound to nitrocellulose, and larger scale, column-based techniques have been reported for the affinity purification of antibodies for immunological methods in yeast cells. Another stage of purification that can be useful is to pre-absorb the antiserum against fixed yeast cells or a yeast acetone powder (Protocol 5) in which the gene encoding the protein of interest has been deleted. If this is not possible then simply re-using an antibody can lead to a marked improvement on the second use, as the non-specific background antibodies are absorbed in the first application.

It should also be remembered that animals in which antibodies are raised will often have had fungal infections and this can seriously affect the quality of antisera that can be used for studies in yeast. However, there is a marked variation in the response of animals to yeast infection and often generation of antibodies in 2–3 animals should give at least one antiserum that is of suitable quality.

B. Commercial Antibodies and Protein Tagging

1. Commercial antibodies

There are an increasing number of antibodies available commercially. Some of these have been tried and tested for use in yeast immunological methods but most have not. In addition, many have been shown to recognise proteins by western blotting and only in some cases by immunoprecipitation. It is important to note that success in one procedure by no means guarantees success for other purposes.

Use of antibodies generated to homologues of your protein from another organism can offer a real alternative to tagging, but care must be taken to demonstrate specificity. Examples of such antibodies include anti-actin, which is most often available as a peptide antiserum raised to the highly conserved C-terminal region of the protein. In addition, the phosphorylated form of the yeast MAP kinase, Slt2p is recognised by antibody to phospho-p44/42 MAP kinase (New England Biolabs; see Martin *et al.*, 2000) but it is always important to consider the specificity of signals in such cases.

In addition, commercial antibodies can be good for positive controls as they are usually a more readily replenishable supply than your own.

2. Protein tagging

A more recent development in studies that use antibodies has been the advent of numerous peptide tags that can be attached via recombinant DNA manipulations to the protein of interest. For studies in yeast, genes can be tagged on a plasmid, or within the genome (see Chapter 4). The most commonly used tags that may require subsequent detection with antisera are HA, *myc* and His_6. Following tagging, proteins can then be detected using readily available antibodies to these epitopes.

There are a number of problems associated with tagging that should be considered, which are primarily concerned with the altered functionality of the protein. For example, N-terminal tags could affect membrane insertion and are usually under control of a different, often higher level promoter. Inappropriate levels of protein could affect localisation and protein–protein interactions. C-terminal tags are often considered preferable, but still could affect interactions with other proteins or localisation of the tagged protein. If possible, it is necessary to show that the tag does not affect the normal function of the protein. If there is a phenotype caused by deletion or a mutation in the gene for the protein then rescue of this phenotype is a good indication that the tag is not interfering with this function of the protein. It is essential to recognise though, that even if a tagged protein is capable of rescuing such a phenotype, it neither guarantees that all interactions are maintained, nor that they are as strong as in the wild-type situation. This caveat is probably the strongest argument against protein tagging and the one that should be most carefully addressed. As with many other techniques the use of complementary approaches can strengthen any data obtained.

♦♦♦♦♦♦ II. WESTERN BLOTTING

SDS polyacrylamide gel electrophoresis (SDS-PAGE), followed by western blotting or immunoblotting is the most commonly used immunological method for detecting a specific protein in a complex protein mixture. Western blotting is used to determine the relative expression level and molecular weight of endogenous proteins and it is used to test expression of mutants or epitope-tagged proteins. The method is sensitive and can be quantitative, so it can be used to compare relative steady state levels of a protein between different samples. This technique is frequently the readout of other techniques, such as subcellular fractionation or *in vitro* and *in vivo* binding assays. Of particular importance, it allows analysis of proteins that are difficult to solubilise or easily degraded as a denaturing lysate can be analysed.

The basic method can be considered in distinct stages. First, a complex protein mixture is prepared, for instance a cell lysate, which is then separated usually using denaturing SDS-PAGE. Subsequently, the proteins are transferred electrophoretically to a membrane. The membrane is then incubated with a specific primary antibody directed against the protein of interest. This immuno-complex is then detected by, for example, an enzyme-conjugated second antibody that is directed against the first antibody. A chemiluminescent or colourigenic assay will subsequently detect the position of the protein on the membrane.

A. Practical Considerations for Western Blotting in Yeast

1. Preparation of yeast extracts

Although western blotting is a method used generally throughout molecular and cell biology labs, one of the stages that needs special attention when working with yeast is extract preparation.

The presence of the cell wall and the possibility of extensive protein degradation are important considerations when preparing extracts. In addition, many proteins require the use of detergents for efficient extraction. In general, if denaturing extracts are being prepared then use of SDS (at 1–2%) is possible. Otherwise, use of non-ionic or zwitterionic detergents is preferable, for which we most often use Triton-X100 or CHAPS (3-[(3-cholamidopropyl)dimethylammonio]-1-propanesulfonate). In this section, we describe two protocols to prepare denatured protein extracts (Protocols 1A, B). These are both good for a relatively quick check as to whether a protein of interest is expressed and at what level. Additionally, we will describe three ways of preparing native lysates that are also compatible with, for instance, immune precipitations, co-immune precipitations, affinity chromatography and sucrose gradient centrifugation (Protocols 2A, B and 3).

2. Gel electrophoresis

There are no unusual requirements for gel electrophoresis when running yeast extracts compared with those from other cell types. Most laboratories continue to use the standard SDS-polyacrylamide gels containing a Tris-glycine discontinuous buffer system. However, other buffer systems are compatible, and many commercial companies sell pre-cast gels with appropriate buffer systems. These commercial gels can be particularly useful if one needs a high level of comparison between gels and blots; if one wants to carry out mass spectrometric analysis at later stages (more commonly after immunoprecipitation type experiments); or if one wants to run a low percentage gradient gel that is relatively difficult to handle. In addition, extracts can be made in buffers appropriate for isoelectric focussing (IEF) or a combination of IEF and SDS PAGE as in 2D electrophoresis.

3. Electrophoretic transfer from gel to membrane

Transfer of proteins from the gel to a membrane is done by electrophoresis. Both nitrocellulose and PVDF membranes are commonly used. The PVDF membrane is more robust, and so useful where it is desired to probe the membrane and then strip off

Protocol 1. Preparation of denatured whole yeast cell lysates.

(A) *Method using TCA/Acetone*

1. Grow cells to OD_{600} 0.5, harvest 2 ml of cells in a 2 ml microfuge tube and centrifuge for 1 min at 4 °C at top speed of microfuge.
2. Resuspend pellet in 300 µl ice-cold Millipore water and add 50 µl Lysis Buffer (1.85 M NaOH, 7.4% 2-mercaptoethanol). Mix well and incubate on ice for 10–15 min.
3. Centrifuge lysate for 5 min at 4 °C at top speed in a microfuge. Collect supernatant and add 42 µl of 50% TCA (Trichloracetic acid). Mix well and incubate on ice for 10–15 min.
4. Centrifuge lysate for 5 min at 4 °C at top speed of microcentrifuge and remove as much of the supernatant as possible.
5. Wash pellet with 1 ml ice-cold acetone to remove the last traces of TCA.
6. Centrifuge lysate for 5 min at 4 °C at top speed of microcentrifuge, remove the supernatant and dry the pellet in a vacuum centrifuge (or leave tube open on the bench to air dry).
7. The pellet can now be resuspended in sample buffer (SB) for SDS-PAGE analysis. If SB turns yellow, add 1 M Tris that has not been adjusted for pH until SB turns blue. 4 × Sample Buffer: 250 mM Tris-HCl pH 6.8, 9.2% SDS, 40% glycerol, 0.2% Bromophenol Blue, 100 mM dithiothreitol (added just before use).

(B) *Small scale glass bead lysis*

1. Pellet about 2.0 OD_{600} units of cells in 1.5 ml microfuge tube.
2. Wash twice in ultra pure water. Pellet and aspirate away excess liquid.
3. Add about 100 µl acid-washed glass beads and mix (Sigma, 425–600 µm).
4. Add 25 µl 2 × Sample Buffer (For stock 2 × SB: 2.5 ml 0.5 M Tris pH 6.8; 2.0 ml 10% SDS; 2.0 ml glycerol; 2.0 ml ultra pure water; 1.0 ml β-mercaptoethanol; 0.1% Bromophenol Blue solution.
5. Boil immediately for 3 min.
6. Vortex full speed for 2 min.
7. Add 100 µl more 2 × SB.
8. Vortex briefly and draw off Sample Buffer. Less than 100 µl is retrievable. Use 2–4 µl/lane on minigels. Either load immediately, or store at –20 °C and boil again before loading.

Protocol 2. Native lysates.

(A) *From spheroplasts*

1. Grow cells to an OD_{600} of 0.5. Harvest cells and resuspend in 100 mM Tris–HCl pH 9.4, 10 mM DTT (dithiothreitol). Use 10 ml buffer for 100 OD units. Incubate at room temperature for 5 min. Harvest cells by centrifugation (3 min, 800*g*; about 2500–3000 rpm in a benchtop clinical centrifuge).
2. Resuspend cells in Spheroplasting Buffer (1.2 M sorbitol, 20 mM potassium phosphate buffer pH 7.4). Add 5 mg of Zymolyase 20T per gram of cells (wet weight) and incubate at 30 °C for 30 min with gentle agitation. To check for spheroplast formation, dilute spheroplast solution in 1% Triton. If the OD_{600} is about one tenth of that of the spheroplasts in sorbitol-containing buffer then spheroplasting is complete.
3. Spin down spheroplasts (3 min at 3000 rpm) and carefully resuspend in spheroplasting buffer. Spin down spheroplasts again as above, remove supernatant and snap freeze pellet in liquid nitrogen and store at –80°C, or proceed directly with the rest of the protocol.
4. If using frozen spheroplasts, thaw on ice. Resuspend spheroplasts in ice-cold Lysis Buffer (50 mM HEPES-KOH pH7.6, 5 mM $MgCl_2$, 150 mM KCl containing 1 mM PMSF and Roche EDTA-free protease inhibitor cocktail). We use 5 ml of Lysis Buffer per gram spheroplasts. Lyse the cell suspension in a Dounce Homogeniser using 10 full strokes. If detergent is required, for instance 1% Triton X-100 or CHAPS, add it at this stage and leave the homogenate on ice for 10 min.
5. Centrifuge lysate for 10 min at 4 °C at top speed of microfuge (13 000*g*).
6. An optional step when intending co-immunoprecipitation, sucrose gradients or affinity chromatography, an extra spin at 100 000*g* for 30 min at 4°C in Beckman clinical centrifuge can be included. Collect the high-speed supernatant and use that for further experimentation.

Always check whether your protein of interest is well extracted with this method by monitoring the 13 000*g* and 100 000*g* pellets for its presence/absence.

(B) *Small scale native glass-bead method*

1. Grow cells to an OD_{600} of 0.5. Harvest 1.5 ml cells by centrifugation for 1 min at 4°C at full speed in a microfuge.
2. Resuspend pellet in ice-cold Lysis Buffer to about 200 µl (for 25 ml stock Lysis Buffer: PBS containing 1% Triton X-100, 1 mM PMSF freshly added, 1 Roche EDTA-free complete protease inhibitor cocktail tablet) and add 200 µl of ice-cold acid-washed Glass beads (425–600 microns, Sigma).

3. Lyse cells using the Mini Bead Beater (Biospec Products) with a 45 s pulse at near maximal speed.
4. Put tubes in ice water immediately. Add an extra ice-cold 200 μl Lysis Buffer, mix and pipette off the liquid from the beads and transfer it to a new tube.
5. Spin for 15 min at 4°C at full speed in a microfuge. Collect the supernatant.

antibodies for a further analysis. Two main systems are used for the transfer, either semi-dry blotting or submerged (wet) blotting. Semi-dry blotting has the advantage of being economical in use of blot buffer and it is relatively fast. However, proteins of high molecular weight tend to give low transfer efficiency. To check effectiveness of transfer the blot can be stained with Ponceau S (see Protocol 4). The buffer most commonly used for both wet and semi-dry blotting is Tris-Glycine based. If problems with transfer are encountered and particularly if your protein is relatively basic, a good alternative is CAPS buffer (10 mM CAPS (3-[cyclohexylamino]-1-1propanasulfonic acid), 10% methanol, pH 11.0; dissolve 8.8 g CAPS in 3 l of water, adjust pH to 11.0 with 10 M NaOH solution, add 400 ml methanol and make up to 4 l with water). This buffer is most often used cold for wet blotting approaches.

4. Detection of antigen

As SDS-PAGE involves denaturing of the protein sample, the antibodies need to be able to recognise denatured protein. Polyclonal antisera normally work well but monoclonal antibodies are frequently conformation-specific, recognising the three-dimensional structure of an epitope, and may therefore be less useful.

As outlined in Protocol 5, prior to incubation of blots with antibody, the blot needs to be incubated with a blocking agent that prevents the antibody to stick non-specifically to the membrane. We normally block with 5% dried skimmed milk (less than 1% fat) or with 2% BSA fraction V in the presence or absence of detergent (usually Tween 20). If there is a high background on the developed blot the level of Tween can be raised to about 0.2%.

Most considerations about primary antibodies are discussed in the introductory section. However, relevant to western blotting is the need to determine the appropriate dilution to use. If the antibody is commercial, this will be the one method that will usually have been tested and the relevant dilution will normally be suggested. If you have raised your own antibody, or if you are using an antibody with unknown association with a yeast protein, it would be best to run a gel of wild-type yeast extract and to cut it into strips and test each with different antibody titrations.

Protocol 3. Small scale liquid nitrogen grinding lysates.

1. This method works best using 2×10^9–1.5×10^{10} cells. Both log-phase and stationary phase cells are broken efficiently. For log-phase cells this should be about 50–100 ml cells.
2. Spin down in 50 ml conical tubes in bench top centrifuge (about 800–1000*g*).
3. Wash with 50 ml distilled water and then with 50 ml breakage buffer. Breakage buffer should include buffer, salt, glycerol to 10% and possibly detergent (for example, 50 mM 4-(2-hydroxyethyl)piperazine-1-ethanesulfonic acid (HEPES) to pH 7.4 using KOH, 20–50 mM KCl, 1 mM $MgCl_2$, 0.5% NP-40, 10% glycerol).
4. Resuspend in a minimal volume of breakage buffer. This usually involves adding a volume of buffer equal to the volume of the fully drained pellet and is about 0.7 ml for 1×10^{10} cells. Final volume should not exceed about 1.5 ml.
5. Add protease inhibitors (EDTA-free protease inhibitor cocktail; Roche) and 1 mM PMSF).
6. Freeze cells by taking up in a pipette and slowly dropping into a 50 ml tube filled with liquid nitrogen. Cap with a lid that has been pierced to allow N_2 to escape. Store at –80°C.
7. Grind cells in a mortar. This should be pre-chilled extensively using liquid N_2. Add the frozen cell pellet and coarsely fragment it using a cooled pestle. Then add more nitrogen and grind for about 100 strokes. More nitrogen may need to be added since it is essential to keep everything very cold. Breakage can be monitored by withdrawing small samples, spinning 2 min in microcentrifuge and checking protein concentration. It should plateau after about 75 strokes.
8. Scrape cell powder out using a cooled spatula. Lysate can be returned to –80°C if desired.
9. Thaw samples slowly on ice when required.
10. Spin 15 min at full speed in microfuge at 4°C, withdraw supernatant and re-spin this for 15 min (alternatively spin at 100 000*g* in an ultracentrifuge to generate a high-speed supernatant). Freeze in small aliquots in liquid N_2.
11. Protein concentrations of 50–75 mg/ml are typical but up to 120 mg/ml can be achieved by minimising resuspension volume.

The best control for antibody specificity is to use an extract in which the gene encoding the protein of interest is deleted. If this is not possible because the gene is essential, showing a size shift in a band when the protein is tagged, or showing an increase in level when the gene is expressed from an over-expressing plasmid, are additional ways of demonstrating that an antibody is recognising

Protocol 4. Transfer of protein to membranes.

(A) *Semi-dry blotting*

1. Run protein samples on an SDS-PAGE gel. It is useful to include a pre-stained marker as this will indicate the efficiency of transfer and it is helpful for the orientation of the filter.
2. Cut six sheets of Whatman 3 MM paper at the same size of the gel. Cut one sheet of membrane the same size as the gel.
3. Immerse filter papers one at a time in Transfer Buffer (20% methanol, 48 mM Tris, 390 mM glycine, 0.1% SDS (w/v) prepared in ultra pure water) and put on the flat electrode. Take care not to trap air bubbles between electrode and filter paper. Put the next two sheets of wetted filter paper on top of the first filter paper followed by the nitrocellulose or PVDF membrane (pre-wetted in ultra pure water or methanol, respectively, before soaking in blotting buffer).
4. Take the gel, soak in Transfer Buffer for 5 min and put on top of the membrane. Then put three layers of filter paper wetted in blot buffer on top of the gel, remove air bubbles if necessary and place the cathode on top. Transfer at 1 mA/cm^2 for 1 h.
5. Disassemble the sandwich and mark the blot for orientation.
6. Check transfer by inspecting the amount of pre-stained marker present on the membrane and left in the gel after transfer. Alternatively, the gel can be stained with Coomassie Brilliant Blue R-250 and the blot can be reversibly stained with Ponceau S (0.2% Ponceau S (w/v) in 1% acetic acid (v/v): incubate with stain for 5 min and then wash off with water until blot is clear but bands are visible. Stain can be completely washed off with 0.1 M NaOH).

(B) *Submerged/wet blotting*

As discussed above, for larger proteins, or proteins that appear to have problems transferring using the above method it can be preferable to use submerged blotting. The same transfer buffer can be used as for semi-dry blotting (20% methanol, 48 mM Tris, 390 mM glycine, 0.1% SDS (w/v) prepared in ultra pure water) and the 'sandwich' of filter paper, gel and membrane is assembled in the same way before placing in an appropriate holder between filter pads in the blotter. The blot can be run at 75 V (9 cm × 18 cm gel) in the cold room overnight or for 360 mA for 1 h (though a cooler block or cooling system should be used in this case).

If you want to use submerged blotting for proteins of a molecular weight below 80 kDa, omit the 0.1% SDS from the transfer buffer.

Protocol 5. Immunological detection of proteins on membranes.

1. After transfer, membranes are washed in phosphate buffered saline, PBS (137 mM NaCl, 2.7 mM KCl, 10 mM Na_2HPO_4, 2 mM KH_2PO_4) for 5 min (change PBS once).
2. Non-specific binding sites are blocked by incubating the membrane in 5% milk in PBST (0.1% Tween 20 in PBS). Incubate at room temperature for at least half an hour with agitation, or overnight in the fridge. If left in the fridge, replace the blocking buffer next morning and leave at room temperature for 10 min with agitation.
3. Dilute antibody in fresh blocking solution and add to membrane. If antibody is scarce, incubations can be performed in sealed bags only slightly larger than the membrane – this then only requires 3–5 ml diluted antibody. If using a new antibody, it must first be titrated to assess the best ratio between specific signal and background. Incubate for at least 1 h at room temperature with agitation
4. Wash blot three times with PBST for 5 min each.
5. Incubate for 1 h at room temperature with secondary antibody in the blocking buffer.
6. Wash blot four times with PBST for 5 min each.
7. *For chemiluminescence detection* – a proprietary method is usually supplied by the manufacturer. Usually 2 reagents are mixed together and added immediately to the blot for 1 min. The blot should then be taken out of the solution, excess solution allowed to drip off and wrapped in plastic wrap. The membrane is then exposed to film for various times usually from a few seconds to half an hour.
8. *For alkaline phosphatase detection* – after probing with an alkaline phosphatase conjugated secondary antibody, the blot is washed with PBST and then incubated with NBT/BCIP developing solution (NBT (nitro blue tetrazolium); 0.5 g in 10 ml 100% dimethyl formamide, store at 4 °C. BCIP (5-bromo-4-chloro-3-indolyl phosphate *p*-toluidine salt); 0.25 g in 10 ml dimethyl formamide, store at room temp. Alkaline phosphatase buffer: 100 mM NaCl, 5 mM $MgCl_2$, 100 mM Tris–HCl pH 9.5). Mix 66 µl NBT stock in 10 ml alkaline phosphatase buffer and then add 66 µl BCIP stock, mix and add to the blot. After sufficient colour development, wash off the chromogenic reagent in water. The reaction can be stopped by addition of EDTA.

Additional notes

(i) If blot is to be reprobed, unless the second protein to be detected is of a significantly different size, stripping the blot may be necessary. This procedure does remove some protein so it is best to probe with the weakest antibody first. In brief, incubate the blot in Stripping Buffer (1.5% glycine, 1% SDS in ultra pure water, adjusted to pH 2.5 using HCl) at room temperature for 1 h, then wash in PBST and re-probe.

(ii) If the primary antibody appears non-specific, some improvement can be gained by incubating with a yeast acetone powder. In brief, take a small volume of spheroplast lysates, add 4 volumes of acetone and mix vigorously. Leave on ice 30 min with occasional mixing and then spin at 10 000*g* for 10 min. Keep pellet, resuspend in acetone (−20°C) and incubate on ice for 10 min with occasional mixing. Centrifuge at 10 000*g*, remove supernatant and air dry pellet. This dried powder can then be used as a 1% solution in antiserum. Incubate at 4°C for 15 min, then spin and use antiserum supernatant.

(iii) If blots are to be reprobed at a later date they are best stored dry at –20°C. You must remember to wet a PVDF blot with methanol before attempting to reuse.

(iv) Throughout this protocol, use of PBST can be replaced with Tris-buffered saline+Tween (TBST: 20 mM Tris–HCl (pH 7.6), 137 mM NaCl, 0.5%Tween 20) – most labs adopt one buffer system or the other.

the appropriate protein. It should be noted that many proteins do not run true to size even on a denaturing gel and so size alone should not be considered the only indicator of specificity.

Finally the antigen–primary antibody interaction is detected by using a secondary antibody that recognises the primary antibody. The secondary antibody is usually conjugated to an enzyme that allows detection using either luminescence approaches (if coupled to horse radish peroxidase) or colorimetric methods (if using alkaline phosphatase). Radiolabelled antibodies are also available, but as other methods are now considered as sensitive, these are not so often used.

There is an extensive range of commercial secondary antibodies, and primary factors in choosing between them include (i) species specificity (if one wants to probe with 2 antibodies the cross reactivity of these must also be considered); (ii) the detection method available – detection of luminescence requires access to a luminometer/phosphorimager or to an X-ray developing machine if exposing to film; (iii) whether quantitation of the results is required. Detection using light-based emission onto X-ray film requires great care in analysis and sometimes a simpler, colour-based detection can allow the data to be quantified more readily. One must remember however, that after a blot has been exposed to detection reagents for alkaline phosphatase it is no longer possible to reprobe with other antibodies. Those membranes exposed to horseradish peroxidase-conjugated secondary antibodies coupled with luminescence detection can be readily reprobed to detect other proteins of interest.

5. Troubleshooting

A major problem in western blotting is that an antibody may recognise other proteins on the blot. If this is caused by the primary antiserum, the antiserum can be absorbed either against fixed whole cells, or with an acetone powder of a yeast strain that does not express the protein of interest (see Protocol 5). If it is the secondary antibody that gives the background, a first step would be simply to try one from a different source. Alternatively, one could clean up the antibody by incubating against fixed whole yeast cells, as many animals suffer fungal infections that would generate non-specific antibodies in any antiserum. If there is a diffuse background staining this indicates either insufficient blocking or that the antibody reacts with the blocking agent. In this case another blocking agent could be used or less of either primary or secondary antibody (depending on which is causing the background) could be used, or the antibody incubation periods shortened. Finally, the stringency of the washes can be increased by adding more Tween 20 or Triton X-100 or 0.1% SDS. Careful titration of antiserum can often help in cleaning up blots. Finally, note that sodium azide should not be added to PBS (or TBS) or blocking solution as it inhibits peroxidase activity.

◆◆◆◆◆◆ III. IMMUNOPRECIPITATION

Immunoprecipitation (IP) is the technique whereby a protein interacts with a specific antibody in solution followed by separation of this immune complex from other components of this protein mixture. Again, the main reason for using IP rather than a GST tag or TAP tag pull-down approach concerns functionality issues regarding the tag addition. IP allows a useful alternative to detect proteins that fail to be expressed, are unstable or are inappropriately localised when tagged.

In essence, a lysate is prepared that is then incubated with a high-affinity antibody. Addition of antibody-binding proteins (Protein A or Protein G) bound to agarose or Sepharose beads allows harvesting of immune complexes by centrifugation followed by elution from the beads.

Immunoprecipitation can be used for many purposes including:

(i) Analysis of the level of expression, molecular weight and isoelectric point of a protein of interest.
(ii) Determination of post-translational modifications, for example, glycosylation or phosphorylation.
(iii) Quantification of the rate at which a protein is synthesised. This can be done by measuring the amount of incorporated radiolabel into the protein of interest during a fixed period of labelling time.
(iv) Analysis of precursor–product relationships using pulse-chase labelling experiments followed by immunoprecipitation.

(v) If the lysis and precipitation are performed under mild conditions, this technique can also be used for the identification of proteins that co-precipitate with the immuno-precipitated protein.

Protocol 6 summarises how to carry out an immunoprecipitation experiment.

A. Practical Considerations for Immunoprecipitation in Yeast

1. Lysate preparation for immunoprecipitation

Lysates can be prepared in a number of ways (see Protocols 2 and 3). Normally we use a spheroplast lysate or liquid nitrogen powder lysate prepared in a buffer containing non-ionic detergent. For some IPs we denature the lysate with 0.1% SDS, which can subsequently be quenched by addition of Triton X-100 to 1%. This of course is only compatible with antibodies that recognise denatured proteins. A major constraint of IPs is that the antigen needs to be soluble as insoluble proteins, or proteins that are present in large protein assemblies, might be pelleted during the centrifugation steps of lysate.

2. Pre-clearing

To reduce non-specific background, the lysate is incubated with an antibody that does not recognise the antigen of interest. The pre-immune serum is commonly used for this purpose. The antibody is subsequently removed by the addition of Protein A or Protein G bound to beads. If no pre-immune serum is available, any other antibody that is efficiently precipitated by Protein A or Protein G can be used.

Washing of the immune complex bound to Protein A or Protein G beads is an important step to get rid of non-specific binding. Different salt concentrations or amount of detergent can be used during washes to increase specificity.

3. Antibodies

Both monoclonal antibodies and polyclonal antibodies can be used for immunoprecipitations (IPs). The affinity of an antibody for its antigen is very important for a successful IP. With a high affinity antibody, purification of 10 000-fold can be achieved on an analytical scale. Combined with SDS-PAGE, a further 10- to 100-fold purification can be achieved (Harlow and Lane, 1998). This means that rare antigens can be studied using this technique, making it much more sensitive than western blotting. A common problem intrinsic to working with antisera is that some antibodies cross-react with other proteins. This can be difficult to resolve but with the proper controls, the cross-reacting proteins can be distinguished from the

Protocol 6. Immunoprecipitation of radiolabelled cells.

Because radiolabelling is less commonly performed we have included a full protocol here. The following IP method is, however, appropriate for labelled or non-labelled cells.

Radiolabelling cells

1. Dilute log phase cells in 25 ml depletion medium (minimal medium containing the required amino acids) at OD_{600} 0.3 and grown for another 2 h.
2. Harvest cells by centrifugation and resuspend in 2 ml depletion medium containing 50 µCi ^{35}S-labelled methionine (we use 15 ml polypropylene tubes).
3. Label cells for 1 h in shaking water bath at 30°C.
4. Transfer tubes to ice slurry for at least 2 min and
5. Harvest cells by centrifugation.

Immunoprecipitation

1. Radiolabelled cells (10 OD_{600} units) are washed in ice-cold PBS and lysed according to the native glass bead method (Protocol 2B).
2. 5 µl antiserum is coupled to 10% slurry of Protein A Sepharose beads (GE Healthcare, UK) in lysis buffer containing 0.2% BSA fraction V (Sigma) and 0.05% Tween 20. BSA prevents non-specific sticking of proteins to beads and the Tween 20 prevents beads sticking to the tube and pipette tips. At the same time we bind pre-immune serum to protein A beads (treated in exactly the same way as the antiserum).
3. Incubate for 30 min at 4°C on rotator wheel. Protein A on beads should be in excess of antibodies.
4. Wash preincubated beads with 500 µl Lysis Buffer and resuspend to make up a 10% slurry of beads in Lysis Buffer.
5. To 300 µl lysate add 100 µl of pre-immune serum-Protein A beads slurry and incubate for 1 h at 4°C on rotator wheel. This can also be left overnight.
6. Spin down beads for 1 min at 2500 rpm in microcentrifuge. Transfer pre-cleared supernatant to new tube.
7. Add 100 µl of antibody-Protein A beads slurry and incubate for 1 h at 4°C on rotator wheel.
8. Spin down beads for 1 min at 2500 rpm in a microfuge. Aspirate supernatant and wash twice with 1 ml 50 mM Tris–Cl pH 8.0, 150 mM NaCl. If there is a lot of non-specific binding, vary the salt concentration in washes or add detergent (for instance 50 mM Tris–Cl pH 8.5, 300 mM NaCl, 0.05% Triton X-100, 0.05% SDS).
9. Resuspend pellet in 50 µl 1 × SDS-PAGE Sample Buffer.

protein of interest. Another common problem is the high amount of non-specific background. Titrating the antibody to enable precipitation of the maximal amount of antigen using the lowest possible antibody input will reduce non-specific background.

Whether Protein A- or Protein G-beads are used to bind the antibodies depends on the origin of the antibody and its IgG class. Briefly, Protein A is used for binding of rabbit polyclonal antisera and mouse monoclonals with the isotype IgG2a, IgG2b and IgG3. Protein G is used for mouse, rat and goat polyclonal antisera and mouse monoclonal IgG1. For more details on Protein A and G binding specificities see http://www.sigmaaldrich.com/img/assets/8181/protein_agl_A.pdf.

4. Detection of the precipitated protein

This can be done in a variety of ways. If using radiolabelled cells, the antigen can be visualised by combining IP with SDS-PAGE and subsequent autoradiography of the gel. If precipitating an enzyme, one could measure its activity if an appropriate assay is available. IPs combined with western blotting or SDS-PAGE followed by Coomassie- or Silver staining can also be used to detect the antigen.

5. Co-immunoprecipitation experiments

Co-immunoprecipitation is performed in fundamentally the same way except that extra care is taken during lysate preparation to maintain protein–protein interactions. This could mean use of lower salt concentrations and milder washing conditions. If the use of detergent is required non-ionic detergents such as Triton X-100 would be recommended. For all IPs and co-IPs we include a control sample that does not contain the antigen to be precipitated to ensure the validity of any results obtained.

◆◆◆◆◆◆ IV. IMMUNOFLUORESCENCE METHODS

Immunofluorescence microscopy has for many years been an invaluable tool for localising a protein of interest within its cellular context. While the advent of GFP tagging and live cell imaging has hugely facilitated our understanding of the dynamics of proteins within their biological environment, it remains critical to have complementary methods either to localise proteins when GFP tagging cannot be successfully achieved, or to add credence to GFP localisation data. There are, however, a number of problems associated with protein immuno-localisation in yeast that are rarely encountered in the more commonly studied higher organisms. One of the most obvious differences is the presence of a cell wall. This generates a barrier to fixation and perturbs some of the standard manipulations that are routinely used with many other cell types. The second feature of yeast that affects immuno-localisation technology is their dense cytoplasm. This presents two problems; first, the fixative must penetrate this dense matrix and second there must be enough room for the antibodies to diffuse for specific labelling.

While it is possible to generalise the cell biology of yeast, the processing methods used in different yeasts for immunofluorescence microscopy vary almost as much as in comparison with higher systems. For example, techniques that preserve microtubules perfectly in *Saccharomyces cerevisiae* (Adams and Pringle, 1984) do not preserve the full array of *Schizosaccharomyces pombe* microtubules (Hagan and Hyams, 1988). As the approach has been developed and used most extensively in *S. cerevisiae*, this is the organism that we refer to in the majority of this section.

A. Practical Considerations for Immunofluorescence in Yeast

1. Growing yeast

The growth state of the cells and the type of media in which they are cultured can both have considerable bearing upon the quality of fixation for immunofluorescence. Although this effect is often negligible, rich media generally seem to give more reliable and better fixation than minimal media. If minimal medium must be used because selection is required to maintain a plasmid, cells can be pelleted from the selective medium and then resuspended in rich medium for a period (often 2–3 h) prior to fixation. It should also be noted that the particular stage of the life cycle may also influence staining. There are reported difficulties in staining yeast cells in stationary phase or during sexual differentiation compared with the same strain with the same antibody during vegetative growth. It is not clear whether this effect is due to alterations in internal protein composition that occurs in these phases or the altered cell wall structure.

2. Fixing cells

The aim of fixation is to "freeze" cell structure permanently in the *in vivo* state. Generally this is achieved either by the use of a molecular "glue" to cross-link the cell's components, or by rapid dehydration to induce precipitation of proteins. The chemistry of fixation by cross-linking is complex but essentially it involves chemical reactions of the fixative with specific amino acids on different proteins. If this amino acid forms part of the epitope one wishes to see, then immunofluorescence will be difficult. A further problem arises if the antigen is obscured by being buried deep within a complex. In this context, the strongest fixation is sometimes not the best way to see the protein of interest, as epitopes may be more exposed on a weakly fixed sample that is loosely held in place than in well-fixed samples in which structural integrity is totally preserved. There are inherent advantages and disadvantages with the use of cross-linking or precipitation in yeast immunofluorescence so each approach will be discussed in turn.

(a) Chemical fixation

This is the most popular method for fixation in use in yeast cell biology at present. Formaldehyde is the reagent of choice and generally most budding yeast researchers buy ready-made formaldehyde solutions. One of the problems experienced with aldehyde fixation is that of poor penetration of the fixative into the sample; the fixative needs to cross-link efficiently but not so efficiently that it impedes the penetration of further aldehyde. In situations where strong fixation is of paramount importance, the problem can be overcome through the use of both glutaraldehyde and formaldehyde fixatives. This has been a widely used approach in immunofluorescence studies in *S. pombe*.

The time that a cell spends in fixative is also a crucial factor in protein localisation. For many fusions with the HA epitope, staining is completely abolished by incubations in fixative for more than 10 min. In addition, extended fixation periods can alter protein localization. Spc110p was found at the SPB if fixation in 3% formaldehyde was for 1–2 min (Kilmartin *et al.*, 1993). Longer fixation resulted in general nuclear fluorescence. Data from genetic and biochemical analysis of the spindle pole body (SPB) indicate that Spc110p is an SPB protein, showing that the general nuclear fluorescence is likely to be a fixation artefact. The dangers of artefacts during sample preparation should always be considered.

(b) Solvent fixation

Cold solvents rapidly dehydrate samples, resulting in the precipitation of the proteins *in situ*. Solvent fixation has the disadvantage that cells shrink as they are dehydrated, but this has actually been put to good use in several protocols as it can help expose epitopes that would be hidden by chemical fixation. Furthermore, because the epitope is not modified by reaction with the fixative, antigen preservation can be much better during solvent rather than aldehyde fixation. The most commonly used budding yeast protocols use a solvent fixation after an initial chemical fixation with formaldehyde (see Protocols 7 and 8). This second step helps to open up the fixed cell structure to the antibody. We have also included a method that uses only solvent fixation (Protocol 9). This is not widely used, but is sometimes the only available method if the major epitope is modified by aldehyde fixation. The method included here is adapted for more general use from one published to localise carbonyl-derivatised proteins in *S. cerevisiae* (Aguilaniu *et al.*, 2003).

One problem that is associated with simple solvent fixation is that while the cytoplasm generally shows excellent preservation, nuclear preservation is highly variable. This may be explained by the

Protocol 7. Immunofluorescence in *S. cerevisiae* using formaldehyde fixation followed by methanol/acetone treatment.

1. Grow a 5 ml culture of *S. cerevisiae* in rich medium to log phase (for wild-type cells this corresponds to an OD_{600} of about 0.2–0.5).
2. Fix the actively growing population of cells by addition of 0.67 ml of 37% formaldehyde and allow to stand at room temperature for 1 h.
3. Spin the cells for 2 min at 1000*g* to pellet, and wash twice with 2.5 ml phosphate/sorbitol buffer (filter sterilised 0.1 M potassium phosphate buffer pH 7.5, 1.2 M sorbitol).
4. Resuspend cells in 0.5 ml phosphate/sorbitol buffer and add 20 µl of 1 mg/ml Zymolyase 100T stock (1 mg/ml stock in potassium phosphate buffer pH 7.5. Store at –20°C as 1 ml aliquots) and 1 µl β-mercaptoethanol. Incubate for 30–35 min at 37°C.
5. During the cell wall digestion incubation time the slides can be prepared for cell mounting. Place 15 µl of 1 mg/ml poly-L-lysine (make up and store at –20°C as a 1 mg/ml stock in ultra pure) on each well of a multi-well slide (pre-clean slide by immersing in ultra pure then in 95% ethanol and air drying). Allow to sit for 5 min then wash each well 3 times with distilled water and air dry.
6. After cell wall digestion put 15 µl of cell suspension on each well. Allow to settle for 5–10 min then aspirate off gently. If obtaining enough cells for observation is a problem, the suspension can be spun down and resuspended in a smaller volume just before placing on the slide.
7. Place the slide in –20°C methanol for 6 min, then in –20°C acetone for 30 s. This step flattens the cells, permeabilises them and can aid antibody penetration.
8. After air-drying the slides, wash each well 10 times with blocking buffer (1% BSA – Fraction V in 1 × Phosphate-buffered saline, PBS) by gently aspirating off the wash buffer and then reapplying from a pipette. From this step onward it is important not to let the cells become dry. Keep the slides in a humid environment, for example, in a covered Petri dish next to a damp tissue.
9. Remove the final blocking buffer wash from the well and place 15 µl of the diluted primary antibody[a] onto the cells. Incubate for at least 1 h.
10. Wash each well 10 times with blocking buffer and add 15 µl of diluted secondary antibody.[b] Incubate for 1 h in a dark place to prevent bleaching of the fluorophore.
11. Wash again 10 times with blocking buffer, aspirate the buffer and put 5 µl of mounting solution (Protocol 10) onto each well. Cover with coverslip, then seal around the edges

with nail polish. Slides can be viewed immediately or stored at −20°C in the dark for several months.

Notes

[a]The antibody should be diluted as necessary in blocking buffer. In general antibodies are used at a greater concentration (about 10-fold greater) for immunofluorescence than for western blotting.

[b]The secondary antibody should be diluted using blocking buffer. For many secondary antibodies we find that a 1/500–1/1000 dilution is suitable.

Protocol 8. Immunofluorescence in *S. cerevisiae* using formaldehyde fixation followed by SDS treatment.

1. Protocol 1 should be followed through to step 6.
2. Following mounting of the cells on the slide, aspirate excess cells and add 10 μl of 0.1% SDS in PBS for 30–60 s. Owing to the presence of SDS the solution loses surface tension. To avoid spreading of solution from the wells, the volume of solution used in this permeabilisation stage and in all subsequent wash steps and incubation procedures should be reduced to 10–2 μl rather than 15 μl.
3. Gently aspirate the SDS solution and wash ten times with blocking buffer. Do not leave the slide to dry in between the SDS step and washes.
4. Continue to process the cells for immunofluorescence as in Protocol 7 from step 9.

fixation being insufficiently rapid to preserve the central structures before they are destroyed by the other changes in the cell. One possible way around the problem of loss of structure in the middle of the cell is to put very small samples into extremely cold liquids with low heat capacity, such as liquid helium and liquid propane. However, this technique is so specialised that it is not recommended for routine fluorescence microscopy, rather for electron microscopy where precise preservation is a necessity. The temperature of the solvent is important as solvents at –80°C often give better preservation than the same solvent at –20°C.

3. Harvesting cells

It has been noted in *S. cerevisiae* that harvesting cells by centrifugation can disrupt the actin cytoskeleton and organelle morphology (Pringle *et al.*, 1991). If chemical fixation is used, this is not generally a problem as cells are directly fixed in culture by the rapid addition

Protocol 9. Solvent fixation for immunofluorescence.

1. Grow a 5 ml culture of *S. cerevisiae* in rich medium to log phase.
2. Harvest cells onto a glass fibre filter, wash twice with ultra pure water and then plunge filter into a tube of cold 70% ethanol (−20°C) for 45 min.
3. Scrape cells from filter into the ethanol, spin briefly and resuspend pellet in 2.5 ml phosphate/sorbitol buffer (filter-sterilised 0.1 M potassium phosphate buffer pH 7.5, 1.2 M sorbitol). Wash pellet twice in 2.5 ml buffer.
4. Resuspend cells in 0.5 ml phosphate/sorbitol buffer and add 20 µl of 1 mg/ml Zymolyase 100T stock (1 mg/ml stock in potassium phosphate buffer pH 7.5. Store at –20°C as 1 ml aliquots) and 1 µl β-mercaptoethanol. Incubate for 20 min at 37°C.
5. Prepare the slides for cell mounting. Place 15 µl of 1 mg/ml poly-L-lysine (make up and store at –20°C as a 1 mg/ml stock in ultra pure water) on each well of a multi-well slide (pre-clean slide by immersing in ultra pure water then in 95% ethanol and air drying). Allow to sit for 5 min, then wash each well 3 times with ultra pure water and air dry.
6. After cell wall digestion put 15 µl of cell suspension on each well. Allow to settle for 30 min, then aspirate off gently. If obtaining enough cells for observation is a problem, the suspension can be gently spun down and resuspended in a smaller volume just before placing on the slide.
7. Wash each well 10 times with Blocking Buffer (1% BSA in PBS) by gently aspirating off the wash buffer and then re-applying from a pipette. Keep the slides in a humid environment, for example, in a covered Petri dish next to a damp tissue.
8. Remove the final blocking buffer wash from the well and place 15 µl of the diluted primary antibody onto the cells. Incubate for at least 1 h.
9. Wash each well 10 times with blocking buffer and add 15 µl of diluted secondary antibody. Incubate for 1 h in a dark place to prevent bleaching of the fluorophore.
10. Wash again 10 times with blocking buffer, aspirate the buffer and put 5 µl of mounting solution (Protocol 10) onto each well. Cover with coverslip, then seal around the edges with nail polish. Slides can be viewed immediately or stored at −20°C in the dark for several months.

of about 1/10th volume of culture. For solvent fixation however, fixing in culture is not practical and cells are preferentially harvested by filtration onto glass fibre filters. With filtration it is important not to harvest too many cells at one time as the filter quickly gets clogged. Fixation is achieved by simple plunging of the filter into cold solvent in a tube. Because the filter warms up the solvent, and because temperature affects the fidelity of solvent fixation, the quality of fixation may be reduced if more than one filter is put into a tube. It is therefore advisable to use a large volume of solvent, and to use the solvent at as low a temperature as possible, i.e. at –80 °C. A final form of harvesting is to scrape cells from the surface of an agar plate with a coverslip. However, for most the interest in protein localisation is in cells during vegetative growth and the physiology of cells taken from plates is likely to be highly variable, making this unlikely to be a method of first choice.

4. Cell wall digestion

A critical stage in the immunofluorescence microscopy of yeast is the removal of the cell wall. In most cases this is done after the cells have been fixed. It should be noted, however, that there are occasions when removal of the cell wall to make spheroplasts has been performed prior to fixing. This has been used for studies on spindle pole body components in *S. cerevisiae* (Rout and Kilmartin, 1990) when a minimal fixation is used (5–10 min in formaldehyde). In this case the cell wall was digested and the cells allowed to recover for about 30 min before addition of fixative to the medium.

There are several enzyme preparations available, of varying purities. The most common enzyme preparation used for cell wall digestion in *S. cerevisiae* is Zymolyase-20T or -100T (MP Biomedicals). This is a relatively pure preparation and is suitable for digestion of cell walls for yeast that have been growing in rich media. Growth in minimal media to beyond about 0.5 OD_{600}, and certain mutations, directly affect the composition of the cell wall such that digestion with this relatively pure enzyme preparation is not always sufficient. If growth in minimal media cannot be avoided due, for example, to the need to preserve selection for a plasmid, then two main approaches are possible. One is to grow the cells with selection until the final generation time. At this point the cells can be spun down and resuspended in rich medium and grown for a further 2–3 h. Alternatively, more crude enzyme preparations can be used. Such crude enzyme mixtures are likely to contain enzymes capable of digesting a wider range of sugar modifications and are likely therefore to result in a more complete removal of the cell wall. However, these mixtures tend to be avoided by many researchers since they have been known to contain significant levels of protease contamination, which could be detrimental to later procedures.

5. Mounting cells

For most standard budding yeast procedures, cells are generally mounted in wells on poly-L-lysine-coated slides after digestion of the cell wall. The remainder of the processing is done on the slide. The advantage of this procedure is that many samples can be easily processed by aspiration of the wash solution from the slide, and that relatively small volumes (10–15 µl) of antibodies are required for each sample. It can be noted that the *S. pombe* community mostly mounts samples after incubation with antibodies. While this approach uses larger amounts of antibodies it is possible to keep labelled cells for long periods of time at 4°C.

6. Epitope masking and further processing procedures

One problem frequently encountered with protein localization is that the epitope recognised by the antibody is obscured by other proteins. This can be circumvented by several approaches. Most commonly, after mounting on the slide, cells can be plunged into cold methanol and acetone. A particularly useful step is a 30 s to 2 min incubation with 0.1% SDS solution prior to addition of primary antibody (see Protocol 8).

7. Considerations for use of antibodies in immunofluorescence microscopy

(a) Primary antibodies

There are an increasing number of commercially available antibodies to specific proteins of interest. However, many of these have only been shown to recognise proteins by western blotting and in some cases by immunoprecipitation. One should note that success in one procedure by no means guarantees success for other purposes. Commonly used, commercially available antibodies for immunofluorescence include those for tubulin (YOL1/34, Serotec), for nuclear pores (MAb414, Covance Research Products), and for actin (N350, GE Healthcare). These antibodies can often be used as positive control antibodies that can be very valuable when attempting immunofluorescence for a protein of unknown cell localization. Use of these antibodies will allow you to determine whether the technique has at least been successful for proteins that should be localised in a specific way. As well as commercial antibodies, colleagues working in the field are often a source of antibodies. However, this source is often relatively limited, so for localising a protein over a series of experiments it will most often be the case that antibodies will not be readily available. The decision can then be made as to whether to raise your own antibodies, get a company to raise them for you, or to tag your protein and use the more readily available commercial antibodies that recognise protein tags.

For most immunofluorescence procedures, purified polyclonal antibodies are as good as, or superior to, monoclonal antibodies. A preference for raising polyclonal antibodies derives from the probability that the serum will contain multiple antibody species that recognise different epitopes on the protein of interest. Thus, if a particular epitope is sensitive to fixation or deeply embedded in the native protein, it is also probable that other epitopes will not be. In addition, if an antiserum is raised to a large part of the protein it is possible that several antibodies could bind simultaneously, so increasing the strength of the immunofluorescence signal. When generating a polyclonal antiserum it is important to recognise that the serum is likely to contain many antibodies, some of which might react with other antigens in yeast cells. To ensure specificity, antibodies should be affinity purified (see above). Both micro-affinity purification and larger scale, column-based approaches have been successfully used in producing antibodies for immunofluorescence studies in yeast cells. Another stage of purification that is relatively simple and has been particularly useful in immunofluorescence studies is to pre-absorb the antiserum using fixed yeast cells or an acetone powder of a yeast lysate (Protocol 5) from a strain in which the gene encoding the protein of interest has been deleted.

The most significant recent impact on the study of protein localization in yeast has come from advances in protein tagging techniques. Tags used for immunofluorescence studies are coding sequences that are fused to the gene encoding a protein of interest which constitute epitopes recognised by specific antibodies that are readily available. Proteins have been successfully localised in *S. cerevisiae* using a range of tags including the *myc* tag (Gourlay *et al.*, 2003) and the HA- tag (Kilmartin *et al.*, 1993). There are clear advantages to using tagging as an approach to localise proteins, including the relatively short timescale needed to manipulate the gene and express it in cells, compared with raising antibodies. In addition, providing the antibody preparation is pure, only the protein carrying the tag will be detected by immunofluorescence. This is particularly relevant when studying one member of a family of highly related proteins. Furthermore, being able to use the same antibodies for successive localisations instead of having to raise antibodies for each protein of interest can be a considerable monetary saving. Limitations to use of tags of particular relevance to immunofluorescence studies are the problems with tag preservation, especially for HA; the need to use multiple tags that may confer different properties on the protein, for example, *myc* is a negatively charged epitope and the commonly used 9 and 13 repeat motifs could have a significant effect; the presence of the tag could interfere with localisation directly by obscuring normal interactions with other proteins or membranes; finally, the expression levels should be considered especially if the tagged protein is expressed from a plasmid, although the ability to integrate tag sequences directly into the genome (Chapter 4) has to an extent circumvented this problem.

Once a staining pattern has been observed it is imperative that the localization is then demonstrated to be specific and preferably seen when several different fixation procedures are used. In this regard, showing a lack of staining in a strain in which the gene encoding the protein of interest has been deleted is a suitable control. Another good control when using affinity-purified antiserum is to pre-absorb the serum against the antigen and then show that a specific staining pattern is no longer obtained. GFP fusions now offer one of the best controls as the distribution in living cells should offer relatively artefact-free results as long as expression levels are appropriate. Ultimately, there is no perfect control to show that the staining pattern reflects normal localization and it is important to confirm results by complementary approaches. These approaches might include co-immunoprecipitation to show binding to another component of the structure to which it is localised, and co-fractionation *in vitro*.

(b) Secondary antibodies

The choice of secondary antibody is critical for good immunofluorescence staining. Two major considerations are the choice of fluorophore and the quality of the antibody preparations. The last five years have seen a plethora of new fluorophores available conjugated to a wide range of secondary antibodies. Advantages of some of the new fluorophores such as Cy3 (Sigma) or the Alexa™ dyes (Molecular Probes) are their brightness, their increased photostability, and sometimes much tighter excitation and emission spectra such that the fluorescence is therefore less likely to "bleed" into other channels, which is of particular importance for double labelling experiments. It should be remembered, however, that a brighter secondary is not going to overcome problems of a poor primary antibody. Rather it is of particular use when the antigen is present at low levels within the cells or if the primary antibody binds very specifically but weakly to its antigen. If the primary antibody is not sufficiently specific for the protein of interest, then a brighter secondary will often result in an increase in general background staining but give no additional information. Often the increased fluorescence might even mask an otherwise weak signal.

When considering buying secondary antibodies it is often a good idea to try similar products from several companies to allow evaluation of brightness, often balanced against an increased background of fluorescence. The level of cross reactivity might also be important for particular experiments. Once a suitable antibody has been identified, it is usually a good idea to note the batch number and to order a number of vials (they can be stored at –80°C for several years without problems).

(c) Double labelling

To demonstrate the spatial relationships of two different proteins in the same cell, double labelling is often valuable. This can be relatively straightforward if primary antibodies from different types of animals are available. Appropriate secondary antibodies can then discriminate the antibody types to allow the staining patterns of each protein to be observed. It should be noted that the primary and secondary antibodies can be incubated with cells either sequentially or simultaneously. If the proteins show co-localization it is important to demonstrate that this is not due to cross-species reactivity of the secondary antibodies used. Thus, control samples should be set up in which one or the other primary antibody is left out of the incubation. Both secondary antibodies should then be added. A specific signal in the absence of the relevant primary antibody would indicate cross-reactivity. To remedy this, cross-reacting antibodies can be pre-absorbed against immobilised antibodies from the second class. A second concern with regard to double labelling is that of crossover fluorescence. For example, sometimes illumination for FITC can result in some rhodamine fluorescence from a double-labelled sample. This can usually be circumvented by the use of different fluorophores with different excitation spectra further away from that of FITC (e.g. Cy3 or Texas Red) or by using different filter sets on the microscope.

8. Use of anti-fade and DNA co-staining solutions

A final step before the cells are to be visualised requires the addition of mounting medium. This is usually a glycerol-based solution that often contains a dye to permit the additional visualisation of DNA (Protocol 10). There are various anti-fade mounting solutions commercially available, for example, Citifluor (Agar Scientific) and Vectashield (Vector Laboratories). It is however possible and relatively straightforward to make mounting solution in the lab using an anti-fade reagent *p*-phenylenediamine. As this chemical is toxic and reported to be carcinogenic, it is preferable to weigh it out infrequently and make large batches that can be stored at –80 °C for long periods of time.

Protocol 10. Mounting medium for immunofluorescence procedures.

1. Add 100 mg *p*-phenylenediamine to 10 ml PBS (Phosphate buffered saline: dissolve 8 g NaCl, 0.2 g KCl, 1.44 g Na_2HPO_4, 0.24 g KH_2PO_4 in 800 ml ultra pure water. Adjust the pH to 7.2 and make up to 1 l and autoclave). If the pH is below 9.0, bring it to pH 9.0 by adding NaOH while stirring.
2. Add 90 ml glycerol and stir until homogeneous.
3. Add 2.25 μl of DAPI (4′,6′-diamidino 2-phenylindole dihydrochloride-1 mg/ml in water) if you wish nuclei/mitochondria to be visualised.
4. Store mounting medium at –80°C in the dark. The solution should be discarded when it loses its clear colour and turns brown.

In addition to mounting cells in anti-fade solution, it is often useful to co-stain with a DNA binding compound. DNA staining is often bright and can help when trying to visualise cells. It also can give an idea of the preservation state of the cell. The immunofluorescence protocols outlined here suggest 4′,6′ diamidino 2-phenylindole dihydrochloride (DAPI) for DNA staining although propidium iodide or ethidium bromide can be used following RNase digestion. This becomes a particularly attractive option when it is necessary to conduct double labelling with another stain that emits in the blue wavelengths, such as calcofluor, or when it is necessary to use a confocal microscope that does not have a UV laser.

9. Visualising cells

The small size of yeast cells means that the intensity of the signal is a major concern. It is important to have access to a good quality microscope that has high magnification objectives. The brighter the light source the better. Some fluorescence microscopes have a 50 W mercury lamp fitted as standard. However, if a 100 W light source is used there is a very significant difference in the image. Different objectives are designed for different purposes. Some objectives have extra lenses in order to generate a flat field in which the entire image is in focus. The more glass a lens has the more light is lost, so it is often preferable with dim samples to use lenses that have been optimised for light transmission rather than generating a flat field. The numerical aperture of a lens gives an indication of the amount of light transmitted by the lens, the higher the value, the more light is transmitted. The numerical aperture is usually written alongside the magnification power on the side of the objective. Some microscopes offer internal magnification in the microscope turret. The temptation is to use the greatest magnification possible. After a certain point however, better images are generated by capture with a low magnification and increasing the image size in subsequent image or film processing steps.

10. Setting about localising a novel protein

As with many methods, each lab uses slight variations to optimise staining for particular antigens. It should be remembered, however, that artefactual staining is possible and varying the fixation procedures or the use of antibodies combined with GFP-fusion protein localization will give credence to any localisation data. The first immunofluorescence method given here is the procedure that has been most widely used and is very similar to that described by Pringle and colleagues (1991). It has been used for visualising the actin cytoskeleton and several actin-binding proteins, for neck filament proteins, for proteins involved in secretion and for microtubules.

While this first protocol serves as a useful starting point, many antigens are better observed when this protocol is modified in some way. One of the most useful adaptations has been the replacement of the methanol/acetone step with a short incubation in a dilute SDS solution (Protocol 8). This has been particularly useful in *S. cerevisiae* for localising many proteins that are found at the presumptive bud site, such as Cdc42p and Sec4p, or within the dense complex of the endocytic patch. As discussed earlier in this section, other variations to optimise staining might include changes in growth conditions, the duration of formaldehyde fixation time or increasing the time of digestion of the cell wall.

♦♦♦♦♦♦ V. SUMMARY

In summary, methods for tagging proteins of interest and use of GFP in particular has revolutionised the approaches available to yeast researchers for undertaking studies on localisation and interactions of specific proteins. However, an inability to tag a protein and maintain functionality should not preclude important studies and for this reason there remains an important role for immunological methods in our standard battery of techniques. The methods outlined here should, we hope, provide a useful start point for many of the more commonly used approaches.

References

Adams, A. E. and Pringle, J. R. (1984). Relationship of actin and tubulin distribution to bud growth in wild-type and morphogenetic-mutant *Saccharomyces cerevisiae*. *J. Cell Biol.* **98**, 934–945.

Aguilaniu, H., Gustafsson, L., Rigoulet, M. and Nystrom, T. (2003). Asymmetric inheritance of oxidatively damaged proteins during cytokinesis. *Science* **299**, 1751–1753.

Gourlay, C. W., Dewar, H., Warren, D. T., Costa, R., Satish, N. and Ayscough, K. R. (2003). An interaction between Sla1p and Sla2p plays a role in regulating actin dynamics and endocytosis in budding yeast. *J. Cell Sci.* **116**, 2551–2564.

Hagan, I. M. and Hyams, J. S. (1988). The use of cell division cycle mutants to investigate the control of microtubule distribution in the fission yeast *Schizosaccharomyces pombe*. *J. Cell Sci.* **89**, 343–357.

Harlow, E. and Lane, D. (1988). *Antibodies: A laboratory manual*. Cold Spring Harbor Laboratory Press, New York.

Harlow, E. and Lane, D. (1998). *Using antibodies: A laboratory manual*. Cold Spring Harbor Laboratory Press, New York.

Kilmartin, J. V., Dyos, S. L., Kershaw, D. and Finch, J. T. (1993). A spacer protein in the *Saccharomyces cerevisiae* spindle pole body whose transcript is cell cycle-regulated. *J. Cell Biol.* **123**, 1175–1184.

Martin, H., Rodriguez-Pachon, J. M., Ruiz, C., Nombela, C. and Molina, M. (2000). Regulatory mechanisms for modulation of signaling through the cell integrity Slt2-mediated pathway in *Saccharomyces cerevisiae*. *J. Biol. Chem.* **275**, 1511–1519.

Pringle, J. R., Adams, A. E., Drubin, D. G. and Haarer, B. K. (1991). Immunofluorescence methods for yeast. *Methods Enzymol.* **194**, 565–602.

Rout, M. P. and Kilmartin, J. V. (1990). Components of the yeast spindle and spindle pole body. *J. Cell. Biol.* **111**, 1913–1927.

12 Measuring the Proximity of Proteins in Living Cells by Fluorescence Resonance Energy Transfer between CFP and YFP

Trisha N Davis and Eric GD Muller
Department of Biochemistry, University of Washington, Seattle, WA 98195, USA

♦♦

Measuring the Proximity of Proteins in Living Cells

CONTENTS

Introduction
Fluorescent proteins
Tagging the genes of interest
Preparation of the cells for imaging
Preparing the slide
Image acquisition

List of Abbreviations

CFP	cyan fluorescent protein
FRET	fluorescence resonance energy transfer
PAGE	polyacrylamide gel electrophoresis
SPB	spindle pole body
YFP	yellow fluorescent protein
YRC	Yeast Resource Center (http://depts.washington.edu/~yeastrc/index.html)

♦♦♦♦♦♦ I. INTRODUCTION

Protein localization via live cell microscopy offers unique advantages for exploring a proteome. The availability of fluorescent proteins of a wide range of colors and the simple methods available for tagging proteins in yeast make localization of proteins in living cells a simple technique for initial characterization of a new protein. Certain localizations (e.g. the kinetochore or bud tip) give strong predictions of the function of a protein (Hazbun *et al.*, 2003).

METHODS IN MICROBIOLOGY, VOLUME 36
0580-9517 DOI:10.1016/S0580-9517(06)36012-6

Fluorescence microscopy can determine whether two tagged proteins localize to the same compartment, but because of the limit of resolution of light microscopy (~200 nm), it cannot demonstrate interactions between a protein pair. Fluorescence resonance energy transfer (FRET) overcomes this limitation because typically FRET can only occur if two fluorophores are at least within 7 nm. FRET can be measured in living cells and therefore provides direct information about the interactions of proteins in their native context.

FRET is the transfer of excitation energy from one fluorophore, referred to as the donor, to a second fluorophore, the acceptor. It can occur when the wave properties of the electrons in the excited state of the donor overlap the wave properties of the ground state electrons of the acceptor. The two electronic states resonate with the excitation energy moving to the acceptor chromophores. Some of the energy transferred to the acceptor decays to the ground state, emitting light. The donor, by transferring its energy to the acceptor, returns to the ground state without the radiation of light.

FRET has a strong dependence on the intermolecular distance, r, between the chromophores. As described by Förster,

$$E = \frac{R_0^6}{R_0^6 + r^6}$$

where E is the efficiency of transfer, and R_0 is the distance at which the efficiency is 50%, referred to as the Förster distance (Stryer, 1978). R_0 is a function of the quantum yield of the donor in the absence of the acceptor, the overlap of the emission spectrum of the donor and the excitation spectrum of the acceptor, the refractive index and the relative orientation of the donor and acceptor. For the CFP–YFP pair, R_0 is estimated to be 49 Å (Heim, 1999). Thus, FRET between cyan fluorescent protein (CFP) and yellow fluorescent protein (YFP) is most sensitive to changes in distance between 35 and 65 Å.

The efficiency of energy transfer is also a function of the orientation of the two chromophores. The flexibility and many vibrational and rotational states of small biological molecules led to the conclusion, supported experimentally, that the orientation effectively randomizes during the fluorescence lifetime of the donor (Stryer, 1978). However, CFP is relatively large, with a rotational correlation time of approximately 15 ns compared to lifetime of 1–4 ns (Borst *et al.*, 2005). Thus any given molecule of CFP will not change orientation between the time it is excited and the time it transfers energy to a nearby YFP. When measuring FRET *in vivo* by epifluorescence microscopy, the FRET signal is an average of many transfer events during the time of exposure. During this time the population of CFP and YFP molecules present a collection of orientations. Since CFP and YFP are tethered to the surface of the tagged proteins via flexible linkers, it is unlikely that their orientations remain fixed in a position that significantly distorts the transfer efficiency. Therefore we maintain that the magnitude of FRET is still largely a function of the distance between the two fluorophores. In our application of FRET to study the organization of the yeast spindle pole body (SPB), we found that the strength of FRET signals between CFP- and YFP-labeled proteins were consistent with a model that interpreted FRET solely as a measure of distance without regard to orientation (Muller *et al.*, 2005).

Here the complete process of a FRET-based analysis of the organization of a protein complex is described. The properties of the two versions of YFP are compared for their value as FRET donors in yeast. Protocols for tagging proteins are presented. Methods for imaging cells are described. FRET is measured by digital epifluorescence microscopy. We demonstrate use of our new metric, $FRET_R$, to measure sensitized emission. $FRET_R$ is both intuitive, easy to implement and simplifies comparisons across projects and labs (Muller *et al.*, 2005). Additional information is available in the recent review of Muller and Davis (2006).

♦♦♦♦♦♦ II. FLUORESCENT PROTEINS

For FRET measurements *in vivo* only CFP and YFP are well established for work in yeast. BFP and GFP are possible FRET partners, but BFP expresses poorly in yeast. The new mOrange is a recommended FRET partner for T-Sapphire, but the slow folding rate of mOrange ($t_{1/2}$ of 2.5 h) may preclude its use in yeast (Shaner *et al.*, 2004). Indeed when labeling SPB components, we find that the mOrange tag interferes with the function of the tagged protein.

Two versions of YFP, YFP 10C Q9 K (chosen because of its reduced pKa; Miyawaki *et al.*, 1999) and Venus (Table 1), are both viable FRET partners for CFP. Both Venus and YFP confer a similar level of fluorescence at 30°C. However, the signal from Venus is about 20% better at 37°C and therefore is useful for analyzing the localization of proteins in temperature-sensitive mutants. One test we have for a fast maturation rate of a fluorescent protein *in vivo* is to determine whether both spindle pole bodies are equally labeled when Spc42 is tagged. For example, DsRed, with a maturation time of hours, strongly labels the old SPB, which goes into the daughter cell, whereas GFP equally labels each SPB (Pereira *et al.*, 2001). We find that YFP also equally labels each SPB (Table 1). However, Venus slightly labels the old SPB better at 30°C. For this reason YFP is recommended for all FRET experiments performed at 30°C.

♦♦♦♦♦♦ III. TAGGING THE GENES OF INTEREST

The first step toward measuring FRET between two proteins of interest is to tag the corresponding genes with CFP and YFP. There

Table 1. Comparison of YFP and Venus fluorescent proteins[a]

	30°C	30°C	37°C	37°C
Fluorescent protein	Intensity[b]	Percentage of cells with brightest SPB going to daughter[c]	Intensity[b]	Percentage of cells with brightest SPB going to daughter[c]
YFP	821±176	59%	574±112	59%
Venus Q69M	802±174	85% (1.28)[d]	694±122	47%

[a]The spindle pole body component Spc42 was tagged with either YFP 10C, Q69K, (S65G, V68L, Q69K, S72A, T203Y) or Venus Q69M (F46L, F64L, S65G, V68L, Q69M, S72A, M153T, V163A, S175G, T203Y: Nagai *et al.*, 2002) in diploid strains of the W303 background (BESY114 and BESY116). Cells were grown at the temperature shown, mounted and imaged as described in the text.
[b]Intensity was measured in 5 × 5 pixel box surrounding the SPB using MatLab, after conversion of files to 8-bit grayscale using graphic converter. Background was measured in a 5 × 5 pixel box adjacent to the SPB and subtracted from the values given. Values are shown ± the standard deviation. At least 60 SPBs in cells with short spindles were measured for each condition.
[c]This is a measure of the maturation rate of the fluorescent protein. The SPB closest to the neck is the older SPB. If the fluorescent protein folds slowly, the older SPB will have more mature fluorescent protein and will be significantly brighter than the SPB in the mother cell, which is the new SPB.
[d]Average ratio of intensity of SPB going to daughter/intensity of SPB staying in mother. This result suggests that Venus may show a delay in maturation at 30°C.

Table 2. Plasmids available for FRET[a]

Plasmid name	Selectable marker	Fluorescent protein	*N*-terminal or *C*-terminal tag
pDH5	His	YFP	*C*-terminal
pDH6	G418	YFP	*C*-terminal
pDH3	G418	CFP	*C*-terminal
pBS4	Hygromycin-B	CFP	*C*-terminal
pDH18	His	YFP-CFP	*C*-terminal
pDH22	G418	YFP	*N*-terminal
pBS5	G418	CFP	*N*-terminal

[a]Plasmids can be requested via our web site: http://depts.washington.edu/yeastrc

are now standard techniques to tag the 5′ end (Prein *et al.*, 2000) or the 3′ end (Wach *et al.*, 1997) of an open reading frame at the endogenous locus as described in Chapter 4, resulting in the tagged gene being the only copy of the gene and being expressed under the control of its own promoter. The Yeast Resource Center (YRC) (http://depts.washington.edu/yeastrc) has created vectors for tagging any protein on either end with either CFP or YFP using PCR amplification with customized primers (Table 2).

To maximize the probability of the tagged protein functioning properly, several steps are typically taken. A 10 amino acid linker of Gly-Ala repeats between the tagged protein and the fluorescent protein provides the greatest degree of flexibility to the linkage. Amplification of the cassette is carried out using a high-fidelity DNA polymerase, such as Expand or Phusion. Although time consuming, the cassette is transformed into a diploid to integrate into one copy of the gene and leave the other copy wild type. This reduces any selective pressure against the tagged copy of the gene during the integration step, a problem that can lead to the selection of frameshifts that prevent the expression of the fluorescent protein even though the cassette is integrated. Usually a dozen transformants are saved for analysis.

Correct integration is checked by PCR. For C-terminal integrations, a 5′ primer is designed that is about 500 base pairs (bp) upstream of the stop codon and a 3′ primer that is just downstream of the stop codon. This gives a 500-bp fragment if integration has not occurred in the correct place and about a 2.5 kb fragment (for YRC vectors) if integration has occurred correctly. In the heterozygous diploid, both fragments should appear and the 500-bp fragment is then a good control for the PCR. The diploids can be checked for a fluorescence signal by microscopy; however, the signal may be faint in the presence of the wild type gene. For N-terminal integrations the 5′ primer is designed to hybridize 500 bp upstream of the start site, and the 3′ primer site is just downstream of the start site. Correct integration leads to a 2.7 kb fragment before the induction of the *cre* recombinase to loop out the selectable marker.

Haploids are obtained by sporulation and tetrad dissection of the diploids (see Chapter 2), and again checked by PCR for correct integration using the same primers as before. In the haploids, only the 2.5-kb fragment should be present for C-terminal tags, and a 1.2-kb fragment for *N*-terminal tags. A common incorrect integration shows the presence of both the tagged gene and the untagged gene even in the haploid. Frequently a correct haploid can be obtained from a different diploid integrant from the original transformation. If the haploid strains are correct as indicated by PCR, we then examine them in the microscope to be sure the fluorescent tag is expressed. If the integration was correct by PCR, but show no fluorescence this may be because the protein is of very low abundance, but can also be due to an error in the sequence such that a frameshift or stop codon is introduced between the gene and the tag at the point of integration. To minimize this problem, the primer at the joint between the fluorescent protein and the target should be purified by polyacrylamide gel electrophoresis (PAGE). To save expense the opposing primer need not be purified.

We construct strains with two tagged proteins by crossing two haploids each with one gene correctly tagged, sporulating the ensuing diploid and then dissecting and identifying the progeny with both tags (see Chapter 2 for details). Although it is possible to transform a second tag into a strain with one protein already tagged, most of the transformants are incorrect because of the extensive homology between the cassettes. For example, if the first protein is tagged with CFP, then after transformation with a YFP cassette intended for a second gene, we commonly find that the tag on the first protein is switched such that the CFP-tagged protein is now tagged with YFP. This is true even in transformants that are resistant to both the drug marking the CFP cassette and the drug marking the YFP cassette.

♦♦♦♦♦♦ IV. PREPARATION OF THE CELLS FOR IMAGING

A. Recipes

YPD 3 × Ade Plates (per liter). Add 10 g yeast extract, 30 g Bacto peptone, 15 ml of 5 mg/ml adenine, 5 ml 2.5 mg/ml uracil and 20 g Agar to 930 ml deionized water, autoclave and then add 50 ml 40% glucose after autoclaving.

S medium (per liter). Add 1.7 g yeast nitrogen base without amino acids or ammonium sulfate and 5 g ammonium sulfate to 950 ml deionized water. Typically divide into 95 ml aliquots in 100-ml bottles and autoclave.

SD medium. Add 5 ml of 40% glucose per 100 ml S medium.

Seakem GTG agarose. Add 1 g Seakem GTG agarose (Cambrex) to 100 ml S medium, melt the agarose and aliquot into 0.9 ml aliquots in 1.5-ml microfuge tubes.

10 × nutrients. Mix 500 μl of 40% glucose, 100 μl of 5 mg/ml adenine, 100 μl of 2.5 mg/ml uracil, 100 μl of filter sterilized 10% casamino acids (Difco), 100 μl of 1% filter sterilized tryptophan (store refrigerated).

SDComplete agarose. Melt 0.9 ml 1% Seakem GTG agarose at 100°C for 5 min and cool to 65°C), then add 90 μl 10 × nutrients.

B. Growing the Cells

On the day before image acquisition, fresh cells are plated on YPD supplemented with 3 × adenine and grown at 30°C. The addition of extra adenine eliminates the red pigment that would otherwise accumulate in our *ade2* strains (Smirnov et al., 1967) and interfere with fluorescence microscopy. The next day colonies the size of pin heads are scraped up together and resuspended in 10–30 μl of SD medium. There is no need for sonication before visualization since cell separation is complete under these conditions. A 3 μl aliquot is placed on a 1% SDC agarose pad on a microscope slide prepared as described below.

♦♦♦♦♦♦ C. PREPARING THE SLIDE

To minimize aberrations in the acquired image the yeast must be mounted adjacent to the coverslip (Nikon). We accomplish this by preparing an agarose pad, depositing the yeast on the pad and then putting down the coverslip. In the next few paragraphs the preparation of the slide is described.

Two strips of scotch tape are placed on either end of a standard microscope slide, such as the pre-cleaned Gold Seal microslides, cat. no. 3010. The slide is wiped with a tissue before the tape is applied.

Then a 30 μl aliquot of the SDC agarose is pipetted into the center of the slide.

At this time a second slide, also wiped with a kimwipe, is gently placed on top of the taped slide to spread the agarose without trapping bubbles. Scotch Tape is 62.5-nm thick and so the agarose forms a pad of similar thickness. The slide sandwich is then placed on a −20°C block that was just removed from a freezer. The agarose is allowed to harden on the block for about 10 s. The key is to harden the agarose but not to let it freeze. Freezing leads to high background in the images and frozen pads should be discarded.

The two slides are slid apart, like the technique used to separate the two halves of an Oreo cookie to get to the filling except here one uses a constant, gentle sliding action. The pad usually sticks to one slide or the other. If you are preparing multiple slides, then reuse the slide that does not get the pad by cleaning it with a kimwipe and warming to room temperature. The technique can be frustrating at first, as the pad is fragile and can easily rip apart. After a while you can expect 95% success rate.

The pad is left out to dry for approximately a minute, the length of time depending on the humidity. The tube with the yeast is briefly vortexed, and a 3 μl sample is dropped onto the center of the pad. A standard No. 1 1/2, 22-mm square coverslip (e.g. from Corning) is placed on top of the cells on top of the pad. The coverslip is not sealed since the pad will not dry out for a few hours, plenty of time to image the cells.

◆◆◆◆◆◆ V. IMAGE ACQUISITION

Before proceeding with our general acquisition parameters, several properties of YFP and CFP need to be introduced. First, CFP and YFP photobleach much more rapidly than GFP. On a DeltaVision microscope equipped with a HBO 100 W mercury short arc lamp for illumination, six repetitions of 0.25 s exposures (a total of just 1.5 s!) reduced CFP fluorescence by 50% (Figure 1). YFP is more photostable, as it takes 6.5 s of total exposure to diminish the signal intensity to 50%. The rapid photobleaching of CFP requires that exposure times be kept to a minimum. In practice exposure times do not exceed 0.4 s for FRET experiments. In addition, to protect against photobleaching the image is focused by observation of the DIC image and fluorescence is never examined by eye before acquisition by the camera.

One other property of YFP dictates the order of image acquisition. YFP is rapidly photobleached when excited with the CFP excitation energy (Figure 2). Thus YFP must be imaged first, followed by the FRET channel (CFP excitation, YFP emission) and finally the CFP channel.

The precise details for acquisition will depend on the microscope. We capture images using the SoftWorX software on our DeltaVision

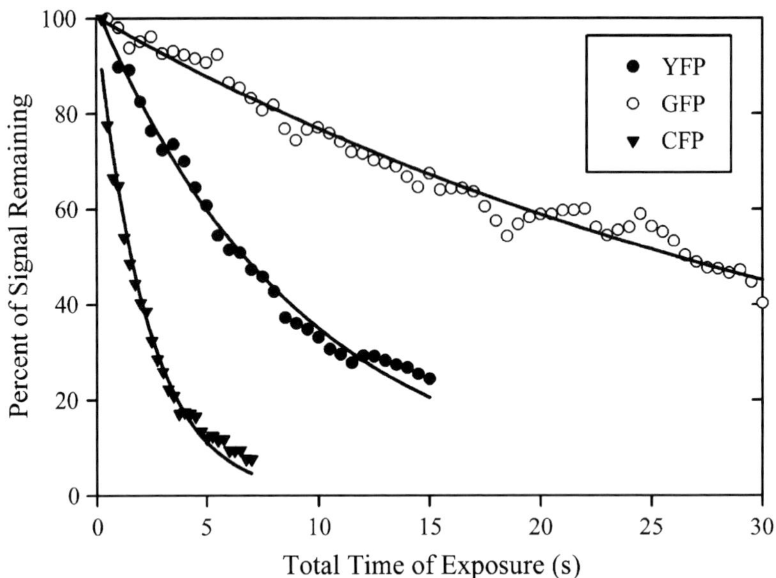

Figure 1. Decay of fluorescence from photobleaching. Yeast cells carrying Spc110 tagged with CFP, GFP or YFP were prepared as described in the text. Cells were repetitively imaged using the appropriate filter set (e.g. YFP filter set for Spc110-YFP) and short exposure times, 0.25 s for Spc110-CFP, and 0.5 s for Spc110-GFP and Spc110-YFP. The signal intensity in a 5 × 5 pixel box surrounding the SPB was measured and background from an adjacent 5 × 5 pixel box was subtracted. (Reprinted from Muller and Davis (2006) by permission.)

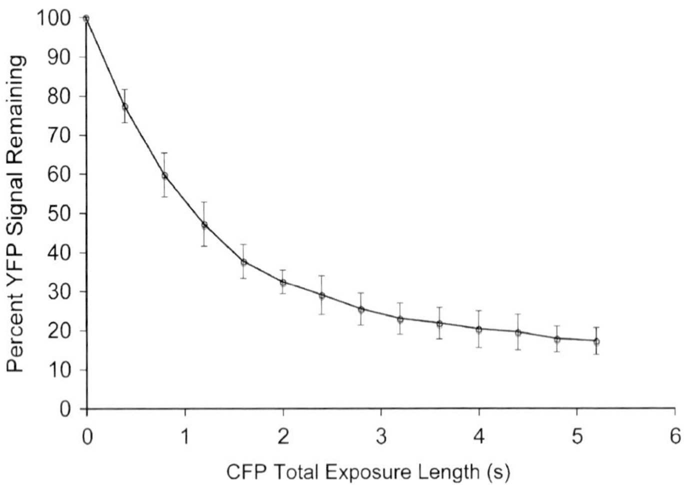

Figure 2. Decay of YFP fluorescence from photobleaching by the CFP filter set. Yeast cells carrying Spc110 tagged with YFP were prepared as described in the text. Cells were repetitively imaged with the CFP filter set using 0.4 s exposures. The signal intensity in a 5 × 5 pixel box surrounding the SPB was measured and background from an adjacent 5 × 5 pixel box was subtracted.

microscope. We image with an Olympus UPlanApo/100 × 1.35 Oil objective. The Photometric Coolsnap HQ camera is set for 2 × 2 binning; 3 sequential 0.4 s exposures capture the YFP, FRET and CFP channels. Finally, a 0.05 s DIC image is captured.

A. Image Analysis

There are two problems that are encountered when using CFP and YFP as FRET partners. First the FRET channel is polluted with fluorescence spilling over from CFP. Even when CFP is expressed alone a certain amount of fluorescence is observed in the FRET channel. We quantify the amount of fluorescence in the FRET channel when CFP is expressed alone and define a spillover factor:

$$Spillover_{CFP} = \frac{FRETchannel}{CFPchannel}$$

where *FRET channel* is the fluorescence intensity in the FRET channel minus background and *CFP channel* is the fluorescence intensity in the CFP channel minus background.

In addition the FRET channel is polluted by some direct excitation of YFP. Just as in the case with CFP, even when YFP is expressed alone a certain amount of fluorescence is observed in the FRET channel. Again, we quantify the amount of fluorescence in the FRET channel when YFP is expressed alone and define a spillover factor:

$$Spillover_{YFP} = \frac{FRETchannel}{YFPchannel}$$

where *YFP channel* is the fluorescence intensity in the YFP channel minus background.

For measuring FRET, the amount of CFP and YFP is measured in each image and the sum of the spillover from CFP and YFP yields the total expected baseline fluorescence.

$$Spillover_{total} = (Spillover_{CFP} \times CFPchannel) + (Spillover_{YFP} \times YFPchannel)$$

Previously, $Spillover_{total}$ was treated as background and subtracted from the FRET channel (Berney and Danuser, 2003). In contrast, we defined a new metric to measure FRET, $FRET_R$, in which the intensity in the FRET channel is divided by the $Spillover_{total}$:

$$FRET_R = \frac{FRETchannel}{Spillover_{total}}$$

$FRET_R$ represents the relative increase in the FRET signal above a baseline defined by $Spillover_{total}$. Thus in the absence of energy transfer, $FRET_R$ has a predicted value of 1. The maximum $FRET_R$ we have observed is approximately 2.5 for positive controls in which a protein is tagged with CFP and YFP in tandem using plasmid pDH18 (Muller *et al.*, 2005).

B. Example

We give an example of how to determine if two hypothetical proteins, Hyp1 and Hyp2, interact. To begin we would recommend constructing the following strains:

Hyp1-CFP
Hyp1-YFP
Hyp2-CFP
Hyp2-YFP
Hyp1-CFP, Hyp2-YFP
Hyp1-YFP, Hyp2-CFP
Hyp1-CFP-YFP (positive control).

Ideally, a negative control where Hyp1 is tagged and another protein that colocalizes but does not interact also is tagged. Alternatively, one could image one of the negative controls from the FRET study of the SPB (Muller *et al.*, 2005). The negative controls check the accuracy of the spillover factors, since they should have $FRET_R$ values of 1.00 ± 0.05.

Three criteria should be met before pursuing a FRET analysis. First and obvious, the localization of Hyp1 and Hyp2 must overlap if there is any chance for a FRET to occur. Second, the fluorescent signal from both Hyp1-YFP and Hyp2-YFP should be more than twofold above background. Finally, the levels of Hyp1 and Hyp2 should be within threefold of each other, otherwise fluorescence spillover will mask the FRET signal. The latter can be readily determined from measuring the fluorescence intensity in the singly tagged strains. If these three criteria are met, but no FRET signal is observed with the listed combinations, then one may consider making additional strains tagging the *N*-termini of the proteins, or using recombinant techniques to tag internal regions.

Spillover factors are measured using the singly tagged strains. Four images are taken as described above (YFP, FRET, CFP, DIC) and the intensity of the region containing fluorescence is measured. Approximately, 60–100 cells should be analyzed for each $FRET_R$ measurement. For SPBs we measure the fluorescence intensity in a 5×5 pixel box around the SPB and subtract background measured in an adjacent 5×5 pixel box. It is important to measure background within the cell as yeast cells autofluoresce in all three channels, YFP, FRET and CFP.

If experiments extend over the course of several months, spillover factors should be measured several times throughout the course of the experiments. Spillover factors change as the microscope lamp ages, and the lamp should be replaced after 150 h of use. If the spillover factors change dramatically, the microscope optics may have fallen out of alignment or the filter sets may be damaged. Accurate FRET measurements require a microscope that is well maintained.

FRET is measured in the experimental strains in which both Hyp1 and Hyp2 are tagged and in the positive and negative controls. We recommend measuring FRET in two experimental strains, one where Hyp1 is tagged with CFP and Hyp2 tagged with YFP and one where Hyp1 is tagged with YFP and Hyp2 tagged with CFP. Any bias suggests that the stoichiometry in the complex is not 1:1. Total fluorescence intensity and background is measured in the YFP channel, FRET channel and CFP channel for the fluorescent region of each cell. From these values $FRET_R$ is calculated as described above. If a good negative control is available and yields a $FRET_R$ value of 1.0, then $FRET_R$ values in the experimental strains as low as 1.1 can be significant. If a good negative control is not available then $FRET_R$ of 1.3 or above indicate the CFP and YFP tags on Hyp1 and Hyp2 are within 70 Å of each other and therefore Hyp1 and Hyp2 are within close proximity.

References

Berney, C. and Danuser, G. (2003). FRET or no FRET: a quantitative comparison. *Biophys. J.* **84**, 3992–4010.

Borst, J. W., Hink, M. A., van Hoek, A. and Visser, A. J. (2005). Effects of refractive index and viscosity on fluorescence and anisotropy decays of enhanced cyan and yellow fluorescent proteins. *J. Fluorescence* **15**, 153–160.

Hazbun, T. R., Malmström, L., Anderson, S., Graczyk, B. J., Fox, B., Riffle, M., Sundin, B. A., Aranda, J. D., McDonald, W. H., Chiu, C. H., Snydsman, B. E., Bradley, P., Muller, E. G. D., Fields, S., Baker, D., Yates, J. R., III and Davis, T. N. (2003). Assigning function to yeast proteins by integration of technologies. *Mol. Cell* **12**, 1353–1365.

Heim, R. (1999). Green fluorescent protein forms for energy transfer. *Methods Enzymol.* **302**, 408–423.

Miyawaki, A., Griesbeck, O., Heim, R. and Tsien, R. Y. (1999). Dynamic and quantitative Ca^{2+} measurements using improved cameleons. *Proc. Natl. Acad. Sci. USA* **96**, 2135–2140.

Muller, E. G. D. and Davis, T. N. (2006). Protein localization by cell imaging. In: *Proteomics for Biological Discovery* (T. D. Veenstra and J. R. Yates, eds), Wiley, New York.

Muller, E. G. D., Snydsman, B. E., Novik, I., Hailey, D. W., Gestaut, D. R., Niemann, C. A., O'Toole, E. T., Giddings, T. H., Jr. Sundin, B. A. and Davis, T. N. (2005). The organization of the core proteins of the yeast spindle pole body. *Mol. Biol. Cell* **16**, 3341–3352.

Nagai, T., Ibata, K., Park, E. S., Kubota, M., Mikoshiba, K. and Miyawaki, A. (2002). A variant of yellow fluorescent protein with fast and efficient maturation for cell-biological applications. *Nat. Biotechnol.* **20**, 87–90.

Pereira, G., Tanaka, T. U., Nasmyth, K. and Schiebel, E. (2001). Modes of spindle pole body inheritance and segregation of the Bfa1p-Bub2p checkpoint protein complex. *EMBO J.* **20**, 6359–6370.

Prein, B., Natter, K. and Kohlwein, S. D. (2000). A novel strategy for constructing *N*-terminal chromosomal fusions to green fluorescent protein in the yeast *Saccharomyces cerevisiae*. *FEBS Lett.* **485**, 29–34.
Shaner, N. C., Campbell, R. E., Steinbach, P. A., Giepmans, B. N., Palmer, A. E. and Tsien, R. Y. (2004). Improved monomeric red, orange and yellow fluorescent proteins derived from *Discosoma* sp. red fluorescent protein. *Nat. Biotechnol.* **22**, 1567–1572.
Smirnov, M. N., Smirnov, V. N., Budowsky, E. I., Inge-Vechtomov, S. G. and Serebrjakov, N. G. (1967). Red pigment of adenine-deficient yeast *Saccharomyces cerevisiae*. *Biochem. Biophys. Res. Commun.* **27**, 299–304.
Stryer, L. (1978). Fluorescence energy transfer as a spectroscopic ruler. *Ann. Rev. Biochem.* **47**, 819–846.
Wach, A., Brachat, A., Alberti-Segui, C., Rebischung, C. and Philippsen, P. (1997). Heterologous *HIS3* marker and GFP reporter modules for PCR-targeting in *Saccharomyces cerevisiae*. *Yeast* **13**, 1065–1075.

13 Identification, Characterization, and Phenotypic Analysis of Covalently Linked Cell Wall Proteins

Frans M Klis, Piet De Groot and Stanley Brul

Swammerdam Institute for Life Sciences, University of Amsterdam, BioCentrum Amsterdam, Nieuwe Achtergracht 166, 1018 WV Amsterdam, The Netherlands

♦♦

CONTENTS

Cell wall protein–polysaccharide complexes
Release and identification of CWPs and CWP–polysaccharide complexes
Identification of cell wall proteins
Glycosylation of cell wall proteins
Phenotypical analysis of cell wall proteins
Cell surface display of heterologous proteins

List of Abbreviations

AA_{ω}	C-terminal amino acid of a protein linked to a GPI-anchor
ASL	alkali-sensitive linkage
BCA	bicinchoninic acid
BSA	bovine serum albumin
ConA	concanavalin A
CWP	cell wall protein
DMSO	dimethylsulfoxide
ECL	enhanced chemiluminescence
ER	endoplasmic reticulum
fr	fragment
GPI	glycosylphosphatidylinositol
HF	hydrofluoric acid
MALDI-TOF MS	matrix-assisted laser desorption/ionizaton time-of-flight mass spectrometry
m/z	mass-to-charge ratio
Pir	proteins with internal repeats
PBS	phosphate-buffered saline
PMSF	phenylmethylsulfonyl fluoride
PVDF	polyvinylidene fluoride

METHODS IN MICROBIOLOGY, VOLUME 36
0580-9517 DOI:10.1016/S0580-9517(06)36013-8

RT	room temperature
SGD	*Saccharomyces* Genome Database
Q-TOF MS	quadrupole time-of-flight mass spectrometry
WGA	wheat germ agglutinin
YPD	rich medium consisting of yeast extract, peptone, and dextrose

♦♦♦♦♦♦ I. CELL WALL PROTEIN–POLYSACCHARIDE COMPLEXES

Proteins that are destined for the cell wall and become covalently attached to the cell wall polysaccharide network first have to traverse the secretory pathway, where they may undergo various posttranslational modifications. In the endoplasmic reticulum (ER) their N-terminal signal peptide is removed, pre-assembled N-linked carbohydrate side-chains are attached to asparagine residues in the amino sequence Asn-Xxx-Ser/Thr, and O-glycosylation of serine and threonine residues is initiated (reviewed by Orlean, 1997). In *Saccharomyces* and *Candida* species and other ascomycetous yeasts, N- and O-linked carbohydrate side-chains consist mainly of α-linked mannose residues, but for example in *Schizosaccharomyces pombe*, N- and O-chains have been observed that consist of a core of α-linked mannose residues decorated with galactose residues (Gemmill and Trimble, 1999). GPI (glycosylphosphatidylinositol) proteins undergo an additional processing step in the ER, where their C-terminal GPI-anchor addition signal is replaced by a pre-assembled glycolipid (GPI) anchor that associates them with the ER membrane. N- and O-linked carbohydrate side-chains and the GPI-anchor are subsequently extended and processed in later compartments of the secretory pathway. The covalent attachment of cell wall proteins (CWPs) to cell wall polysaccharides, including 1,6-β-glucan, takes place at the cell surface (Lu *et al.*, 1995; Montijn *et al.*, 1999).

Cell wall proteins of *Saccharomyces cerevisiae* and other ascomycetous yeasts form an external layer surrounding an internal layer of stress-bearing polysaccharides (reviewed in Klis *et al.*, 2002; De Groot *et al.*, 2005). The internal layer is composed of a continuous three-dimensional network of moderately branched 1,3-β-glucan molecules that is kept together by hydrogen bonding between locally aligned chains. This network is highly elastic and may shrink or extend depending on the osmotic strength of the medium. The nonreducing ends of the 1,3-β-glucan molecules may function as acceptor sites for the attachment of chitin chains, which stiffens the network, and of 1,6-β-glucan molecules. The 1,6-β-glucan molecules are highly branched and thus water soluble. They may be tethered at one of their nonreducing ends to GPI-modified cell wall proteins (GPI-CWPs), forming the CWP–polysaccharide complex GPI-CWP → 1,6-β-glucan → 1,3-β-glucan. The GPI-CWPs represent the main class of CWPs and form the bulk of the external protein layer. As GPI-CWPs are linked to 1,6-β-glucan through a lipidless GPI-anchor remnant at their C-terminal end, their N-terminal region, which in GPI-CWPs tends to contain the functional domain, extends into the medium (De Groot *et al.*, 2005). The ASL-CWPs form the second class of CWPs; they are directly linked to 1,3-β-glucan through an alkali-sensitive linkage (ASL), which involves a glutamine residue and a glucosyl hydroxyl group (Ecker *et al.*, 2006), and include the Pir-proteins (see below). This corresponds to the CWP–polysaccharide complex ASL-CWP–1,3-β-glucan. A small group of GPI-CWPs may contain in addition to their normal GPI-linkage an ASL as well, resulting in the CWP–polysaccharide complexes 1,3-β-glucan – GPI-CWP → 1,6-β-glucan → 1,3-β-glucan and GPI-CWP – 1,3-β-glucan. In case of cell wall stress, an otherwise rare CWP–polysaccharide complex becomes much more prominent (GPI-CWP → 1,6-β-glucan ← chitin). In total, five CWP–polysaccharide complexes have been identified (Table 1).

Importantly, similar CWP–polysaccharide complexes have also been found in other ascomycetous yeasts such as *Candida albicans* and *Candida glabrata*, *Exophiala dermatitidis*, and *Yarrowia lipolytica*, and in ascomycetous mycelial species such as *Aspergillus niger*, *Fusarium oxysporum*, and *Paecilomyces variottii* (Schoffelmeer *et al.*, 1996, 2001; Brul *et al.*, 1997; Montijn *et al.*, 1997; Kapteyn *et al.*, 2000; Klis *et al.*, 2001; Frieman *et al.*, 2002; Jaafar and Zueco, 2004; Weig *et al.*, 2004; Damveld *et al.*, 2005b). A minority of proteins are bound indirectly to the cell wall polysaccharide network through a disulfide bond to other proteins (Cappellaro *et al.*, 1994, 1998; Moukadiri *et al.*, 1999; Jaafar *et al.*, 2003). Finally, because of the

Table 1. CWP–polysaccharide complexes found in *S. cerevisiae* and *C. albicans*

A	GPI-CWP → 1,6-β-glucan → 1,3-β-glucan
B	GPI-CWP → 1,6-β-glucan ← chitin
C	ASL-CWP — 1,3-β-glucan
D	1,3-β-glucan — GPI-CWP → 1,6-β-glucan → 1,3-β-glucan
E	GPI-CWP — 1,3-β-glucan

Note: The arrows represent glycosidic linkages and point to a nonreducing end of the acceptor polysaccharide. Note that branched polysaccharides have a single reducing end and multiple nonreducing ends. GPI-CWPs are linked through a trimmed lipidless form of their original GPI-anchor to 1,6-β-glucan (Kollar *et al.*, 1997) and as a result the N-terminal region of GPI-CWPs extends into the medium. The alkali-sensitive linkage (ASL) involves an ester linkage between a glutamine residue and a glucosyl hydroxyl group (Ecker *et al.*, 2006). Complex B becomes much more abundant in response to cell wall stress (Kapteyn *et al.*, 1997). Complexes D and E are relatively rare. Note that ASL-CWPs include the Pir-proteins (Yin *et al.*, 2005). CWP, cell wall protein; ASL-CWP, alkali-sensitive linkage-CWP; GPI-CWP, glycosylphosphatidylinositol-CWP.

similarities between the cell wall organization of *S. cerevisiae* and other ascomycetous fungi including mycelial species, the techniques developed for *S. cerevisiae* are often equally effective in other ascomycetous fungi (Brul *et al.*, 1997; Schoffelmeer *et al.*, 1999; Kapteyn *et al.*, 2000; Frieman *et al.*, 2002; Jaafar and Zueco, 2004; Perez and Ribas, 2004; Weig *et al.*, 2004; Damveld *et al.*, 2005b).

♦♦♦♦♦♦ II. RELEASE AND IDENTIFICATION OF CWPS AND CWP–POLYSACCHARIDE COMPLEXES

A. Release of CWPs and CWP–Polysaccharide Complexes

Generally, isolated cell walls are used as starting material for the release and isolation of covalently linked CWPs. To avoid heavy contamination with cytosolic proteins, cell wall preparations should be completely free from unbroken cells. The use of a FastPrep Instrument is recommended for full cell breakage. To prevent contamination by ionically bound cytosolic proteins and entrapped membrane proteins, isolated walls are washed with concentrated salt solutions and extracted with hot detergent containing a reducing agent. The procedure for cell breakage and isolation of cell walls is given in Protocol 1.

Purified cell walls usually have a protein content of about 3%. Isolated walls are treated with hot alkali and the extract is used for assaying the protein content as described in Protocol 2.

Cell wall proteins can be liberated from isolated cell walls either chemically or enzymatically. As the GPI-remnant through which GPI-CWPs are linked to 1,6-β-glucan (protein – AA_{ω} – **ethanolamine – P_i – tetramannoside** → 1,6-β-glucan; the GPI-remnant is in bold) contains a phosphodiester bridge, GPI-CWPs can be released using HF-pyridine (Protocol 3), which under the right conditions specifically cleaves phosphodiester bridges as described below (De Groot *et al.*, 2004). ASL-CWPs can be released by incubating isolated walls in the presence of 30 mM NaOH overnight in the cold using Protocol 4 (Mrsa *et al.*, 1997; Kapteyn *et al.*, 1999; De Groot *et al.*, 2004; Weig *et al.*, 2004).

Various enzymes allow isolation of CWPs and specific CWP–polysaccharide complexes. Phosphodiesterases have been used to release GPI-CWPs (Kapteyn *et al.*, 1996). 1,6-β-Glucanase can be used to release GPI-CWPs, resulting in GPI-CWPs connected to a 1,6-β-glucan fragment (GPI-CWP → 1,6-β-$glucan_{fr}$; Kapteyn *et al.*, 1997, 2000). A simple procedure has been developed for the purification of recombinant 1,6-β-glucanase (Bom *et al.*, 1998). 1,3-β-Glucanase, which is commercially available as a recombinant enzyme (Quantazyme ylg^{TM}), can be used to release both GPI-CWPs and ALS-CWPs (Kapteyn *et al.*, 2001). This results in complexes consisting of a GPI-CWP connected to 1,6-β-glucan, which in turn is

Protocol 1. Cell wall isolation.

1. Harvest the cells and wash them with 10 mM Tris–HCl, pH 7.5.
2. Transfer the cells to screw-capped Eppendorf tubes (Sarstedt; $\sim 2 \times 10^9$ cells per tube) and spin the cells down at 7000 rpm for 1 min.
3. Resuspend the cells in 200 μl 10 mM Tris–HCl, pH 7.5. Add glass beads (0.25–0.5 mm in diameter) and tip off excess dry beads. Add 10 μl protease inhibitor cocktail (SIGMA).
4. Break the cells with a FastPrep instrument (Qbiogene) by two runs at speed 6 for 20 s with intermediate cooling on ice.
5. Check cell rupture with a light microscope, and repeat the breaking step if necessary.
6. Collect the cell lysate by repeatedly washing the glass beads with 1 M NaCl. Centrifuge the lysate at 3000 rpm for 5 min to isolate the walls.
7. Wash the cell walls repeatedly with 1 M NaCl until the supernatant is clear. Repeat the washing step once with water. During steps 1–7, the samples should be kept cold (4 °C) where possible.
8. Clean the walls by heating them in SDS-extraction buffer (100 mM Na–EDTA, 50 mM Tris–HCl, 2% (w/v) SDS, 50 mM β-mercaptoethanol, at pH 7.8) at 100 °C for 5 min. Spin the walls down and repeat this step once.
9. Remove SDS-extraction buffer by washing three times with water.
10. Cell walls may be lyophilized depending on experimental purposes.

linked to a 1,3-β-glucan fragment (GPI-CWP → 1,6-β-glucan → 1,3-β-glucan$_{fr}$), and complexes of ASL-CWPs connected to a 1,3-β-glucan fragment (ASL-CWP – 1,3-β-glucan$_{fr}$). The procedure based on the use of Quantazyme is given in Protocol 5.

In combination with 1,3-β-glucanase, chitinase can be used to release chitin-linked GPI-CWPs, resulting in a complex consisting of a GPI-CWP linked through 1,6-β-glucan to a chitin fragment (GPI-CWP → 1,6-β-glucan → chitin$_{fr}$; Kapteyn *et al.*, 1997). As mentioned above, some CWPs can be released using reducing agents, both from isolated walls and from intact cells. For intact cells, we recommend the extraction method introduced by Cappellaro and co-workers (Cappellaro *et al.*, 1998). According to their protocol, the cells are shaken in 2 mM dithiothreitol, 25 mM Tris–HCl (pH 8.5) at 4 °C for 2 h. These conditions are much more gentle than those used in other methods and are thus less likely to introduce artefacts resulting from limited cell lysis. To distinguish between authentic CWPs and adventitious proteins, cell surface proteins may be

Protocol 2. Hot alkali extraction of CWPs for quantitation.

1. Resuspend 4 mg hot detergent-extracted, freeze-dried walls in 100 µl 1 N NaOH.
2. Boil the suspension for 10 min.
3. Neutralize the suspension by adding 100 µl 1 N HCl.
4. Spin the cell walls down at 10 000 g for 5 min.
5. Take a 20 µl sample from the supernatant.
6. Add 1 ml of BCA protein reagent (Pierce).
7. Incubate the mixture at 37°C for 30 min.
8. Measure the OD at 562 nm and estimate the protein content from a calibration curve generated with BSA samples treated in the same way as isolated walls.

Protocol 3. Release of GPI-CWPs by HF-pyridine cleavage.

1. Incubate 4 mg lyophilized cell walls in 300 µl HF-pyridine commercially available for 3 h on ice.
2. Quench the reaction by adding a similar volume of ice-cold H_2O.
3. Remove HF-pyridine by dialysis overnight against cold water.
4. Centrifuge to separate released CWPs from remaining cell wall matrix. If the cell wall pellet is to be used for further analysis, it is advisable to wash the walls with SDS-extraction buffer (as described under cell wall isolation) to remove any noncovalently adsorbed GPI-CWPs.
5. Concentrate extracted GPI-CWPs by lyophilization.

Protocol 4. Release of ASL-CWPs using alkali.

1. Incubate 4 mg lyophilized cell walls in 91 µl 30 mM NaOH overnight with gentle shaking and at 4°C.
2. Neutralize by adding 109 µl 30 mM acetic acid.
3. Spin the reaction tube and collect the supernatant containing ASL-proteins. If needed for further analysis, walls can be washed with SDS-extraction buffer (as described in Protocol 1).

biotinylated before extraction; for this, the labeling procedure that is developed by Mrsa and co-workers and is carried out at 0°C is recommended (Mrsa *et al.*, 1997).

DMSO has been used as a solvent for size-exclusion chromatography for 1,3-β-glucans (Williams *et al.*, 1994). This approach has

Protocol 5. Release of GPI-CWPs and ALS-CWPs using Quantazyme.

1. Resuspend 4 mg lyophilized cell walls in 200 µl 50 mM Tris–HCl, 50 mM β-mercaptoethanol, pH 7.4.
2. Add 6 µl Quantazyme and incubate overnight at 37°C with gentle shaking.
3. Spin the reaction tube and collect the supernatant containing CWP. If needed for further analysis, walls can be washed with SDS-extraction buffer (as described in Protocol 1).

been extended by Grün *et al.* (2005). They were able to completely dissolve isolated walls of *S. pombe* in hot DMSO, and to distinguish three subpopulations of macromolecules or macromolecular complexes using high-performance size-exclusion chromatography. In principle, this technique should allow the isolation of intact CWP–polysaccharide complexes. Because this technique does not require any prior degradation of cell wall components, it promises to open new avenues not only for the study of CWP–polysaccharide complexes, but also of other cell wall components.

B. Identification of CWPs and CWP–Polysaccharide Complexes

First, a note of caution. Many gel systems allow separation of proteins up to about 200 kDa only. Because CWPs may possess long peptide backbones and are often heavily glycosylated or are present as CWP–polysaccharide complexes, it is advisable to use SDS-PAGE gradient gels that allow separation of proteins with an apparent mass of up to 600 kDa such as 3–8% Tris–Acetate gels (Invitrogen) or 2.2–20% SDS-PAGE gels. Otherwise, CWPs and CWP–polysaccharide complexes may not enter the gel and will be overlooked. As CWPs are mannosylated, they may be sensitively detected using lectin blotting with peroxidase-conjugated ConA as described in Protocol 6. CWPs, which are often heavily glycosylated, may also be sensitively detected using silver staining in combination with periodate oxidation. After protein fixation the gels are soaked for 30 min in a 50 mM solution of periodic acid in 100 mM acetic acid at room temperature to generate (reactive) aldehyde groups in the sugar residues of the carbohydrate side-chains of the separated proteins, resulting in strongly enhanced staining (De Nobel *et al.*, 1989). It may also be convenient specifically to label cell surface proteins (including CWPs) through their lysine residues, using a negatively charged and thus plasma membrane-impermeable biotinylation reagent, which allows sensitive detection of labeled proteins (Mrsa *et al.*, 1997).

GPI-CWPs commonly possess a modular structure (signal peptide – functional domain – spacer domain – GPI-anchor addition signal; De Groot *et al.*, 2005; see Figure 1). GPI-CWPs have been

Protocol 6. Detection of CWPs and CWP–polysaccharide complexes by lectin blotting.

1. Separate proteins by SDS-PAGE and transfer them onto a PVDF-filter. Note that ConA staining is very sensitive. Resolution of protein bands may improve by limiting the amount of protein loaded on the gel (protein from 80 μg dry weight of walls is sufficient).
2. Block the filter with 6% (w/v) BSA in PBS (BSA/PBS) at room temperature with gentle shaking for at least 1 h.
3. Rinse the filter twice with PBS for 5 min.
4. Incubate the filter with 0.5 μg/ml peroxidase-conjugated ConA (SIGMA) in 3% (w/v) BSA/PBS, containing 2.5 mM $CaCl_2$ and 2.5 mM $MgCl_2$ at room temperature for 1 h.
5. Rinse the filter twice in PBS for 5 min and once for 30 min.
6. Develop the filter using ECL-detection reagents GE Healthcare. Longer exposure times may help to detect proteins with a relatively low level of glycosylation.

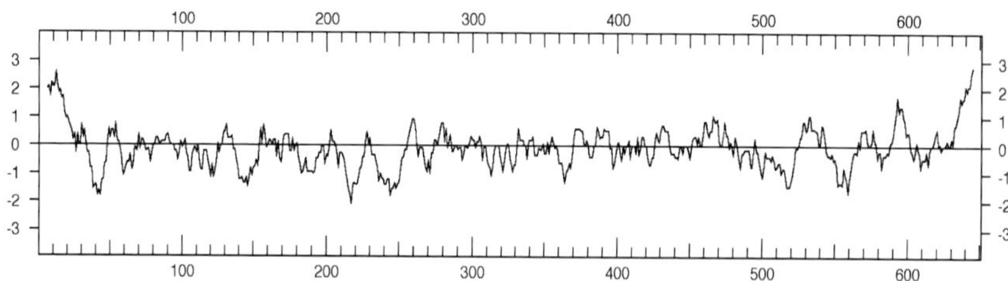

Figure 1. Hydropathy plot of Sag1p. Sag1p or α-agglutinin is a well-characterized GPI-modified cell wall protein (GPI-CWP) in *MATα* cells that is involved in sexual agglutination and is both N- and O-glycosylated (Chen *et al.*, 1995). Note the hydrophobic regions at both ends of the protein. Sag1p displays the general organization of GPI-CWPs: SP – functional domain – spacer domain – GPI-anchor addition signal. The N-terminal signal peptide is predicted to consist of the first 19 amino acid residues (Bendtsen *et al.*, 2004; http://www.cbs.dtu.dk/services/SignalP/). The functional domain is predicted to comprise at least residues 45–264 and probably some more (Chen *et al.*, 1995; Marchler-Bauer *et al.*, 2003), and the hydroxy-amino acid-rich region extends from about residue 300 to residue 620, close to the predicted C-terminal amino acid or AA_{ω} of the mature protein, residue 627, to which the GPI-anchor is expected to be attached (Eisenhaber *et al.*, 2004). As expected, C-terminally truncated forms of Sag1p, which lack the GPI-anchor addition signal, behave like normal secretory proteins and are secreted into the medium (Lu *et al.*, 1995). This is generally the case with C-terminally truncated forms of GPI-CWPs.

successfully tagged by inserting the tag directly after the postulated signal peptide (Ram *et al.*, 1998; Mao *et al.*, 2003; Smits *et al.*, 2006). Because some GPI-CWPs such as Crh1p possess a Kex2 protease cleavage site (–KR–), it is probably safer in these cases to insert the tag after this cleavage site or in the proximity of the GPI-anchor addition site. In the latter case, a reliable position for insertion seems to be about 5–10 amino acids before the C-terminal amino acid of the mature protein; this position corresponds to about 30–40 amino acids before the C-terminus of the predicted polypeptide chain. As also mentioned below, a simple algorithm is available to identify the most likely amino acid for GPI-anchor attachment (Eisenhaber *et al.*, 2004; http://mendel.imp.univie.ac.at/gpi/fungi_server.html). Note that insertion after the GPI-anchor addition site may cause mis-localization of the tagged protein (or possibly only of the tag) to the ER (Huh *et al.*, 2003; SGD; http://www.yeastgenome.org/).

Antisera against neoglycoproteins consisting of 1,6-β-glucan and 1,3-β-glucan oligosaccharides conjugated to BSA have been raised by Montijn *et al.* (1994). Hepta-oligosaccharides of 1,3-β-glucan are commercially available. The specificities of these antisera for β-glucans can be tested by carrying out competition experiments using pustulan (1,6-β-glucan) or periodate-treated laminarin (1,3-β-glucan; Montijn *et al.*, 1994). These sera efficiently recognize 1,6-β-glucan and 1,3-β-glucan, allowing the characterization of various CWP–polysaccharide complexes (Kapteyn *et al.*, 1996, 2000). Monoclonal antibodies directed against 1,3-β-glucan are also commercially available (Biosupplies). In addition, the lectin WGA specifically binds to chitin and can thus be used to identify the chitin-containing CWP–polysaccharide complex GPI-CWP → 1,6-β-glucan ← $chitin_{fr}$ (Kapteyn *et al.*, 1997). When raising antisera against specific GPI-CWPs, it is probably best to target their functional domain, because that domain is the most characteristic part of the protein, and also because it is located in the N-terminal half of the protein and therefore more accessible to antibodies in intact cells; it is also likely to be less glycosylated than the C-terminal region.

♦♦♦♦♦♦ III. IDENTIFICATION OF CELL WALL PROTEINS

A. *In Silico* Identification of Potential Yeast CWPs

As already mentioned, the predicted amino acid sequence of a GPI-CWP starts with an N-terminal signal peptide, which is generally followed by the functional domain and then a spacer domain, and terminates in a GPI-anchor addition signal. The spacer domain often contains repeats and is usually rich in hydroxyamino acid residues, which may function as attachment sites for short O-chains, resulting in a rigid rod-like structure (Jentoft, 1990). Several algorithms are available to identify potential GPI-proteins in fungi. De Groot *et al.* (2004) have developed a genomic algorithm for genome-wide identification of proteins with a potential GPI-anchor addition signal. Selected proteins are further analyzed for the presence of an N-terminal signal peptide and the absence of an internal transmembrane sequence. For individual proteins, another useful algorithm has been developed, which also predicts which amino acid is most likely to receive a GPI-anchor (Eisenhaber *et al.*, 2004). GPI-proteins may be largely retained in the plasma membrane such as Gas1p or incorporated in the cell wall (Hamada *et al.*, 1999). Often, GPI-proteins that are retained in the plasma membrane contain a dibasic motif close to the C-terminus (Caro *et al.*, 1997; Frieman and Cormack, 2003). Hydrophobic amino acids in this region may also affect the final destination of GPI-proteins (Hamada *et al.*, 1998, 1999). In addition, long regions that are rich in serine and threonine residues tend to promote incorporation of GPI-proteins into the cell wall (Frieman and Cormack, 2004). An algorithm to identify all ASL-CWPs *in silico* is still lacking, but it is possible to recognize putative Pir-CWPs, an important group of ASL-CWPs. Pir-CWPs also possess a common organization (SP – pro-peptide – repeats – functional domain). Their predicted amino acid sequence begins with an N-terminal signal sequence and is followed by a propeptide terminating in a putative Kex2 cleavage site (–KR–). Beyond the Kex2 site, one or more repeats occur that contain the sequence DGQJQ – at least in *S. cerevisiae*, *C. albicans*, *C. glabrata*, and *Yarrowia lipolytica* – in which J represents a hydrophobic amino acid. The C-terminal part of the protein is highly conserved and includes a characteristic cysteine pattern; it probably represents the functional domain of the protein (De Groot *et al.*, 2005).

Conveniently, the *Saccharomyces* Genome Database allows a rapid search for fungal homologs of all proteins of baker's yeast through a tool called "Comparison Resources", which can be accessed in all gene entries (SGD; http://www.yeastgenome.org/).

B. Mass Spectrometric Identification of CWPs

Mass spectrometric identification of proteins is usually based on protein separation and visualization on 1-D or 2-D gels, followed by tryptic digestion of excised protein bands. Identification of covalently bound CWPs based on this approach is complicated by the fact that CWPs are usually heavily glycosylated, carrying numerous O-linked carbohydrate side-chains and some N-linked carbohydrate side-chains, which may be very long. As the number and size of these side-chains vary between individual protein molecules (Reddy *et al.*, 1988) resulting in numerous glycoforms, separation of CWPs on SDS-PAGE gels is often poor. Generally, proteins appear as broad bands or even smears on 1-D gels (see also next section). Actually, when sharp, distinctive bands are observed on 1-D gels for proteins released from cell wall preparations, this raises the suspicion that the sample might be contaminated with cytosolic (unglycosylated) proteins. A second complication is caused by the presence of phosphodiester groups in carbohydrate side-chains of fungal glycoproteins, resulting in multiple isoforms of the protein with different isoelectric points (see also next section). Thus, on 2-D gels CWPs often appear as multiple bands of similar mass but with different p*I* values (see e.g. Weig *et al.*, 2004). Removal of N-linked carbohydrate side-chains from CWPs (see below) may improve their resolution on gel. However, in contrast to mammalian glycoproteins there are as yet no suitable enzyme preparations available that can efficiently remove O-linked carbohydrate side-chains from fungal glycoproteins.

As an alternative to separation of CWPs by 1-D or 2-D gel electrophoresis and proteolytic digestion of separate bands, CWPs can also be directly digested with suitable endoproteases to generate peptide fragments for mass spectrometric analysis. This can be achieved either by direct digestion of isolated cell walls that are kept in suspension or by digestion of either chemically or enzymatically released CWP mixtures (see Protocols 3–5). CWP identification by direct proteolytic digestion has been successfully applied in *S. cerevisiae* (Yin *et al.*, 2005), *C. albicans* (De Groot *et al.*, 2004), and *C. glabrata* (our unpublished results), using a nano-LC system coupled online to a quadrupole time-of-flight (Q-TOF) mass spectrometer. A procedure for direct proteolytic digestion of isolated cell walls is presented in Protocol 7.

Direct digestion of proteins from intact cell wall preparations solubilizes about 50% of the protein present in yeast walls (our unpublished results). This incomplete digestion is probably due to the high degree of glycosylation that is often observed in the C-terminal region of GPI-CWPs. This region may also be less accessible to endoproteases because it is further away from the

Protocol 7. Proteolytic digestion of isolated cell walls.

1. Resuspend 4 mg freeze-dried cell walls in 100 mM NH_4HCO_3/10 mM dithiothreitol and incubate them at 56°C for 1 h.
2. Spin the walls down (3000 rpm for 5 min) and *S*-alkylate them in 100 mM NH_4HCO_3/55 mM iodoacetamide in the dark at room temperature for 45 min.
3. Wash the cell wall pellet three times with 50 mM NH_4HCO_3 and dry the walls under vacuum.
4. For proteolytic cleavage, incubate the cell walls in 50 mM NH_4HCO_3 overnight at 37°C with gentle shaking in the presence of sequencing grade trypsin (Roche, Basel, Switzerland), or at 25°C in the presence of endoprotease Glu-C (Sigma, St. Louis, MA), using a CWP/enzyme ratio of 50:1. Protein accounts for approximately 3% (w/w) of the cell wall dry weight.
5. Centrifuge the digested samples. The supernatants contain the solubilized peptides that can be analyzed by mass spectrometry.

surface. Importantly, due to the numerous O-linked side-chains present in CWPs, CWP-derived tryptic peptides may have higher masses than predicted. This severely hampers identification of CWPs by peptide mass fingerprinting. For reliable identification it is recommended to use peptide sequencing which can for instance be achieved by electrospray ionization time-of-flight tandem mass spectrometry (ESI-TOF-MS/MS). Collision-induced fragmentation of an ionized peptide may result in a series of fragment ions, from which the peptide sequence, and thus the identity of the corresponding protein, can be elucidated. An important advantage of this method is that a single peptide may be sufficient for protein identification. As mentioned earlier, the masses of many peptides are increased due to glycan additions, causing them to fall out of the range that can be analyzed by ESI-TOF-MS (upper *m/z* limit ~3500). For most GPI-CWPs, the N-terminal half, which specifies the functional domain and usually has a relatively low level of glycosylation, seems most amenable to mass spectrometry (Yin *et al.*, 2005). As the number of covalently bound CWPs in ascomycetous yeasts, in any given growth condition, is limited to ~20 different proteins, the vast majority of unglycosylated peptides generated by direct digestion can be readily identified in a single mass spectrometric experiment. As all CWPs receive an N-terminal signal peptide, which is cleaved off in the ER, the N-terminal tryptic peptide of the mature protein may have a lower mass than calculated by the standard algorithms for predicting the masses of tryptic peptides. Algorithms that predict signal peptidase cleavage sites such as SignalP (Bendtsen *et al.*, 2004; http://www.cbs.dtu.dk/services/SignalP/) may help to calculate the correct mass of the N-terminal tryptic peptide of CWPs.

◆◆◆◆◆◆ IV. GLYCOSYLATION OF CELL WALL PROTEINS

Cell wall mannoproteins are often heavily glycosylated carrying both N- and O-linked carbohydrate side-chains, allowing their sensitive detection using lectin blotting based on ConA, which recognizes α-linked mannose residues as described above. As the number and size of these side-chains vary between individual protein molecules, polydispersed bands are observed. In case of CWP–polysaccharide complexes even extended smears may appear on the gels (Kapteyn *et al.*, 1996). To determine if CWPs and other secretory proteins are N-glycosylated, digestion with Endo-H, which cleaves after the first *N*-acetylglucosamine residue thereby removing almost the entire N-chain is an obvious choice; alternatively, *N*-glycanase, which removes the entire N-chain, may be used. The protocol for Endo-H digestion is presented in Protocol 8. Alternatively,

Protocol 8. Endo-H digestion.

1. Mix 1 mg of mannoprotein (CWPs released from 2.5 mg dry weight walls) in 40 µl water, 10 µl 2% SDS, and 1.5 µl β-mercaptoethanol.
2. Denature the proteins by heating at 100°C for 5 min.
3. Add 150 µl of 50 mM Na-acetate, pH 5.5, containing 2 mM EDTA and 1 mM PMSF.
4. Add 1 µl pepstatin (0.7 mg/ml in methanol) and 1 µl leupeptin (0.7 mg/ml in water).
5. Add 50 mU recombinant Endo-H (Roche) and incubate overnight at 37°C.

tunicamycin may be used, because this drug inhibits the addition of N-chains to the protein (Orlean *et al.*, 1991). Interestingly, underglycosylation of secretory proteins often results in increased resistance to vanadate and decreased resistance to hygromycin. This has not only been observed in *S. cerevisiae* but also in other species such as *C. albicans*, *K. lactis*, and *Y. lipolytica*, indicating that this approach is generally useful in fungal research (Ballou *et al.*, 1991; Dean, 1995; Uccelletti *et al.*, 2000; Jaafar *et al.*, 2003). Both types of protein-linked carbohydrate side-chains may be phosphorylated in the form of phosphodiester bridges connecting two mannose residues. As a result, yeast cell walls contain numerous negative charges at physiological pHs. This can be both visualized and quantified by staining cells with the cationic dye Alcian blue (Ballou, 1990; Conde *et al.*, 2003). The presence of so many negative charges at pHs ≥ 3 in the external, protein-enriched wall layer explains why yeast cell walls avidly bind soluble, positively charged proteins, for example, released as a result of cell lysis.

♦♦♦♦♦♦ V. PHENOTYPICAL ANALYSIS OF CELL WALL PROTEINS

The functions of fungal CWPs are manifold (reviewed in De Groot *et al.*, 2005) and include both collective functions such as cell wall permeability for macromolecules and individual functions such as adhesiveness. Cell wall permeability depends on the external protein layer of the wall and can be easily assayed by measuring the relative sensitivity of washed, intact cells to polycations, which cause cell leakage and the release of UV-absorbing compounds (De Nobel *et al.*, 1990a). As the thickness and composition of the CWP layer are highly dependent on growth conditions, cell wall permeability will vary correspondingly and is thus a sensitive way of detecting changes in cell wall structure (De Nobel *et al.*, 1990a,b).

As expected, this assay is also functional in other ascomycetous yeasts (De Nobel *et al.*, 1990a). Resistance of intact cells to the lytic activity of 1,3-β-glucanase, which depends on both the permeability of the outer protein layer and the composition and thickness of the internal skeletal layer, also sensitively reflects cell wall structure (De Nobel *et al.*, 1990b, 2000; Ovalle *et al.*, 1998; Boorsma *et al.*, 2004). Protein sulfhydryl groups in cell walls can be quantified using Ellman's reagent (De Nobel *et al.*, 1990a). The protocol for measuring the resistance of intact cells to 1,3-β-glucanase by monitoring the optical density of the cell suspension time in time is presented in Protocol 9.

Deletion of individual CWP may affect cell wall integrity, resulting in altered resistance of intact cells to cell wall-perturbing compounds such as Calcofluor white and Congo red and to the detergent SDS (Van der Vaart *et al.*, 1995; Klis *et al.*, 1998; De Groot *et al.*, 2001; Jaafar *et al.*, 2003). The protocol for determining loss of cell wall integrity by measuring the relative resistance to cell wall-perturbing compounds is described in Protocol 10. For Calcofluor white, Congo red, and SDS, concentrations in the order of 10–100 μg/ml are used, but the actual concentration needed may depend on the genetic background. As cells are often much more sensitive to cell wall-perturbing compounds at elevated temperatures, it may be necessary to use higher temperatures to see an effect. Recently, it was shown that deletion of a GPI-CWP in *A. niger* also results in a decrease in Calcofluor resistance, confirming that mycelial fungi can also be analyzed using this type of assay (Damveld *et al.*, 2005a).

More specific functions of individual CWP may be inferred by *in silico* analysis. The CAZy database indicates that many CWPs

Protocol 9. Measuring the resistance of cells to 1,3-β-glucanase.

1. Grow cells overnight in YPD.
2. Inoculate cells in fresh YPD at a starting OD_{600} of 0.1. Culture at 30°C to $OD_{600} = 0.5–1.0$.
3. Spin the cells down. Resuspend them in 1 ml 50 mM Tris–HCl, 40 mM β-mercaptoethanol, pH 7.4, at $OD_{600} = 1.3$, and incubate them at room temperature for 1 h. During this pre-incubation, which will open up disulfide bridges, the OD_{600} will drop to ~1.
4. Measure OD_{600} and add 60–100 Units Quantazyme. Sensitivity to 1,3-β-glucanase treatment is monitored by measuring the decrease of OD_{600} in time. The amount of Quantazyme to be added may vary depending on the cell wall organization of the strain or organism to be analyzed. Alternatively, Zymolyase may be used. As Quantazyme is a recombinant 1,3-β-glucanase and Zymolyase contains in addition to 1,3-β-glucanase also a protease, the results for the two enzyme preparations may differ slightly.

Protocol 10. Monitoring resistance to cell wall-perturbing compounds.

1. Grow cells overnight in YPD.
2. Prepare a 1% (w/v) filter-sterilized stock solution of the compound to be tested. The commonly used Calcofluor white from Sigma (Fluorescent Brightener 28) should be dissolved under alkaline conditions (25 mM NaOH).
3. Prepare YPD with and without the compound to be tested. Plates containing Fluorescent Brightener 28 are buffered to pH 6.0 with 150 mM MES-NaOH to prevent Calcofluor precipitation.
4. Prepare 10-fold serial dilutions of cells. Spot 4 μl containing 10^5–10^1 cells on the plates. Monitor growth after 2 and 3 days at 30 or 37°C.

possess carbohydrate-processing activity (Coutinho and Henrissat, 1999). In addition, various simple bioassays are available to determine if specific proteins may contribute to cell surface hydrophobicity (Straver and Kijne, 1996), flocculation (Straver and Kijne, 1996), invasive growth (Cullen and Sprague, 2000), and biofilm formation (Reynolds and Fink, 2001). As mentioned before, the fungal cell wall is a dynamic entity strongly depending in composition and molecular organization on environmental conditions. To further the study of cell wall dynamics, Rodriguez-Pena *et al.* (2005) have developed a "yeast cell wall chip", a simplified DNA microarray containing 390 cell wall-related genes, allowing sensitive and focused transcript analysis of cell wall-related genes including CWPs.

♦♦♦♦♦♦ VI. CELL SURFACE DISPLAY OF HETEROLOGOUS PROTEINS

As the external layer of the cell wall of *S. cerevisiae* consists of proteins emanating into the medium, various approaches have been developed to target heterologous proteins to that layer to create strains that may function as renewable and immobilized biocatalysts, immunoadsorbent, heavy metal-adsorbent, etc. Both GPI-CWPs and Pir-CWPs have been successfully used (Schreuder *et al.*, 1996; Kondo and Ueda, 2004; Andres *et al.*, 2005). Also co-expression of heterologous proteins has been successful, allowing for example the efficient production of ethanol by yeast cells growing on cellulose as the sole carbon source (Fujita *et al.*, 2002). Surface display of llama single-domain antibody fragments also seems to be a very attractive option, not only because *S. cerevisiae* can produce them in large quantities, but also because they are often exceptionally stable (Van der Linden *et al.*, 1999; Dolk *et al.*, 2005).

Based on the general organization of GPI-CWPs (SP – functional domain – spacer domain – GPI-anchor addition signal), it is probably best to replace the functional domain of a long, homologous GPI-protein such as a flocculin by the protein (domain) of interest. The presence of a spacer domain is important for the accessibility and functionality of the incorporated protein (Frieman *et al.*, 2002; Sato *et al.*, 2002). Boder and Wittrup (1997) introduced the **a**-agglutinin complex for cell surface display of heterologous proteins. The **a**-agglutinin complex consists of a GPI-CWP (Aga1p), which functions as a carrier for the adhesion subunit of the complex (Aga2p). The adhesion subunit of the complex is a short secretory peptide that is covalently linked to Aga1p by a disulfide bond (Cappellaro *et al.*, 1994). The Aga2p peptide can be C-terminally extended with a protein of choice, which is then displayed at the cell surface.

As expected in view of the similarities between the molecular architecture of the cell wall of *S. cerevisiae* and other ascomycetous yeasts, other ascomycetous yeasts such as *Hansenula*, *Pichia pastoris*, and *Y. lipolytica* can also be used as platforms for cell surface engineering. In view of the growing evidence that similar CWP–polysaccharide complexes as identified in *S. cerevisiae* also occur in mycelial Ascomycetes, there is thus every reason to believe that many mycelial Ascomycetes including such well-known production organisms as *A. niger* will prove to be suitable hosts for cell surface engineering.

Acknowledgements

We would like to acknowledge all present and former members of the Klis lab for their contributions. This work was financially supported by the EU programs GALAR FUNGAIL I and II and FUNGWALL.

References

Andres, I., Gallardo, O., Parascandola, P., Javier Pastor, F. I. and Zueco, J. (2005). Use of the cell wall protein Pir4 as a fusion partner for the expression of *Bacillus* sp. BP-7 xylanase A in *Saccharomyces cerevisiae*. *Biotechnol. Bioeng.* **89**, 690–697.

Ballou, C. E. (1990). Isolation, characterization, and properties of *Saccharomyces cerevisiae mnn* mutants with nonconditional protein glycosylation defects. *Methods Enzymol.* **185**, 440–470.

Ballou, L., Hitzeman, R. A., Lewis, M. S. and Ballou, C. E. (1991). Vanadate-resistant yeast mutants are defective in protein glycosylation. *Proc. Natl. Acad. Sci. USA* **88**, 3209–3212.

Bendtsen, J. D., Nielsen, H., Von Heijne, G. and Brunak, S. (2004). Improved prediction of signal peptides: SignalP 3.0. *J. Mol. Biol.* **340**, 783–795.

Boder, E. T. and Wittrup, K. D. (1997). Yeast surface display for screening combinatorial polypeptide libraries. *Nat. Biotechnol.* **15**, 553–557.

Bom, I. J., Dielbandhoesing, S. K., Harvey, K. N., Oomes, S. J., Klis, F. M. and Brul, S. (1998). A new tool for studying the molecular architecture of the fungal cell wall: one-step purification of recombinant *Trichoderma* β-(1-6)-glucanase expressed in *Pichia pastoris*. *Biochim. Biophys. Acta* **1425**, 419–424.

Boorsma, A., De Nobel, H., Ter Riet, B., Bargmann, B., Brul, S., Hellingwerf, K. J. and Klis, F. M. (2004). Characterization of the transcriptional response to cell wall stress in *Saccharomyces cerevisiae*. *Yeast* **21**, 413–427.

Brul, S., King, A., Van der Vaart, J. M., Chapman, J., Klis, F. and Verrips, C. T. (1997). The incorporation of mannoproteins in the cell wall of *S. cerevisiae* and filamentous *Ascomycetes*. *Antonie Van Leeuwenhoek* **72**, 229–237.

Cappellaro, C., Baldermann, C., Rachel, R. and Tanner, W. (1994). Mating type-specific cell–cell recognition of *Saccharomyces cerevisiae*: cell wall attachment and active sites of **a**- and α-agglutinin. *EMBO J.* **13**, 4737–4744.

Cappellaro, C., Mrsa, V. and Tanner, W. (1998). New potential cell wall glucanases of *Saccharomyces cerevisiae* and their involvement in mating. *J. Bacteriol.* **180**, 5030–5037.

Caro, L. H. P., Tettelin, H., Vossen, J. H., Ram, A. F. J., Van den Ende, H. and Klis, F. M. (1997). *In silicio* identification of glycosyl-phosphatidylinositol-anchored plasma-membrane and cell wall proteins of *Saccharomyces cerevisiae*. *Yeast* **13**, 1477–1489.

Chen, M. H., Shen, Z. M., Bobin, S., Kahn, P. C. and Lipke, P. N. (1995). Structure of *Saccharomyces cerevisiae* α-agglutinin. Evidence for a yeast cell wall protein with multiple immunoglobulin-like domains with atypical disulfides. *J. Biol. Chem.* **270**, 26168–26177.

Conde, R., Pablo, G., Cueva, R. and Larriba, G. (2003). Screening for new yeast mutants affected in mannosylphosphorylation of cell wall mannoproteins. *Yeast* **20**, 1189–1211.

Coutinho, P. M. and Henrissat, B. (1999). Carbohydrate-active enzymes: an integrated database approach. In: *Recent Advances in Carbohydrate Bioengineering* (H. J. Gilbert, G. Davies, B. Henrissat and B. Svensson, eds), pp. 3–12. The Royal Society of Chemistry, Cambridge.

Cullen, P. J. and Sprague, G. F., Jr. (2000). Glucose depletion causes haploid invasive growth in yeast. *Proc. Natl. Acad. Sci. USA* **97**, 13619–13624.

Damveld, R. A., Arentshorst, M., Vankuyk, P. A., Klis, F. M., Van den Hondel, C. A. M. J. J. and Ram, A. F. J. (2005a). Characterisation of CwpA, a putative glycosylphosphatidylinositol anchored cell wall mannoprotein in the filamentous fungus *Aspergillus niger*. *Fungal Genet. Biol.* **42**, 873–885.

Damveld, R. A., Vankuyk, P. A., Arentshorst, M., Klis, F. M., van den Hondel, C. A. and Ram, A. F. J. (2005b). Expression of *agsA*, one of five 1,3-α-d-glucan synthase-encoding genes in *Aspergillus niger*, is induced in response to cell wall stress. *Fungal Genet. Biol.* **42**, 165–177.

De Groot, P. W. J., De Boer, A. D., Cunningham, J., Dekker, H. L., De Jong, L., Hellingwerf, K. J., De Koster, C. and Klis, F. M. (2004). Proteomic analysis of *Candida albicans* cell walls reveals covalently bound carbohydrate-active enzymes and adhesins. *Eukaryot. Cell* **3**, 955–965.

De Groot, P. W. J., Ram, A. F. and Klis, F. M. (2005). Features and functions of covalently linked proteins in fungal cell walls. *Fungal Genet. Biol.* **42**, 657–675.

De Groot, P. W. J., Ruiz, C., Vázquez de Aldana, C. R., Dueñas, E., Cid, V. J., Del Rey, F., Rodríguez-Peña, J. M., Pérez, P., Andel, A., Caubín, J., Arroyo, J., García, J. C., Gil, C., Molina, M., García, L. J., Nombela, C. and Klis, F. M. (2001). A genomic approach for the identification and classification of genes involved in cell wall formation and its regulation in *Saccharomyces cerevisiae*. *Comp. Funct. Genom.* **2**, 124–142.

De Nobel, H., Ruiz, C., Martin, H., Morris, W., Brul, S., Molina, M. and Klis, F. M. (2000). Cell wall perturbation in yeast results in dual phosphorylation of the Slt2/Mpk1 MAP kinase and in an Slt2-mediated increase in *FKS2-lacZ* expression, glucanase resistance and thermotolerance. *Microbiology* **146**, 2121–2132.

De Nobel, J. G., Dijkers, C., Hooijberg, E. and Klis, F. M. (1989). Increased cell wall porosity in *Saccharomyces cerevisiae* after treatment with dithiothreitol or EDTA. *J. Gen. Microbiol.* **135**, 2077–2084.

De Nobel, J. G., Klis, F. M., Munnik, T., Priem, J. and Van den Ende, H. (1990a). An assay of relative cell wall porosity in *Saccharomyces cerevisiae, Kluyveromyces lactis* and *Schizosaccharomyces pombe*. *Yeast* **6**, 483–490.

De Nobel, J. G., Klis, F. M., Priem, J., Munnik, T. and Van den Ende, H. (1990b). The glucanase-soluble mannoproteins limit cell wall porosity in *Saccharomyces cerevisiae*. *Yeast* **6**, 491–499.

Dean, N. (1995). Yeast glycosylation mutants are sensitive to aminoglycosides. *Proc. Natl. Acad. Sci. USA* **92**, 1287–1291.

Dolk, E., Van der Vaart, M., Lutje Hulsik, D., Vriend, G., De Haard, H., Spinelli, S., Cambillau, C., Frenken, L. and Verrips, T. (2005). Isolation of llama antibody fragments for prevention of dandruff by phage display in shampoo. *Appl. Environ. Microbiol.* **71**, 442–450.

Ecker, M., Deutzmann, R., Lehle, L., Mrsa, V. and Tanner, W. (2006). Pir proteins of *Saccharomyces cerevisiae* are attached to beta-1,3-glucan by a new protein–carbohydrate linkage. *J. Biol. Chem.* **281**, 11523–11529.

Eisenhaber, B., Schneider, G., Wildpaner, M. and Eisenhaber, F. (2004). A sensitive predictor for potential GPI lipid modification sites in fungal protein sequences and its application to genome-wide studies for *Aspergillus nidulans, Candida albicans, Neurospora crassa, Saccharomyces cerevisiae* and *Schizosaccharomyces pombe*. *J. Mol. Biol.* **337**, 243–253.

Frieman, M. B. and Cormack, B. P. (2003). The omega-site sequence of glycosylphosphatidylinositol-anchored proteins in *Saccharomyces cerevisiae* can determine distribution between the membrane and the cell wall. *Mol. Microbiol.* **50**, 883–896.

Frieman, M. B. and Cormack, B. P. (2004). Multiple sequence signals determine the distribution of glycosylphosphatidylinositol proteins between the plasma membrane and cell wall in *Saccharomyces cerevisiae*. *Microbiology* **150**, 3105–3114.

Frieman, M. B., McCaffery, J. M. and Cormack, B. P. (2002). Modular domain structure in the *Candida glabrata* adhesin Epa1p, a β1,6 glucan-cross-linked cell wall protein. *Mol. Microbiol.* **46**, 479–492.

Fujita, Y., Takahashi, S., Ueda, M., Tanaka, A., Okada, H., Morikawa, Y., Kawaguchi, T., Arai, M., Fukuda, H. and Kondo, A. (2002). Direct and efficient production of ethanol from cellulosic material with a yeast strain displaying cellulolytic enzymes. *Appl. Environ. Microbiol.* **68**, 5136–5141.

Gemmill, T. R. and Trimble, R. B. (1999). Overview of N- and O-linked oligosaccharide structures found in various yeast species. *Biochim. Biophys. Acta* **1426**, 227–237.

Grün, C. H., Hochstenbach, F., Humbel, B. M., Verkleij, A. J., Sietsma, J. H., Klis, F. M., Kamerling, J. P. and Vliegenthart, J. F. (2005). The structure of cell wall α-glucan from fission yeast. *Glycobiology* **15**, 245–257.

Hamada, K., Terashima, H., Arisawa, M. and Kitada, K. (1998). Amino acid sequence requirement for efficient incorporation of glycosylphosphatidylinositol-associated proteins into the cell wall of *Saccharomyces cerevisiae*. *J. Biol. Chem.* **273**, 26946–26953.

Hamada, K., Terashima, H., Arisawa, M., Yabuki, N. and Kitada, K. (1999). Amino acid residues in the omega-minus region participate in cellular localization of yeast glycosylphosphatidylinositol-attached proteins. *J. Bacteriol.* **181**, 3886–3889.

Huh, W. K., Falvo, J. V., Gerke, L. C., Carroll, A. S., Howson, R. W., Weissman, J. S. and O'Shea, E. K. (2003). Global analysis of protein localization in budding yeast. *Nature* **425**, 686–691.

Jaafar, L., Moukadiri, I. and Zueco, J. (2003). Characterization of a disulphide-bound Pir-cell wall protein (Pir-CWP) of *Yarrowia lipolytica*. *Yeast* **20**, 417–426.

Jaafar, L. and Zueco, J. (2004). Characterization of a glycosylphosphatidylinositol-bound cell-wall protein (GPI-CWP) in *Yarrowia lipolytica*. *Microbiology* **150**, 53–60.

Jentoft, N. (1990). Why are proteins O-glycosylated?. *Trends Biochem. Sci.* **15**, 291–294.

Kapteyn, J. C., Hoyer, L. L., Hecht, J. E., Muller, W. H., Andel, A., Verkleij, A. J., Makarow, M., Van den Ende, H. and Klis, F. M. (2000). The cell wall architecture of *Candida albicans* wild-type cells and cell wall-defective mutants. *Mol. Microbiol.* **35**, 601–611.

Kapteyn, J. C., Montijn, R. C., Vink, E., De la Cruz, J., Llobell, A., Douwes, J. E., Shimoi, H., Lipke, P. N. and Klis, F. M. (1996). Retention of *Saccharomyces cerevisiae* cell wall proteins through a phosphodiester-linked β-1,3-/β-1,6-glucan heteropolymer. *Glycobiology* **6**, 337–345.

Kapteyn, J. C., Ram, A. F. J., Groos, E. M., Kollar, R., Montijn, R. C., Van den Ende, H., Llobell, A., Cabib, E. and Klis, F. M. (1997). Altered extent of cross-linking of β1,6-glucosylated mannoproteins to chitin in *Saccharomyces cerevisiae* mutants with reduced cell wall β1,3-glucan content. *J. Bacteriol.* **179**, 6279–6284.

Kapteyn, J. C., Ter Riet, B., Vink, E., Blad, S., De Nobel, H., Van den Ende, H. and Klis, F. M. (2001). Low external pH induces *HOG1*-dependent changes in the organization of the *Saccharomyces cerevisiae* cell wall. *Mol. Microbiol.* **39**, 469–479.

Kapteyn, J. C., Van Egmond, P., Sievi, E., Van den Ende, H., Makarow, M. and Klis, F. M. (1999). The contribution of the O-glycosylated protein Pir2p/Hsp150 to the construction of the yeast cell wall in wild-type cells and β1,6-glucan-deficient mutants. *Mol. Microbiol.* **31**, 1835–1844.

Klis, F. M., De Groot, P. and Hellingwerf, K. (2001). Molecular organization of the cell wall of *Candida albicans*. *Med. Mycol.* **39**(Suppl. 1), 1–8.

Klis, F. M., Mol, P., Hellingwerf, K. and Brul, S. (2002). Dynamics of cell wall structure in *Saccharomyces cerevisiae*. *FEMS Microbiol. Rev.* **26**, 239–256.

Klis, F. M., Ram, A. F. J., Montijn, R. C., Kapteyn, J. C., Caro, L. H. P., Vossen, J. H., Van Berkel, M. A. A., Brekelmans, S. S. C. and Van den Ende, H. (1998). Posttranslational modifications of secretory proteins. *Methods Microbiol.* **26**, 223–238.

Kollar, R., Reinhold, B. B., Petrakova, E., Yeh, H. J., Ashwell, G., Drgonova, J., Kapteyn, J. C., Klis, F. M. and Cabib, E. (1997). Architecture of the yeast cell wall. β(1→6)-glucan interconnects mannoprotein, β(1→3)-glucan, and chitin. *J. Biol. Chem.* **272**, 17762–17775.

Kondo, A. and Ueda, M. (2004). Yeast cell-surface display – applications of molecular display. *Appl. Microbiol. Biotechnol.* **64**, 28–40.

Lu, C. F., Montijn, R. C., Brown, J. L., Klis, F., Kurjan, J., Bussey, H. and Lipke, P. N. (1995). Glycosylphosphatidylinositol-dependent cross-linking of α-agglutinin and β-1,6-glucan in the *Saccharomyces cerevisiae* cell wall. *J. Cell Biol.* **128**, 333–340.

Mao, Y., Zhang, Z. and Wong, B. (2003). Use of green fluorescent protein fusions to analyse the N- and C-terminal signal peptides of GPI-anchored cell wall proteins in *Candida albicans*. *Mol. Microbiol.* **50**, 1617–1628.

Marchler-Bauer, A., Anderson, J. B., DeWeese-Scott, C., Fedorova, N. D., Geer, L. Y., He, S., Hurwitz, D. I., Jackson, J. D., Jacobs, A. R., Lanczycki, C. J., Liebert, C. A., Liu, C., Madej, T., Marchler, G. H., Mazumder, R., Nikolskaya, A. N., Panchenko, A. R., Rao, B. S., Shoemaker, B. A., Simonyan, V., Song, J. S., Thiessen, P. A., Vasudevan, S., Wang, Y., Yamashita, R. A., Yin, J. J. and Bryant, S. H. (2003). CDD: A curated Entrez database of conserved domain alignments. *Nucleic Acids Res.* **31**, 383–387.

Montijn, R. C., van Rinsum, J., van Schagen, F. A. and Klis, F. M. (1994). Glucomannoproteins in the cell wall of *Saccharomyces cerevisiae* contain a novel type of carbohydrate side chain. *J. Biol. Chem.* **269**, 19338–19342.

Montijn, R. C., Van Wolven, P., De Hoog, S. and Klis, F. M. (1997). β-Glucosylated proteins in the cell wall of the black yeast *Exophiala* (*Wangiella*) *dermatitidis*. *Microbiology* **143**, 1673–1680.

Montijn, R. C., Vink, E., Muller, W. H., Verkleij, A. J., Van Den Ende, H., Henrissat, B. and Klis, F. M. (1999). Localization of synthesis of β1,6-glucan in *Saccharomyces cerevisiae*. *J. Bacteriol.* **181**, 7414–74120.

Moukadiri, I., Jaafar, L. and Zueco, J. (1999). Identification of two mannoproteins released from cell walls of a *Saccharomyces cerevisiae mnn1 mnn9* double mutant by reducing agents. *J. Bacteriol.* **181**, 4741–4745.

Mrsa, V., Seidl, T., Gentzsch, M. and Tanner, W. (1997). Specific labelling of cell wall proteins by biotinylation. Identification of four covalently linked O-mannosylated proteins of *Saccharomyces cerevisiae*. *Yeast* **13**, 1145–1154.

Orlean, P. (1997). Biogenesis of yeast wall and surface components. In: *The Molecular and Cellular Biology of the Yeast Saccharomyces. Cell Cycle and Cell Biology*, vol. 3 (J. R. Pringle, J. R. Broach and E. W. Jones, eds), pp. 229–362. Cold Spring Harbor Laboratory Press, New York.

Orlean, P., Kuranda, M. J. and Albright, C. F. (1991). Analysis of glycoproteins from *Saccharomyces cerevisiae*. *Methods Enzymol.* **194**, 682–697.

Ovalle, R., Lim, S. T., Holder, B., Jue, C. K., Moore, C. W. and Lipke, P. N. (1998). A spheroplast rate assay for determination of cell wall integrity in yeast. *Yeast* **14**, 1159–1166.

Perez, P. and Ribas, J. C. (2004). Cell wall analysis. *Methods* **33**, 245–251.

Ram, A. F. J., Van den Ende, H. and Klis, F. M. (1998). Green fluorescent protein–cell wall fusion proteins are covalently incorporated into the cell wall of *Saccharomyces cerevisiae*. *FEMS Microbiol. Lett.* **162**, 249–255.

Reddy, V. A., Johnson, R. S., Biemann, K., Williams, R. S., Ziegler, F. D., Trimble, R. B. and Maley, F. (1988). Characterization of the glycosylation sites in yeast external invertase. I. N-linked oligosaccharide content of the individual sequons. *J. Biol. Chem.* **263**, 6978–6985.

Reynolds, T. B. and Fink, G. R. (2001). Bakers' yeast, a model for fungal biofilm formation. *Science* **291**, 878–881.

Rodriguez-Pena, J. M., Perez-Diaz, R. M., Alvarez, S., Bermejo, C., Garcia, R., Santiago, C., Nombela, C. and Arroyo, J. (2005). The 'yeast cell wall chip' – a tool to analyse the regulation of cell wall biogenesis in *Saccharomyces cerevisiae*. *Microbiology* **151**, 2241–2249.

Sato, N., Matsumoto, T., Ueda, M., Tanaka, A., Fukuda, H. and Kondo, A. (2002). Long anchor using Flo1 protein enhances reactivity of cell surface-displayed glucoamylase to polymer substrates. *Appl. Microbiol. Biotechnol.* **60**, 469–474.

Schoffelmeer, E. A. M., Kapteyn, J. C., Montijn, R. C., Cornelissen, B. C. and Klis, F. M. (1996). Glucosylation of fungal cell wall proteins as a potential target for novel antifungal agents. In: *Modern Fungicides and Antifungal Compounds* (H. Lyr, P. E. Russel and H. D. Sisler, eds), pp. 157–162. Intercept Ltd, Andover.

Schoffelmeer, E. A. M., Klis, F. M., Sietsma, J. H. and Cornelissen, B. J. (1999). The cell wall of *Fusarium oxysporum*. *Fungal Genet. Biol.* **27**, 275–282.

Schoffelmeer, E. A. M., Vossen, J. H., Van Doorn, A. A., Cornelissen, B. J. and Haring, M. A. (2001). FEM1, a *Fusarium oxysporum* glycoprotein that is covalently linked to the cell wall matrix and is conserved in filamentous fungi. *Mol. Genet. Genom.* **265**, 143–152.

Schreuder, M. P., Mooren, A. T., Toschka, H. Y., Verrips, C. T. and Klis, F. M. (1996). Immobilizing proteins on the surface of yeast cells. *Trends Biotechnol.* **14**, 115–120.

Smits, G. J., Schenkman, L. R., Brul, S., Pringle, J. R. and Klis, F. M. (2006). Role of cell cycle-regulated expression in the localized incorporation of cell wall proteins in yeast. *Mol. Biol. Cell* **17**, 3267–3280.

Straver, M. H. and Kijne, J. W. (1996). A rapid and selective assay for measuring cell surface hydrophobicity of brewer's yeast cells. *Yeast* **12**, 207–213.

Uccelletti, D., Pacelli, V., Mancini, P. and Palleschi, C. (2000). *vga* Mutants of *Kluyveromyces lactis* show cell integrity defects. *Yeast* **16**, 1161–1171.

Van der Linden, R. H., Frenken, L. G., de Geus, B., Harmsen, M. M., Ruuls, R. C., Stok, W., De Ron, L., Wilson, S., Davis, P. and Verrips, C. T. (1999). Comparison of physical chemical properties of llama VHH antibody fragments and mouse monoclonal antibodies. *Biochim. Biophys. Acta* **1431**, 37–46.

Van der Vaart, J. M., Caro, L. H. P., Chapman, J. W., Klis, F. M. and Verrips, C. T. (1995). Identification of three mannoproteins in the cell wall of *Saccharomyces cerevisiae*. *J. Bacteriol.* **177**, 3104–3110.

Weig, M., Jansch, L., Gross, U., De Koster, C. G., Klis, F. M. and De Groot, P. W. J. (2004). Systematic identification *in silico* of covalently bound cell wall proteins and analysis of protein–polysaccharide linkages of the human pathogen *Candida glabrata*. *Microbiology* **150**, 3129–3144.

Williams, D. L., Pretus, H. A., Ensley, H. E. and Browder, I. W. (1994). Molecular weight analysis of a water-insoluble, yeast-derived $(1\rightarrow 3)$-β-d-glucan by organic-phase size-exclusion chromatography. *Carbohydr. Res.* **253**, 293–298.

Yin, Q. Y., de Groot, P. W. J., Dekker, H. L., de Jong, L., Klis, F. M. and de Koster, C. G. (2005). Comprehensive proteomic analysis of *Saccharomyces cerevisiae* cell walls: identification of proteins covalently attached via glycosylphosphatidylinositol remnants or mild alkali-sensitive linkages. *J. Biol. Chem.* **280**, 20894–20901.

14 Yeast Protein Microarrays

Jason Ptacek[1] and Michael Snyder[1,2]
[1] *Department of Molecular Biophysics & Biochemistry, Yale University, New Haven, CT, USA;*
[2] *Molecular, Cellular & Developmental Biology, Yale University, New Haven, CT, USA*

♦♦♦

CONTENTS

Introduction
Development of the microarray
Expression libraries for protein microarrays
Technical aspects of protein microarrays
Current applications of protein microarrays

List of Abbreviations

GST	glutathione-S-transferase
ORF	open reading frame
SELDI	surface enhanced laser desorption/ionization
SPR	surface plasmon resonance
PPI	protein–protein interaction

♦♦♦♦♦♦ I. INTRODUCTION

Since the publication of the sequence of *S. cerevisiae* in 1996 (Goffeau *et al.*, 1996), over 100 genomes have been sequenced. This sequencing effort has led to the systematic identification of genes and allowed targeted deletions to be made for functional analysis. Complete genome sequences have also led to the creation of DNA microarrays, arrays of oligonucleotides printed at high spatial density, that allow the identification of transcribed mRNA in cells. Studies using DNA microarrays have generated expression profiles of yeast genes under a variety of conditions. However, these changes in gene expression and linkage of gene deletions to phenotypes provide only limited answers into how a cell is able to perform all the functions necessary for life in a dynamic environment. Answers in more detail will be provided once the proteome, the complement to the genome, is defined and all the biological activities and interactions of each protein have been identified.

METHODS IN MICROBIOLOGY, VOLUME 36
0580-9517 DOI:10.1016/S0580-9517(06)36014-X

Proteins are the players in the cell, interacting together to maintain and propagate life, and the proteome is the set of all the proteins in an organism. Proteomics seeks to determine what each protein does in the context of the entire proteome, providing in much more detail how a cell functions. This is undoubtedly a larger task than sequencing a genome, considering that the sheer number of proteins will be significantly more than the number of genes due to all the isoforms of all the different cell types at different points in development. It is likely that the number of unique proteins in humans (30 000–40 000 genes) will exceed 1–2 million. Humans are not the exception in proteome complexity; even in yeast the exact number of different proteins is unknown. Despite a relatively low number of splice variants (<300; Davis *et al.*, 2000), *S. cerevisiae* still has a complex proteome when one considers that 30% of the yeast proteome is predicted to be phosphorylated (Cohen, 2000; Ficarro *et al.*, 2002) and 20–50% is predicted to be glycosylated (Apweiler *et al.*, 1999) in addition to other modifications. Protein microarrays are arrays of protein, or in the case of yeast nearly the entire proteome, which will expedite our study of the proteome by providing a platform to elucidate a protein's function and how it relates to other proteins on a global scale. This chapter seeks to address the many challenges and options that exist in designing a yeast protein array and the many questions that have been addressed using this technology, predominantly in the form of functional protein microarrays.

To better appreciate the utility of protein microarrays, it is useful to understand its limits and advantages in the context of competing technologies. All technologies involving proteins are challenged by the large scale of the proteome and the difficulty in working with proteins given that their chemistry and solubility are much more variable. The goals of proteomics and a sample of the common technologies applied to each are listed in Table 1.

Many techniques have been used to address different aspects of these goals. Mass spectrometry has been used to identify protein complexes (Gavin *et al.*, 2002; Ho *et al.*, 2002), components of the yeast nuclear pore complex (Rout *et al.*, 2000), and to catalogue 1484 proteins from yeast in log-phase (Washburn *et al.*, 2001). Using this technique it is difficult, but not impossible, to determine if a protein interaction is a direct, or binary, interaction or if it is an indirect interaction, mediated by other components of the complex (Ranish *et al.*, 2003). Additionally, multiple experiments with different bait proteins from the same identified complex are needed to confirm if a protein is part of a one large complex or rather is a member of multiple, smaller complexes. Nonetheless, mass spectrometry has a high success rate at identifying *in vivo* interactions and simply knowing that two proteins interact in a complex, even if indirectly, provides much information concerning the role of these proteins. Another utility of mass spectrometry is its ability to catalogue proteins in a proteome and has been used to identify many posttranslational modifications in yeast (Ficarro *et al.*, 2002; Gruhler *et al.*,

Table 1. The goals and technology of proteomics

Goal	Methods
Catalogue all proteins and isoforms in each cell type, at each developmental stage, and in different disease states (phosphorylation, methylation, splicing variants, etc.)	• 2-dimensional electrophoresis • Mass spectrometry • Fluorescence microscopy • Protein microarrays
Identify the protein-biomolecule (protein, DNA, lipid) interactions and complexes within the proteome	• Two-hybrid • Chromatin immunoprecipitation (ChIP-chip) • Mass spectrometry • Protein microarrays
Determine the enzymatic activities of the proteins and identify their substrates	• Mutant *as*-alleles for kinase-substrate identity • Solution enzymatic assays • Protein microarrays
Construct pathways of the proteins and determine how these pathways are regulated	• Integration of multiple methods

2005). The limitations of this technique are that it is relatively low-throughput and that it is difficult to get complete coverage of a single protein, let alone an entire proteome. Undoubtedly many more protein isoforms wait to be identified. However, knowing that a protein is modified and the site of modification has allowed mutational analysis into the significance of the modifications. As discussed later, protein microarrays can complement mass spectrometry studies by identifying the enzyme responsible for the modification and help fill in the links between the proteins in an organism.

Another technique for the identification of protein–protein interactions is the yeast 2-hybrid genetic screen (Fields and Song, 1989; Uetz *et al.*, 2000; Ito *et al.*, 2001; see Chapters 6 and 7). The technique is high-throughput but is less accurate in identifying valid interactions, partly because the interactions take place in the nucleus rather than their native environment. The assay is also skewed by transcriptional activators. Different two-hybrid studies have not shown a high degree of overlap with each other or with mass spectrometry studies. Of the 80 000 binary interactions suggested for yeast proteins, only around 2400 interactions are supported by at least two lines of evidence (von Mering *et al.*, 2002). Undoubtedly multiple techniques will be useful in identifying valid interactions within the cell.

♦♦♦♦♦♦ II. DEVELOPMENT OF THE MICROARRAY

The concept of protein microarrays can be attributed to Dr. Roger Ekins, who first spotted antibodies in a microarray format that could be used to detect antigens (Ekins *et al.*, 1990). Protein microarrays are sets of proteins or, in the case of yeast, nearly the whole proteome (Zhu *et al.*, 2001; Gelperin *et al.*, 2005), printed in an addressable, high-density format onto a solid surface. DNA microarrays, first produced in 1995 (Schena *et al.*, 1995), established the technology needed to produce complex microarrays of thousands of features. Protein microarrays use similar spotting techniques to fabricate the arrays. However, unlike DNA, which is biochemically uniform and stable, protein microarrays present unique challenges. These include the sensitivity of proteins, the variability in structure, size, and modifications that proteins have, and the difficult task of producing each individual protein. Today, protein microarrays are largely of two types: antibody microarrays and functional protein microarrays.

A. Antibody Microarrays

Antibody microarrays usually consist of antibodies of known specificity spotted onto a microscope slide. A complex mixture of antigens can be labeled and passed over the chip, the microarray washed, and captured antigens detected by their label. Many antibody microarrays are available for mammalian studies from more than a dozen companies, including the 224 antibody Panorama Ab microarray kit (Sigma-Aldrich) or the Signal Transduction AntibodyArray (Hypromatrix) that contains 400 antibodies to popular proteins in cell signaling. Antibody microarrays can be used to screen for proteins of interest from different cell states or cellular compartments, to identify changes in modifications (e.g. phosphorylation), and can be used to identify protein–protein interactions. This latter assay works by adding a lysate to an antibody microarray, washing the array, and then probing with an antibody to the protein of interest. The binding partner to the protein of interest will be bound to the array by a known antibody, providing the identity of the interacting partner. Despite the promise of the technology, current microarrays are limited to sets of well-characterized antibodies, such as antibodies to cytokines.

The greatest challenge with antibody microarrays is the inadequate number of antibodies of high specificity (Haab *et al.*, 2001; Michaud *et al.*, 2003). This is needed in part due to the wide range of protein concentrations (<50–10^6 copies per yeast cell; Ghaemmaghami *et al.*, 2003) that results in an antibody with low affinity for a protein binding to the protein if it is presented to the antibody at a high enough concentration. Polyclonal antibodies often lack high-specificity, while producing monoclonal antibodies is an arduous task; both are costly endeavors. To accelerate the screening of specific antibodies, functional protein microarrays could be used. This was done for 11 antibodies using a yeast protein microarray containing approximately 5000 yeast proteins (Michaud *et al.*, 2003). Six antibodies recognized multiple proteins on the array; the anti-Nap1 antibody recognized over 1700 proteins. As Haab *et al.* (2001) had found earlier, research-grade antibodies of high-specificity are the exception rather than the norm.

A sandwich assay can be used to achieve better specificity by using two antibodies to different epitopes of the same antigen (Silzel *et al.*, 1998; Delehanty and Ligler, 2002). The antigen is recognized first by an antibody arrayed onto a chip and then by a second labeled antibody applied after the lysate. This of course requires two antibodies for each antigen, currently not feasible for most proteins. Using proteome arrays to screen antibodies could likely expedite the manufacture of high-quality antibodies, leading to better medicines and diagnostic tools, such as improved antibody microarrays. A microarray with antibodies of high-specificity to the majority of proteins could provide a more accurate snapshot of the changes to the cell's proteome than current expression analysis performed using DNA microarrays (Ideker *et al.*, 2001). Currently though, greater potential for antibody microarrays lies in their use in medical diagnostics in which the microarrays quickly identify antigens of interest, such as cancer markers (Knezevic, *et al.*, 2001; Sreekumar *et al.*, 2001).

B. Functional Protein Microarrays

Functional protein microarrays provide a platform to screen tens to thousands of proteins, and can be used for detecting protein–protein, protein–lipid, protein–DNA, and protein–small molecule interactions, for cataloguing the many posttranslational modifications to proteins, and for the identification of enzyme substrates (Figure 1). Protein microarrays enable one to study the big picture by determining a protein's relationship to not just one other protein but to all its cellular counterparts simultaneously.

To make a protein microarray, proteins must first be produced, requiring a high-quality, addressable collection of ORFs in an expression library. Proteins are then expressed and purified, and arrayed on a surface such as a modified glass slide. Assays have been developed that allow the screening of the majority of the yeast proteome in an unbiased, flexible, and high-throughput process. The speed at which data is generated can be almost overwhelming – we routinely print 80 yeast proteome microarrays in a day and can do 80 assays the next day, generating hundreds of thousands of data points on ~5800 proteins. Often the bottleneck is in data analysis, where bioinformatic methodologies are used to normalize the background and signal from each slide and computer algorithms identify hits that are above a certain threshold. With the data in hand providing a list of candidate interactions, well-defined *in vivo* experiments can be performed to validate the results, and the intricacies of biological pathways better defined.

C. Advantages and Limitations of Protein Chips

Protein microarrays can be applied to many important biological questions in a high-throughput, discovery-orientated process. Protein microarrays allow biochemical assays to be completed on

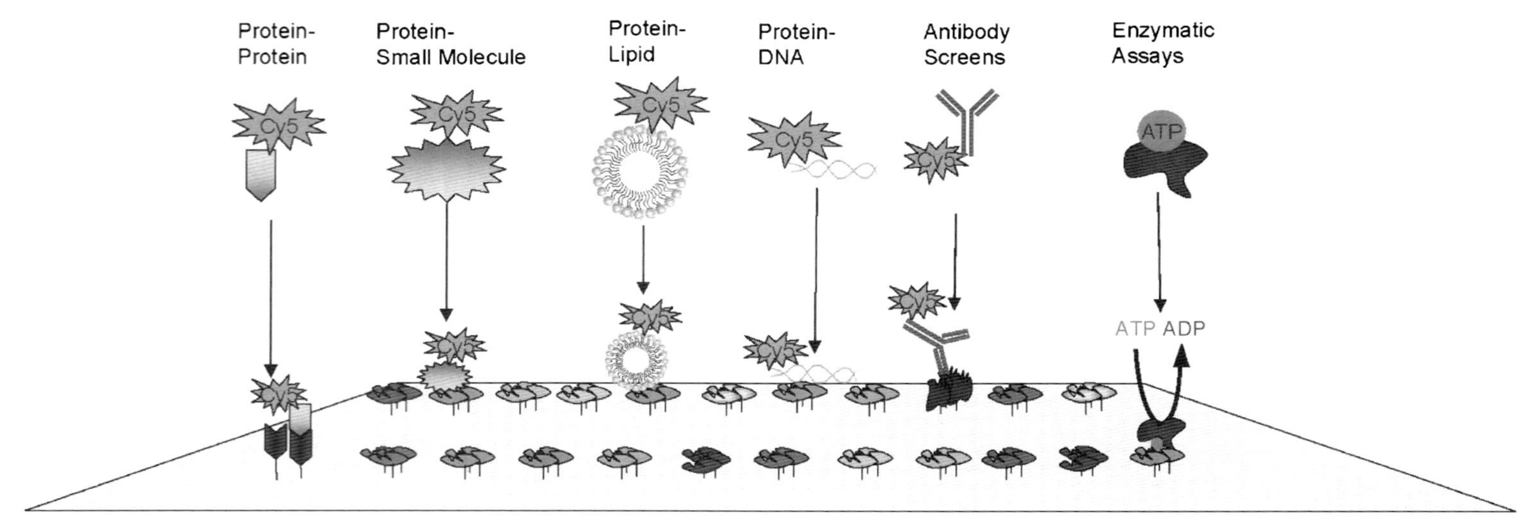

Figure 1. Different assays that have been performed on yeast protein microarrays. While Cy5 is the fluorophore shown, many other fluorophores can be used for detection. (See color plate section).

the entire proteome simultaneously, require less reagents and proteins, and are more sensitive. Better motifs (e.g. binding or phosphorylation motifs) using unbiased data can be generated. Using protein microarrays, the experimenter has control over the assay conditions, such as the buffer used and the cofactors present, and therefore protein microarrays can be tailored towards a variety of assays. Protein interactions between protein, lipid, DNA, and small molecules can easily be done in a robust fashion, posttranslational modifications identified, substrates determined, and cellular pathways delineated.

Despite the strengths microarrays offer, they do at best provide a picture of what is possible biologically. Integration of protein chip data with other high-throughput data, such as expression (Horak *et al.*, 2002; Lee *et al.*, 2002) and protein-binding data (Bader and Hogue, 2000; Uetz *et al.*, 2000; Xenarios *et al.*, 2000; Ito *et al.*, 2001), can provide multiple lines of evidence suggesting that an interaction is likely to occur *in vivo*. Ideally, with a small list of candidate interactions identified from the microarrays, *in vivo* validation can be performed to confirm that an enzyme acts upon an identified substrate or that a drug does bind a specific protein. The ability of a protein microarray with thousands of features to distil into a manageable list of candidate proteins for further analysis will be even more important as larger, more complex proteomes are studied.

♦♦♦♦♦♦ III. EXPRESSION LIBRARIES FOR PROTEIN MICROARRAYS

Expression libraries of *S. cerevisiae* ORFs have been constructed and used in the development of yeast protein microarrays. An N-terminally tagged library consisting of 5800 ORFs was created by cloning into a high-copy plasmid in which the NH_2-termini of each protein is fused to glutathione *S*-transferase polyhistidine (GST-His_6) using homologous recombination (Zhu *et al.*, 2001). DNA sequencing ensured that only plasmids containing the ORF fused in-frame to the N-terminal tag were used. Proteins from this library are expressed under the strongly inducible *GAL1* promoter and purified from yeast (Protocol 1, adapted from Zhu *et al.*, 2001). An advantage of producing proteins in yeast rather than in *E. coli* is that the proteins are more likely to be properly folded and modified; expression in *E. coli* has a higher tendency to produce protein aggregates. High-throughput expression and purification is done in a 96-well format allowing 1152 proteins to be purified per day. Briefly, yeast strains expressing one fusion protein each are grown in SC-Ura containing 2% raffinose to log phase and protein expression induced in the presence of 2% galactose. A combined 3 ml culture of each strain is pelleted and lysed by bead-beating. Proteins are purified using glutathione–agarose beads, the bound beads are washed with high-salt buffers, and eluted

Protocol 1. High-throughput growth and purification of GST-tagged proteins. See Amberg *et al.* (2005) for composition of growth media.

1. Inoculate yeast glycerol stocks (stored at −80°C in 96-well plates) onto SD-Ura agar plates using a 96-pronger and grow at 30°C for two days.
2. Use a 96-pronger to inoculate yeast cells from agar plates to uniboxes with 2 ml wells (Fisher) in which every well contains 300 μl SC-Ura liquid medium containing 2% raffinose and a 2 mm diameter glass ball for aeration.
3. Grow the starter cultures at 30°C with vigorous shaking (300 rpm) to an OD_{600} 4.0 (~16 h).
4. From these starter cultures, inoculate 15 μl into four different uniboxes containing a 2 mm glass ball and 750 μl of SC-Ura liquid medium containing 2% raffinose, giving a total of 3 ml of culture.
5. Grow cultures at 30°C with vigorous shaking to an OD_{600} 0.6–0.8 (~15 h) and induce protein expression by the addition of 40% galactose stock to achieve a final concentration of 2%, using an automated plate filling device (Q-Fill). Incubate cultures at 30°C for 4 h with shaking to allow expression.
6. Harvest the cells by spinning at 3000 rpm for 4 min, and wash the cell pellets once with cold water. Combine the 4 uniboxes containing the same cultures into 1 unibox, and then wash once with cold Lysis Buffer (50 mM Tris pH 7.5, 100 mM NaCl, 1 mM EGTA, 0.1% Triton X-100, 0.1% β-mercaptoethanol (BME), 0.5 mM PMSF, Roche Protease inhibitor tablets containing EDTA). The washed semi-dry culture is immediately stored in a –80°C freezer. The culture can be kept for weeks.
7. Transfer 12 uniboxes containing frozen cell pellets to ice and to the wells of the uniboxes add 100 μl of zirconia beads (0.5 mm diameter; from BSP, Germany) and 300 μl of Lysis Buffer. A cap mat is used to seal each well.
8. Lyse the cells by shaking in a paint-shaker for 4 × 1 min with 2 min intervals on ice.
9. Spin the lysates at 3000 rpm for 5 min at 4°C, collect the supernatants using either a multipipetor or a Hydra 96 microdispenser (Robbins Scientific, Sunnyvale, CA), and transfer them into a 96-well filter plate (Whatman) placed on top of a 96-well box. Care should be taken to collect only the supernatant and not disturb cellular debris as it will clog the filter plates, reducing the clarified lysate collected.
10. Perform a second lysis with 300 μl of Lysis Buffer and repeat Steps 8 and 9.
11. Spin the combined cell lysates through the filter plate (Whatman, #7700–2806) into a clean 96-well box for 15 min at 3000 rpm at 4°C, with aluminium foil serving as a lid for

the filter plates. The volume of filtered lysate in each well should be roughly 500 µl.

12. Wash glutathione beads (Amersham) with Lysis Buffer and resuspend in 5 × their volume of Lysis Buffer. Add 100 µl of this suspension to each well (i.e. 20 µl beads per well) and seal tightly with a cap mat. The beads are incubated with the lysate by rotating 360 degrees on a roller drum at 4°C for 1 h.
13. Collect the beads containing bound fusion protein by spinning at 3000 rpm for 10 s and remove the supernatant. Wash the beads 4 × with 400 µl of Wash Buffer I (Lysis Buffer containing 500 mM NaCl).
14. Wash the beads 2 × times with 400 µl of Wash Buffer II (50 mM HEPES pH 7.5, 100 mM NaCl, 10% glycerol). A last wash is done with 150 µl of Wash Buffer II and the bead slurry was transferred to a cold filter plate (Millipore, # MHVBN4550) and wash buffer removed by centrifugation for 1 min at 3000 rpm.
15. With the beads on the filter plate, add 30 µl of Elution Buffer (50 mM HEPES pH 7.5, 100 mM NaCl, 40% glycerol, 25 mM glutathione) to each well. Rock the beads for 1 h at 4°C. Different elution buffers can be used and should be tested in a pilot experiment for compatibility with slide chemistry and assay conditions.
16. Collect the eluate in a 96-well PCR plate by spinning through the filter plate for 1 min at 3000 rpm with the filter plate held on top of the PCR plate by the use of two rubber bands.
17. Aliquot each purified protein into three 96-well PCR plates and immediately store at –80°C.

using buffer containing glutathione. A non-Tris containing buffer is used for elution since primary amines react with many surface chemistries. Over 80% of the 5800 strains produced fusion protein at the expected molecular weight as calculated from immunoblots of 60 random samples (Zhu *et al.*, 2001).

A new collection of movable ORFs (the MORF collection) consists of ORFs cloned into a vector that allows easy transfer of the ORF into a variety of vectors using the Gateway® system (Invitrogen; Gelperin *et al.*, 2005). The MORF collection is the most comprehensive collection of cloned ORFs from a eukaryote constructed to date. Each ORF is expressed in yeast from the *GAL1* promoter and has His_6, an HA epitope, a protease 3C cleavage site, and the IgG binding domain from Protein A fused in-frame to the protein's C-terminus. Purification is also done in a 96-well format (Protocol 2) but unlike Protocol 2, 30 ml cultures of each strain are grown to obtain more protein at only a minor cost in effort. Both protocols can be scaled up or down depending upon the needs of the researcher. A challenge to

Protocol 2. Purification of Protein A fusion-proteins (Gelperin *et al.*, 2005). See Amberg *et al.* (2005) for composition of growth media.

1. Yeast glycerol stocks of MORF transformants are stored in 96-well plates at –80°C. Use a 96-pronger to transfer yeast transformants (Y258+MORF expression plasmid) from frozen stocks onto SD-Ura plates and grown at 30°C for two days.
2. Pick the yeast using a 96-pronger into 0.8 ml of liquid SCD-Ura in a unibox with 2 ml wells (Fisher) and grow overnight at 30°C.
3. Pellet these starter cultures and wash once with SC-Ura containing 2% raffinose, before adding 1/6th of the cells from each well to individual snap-cap 50 ml tubes (Nunc) containing 20 ml of SC-Ura/2% raffinose and a sterile 8 mm glass bead (PGC Scientific) to yield a starting OD_{600} of ~0.05.
4. Grow the SC-Ura/2% raffinose cultures at 30°C with shaking at 250 rpm for 15 h until the OD_{600} reaches 0.6–0.8, at which point protein expression is induced by adding 10 ml of 3 × YEP-Gal (3% yeast extract, 6% peptone, 6% galactose) to each tube.
5. After 6 h of induction harvest the cells by spinning each tube at 2500 × *g* and transfer the cells to a 96-well unibox. Wash the cell pellets once with ice-cold water before freezing at –80°C.
6. To purify proteins, resuspend the cells in 200 μl of ice-cold Lysis Buffer 150 (50 mM Tris–HCl pH 7.5, 150 mM NaCl, 1 mM EGTA, 10% glycerol, 0.1% Triton X-100, 0.5 mM DTT, 1 mM PMSF) containing 1 × Complete protease inhibitors (Roche) and transfer to a deepwell box with 1 ml wells (Fisher) containing 250 μl of acid-washed glass beads.
7. Break the cells by shaking for 6 min in a paint-shaker at 4°C.
8. Spin the crude lysates at 2500 × *g* for 5 min at 4°C and remove the lysate to a 1.2 micron PVDF filter-plate (Millipore, #MABVN1250).
9. Resuspend the broken cell pellets in 200 μl of ice-cold Lysis Buffer 650 (50 mM Tris–HCl pH 7.5, 650 mM NaCl, 1 mM EGTA, 10% glycerol, 0.1% Triton X-100, 0.5 mM DTT, 1 mM PMSF) containing 1 × Complete protease inhibitors (Roche) and lyse a second time in the paint-shaker, before spinning and transferring the lysates to a 1.2 micron filterplate as before.
10. Spin both 1.2 micron filterplates and transfer the clarified lysates to a deepwell box containing 40 μl of IgG-Sepharose (20 μl beads with 20 μl Lysis Buffer, Amersham) and 400 μl of Lysis Buffer (50 mM Tris–HCl pH 7.5, 1 mM EGTA, 10% glycerol, 0.1% Triton X-100, 0.5 mM DTT, 1 mM PMSF)

containing 1 × Complete protease inhibitors (Roche) in each well.

11. Rotate the lysates end-over-end at 4°C for 2 h to allow binding of the fusion proteins to the IgG beads.
12. Wash the beads 5 × with 800 μl of Wash Buffer (50 mM Tris–HCl pH 7.5, 150 mM NaCl, 10% glycerol, 0.1% Triton X-100) and rock for 5 min.
13. After washing, transfer the beads to a 3.0 micron polycarbonate filterplate (Millipore, #MAPB MN3) and incubate with 40 μl of GST-3C protease in Elution Buffer (50 mM Tris–HCl pH 7.5, 150 mM NaCl, 30% glycerol, 0.1% Triton X-100) overnight at 4°C to cleave the purified proteins off the beads. Triton X-100 should only be used if printing on nitrocellulose-coated slides.
14. In order to remove the GST-3C from each purified protein, add 20 μl of glutathione Sepharose 4B (Amersham) to each well and incubate for 2 h at 4°C with shaking.
15. Finally, recover the purified cleaved proteins by spinning the filterplate at 2500 rpm for 5 min at 4°C, array the proteins into 384-well plates using a Biomek FX arrayer (Beckman Coulter) and freeze at −80°C.

producing microarrays from purified proteins is the variability in expression levels. Protein expression from the MORF collection resulted in 63% of fusion proteins being expressed at medium (~0.1 mg/l) or high levels (~1+ mg/l) with the remaining being low expressers (~0.01 mg/l). The MORF collection is of higher quality than the earlier N-terminal GST collection (Zhu *et al.*, 2001) due to more extensive sequence verification and is made from a recent annotation of the yeast genome. While the C-terminal collection has the advantage of proteins with signal peptides being properly processed, both will complement each other in future proteomic endeavors.

Recombinant proteins from different organisms can be produced from bacteria, yeast, or insect cells. Expression of eukaryotic proteins in bacteria does lead to a significant number of proteins being insoluble or inactive (Braun *et al.*, 2002). Baculovirus-transfected cells provide mammalian proteins that are processed similarly to native mammalian processing and provide greater protein expression (Phizicky *et al.*, 2003). Yeast ORFs can be cloned using homologous recombination by PCR amplification of the ORF using primers containing the recombination sequences to the vector of interest. Alternatively, the Gateway® system (Invitrogen) uses the lambda *int* recombinase and the Creator system (BD Biosciences) uses Cre to clone ORFs into expression vectors.

We have used Protocols 1 and 2 for expressing protein under a *GAL1* promoter and purification of 1152 GST-tagged proteins or Protein A-tagged proteins per day, respectively. The composition of the elution buffer, particularly the concentration of glycerol and detergents, should be tested beforehand on the expected slide to be used, as the elution buffer and surface chemistry can alter spot morphology and limit the density at which proteins can be printed.

♦♦♦♦♦♦ IV. TECHNICAL ASPECTS OF PROTEIN MICROARRAYS

A. Surface Chemistry

The surface to print on depends upon the assays to be completed, and currently no attachment method can be defined as the clear choice (Table 2; see Angenendt *et al.*, 2002; Kusnezow *et al.*, 2003; Peluso *et al.*, 2003; Zhu and Snyder, 2003). In developing an assay using microarrays, several slide chemistries should be tested. Aldehyde- or epoxy-coated slides covalently attach protein by cross-linking to the primary amines of protein. This results in random attachment, reducing the possibility of attachment preventing a certain epitope from being exposed to the probe. Also used are nitrocellulose-coated slides that bind protein by adsorption and absorption. This also produces proteins randomly attached. Owing to the white surface, these slides can give higher background in microarray scanners; scanning at a lower voltage minimizes this.

Table 2. Popular surface chemistries used in protein microarrays

Surface chemistry	Attachment	Sample of assays completed
Aldehyde	Random, covalent cross-linking	Enzymatic, PPI (MacBeath and Schreiber, 2000) Antibody (Sreekumar *et al.*, 2001)
Epoxy	Random, covalent cross-linking	Enzymatic (Zhu *et al.*, 2000)
Nitrocellulose	Adsorption and absorption	Antigen (Joos *et al.*, 2000) PPI (Ge, 2000) Protein–DNA (Ge, 2000; Ho *et al.*, 2006) PTM (Gelperin *et al.*, 2005)
Ni-NTA coated	Affinity binding via histidine tag	PPI, protein–lipid (Zhu *et al.*, 2001) Protein–DNA (Hall *et al.*, 2004) Protein–small molecule (Huang *et al.*, 2004)
Hydrogel	Diffusion, adsorption	Antibody (Miller *et al.*, 2003)
Gold-coated silicon	Random, covalent cross-linking	Antigen arrays (Kanda *et al.*, 2004)

PPI = protein–protein interaction, PTM = posttranslational modification.

Proteins with affinity tags, including His_6 or biotin, can bind to nickel-coated slides or streptavidin-coated slides, respectively, by electrostatic interactions. For the N-terminally tagged GST-His_6 ORF collection proteins can be bound by the tag on nickel-coated slides, likely resulting in an orientation of the protein away from the slide. A study showed that attaching antibodies to a streptavidin-coated slide by a biotinylated Fc region of the antibodies increased the amount of antibody attached and led to a 10-fold increase in analyte bound (Peluso *et al.*, 2003) while other studies did not show an advantage over random attachment (Vijayendran and Leckband, 2001; Thulasiraman *et al.*, 2004). Attachment of the protein by a tag (His_6, biotin, etc.) also reduces the possibility that direct attachment of the protein to the slide causes misfolding.

Surfaces are also being developed that may better preserve the native protein confirmation and have a higher binding capacity, such as the polyacrylamide-based HydroGels (PerkinElmer; Rubina *et al.*, 2003). Interestingly, there are efforts to produce protein chips with surfaces, such as gold-coated, that allow for mass spectrometry (Davies *et al.*, 1999; Tang *et al.*, 2004; Thulasiraman *et al.*, 2004) or surface plasmon resonance (SPR) analysis (Bieri *et al.*, 1999; Rich *et al.*, 2001; Houseman *et al.*, 2002; Yuk and Ha, 2005). Further innovations into surface chemistries will continue with the goal of increasing the amount of protein that can be attached, providing attachment via linkers that expose more of the bound protein to the probe environment and minimize misfolding, and that allow quantitative measurements of thousands of features.

B. Printing Protein Microarrays

With the purified proteins of interest arrayed in a 384-well format, the proteins are printed on slides, usually using a 48-pin contact printer (MacBeath and Schreiber, 2000; Zhu *et al.*, 2001; Gelperin *et al.*, 2005; Ptacek *et al.*, 2005) as described in Protocol 3. An example of a yeast proteome array is shown in Figure 2A, with ~10 000 features probed with anti-GST antibodies followed by Cy5-labeled anti-rabbit IgG. In addition to the yeast proteins, many control proteins are also spotted and will vary according to the assay. Protein-binding assays often have dilutions of a labeled protein (Figure 2B) and GST or Protein A standards to aid in the data analysis. Kinase assays (Figure 2C) benefit from having kinases that autophosphorylate spotted at the corners of each of the 48 blocks on the slide. These kinase controls provide landmarks for identification of phosphorylation signals.

Amounts of 0.5–12.5 nl can be delivered with spot size ranging from 62.5 to 600 μm depending on the pin characteristics (solid or capillary; Telechem), the elution buffer used, and the temperature and relative humidity at the time of printing. Alternatively, using a piezoelectric arrayer (PerkinElmer) the volume, and therefore the amount, of protein deposited can be controlled (Avseenko *et al.*, 2002). The tradeoff for this flexibility is the lengthy time required to

Protocol 3. Printing protein microarrays

**The values we print with are shown in parentheses*

1. Prepare a program to run the arrayer. Consult the arrayer manufacturer's instructions; factors to be aware of include:
   a. Spot-to-spot distance, adjusted according to the pin size (spot size is 150 µm with spot-to-spot distance set at 250 µm with SMP3 pins from Telechem)*
   b. Number of times each sample will be printed (we double-spot our proteins)
   c. Quantity and orientation of the slides to be printed
   d. Number of pin washes (3 ×) and preprinting of pins (10 ×) to remove excess protein solution
2. Ensure that the relative humidity in the room is below 50%, ideally 20–30%, as spot morphology is affected by temperature and humidity.
3. Rearray proteins into a 384-well format with 10 µl per well; centrifuge briefly to ensure sample is at the bottom of the well.
4. Print proteins onto slides.
5. Store printed slides at 4°C under low humidity or for long-term storage slides can be kept at –20°C.

print many proteins, and therefore this approach precludes its use in microarrays of more than a couple of hundred features. The quality of the printing, measured in terms of spot quality and reproducibility of spot-to-spot quantity, are critical if the results are to be quantitated. Often the signal will be normalised to protein quantity per spot to provide a more informed ranking of hits.

An alternative to producing protein *in vivo* is the technique known as Nucleic Acid Programmable Protein Array (NAPPA) (Ramachandran *et al.*, 2004). Protein slides are printed with cDNA GST expression plasmids and anti-GST antibodies, which can then be stored until a protein microarray is needed. When needed, the slide can be overlayed with rabbit reticulocyte lysates containing T7 polymerase or another cell-free transcription/translation system. GST-tagged proteins are expressed *in situ* and bound to the microarray by the GST antibodies. This has been done as a proof-of-principle by creating a microarray of 29 human DNA replication proteins and testing for protein–protein interactions (Ramachandran *et al.*, 2004). Of known protein–protein interactions, 85% were observed from their assays.

C. Detection Methods

Today, most protein microarray assays involve labeling the probe of interest, usually fluorescently. The probes are detected with a microarray scanner, such as a Genepix 4200A (Axon Instruments).

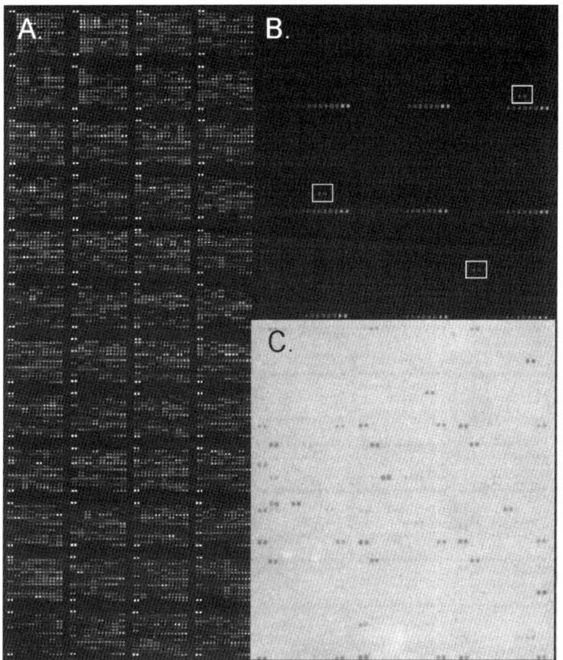

Figure 2. Yeast protein microarrays in use. (A) A yeast proteome microarray probed with anti-GST antibodies followed by Cy5-labeled anti-rabbit antibodies. All 48 blocks are shown. (B) Biotinylated Cpr7p probed against the yeast proteome array, followed by Cy5-labeled streptavidin. A dilution of Cy5 labeled controls can also be seen in each of the 9 blocks shown. (C) A kinase assay with [γ-^{33}P]ATP and active kinase on a yeast proteome microarray. Dark spots represent radiolabeled phosphorylated substrates. Kinases that autophosphorylate are printed in each of the nine blocks shown and serve as reference points on the slide. (See color plate section).

Proteins and antibodies, DNA, lipids, and other biomolecules can be labeled either with biotin and further detected with labeled streptavidin or directly labeled with Cy3 or Cy5 dyes (Lueking *et al.*, 1999; MacBeath and Schreiber, 2000; Zhu *et al.*, 2001; Michaud *et al.*, 2003; Hall *et al.*, 2004; Huang *et al.*, 2004; Gelperin *et al.*, 2005). In addition to cyanines, many other fluorophores can be used, including rhodamine and Alexa Fluor dyes (Invitrogen). While detection of femtogram quantity of protein is possible (Ekins, 1998), real-world conditions limit fluorescence detection of proteins of very low concentration ($\geq$1 ng/ml) or in a background of more abundant proteins ($\geq$5 ng/ml; Sreekumar *et al.*, 2001). Methods that offer more sensitivity have been described. One technique is the rolling circle amplification (RCA) that provides a very sensitive, yet more time-consuming, method to detect protein captured onto a microarray (Schweitzer *et al.*, 2000; Nallur *et al.*, 2001; Kingsmore and Patel, 2003). The technique involves replicating a circular DNA tag using DNA polymerase and is often detected using fluorescent nucleotides, thereby producing an amplified signal. Screening against the prostate marker, PSA, it was shown that RCA is 3 orders of magnitude more sensitive than other detection techniques with the ability to detect PSA at 0.1 pg/ml or 300 zeptomoles (Schweitzer *et al.*, 2000). Quantum dots, which do not suffer from photobleaching and are very sensitive, can potentially be used for single-molecule detection (Lian *et al.*, 2004; Geho *et al.*, 2005; Michalet *et al.*, 2005). Quantum dots (Invitrogen) are fluorescent semiconductor nanocrystals that currently are offered in eight unique emissions allowing multiplexed assays to be completed on a single slide. Direct comparison of different probes can be performed simultaneously allowing relativistic quantification of the results.

Labeling a protein can affect the protein's characteristics and activities. Proteins are most often labeled by their primary amine groups or by cysteine residues, though 8% of yeast proteins lack a cysteine (Gygi *et al.*, 1999). Protocols are available from the manufacturer of the fluorophores but it is often useful to perform the labeling reaction in triplicate, using three dilutions of label. This way a sample that is not over-labeled with altered binding is produced. Besides the residue labeled, the steric hindrance caused by a label may be minimized by testing different size linkers for changes in binding behaviour or enzymatic activity. At least for PPIs, it is possible to perform the reciprocal PPI assay by testing the protein pairs in both surface-bound and solution-phase configurations.

Alternatively, protein-binders could be identified without labeling using SPR or Surface Enhanced Laser Desorption/Ionization (SELDI) mass spectrometry. SPR allows quantitative measurements of binding constants to be made in real-time; the current capacity for this assay has recently risen to 400 proteins per assay using a Flexchip (Biacore, Uppsala, Sweden). SELDI ionizes and analyzes bound biomolecules from a protein microarray by mass spectrometry (Davies *et al.*, 1999; Tang *et al.*, 2004; Thulasiraman *et al.*, 2004). SELDI would circumvent possible artifacts of labeling but while use of SELDI is possible with a limited number of features, large-scale application of SELDI to proteome microarrays is still in the future.

Radiolabeling is another detection technique that does not structurally alter the labeled protein, is very sensitive, and can be used to identify enzymes' substrates. Owing to the small feature size (~150 μm) and high-spatial density of the protein microarrays we used, [γ-^{33}P]ATP with its lower emission energy that prevents phosphorylated features from overwhelming the signal from neighbouring spots (MacBeath and Schreiber, 2000; Zhu *et al.*, 2000; Ptacek *et al.*, 2005). Radiolabeled small molecules could also be used in screening for drug targets. While this minimizes effects caused by larger-fluorescent labels, the signal for small molecules would be dependent upon the isotope incorporated into the molecule. Detection of phosphorylation events is readily done, while detecting methylation by radiolabeled *S*-adenosyl-L-methionine would be significantly more challenging. However, detection methodology is continually improving and will only increase the utility of protein microarrays.

♦♦♦♦♦♦ V. CURRENT APPLICATIONS OF PROTEIN MICROARRAYS

A. Protein-Binding Assays

Protein microarrays have been used to identify protein interactors of calmodulin and different lipids (Zhu *et al.*, 2001). In another study, novel DNA binding proteins were identified by probing N-terminally tagged yeast proteome chip with labeled genomic DNA (Hall *et al.*, 2004). Recently, a protein microarray containing 282 known or putative yeast transcription factors was used to identify *cis*-regulatory sequences recognized by the transcription factors (Ho *et al.*, 2006). The array was probed with 40 Cy3-labeled DNA probes containing 75 novel DNA sequence motifs that are evolutionarily conserved. As controls, the authors used Cy5-labeled probes that contained the same 75 sequence motifs, except with 1 or 2 mutations to test the stringency of the interactions. From their study, 214 specific DNA–protein interactions were identified (Ho *et al.*, 2006). Many proteins bound their known or predicted sequence while potential binding sites were identified for many of the roughly half of the yeast DNA-binding proteins that lack a sequence binding assignment. Electrophoretic mobility shift assays were performed to validate several of the results.

Another aspect of protein chips is their emerging utility in small molecule screens. Huang *et al.* (2004) screened 16 320 small molecules (Diverset E, ChemBridge, San Diego, CA) to find inhibitors or enhancers to rapamycin's growth-inhibiting effect in yeast. From this set 26 compounds were able to counter rapamycin's effects and promote the growth of rapamycin-exposed yeast; using whole-genome expression profiling one molecule termed SMIR4 (small molecule inhibitor of rapamycin) is able to restore expression of rapamycin-treated cells to a near normal profile. To understand how SMIR4 is able to achieve this, they probed a yeast N-terminally tagged proteome chip with biotinylated SMIR4, washed, and then identified SMIR4-binding proteins with Cy3-labeled streptavidin. This resulted in 30 candidate proteins that may be involved in SMIR4's ability to undo rapamycin's growth-inhibitory action. The authors made genetic knockouts of each of the 30 proteins and looked for a reduction in SMIR4's effects and thereby a restoration of rapamycin's effects. They found one strain, *ybr077cΔ*, that was susceptible to rapamycin and SMIR4 was unable to restore growth. These experiments, from a phenotype-based screen to expression analysis to protein-small-molecule screens performed on protein microarrays, demonstrate the success of complementary high-throughput techniques. Drugs can be selected for their phenotypic action, their targets determined from DNA and protein arrays, and a reiteration of the process can be completed to find additional small molecules that modulate specific aspects of the drug response by either increasing the efficacy of the drug or diminishing unwanted effects of the drug.

B. Detection of Posttranslational Modifications

Recently, assays performed in our lab measured the specificity of 87 kinases on a version of the N-terminally tagged proteome chip, the Protoarray (Invitrogen, Carlsbad, CA, USA; Ptacek *et al.*, 2005). The microarray included ~4400 yeast proteins verified for correct size by western blot, in addition to controls that included the kinases Pka2, Pkc-δ and/or calmodulin-dependent kinase Cmk1 at defined locations to aid in the protein identification of phosphorylated spots (Figure 2C). Most of the kinases were purified as GST fusions and purified out of yeast; purification of yeast kinases out of insect cells also provides active kinase (Moffat and Andrews, 2004). Protocol 4 outlines the procedures used.

From the 87 kinase assays, 4192 phosphorylation events involving 1325 proteins were identified as potential substrates for specific kinases and provide the first observation of the phosphorylome of an organism. A third of these interactions occur between kinases and substrates that are in the same localization category (Huh *et al.* 2003). On the basis of functional data from MIPS (Mewes *et al.*, 1997), 18.4% of the kinase-substrate pairs, or "kinates", occur between proteins of the same functional category. This is significant ($P<10^{-99}$), especially when one considers that at least a third of yeast proteins still have no known function. Furthermore, proteins with known function may have novel functions not yet identified. These results were validated *in vivo* for 15 cases by deleting the kinase from TAP-tagged strains (Ghaemmaghami *et al.*, 2003) of the substrate and looking for a mobility shift by gel electrophoresis or loss of phosphorylation using phospho-specific antibodies. Related kinases, such as the yeast protein kinase A homologues Tpk1-3p, were analyzed on a proteome-wide level and showed striking differences in the substrates they recognize. Only six substrates were common between Tpk1, Tpk2, and Tpk3; most substrates (87.7%) were recognized by only one of the Tpks. These experiments using protein microarrays provide possible roles for many of the proteins lacking an established function based on "guilt by association" with kinases of known function, as well as providing additional roles for proteins of known function.

There are many other posttranslational modifications in yeast, such as ubiquitination, methylation, and glycosylation. The study of glycosylation is difficult due to the challenge in making properly modified protein and also by the many forms glycosylation takes. Of the yeast proteome, 20–50% is expected to be glycosylated (Apweiler *et al.*, 1999). Since proper targeting of proteins is largely dependent upon the N-terminal signal peptide, a new high quality moveable ORF (MORF) collection has recently been constructed. Placement of the tag at the C-termini allows for maintenance of N-terminal signal peptides and proper processing of the proteins. This is especially important for the 20–30% of eukaryotic proteins that are predicted to be membrane proteins (Krogh *et al.*, 2001).

Protocol 4. Kinase assays on protein microarrays (Ptacek *et al.*, 2005)

1. Purify active kinase, preferably from yeast to ensure proper modifications (see Protocols 1 and 2) and elute into Kinase Buffer (as listed in Step 5 below but without the ATP).
2. Verify purity of kinase by gel electrophoresis.
3. Determine the optimal amount of kinase to use on the microarray that will produce desirable signal:noise ratios by performing kinase assays on test protein chips (slides with ~100 proteins, including common kinase substrates). This is performed as detailed below for the ~4,400 protein microarrays.
4. Two slides for every kinase to be assayed and two slides for a negative control are blocked using Superblock (Pierce) with 0.1% Triton X-100 for 1 h at 4°C.
5. The amount of kinase as determined in Step 3 is diluted into kinase buffer containing 100 mM Tris-HCl (pH 8.0), 100 mM NaCl, 10 mM $MgCl_2$, 1 mM DTT, 0.5 mg/ml BSA, 0.1% Triton X-100, and 33.3 nM [γ-^{33}P]ATP (GE Healthcare Bio-Sciences Corp, Piscataway, NJ, USA).
6. The microarrays are incubated at 30°C for 1 h in a humidified chamber.
7. The kinase and free label are removed by washing twice with 0.5% SDS in 10 mM Tris-HCl (pH 7.4) and then once in ultrapure water.
8. The slides are spun dry at 1,500 rpm and then exposed to film (a phosphoimager with high-resolution may also be used).
9. From the film the resulting spots are analyzed using Genepix software. We have also used an algorithm written in our lab to normalize the signal and background and identify hits that were 2 standard deviations above background in 3 of the 4 spots.
10. For each set of assays, two slides are incubated in the absence of kinase, serving as a negative control and identifying proteins on the array that autophosphorylate.

Purified proteins from the MORF collection were printed onto nitrocellulose-coated FAST slides (Schleicher and Schuell). The slides were blocked with Superblock (Pierce) for 1 h at 4°C and probed then with a polyclonal antibody that recognizes yeast glycans at 1:10 000 dilution for 1 h at 4°C. The slides were washed extensively, probed with Alexa 647-coupled secondary antibody for an hour, and washed again. The authors identified 509 potential glycosylated proteins, including 55 of the 136 previously known glycoproteins and 20 of 30 of known GPI-anchored proteins. These results are significantly enriched for known glycoproteins, suggesting that many of the other putative glycoproteins are likely to be glycosylated *in vivo*. To validate the results, the mobilities of potential glycoproteins were tested with or without treatment by Endo H and PNGase F, two enzymes used to remove N-linked glycans. Of 25 known N-linked glycans tested, 21 showed a mobility shift while none of the 19 negative control proteins showed a change. Using this same strategy with previously unknown glycoproteins identified in their screen, their results confirm 109 glycosylated proteins and suggest that about half of the potential glycoproteins may indeed be glycosylated. Microarrays have increased the number of known glycoproteins from 136 to 245, and will lead to many more experiments to determine the significance of these modifications.

C. Enzymatic Assay Using Protein Microarrays

Jung and Stephanopoulos (2004) produced a protein chip using an *in vitro* translation system from rabbit reticulocyte lysate to express mRNA–protein fusions. The proteins were bound to the slide by capture DNA molecules crosslinked to the slide that were complementary to the mRNA–protein fusions. These "self-assembled" microarrays allowed them to strictly control the amount of protein bound to the slide since protein quantity was dependent upon the amount of captured DNA. Assays were completed in a well format that allowed the controlled variation of the quantities of seven enzymes involved in the 5-step trehalose pathway in yeast. Trehalose is a possible energy source, has the ability to protect proteins from misfolding, and has industrial uses. The authors showed that all five chimeric molecules retained activity while bound to a solid surface (Jung and Stephanopoulos, 2004). The optimal amounts or ratios of enzymes (3:2 of PGM to OtsA) to produce maximal trehalose accumulation rate was determined by capturing different quantities of the five enzymes in each well and measuring trehalose production. The authors showed a novel method in which to dissect a pathway to determine its optimal configuration using protein microarrays.

D. Successes in Other Organisms Using Protein Microarrays

In humans, assays have been performed to test the functionality of wild-type p53 and 45 of its mutants identified in cancers (Boutell *et al.*, 2004). The authors were able to detect relative affinity constants (K_d) of binding of the radiolabeled GADD45 promoter element to the p53 variants. Phosphorylation of the p53 proteins by casein kinase (CKII) was also tested by antibodies to a specific phosphoserine residue (Boutell *et al.*, 2004). In a different study, the predicted coiled-coil domains of 49 of the ~55 human bZIP proteins were purified by HPLC and arrayed onto aldehyde slides (Newman and Keating, 2003). All pairwise interactions were tested with each bZIP domain with many known interactors identified; a false-positive rate of $<6\%$ was calculated from their results (Newman and Keating, 2003).

Many reports have shown that enzymes maintain activity while arrayed onto a solid surface (Zhu *et al.*, 2000; Jung and

Stephanopoulos, 2004; Lee *et al.*, 2005; Merkel *et al.*, 2005). A novel toxicology screened using a "Metachip" protein array of two P450 isozymes, CYP3A4 and CYP2B6, was able to identify drugs that are metabolized by the P450 isozymes into toxic species (Lee *et al.*, 2005). A slide with a monolayer of cells was placed over the Metachip slide with the toxic metabolites and incubated up to 6 h. Cell staining was performed to determine live/death ratio for each drug. This extension of protein microarrays to screen for toxic metabolites produced from enzymes arrayed onto a slide could accelerate the screening process of drugs. An additional application of protein microarrays will be in detecting immune responses to certain antigens. This has been done for certain autoimmune conditions and for detecting serum antibodies to cancer markers (Robinson *et al.*, 2002, 2003; Wang *et al.*, 2005).

E. Future Directions

Protein microarrays have been used in multiple organisms and have addressed a variety of questions. The technology allows many biochemical assays to be completed on the entire proteome simultaneously. The cataloguing of protein–protein interactions is likely to push forward and it will be interesting to see the overlap between the datasets produced by protein microarrays, yeast two-hybrid (Uetz *et al.*, 2000; Ito *et al.*, 2001), and mass spectrometry (Gavin *et al.*, 2002; Ho *et al.*, 2002), the latter two of which show little overlap with one another (von Mering *et al.*, 2002). All three techniques have false-positive and false-negative concerns. An advantage of protein microarrays is that it allows direct interactions to be detected simultaneously minimizing experimental variations; the environment or buffer used can be adjusted, and cofactors and scaffold proteins can be added. The relative binding affinities can be determined to provide a better idea as to which interactions are most likely to occur *in vivo*. Protein microarrays have been used to identify binary interactions but complexes could also be tested.

Future uses of protein microarrays may be to analyze cell lysates. For example, it would be interesting to label a lysate from yeast grown to log-phase with one dye, such as Cy3, and label a lysate that is nitrogen-starved with another dye, like Cy5. As is done with DNA microarrays, relativistic ratios for binding can be determined. DNA microarrays have determined the expression differences for yeast with limited nitrogen (Gasch *et al.*, 2000), but the protein chip works at the proteome level and could identify changes in protein abundance not seen at an mRNA expression level and elucidate the necessary interactions for a cellular response to nitrogen starvation.

The cataloguing of small-molecule binding is another avenue of research ripe for discovery. For example, the targets of cAMP, GTP, or other small molecules could be identified using protein microarrays. The second messenger cAMP is important cell growth, stress resistance, and metabolism (Rolland *et al.*, 2001). Radiolabeled cAMP could be used to identify all the proteins capable of cAMP binding, such as the cAMP-dependent protein kinase, PKA. In addition to cAMP, protein microarrays provide an excellent platform to identify the proteins with which drugs and inhibitors interact to produce their effects (Huang *et al.*, 2004; Lee *et al.*, 2005).

Protein and DNA microarrays share much of the arraying and detection technology and should allow the transition from genomic studies to proteomic studies to continue to accelerate. Yet protein microarrays are years behind DNA microarrays, largely due to the difficult task of cloning, expressing, and purifying all the proteins in an organism. However, more and more researchers are entering the protein microarray field, and more uses for microarrays are being found. The mountains of data being generated are pushing the limits of bioinformatics but undoubtedly the analysis will identify many new cellular connections and allow biological pathways to be integrated with one another. Protein microarrays are already making many lasting contributions to our understanding of the cell and are poised to improve medicine by providing better diagnostics and therapeutics.

Acknowledgements

We thank Daniel Gelperin and Li Kung for comments on this work.

References

Amberg, D., Burke, D. and Strathern, J. (2005). *Methods in Yeast Genetics. A Cold Spring Harbor Laboratory Course Manual*. Cold Spring Harbor Laboratory Press, New York.

Angenendt, P., Glokler, J., Murphy, D., Lehrach, H. and Cahill, D. J. (2002). Toward optimized antibody microarrays: a comparison of current microarray support materials. *Anal. Biochem.* **309**, 253–260.

Apweiler, R., Hermjakob, H. and Sharon, N. (1999). On the frequency of protein glycosylation, as deduced from analysis of the SWISS-PROT database. *Biochim. Biophys. Acta* **1473**, 4–8.

Avseenko, N. V., Morozova, T. Y., Ataullakhanov, F. I. and Morozov, V. N. (2002). Immunoassay with multicomponent protein microarrays fabricated by electrospray deposition. *Anal. Chem.* **74**, 927–933.

Bader, G. D. and Hogue, C. W. (2000). BIND – a data specification for storing and describing biomolecular interactions, molecular complexes and pathways. *Bioinformatics* **16**, 465–477.

Bieri, C., Ernst, O. P., Heyse, S., Hofmann, K. P. and Vogel, H. (1999). Micropatterned immobilization of a G protein-coupled receptor and direct detection of G protein activation. *Nat. Biotechnol.* **17**, 1105–1108.

Boutell, J. M., Hart, D. J., Godber, B. L., Kozlowski, R. Z. and Blackburn, J. M. (2004). Functional protein microarrays for parallel characterisation of p53 mutants. *Proteomics* **4**, 1950–1958.

Braun, P., Hu, Y., Shen, B., Halleck, A., Koundinya, M., Harlow, E. and LaBaer, J. (2002). Proteome-scale purification of human proteins from bacteria. *Proc. Natl. Acad. Sci. USA* **99**, 2654–2659.

Cohen, P. (2000). The regulation of protein function by multisite phosphorylation – a 25 year update. *Trends Biochem. Sci.* **25**, 596–601.

Davies, H., Lomas, L. and Austen, B. (1999). Profiling of amyloid beta peptide variants using SELDI Protein Chip arrays. *Biotechniques* **27**, 1258–1261.

Davis, C. A., Grate, L., Spingola, M. and Ares, M., Jr. (2000). Test of intron predictions reveals novel splice sites, alternatively spliced mRNAs and new introns in meiotically regulated genes of yeast. *Nucl. Acids Res.* **2**, 1700–1706.

Delehanty, J. B. and Ligler, F. S. (2002). A microarray immunoassay for simultaneous detection of proteins and bacteria. *Anal. Chem.* **74**, 5681–5687.

Ekins, R., Chu, F. and Biggart, E. (1990). Multispot, multianalyte, immunoassay. *Ann. Biol. Clin. (Paris)* **48**, 655–666.

Ekins, R. P. (1998). Ligand assays: from electrophoresis to miniaturized microarrays. *Clin. Chem.* **44**, 2015–2030.

Ficarro, S. B., McCleland, M. L., Stukenberg, P. T., Burke, D. J., Ross, M. M., Shabanowitz, J., Hunt, D. F. and White, F. M. (2002). Phosphoproteome analysis by mass spectrometry and its application to *Saccharomyces cerevisiae*. *Nat. Biotechnol.* **20**, 301–305.

Fields, S. and Song, O. (1989). A novel genetic system to detect protein–protein interactions. *Nature* **340**, 245–246.

Gasch, A. P., Spellman, P. T., Kao, C. M., Carmel-Harel, O., Eisen, M. B., Storz, G., Botstein, D. and Brown, P. O. (2000). Genomic expression programs in the response of yeast cells to environmental changes. *Mol. Biol. Cell* **11**, 4241–4257.

Gavin, A. C., Bosche, M., Krause, R., Grandi, P., Marzioch, M., Bauer, A., Schultz, J., Rick, J. M., Michon, A. M., Cruciat, C. M. *et al.* (2002). Functional organization of the yeast proteome by systematic analysis of protein complexes. *Nature* **415**, 141–147.

Ge, H. (2000). UPA, a universal protein array system for quantitative detection of protein–protein, protein–DNA, protein–RNA and protein–ligand interactions. *Nucl. Acids Res.* **28**, e3.

Geho, D., Lahar, N., Gurnani, P., Huebschman, M., Herrmann, P., Espina, V., Shi, A., Wulfkuhle, J., Garner, H., Petricoin, E., 3rd *et al.* (2005). Pegylated, steptavidin-conjugated quantum dots are effective detection elements for reverse-phase protein microarrays. *Bioconjug. Chem.* **16**, 559–566.

Gelperin, D. M., White, M. A., Wilkinson, M. L., Kon, Y., Kung, L. A., Wise, K. J., Lopez-Hoyo, N., Jiang, L., Piccirillo, S., Yu, H. *et al.* (2005). Biochemical and genetic analysis of the yeast proteome with a movable ORF collection. *Genes Dev.* **19**, 2816–2826.

Ghaemmaghami, S., Huh, W. K., Bower, K., Howson, R. W., Belle, A., Dephoure, N., O'Shea, E. K. and Weissman, J. S. (2003). Global analysis of protein expression in yeast. *Nature* **425**, 737–741.

Goffeau, A., Barrell, B. G., Bussey, H., Davis, R. W., Dujon, B., Feldmann, H., Galibert, F., Hoheisel, J. D., Jacq, C., Johnston, M., *et al.* (1996). Life with 6000 genes. *Science* **274**, 546, 563–547.

Gruhler, A., Olsen, J. V., Mohammed, S., Mortensen, P., Faergeman, N. J., Mann, M. and Jensen, O. N. (2005). Quantitative phosphoproteomics applied to the yeast pheromone signaling pathway. *Mol. Cell Proteom.* **4**, 310–327.

Gygi, S. P., Rist, B., Gerber, S. A., Turecek, F., Gelb, M. H. and Aebersold, R. (1999). Quantitative analysis of complex protein mixtures using isotope-coded affinity tags. *Nat. Biotechnol.* **17**, 994–999.

Haab, B. B., Dunham, M. J. and Brown, P. O. (2001). Protein microarrays for highly parallel detection and quantitation of specific proteins and antibodies in complex solutions. *Genome Biol.* **2** RESEARCH0004.

Hall, D. A., Zhu, H., Zhu, X., Royce, T., Gerstein, M. and Snyder, M. (2004). Regulation of gene expression by a metabolic enzyme. *Science* **306**, 482–484.

Ho, S.-W., Jona, X., Chen, C. T., Johnston, M. and Snyder, M. (2006). Linking DNA-binding proteins to their recognition sequences using protein microarrays. *Proc. Natl. Acad. Sci. USA*, **103**, 9940–9945.

Ho, Y., Gruhler, A., Heilbut, A., Bader, G. D., Moore, L., Adams, S. L., Millar, A., Taylor, P., Bennett, K., Boutilier, K. *et al.* (2002). Systematic identification of protein complexes in *Saccharomyces cerevisiae* by mass spectrometry. *Nature* **415**, 180–183.

Horak, C. E., Luscombe, N. M., Qian, J., Bertone, P., Piccirrillo, S., Gerstein, M. and Snyder, M. (2002). Complex transcriptional circuitry at the G1/S transition in *Saccharomyces cerevisiae*. *Genes Dev.* **16**, 3017–3033.

Houseman, B. T., Huh, J. H., Kron, S. J. and Mrksich, M. (2002). Peptide chips for the quantitative evaluation of protein kinase activity. *Nat. Biotechnol.* **20**, 270–274.

Huang, J., Zhu, H., Haggarty, S. J., Spring, D. R., Hwang, H., Jin, F., Snyder, M. and Schreiber, S. L. (2004). Finding new components of the target of rapamycin (TOR) signaling network through chemical genetics and proteome chips. *Proc. Natl. Acad. Sci. USA* **101**, 16594–16599.

Huh, W. K., Falvo, J. V., Gerke, L. C., Carroll, A. S., Howson, R. W., Weissman, J. S. and O'Shea, E. K. (2003). Global analysis of protein localization in budding yeast. *Nature* **425**, 686–691.

Ideker, T., Thorsson, V., Ranish, J. A., Christmas, R., Buhler, J., Eng, J. K., Bumgarner, R., Goodlett, D. R., Aebersold, R. and Hood, L. (2001). Integrated genomic and proteomic analyses of a systematically perturbed metabolic network. *Science* **292**, 929–934.

Ito, T., Chiba, T., Ozawa, R., Yoshida, M., Hattori, M. and Sakaki, Y. (2001). A comprehensive two-hybrid analysis to explore the yeast protein interactome. *Proc. Natl. Acad. Sci. USA* **98**, 4569–4574.

Joos, T. O., Schrenk, M., Hopfl, P., Kroger, K., Chowdhury, U., Stoll, D., Schorner, D., Durr, M., Herick, K., Rupp, S. *et al.* (2000). A microarray enzyme-linked immunosorbent assay for autoimmune diagnostics. *Electrophoresis* **21**, 2641–2650.

Jung, G. Y. and Stephanopoulos, G. (2004). A functional protein chip for pathway optimization and in vitro metabolic engineering. *Science* **304**, 428–431.

Kanda, V., Kariuki, J. K., Harrison, D. J. and McDermott, M. T. (2004). Label-free reading of microarray-based immunoassays with surface plasmon resonance imaging. *Anal. Chem.* **76**, 7257–7262.

Kingsmore, S. F. and Patel, D. D. (2003). Multiplexed protein profiling on antibody-based microarrays by rolling circle amplification. *Curr. Opin. Biotechnol.* **14**, 74–81.

Knezevic, V., Leethanakul, C., Bichsel, V. E., Worth, J. M., Prabhu, V. V., Gutkind, J. S., Liotta, L. A., Munson, P. J., Petricoin, E. F., 3rd and Krizman, D. B. (2001). Proteomic profiling of the cancer microenvironment by antibody arrays. *Proteomics* **1**, 1271–1278.

Krogh, A., Larsson, B., von Heijne, G. and Sonnhammer, E. L. (2001). Predicting transmembrane protein topology with a hidden Markov model: application to complete genomes. *J. Mol. Biol.* **305**, 567–580.

Kusnezow, W., Jacob, A., Walijew, A., Diehl, F. and Hoheisel, J. D. (2003). Antibody microarrays: an evaluation of production parameters. *Proteomics* **3**, 254–264.

Lee, M. Y., Park, C. B., Dordick, J. S. and Clark, D. S. (2005). Metabolizing enzyme toxicology assay chip (MetaChip) for high-throughput micro-scale toxicity analyses. *Proc. Natl. Acad. Sci. USA* **102**, 983–987.

Lee, T. I., Rinaldi, N. J., Robert, F., Odom, D. T., Bar-Joseph, Z., Gerber, G. K., Hannett, N. M., Harbison, C. T., Thompson, C. M., Simon, I. *et al.* (2002). Transcriptional regulatory networks in *Saccharomyces cerevisiae*. *Science* **298**, 799–804.

Lian, W., Litherland, S. A., Badrane, H., Tan, W., Wu, D., Baker, H. V., Gulig, P. A., Lim, D. V. and Jin, S. (2004). Ultrasensitive detection of biomolecules with fluorescent dye-doped nanoparticles. *Anal. Biochem.* **334**, 135–144.

Lueking, A., Horn, M., Eickhoff, H., Bussow, K., Lehrach, H. and Walter, G. (1999). Protein microarrays for gene expression and antibody screening. *Anal. Biochem.* **270**, 103–111.

MacBeath, G. and Schreiber, S. L. (2000). Printing proteins as microarrays for high-throughput function determination. *Science* **289**, 1760–1763.

Merkel, J. S., Michaud, G. A., Salcius, M., Schweitzer, B. and Predki, P. F. (2005). Functional protein microarrays: just how functional are they? *Curr. Opin. Biotechnol.* **16**, 447–452.

Mewes, H. W., Albermann, K., Heumann, K., Liebl, S. and Pfeiffer, F. (1997). MIPS: a database for protein sequences, homology data and yeast genome information. *Nucl. Acids Res.* **25**, 28–30.

Michalet, X., Pinaud, F. F., Bentolila, L. A., Tsay, J. M., Doose, S., Li, J. J., Sundaresan, G., Wu, A. M., Gambhir, S. S. and Weiss, S. (2005). Quantum dots for live cells, *in vivo* imaging, and diagnostics. *Science* **307**, 538–544.

Michaud, G. A., Salcius, M., Zhou, F., Bangham, R., Bonin, J., Guo, H., Snyder, M., Predki, P. F. and Schweitzer, B. I. (2003). Analyzing antibody specificity with whole proteome microarrays. *Nat. Biotechnol.* **21**, 1509–1512.

Miller, J. C., Zhou, H., Kwekel, J., Cavallo, R., Burke, J., Butler, E. B., Teh, B. S. and Haab, B. B. (2003). Antibody microarray profiling of human prostate cancer sera: antibody screening and identification of potential biomarkers. *Proteomics* **3**, 56–63.

Moffat, J. and Andrews, B. (2004). Late-G1 cyclin-CDK activity is essential for control of cell morphogenesis in budding yeast. *Nat. Cell Biol.* **6**, 59–66.

Nallur, G., Luo, C., Fang, L., Cooley, S., Dave, V., Lambert, J., Kukanskis, K., Kingsmore, S., Lasken, R. and Schweitzer, B. (2001). Signal amplification by rolling circle amplification on DNA microarrays. *Nucl. Acids Res.* **29**, E118.

Newman, J. R. and Keating, A. E. (2003). Comprehensive identification of human bZIP interactions with coiled-coil arrays. *Science* **300**, 2097–2101.

Peluso, P., Wilson, D. S., Do, D., Tran, H., Venkatasubbaiah, M., Quincy, D., Heidecker, B., Poindexter, K., Tolani, N., Phelan, M. *et al.* (2003). Optimizing antibody immobilization strategies for the construction of protein microarrays. *Anal Biochem.* **312**, 113–124.

Phizicky, E., Bastiaens, P. I., Zhu, H., Snyder, M. and Fields, S. (2003). Protein analysis on a proteomic scale. *Nature* **422**, 208–215.

Ptacek, J., Devgan, G., Michaud, G., Zhu, H., Zhu, X., Fasolo, J., Guo, H., Jona, G., Breitkreutz, A., Sopko, R. *et al.* (2005). Global analysis of protein phosphorylation in yeast. *Nature* **438**, 679–684.

Ramachandran, N., Hainsworth, E., Bhullar, B., Eisenstein, S., Rosen, B., Lau, A. Y., Walter, J. C. and LaBaer, J. (2004). Self-assembling protein microarrays. *Science* **305**, 86–90.

Ranish, J. A., Yi, E. C., Leslie, D. M., Purvine, S. O., Goodlett, D. R., Eng, J. and Aebersold, R. (2003). The study of macromolecular complexes by quantitative proteomics. *Nat. Genet.* **33**, 349–355.

Rich, R. L., Day, Y. S., Morton, T. A. and Myszka, D. G. (2001). High-resolution and high-throughput protocols for measuring drug/human serum albumin interactions using BIACORE. *Anal. Biochem.* **296**, 197–207.

Robinson, W. H., DiGennaro, C., Hueber, W., Haab, B. B., Kamachi, M., Dean, E. J., Fournel, S., Fong, D., Genovese, M. C., de Vegvar, H. E. *et al.* (2002). Autoantigen microarrays for multiplex characterization of autoantibody responses. *Nat. Med.* **8**, 295–301.

Robinson, W. H., Fontoura, P., Lee, B. J., de Vegvar, H. E., Tom, J., Pedotti, R., DiGennaro, C. D., Mitchell, D. J., Fong, D., Ho, P. P. *et al.* (2003). Protein microarrays guide tolerizing DNA vaccine treatment of autoimmune encephalomyelitis. *Nat. Biotechnol.* **21**, 1033–1039.

Rolland, F., Winderickx, J. and Thevelein, J. M. (2001). Glucose-sensing mechanisms in eukaryotic cells. *Trends Biochem. Sci.* **26**, 310–317.

Rout, M. P., Aitchison, J. D., Suprapto, A., Hjertaas, K., Zhao, Y. and Chait, B. T. (2000). The yeast nuclear pore complex: composition, architecture, and transport mechanism. *J. Cell Biol.* **148**, 635–651.

Rubina, A. Y., Dementieva, E. I., Stomakhin, A. A., Darii, E. L., Pan'kov, S. V., Barsky, V. E., Ivanov, S. M., Konovalova, E. V. and Mirzabekov, A. D. (2003). Hydrogel-based protein microchips: manufacturing, properties, and applications. *Biotechniques* **34**, 1008–1014, 1016–1020, 1022.

Schena, M., Shalon, D., Davis, R. W. and Brown, P. O. (1995). Quantitative monitoring of gene expression patterns with a complementary DNA microarray. *Science* **270**, 467–470.

Schweitzer, B., Wiltshire, S., Lambert, J., O'Malley, S., Kukanskis, K., Zhu, Z., Kingsmore, S. F., Lizardi, P. M. and Ward, D. C. (2000). Inaugural article: immunoassays with rolling circle DNA amplification: a versatile platform for ultrasensitive antigen detection. *Proc. Natl. Acad. Sci. USA* **97**, 10113–10119.

Silzel, J. W., Cercek, B., Dodson, C., Tsay, T. and Obremski, R. J. (1998). Mass-sensing, multianalyte microarray immunoassay with imaging detection. *Clin. Chem.* **44**, 2036–2043.

Sreekumar, A., Nyati, M. K., Varambally, S., Barrette, T. R., Ghosh, D., Lawrence, T. S. and Chinnaiyan, A. M. (2001). Profiling of cancer cells using protein microarrays: discovery of novel radiation-regulated proteins. *Cancer Res.* **61**, 7585–7593.

Tang, N., Tornatore, P. and Weinberger, S. R. (2004). Current developments in SELDI affinity technology. *Mass Spectrom. Rev.* **23**, 34–44.

Thulasiraman, V., Wang, Z., Katrekar, A., Lomas, L. and Yip, T. T. (2004). Simultaneous monitoring of multiple kinase activities by SELDI-TOF mass spectrometry. *Methods Mol. Biol.* **264**, 205–214.

Uetz, P., Giot, L., Cagney, G., Mansfield, T. A., Judson, R. S., Knight, J. R., Lockshon, D., Narayan, V., Srinivasan, M., Pochart, P. *et al.* (2000). A comprehensive analysis of protein–protein interactions in *Saccharomyces cerevisiae*. *Nature* **403**, 623–627.

Vijayendran, R. A. and Leckband, D. E. (2001). A quantitative assessment of heterogeneity for surface-immobilized proteins. *Anal. Chem.* **73**, 471–480.

von Mering, C., Krause, R., Snel, B., Cornell, M., Oliver, S. G., Fields, S. and Bork, P. (2002). Comparative assessment of large-scale data sets of protein–protein interactions. *Nature* **417**, 399–403.
Wang, X., Yu, J., Sreekumar, A., Varambally, S., Shen, R., Giacherio, D., Mehra, R., Montie, J. E., Pienta, K. J., Sanda, M. G. *et al.* (2005). Autoantibody signatures in prostate cancer. *N. Engl. J. Med.* **353**, 1224–1235.
Washburn, M. P., Wolters, D. and Yates, J. R., 3rd. (2001). Large-scale analysis of the yeast proteome by multidimensional protein identification technology. *Nat. Biotechnol.* **19**, 242–247.
Xenarios, I., Rice, D. W., Salwinski, L., Baron, M. K., Marcotte, E. M. and Eisenberg, D. (2000). DIP: the database of interacting proteins. *Nucl. Acids Res.* **28**, 289–291.
Yuk, J. S. and Ha, K. S. (2005). Proteomic applications of surface plasmon resonance biosensors: analysis of protein arrays. *Exp. Mol. Med.* **37**, 1–10.
Zhu, H., Bilgin, M., Bangham, R., Hall, D., Casamayor, A., Bertone, P., Lan, N., Jansen, R., Bidlingmaier, S., Houfek, T. *et al.* (2001). Global analysis of protein activities using proteome chips. *Science* **293**, 2101–2105.
Zhu, H., Klemic, J. F., Chang, S., Bertone, P., Casamayor, A., Klemic, K. G., Smith, D., Gerstein, M., Reed, M. A. and Snyder, M. (2000). Analysis of yeast protein kinases using protein chips. *Nat. Genet.* **26**, 283–289.
Zhu, H. and Snyder, M. (2003). Protein chip technology. *Curr. Opin. Chem. Biol.* **7**, 55–63.

15 Smart Genetic Screens

Michael Breitenbach[1], J Richard Dickinson[2] and Peter Laun[1]
[1] *Department of Cell Biology, Division of Genetics, University of Salzburg, Hellbrunnerstrasse 34, 5020 Salzburg, Austria;* [2] *Cardiff School of Biosciences, Main Building, Museum Avenue, Cardiff CF10 3TL, UK*

♦♦♦

CONTENTS

Introduction
Chemical screens, screens using lectins/antibodies and FACS and screens based on cell size and density
How to screen clone bank transformants and how to identify the gene corresponding to a mutation
Screens for conditional mutants including temperature-sensitive (ts) mutants and using counter selection with Fluoroorotic-acid (FOA)
Screening for extragenic suppressors of conditional mutants. Screening for suppressees of a given multicopy suppressor
Synthetic lethality and dosage lethality screens
Concluding remarks

Abbreviations

ARS	Autonomously replicating sequence
CRE	Cyclization recombination
FACS	Fluorescence activated cell sorting
FITC	Fluorescein isothiocyanate
FOA	5-Fluoroorotic acid
GFP	Green fluorescence protein
HPLC	High Performance Liquid Chromatography
LOX	Locus of X-over P1
NUP	Nuclear pore
OD	Optical density
PCR	Polymerase chain reaction
SGA	Synthetic genetic array
TOR	Target of rapamycin
ts	Temperature sensitive
FTL	Ferritin light chain

Smart Genetic Screens

METHODS IN MICROBIOLOGY, VOLUME 36
0580-9517 DOI:10.1016/S0580-9517(06)36015-1

♦♦♦♦♦♦ I. INTRODUCTION

In the 10 years since the first edition of this book (de la Cruz *et al.*, 1998), enormous progress has been made in developing "genomic" methods for functional analysis of yeast genes. Although the two chapters following this cover very important genomic methods, it is impossible for us to avoid discussing the impact of genomics on the development of new genetic screening methods in yeast. In fact, these methods have revolutionized and enormously speeded up the discovery of gene function. It goes without saying that the developments described in this chapter depend heavily on the bioinformatic analysis of whole genome sequences, including most importantly, the genome sequence of *Saccharomyces cerevisiae* that was completed in 1996 (Goffeau *et al.*, 1996). The genome sequences of closely related taxa also proved to be invaluable, among other things for identification of functionally important upstream regulatory sequences, since these generally evolve much more rapidly than the protein coding parts of gene sequences (Dujon *et al.*, 2004, http://cbi.labri.fr/Genolevures). The limited number of about 6000 yeast genes and the development of new techniques like DNA-microarrays, protein arrays, biological mass spectrometry, robotics and others have made possible the use of high throughput "brute force" screening methods. Some of the techniques just mentioned are covered elsewhere in this book (see Chapters 7, 9, 14, 16 and 17). The actual number and annotation of yeast genes was corrected several times after the first annotation based on refined methods of genome-sequence analysis and on experimental data (Mewes *et al.*, 1997; Wood *et al.*, 2001). It turned out that classical yeast genetic screens could uncover less than half of the genes that actually exist in a yeast cell. The reasons for this surprising fact cannot be discussed in detail here, but they obviously have to do (i) with the probability of hitting a specific gene with a mutagenic agent, and (ii) with our limited knowledge of cellular physiology. To make this clearer, for instance *BPH1* (YCR032W), the gene with the largest open reading frame in the yeast genome, was never discovered, because its only known function is to pump out acetic acid when yeast cells grow in a very acidic medium, a physiological situation that was never investigated. Even now, depending on the definition of what is a "known function", a large part of the known yeast genes are functionally unknown or partially unknown. So, the purpose of genetic screens has changed: the aim is no longer to discover new genes, which in classical times could only be done via discovering new mutants. On the contrary, the genes as such are now very well known through genome sequencing and bioinformatic analysis, and the aim of our research is to discover the biological (physiological and biochemical) function of genes whose function is not known or not known in sufficient detail (Oliver, 1997). In many cases, this involves the discovery of new interactors for a given gene, in other terms, genetic networks, a key term in modern biology, which might clarify relationships between and within groups of genes.

What are the typical numbers of interactors that are found for a given yeast gene? Very few exhaustive screens have been published (referenced in Hartman *et al.*, 2001) and it turns out that between four and eight functionally interacting genes were found by classic synthetic lethality methods for the query genes that were investigated. A very informative set of interacting genes obtained with deletion mutants was published more recently (Tong *et al.*, 2001, 2004). This so-called Synthetic Genetic Array (SGA) analysis (see Chapter 18) showed a record of 31 interactors for one query gene and a large majority of genes that had only one interactor.

Only about 1200 yeast genes are "essential" in the sense that a haploid cell in which the gene has been deleted cannot grow (and so dies) on rich media. The remaining approximately 4800 genes are non-essential in the sense that each individual gene deletion strain still grows on full media under simple laboratory conditions, but may have defects under special environmental or physiological conditions. For example, such a strain could grow perfectly well on full media, but the homozygous diploid could be completely defective in sporulation (or other special situations).

Haplo-insufficiency on full media is extremely rare in a simple unicellular organism like yeast. This would mean that the heterozygous diploid deletion strain has a defect even on full media. However, in higher organisms like humans, such a situation is frequently incompatible with normal functioning of the organism, or even at the single-cell level. Also in yeast, haplo-insufficiency can be used as means for genetic screening (see Chapter 17) because a large number of chemical inhibitors can be screened using heterozygous diploids. This opens one door for the analysis of essential genes in yeast because the heterozygotes are viable. Other methods for screening essential genes will be described here.

Some of the genetic screens that we will discuss depend on the availability of genome-wide collections of artificially created yeast mutants. We name here the collection of deletion strains which was constructed by an international consortium (EUROSCARF, Chapter 25), the collection of mutants consisting of doxycyclin-regulatable chromosomally integrated versions of nearly all essential genes, and the collection of mutant strains containing N-terminal and C-terminal fusion constructs of all yeast genes with GFP. Direct phenotypic screens can now be done not only with the deletion mutants corresponding to the non-essential genes, but also with the doxycyclin-regulated genes and proteins and their location in the cell can be routinely tested, with strains that are ready for use. What is available in the deletion collection is by definition a null allele and this means that point mutations, which could be hypomorphs or could be dominant with respect to certain phenotypes, cannot be found in this way. Nevertheless these mutant collections are enormously valuable for functional analysis. In addition to the mutant collections just mentioned, the growing heuristic databases and bioinformatics tools are enormously helpful for functional analysis.

The "problem space" defined by the 6000 genes and by the different cell differentiation and other physiological situations with which a yeast cell can cope is enormous (e.g. 36 million possible binary gene interactions), but finite, and therefore accessible with modern high-throughput methods.

The literature on yeast genetics and molecular biology over the last 30 years clearly shows that the main driving force of research was at the beginning the wish to discover and analyze new genes. With the advent of genomics there was a shift to functional analysis of genes that were already known in principle. The screening methods were now devised mainly to discover new genetic interactions (networks of interactions) and new genotype/phenotype relationships. The starting point in every case was a biological or physiological question. The "smart screening procedures" that were published in considerable number were always designed to meet the needs of a specific biological problem. Therefore the published screening procedures are all different from each other; however, a relatively small number of technical tricks tend to reappear in those reports, for instance the colony sectoring assay, the creation and use of temperature-sensitive mutants, the counterselection based on FOA (fluoroorotic acid) and others. We describe these tools in detail. The main emphasis is on genetic (physiologic) interaction screens, but we start with basic procedures for mutant isolation, since the mutants are the necessary raw material for the more sophisticated screens that follow.

In this chapter, we present the reader with a number of screening methods of which we personally have experience and/or we consider to be of special importance. We will discuss these methods by giving concrete examples, which we think have a special importance for the yeast molecular biology and genetics community. However, it is impossible to give a comprehensive description of all yeast mutant-screening methods.

♦♦♦♦♦♦ II. CHEMICAL SCREENS, SCREENS USING LECTINS/ANTIBODIES AND FACS AND SCREENS BASED ON CELL SIZE AND DENSITY

A. Example I: The Dityrosine Screen

Dityrosine was discovered by chemical analysis of yeast spore walls as a sporulation-specific and abundant component of ascospore walls which is located in the outermost spore wall layer and confers resistance of the spore wall to enzymatic attack (Briza *et al.*, 1986, 1988). In order to use the natural strong fluorescence of dityrosine for devising a mutant screen that would recognize recessive mutations of meiosis, sporulation and spore-wall synthesis, it was necessary to construct a strain that sporulated as a haploid. The "haploid meiosis" strain was mutagenized, plated out on complete medium, replica plated to sporulation medium and a few thousand single colonies were screened for the presence of dityrosine under UV light (Briza *et al.*, 1990; Esposito *et al.*, 1991). This simple and efficient method allowed the isolation of spore-wall mutants. Those were mutants that on the one hand lacked fluorescence and on the other did show asci on microscopic examination. The mutants were used to isolate the corresponding genes from a standard genomic clone bank, again using the fluorescence of dityrosine as a chemical marker. Two genes, *DIT1* and *DIT2,* were discovered coding for proteins which catalyze the final steps of the biosynthesis of dityrosine, and are regulated in coordinate fashion from a common intergenic region of about 900 bp. These genes turned out to be transcribed only in a short time window during mid-late sporulation. A further gene was discovered in the same screen, which codes for chitin synthase III (*CHS3*). *CHS3* is not transcribed in a sporulation-specific way and is needed also for vegetative growth (Pammer *et al.*, 1992). The authors extended the screening method described so far to screening the EUROSCARF deletion collection of about 4800 homozygous deletion mutants in non-essential genes. The strains were sporulated, hydrolyzed and analyzed using reversed phase HPLC and a fluorescence detector set to the optimal excitation and emission wavelengths of dityrosine. The wild type showed nearly equal amounts of 60% LL- and 40% DL-dityrosine as relatively abundant amino acid peaks. The occurrence of the DL form is believed to contribute to the spore's resistance and probably is due to a sporulation-specific dityrosine epimerase. Five classes of mutants deviating from this pattern were further analyzed (Briza *et al.*, 2002).

Genes coding for two chitin deacetylases (*CDA1, CDA2*), one dityrosine-specific membrane transporter of the MDR family (*DTR1*) and one sporulation-specific chitinase (*CTS2*), were identified. The two chitin deacetylases are both contributing to spore-wall synthesis and are both transcribed in a sporulation-specific manner (Christodoulidou *et al.*, 1996, 1999). *CHS3*, *CDA1* and *CDA2* cooperate in the biosynthesis of the second outer layer of the spore wall, which is chitosan (Briza *et al.*, 1988). The lack of the chitosan layer indirectly leads to a failure to assemble the outermost layer of the spore wall and to a lack of dityrosine. *DTR1* (transcribed only in sporulation) is needed for the transport of a dityrosine-containing spore wall precursor across the plasma membrane of the prospore (Felder *et al.*, 2002). *CTS2* (also transcribed only in sporulation) apparently is needed for the morphogenesis of the chitin layer that is subsequently deacetylated to form the mature spore wall chitosan layer (Coluccio *et al.*, 2004). The function of the only other yeast chitinase (*CTS1*) is not related to the spore wall but to restructuring of the cell wall in the budding process, and was consequently not found in the dityrosine screen. The example shows how the genes and mutants obtained with a simple dedicated chemical screen can contribute to the elucidation of a whole metabolic pathway, and furthermore, how powerful brute force screens employing the yeast deletion collection can be.

During the same screening procedure a large set of mutants were discovered which are defective in sporulation in any of the sequential steps from initiation of meiosis to maturation of the spore wall (Briza *et al.*, 2002) further underlining the power of this screen. From a sample of 624 deletion mutants of genes functionally unknown at the time, 54 showed a well-defined sporulation defect. Only 19 of those (35%) are transcribed in a sporulation-specific manner. Extrapolating from this sample to the whole genome predicts that about 500 yeast gene deletions confer a well-defined sporulation defect, which was confirmed in a subsequent paper. Using 4323 single gene deletion mutants, 334 were identified which confer a complete sporulation deficiency (Enyenihi and Saunders, 2003).

One can easily imagine that similar chemical screens can be devised for any metabolite of interest, given that a sensitive analytical detection system for the compound exists or can be invented.

B. Example 2: Screening for Mutants Extending the Mother Cell-Specific Life span

We choose this example to demonstrate how in spite of enormous technical difficulties the right combination of methods can solve the problem of devising a powerful screening system.

Yeast mother cell-specific life span is defined by the number of buds an individual mother cell can produce. In a population of yeast mother cells, the life span distribution follows the Gompertz law (Finch, 1994). In an exponentially growing yeast culture, most cells are of the zeroth or first generation and the fraction of very old mothers is extremely small. Therefore, the isolation of a large set of long-lived mutants by direct selection, which would be the best way to achieve comprehensive genetic analysis of the phenomenon of mother cell-specific aging, has not been possible up to now. It would require a powerful physical selection for old mother cells (discussed in Breitenbach *et al.*, 2004). Old cells are much larger than daughter cells, and the same is true for old human cells. Methods are available for separation of cells by size on sucrose density gradients (Egilmez *et al.*, 1990) and by elutriation centrifugation (Laun *et al.*, 2001). These methods were shown to be sufficient for isolating reasonably pure or enriched old cell populations for biochemical analysis of those cells. However, they still contain young cells as well as attached daughter cells and could not be used for direct selection of mutants. They were used to examine the mRNA complement of old cells either by library subtraction or differential screening methods (D'Mello N *et al.*, 1994) or on microarrays (Laun *et al.*, 2005) and candidate genes were identified which are differentially transcribed in old vs. young cells. The corresponding deletion mutants in some cases showed substantial elongation of life span. Examples are *LAG1* (D'Mello *et al.*, 1994; Jazwinski and Conzelmann, 2002) and the gene coding for the mitochondrial ribosomal protein, *YGR076c* (Heeren *et al.*, in preparation). Generally, differential transcription of genes was not found to be a powerful pre-screen for genes functionally relevant for the process in question. Other indirect mutant screens were used, for instance, pre-screening for resistance to oxidative stress or other stresses. Such methods have not yet led to a comprehensive picture of the genetics of aging in yeast.

The problem was solved in principle by combining two different methods for mild chemical marking of the surface of yeast cells so that selection and counter-selection of cells by fluorescence activated cell sorting could be used (Chen *et al.*, 2003). By using this method, up to now, no yeast mutant collection was produced, but human cDNAs were selected in yeast cells that could substantially elongate the yeast life span. One of those cDNAs codes for ferritin light chain (FTL) which could also elongate the life span of a metazoan aging model organism, *Caenorhabditis elegans*. The mechanism for life span elongation in this case is based on the antioxidative properties of FTL by binding of iron ions and preventing superoxide production, which is not surprising in view of the proven role of oxygen radicals in the aging process in practically all model systems of aging.

Very briefly, the critical enrichment of old mother cells was achieved by first labeling the surface of the starting cells with biotin and separating mothers from daughters on magnetic beads after seven generations (Smeal *et al.*, 1996; Laun, 1997). After regrowth of mother cells for more generations, the second step was performed. The cells were now decorated with streptavidin-phycoerythrin creating an orange fluorescence signal discriminating all mothers from contaminating daughters. In addition, cells were stained with wheat germ agglutinin-FITC creating a green fluorescence signal which depends on the number of bud scars indicating the number of generations an individual cell has undergone, in other words on the age. The cells were sorted on a FACS and the fraction positive for the orange signal and showing the highest FITC fluorescence was plated to single colonies. Plasmids isolated from these colonies were retransformed into yeast and consistently caused an increase in longevity, thus showing the validity of the method.

This example shows two important facts: First, depending on the specific problem, decorating the surface of yeast cells (for instance with antibodies recognizing surface structures) and separating the cells either on magnetic beads or with a FACS can be employed successfully for mutant screens.

Second, using yeast for screening human cDNAs, like in this example, is a way to learn something about human cells, where genetic experiments cannot be performed so easily.

This latter aspect, employment of yeast screening methods to solve problems of human biology/genetics is important in several other examples. For instance, consider the screening for human cDNAs and also for yeast mutants which render the yeast cells resistant to killing by mammalian Bax (Huckelhoven, 2004) or the screening of pharmaceutically relevant drug candidates against a library of yeast mutants (Giaever *et al.*, 1999) or the study of human inherited disease by investigating an orthologous yeast gene (Breitenbach *et al.*, 2003; Kellermayer, 2005).

The problem of screening directly for yeast mutants displaying an increased mother cell-specific life span has also been attacked by genetic methods. A strain was constructed in which all daughter cells die while the mother cell lives and can undergo a number of cell divisions according to her individual life span. Such a strain shows linear growth characteristics instead of the normal logarithmic growth of wild-type strains. This was accomplished by expressing the only copy of the essential gene, *CDC6*, under control of the mother cell-specific *HO*-promoter (Bobola *et al.*, 1996). The strain was successfully used to isolate mutants that abolish mother cell-specificity thus elucidating the mechanism of mother cell-specific regulation of the *HO*-promoter. The strain was also used to devise a simple test for the life span of yeast mutants (Jarolim *et al.*, 2004) which consists of only cell density measurements or OD measurements instead of determination of life span by micromanipulation. Due to the relatively high frequency of extragenic revertants in genes regulating the *HO*-promoter the method was not sufficient for direct screening of mutants.

An alternative was presented recently (Gottschling-DE, personal communication) using a daughter cell-specific promoter controlling the *cre* recombinase which only in daughter cells inactivates an essential gene constructed with the appropriate *lox* flanking sequences. Apparently, this is less prone to extragenic reversion, perhaps, because the *cre-lox* system is prokaryotic and not easily modified by yeast mutations.

C. Example 3: Selectively Screening for Small Size Mutants

The biochemical mechanism by which cell size is controlled at the START event of the cell division cycle might be discovered through the analysis of small-sized mutants. These were isolated in *Schizosaccharomyces pombe* by velocity separation through a sucrose gradient (Thuriaux *et al.*, 1978). A similar technique applied to *S. cerevisiae* proved to be more difficult, probably due to the differences in cell morphology and cell cycle control of this yeast. Cell-size mutants were isolated by preparing small daughter cells on sucrose gradients, growing them synchronously, and applying α-factor, which arrested wild-type cells but not the sought-for small-sized mutants, which would divide earlier than wild type. After another cell cycle the size difference between the arrested cells (shmoos) of wild type and mutant cells would be even bigger and now the smaller cells were separated by zonal centrifugation, plated and analyzed genetically (Sudbery *et al.*, 1980). The genetic analysis of one of these mutants (*whi1-1*) contributed to the elucidation of the mechanism of cell cycle control through the cyclin-dependent kinase (CDK) *CDC28*. *WHI1* was later shown to be identical to *CLN3* (one of the three G1 cyclins), which was also found as *DAF1* (Cross, 1988) by a different screening method (resistance to α-factor). Only dominant gain of function mutations could be isolated because of the functional redundancy of the three G1 cyclins and because they as a group are essential for survival (Richardson *et al.*, 1989). We will come back to this problem when describing the hunt for cell cycle mutants in yeast. The dominant phenotype of the *CLN3* mutants is caused by C-terminal truncations leading to Cln3 protein that is not degraded by the proteasome.

The biological question behind this result is: how is cell size determined in yeast (and for that matter in any eukaryotic cell)? Surprisingly, this question is still not completely answered. In *S. cerevisiae* it must have to do with the transition from the nutrient-controlled step in the START of the cell cycle to the CDK-controlled step that allows growth, but not initiation of a new cycle. We know now that *CLN3* is somehow controlled by protein kinase A and the TOR pathway (which are under nutrient control), and thus probably transmit a signal to CDK (Mendenhall and Hodge, 1998). This picture is far from complete, but explains why a dominant gain of function in *CLN3* leads to small cell size.

◆◆◆◆◆◆ III. HOW TO SCREEN CLONE BANK TRANSFORMANTS AND HOW TO IDENTIFY THE GENE CORRESPONDING TO A MUTATION

A. Cloning by Functional Complementation

Yeast genomic clone banks, which are most commonly used have been prepared in simple episomal vectors like Yep13 (Rose and Broach, 1990) or in centromeric vectors, like Ycp50 (Johnston and Davis, 1984). All vectors are shuttle vectors that can be propagated in *Escherichia coli* as well as in yeast. They comprise sequences enabling replication and selection in *E. coli* and in yeast, namely the pMB1 origin of replication and ampicillin or tetracycline resistance genes for *E. coli* and a wild-type gene to allow plasmid selection in yeast (*URA3, LEU2, TRP1* or *HIS3* are commonly used). In addition, shuttle vectors have either the origin of replication derived from the 2μ episomal plasmid of yeast (conferring multicopy properties), or alternatively, chromosomal ARS and centromeric CEN) sequences (conferring single-copy character). These genomic clone banks have the advantage (compared to the more common cDNA clone banks often used in higher organisms) that they contain every part of the genome and not only the genes transcribed under specific conditions. However, great care must be taken to efficiently work with partial restriction digests (for instance with *Sau*3A) when constructing the bank in order to get a large enough number of random fragments of the genome so that every gene (even large ones) is represented in a complete and expressible form. Two very different strategies are necessary for isolation of the gene corresponding for a given mutation, depending on the recessivity or dominance of the mutation in question.

1. Recessive mutations

The mutation must confer a phenotype for which a positive selection can be found and the strain must carry an auxotrophic marker corresponding to the above-mentioned plasmid selection genes, e.g. *LEU2*. Taking the example of temperature-sensitive (*ts*) *cdc* mutants described in the following section, the selectable phenotype is a growth defect at 37°C. Clone bank DNA is transformed by the lithium-acetate method (Schiestl and Gietz, 1989; Gietz *et al.*, 1992, and this volume, Chapter 3) into the mutant cells and the transformants are plated out on –Leu plates at the permissive temperature (24°C). The number of colonies obtained must reach a few thousand (depending on the quality of the clone bank, insert size and on the number of genes on the yeast genome). The leucine-prototrophic and therefore plasmid-containing colonies are scraped off the plates and the cells are re-plated at the restrictive temperature (37°C). Colonies growing at 37°C are isolated individually and plasmids are prepared from the yeast cells and retransformed into *E. coli*. Note that ideally a yeast colony is derived clonally from a single cell. Due to the multicopy properties of episomal plasmids, more than one type of clone bank plasmid can be contained in a yeast cell. Therefore, plasmids are prepared from several *E. coli* transformant colonies. Note that an *E. coli* cell usually contains only one type of plasmid due to the properties of the replication origin used. These plasmid preparations are analyzed by restriction digest and electrophoresis to establish how many different plasmids were present in the original yeast transformant. Every plasmid type is now retransformed into the *cdc*-ts yeast mutants and tested at 37°C. Two possible outcomes of this experiment are considered: (i) only one or a few plasmids confer functional complementation (growth at 37°C) and these all contain overlapping inserts and a common region, which probably contains the gene of interest; and (ii) two or more unrelated plasmids allow the mutation to grow at 37°C. In the first case, the relevant region of the insert is sequenced and the gene is identified by comparison with the published genome sequence (http://www.yeastgenome.org). However, it is still necessary to show that the cloned gene is identical with the gene that carries the original ts mutation. This is done by integrative transformation of the cloned gene (which automatically leads to co-integration of the *LEU2* gene), performing a genetic cross with the *cdc*-ts mutant, sporulation and tetrad analysis. If all tetrads are parental di-types (see Chapter 2) with respect to *LEU2* and ts, absence of recombination and hence genetic identity are shown.

In the second case, where two or multiple different plasmids are recovered which all restore growth (perhaps in some cases slow growth) of the mutant, the integration and recombination test just described is also performed, in order to identify the *CDC* gene. The other plasmids must encode extragenic multicopy suppressors. It is possible to isolate such multicopy suppressors from clone banks constructed in episomal plasmids, but less frequently, also from centromeric clone banks since these plasmids are also not strictly single copy and, moreover, gene expression may be different for a gene on a plasmid compared to the chromosomal gene. These multicopy suppressors are not irrelevant genes but are in most cases readily understood and can clarify the function of the mutant gene because they encode genes which functionally interact with the gene in question, as we discuss later in this chapter.

2. Dominant mutations

In the case of dominant mutations cloning of the corresponding gene is impossible from a standard clone bank constructed with wild-type DNA. A prerequisite for the cloning procedure is that the dominant mutation confers a selectable phenotype. Therefore, the only way is to construct a new clone bank from the mutant strain, to transform this bank into wild-type cells and to select (in the same manner as described above) a yeast colony which shows the phenotype searched for. To make this clearer, we describe an example in the section on extragenic suppressors. These suppressor mutations are frequently dominant and are cloned in the way just described.

B. Positional Cloning

Finally, in very rare cases, molecular cloning is not possible with the methods described so far. The reasons for such rare cases may be that a mutation's phenotype is not easily selectable or perhaps in the rare occurrence (for technical reasons) of the sought for gene in the clone bank. In these cases, still another possibility exists: the mutant can be mapped by classical methods and the region of the chromosome corresponding to the map location can be analyzed on the genome sequence of yeast. This so-called positional cloning is very labor intensive, but is a method that works well and is completely independent of cloning by functional complementation.

♦♦♦♦♦♦ IV. SCREENS FOR CONDITIONAL MUTANTS INCLUDING TEMPERATURE-SENSITIVE (ts) MUTANTS AND USING COUNTER SELECTION WITH FLUOROOROTIC-ACID (FOA)

As will become clear in the remaining part of this chapter, conditional mutants, including temperature-sensitive mutants, are one of the preferred starting points for screening for interactors of essential yeast genes. Classical genetic methods have not yielded ts mutants in the majority of the approximately 1200 essential genes. Therefore, it is often necessary to start from a DNA clone of a given essential gene, and to devise methods of making conditional mutants, preferably ts mutants in that gene. To this end, *in vitro* mutagenesis and various ''plasmid shuffling'' and ''counter-selection'' methods (for instance on FOA) are useful, which are described here. A typical cloning experiment was described in the preceding section for using yeast genomic clone banks in order to isolate a gene corresponding to a given mutant. However, we start with a classic set of ts mutants, which originates in the early 1970s before the advent of DNA cloning.

A. Example 4: Screening for ts Cell Division Cycle-Specific Mutants

This set of mutants that goes back to the early 1970s has unveiled the basic ''mitotic engine'', has revolutionized our thinking about the cell cycle and about the biological basis of cancer, and is useful here to show the power and the limits of such a mutant screen. The original idea was to separate by appropriate selection criteria the metabolic functions necessary for life (for instance glycolysis, protein synthesis or any metabolic pathways leading to metabolites necessary for survival on complex media) from cell cycle regulatory functions, which were unknown at the time.

In yeast, ts mutants are usually screened as colonies that grow at 24°C (permissive temperature), but after replica plating fail to grow at 35°C or 37°C (restrictive temperature). This is only possible for an essential gene. By definition, ''essential'' genes are present in a single copy in the haploid genome, and are not redundant with any other gene for the given essential function. Given the high degree of ''buffering'' in the yeast genome (Hartman *et al.*, 2001) this sets strict limits to the number of genes defined in this way. Growth at the permissive temperature is usually somewhat slower than in the wild type or leads to some recognizable additional phenotype. The biochemical reasons for a ts mutant phenotype can be very different: The structure of the protein can be rendered ts by an appropriate amino acid change, but also the synthesis of the protein or its degradation or assembly can be ts. Hartwell and his colleagues in a series of papers in the early 1970s (for review see Mendenhall and Hodge, 1998) argued that the loss of a metabolic function would lead to growth arrest at the high temperature which could be immediate, but would be independent of cell division cycle stage. Depending on the half-life of the protein, cells at the restrictive temperature sometimes undergo a small number of cell cycles and form microcolonies. Conversely, the loss of a cell cycle regulatory function should result in quite distinct phenotypic traits at the restrictive temperature, in particular a cell cycle stage-specific arrest, first cycle arrest, continued growth of the arrested cells, recessivity and reversibility of the arrest after a shift-back to the permissive temperature. The last two criteria were included to enable genetic analysis and reciprocal shift experiments in mutants and double mutants, which led to the ordering of these mutants in pathways of a cell cycle scheme, and are not discussed here. The power of these ideas is shown by the fact that many of these mutants later were shown to be in highly conserved eukaryotic genes, which indeed regulate the cell cycle. Above all, we mention the regulation of START and of other points in the cell cycle by the universal cyclin-dependent kinases (CDKs) and the discovery of the numerous checkpoints of the cell cycle which ensure coordination between the different parallel regulatory pathways and are also highly conserved in all eukaryotes (Weinert and Hartwell, 1988; Hartwell and Weinert, 1989; Mendenhall and Hodge, 1998). Both concepts are of overriding importance for modern cancer research.

It goes without saying that the classical methods just described cannot give a complete picture of cell cycle regulation, for which the screen was originally devised. Many other methods of molecular genetics and biochemistry, including *in vitro* experiments had to be employed to arrive at the picture of the cell cycle that we now enjoy. Many of the important genes for cell cycle regulation could not be found as *cdc* mutants because of functional redundancy, for instance the cyclins (please compare the discussion of *CLN3* in Example 3.) Some others might, due to their 3D structure not be particularly prone to produce ts mutants. This is a major problem that can only be overcome by *in vitro* mutagenesis of a cloned gene. Some of the genes recovered as complementing the *cdc* mutations are still not understood at present, pointing to, perhaps, complex interactions or "double" functions of some of the metabolic housekeeping genes of yeast, and perhaps also higher organisms. For example, *cdc19*-ts is a START 'A' mutation affecting the same step as the RAS/cAMP pathway and is located in the *PYK1* gene, coding for pyruvate kinase. This possibly makes sense because in the non-growing G1 phase basic metabolic functions must be turned down and we can assume that defects in some of these functions should arrest the cell cycle at the same step, thus pointing to a second (regulatory) function of pyruvate kinase. The strict distinction between "metabolic" and "regulatory" that was assumed to exist 30 years ago may not actually exist in such an exclusive fashion.

B. Example 5: *In vitro* Mutagenesis and Screening of a Library of Point Mutations in a Cloned Gene

Over the years, several methods have been developed to create small libraries of mutated genes starting from a specific cloned gene. Some of these methods have mostly historical interest and will not be discussed in detail, such as mutagenizing a gapped plasmid *in vitro* with $NaHSO_3$ (Shortle *et al.*, 1984). Another approach is to use PCR and to induce errors in the polymerization step by, for instance, adding unphysiological concentrations of Mn^{2+} ions or unbalanced concentrations of nucleoside triphosphates (Clontech: Diversify™ PCR Random Mutagenesis Kit). Chapter 5 describes a detailed protocol for such PCR-induced mutagenesis. In our view, the best and up-to-date method is to introduce the gene of interest (from now on called gene X) in a shuttle vector into an *E. coli* mutant strain which is severely deficient in DNA repair. Commercial kits including such *E. coli* strains exist and are easy to use, e.g., Stratagene: XL1-Red competent cells for random mutagenesis (Cline *et al.*, 1996). In principle, the mutagenized plasmid (actually a library of plasmids) is recovered from *E. coli*, the gene of interest is excised and recloned in a non-mutagenized but otherwise identical plasmid (selectable by *LEU2*) and introduced into haploid yeast cells. The yeast cells are deleted for the essential gene X and carry a wild-type copy of gene X on an episomal plasmid selectable by *URA3*. After introducing the library, haploid cells result that contain versions of gene X on two plasmids, the wild-type copy on a *URA3*-bearing plasmid and the library on a *LEU2*-bearing plasmid. The method is also known as "plasmid shuffling". In one version of the screen, a few thousand colonies containing both plasmids and therefore growing on media lacking both leucine and uracil at 24°C are investigated by replica plating. Replica plating in parallel on media which contain uracil and FOA (fluoroorotic acid) at the two temperatures shows three types of colonies. The majority of the colonies either grow at both temperatures because the library plasmids contain active versions of gene X or don't grow at both temperatures because the library version of gene X is inactive at both temperatures. However, those colonies which grow on FOA plates at 24°C but fail to grow on FOA plates at 37°C, contain mutant versions of gene X, which code for a ts protein. FOA selects for the loss of any plasmid containing *URA3* and therefore, for the loss of the wild type copy of gene X. The biochemical basis for this counter-selection (selection against *URA3*$^+$) is that *URA3* encodes orotic acid decarboxylase catalyzing the last step of uracil biosynthesis leading to biosynthesis of fluoro-uracil which is highly toxic. If a cell is *URA3*$^+$ it will be killed by FOA. If it is *ura3*$^-$, it can survive in the presence of uracil in the medium. Therefore, FOA selects against the presence of the *URA3*$^+$ plasmids.

This method was successfully used to isolate ts and cold-sensitive (cs) mutants in *TUB1*, the yeast gene coding for alpha-tubulin

(Schatz *et al.*, 1988). The cs mutants were screened on FOA medium at 14°C. In this example, the haploid starting strain had to be constructed with a double deletion of *TUB1* and *TUB3* because the two genes form an essential gene pair, a situation often encountered with yeast. A scheme of this screening procedure is shown in Figure 1. In a similar procedure, colony sectoring is used as a screening phenotype. This is described in detail in the section dealing with synthetic lethality screens (this chapter, and Chapter 17), but can in principle also be used for screening of ts and cs mutants.

It is perhaps appropriate at this point to speculate about the possibility of obtaining cs as well as ts mutations in a protein. In a series of publications dealing with the genetic analysis of the cytoskeleton (for review see Solomon, 1991), it was found that proteins such as tubulin and actin which *in vivo* form large functional complexes by homotypic and heterotypic interaction, are particularly prone to yielding cs mutants. This was expected, because it had been found by the same authors earlier that cs mutants can also be recovered in phage coat proteins which also form large aggregates. Similar results have also been found for ribosomal proteins. What is common to all of these proteins is that they spontaneously assemble into large structures and we speculate that this may be a general rule.

What about essential proteins that are particularly resistant to the forming of ts (or cs) mutations? Systematic studies employing *in vitro* mutagenesis indicate that in any of those genes it is indeed possible to obtain ts mutations; however, they are often double of even multiple point mutations within the same gene, a situation which cannot be mimicked by mutagenizing living cells with a chemical mutagen.

A very interesting possibility arises in the case of a protein whose three-dimensional structure is known. In one case (Wertman *et al.*, 1992), representing yet another powerful method for *in vitro* mutagenesis, the yeast actin gene, *ACT1*, was systematically mutagenized with mutagenic primers and PCR at all points where two or more consecutive charged residues reside at the surface of the protein. These residues were replaced by alanine. Among 34 mutants, 11 were recessive lethals, 16 were conditional-lethal (either ts or salt-sensitive) and 7 were without recognizable phenotype. Two mutations may have had a dominant lethal effect.

It should be mentioned that it has been attempted to predict ts mutations in a globular protein based solely on the amino acid sequence (Varadarajan *et al.*, 1996). The reasoning used here is quite different from the one used by Wertman *et al.* (1992). Buried hydrophobic residues are believed to contribute substantially to the stability of the protein, hence exchanging them for other residues often results in a weakened, temperature-sensitive structure. The core of the method is an algorithm to predict which hydrophobic residues are buried in the structure. Comparison of the results of this calculation with a set of known structures and of substitutions that make the structure ts are presented.

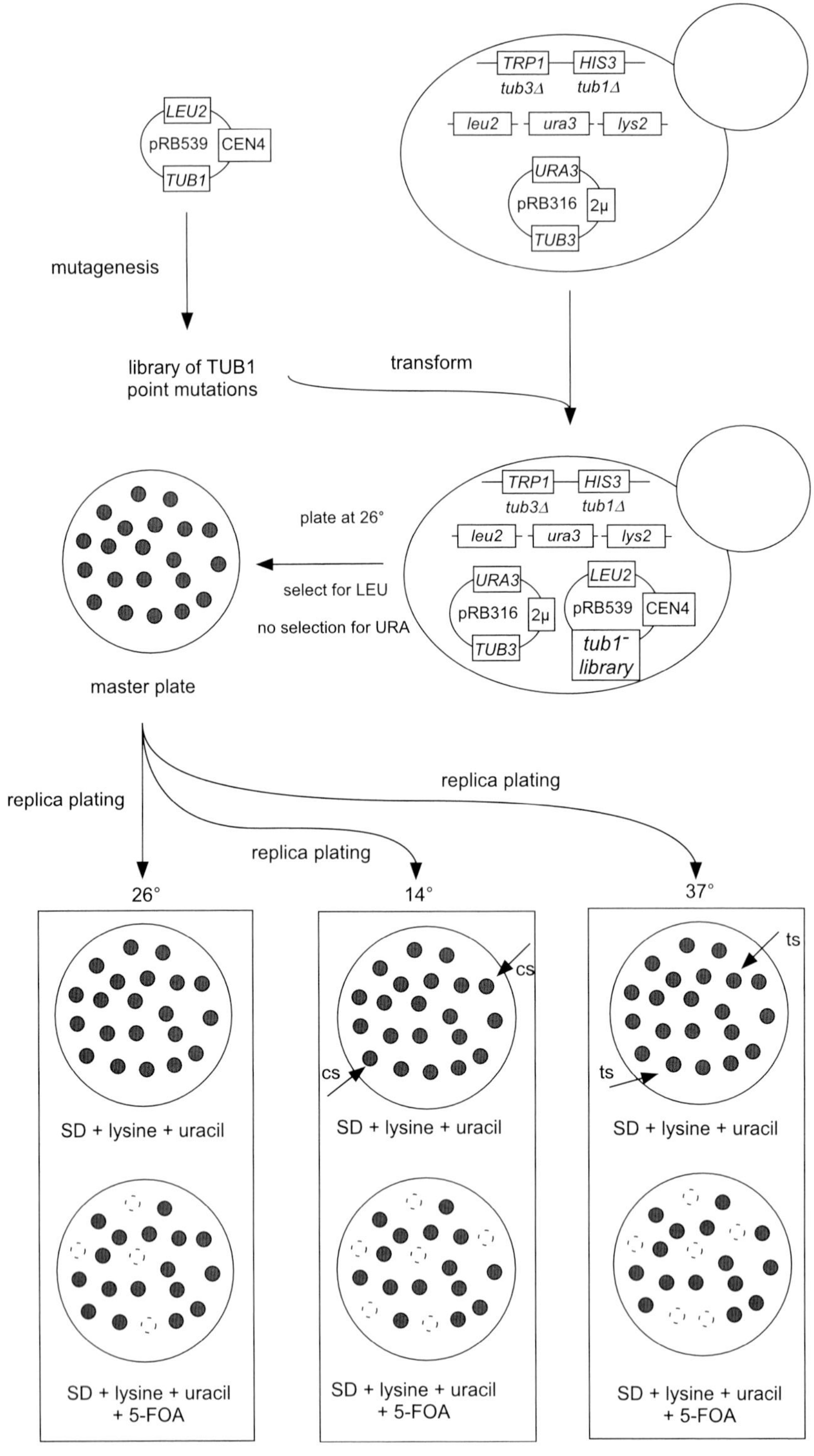

TRP1
HIS3
tub3Δ
tub1Δ
leu2
ura3
lys2
URA3
pRB316
2μ
TUB3
LEU2
pRB539
CEN4
TUB1
mutagenesis
library of TUB1
point mutations
transform
plate at 26°
select for LEU
no selection for URA
tub1⁻
library
master plate
replica plating
replica plating
replica plating
26°
14°
37°
cs
cs
ts
ts
SD + lysine + uracil
SD + lysine + uracil
SD + lysine + uracil
SD + lysine + uracil
+ 5-FOA
SD + lysine + uracil
+ 5-FOA
SD + lysine + uracil
+ 5-FOA

C. Example 6: Constructing Conditional Mutants using Regulatable Promoters and Protein Destabilization.

Cloning a yeast gene under control through the *GAL1/10* regulatory region is a powerful way to both increase (on galactose) and very substantially decrease (on glucose as sole carbon source) the expression level of a gene. The method has been used many times for functional characterization of a functionally unknown essential yeast gene to determine the terminal phenotype of the cells after shutting off transcription. On galactose, the consequences of severe overexpression of the gene were analyzed. The method is easy and useful, but it has some drawbacks, because on galactose rather large changes of cellular metabolism occur as compared to glucose, and these might influence the outcome of the experiment. Furthermore, it is apparently not possible to fine tune gene expression by varying the concentration of galactose.

The construction of a yeast cDNA library in a centromeric yeast vector under control of the *GAL1* promoter has been described (Liu *et al.*, 1992). By transforming the library into yeast, plating out on glucose and replica plating onto galactose media, clones were identified which on overexpression are lethal. As expected, several components of tightly regulated multiprotein complexes were identified which are detrimental to the cell if production in stoichiometric amounts is disturbed, among them actin and beta-tubulin. Other genes whose over expression is lethal were also identified, among them *GCL7* (coding for Type I phosphoprotein phosphatase) and some genes which are not yet characterized.

To overcome the inherent problems of the *GAL1/10* system, a prokaryotic regulatable promoter element was introduced into eukaryotic cells together with the binding protein tTA (tetracyclin-dependent transactivator protein) which depends on the addition of tetracycline (Gossen and Bujard, 1992). Then, the same system was adapted to yeast by chromosomally integrating in front of the gene in question the basic minimal *CYC1* promoter together with the tTA binding sequence (Gari *et al.*, 1997; Nagahashi *et al.*, 1997; Belli *et al.*, 1998). By site-directed mutagenesis of the binding sequence as well as the tTA protein and by using fusions of tTA with endogenous yeast repressor proteins, the system could be tuned to repression as

Figure 1. Isolation of conditional (ts and cs) mutants in *TUB1* by "plasmid shuffling". The starting strain is deleted for *tub1* and *tub3* and carries a 2μ *URA3* plasmid with *TUB3*. A library with mutations in *TUB1* is introduced on a centromeric *LEU2* plasmid. A master plate is created at the permissive temperature without selection for uracil prototrophy. The master plate is replicated on synthetic medium with lysine and uracil at three temperatures and on the same media with FOA also at three temperatures. Colonies not growing at the FOA plates at 14°C, but growing in every other condition tested carry a cs mutant allele from the plasmid library. Likewise, colonies not growing on FOA plates at 37°C, but growing in every other condition, carry a ts mutant allele of *TUB1* from the plasmid library. (after Schatz *et al.*, 1988, with permission).

well as activation of the downstream gene. A systematic study of yeast essential genes (Hegemann, J. H., personal communications) in the system engineered to repression by addition of tetracycline (doxycycline) showed that in all cases a very substantial down-regulation occurred, but that the majority of these strains were still growing although very slowly. This is an advantage because the repressed strains can still be used for genetic experiments (for instance, screening for synthetic lethality). If a binary system is used where repression is through a fusion protein of tTA with the yeast repressor Ssn6, complete inhibition of growth can often be achieved (Gari *et al.*, 1997). The big advantage of the system is that the prokaryotic binding sequence is not recognized by yeast proteins and that doxycycline, at the dose used for repression, has no strong effect on yeast physiology. Finally, it has been shown that gene expression can be titrated by adding increasing amounts of doxycycline. The target genes tested in this case were the four members of the Brix gene family involved in ribosome biosynthesis in yeast (Bogengruber *et al.*, 2003). A commercially available yeast strain collection, the 'Yeast Tet-Promoter Hughes collection' (yTHC), is now available which contains the simpler version (i.e. not with tTA-SSN6) of the system in front of nearly every essential yeast gene (Mnaimneh *et al.*, 2004).

Another addition to the toolbox for regulating gene expression concerns not transcription *per se*, but in addition, protein degradation. *ARD1* codes for N-terminal acetyl transferase, an essential gene of yeast. After shutoff on glucose of the *ARD1* gene, the strain grows normally with only marginal phenotypic consequences such as the inability to arrest the cell cycle after addition of alpha factor. This is explained by the necessity for N-terminal acetylation for complete repression of the silent mating type loci. Otherwise, Ard1p seems to be required only in very small amounts and the protein displays a relatively long half-life. It is possible to achieve a strong mutant phenotype by decreasing the half-life of the protein in addition to using the *GAL10* promoter. This was done by a combination of transcriptional control though the *GAL10* promoter with an N-terminal fusion of *ARD1* with the (monomeric) ubiquitin sequence so that after cleavage of ubiquitin, an N-terminal arginine remains which makes the protein extremely unstable (Park *et al.*, 1992). The fusion part additionally contained part of the *lac* repressor and an epitope tag. The fusion protein was enzymatically active, but would support growth only on galactose, not on glucose. The protein disappeared within one cell generation, but presumably due to stability of the acetylated substrates of Ard1p, several generations were needed before growth stopped completely.

Summarizing this part of the chapter, we can say that a large number of methods exist to efficiently make conditional mutants of any essential yeast gene, thus creating a possibility for genetic analysis.

♦♦♦♦♦♦ V. SCREENING FOR EXTRAGENIC SUPPRESSORS OF CONDITIONAL MUTANTS. SCREENING FOR SUPPRESSEES OF A GIVEN MULTICOPY SUPPRESSOR

Revertants are secondary mutations that restore the original wild-type phenotype of a mutant. Revertants can be identical revertants (restoring the original gene sequence), or intragenic second site mutations, which compensate the mutant phenotype of (mostly) a missense mutation through a change in amino acid sequence occurring at a second site in the same gene. This effect is explained through changes in the three-dimensional structure of the protein. However, still another class of revertants occurs, which is the one that interests us here. These are extragenic revertants, which are often called suppressors. The suppressed strain is therefore a double mutant. As we have mentioned already, the phenotype of the suppressed strains can be exactly like wild type, but very often the suppressor mutations confer additional, sometimes selectable, phenotypes. One example which played a very important role in the history of molecular genetics is, of course, the case of the nonsense suppressors, which restore the mutant phenotype of a nonsense (STOP) mutation in a protein-coding reading frame. Another example occurs, when a missense mutation in a protein, which influences the interaction with a partner protein, is compensated by a second mutation within that partner protein. Here, the three-dimensional structures of the two partner proteins fit together because of compensatory structural changes in both mutant proteins. It is readily understood that in such a case the suppressors are highly allele-specific as well as gene-specific. In such a case, the partner protein would also be found by various other screening techniques besides suppressor analysis, for instance through the yeast two-hybrid system or through mass spectrometric analysis of protein complexes. The cases where a suppressor mutation occurs in a partner protein directly interacting (binding) to the query protein, are only a minority. In many other cases physiological suppressors are found, which, for instance, compensate through influencing signal strength in a signal transduction pathway, which encompasses both the query and the suppressor gene. Another possibility is by influencing metabolism in a great number of possible ways. Cloning the suppressor genes obviously can reveal previously unknown physiological interaction partners, which are often not directly binding to each other. Dominant as well as recessive suppressor mutations have been found.

A. Example 7: *EFL1* and its Suppressor *TIF6* in Ribosome Biogenesis

Efl1p is a cytosplasmic GTPase showing sequence homology to the ribosomal translocase EF-2/EF-G. It is not a component of ribosomes, and originally its function was unknown. Suppressor analysis revealed a function in ribosome biogenesis, which could be further corroborated with biochemical techniques (Senger *et al.*, 2001). The deletion of *EFL1* causes extremely slow growth. This is a phenotype which is easy to use for suppressor analysis. Plating many millions of cells without mutagenesis allowed the isolation of clones of suppressed strains that grew with near wild-type velocity and formed normal-sized colonies after two days. One advantage of using spontaneous mutants is that the probability for isolating double mutants is extremely low. Genetic analysis showed that the mutations were dominant. Crossing the suppressed strain to the wild type revealed that the suppressor was in a gene different from *EFL1*, because most of the tetrads from the cross were tetratypes, indicating that the suppressor was not linked to *EFL1*. Crossing several different suppressed mutants among each other showed that they all carried a suppressor in the same gene, because they did not recombine with each other. Note that this rather laborious technique was necessary because of the dominant nature of the suppressor. The next step was the construction of a clone bank from the suppressed strain, transforming this clone bank into the slow growing *efl1* deletion mutant, and isolating strains which regained the ability to grow with near wild-type velocity. Since the clone had been constructed from the deletion mutant *efl1*, it was impossible to isolate the wild-type *EFL1* gene from that clone bank. All the suppressing plasmids contained a common region harboring the *TIF6* gene. This was surprising since Tif6p was a nucleolar protein with a known implication in 60S ribosomal subunit synthesis. Testing the *efl1* deleted strain by polysome analysis on sucrose gradients and analyzing the processing of ribosomal RNA showed a defect in 60S ribosomal subunit synthesis and, as a secondary effect, a defect of ribosomal RNA processing. Further biochemical experiments suggested that Tif6p binds to the pre-60S subunits during nucleolar assembly and transport to the cytoplasm and is released and re-imported to the nucleus only after a structural change of the pre-60S ribosome triggered by Efl1p. The suppressor mutant protein apparently becomes independent of the Efl1p-catalyzed step for re-import. Screening the *efl1* deletion mutants for large colonies after transformation with a wild-type multicopy clone bank also led to the same single multicopy suppressor, *TIF6*. This means that the effect of the Efl1p-catalyzed step can also be mimicked by an over-supply of the Tif6p factor. The example is a typical one and shows very nicely how a gain of function of the suppressor gene can compensate for a defect of a different gene.

B. Example 8: Suppressor Analysis of the *RAS*/cAMP Signaling Pathway

This example is about 20 years old, but still very instructive about the possibilities and limits of suppressor analysis. We are presenting the now classic analysis performed by the group of Kelly Tatchell starting around 1984 and supplementing this picture with some more modern subsequent investigations of the pathway. The haploid yeast genome harbors two genes which in the N-terminal part of the encoded proteins are closely related to mammalian ras protooncogenes (Kataoka *et al.*, 1984; Tatchell *et al.*, 1984). This caused a lot of excitement because it held promise for the elucidation of signaling mechanisms in cancer cells with the help of yeast genetics. The two yeast *RAS* genes (*RAS1* and *RAS2*) form an essential gene pair: single gene deletions are viable but the double deletion is inviable. As has often been found for yeast gene pairs, the known functions of the two genes are nearly the same, but their transcriptional regulation is different and depends on the carbon source (Breviario *et al.*, 1986). Deletion of *RAS1* has nearly no effect while the deletion of *RAS2* leads to lack of growth on nonfermentable carbon sources (Tatchell *et al.*, 1984). This property of *ras2*::*LEU2* strains was used to perform a selective screen for genetic suppressors of the growth defect (Tatchell, 1986; Cannon and Tatchell, 1987). Both recessive and dominant suppressors were found which significantly contributed to our present understanding of the *RAS*/cAMP signaling pathway. The corresponding genes were cloned as described above in the Section III of this chapter. Dominant suppressors (called *SRA4* by the authors) were found in the START gene, *CDC35*, coding for yeast adenylate cyclase. The mechanism of suppression is easy to understand: the mutant adenylate cyclase is independent of *RAS*, a small GTPase which is an activator of adenylate cyclase in the GTP bound conformation, but not in the GDP bound confirmation. Both the yeast *ras* proteins and the adenylate cyclase are located at the inner face of the plasma membrane and the regulatory interaction between the two proteins was also shown by *in vitro* biochemical experiments (Toda *et al.*, 1985). Two side effects of the dominant *CDC35* mutation were predicted and shown to be true: (i) the suppressor mutant should be viable even when both *ras* genes are deleted (it is independent of *ras*), and (ii) the proper response to nutrient limitation which is in the wild type mediated by *ras*, should be lost, leading to a defect in survival of starvation. Another dominant suppressor mutation (*SRA3*) was found in *TPK1*, one of the three genes coding for the catalytic subunits of yeast protein kinase A. The properties of this mutant were similar to the adenylate cyclase mutant, leading to *ras*-independent signaling and the side effects just described. Recessive suppressor mutations were found in *BCY1* (*SRA1*), the gene coding for the regulatory subunit of protein kinase A. A loss of function of this relatively small protein leads to constitutive activity of protein kinase A and to similar properties as just described. Another recessive suppressor mutation was identified in *PDE2*, the gene coding for the high affinity cAMP phosphodiesterase, another negative regulator of the pathway. A further suppressor (*SRA6*) was not followed up closely because it was shown to regulate transcription of *RAS1*, thereby leading to the restoration of growth on non-fermentable carbon sources.

Key features of the pathway were elucidated by this simple genetic investigation, but additional important genes could not be found. For instance, genes acting upstream of *ras* in the same pathway (*CDC25, IRA1, IRA2*) cannot give rise to mutations compensating for a deletion of *ras2*. Other genes (proteins) that were later discovered to interact or to act in parallel with the *RAS*/cAMP pathway, like *CAP1* (*SRV2*) or *SCH9*, were also not found. *SCH9* was discovered as a multicopy suppressor of a temperature-sensitive growth defect of *cdc25* mutant. *SCH9* codes for a kinase which has partially overlapping function with protein kinase A (Toda *et al.*, 1988). Other questions which are completely open at present concern the relation of the *RAS*/cAMP pathway to the TOR pathway (which is also active in nutrient dependent growth control and control of ribosome synthesis in yeast) and, for instance, the role and regulation of the pathway in control of the cell's actin cytoskeleton in response to several kinds of stresses (Thevelein and de Winde, 1999; Ho and Bretscher, 2001). Note that the conserved eukaryotic *ras* genes are involved in cell cycle and growth control in both yeast and human cells, but the signaling pathways of which they are part, are different in the two systems (Ory and Morrison, 2004).

C. Example 9: Screening for Suppressees of a given Multicopy Suppressor

As a technique, multicopy suppression, used and discussed in both the examples just described, is frequently successful. However, this technique can be "turned around" by screening for mutants or gene deletions from the deletion collection that require a certain gene on a multicopy plasmid (or the same gene overexpressed under control of the *GAL1* or other regulatable promoter) in order to survive. In the example given, a foreign gene (*Drosophila* topoisomerase II) was used in this way to isolate mutants in yeast toposiomerase II (Kranz and Holm, 1990). This paper, which is now considered a classic, served as a proof of principle for the method. This approach is still useful in the postgenomic era, because there are still many genes which are functionally unknown (even essential ones) and sequence comparisons often do not reveal the true yeast homolog of a given gene from higher cells. This was the case, for instance, for the yeast caspase, *YCA1* (Madeo *et al.*, 2002). Furthermore, the method can also detect functional interaction between non-homologous yeast genes. We are also using this example to discuss the advantage of the colony-sectoring assay when mutations in essential genes are sought.

The authors (Kranz and Holm, 1990) placed the cDNA coding for *Drosophila* topoisomerase II on a yeast expression plasmid harboring also the yeast *ADE3* and *URA3* genes. This plasmid was transformed into a haploid yeast strain, which was marked *ade2*, *ade3*, and *ura3*. See Figure 2 for a scheme of the procedure. As the plasmid could be lost mitotically (Figure 3), red/white sectored colonies resulted (Figure 4). This sectoring assay was based on earlier work by the Hartwell group (Koshland *et al.*, 1985). The red sectors still contained the plasmid which covered the *ade3* defect resulting in the color development because of the *ade2* mutation. The white sectors had lost the plasmid and the *ade3* mutation prevented color development. The strain was mutagenized and plated out. Colonies that were entirely red carried a mutation which caused complete dependency on the foreign *Drosophila* gene. It was shown that transforming such a mutant with a wild-type standard clone bank, and screening for colonies that regain the sectoring phenotype, can be used to clone the yeast gene (*TOP2*) corresponding to the mutant. Moreover, if the first plating is performed at 35°C and the red colonies are retested at 24°C, temperature-sensitive mutants of the gene in question can be directly isolated. The colony sectoring assay as described here has two main advantages: (i) the sectoring phenotype avoids replica plating after the primary screen, and (ii) the candidates isolated by the sectoring assay can be retested by an independent method using FOA for the ability to lose the *URA3* carrying plasmid. This is important because non-sectoring colonies can arise infrequently through artifacts like mutations in the chromosomal background, for instance reversions of *ade3* through recombination with the *ADE3* gene on the plasmid. These false positives can be recognized by selection on FOA (they are FOA^+). Finally, the method can be adapted to make in yeast cells temperature-sensitive mutants of genes from higher organisms, which are important for *in vitro* biochemical tests and also for making partial loss of function mutants, for instance of mouse genes. Such mouse mutants could not be obtained *in vivo* because of the constant body temperature of the animal.

♦♦♦♦♦♦ VI. SYNTHETIC LETHALITY AND DOSAGE LETHALITY SCREENS

These techniques are in a sense opposite to the suppressor screens. This is because a suppressor compensates for a defect in the query gene, while a mutation leading to synthetic lethality enhances the defect in the query gene. Therefore, starting from the same query gene it is not expected to recover the same set of genes when these two alternative methods are applied. In cases where both methods have been tried, it was sometimes found that synthetic lethality screens were the more successful and more generally applicable method (Bender and Pringle, 1991; Doye and Hurt, 1995). However, no single screening method will detect all genes involved in the process in question and the greatest power of detection will come from the combination of several complementary screening techniques. Like in the last-mentioned example (Kranz and Holm, 1990), the colony sectoring phenotype is often used for screening synthetic lethal and dosage lethal mutants.

A. Example 10: The Nuclear Pore Complex

Synthetic lethality screens were successful to find a large number of genes coding for proteins of the nuclear pore complex, the largest

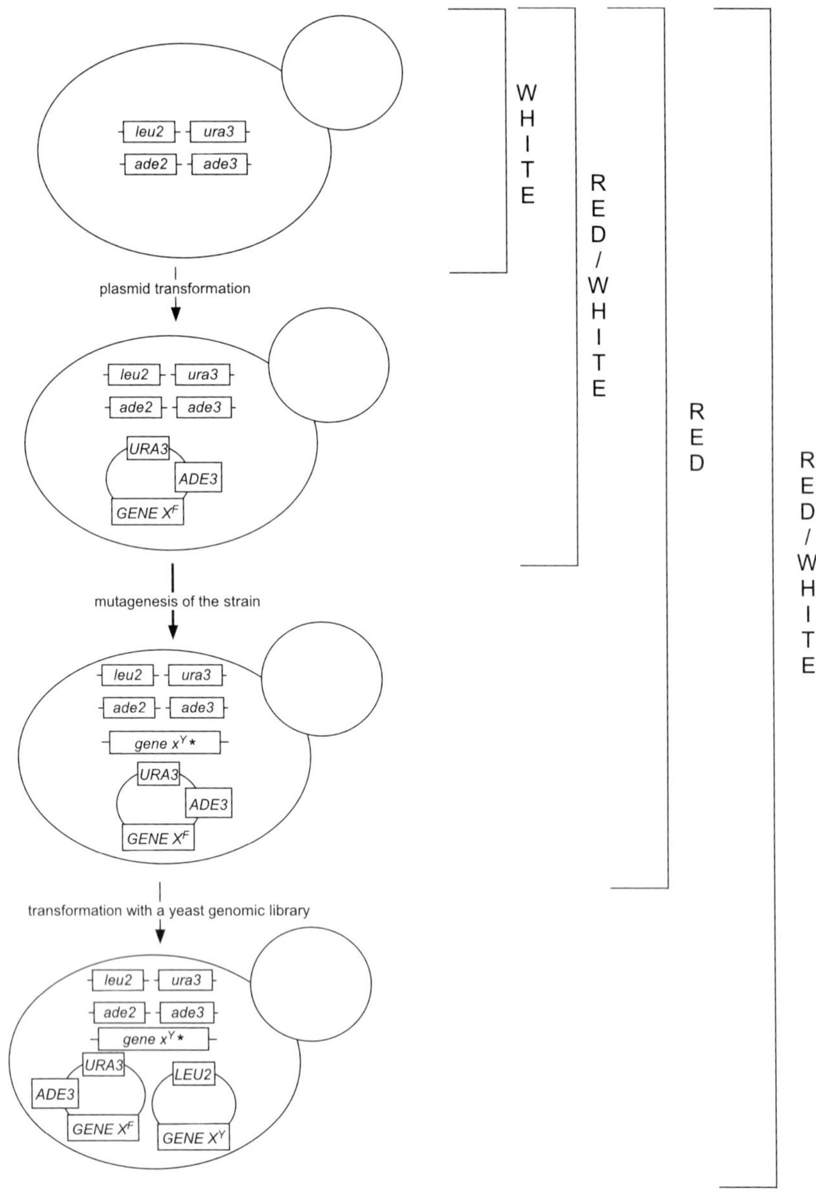

protein complex of the eukaryotic cell which is highly conserved between yeast and human cells (Doye and Hurt, 1995). Because the query gene, *NSP1* is essential (and many of the nuclear pore (NUP) genes are also essential), *NSP1* was introduced both as the wild type and as a temperature-sensitive allele on two separate plasmids in a haploid strain deleted for *NSP1* on the chromosome and marked *ade2* and *ade3* (Figure 5). Plasmid 1 carried the wild-type allele of *NSP1* and *ADE3*. Plasmid 2 carried the temperature-sensitive allele, *nsp1-ts*. All tests were performed at the permissive temperature, where *nsp1-ts* allowed growth but had a reduced biochemical activity. The strain was mutagenized and plated out at the permissive temperature. Sectoring colonies (see above, Example 9) indicated that *nsp1-ts* was still sufficient for growth in that colony. Solid red colonies indicated that the wild-type allele of *NSP1* could no longer be lost, because in that mutant colony the *nsp1-ts* allele was no longer sufficient for growth due to synthetic lethality with a second mutation (*nup-X*) in the chromosomal background. Candidates for synthetically lethal mutants were tested further to exclude other causes for non-sectoring like back mutations in *ade3*. The synthetically lethal genes were cloned by screening for the restoration of the sectoring phenotype after transformation with a standard clone bank. Note that practically all synthetically lethal mutations were recessive. Finally, synthetic lethality was confirmed in crosses of the single mutations (where it was necessary in the presence of a plasmid carrying the wild type gene). In those crosses, both mutations segregated independently and 1/4 of the progeny was double mutant in the tetrads and did not grow.

Four mechanisms for synthetic lethality are discussed (Figure 6):

(i) Physical interaction between the two proteins. Mutant protein *nsp1-ts* can still interact with *NUP-X*, but not with the *nup-X* mutant protein.
(ii) *NSP1* and *NUP-X* have redundant functions and weakening of both proteins is not sufficient for growth.

Figure 2. Proof of principle for the isolation of a yeast gene (topoisomerase II) functionally equivalent to a non-yeast gene. X^F is foreign gene X (*topoisomerase* II from *Drosophila*). X^Y is the yeast orthologue of X^F. * denotes the inactivating mutation in gene X^Y. The starting strain is homogeneously white because both auxotrophic markers *ade2* and *ade3* are present. This strain is transformed with gene X^F on an episomal plasmid carrying both *ADE3* and *URA3* resulting in sectored colonies and mutagenized. Inactivating mutants in gene X^Y render the strain homogeneously red because the plasmid has now become essential for life. In a final step, the wild-type yeast gene complementing the mutant X^Y is being cloned by functional complementation from a genomic yeast library. Transformants from the library which contain the WT gene X^Y are again sectoring because now the gene X^F has become dispensable. This is a model experiment which of course depends on the assumption that gene X^F is functional in yeast and can replace gene X^Y. Nowadays a similar approach could be used for isolating an orthologue of a given yeast gene from a cDNA bank of a higher organism. (after Kranz and Holm, 1990, with permission).

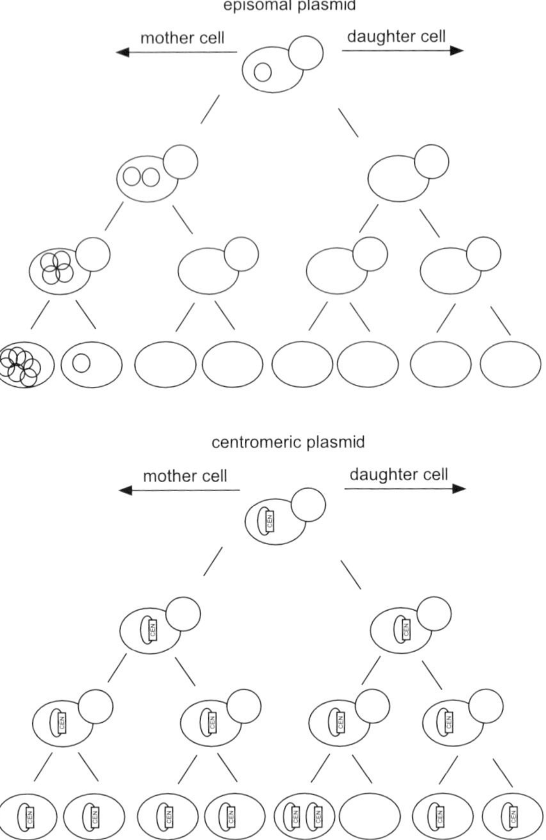

Figure 3. Inheritance of episomal (upper portion) and centromeric (lower portion) plasmids in yeast. Episomal plasmids accumulate in mother cells and lead to increased copy number. Centromeric plasmids are evenly distributed between mother and daughter cells.

(iii) *NSP1* and *NUP-X* are in the same pathway, for instance NUP-X could be a factor necessary for maturation of *NSP1*. Weakening both the structure and the efficiency of synthesis (maturation) of *NSP1* could lead to an amount of *NSP1* which is insufficient for growth.

(iv) *NSP1* and *NUP-X* act in two different dependent pathways, for instance one could be necessary for nuclear export and the other for nuclear import, and they could be mechanistically coupled, which would then lead to a weakening of the nuclear pore function which is insufficient for growth.

These possibilities are schematically depicted in Figure 6. Systematic application of this technique resulted in a large network of interactions between the NUPs, but also with further genes that are not directly involved in the architecture of the nuclear pore complex, like for instance, genes involved in pre-mRNA processing.

Of course, screening for synthetic lethality with a deletion mutant of a non-essential gene can be performed in a very similar way, but

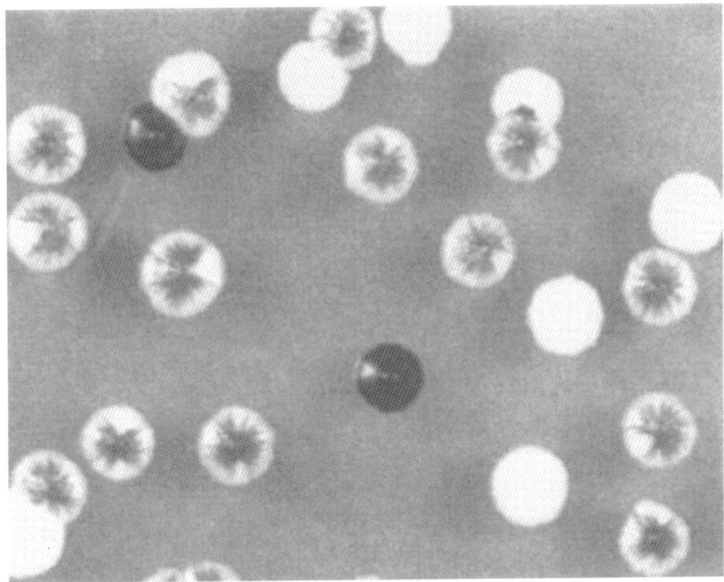

Figure 4. Example for the sectored colony phenotype. The red (dark tone) colonies are mutants for X^Y (see Figure 2) and therefore cannot lose the plasmid carrying X^F. The sectored colonies carry the X^Y wild-type gene and therefore statistically lose the plasmid. The homogenously white colonies have lost the plasmid even in the cell starting the colony (after Kranz and Holm, 1990, with permission).

is technically easier, because a TS mutant of the query gene is not necessary for the screen.

B. Example 11: Cell Polarity and Bud Formation

Establishment of cell polarity and determination of the future budding site at first seems to be a very yeast-specific topic. However, also in this case a very general eukaryotic problem of cellular morphogenesis lies behind the surface. We are discussing a classic paper (Bender and Pringle, 1991) which was first to implement a systematic synthetic lethality screen and to use the colony-sectoring assay to this end. The methods employed are still very important for functional analysis of yeast genes.

The cell division cycle genes, *CDC24* and *CDC42*, were discovered through their respective ts mutants and shown by temperature shift and double mutant experiments to constitute the so-called budding branch of the cell cycle regulatory network. In order to discover additional genes and functions involved in the process of cell polarity establishment and budding which are directly interacting with *CDC24* and *CDC42*, the authors started out with a multicopy suppressor screen of *cdc24*-ts. This screen revealed the gene *MSB1*; however, deletion of *msb1* showed no phenotype by itself (this was, of course, the reason why it was not included in the original collection of *cdc* mutants). Performing a synthetic lethality screen

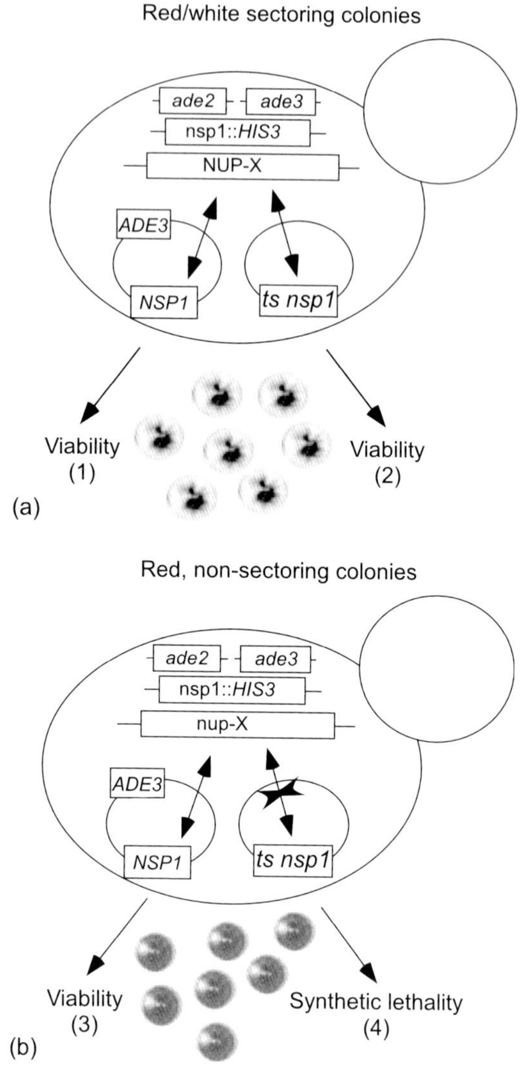

Red/white sectoring colonies

ade2 ade3

nsp1::*HIS3*

NUP-X

ADE3

NSP1

ts nsp1

Viability (1)

Viability (2)

(a)

Red, non-sectoring colonies

ade2 ade3

nsp1::*HIS3*

nup-X

ADE3

NSP1

ts nsp1

Viability (3)

Synthetic lethality (4)

(b)

Figure 5. Screening for synthetically lethal mutations in genes functionally interacting with *NSP1*. These genes are called here *NUP-X*. The ts *nsp1* is functional at 25°C but non-functional at 37°C. Cells carrying both plasmids are mutagenized and plated out at the permissive temperature (25°C). a) No mutations on the genome (1) or only irrelevant mutations not interacting with *NSP1* (2) leading to sectored colonies. b) Mutations in *nup-x*: Cell is only viable because wild-type *NSP1* is present (3). *Nup-x* and *ts nsp1* are synthetically lethal at 25°C (4). (After Doye and Hurt, 1995, with permission).

based on the deletion of *msb1*, carried out in much the same way as described above, revealed seven complementation groups of synthetically lethal mutants. It is interesting that all of those (non-sectoring) mutants were recessive, as expected, and colony sectoring could easily be applied to a complete complementation analysis. Classical methods that have been discussed already led to cloning of the genes, again using colony sectoring as a screenable phenotype.

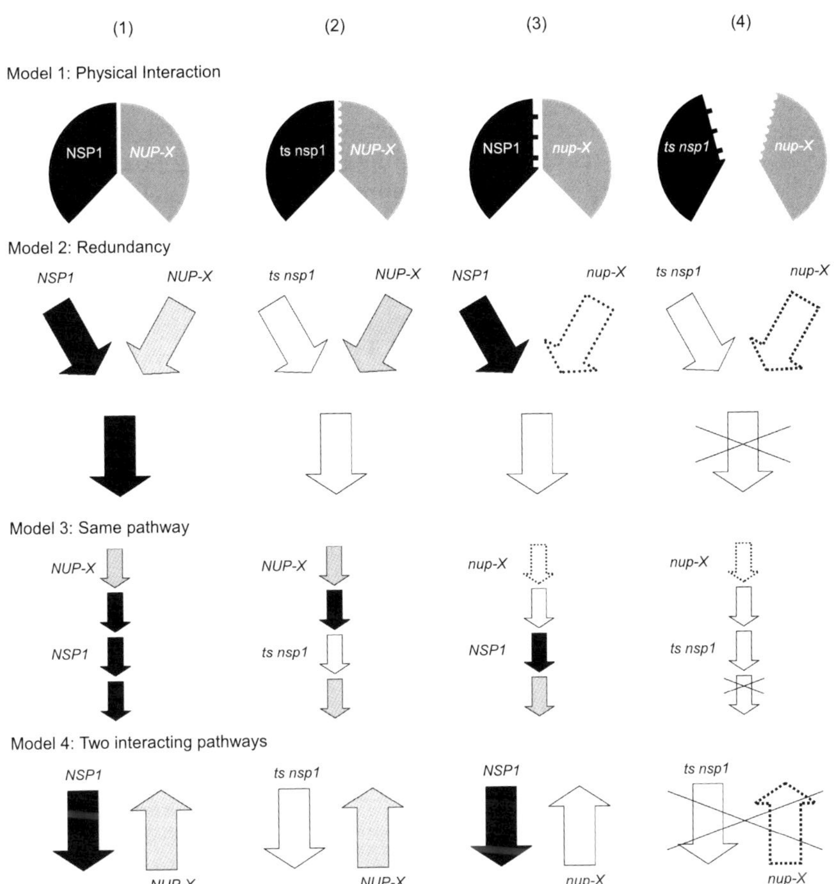

Figure 6. Four possible mechanisms for synthetical lethality. Numbers in brackets refer to Figure 5. *Model* 1: *NSP1* and *NUP-X* form a heterodimer. The *ts nsp1* mutant as well as a *nup-x* mutant are viable as long as they have a wild-type interacting partner. The *ts nsp1/nup-x* heterodimer is non-functional. *Model* 2: No physical interaction between *NSP1* and *NUP-X* protein. The two proteins have similar functions. Only if both of them are non-functional synthetic lethality occurs. *Model* 3: Both genes (proteins) are acting in the same dependent pathway. Both mutations inactivate the respective protein only partially. Only if both consecutively acting proteins are mutated the flux through the pathway remains below the threshold for viability. *Model* 4: *NUP-X* and *NSP1* are acting in two different but dependent pathways (for example, nuclear export and nuclear import). Only if both proteins are mutant the pathways are blocked (after Doye and Hurt, 1995, with permission).

One of those complementation groups was *CDC24*, which was reassuring because it meant that the screen revealed true functionally interacting genes. Two of the mutants recovered turned out, unexpectedly, to be ts mutants in essential genes, *bem1-ts* and *bem2-ts*. Their ultrastructural phenotype was very similar to the original *cdc24-ts* phenotype (abnormal elongated buds etc.). The *bem2-ts* mutant could be suppressed by *MSB1* on a multicopy plasmid. Taken together, these elegant experiments showed that the genes, *CDC24*, *CDC42*, *MSB1*, *MSB2*, *BEM1* and *BEM2*, form a network of interactions in cell polarity determination and bud formation. In the 15 years since this work was done, many details of the biochemical mechanisms of these processes were discovered based on the founding work of Bender and Pringle (Pruyne *et al.*, 2004).

It is also interesting that of the seven genes found in the synthetic lethality screen, four were unrelated to cell polarity establishment and budding, but were rather technical artifacts of the method. However, as the authors say, these can easily be sorted out. They were due to the fact that some elements on the plasmid carrying *MSB1* became essential through mutations on the chromosome, mostly this was the *ADE3* gene, as has been discussed already. They could be recognized as *MSB1* or *CDC24* on another plasmid did not rescue the synthetic lethality. However, the ultimate test for proving beyond reasonable doubt that a synthetic lethality relationship exists between the query gene and the gene recovered was a cross between the *msb1* deletion strain and the new mutation. In regular tetrads all double mutants were shown to be inviable.

C. Example 12: Synthetic Dosage Lethality Screens

This method is similar to the synthetic lethality screen discussed above. Synthetic lethality can be screened when a query gene is turned down with a regulatable promoter. This is useful because in most cases the down regulation by means of the *GAL1* promoter or the tTA/tetracycline system was shown to lead to slow growth but not to death. However, of greater importance is synthetic lethality when a query gene is overexpressed, which was discussed in detail in recent years (Measday and Hieter, 2002) and shown to reveal novel genes functionally interacting with the yeast kinetochore (Measday *et al.*, 2005). In the latter communication it was shown that, as expected, different sets of genes are recovered in a synthetic lethal screen and in a synthetic dosage lethality screen. In the first case, mutants are recovered that cannot tolerate the loss of function of the query gene; in the second case mutants are recovered that cannot tolerate the overexpression of the same query gene. Both cases point to functional interaction of the two genes.

We are discussing here the second case. A prerequisite is that overexpression of most yeast genes is not toxic to wild-type strains (Liu *et al.*, 1992). One of the query genes was *CTF13* (a structural component of the yeast kinetochore) and the aim was to identify further components of the machinery necessary for faithful chromosome transmission. *CTF13* was placed on a centromeric plasmid under *GAL1* control, which on galactose did not cause a recognizable phenotype of the yeast cells. Mutants that grow on glucose but fail to grow on galactose were isolated. If this is performed at several different temperatures, say 25°C, 30°C and 35°C, conditional mutants can be isolated which are sensitive to overexpression of the query gene. Conversely, the method is used to start from a ts mutant, a clone bank constructed in the centromeric galactose-inducible vector is transformed into the mutant, and colonies are screened which at the permissive temperature are synthetically lethal on galactose. In the same experiment colonies also can be screened which regain the ability to grow at the non-permissive temperature, so that synthetic dosage lethal as well dosage suppressor plasmids can be isolated (Figure 7). For example, overexpression of Ctf13p was synthetically lethal with the *ctf14-42*-ts mutation at the permissive temperature. As controls it was tested that overexpression of Ctf13p did not prohibit growth of the wild type at any temperature and that empty vector on galactose likewise did not have any effect on wild type or on the *ctf14-42*-ts mutant. The authors then screened a collection of *ctf* (chromosome

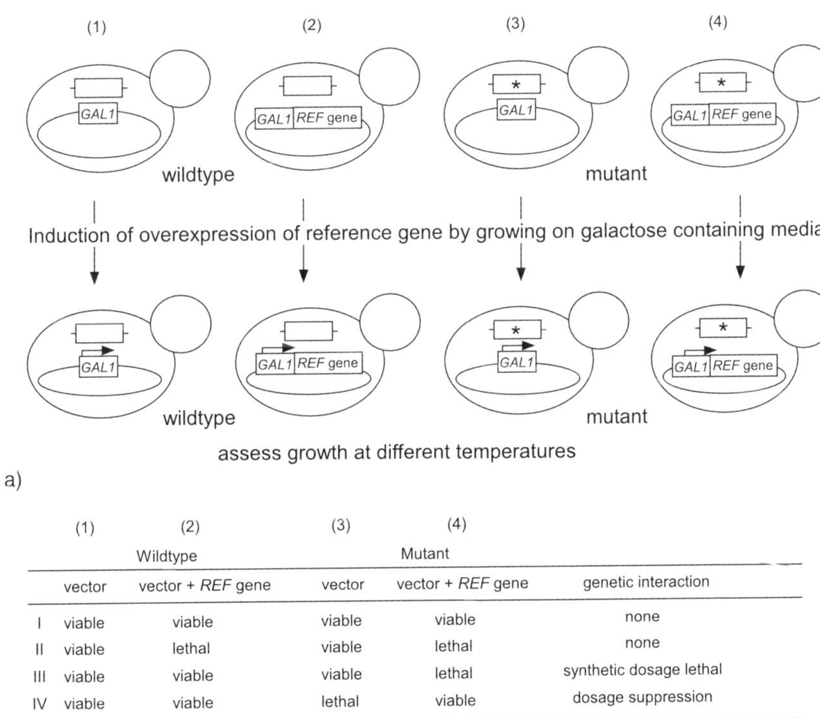

	(1)	(2)	(3)	(4)	
	Wildtype		Mutant		
	vector	vector + *REF* gene	vector	vector + *REF* gene	genetic interaction
I	viable	viable	viable	viable	none
II	viable	lethal	viable	lethal	none
III	viable	viable	viable	lethal	synthetic dosage lethal
IV	viable	viable	lethal	viable	dosage suppression

Figure 7. Synthetic dosage and dosage suppression. (a) Yeast is transformed with a reference gene under a *GAL1* inducible promoter for which interactions are being sought. The strain is mutagenized and plated on glucose on a reference plate and replica plated on glucose and galactose at two different temperatures. (1) and (3) are control experiments without reference gene. (2) and (4) are to test the effect of the overexpression of the *REF* gene at different temperatures. (b) Four different situations can be discriminated: (I) no synthetic lethality, no genetic interaction. (II) overexpression of the *REF* gene is lethal irrespective of additional mutation, no genetic interaction can be seen. (III) a mutant is synthetically lethal with overexpression of *REF* gene. (IV) Mutation is a ts mutation which is suppressed by overexpression of *REF* gene. The procedure can recognize synthetic dosage lethality as well as dosage suppression (after Measday and Hieter, 2002, with permission).

transmission fidelity) mutants by the assay just described using as query genes *CTF13*, *CTF14* and *CTF19*, which had all been previously identified as kinetochore protein encoding genes. The purpose of this screen was to obtain candidates for additional kinetochore protein encoding genes from this large collection of mutants. *CTF5* was shown to be a newly recognized kinetochore protein and the synthetic dosage lethality was shown to depend on the specific allele of *ctf5*. Cloning of the *CTF5* gene showed that it is identical with *MCM21* that had been previously shown to be a kinetochore protein. Three years later (Measday *et al.*, 2005), the synthetic dosage lethality screening method was applied to the now available yeast deletion collection and a large number of new candidates for structural proteins of the kinetochore were identified.

♦♦♦♦♦♦ VII. CONCLUDING REMARKS

In this chapter, we have shown that a large number of "smart screening" procedures have been developed in the field of yeast genetics which increase the power of this genetic system and are made even more powerful by the advent of the genomics-based tools that are now available. At the same time there was a shift of paradigm from the main aim of "discovery of new genes" to the new aim of "discovery of new gene interactions" leading ultimately to the identification of large genetic networks. In several places, we have shown that increasingly this most advanced of all genetic systems is also used to attack problems of human genetics. This can be done in the yeast cell as a little laboratory, but also in the human cell. For instance, synthetic lethality screens have been started to be applied even to human cell cultures (Simons *et al.*, 2001). The future looks bright for yeast genetics.

Acknowledgements

Financial support by FWF (Vienna, Austria) for project S9302-B05 (to M.B.), and by the EC (Brussels, Europe) for project MIMAGE (contract no. 512020; to M.B.) is gratefully acknowledged.

We are grateful to Patrick Linder, who wrote the chapter on Smart Genetic Screens in the first edition of this book for his ideas and for his support and encouragement and to Ian Stansfield for his suggestions for improvement of the text.

References

Belli, G., Gari, E., Piedrafita, L., Aldea, M. and Herrero, E. (1998). An activator/repressor dual system allows tight tetracycline-regulated gene expression in budding yeast. *Nucleic Acids Res.* **26**, 942–947.

Bender, A. and Pringle, J. R. (1991). Use of a screen for synthetic lethal and multicopy suppressee mutants to identify two new genes involved in morphogenesis in *Saccharomyces cerevisiae*. *Mol. Cell Biol.* **11**, 1295–1305.

Bobola, N., Jansen, R. P., Shin, T. H. and Nasmyth, K. (1996). Asymmetric accumulation of Ash1p in postanaphase nuclei depends on a myosin and restricts yeast mating-type switching to mother cells. *Cell* **84**, 699–709.

Bogengruber, E., Briza, P., Doppler, E., Wimmer, H., Koller, L., Fasiolo, F., Senger, B., Hegemann, J. H. and Breitenbach, M. (2003). Functional analysis in yeast of the Brix protein superfamily involved in the biogenesis of ribosomes. *FEMS Yeast Res.* **3**, 35–43.

Breitenbach, M., Laun, P., Heeren, G., Jarolim, S. and Pichová, A. (2004). Mother cell-specific aging in *Saccharomyces cerevisiae*. In: *Metabolism and Molecular Physiology of Saccharomyces cerevisiae* (J. R. Dickinson and M. Schweizer, eds), pp. 20–41. CRC Press, London.

Breitenbach, M., Madeo, F., Laun, P., Heeren, G., Jarolim, S., Frohlich, K.-U., Wissing, S. and Pichova, A. (2003). Yeast as a model for ageing and apoptosis research. In: *Topics in Current Genetics 3 Model Systems in Aging* (H. Osiewacz and T. Nystrom, eds), pp. 61–97. Springer Verlag, Berlin.

Breviario, D., Hinnebusch, A., Cannon, J., Tatchell, K. and Dhar, R. (1986). Carbon source regulation of RAS1 expression in *Saccharomyces cerevisiae* and the phenotypes of ras2- cells. *Proc. Natl. Acad. Sci. USA* **83**, 4152–4156.

Briza, P., Bogengruber, E., Thur, A., Rutzler, M., Munsterkotter, M., Dawes, I. W. and Breitenbach, M. (2002). Systematic analysis of sporulation phenotypes in 624 non-lethal homozygous deletion strains of *Saccharomyces cerevisiae*. *Yeast* **19**, 403–422.

Briza, P., Breitenbach, M., Ellinger, A. and Segall, J. (1990). Isolation of two developmentally regulated genes involved in spore wall maturation in *Saccharomyces cerevisiae*. *Genes Dev.* **4**, 1775–1789.

Briza, P., Ellinger, A., Winkler, G. and Breitenbach, M. (1988). Chemical composition of the yeast ascospore wall. The second outer layer consists of chitosan. *J. Biol. Chem.* **263**, 11569–11574.

Briza, P., Winkler, G., Kalchhauser, H. and Breitenbach, M. (1986). Dityrosine is a prominent component of the yeast ascospore wall. A proof of its structure. *J. Biol. Chem.* **261**, 4288–4294.

Cannon, J. F. and Tatchell, K. (1987). Characterization of *Saccharomyces cerevisiae* genes encoding subunits of cyclic AMP-dependent protein kinase. *Mol. Cell Biol.* **7**, 2653–2663.

Chen, C., Dewaele, S., Braeckman, B., Desmyter, L., Verstraelen, J., Borgonie, G., Vanfleteren, J. and Contreras, R. (2003). A high-throughput screening system for genes extending life-span. *Exp. Gerontol.* **38**, 1051–1063.

Christodoulidou, A., Bouriotis, V. and Thireos, G. (1996). Two sporulation-specific chitin deacetylase-encoding genes are required for the ascospore wall rigidity of *Saccharomyces cerevisiae*. *J. Biol. Chem.* **271**, 31420–31425.

Christodoulidou, A., Briza, P., Ellinger, A. and Bouriotis, V. (1999). Yeast ascospore wall assembly requires two chitin deacetylase isozymes. *FEBS Lett.* **460**, 275–279.

Cline, J., Braman, J. C. and Hogrefe, H. H. (1996). PCR fidelity of pfu DNA polymerase and other thermostable DNA polymerases. *Nucleic Acids Res.* **24**, 3546–3551.

Coluccio, A., Bogengruber, E., Conrad, M. N., Dresser, M. E., Briza, P. and Neiman, A. M. (2004). Morphogenetic pathway of spore wall assembly in *Saccharomyces cerevisiae*. *Eukaryot. Cell* **3**, 1464–1475.

Cross, F. R. (1988). *DAF1*, a mutant gene affecting size control, pheromone arrest, and cell cycle kinetics of *Saccharomyces cerevisiae*. *Mol. Cell Biol.* **8**, 4675–4684.

D'Mello N. P., Childress, A. M., Franklin, D. S., Kale, S. P., Pinswasdi, C. and Jazwinski, S. M. (1994). Cloning and characterization of *LAG1*, a longevity-assurance gene in yeast. *J. Biol. Chem.* **269**, 15451–15459.

de la Cruz, J., Daugeron, M. C. and Linder, P. (1998). "Smart" genetic screens. In: *Methods in Microbiology 26 Yeast Gene Analysis* (A. J. P. Brown and M. F. Tuite, eds), pp. 269–295. Academic Press, New York, N.Y.

Doye, V. and Hurt, E. C. (1995). Genetic approaches to nuclear pore structure and function. *Trends Genet.* **11**, 235–241.

Dujon, B., Sherman, D., Fischer, G., Durrens, P., Casaregola, S., Lafontaine, I., De Montigny, J., Marck, C., Neuveglise, C., Talla, E., Goffard, N., Frangeul, L., Aigle, M., Anthouard, V., Babour, A., Barbe, V., Barnay, S., Blanchin, S., Beckerich, J. M., Beyne, E., Bleykasten, C., Boisrame, A., Boyer, J., Cattolico, L., Confanioleri, f., De Daruvar, A., Despons, L., Fabre, E., Fairhead, C., Ferry-Dumazet, H., Groppi, A., Hantraye, F., Hennequin, C., Jauniaux, N., Joyet, P., Kachouri, R., Kerrest, A., Koszul, R., Lemaire, M., Lesur, I., Ma, L., Muller, H., Nicaud, J. M., Nikolski, M., Oztas, S., Ozier-Kalogeropoulos, O., Pellenz, S., Potier, S., Richard, G. F., Straub, M. L., Suleau, A., Swennen, D., Tekaia, F., Wesolowski-Louvel, M., Westhof, E., Wirth, B., Zeniou-Meyer, M., Zivanovic, I., Bolotin-Fukuhara, M., Thierry, A., Bouchier, C., Caudron, B., Scarpelli, C., Gaillardin, C., Weissenbach, J., Wincker, P. and Souciet, J. L. (2004). Genome evolution in yeasts. *Nature* **430**, 35–44.

Egilmez, N. K., Chen, J. B. and Jazwinski, S. M. (1990). Preparation and partial characterization of old yeast cells. *J. Gerontol.* **45**, B9–B17.

Enyenihi, A. H. and Saunders, W. S. (2003). Large-scale functional genomic analysis of sporulation and meiosis in *Saccharomyces cerevisiae*. *Genetics* **163**, 47–54.

Esposito, R. E., Dresser, M. and Breitenbach, M. (1991). Identifying sporulation genes, visualizing synaptonemal complexes, and large-scale spore and spore wall purification. *Methods Enzymol.* **194**, 110–131.

EUROSCARF (2006). http://web.uni-frankfurt.de/fb15/mikro/euroscarf/.

Felder, T., Bogengruber, E., Tenreiro, S., Ellinger, A., Sa-Correia, I. and Briza, P. (2002). Dtrlp, a multidrug resistance transporter of the major facilitator superfamily, plays an essential role in spore wall maturation in *Saccharomyces cerevisiae*. *Eukaryot. Cell* **1**, 799–810.

Finch, C. (1994). *Longevity, Senescence and the Genome*. The University of Chicago Press, Chicago, IL, USA.

Gari, E., Piedrafita, L., Aldea, M. and Herrero, E. (1997). A set of vectors with a tetracycline-regulatable promoter system for modulated gene expression in *Saccharomyces cerevisiae*. *Yeast* **13**, 837–848.

Giaever, G., Shoemaker, D. D., Jones, T. W., Liang, H., Winzeler, E. A., Astromoff, A. and Davis, R. W. (1999). Genomic profiling of drug sensitivities via induced haploinsufficiency. *Nat. Genet.* **21**, 278–283.

Gietz, D., St Jean, A., Woods, R. A. and Schiestl, R. H. (1992). Improved method for high efficiency transformation of intact yeast cells. *Nucleic Acids Res.* **20**, 1425.

Goffeau, A., Barrell, B. G., Bussey, H., Davis, R. W., Dujon, B., Feldmann, H., Galibert, F., Hoheisel, J. D., Jacq, C., Johnston, M., Louis, E. J., Mewes, H. W., Murakami, Y., Philippsen, P., Tettelin, H. and Oliver, S. G. (1996). Life with 6000 genes. *Science,* **274**, 546, 563–567.

Gossen, M. and Bujard, H. (1992). Tight control of gene expression in mammalian cells by tetracycline-responsive promoters. *Proc. Natl. Acad. Sci. USA* **89**, 5547–5551.

Hartman, J. L. T., Garvik, B. and Hartwell, L. (2001). Principles for the buffering of genetic variation. *Science* **291**, 1001–1004.

Hartwell, L. H. and Weinert, T. A. (1989). Checkpoints: controls that ensure the order of cell cycle events. *Science* **246**, 629–634.

Ho, J. and Bretscher, A. (2001). Ras regulates the polarity of the yeast actin cytoskeleton through the stress response pathway. *Mol. Biol. Cell* **12**, 1541–1555.

Huckelhoven, R. (2004). BAX Inhibitor-1, an ancient cell death suppressor in animals and plants with prokaryotic relatives. *Apoptosis* **9**, 299–307.

Jarolim, S., Millen, J., Heeren, G., Laun, P., Goldfarb, D. S. and Breitenbach, M. (2004). A novel assay for replicative lifespan in *Saccharomyces cerevisiae*. *FEMS Yeast Res.* **5**, 169–177.

Jazwinski, S. M. and Conzelmann, A. (2002). *LAG1* puts the focus on ceramide signaling. *Int. J. Biochem. Cell Biol.* **34**, 1491–1495.

Johnston, M. and Davis, R. W. (1984). Sequences that regulate the divergent GAL1-GAL10 promoter in *Saccharomyces cerevisiae*. *Mol. Cell Biol.* **4**, 1440–1448.

Kataoka, T., Powers, S., McGill, C., Fasano, O., Strathern, J., Broach, J. and Wigler, M. (1984). Genetic analysis of yeast *RAS1* and *RAS2* genes. *Cell* **37**, 437–445.

Kellermayer, R. (2005). Hailey-Hailey disease as an orthodisease of *PMR1* deficiency in *Saccharomyces cerevisiae*. *FEBS Lett.* **579**, 2021–2025.

Koshland, D., Kent, J. C. and Hartwell, L. H. (1985). Genetic analysis of the mitotic transmission of minichromosomes. *Cell* **40**, 393–403.

Kranz, J. E. and Holm, C. (1990). Cloning by function: an alternative approach for identifying yeast homologs of genes from other organisms. *Proc. Natl. Acad. Sci. USA* **87**, 6629–6633.

Laun, P. (1997). Immobilisierung von Hefezellen durch genetische Derivatisierung der Zelloberfläche, Diploma thesis, University of Salzburg, p. 91.

Laun, P., Pichova, A., Madeo, F., Fuchs, J., Ellinger, A., Kohlwein, S., Dawes, I., Frohlich, K. U. and Breitenbach, M. (2001). Aged mother cells of *Saccharomyces cerevisiae* show markers of oxidative stress and apoptosis. *Mol. Microbiol.* **39**, 1166–1173.

Laun, P., Ramachandran, L., Jarolim, S., Herker, E., Liang, P., Wang, J., Weinberger, M., Burhans, D. T., Suter, B., Madeo, F., Burhans, W. C. and Breitenbach, M. (2005). A comparison of the aging and apoptotic transcriptome of *Saccharomyces cerevisiae*. *FEMS Yeast Res.* **5**, 1261–1272.

Liu, H., Krizek, J. and Bretscher, A. (1992). Construction of a *GAL1*-regulated yeast cDNA expression library and its application to the identification of genes whose overexpression causes lethality in yeast. *Genetics* **132**, 665–673.

Madeo, F., Herker, E., Maldener, C., Wissing, S., Lachelt, S., Herlan, M., Fehr, M., Lauber, K., Sigrist, S. J., Wesselborg, S. and Frohlich, K. U. (2002). A caspase-related protease regulates apoptosis in yeast. *Mol. Cell* **9**, 911–917.

Measday, V., Baetz, K., Guzzo, J., Yuen, K., Kwok, T., Sheikh, B., Ding, H., Ueta, R., Hoac, T., Cheng, B., Pot, I., Tong, A., Yamaguchi-Iwai, Y., Boone, C., Hieter, P. and Andrews, B. (2005). Systematic yeast synthetic lethal and synthetic dosage lethal screens identify genes required for chromosome segregation. *Proc. Natl. Acad. Sci. USA* **102**, 13956–13961.

Measday, V. and Hieter, P. (2002). Synthetic dosage lethality. *Methods Enzymol.* **350**, 316–326.

Mendenhall, M. D. and Hodge, A. E. (1998). Regulation of Cdc28 cyclin-dependent protein kinase activity during the cell cycle of the yeast *Saccharomyces cerevisiae. Microbiol. Mol. Biol. Rev.* **62**, 1191–1243.

Mewes, H. W., Albermann, K., Heumann, K., Liebl, S. and Pfeiffer, F. (1997). MIPS: a database for protein sequences, homology data and yeast genome information. *Nucl. Acids Res.* **25**, 28–30.

Mnaimneh, S., Davierwala, A. P., Haynes, J., Moffat, J., Peng, W. T., Zhang, W., Yang, X., Pootoolal, J., Chua, G., Lopez, A., Trochesset, M., Morse, D., Krogan, N. J., Hiley, S. L., Li, Z., Morris, Q., Grigull, J., Mitsakakis, N., Roberts, C. J., Greenblatt, J. F., Boone, C., Kaiser, C. A., Andrews, B. J. and Hughes, T. R. (2004). Exploration of essential gene functions via titratable promoter alleles. *Cell* **118**, 31–44.

Nagahashi, S., Nakayama, H., Hamada, K., Yang, H., Arisawa, M. and Kitada, K. (1997). Regulation by tetracycline of gene expression in *Saccharomyces cerevisiae. Mol. Gen. Genet.* **255**, 372–375.

Oliver, S. G. (1997). From gene to screen with yeast. *Curr. Opin. Genet. Dev.* **7**, 405–409.

Ory, S. and Morrison, D. K. (2004). Signal transduction: implications for Ras-dependent ERK signaling. *Curr. Biol.* **14**, R277–R278.

Pammer, M., Briza, P., Ellinger, A., Schuster, T., Stucka, R., Feldmann, H. and Breitenbach, M. (1992). *DIT101* (*CSD2, CAL1*), a cell cycle-regulated yeast gene required for synthesis of chitin in cell walls and chitosan in spore walls. *Yeast* **8**, 1089–1099.

Park, E. C., Finley, D. and Szostak, J. W. (1992). A strategy for the generation of conditional mutations by protein destabilization. *Proc. Natl. Acad. Sci. USA* **89**, 1249–1252.

Pruyne, D., Gao, L., Bi, E. and Bretscher, A. (2004). Stable and dynamic axes of polarity use distinct formin isoforms in budding yeast. *Mol. Biol. Cell* **15**, 4971–4989.

Richardson, H. E., Wittenberg, C., Cross, F. and Reed, S. I. (1989). An essential G1 function for cyclin-like proteins in yeast. *Cell* **59**, 1127–1133.

Rose, A. B. and Broach, J. R. (1990). Propagation and expression of cloned genes in yeast: 2-microns circle-based vectors. *Methods Enzymol.* **185**, 234–279.

Schatz, P. J., Solomon, F. and Botstein, D. (1988). Isolation and characterization of conditional-lethal mutations in the *TUB1* alpha-tubulin gene of the yeast *Saccharomyces cerevisiae. Genetics* **120**, 681–695.

Schiestl, R. H. and Gietz, R. D. (1989). High efficiency transformation of intact yeast cells using single stranded nucleic acids as a carrier. *Curr. Genet.* **16**, 339–346.

Senger, B., Lafontaine, D. L., Graindorge, J. S., Gadal, O., Camasses, A., Sanni, A., Garnier, J. M., Breitenbach, M., Hurt, E. and Fasiolo, F. (2001). The nucle(ol)ar Tif6p and Efl1p are required for a late cytoplasmic step of ribosome synthesis. *Mol. Cell* **8**, 1363–1373.

Shortle, D., Novick, P. and Botstein, D. (1984). Construction and genetic characterization of temperature-sensitive mutant alleles of the yeast actin gene. *Proc. Natl. Acad. Sci. USA* **81**, 4889–4893.

Simons, A. H., Dafni, N., Dotan, I., Oron, Y. and Canaani, D. (2001). Genetic synthetic lethality screen at the single gene level in cultured human cells. *Nucleic Acids Res.* **29**, E100.

Smeal, T., Claus, J., Kennedy, B., Cole, F. and Guarente, L. (1996). Loss of transcriptional silencing causes sterility in old mother cells of *S. cerevisiae*. *Cell* **84**, 633–642.

Solomon, F. (1991). Analyses of the cytoskeleton in *Saccharomyces cerevisiae*. *Annu. Rev. Cell. Biol.* **7**, 633–662.

Sudbery, P. E., Goodey, A. R. and Carter, B. L. (1980). Genes which control cell proliferation in the yeast *Saccharomyces cerevisiae*. *Nature* **288**, 401–404.

Tatchell, K. (1986). RAS genes and growth control in *Saccharomyces cerevisiae*. *J. Bacteriol.* **166**, 364–367.

Tatchell, K., Chaleff, D. T., DeFeo-Jones, D. and Scolnick, E. M. (1984). Requirement of either of a pair of ras-related genes of *Saccharomyces cerevisiae* for spore viability. *Nature* **309**, 523–527.

Thevelein, J. M. and de Winde, J. H. (1999). Novel sensing mechanisms and targets for the cAMP-protein kinase A pathway in the yeast *Saccharomyces cerevisiae*. *Mol. Microbiol.* **33**, 904–918.

Thuriaux, P., Nurse, P. and Carter, B. (1978). Mutants altered in the control co-ordinating cell division with cell growth in the fission yeast *Schizosaccharomyces pombe*. *Mol. Gen. Genet.* **161**, 215–220.

Toda, T., Cameron, S., Sass, P. and Wigler, M. (1988). *SCH9*, a gene of *Saccharomyces cerevisiae* that encodes a protein distinct from, but functionally and structurally related to, cAMP-dependent protein kinase catalytic subunits. *Genes. Dev.* **2**, 517–527.

Toda, T., Uno, I., Ishikawa, T., Powers, S., Kataoka, T., Broek, D., Cameron, S., Broach, J., Matsumoto, K. and Wigler, M. (1985). In yeast, RAS proteins are controlling elements of adenylate cyclase. *Cell* **40**, 27–36.

Tong, A. H., Evangelista, M., Parsons, A. B., Xu, H., Bader, G. D., Page, N., Robinson, M., Raghibizadeh, S., Hogue, C. W., Bussey, H., Andrews, B., Tyers, M. and Boone, C. (2001). Systematic genetic analysis with ordered arrays of yeast deletion mutants. *Science* **294**, 2364–2368.

Tong, A. H., Lesage, G., Bader, G. D., Ding, H., Xu, H., Xin, X., Young, J., Berriz, G. F., Brost, R. L., Chang, M., Chen, Y., Cheng, X., Chua, G., Friesen, H., Goldberg, D. S. *et al.* (2004). Global mapping of the yeast genetic interaction network. *Science* **303**, 808–813.

Varadarajan, R., Nagarajaram, H. A. and Ramakrishnan, C. (1996). A procedure for the prediction of temperature-sensitive mutants of a globular protein based solely on the amino acid sequence. *Proc. Natl. Acad. Sci. USA* **93**, 13908–13913.

Weinert, T. A. and Hartwell, L. H. (1988). The *RAD9* gene controls the cell cycle response to DNA damage in *Saccharomyces cerevisiae*. *Science* **241**, 317–322.

Wertman, K. F., Drubin, D. G. and Botstein, D. (1992). Systematic mutational analysis of the yeast *ACT1* gene. *Genetics* **132**, 337–350.

Wood, V., Rutherford, K. M., Ivens, A., Rajandream, M. A. and Barrell, B. (2001). A re-annotation of the *Saccharomyces cerevisiae* genome. *Comparative and Functional Genomics* **2**, 143–154.

16 High-Throughput Strain Construction and Systematic Synthetic Lethal Screening in *Saccharomyces cerevisiae*

Amy Hin Yan Tong and Charles Boone
Banting and Best Department of Medical Research and Department of Medical Genetics and Microbiology, Terrence Donnelly Centre for Cellular and Biomolecular Research, University of Toronto, 160 College Street, Toronto, Canada M5S 3E1

♦♦

CONTENTS

Introduction
Identification of synthetic lethal interactions
Notes

♦♦♦♦♦♦ I. INTRODUCTION

Genetic analysis is a powerful way to assess gene function *in vivo*, identifying new components of specific pathways and ordering gene products within a pathway. Synthetic genetic interactions are usually identified when a second-site mutation, or increased gene dosage, suppresses or enhances the original mutant phenotype. This type of genetic screening approach has been used extensively in yeast, worms, flies, mice, and other model organisms. In particular, a genetic interaction termed "synthetic lethality" occurs when the combination of two otherwise viable mutations results in a lethal phenotype (Hartman *et al.*, 2001; Kaelin, 2005). When two genes show a synthetic lethal interaction, it often reflects that the gene products impinge on the same essential function, such that one pathway functionally compensates for, or buffers, the defects in the other. Thus, large-scale mapping of genetic interactions should provide a global view of functional relationships between genes and pathways (Tong *et al.*, 2004).

In budding yeast *Saccharomyces cerevisiae*, a complete set of gene deletion mutants has been constructed for each of the ~6000

METHODS IN MICROBIOLOGY, VOLUME 36
0580-9517 DOI:10.1016/S0580-9517(06)36016-3

predicted genes in the genome, identifying ~1000 essential genes and creating ~5000 viable deletion mutants (Winzeler *et al.*, 1999; Giaever *et al.*, 2002). The fact that over 80% of the predicted genes are not required for life reflects the robustness of biological circuits and may reflect cellular buffering against genetic variation (Hartwell *et al.*, 1999; Hartman *et al.*, 2001; Hartwell, 2004). Hence, the collection of ~5000 viable deletion mutants represents a valuable resource for systematic genetic analysis, providing the potential to examine 12.5 million different double-mutant combinations for a synthetic lethal or sick phenotype. In this chapter, we focus on an array-based synthetic lethal analysis approach, termed synthetic genetic array (SGA) analysis (Tong *et al.*, 2001, 2004), an automated method for constructing double mutants (or higher order allele combinations) and large-scale mapping of functional relationships among specific genes and pathways in yeast.

♦♦♦♦♦♦ II. IDENTIFICATION OF SYNTHETIC LETHAL INTERACTIONS

A. Classical Synthetic Lethal Screens

The availability of a haploid life cycle in yeast makes it particularly suitable for genetic analysis such as screens to identify synthetic lethal interactions. A classical synthetic lethal screen typically involves mutagenizing a strain carrying a mutation in a "query" gene of interest, and screening for mutants whose growth is dependent upon expression of the query gene, using a plasmid loss/colony-sectoring assay (Bender and Pringle, 1991). Subsequent identification of the synthetic lethal mutations requires complementation cloning with a plasmid-based genomic library. Although this approach has been used successfully to dissect genetic relationships among genes involved in cell polarity, secretion, DNA repair, transcription and many other biological processes, relatively few interactions are usually identified in a single screen (Bender and Pringle, 1991; Wang and Bretscher, 1997; Chen and Graham, 1998; Macpherson *et al.*, 2000; Hartman *et al.*, 2001; Mullen *et al.*, 2001). Saturation is rarely achieved because the genetic analysis of the synthetic lethal double mutants and the subsequent cloning of the identified genes is time consuming.

B. Systematic Synthetic Lethal Screens – Synthetic Genetic Array (SGA) Analysis

We developed a method termed SGA analysis, which offers an efficient approach for the systematic construction of double mutants and enables a global analysis of synthetic lethal genetic interactions (Tong *et al.*, 2001). A typical SGA screen involves crossing a query mutation to an ordered array of ~5000 viable gene deletion mutants, and, through a series of replica-pinning steps, meiotic progeny harboring both mutations can be recovered and scored for fitness defects (Figure 1, see Colour Plate section). This procedure

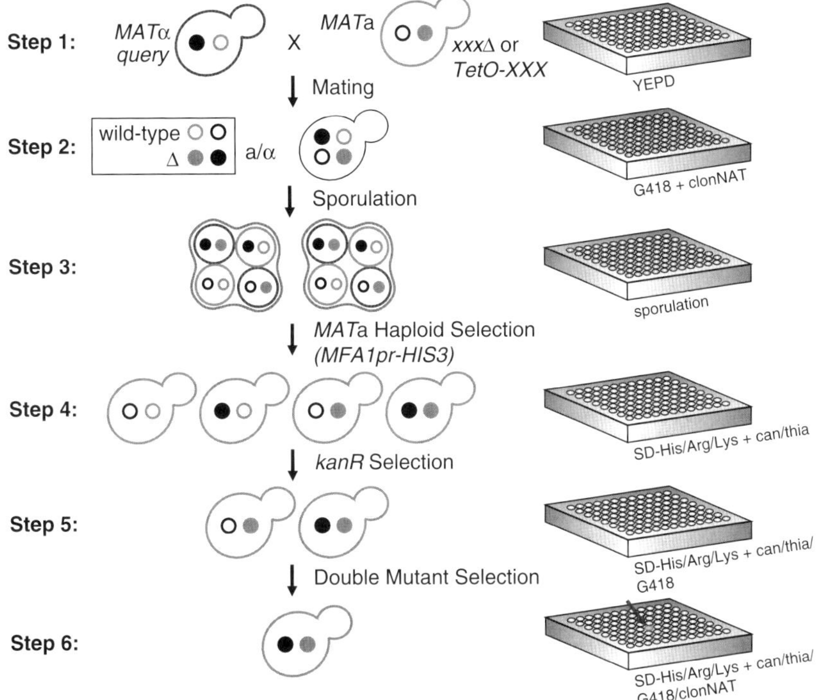

Figure 1. SGA methodology. Step 1, a *MATα* strain carrying a query mutation (e.g. *bni1Δ*) linked to a dominant selectable marker, such as the nourseothricin-resistance marker *natMX* that confers resistance to the antibiotics nourseothricin (clonNAT), and the *MFA1pr-HIS3* reporter, *can1Δ* and *lyp1Δ* reporters is crossed to an ordered array of *MAT***a** viable deletion mutants (*xxxΔ*), each carrying a gene deletion mutation linked to a kanamycin-resistance marker *kanMX* that confers resistance to the antibiotic geneticin (G418). To score genetic interactions amongst essential genes, the query strain can be crossed to an array of conditional yeast mutants. For example, an array in which each mutant carries a different essential gene placed under the control of the conditional Tetracycline-regulated promoter (*TetO-XXX*); however, when screening the conditional array the selection conditions at each step differ from those outlined here as described previously (Mnaimneh *et al.*, 2004; Davierwala *et al.*, 2005). Step 2, growth of resultant zygotes is selected for on medium containing nourseothricin and G418. Step 3, the heterozygous diploids are transferred to medium with reduced levels of carbon and nitrogen to induce sporulation and the formation of haploid meiotic spore progeny. Step 4, spores are transferred to synthetic medium lacking histidine, which allows for selective germination of *MAT***a** meiotic progeny because only these cells express the *MFA1pr-HIS3* reporter, and containing canavanine and thialysine, which allows for selective germination of meiotic progeny that carries the *can1Δ* and *lyp1Δ* markers. Step 5, the *MAT***a** meiotic progeny are then transferred to medium that contains G418, which selects for growth of meiotic progeny that carries the gene deletion mutation (*xxxΔ::kanR*). Finally, the *MAT***a** meiotic progeny are transferred to medium that contains both clonNAT and G418, which then selects for growth of double mutant (*bni1Δ::natR xxxΔ::kanR*). (See color plate section).

can be performed using a colony pinning robot or manually using a hand-held replicator. Here, we outline the genetic logic underlying SGA analysis and describe the most recent version of SGA reagents and methodology. For additional information about the SGA system see Tong and Boone (2005).

1. SGA starting strains and media

(a) MATa-specific SGA reporters

The SGA methodology depends on the germination of *MAT***a** meiotic progeny, specifically, if both *MAT***a** and *MATα* meiotic progeny are germinated then haploid cells can mate with one another and generate diploids that are heterozygous for one or both deletion alleles, thereby leading to false negatives in a synthetic lethal screen. To ensure the germination of a single mating type (Figure 1, Step 4), we linked a haploid mating-type specific promoter to a selectable marker. For example, the *MFA1* promoter (pr) sequence was fused with the *HIS3* open reading frame to create the SGA reporter *MFA1pr-HIS3*, which was then integrated at the *CAN1* locus (*can1Δ::MFA1pr-HIS3*) (Figure 2A). *MAT***a** cells carrying *MFA1pr-HIS3* are able to grow on medium lacking histidine, whereas *MATα* and *MAT***a**/*α* cells carrying *MFA1pr-HIS3* are unable to do so because the expression of *MFA1pr-HIS3* is repressed in these cells.

To investigate which **a**-specific promoter was most productive for SGA analysis, we created six different **a**-specific SGA reporters, derived from the **a**-specific genes listed in Table 1. Each reporter was constructed by fusing a different **a**-specific promoter sequence with the *HIS3* open reading frame, we then examined if appropriate expression of the *HIS3* gene occurs only in *MAT***a** cells but not *MATα* or *MAT***a**/*α* cells, using a selective growth assay on medium lacking histidine (SD-His). We found that all of the reporters showed mating-type specific expression as expected; however, *STE2pr-HIS3* was the most reliable in our experiments for two reasons. First, *MAT***a** cells carrying the reporter were His+ and grew at rates equivalent to that of *HIS3* cells on SD-His. This is in contrast to cells carrying the *ASG7pr-HIS3*, which showed a reduced fitness on SD-His. Second, the *STE2pr-HIS3* appeared to result in the lowest level of inappropriate expression of *HIS3* in *MATα* and *MAT***a**/*α* cells.

Because *can1Δ* is recessive, it can be used as an additional haploid-selectable marker in the SGA procedure (see below) and we therefore often integrate the SGA reporters at the *CAN1* locus (Figure 2B). To facilitate a wide variety of genetic manipulations and improve the SGA selection, we also created a number of SGA reporters in which the **a**-specific promoter was fused to alternative selectable markers. In total, we utilized three selectable markers, the *S. cerevisiae LEU2* and *URA3* genes, as well as the *Schizosaccharomyces pombe his5* gene, which corresponds to the *S. cerevisiae*

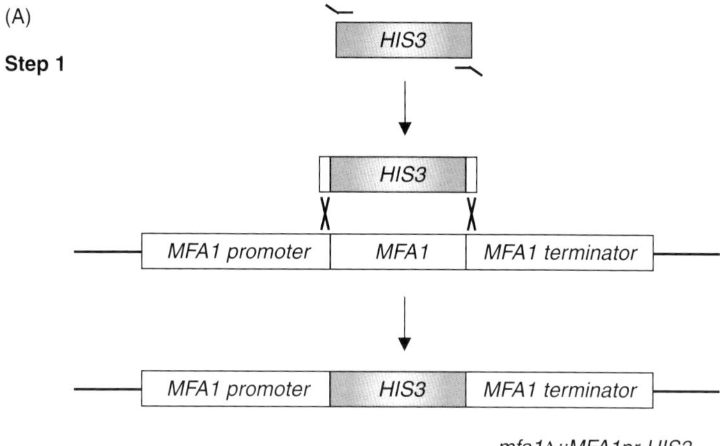

Figure 2. Construction of the SGA reporters. The construction of *can1Δ::MFA1pr HIS3* involves two steps (A) First, the *HIS3* open reading frame (ORF) is integrated at the *MFA1* locus, such that its expression is regulated by the *MFA1* promoter (*MFA1pr*), *mfa1Δ::MFA1pr-HIS3*. (B) Second, *MFA1pr-HIS3* is integrated at the *CAN1* locus, replacing the chromosomal copy of the *CAN1* gene, *can1Δ::MFA1pr-HIS3*.

Table 1. List of **a**-specific genes

Gene	Description
MFA1	**a**-factor mating pheromone precursor
MFA2	**a**-factor mating pheromone precursor
STE2	α-factor receptor
STE6	**a**-factor exporter
BAR1	protease, cleaves and inactivates α-factor
ASG7	**a**-specific gene

HIS3 gene (see SGA reporter genotypes in Table 2). The Burke lab discovered that false negative SGA results may be derived from a gene conversion event in which a *HIS3*-based SGA reporter converts the *his3Δ1* deletion allele carried by the deletion mutant background to *HIS3* within the heterozygous diploids (see Figure 1, Step 2) (Daniel *et al.*, 2005), thereby removing the mating-type specific selection for *MAT***a** meiotic progeny (see Figure 1, Step 4). This gene conversion event is possible because the *his3Δ1* deletion only removes part of the *HIS3* open reading frame (Brachmann *et al.*, 1998). Since *S. pombe his5*$^+$ does not share sequence similarity with *S. cerevisiae HIS3* there is no opportunity for gene conversion to occur. In the case of *LEU2* and *URA3*, the deletion mutant strain background carries a complete deletion of the ORF, corresponding to the *leu2Δ0* and *ura3Δ0* alleles and therefore gene conversion is not an issue.

Table 2. Yeast strains

Strain	Genotype	Source
Y2454	*MATα mfa1Δ::MFA1pr-HIS3 can1Δ ura3Δ0 leu2Δ0 his3Δ1 lys2Δ0*	Tong *et al.* (2001)
Y3068	*MATα can1Δ::MFA1pr-HIS3 ura3Δ0 leu2Δ0 his3Δ1 met15Δ0 lys2Δ0*	Tong *et al.* (2001)
Y3084	*MATα can1Δ::MFA1pr-HIS3 mfα1Δ::MFα1pr-LEU2 ura3Δ0 leu2Δ0 his3Δ1 met15Δ0 lys2Δ0*	Tong *et al.* (2004)
Y3656	*MATα can1Δ::MFA1pr-HIS3-MFα1pr-LEU2 ura3Δ0 leu2Δ0 his3Δ1 met15Δ0 lys2Δ0*	Tong *et al.* (2004)
Y5563	*MATα can1Δ::MFA1pr-HIS3 lyp1Δ ura3Δ0 leu2Δ0 his3Δ1 met15Δ0*	Tong *et al.* (2005)
Y5565	*MATα can1Δ::MFA1pr-HIS3 mfα1Δ::MFα1pr-LEU2 lyp1Δ ura3Δ0 leu2Δ0 his3Δ1 met15Δ0*	Tong *et al.* (2005)
Y6547	*MATα can1Δ::MFA1pr-LEU2 lyp1Δ ura3Δ0 leu2Δ0 his3Δ1 met15Δ0*	Boone Lab
Y7029	*MATα can1Δ::STE2pr-HIS3 lyp1Δ ura3Δ0 leu2Δ0 his3Δ1*	Boone Lab
Y7033	*MATα can1Δ::MFA1pr-his5 lyp1Δ ura3Δ0 leu2Δ0 his3Δ1 met15Δ0*	Boone Lab
Y7039	*MATα can1Δ::STE2pr-LEU2 lyp1Δ ura3Δ0 leu2Δ0 his3Δ1*	Boone Lab
Y7092	*MATα can1Δ::STE2pr-his5 lyp1Δ ura3Δ0 leu2Δ0 his3Δ1 met15Δ0*	Boone Lab
Y8205	*MATα can1Δ::STE2pr-his5 lyp1Δ::STE3pr-LEU2 ura3Δ0 leu2Δ0 his3Δ1*	Boone Lab
Y8835	*MATα can1Δ::STE2pr-his5 lyp1Δ ura3Δ::natR leu2Δ0 his3Δ1 met15Δ0 cyh2*	Boone Lab
Y9230	*MATα can1Δ::STE2pr-URA3 lyp1Δ ura3Δ0 leu2Δ0 his3Δ1 met15Δ0*	Boone Lab

(b) *can1Δ and lyp1Δ markers*

Because mitotic recombination can occur between homologous chromosomes in *MAT***a**/α diploids, a crossover event between the *MAT* locus and the centromere on chromosome III can result in *MAT***a**/**a** or *MAT*α/α diploids. In fact, streaking *MAT***a**/α diploid cells that carry the SGA reporter onto SD-His selects for *MAT***a**/**a** diploids. Because only a fraction (10%) of the heterozygous diploids (see Figure 1, Step 2) sporulate, rare mitotic crossover events within the remaining diploids can contribute to false negative scores, as a *MAT***a**/**a** diploid behaves like a *MAT***a** haploid, expressing *MFA1pr-HIS3*, and carries both deletion alleles. To avoid this complication, we introduced two recessive markers that confer drug resistance, *can1Δ* and *lyp1Δ*, into the query strain. The *CAN1* gene encodes an arginine permease that allows canavanine, a toxic analog for arginine, to enter and kill cells (Kitagawa and Tomiyama, 1929; Sychrova and Chevallier, 1993). Similarly, the *LYP1* gene encodes a lysine permease that allows thialysine, a toxic analog for lysine, to enter and kill cells (Kitagawa, 1929; Sychrova and Chevallier, 1993).

Including *can1Δ* and *lyp1Δ* into the query strain means that *MAT***a**/**a** diploid cells are killed by canavanine and thialysine because they carry a wild-type copy of the *CAN1* and *LYP1* genes. Although it is possible for mitotic recombination to occur in the vicinity of *can1Δ*, *lyp1Δ*, and *MAT* loci, it is unlikely for three independent recombination events (*MAT***a**/**a**, *can1Δ/can1Δ*, and *lyp1Δ/lyp1Δ*) to occur simultaneously within a cell. Hence, by introducing the *can1Δ* and *lyp1Δ* markers, the potential for *MAT***a**/**a** diploids to contribute to false negative SGA scores is reduced substantially.

(c) *SGA starting strains*

All strains are derivatives of BY4741 (*MAT***a** *ura3Δ0 leu2Δ0 his3Δ1 met15Δ0*) or BY4742 (*MAT*α *ura3Δ0 leu2Δ0 his3Δ1 lys2Δ0*) (Brachmann *et al.*, 1998). Among the strains listed in Table 2, six, Y2454, Y3068, Y3084, Y3656, Y5563, and Y5565, were constructed previously and used for SGA analysis (Tong *et al.*, 2001, 2004; Tong and Boone, 2005). Some of these strains, Y3084, Y3656, and Y5565, also carry an *MF*α*1pr-LEU2* reporter, which is activated only in *MAT*α cells, and enables selection of *MAT*α meiotic progeny during SGA analysis. The selection of *MAT*α meiotic progeny is also useful during the construction of *MAT*α SGA query strains by marker replacement of the original deletion mutant alleles, a method that avoids the construction of new alleles and has been outlined in detail previously (Tong and Boone, 2005).

Another seven strains, Y6547, Y7029, Y7033, Y7039, Y7092, Y8205, Y8835, and Y9230 (Table 2), are more recent developments; this set includes strains carrying the **a**-specific SGA reporter based on the *STE2* promoter and a variety of different selectable markers as discussed above. Y7092 (*MAT*α *can1Δ::STE2pr-his5 lyp1Δ ura3Δ0 leu2Δ0 his3Δ1 met15Δ0*) is the starting strain we currently use for the

High-Throughput Strain Construction

construction of SGA query strains. With most of these starting strains, standard protocols for PCR-mediated integration or gene disruption are used to create SGA query strains; however, Y8205 also carries *STE3pr-LEU2* reporter, which is activated only in *MAT*α cells and enables selection of *MAT*α meiotic progeny and the construction of SGA starting strains by marker replacement of the original deletion mutant alleles (see Protocol 1).

Protocol 1. SGA Procedure.

1. Set up cultures for query strain and the deletion mutant array (DMA) as follows:
   (i) Grow the query strain in a 5 ml overnight culture in YEPD.
   (ii) Replicate the 768-density DMA to fresh YEPD+G418. Let cells grow at 30°C for 2 days.
2. Pour the query strain culture over a YEPD plate, use the replicator to transfer liquid culture onto two fresh YEPD plates, generating a source of newly grown query cells for mating to the DMA in the density of 768.[1] Let cells grow at 30°C for 1 day.
3. Mate the query strain with the DMA by first pinning the 768-format query strain onto a fresh YEPD plate, and then pinning the DMA on top of the query cells.[2] Incubate the mating plates at room temperature for 1 day.
4. Pin the resulting *MAT***a**/α zygotes onto YEPD+G418/clonNAT plates. Incubate the diploid-selection plates at 30°C for 2 days.
5. Pin diploid cells to enriched sporulation medium. Incubate the sporulation plates at 22°C for 5 days.[3]
6. Pin spores onto SD – His/Arg/Lys+canavanine/thialysine plates to select for *MAT***a** haploid meiotic progeny. Incubate the haploid-selection plates at 30°C for 2 days.
7. Pin the *MAT***a** meiotic progeny onto SD – His/Arg/Lys+canavanine/thialysine plates for a second round of haploid selection. Incubate the plates at 30°C for 1 day.
8. Pin the *MAT***a** meiotic progeny onto (SD/MSG) – His/Arg/Lys+canavanine/thialysine/G418 plates to select for *MAT***a** meiotic progeny carrying the *kanR* marker. Incubate the *kanR*-selection plates at 30°C for 2 days.
9. Pin the *MAT***a** meiotic progeny onto (SD/MSG) – His/Arg/Lys+canavanine/thialysine/G418/clonNAT plates to select for *MAT***a** meiotic progeny carrying both *kanR* and *natR* markers. Incubate the *kanR*/*natR*-selection plates at 30°C for 2 days.
10. Score double mutants for fitness defects.

(d) Media

Media used in the SGA analysis were described previously (Tong and Boone, 2005). Stock solutions are filtered-sterilized and stored in aliquots at 4°C: canavanine (50 mg/ml, Sigma); thialysine (50 mg/ml, Sigma); clonNAT (100 mg/ml, Werner Bioagents); and G418 (200 mg/ml, Invitrogen Life Technologies), and added to autoclaved medium. Solid medium contains 2% agar.

To minimize contamination on the deletion mutant array (DMA), we propagate it on YEPD+G418 medium. The query strain is mated to the DMA on YEPD. Diploids are selected on YEPD supplemented with 100 mg/l clonNAT and 200 mg/l G418. For efficient sporulation of diploids, the medium is supplemented with an amino-acid powder mixture (20 g/l agar, 10 g/l potassium acetate, 1 g/l yeast extract, 0.5 g/l glucose, 0.1 g/l amino-acids supplement). The amino-acids supplement for sporulation medium contains 2 g histidine, 10 g leucine, 2 g lysine, and 2 g uracil. Because ammonium sulfate impedes the function of G418 and clonNAT, synthetic medium containing these antibiotics are made with monosodium glutamic acid (MSG) as a nitrogen source. For selection of *MAT***a** meiotic progeny carrying *kanR* and, or *natR* markers, (SD/MSG) – His/Arg/Lys+canavanine/thialysine/G418, (SD/MSG) – His/Arg/Lys+canavanine/thialysine/clonNAT, (SD/MSG) – His/Arg/Lys+canavanine/thialysine/G418/clonNAT, the medium lacks histidine (selects for expression of *STE2pr-his5*), arginine, and lysine, and contains 50 mg/l canavanine (selects for *can1Δ*), 50 mg/l thialysine (selects for *lyp1Δ*), and 200 mg/l G418 (selects for *kanR*) and, or 100 mg/l clonNAT (selects for *natR*) [20 g/l agar, 20 g/l glucose, 1.7 g/l yeast nitrogen base w/o ammonium sulfate and amino acids (BD Difco), 1 g/l monosodium glutamic acid (Sigma), 2 g/l amino-acids supplement powder (DO – His/Arg/Lys)]. Tetrad analysis is performed on synthetic dextrose (SD/MSG) complete medium.

2. Yeast cell manipulation

(a) Manual pin tools

An SGA screen can be performed manually using a 96 or 384 floating pin E-clip style manual replicator and registration tools such as a Colony Copier™, or Library Copier™. Hand-held replicator and accessories can be purchased from V & P Scientific, Inc (http://www.vp-scientific.com/floating_e-clip_replicators.htm).

To sterilize the replicator before and between each pinning step, the replicator is first placed in a tray of sterile water for ~1 min, which removes most of the yeast cells from the pins. Next, the replicator is placed in a tray of 10% bleach for 20 s, followed by three sequential rinses in different water baths (5 s/bath). Finally, the replicator is placed in 95% ethanol for 5 s. When excess ethanol drips off the pins, the replicator is flamed and allowed to cool before use.

To ensure the pins are cleaned properly and avoid contamination in the wash procedure, the volume of wash liquids in the cleaning reservoirs is designed to cover the pins sequentially in small increments. For example, in the first step, only the tips of the pins should be submerged in water. As the pins are transferred through the cleaning reservoirs to the final ethanol step, the lower halves of the pins should be covered. To reduce waiting time during the sterilization procedure, it is desirable to have three to four pinning tools such that they can be processed through the sterilization and pinning procedures in rotation.

(b) Robotic pin tools

There are a number of robotic systems available that can be programmed to manipulate yeast cell arrays such as: the BioMatrix (S & P Robtoics Inc., www.sprobotics.com); the VersArray colony arrayer system (BioRad Laboratories, http://www.bio-rad.com); the QBot, QPixXT, MegaPix (Genetix, http://www.genetix.co.uk); and the Singer Rotor HDA bench top robot (Singer Instruments, http://www.singerinst.co.uk).

The Rotor uses disposable plastic replicator pads, whereas most other machines use metal pinning tools, which must be sterilized between each pinning step. Because each robotic system has a different set up for the wash station, the following sterilization procedure is a general outline based on the VersArray colony arrayer system. To clean and sterilize the replicator prior to starting on the robot, the replicator is first placed in the sonicator that is filled with sterile water for 5 min. Next, the sonicator is cleaned and filled with 70% ethanol. The replicator is then placed in the sonicator for 5 min. Finally, the replicator is placed in 95% ethanol for 30 s and allowed to dry over the fan for 30 s.

To sterilize the replicator between each pinning step, the replicator is first placed in a tray of sterile water for 1 min to remove the cells on the pins. Next, the replicator is placed in a second tray of sterile water for 1 min. The replicator is then placed in the sonicator that is filled with 70% ethanol for 2 min. Finally, the replicator is placed in 95% ethanol for 30 s and allowed to dry over the fan for 30 s.

3. Array design

The collection of yeast deletion strains can be purchased from Invitrogen (http://www.resgen.com/products/YEASTD.php3); American Type Culture Collection (http://www.atcc.org/common/special Collections/cydac.cfm); EUROSCARF (http://www.uni-frankfurt.de/fb15/mikro/euroscarf/index.html); and Open Biosystems (http://www.openbiosystems.com/GeneExpression/Yeast/YKO) as stamped 96-well agar plates or frozen stocks in 96-well plates.

The following procedure facilitates the transfer of yeast deletion strains from 96-well frozen stocks to solid agar medium and the building of high-density deletion mutant array (DMA). First, peel off the foil coverings slowly on the frozen 96-well microtiter plates. Second, allow the plates to thaw completely on a flat surface, preferably in a biological safety cabinet. Third, mix the glycerol stocks gently by stirring with a 96-pin hand-held replicator. Fourth, replicate the glycerol stocks from the 96-well plates onto YEPD+G418 agar plates. Take extreme caution that the pins do not drip liquid into neighboring wells. Finally, reseal the 96-well plates with fresh aluminum sealing tape, and return to −80°C. Allow cells to grow at room temperature for ~2 days.

Because fitness is monitored as the output readout in SGA analysis, factors affecting the growth rate of yeast colonies can influence the system sensitivity. Yeast colonies grow faster and become larger in size when they have access to more nutrients in the medium. Hence, colonies surrounding an empty spot or those positioned along the edges of a high-density array, tend to be larger than the ones positioned in a dense area away from the edges (Figure 3A, see Colour Plate section). To minimize the positional effects and ensure a uniform growth rate in a high-density array, four important points need to be considered. First, slow-growing strains can be examined in a less biased manner by removing them from the regular array and creating a special one containing mutants with a slow growth rate. Second, a border can be added around the edges of the plate, i.e. the outermost layer of colonies on four edges of the plate, using a neutral strain carrying all the markers required in the experimental procedure. For example, the *MAT***a** *his3Δ::kanR* deletion strain for SGA analysis. Third, gaps or empty spots can be filled in or removed to make the array more robust for examining subtle differences in fitness amongst the deletion mutants. Fourth, each plate may contain a number of auxotrophic mutants which can be used for plate identification by providing a unique growth pattern or "signature" on medium lacking a specific nutrient (Figure 3B, see Colour Plate section).

A 384-density DMA can be assembled by spotting the strains manually or automatically using a colony arrayer. The collection of 384-density DMA plates can then be maintained as the master plate set for SGA analysis and also as frozen stock at −80°C. The agar plates can be kept at 4°C and propagated as needed, or revived from the frozen stock once every month. The 384-density array is also used as a source to generate working copies of the DMA in density formats such as 768 or 1536.

4. Scoring of putative interactions in an SGA screen

To evaluate the colony sizes of double-mutants generated from a query screen, we compare them to a reference set of wild-type control screens. The control set is generated by crossing *MATα ura3Δ::natR can1Δ::STE2pr-his5 lyp1Δ* to the DMA to create an

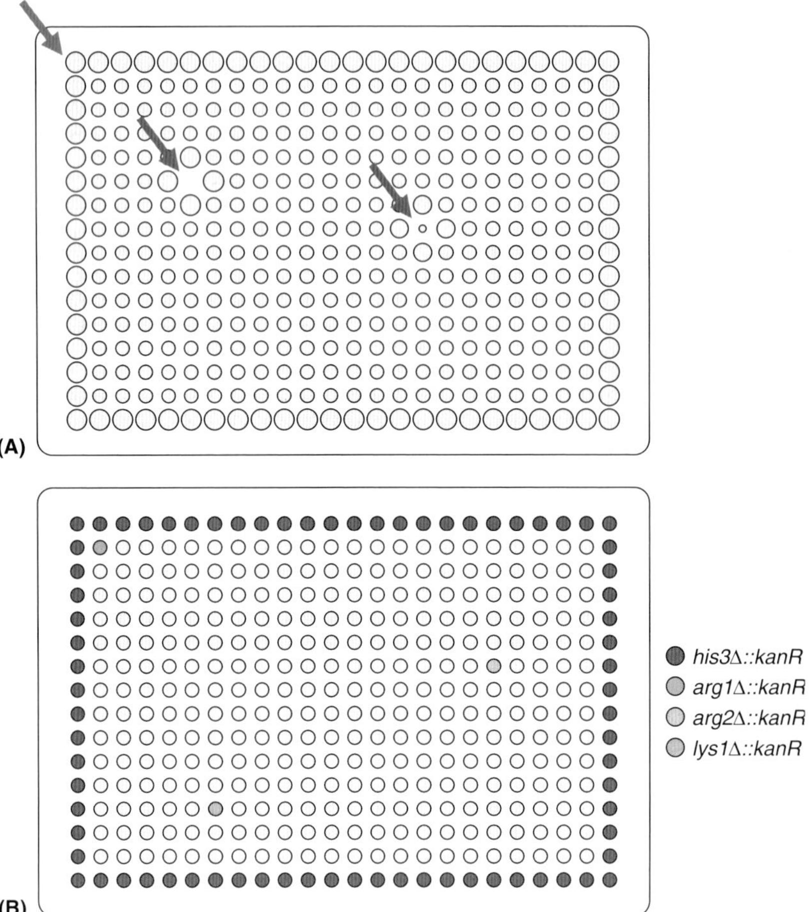

Figure 3. Array Design. Each spot represents a yeast colony growing in a 384-density array. (A) Yeast colonies surrounding an empty spot or a slow-growing strain (red arrows), and those positioned along the edges of the array (blue arrow), have access to more nutrients in the medium and therefore, tend to be larger than the ones positioned in a dense area away from the edges. (B) An ideal array layout for SGA analysis should facilitate accurate output readout and include the following: (i) removal of slow-growing strains from the regular array to a special array containing only mutants with a slow growth rate; (ii) a border around the edges of the plate, i.e. the outermost layer of colonies on four edges of the plate, using a neutral strain carrying all the markers required in the experimental procedure, for example the *MAT***a** *his3Δ::kanR* deletion strain (red colonies); (iii) filled in gaps or empty spots to make the array more robust for examining subtle differences in fitness amongst the deletion mutants; (iv) a number of auxotrophic mutants to be used as a unique plate identification system, for example, the *MAT***a** *ura4Δ::kanR* deletion strain (green colony), the *MAT***a** *trp1Δ::kanR* deletion strain (blue colony), and the *MAT***a** *lys1Δ::kanR* deletion strain (purple colony), are unable to grow on medium lacking uracil, tryptophan, and lysine, respectively. (See color plate section).

output array carrying the SGA markers in every single-deletion mutant background. The double-deletion mutant array can be examined visually and compared to that of the wild-type control array. A synthetic lethal/sick interaction is scored when the colony size

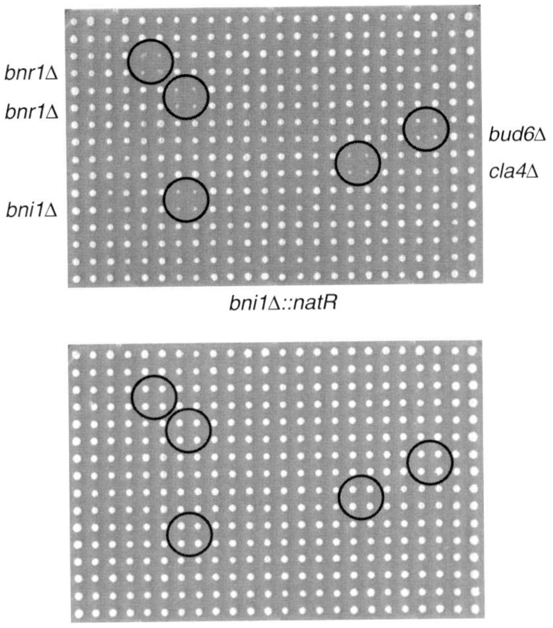

Figure 4. Examples of scoring synthetic lethal/sick interactions in an SGA screen. A *bni1Δ::natR* query strain is crossed to a test array containing 96 deletion mutants, each arrayed in quadruplicate in a square pattern. (note: SGA screens can be carried out at a density of 96, 384, 768 or 1536) *bnr1Δ* is duplicated within the array. The final array that selects for growth of the *bni1Δ* double mutants is shown at the top of the figure. Synthetic lethal/sick interactions lead to the formation of residual colonies (circled) that are smaller than the equivalent colony on the wild-type control plate. Synthetic lethal/sick interactions are scored with *bnr1Δ*, *cla4Δ*, and *bud6Δ*. When the query mutation is identical to one of the gene deletions within the array, double mutants cannot form because haploids carry a single copy of each allele; therefore, *bni1Δ* appeared synthetic lethal with itself.

on the double-deletion mutant array is smaller than that on the wild-type control array (Figure 4). The query mutant is screened two more times, for a total of three independent screens. Screens can be carried out in 96, 384, 768 and 1536 density format with between 2 and 4 replica copies of each deletion mutant on the array.

In general, potential positive hits from three rounds of screening are combined and used to generate an unbiased set of putative interactions, which includes all those that appear two or three times in the three rounds of screening. A biased set of putative interactions is generated by sorting the one-time hits according to the functional annotations such as Gene Ontology (GO) molecular function and biological process, and selecting those that are related functionally to multiple genes within the unbiased set. The programs FunSpec (http://funspec.med.utoronto.ca) and FuncAssociate (http://llama.med.harvard.edu/cgi/func/funcassociate) are used to assign functional annotations in order to assist the sorting of putative interactions. FunSpec takes a list of genes as input and produces a summary of functional annotations from the MIPS and GO databases that are enriched in the list. FuncAssociate takes a list of genes as input and produces a ranked list of the GO annotations as enriched or depleted within the list. Both sets of putative interactions are then combined to create a list of candidates for confirmation.

In addition to visual inspection of the double mutants, we have developed a computer-based scoring system, which generates an estimate of relative growth rates from the area of individual colonies, as measured from digital images of the double-mutant plates (Tong *et al.*, 2004). Following normalization of the images derived from control and double mutant plates, statistical significance can be determined for each strain by comparing the measurements between the mutants and wild-type controls.

5. Confirmation of the putative interactions generated from SGA analysis

To confirm the results obtained from SGA analysis, spores saved from the sporulation step in the SGA procedure (Figure 1, Step 3) can be used. Alternatively, heterozygous diploids of the query mutation and test mutation can also be generated independently by mating the *MAT*α query strain to the *MAT***a** deletion strain of interest (*xxx*Δ*::kanR*). The resulting diploids can then be induced for sporulation and used in random spore analysis (RSA) and tetrad analysis.

(a) Random spore analysis (RSA)

The following procedure facilitates RSA. First, inoculate a small amount of spores (approximately the size of a pinprick) in 1 ml of sterile water, and mix well. Second, plate 20 μl of suspended spores on SD – His/Arg/Lys+canavanine/thialysine medium, 40 μl of suspended spores on (SD/MSG) – His/Arg/Lys+canavanine/thialysine/G418, and (SD/MSG) – His/Arg/Lys+canavanine/thialysine/clonNAT, respectively, and 80 μl of suspended spores on (SD/MSG) – His/Arg/Lys+canavanine/thialysine/G418/clonNAT. Third, incubate the plates at 30°C for ~1.5–2 days. Finally, score the double-drug selection against the single-drug selections (Figure 5).

The expected number of *MAT***a** meiotic progeny on each medium should be roughly equal. SD – His/Arg/Lys+canavanine/thialysine allows germination of the *MAT***a** meiotic progeny that carries the *can1*Δ*::STE2pr-his5* and *lyp1*Δ markers. (SD/MSG) – His/Arg/Lys+canavanine/thialysine/G418 allows the germination of the *MAT***a** meiotic progeny that carries the *can1*Δ*::STE2pr-his5* and *lyp1*Δ markers, and the *kanR*-marked gene deletion. (SD/MSG) – His/Arg/Lys+canavanine/thialysine/clonNAT allows the germination of the *MAT***a** meiotic progeny that carries the *can1*Δ*::STE2pr-his5* and *lyp1*Δ markers, and the *natR*-marked query deletion. (SD/MSG) – His/Arg/Lys+canavanine/thialysine/G418/clonNAT allows the germination of the *MAT***a** meiotic progeny that carries the *can1*Δ*::STE2pr-his5* and

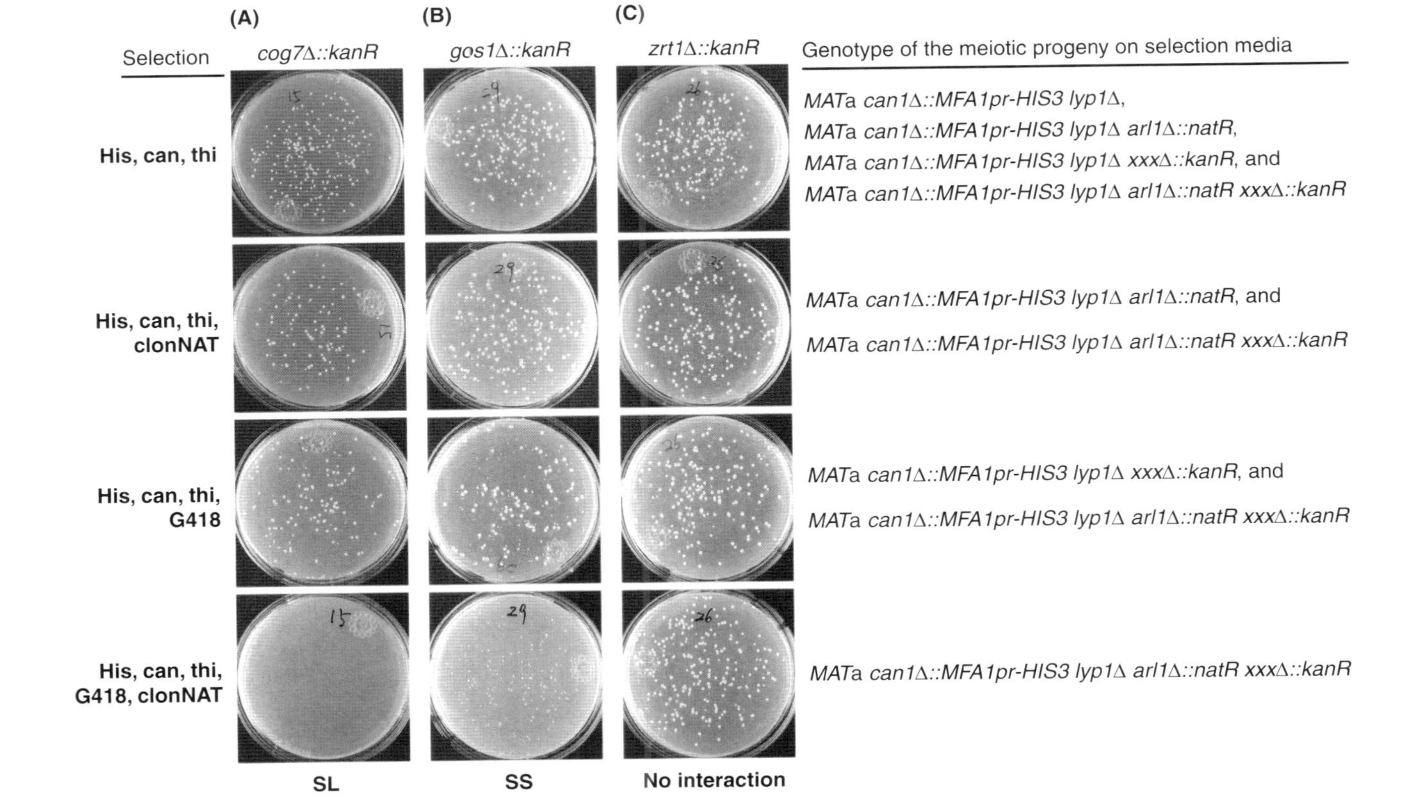

Figure 5. Examples of the random spore analysis (RSA): *MAT***a** meiotic progeny derived from sporulation of heterozygous diploids; *MAT***a**/α *arl1Δ::natR*/– *cog7Δ::kanR*/+ (A), *MAT***a**/α *arl1Δ::natR*/+ *gos1Δ::kanR*/+ (B), and *MAT***a**/α *arl1Δ::natR*/+ *zrt1Δ::kanR*/+ (C), were plated onto media [SD – His/Arg/Lys+canavanine/thialysine], [(SD/MSG) – His/Arg/Lys+canavanine/thialysine/G418], [(SD/MSG) – His/Arg/Lys+canavanine/thialysine/clonNAT], and [(SD/MSG) – His/Arg/Lys+canavanine/thialysine/G418/clonNAT] as indicated. The plates were incubated at 30°C for ~2 days. Cell growth under the four conditions was compared and scored. The *MAT***a** *arl1Δ::natR cog7Δ::kanR* double mutant (A) was scored as having a synthetic lethal (SL) interaction. The *MAT***a** *arl1Δ::natR gos1Δ::kanR* double mutant (B) was scored as having a synthetic sick (SS) interaction. The *MAT***a** *arl1Δ::natR zrt1Δ::kanR* double mutant (C) was scored as having no interaction.

lyp1Δ markers, and the double mutations of the *natR*-marked query and *kanR*-marked gene deletion.

(b) Tetrad analysis

Standard procedure is followed to dissect tetrads except for the medium on which the spores are germinated. Because we cannot add the antibiotics (G418 and clonNAT) into the medium for tetrad analysis, the closest conditions to the double mutant selection step is synthetic dextrose (SD/MSG) complete medium. This medium resembles the final double mutant selection conditions (Figure 1, Step 5), only lacking G418 and clonNAT, and thus is more sensitive than the conventional rich medium in detecting subtle growth defects associated with the double mutant.

6. Applications of the SGA methodology

To examine synthetic genetic interactions with the essential genes, an SGA query strain can be crossed to an array of yeast mutants in which each essential gene has been placed under the control of the conditional Tetracycline-regulated promoter, the Tim Hughes Collection (yTHC; Open Biosystems; Figure 1), double mutants can be selected and scored for growth defects in the presence of doxycycline, which down-regulates the expression of the essential genes (Mnaimneh *et al.*, 2004; Davierwala *et al.*, 2005).

Because double mutants are created by meiotic recombination and since the viable gene deletion alleles represent mapping markers covering all chromosomes in the yeast genome, SGA screens also enable a genome-wide set of two-factor crosses that allow for high-resolution mapping of selectable traits, such as drug-resistant phenotypes or suppressors of temperature-sensitive mutations. In a proof-of-principle study, SGA mapping (SGAM) as applied to identify *ssd1Δ* as a suppressor of the lethality associated with deletion alleles of the RAM pathway (Jorgensen *et al.*, 2002) and further application of SGAM identified *sgs1* mutations as suppressors of the slow growth defect associated with a *rmi1Δ* allele (Chang *et al.*, 2005).

The SGA methodology is versatile because any genetic element (or any number of genetic elements) marked by a selectable marker(s) can be manipulated similarly. This array-based approach automates yeast genetics and can be easily adapted for a number of different screens, including higher order genetic interaction analysis (triple mutant genetic interactions: Tong *et al.*, 2004), dosage lethality (Measday *et al.*, 2005), suppression using high copy plasmid (dosage suppression), or plasmid shuffling. Mutant arrays generated by SGA can also be phenotypically assessed, for example, morphological analysis of genetic arrays using a high-throughput automated imaging system (Saito *et al.*, 2004, 2005) will allow a detailed phenotypic assessment of double mutants. In addition, strain arrays generated by SGA can be used in secondary assays, for example, the *SCB::HIS3* reporter construct (Costanzo *et al.*, 2004) was used to determine transcriptional responses in the ~5000 deletion mutant backgrounds. A yeast overexpression array, in which a wild-type strain was transformed with ~6000 different plasmids, each of which enables the conditional overexpression of a specific gene from the *GAL1* promoter (Zhu *et al.*, 2001), has been assembled and can be used to screen for synthetic dosage lethality and suppression with SGA methodology (Sopko et al., 2006). Other collections of yeast strains such as the green fluorescence protein (GFP) and tag-affinity protein (TAP) fusion libraries can also be integrated with the SGA methodology, allowing systematic examination of protein localization or the assembly of protein complexes in any genetic background.

♦♦♦♦♦♦ NOTES

1. Pinning the query strain and wild-type strain in the 768-format on agar plates is advantageous as cells are evenly transferred to the subsequent mating step.
2. One query plate should contain a sufficient amount of cells for mating with 6–8 plates of the DMA. The DMA can be reused for three to four rounds of mating reactions.
3. It is important to keep the sporulation plates at ~22–24°C for efficient sporulation. The resultant sporulation plates can be stored at 4°C for up to 4 months without significant loss of spore viability, and provide a source of spores for random spore analysis (RSA) and tetrad analysis.

References

Bender, A. and Pringle, J. R. (1991). Use of a screen for synthetic lethal and multicopy suppressee mutants to identify two new genes involved in morphogenesis in *Saccharomyces cerevisiae*. *Mol. Cell Biol.* **11**, 1295–1305.

Brachmann, C. B., Davies, A., Cost, G. J., Caputo, E., Li, J., Hieter, P. and Boeke, J. D. (1998). Designer deletion strains derived from *Saccharomyces cerevisiae* S288C: A useful set of strains and plasmids for PCR-mediated gene disruption and other applications. *Yeast* **14**, 115–132.

Chang, M., Bellaoui, M., Zhang, C., Desai, R., Morozov, P. *et al.* (2005). RMI1/NCE4, a suppressor of genome instability, encodes a member of the RecQ helicase/Topo III complex. *EMBO J.* **24**, 2024–2033.

Chen, C. Y. and Graham, T. R. (1998). An *arf1Δ* synthetic lethal screen identifies a new clathrin heavy chain conditional allele that perturbs vacuolar protein transport in *Saccharomyces cerevisiae*. *Genetics* **150**, 577–589.

Costanzo, M., Nishikawa, J. L., Tang, X., Millman, J. S., Schub, O. *et al.* (2004). CDK activity antagonizes Whi5, an inhibitor of G1/S transcription in yeast. *Cell* **117**, 899–913.

Daniel, J. A., Yoo, J. Y., Bettinger, B. T., Amberg, D. C. and Burke, D. J. (2005). Eliminating gene conversion improves high-throughput genetics in *Saccharomyces cerevisiae*. *Genetics* **172**, 709–711.

Davierwala, A. P., Haynes, J., Li, Z., Brost, R. L., Robinson, M. D. *et al.* (2005). The synthetic genetic interaction spectrum of essential genes. *Nat. Genet.* **37**, 1147–1152.

Giaever, G., Chu, A. M., Ni, L., Connelly, C., Riles, L. *et al.* (2002). Functional profiling of the *Saccharomyces cerevisiae* genome. *Nature* **418**, 387–391.

Hartman, J. L., Garvik, B. and Hartwell, L. (2001). Principles for the buffering of genetic variation. *Science* **291**, 1001–1004.

Hartwell, L. (2004). Genetics. Robust interactions. *Science* **303**, 774–775.

Hartwell, L. H., Hopfield, J. J., Leibler, S. and Murray, A. W. (1999). From molecular to modular cell biology. *Nature* **402**, C47–C52.

Jorgensen, P., Nelson, B., Robinson, M. D., Chen, Y., Andrews, B., Tyers, M. and Boone, C. (2002). High-resolution genetic mapping with ordered arrays of *Saccharomyces cerevisiae* deletion mutants. *Genetics* **162**, 1091–1099.

Kaelin, W. G., Jr. (2005). The concept of synthetic lethality in the context of anticancer therapy. *Nat. Rev. Cancer* **5**, 689–698.

Kitagawa, M. and Tomiyama, T. (1929). A new amino-compound in the jack bean and a corresponding new ferment. *J. Biochem.* **11**, 265–271.

Macpherson, N., Measday, V., Moore, L. and Andrews, B. (2000). A yeast *taf17* mutant requires the Swi6 transcriptional activator for viability and shows defects in cell cycle-regulated transcription. *Genetics* **154**, 1561–1576.

Measday, V., Baetz, K., Guzzo, J., Yuen, K., Kwok, T. *et al.* (2005). Systematic yeast synthetic lethal and synthetic dosage lethal screens identify genes required for chromosome segregation. *Proc. Natl. Acad. Sci. USA* **102**, 13956–13961.

Mnaimneh, S., Davierwala, A. P., Haynes, J., Moffat, J., Peng, W. T. *et al.* (2004). Exploration of essential gene functions via titratable promoter alleles. *Cell* **118**, 31–44.

Mullen, J. R., Kaliraman, V., Ibrahim, S. S. and Brill, S. J. (2001). Requirement for three novel protein complexes in the absence of the Sgs1 DNA helicase in *Saccharomyces cerevisiae*. *Genetics* **157**, 103–118.

Saito, T. L., Ohtani, M., Sawai, H., Sano, F., Saka, A. *et al.* (2004). SCMD: *Saccharomyces cerevisiae* morphological database. *Nucl. Acids Res.* **32**(Database issue), D319–D322.

Saito, T. L., Sese, J., Nakatani, Y., Sano, F., Yukawa, M., Ohya, Y. and Morishita, S. (2005). Data mining tools for the *Saccharomyces cerevisiae* morphological database. *Nucl. Acids Res.* **33**, W753–W757.

Sopko, R., Huang, D., Preston, N., Chua, G., Papp, B., Kafadar, K., Snyder, M., Oliver, S. G., Cyert, M., Hughes, T. R., Boone, C. and Andrews, B. (2006). Mapping pathways and phenotypes by systematic gene overexpression. *Mol. Cell.* **21**, 319–330.

Sychrova, H. and Chevallier, M. R. (1993). Cloning and sequencing of the *Saccharomyces cerevisiae* gene *LYP1* coding for a lysine-specific permease. *Yeast* **9**, 771–782.

Tong, A. H. and Boone, C. (2005). Synthetic genetic array analysis in *Saccharomyces cerevisiae*. *Methods Mol. Biol.* **313**, 171–192.

Tong, A. H., Evangelista, M., Parsons, A. B., Xu, H., Bader, G. D. *et al.* (2001). Systematic genetic analysis with ordered arrays of yeast deletion mutants. *Science* **294**, 2364–2368.

Tong, A. H., Lesage, G., Bader, G. D., Ding, H., Xu, H. *et al.* (2004). Global mapping of the yeast genetic interaction network. *Science* **303**, 808–813.

Wang, T. and Bretscher, A. (1997). Mutations synthetically lethal with *tpm1Δ* lie in genes involved in morphogenesis. *Genetics* **147**, 1595–1607.

Winzeler, E. A., Shoemaker, D. D., Astromoff, A., Liang, H., Anderson, K. *et al.* (1999). Functional characterization of the *S. cerevisiae* genome by gene deletion and parallel analysis. *Science* **285**, 901–906.

Zhu, H., Bilgin, M., Bangham, R., Hall, D., Casamayor, A. *et al.* (2001). Global analysis of protein activities using proteome chips. *Science* **293**, 2101–2105.

17 Chemical Genomic Tools for Understanding Gene Function and Drug Action

Corey Nislow[*] and Guri Giaever[1]

[*] *Banting and Best Department of Medical Research;* [1] *Department of Pharmaceutical Sciences, Donnelley CCBR, University of Toronto, Ontario M5S3E1, Canada*

◆◆◆

CONTENTS

Introduction
The HIP assay: Background and method
The HOP assay: Background and method
The advantages of combining the HIP and HOP assays
Comparison to other technologies
Perspectives and future directions

Abbreviations

YKO	Yeast KnockOut Collection
HIP	HaploInsufficiency Profiling
HOP	Homozygous Profiling
SGA	Synthetic Genetic Array
NER	Nucleotide Excision Repair
PRR	Post Replication Repair
TLS	Translesion Synthesis

◆◆◆◆◆◆ I. INTRODUCTION

As our understanding of biology and physiology increases and the demand for safe effective medicines continues to grow, the means to accomplish these discoveries must also expand. At present there are two broad categories of discovery methods. The traditional approach has been to use a cellular assay where the end result is a discernable phenotype, e.g. cell death. Many widely used therapies have been discovered in this manner. A drawback of the cellular approach is that it cannot identify the cellular target of a therapeutic. This confounds efforts to understand the mechanism

METHODS IN MICROBIOLOGY, VOLUME 36
0580-9517 DOI:10.1016/S0580-9517(06)36017-5

of action of drugs. Such information is essential for characterizing drug efficacy and side-effects prior to designing improvements of the drug of interest. The second discovery approach is to use target-based assays. These assays have been widely adopted in the last several decades. Here, a purified target is assayed for a predefined biochemical activity *in vitro*. Target-based approaches have the benefit of identifying compounds with the selected activity, but suffer from the fact that the target of the identified molecules may act very differently once in the context of all proteins inside a cell. In this short chapter we describe an assay that combines the desired attributes of both types of assays and employs the model eukaryote *Saccharomyces cerevisiae*. Applications of this assay, "HaploInsufficiency Profiling" or HIP (Baetz *et al.*, 2004a; Giaever *et al.*, 1999, 2004; Lum *et al.*, 2004) are presented as well as complementary variants on this method.

Like most cell-based assays, the HIP assay reports whether or not a compound inhibits cell growth. Unlike traditional cell-based-assays, however, HIP identifies the cellular target directly, without the need for time consuming biochemical follow up. The HIP assay therefore identifies the genes most important for growth; these in turn define potential drug targets. Finally, the HIP assay is relevant to basic biological research, as any molecule identified that blocks the activity of a specific protein target can serve as a rapidly reversible, conditional "mutant" (or tool) to study essential gene function for the research biologist (Specht and Shokat, 2002).

♦♦♦♦♦♦ II. THE HIP ASSAY: BACKGROUND AND METHOD

A. The Yeast Deletion Collection and Its Application to Functional Genomics

The 20 000 + Yeast KnockOut (YKO) strains constructed by the International Yeast Deletion Consortium have been described in over 30 papers and several excellent reviews (Brenner, 2004; Giaever, 2003; Grunenfelder and Winzeler, 2002; Pan *et al.*, 2004; Scherens and Goffeau, 2004). For the purposes of the assays described in this chapter, the most noteworthy feature of these strains is the presence of molecular barcodes, or tags, that serve as unique strain identifiers. These 20 bp sequence tags are linked to each gene deletion and therefore allow the strains to be analyzed in parallel (Giaever *et al.*, 2002; Winzeler *et al.*, 1999). Each strain carries two tags, UPTAG and DNTAG. The presence of two tags improves the robustness of the data. In each experiment, a mixed culture containing every deletion mutant is grown, samples are collected at intervals during growth, and the tags PCR amplified from genomic DNA with labeled primers flanking the barcodes. These primers are universal to each deletion strain and allow amplification of the tags from every deletion strain in just two PCR reactions: one for the UPTAGS and one for the DNTAGS (Figure 1; see Colour Plate section). The abundance of each deletion strain is then determined by quantifying the associated molecular bar codes by denaturization and hybridization to an oligonucleotide array carrying the complementary barcode sequences. The intensity of a particular barcode on the array is related to the abundance of the associated deletion strain. The more important a gene is for growth in a particular condition, the slower the growth of the corresponding deletion strain, and therefore the less able it is to compete with the growth of the other strains in the pool, and the more rapidly the molecular barcodes of the strain diminish from the culture. Thus, all genes required for growth can be identified and ranked in order of their relative importance to strain fitness in a single experiment in a particular condition (Birrell *et al.*, 2001; Winzeler *et al.*, 1999). In this way, gene function can be inferred. In the following sections we present results from such fitness profiling of nearly all yeast genes (96%) under diverse experimental conditions.

B. Haploinsufficiency Profiling (HIP)

We have developed a chemical genomics assay that screens all potential targets in parallel, in a robust, high-throughput manner. This assay, HIP, is based on the observation of increased drug sensitivity (as measured by a reduced growth rate) of a heterozygous deletion strain in the presence of a compound that specifically inhibits the gene product of the heterozygous locus (Giaever *et al.*, 1999). This phenotype is known as drug-induced haploinsufficiency (Figure 2). The HIP assay exploits this phenomenon on a genome-wide level to identify drug targets *de novo* in the following way. All molecularly barcoded deletion strains are mixed together and grown in a single culture (Protocol 1). The compound of interest is added at the appropriate concentration (Figure 3 and Protocol 2), and the strains allowed to grow competitively as described in Figure 1. The "appropriate dose" used in these experiments must be determined empirically. In general, a dose that inhibits the wildtype by 10% is a good starting concentration. The samples are processed, and relative strain abundance analyzed via microarray as described in Figure 1 and Protocol 3. The strains that grow most slowly are identified as candidates for the drug target. While most compounds with well-characterized targets interact with one or a few gene products across the genome, other unexpected effects can reveal insights into compound mechanism and potentially uncover off-target/cytotoxic effects. The results of a HIP assay thus provides a comprehensive *in vivo* snapshot of the genome-wide cellular response to small-molecule perturbants (Giaever *et al.*, 2004; Lum *et al.*, 2004). In the following sections we present sample results from such fitness profiling under diverse experimental conditions.

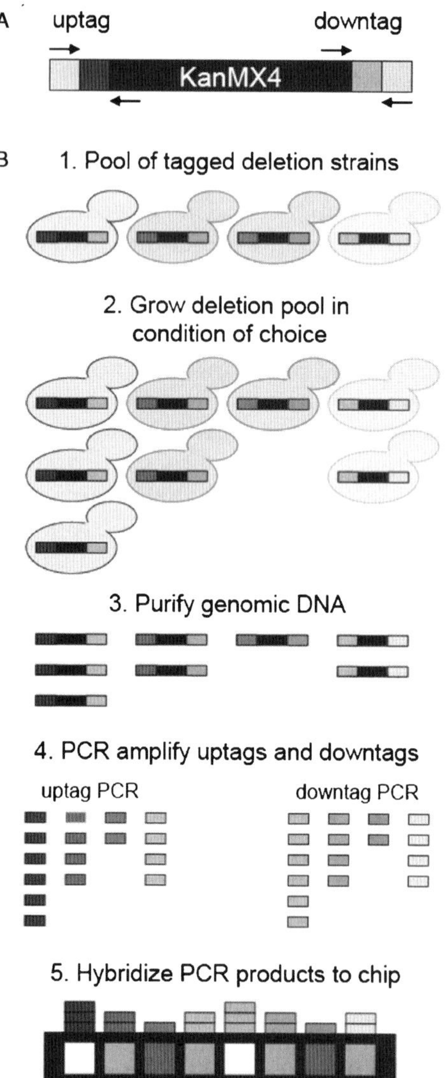

Figure 1. Description of the competitive growth assay. (A) The deletion cassette module is shown. The uptag and dntags (shown in darker blue hues) are flanked by universal primers common to every strain in the collection (shown by arrows). The KanMX4 marker is required for selection of transformants with G418, and 45bp of flanking genomic homology (light blue) to ensure proper insertion by homologous recombination. (B) Cartoon outlining the experimental protocol. (1) Strains are first pooled at approximately equal abundance. (2) The pool is grown competitively under a selection of choice; strains deleted for genes required for growth under a particular condition will grow more slowly and become under-represented in the pool. (3) The pool is sampled over time and genomic DNA isolated. (4) Genomic DNA is PCR amplified using the universal primers in two PCR reactions, one for the uptag and one for the dntag. (5) PCR products are then hybridized to the array carrying the tag complements. Array intensity is related to the amount of each strain present. Those strains that are under-represented are likely deleted for genes important for survival under the condition selected. Modified from Scherens and Goffeau (2004). (See color plate section).

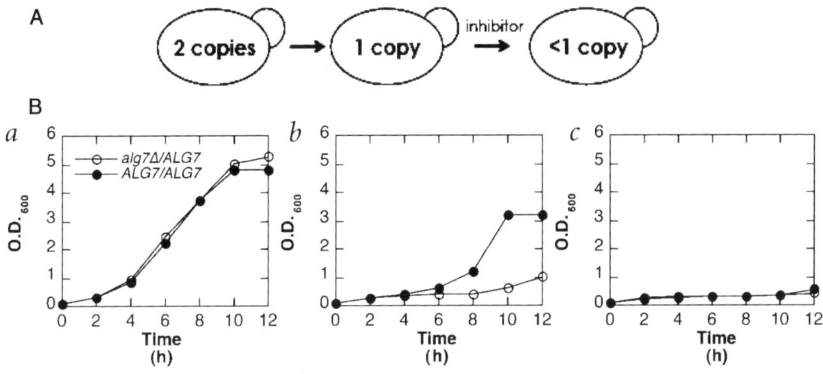

Figure 2. Description of the HIP assay. (A) A schematic outlining the general concept of the HIP assay. The first strain on the left represents a wildtype diploid strain containing two copies of all genes. The strain in the middle represents a heterozygous strain; that is a strain that carries a single gene that has been reduced in copy number from 2 copies to 1 copy. The strain on the right shows the effect of growing a particular heterozygous strain in the presence of a compound that targets the gene product of the heterozygous locus. The concept behind the method is that in the presence of compound the effective gene dosage is further reduced from 1 copy in the heterozygote to less than 1 functional copy of the gene product. If this gene product is important for growth, the result is drug sensitivity as measured by an increased doubling time. This phenotype is called drug-induced haploinsufficiency. (B) An example of drug-induced haploinsufficiency. In the left panel, no drug is present and the heterozygous strain (*ALG7*, the known target of the compound tunicamycin) grows equally well compared to a wildtype strain. In the right panel at high tunicamycin concentrations (2.0 μg/ml), each strain grows equally poorly. It is only in the center panel where drug-induced haploinsufficiency is observed at a tunicamycin concentration of 0.5 μg/ml and the strain heterozygous for the drug target (*ALG7*) has a fitness defect when compared to the wildtype. Taken from Giaever *et al.* (1999).

C. Data Analysis

As for any microarray experiment, there are multiple ways of analyzing the data. In our hands, we find the most important feature of our experimental analysis is that the experimental array be compared with a set of no drug or treatment control arrays. In the set of control arrays, any tag that falls beneath ~3-fold the background array intensity on the control arrays is eliminated from the analysis. The remaining tag intensities are then log transformed, mean normalized and the intensities of the two tags (sense and antisense) associated with each strain are averaged when possible (N.B. not all barcode microarrays carry both strands of the barcode). The mean and standard deviation is calculated for each tag across the set of control arrays. A *z*-score is calculated by taking the difference between the experimental intensity and the control intensity of a particular tag and dividing by the standard deviation for that tag. The *z*-scores for the uptag and dntag are then averaged to define a "fitness defect score". A significance cutoff is determined by calculating the score cutoff required for a 0.05 false discovery

Protocol 1. Pool construction.

1. Obtain the YKO collections as frozen glycerol stocks in 96 well microtiter plates (available at http://www.openbiosystems.com).
2. Convert frozen stocks to solid colonies. Allow plates to thaw completely (cells may have settled prior to being frozen). Insert a 96-well pin tool (V&P Scientific, Inc. catalog # VP407A), into thawed 96-well plates, swirl gently then transfer to a Nunc Omni Tray (VWR catalog # 62409-600) containing 50 ml of YPD-agar including 200 g/ml Geneticin G418 (Agri-Bio catalog # 3000). Allow pin to dwell on agar for 5–10 s.
3. Before each transfer, dip pin tool in water followed by 95% ethanol, then carefully flame the pin tool. Allow pin tool to cool. Make certain that the level of ethanol in the wash container exceeds the level in the water container to ensure all carry-over cells are flamed and removed. Change water frequently.
4. Grow colonies until they reach maximal size at 30°C (2–3 d).
5. After colonies have reached full size, scrape the entire contents of all plates (in a laminar flow hood to avoid contamination) into a 50 ml conical centrifuge tube containing YPD liquid media + 200 g/ml G418.
6. Make note of any strains that are missing or appear as slow-growing colonies. For these strains go to the original frozen stock and streak them out individually using standard yeast procedures.
7. For slow-growing strains add a colony-equivalent of cells using a sterile flat toothpick and add them to the conical tube. Measure the OD_{600} of the pool and adjust to a final 50 OD_{600}/ml (OD_{600} 1.0~2.2×10^7 cells/ml for diploid strains).
8. Add glycerol to 15% or DMSO to 7%, mix well and aliquot into individually capped PCR tubes with 10–25 μl of pool and store at −80°C.

Note: Hybridization to the TAG array will identify any of those strains that are still underrepresented (Deutschbauer *et al.*, 2005). These strains can then be spiked in to the pools individually at the start of each experiment.

rate (Benjamini and Hochberg, 1995; Efron, 2004). An alternative method for analysis of Agilent TAG microarrays is available (Peyser *et al.*, 2005).

D. HIP: Drug Target Identification

To test the feasibility of HIP, we first tested a well-characterized drug with a well-characterized target (Giaever *et al.*, 2004). Methotrexate is

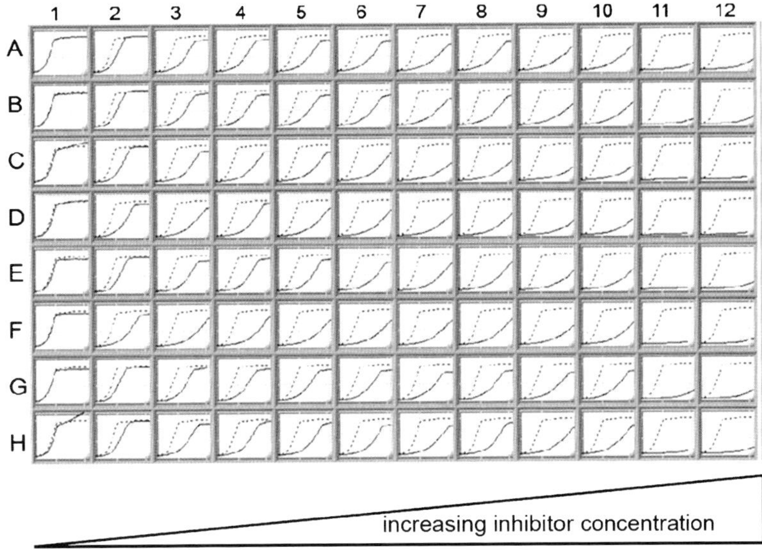

Figure 3. Determining appropriate screening dose. As described in Protocol 2, compounds are applied to wildtype cells, grown overnight in 96-well plates with OD_{600} measured every 15′ for 16 h. Column 1 contains the vehicle (DMSO) control. All compounds are referenced against vehicle (well A1) and this reference line appears as a dotted line in each test well. Each row contains one compound which is serially diluted 1:2 across each row (right to left). Examples of appropriate levels of inhibition (~10%) are seen in wells A2, B2, E2, G2, H2.

a widely used antiproliferative used in oncology (Chabner *et al.*, 2001), reproductive medicine (Hausknecht, 1995) and dermatology (see, for example, Kazlow Stern *et al.*, 2005). In man, dihydrofolate reductase is the known target of methotrexate (Chabner *et al.*, 2001). The yeast homolog (*DFR1*) was identified in the HIP assay as a highly sensitive strain in the optimal concentration window of 250 M methotrexate (Figure 4, for analysis methods see Giaever *et al.*, 2002). Four other strains were identified as significantly sensitive in 8 of 9 replicate experiments. Two of these strains were heterozygous for the essential genes *FOL1* and *FOL2*, which act upstream of *DFR1* and are required for biosynthesis of folic acid in yeast. It is possible that *FOL1* and *FOL2* interact directly with folic acid (and therefore are directly targeted by methotrexate) to control the flux through this essential metabolic pathway. The genome-wide methotrexate results also reveal a common theme; in addition to identifying the potential drug/compound target(s); HIP also identifies those genes involved in compound availability (i.e. membrane transporters, detoxifying enzymes and proteins involved in plasma membrane or cell wall integrity). In the case of methotrexate, the *YBT1* heterozygous deletion strain is highly sensitive to methotrexate. The human homolog of *YBT1* encodes the known methotrexate transporter. Finally, the function of the 5th gene, *YOR072w* is as of yet, unknown.

Protocol 2. Identifying the appropriate screening dose.

Prior to embarking on a genome-wide HIP or HOP (see section below), the appropriate dose of compound must be determined empirically. This is accomplished as follows:

1. Dissolve compound in the appropriate diluent at the highest possible concentration. A good starting point is to prepare 50 mM stocks in DMSO.
2. Set up a dose plate by transferring a small volume of compound to wells A2-A11 of a conical-bottom polypropylene microtiter plate. Wells A1 and A12 should be loaded with diluent alone.
3. With a multichannel pippetor, transfer sufficient diluent to prepare 1:1 dilutions of the compound stock to wells B1-H12.
4. Using a multichannel pippettor, dilute and mix the stock compounds from well 12 to 11, 11 to 10, etc. to finalize the dose plate.
5. Transfer 100 µl of diluted wildtype cells ($OD_{600} = 0.0625$) from a fresh overnight culture into each well of a flat bottom spectrophotometer-compatible clear-bottom 96-well plate using a multichannel pippettor. This is the assay plate.
6. Transfer drug dilutions into the assay plate from the dose plate with a multichannel pippetor or calibrated pin tool to deliver 1 µl of compound to the assay plate.
7. Seal the assay plate with clear plastic adhesive seal (ABgene catalog # ab-0580) and place into a shaking/incubating spectrophotometer (e.g. Tecan Genios).
8. Seal the dose plate with an adhesive foil seal (Fisher catalog # 07-200-684) and store at −80°C.
9. Grow microcultures using the highest possible shaking speed, and measure the OD_{600} of each well of the 96 well assay plate every 15′ at 30°C for a minimum of 12 h in rich media or 36 h in defined media.
10. After the assay is complete, determine the dose of compound that produces a 10–15% inhibition of cell growth relative to wildtype with diluent alone; this defines the screening concentration that will be used on the pooled deletion collection.

Note: Include a barcode on each dose plate and assay plate, and track plates with the use of a handheld scanner.

The HIP assay has been further validated as a means to find protein targets; we have screened ~500 unique compounds (including human therapeutics and bioactives, to be published elsewhere) that inhibit wildtype growth to a measurable degree (approximately 30% of all compounds tested). To the best of our

Protocol 3. Overview of the assay.

A. Genome-wide competitive growth assay (HIP)*

1. Thaw and inoculate frozen aliquot of pool into 100 ml such that on average 1000 cells/strain are present (OD_{600}~0.003).
2. Allow cells to recover overnight (~9 generations in YPD at 30°C with shaking at 250 rpm) without allowing them to saturate ($OD_{600} \leq 2$).
3. Dilute culture into condition of interest to not less than 125 cells/strain ($OD_{600} = 0.0625$).
4. Grow cells logarithmically for 5 generations until $OD_{600} = 2$.
5. Save not less than 1 OD_{600} of cells.
6. Batch dilute culture to not less than 125 cells/strain ($OD_{600} =$ 0.0625).
7. Continue to grow and collect the log culture and dilute and save every 5 generations until cells have reached 20 generations.

*For the HOP assay (see below), cells are grown for 5 generations straight from the frozen stock and collected at 5, 10, 15 and 20 generations. Because 15% of all homozygotes are slow-growing strains, optimum results are usually obtained at the 5 generation time point, minimizing nonspecific strain loss.

B. Prepare genomic DNA from harvested cells
Although any prep for genomic DNA can be used, we use Zymo Research YeaStar kit (catalog # D2002).

C. PCR amplify genomic DNA
Using ~0.2 μg of genomic DNA as template, set up a two PCR reactions, one for the UPTAGs and one for the DNTAGs (using any thermocycler with a heated lid). Cycle as follows in 100 μl with biotinylated primers and PCR master mix.
PCR Mix (100 μl)
PCR buffer ([1 ×], $MgCl_2$ [2.5 mM], dNTPs [0.2 mM], UP or DN mix [1 μM], Taq polymerase [5U], genomic DNA [~0.2 μg]) in a final volume of 100 μl.
UP mix = BUPTAG (5′-GAT GTC CAC GAG GTC TCT-3′) and BUPTAGKANMX4 (5′-GTC GAC CTG CAG CGT ACG-3′) each at 100 pmol/μl, mix in a 1:1 ratio. DN mix = BDNTAG (5′-CGG TGT CGG TCT CGT AG-3′) and BDNKANMX4 (5′-GAA AAC GAG CTC GAA TTC ATC G-3′) each at 100 pmol/μl, mix in a 1:1 ratio. All primers are 5′ biotinylated and desalted.
Cycle as follows
(94°C 3′)(94°C 30″)(55°C 30″)(72°C 30″) go to step 2 29 × then (72°C 3′)(4°C)

D. Hybridize PCR amplified tags to microarray

1. Prewet the probe array by with 140 μl 1 × filter sterilized hybridization buffer (100 mM MES [12 × MES: stock 1.22 M MES,

0.89 M NaCl], 1 M [Na^+], 20 mM EDTA, 0.01% Tween, store at 4°C in the dark) at 42°C for 10′ with rotation at 20 rpm.
2. Separately add 30 μl of uptag PCR and 30 μl of dntag PCR to 90 μl of hybridization mix (hybridization mix: 75 μl 2 × hybridization buffer, 0.5 μl B213 ctrl oligonucleotide [5′ - CTG AAC GGT AGC ATC TTG AC – 3′, (the biotinylated B213 oligonucleotide hybridizes to the border of the microarray)] at 0.2 fmol/μl, 12 μl of mixed oligonucleotides at 12.5 pmol/μl, 3 μl 50 × Denhardts) for a total volume of 150 μl in an 0.5 ml microfuge tube. Mixed oligonucleotides consist of 8 unbiotinylated primers: the 4 amplification primers and their complements at 100 pmol/μl, mixed 1:1 for a final concentration of 12.5 pmol/μl each. These primers act to "soak" up the common regions of the 20 bp tags.
3. Boil each tube for 2′ and then set on ice for 2′.
4. Remove 1 × hybridization buffer from each chip.
5. Add hybridization mix to each chip, place a Tough Spot® over each gasket and hybridize in the oven for 10–16 h at 42°C rotating at 20 rpm.

Post Hybridization Wash

Remove hybridization mix and open the top gasket of the microarray by inserting a disposable pipette tip. Hand wash as follows using a manual pipettor fitted with a 200 μl tip and inserted into the bottom gasket:
Wash A (6 × SSPE + 0.01% Tween 20) at R.T., 2 solution changes, 4 mixes/solution change. Wash B (3 × SSPE + 0.01% Tween 20) at 42°C, 6 solution changes, 4 mixes/solution change. Wash A at RT, 1 solution change, 4 mixes/solution change. One mix = pipetting the fluid up and down into the microarray using 150 μl of liquid.

Biotin staining

6 × SSPE, 1 × Denhardt's, 0.01% Tween 20, 1.7 ng/μl streptavidin-phycoerythrin (Molecular Probes catalog # S-866). Add 140 μl of stain/array, incubate for 10′ at 42°C with 20 rpm rotation.

Post Staining Wash

Wash A, 6 solution changes, 4 mixes. Fill up array with Wash A before scanning. Take care to avoid introducing bubbles into the array chamber.

Scanning

1. Clean glass side of arrays with isopropanol and a cotton swap and scan according to manufacturer's instruction.
2. Analyze as described in the text.

Step-by-step protocols for Agilent TAG arrays are available in Yuan *et al.* (2005) and a copy of the Affymetrix TAG3 protocol can also be found at http://www.chemogenomics.stanford.edu/.

Tips on the procedure:

1. Although we use 125 cells/strain in order to minimize volume (and therefore compound consumption) it is preferable to use a greater number of cells/strain to reduce the sampling error when batch diluting. For example, using 1000 cells/strain gives a theoretical sampling error of ~3%.
2. For the HIP assay, because the growth differences are slight, empirically we have observed the best results at the 20-generation time point. However, for further resolution, hybridization of the earlier time points can be informative in ranking the strains.
3. It is also possible to add the condition of interest without overnight recovery ("thaw and go"). This is particularly important in the homozygous profiling assay due to the fact that there are many more strains that are slow-growers in the absence of condition then during overnight growth.
4. The Affymetrix fluidics station may also be used for the staining/washing steps.

To confirm any strains that appear sensitive in the genome-wide microarray assay

1. Pick appropriate strains from YPD-antibiotic-agar plates.
2. Grow overnight in YPD without compound (N.B. These microcultures are set up similar to those described in Protocol 2).
3. Dilute 1:10 in YPD.
4. Measure OD_{600} in 96-well spectrophotometer.
5. Normalize cell number/well (either manually or robotically – e.g. using a variable span liquid dispenser) to a final OD_{600} of 0.0625. Set aside several wells for wildtype, or other controls.
6. Add compound to every well using a multichannel pippetor or pin tool.
7. Grow and measure in the spectrophotometer as described for prescreening.
8. Calculate each strain's sensitivity relative to wildtype.

knowledge, we have not failed to identify any targets in cases where the target is known and well-characterized. A key feature of the HIP assay is that it allows the cell's physiology to report what gene products are most important for growth, and therefore may be potential drug targets. As mentioned above, the HIP assay identifies both the cellular target and its inhibitor simultaneously. This is in contrast to traditional perturbation strategies, where one must have *a priori* knowledge of the target to design specific siRNAs, antibodies, peptides, aptamers or site-directed mutations against the target.

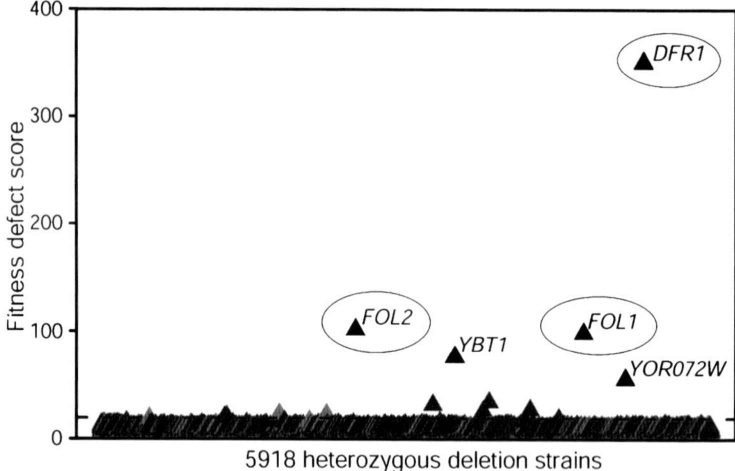

Figure 4. An example of HIP: Identifying the target of methotrexate in yeast. All ~6000 heterozygous strains are plotted along the x-axis. On the y-axis the corresponding fitness defect score of every strain (Giaever *et al.*, 2004) is plotted. Five strains appear significantly sensitive in 8 of 9 experiments. *DFR1* is the most sensitive strain, the human homolog of which is dihydrofolate reductase, the known drug target of methotrexate. The other four sensitive strains are described in the text. Circles indicate that a particular strain is essential. Taken from Giaever *et al.* (2004).

Alternatively, traditional cell-based phenotypic screens define the inhibitory compound but do not directly identify the potential targets. To find the cellular target, time-consuming follow-up characterization is required.

Other laboratories have also demonstrated the ability of the HIP assay to identify drug targets (Baetz *et al.*, 2004a, 2004b; Dorer *et al.*, 2005; Lum *et al.*, 2004). For example, Lum *et al.* (2004) screened 78 compounds against approximately one-third of the yeast genome. Many of their results are in agreement with our own (Giaever *et al.*, 2004). The results with 5-fluorouracil (5-FU) are especially noteworthy. The major mechanism of action of 5-FU is thought to be inhibition of thymidylate synthase (Parker and Cheng, 1990), though there is evidence that it may also be inhibiting RNA metabolism (Engelbrecht *et al.*, 1984; Linke *et al.*, 1996; Longley *et al.*, 2003; Pritchard *et al.*, 1997). The results from both Lum *et al.* (2004) and Giaever *et al.* (2004) however, suggest that the major mode of action is disruption of rRNA processing. Extension of this observation by monitoring rRNA processing directly revealed that 5-FU treatment disrupted exosome-specific rRNA processing (Lum *et al.*, 2004).

E. HIP: Drug Discovery

The HIP assay may be used to identify new compounds that can be used as chemical probes to dissect biological pathways in the lab, as well as to provide a jumping off point for medicinal chemists for drug discovery and development. A useful example of the power of chemical probes is provided by examining mitosis, a process that underlies many oncology therapies. Mitotic poisons include monastrol (Kapoor *et al.*, 2000), and other mitotic kinesin (KSP) inhibitors (Wood *et al.*, 2001). These compounds cause the centrosomes to collapse and result in defective "monopolar" spindles. Colchicine and nocodazole depolymerize spindle microtubules and prevent the formation of a metaphase spindle. Though these compounds proved too toxic in clinical trials, the vinca alkaloids that act by a similar microtubule depolymerizing mechanism are used clinically. The microtubule stabilizers Taxol and Taxotere (as well as the epothilones) are very effective antiproliferatives that act by stabilizing spindle microtubules and freezing the cell in metaphase (for review see Haggarty *et al.*, 2000). Other chemical probes that led to the discovery of potential drug targets include the cytochalasins (actin destabilizers) and blebbistatin (a myosin II inhibitor) (Straight *et al.*, 2003). Over the past 30 years these chemicals have been valuable tools for understanding the molecular details of this key biological process.

A more recent example of the use of HIP for target and drug discovery is the identification of a dual-specificity kinase as a potential drug target. Briefly, a phenotype-based chemical-genetic screen in yeast identified a compound (cincreasin) that abrogated the mitotic spindle checkpoint (Dorer *et al.*, 2005). A standard HIP protocol demonstrated that the target of cincreasin is the kinase Mps1 (Weiss and Winey, 1996), required for checkpoint function. Though cincreasin is structurally very simple and despite the fact that it lacks potency and specificity, the HIP assay had sufficient sensitivity and selectivity to identify the protein target in this case. This study highlights the fact HIP can be used to define the "druggable" fraction of the genome, that is to say all proteins that can be inhibited by a small molecule.

F. HIP: Structure Activity Relationships (SAR)

Another interesting aspect of HIP is the observation that similar compounds can give similar profiles. For example, three therapeutically distinct compounds fenpropimorph (a morpholine antifungal: Lai *et al.*, 1994; Marcireau *et al.*, 1990), alverine citrate (an antispasmodic muscle relaxant: Coelho *et al.*, 2001) and dyclonine (an anesthetic) gave similar profiles. The *ERG24* heterozygous strain is highly sensitive to all three compounds (Figure 5); however, only in the case of fenproprimorph (Lai *et al.*, 1994; Marcireau *et al.*, 1990) was the target known to be *ERG24*. Confirmatory experiments showed that overexpression of the human homolog of *ERG24* confer resistance to all three compounds. This makes *ERG24* the likely target of this class of compounds (Giaever *et al.*, 2004). The structures of these three compounds reveal they share a common chemical core (Figure 5). This example demonstrates that the HIP assay may be used to understand structure–activity relationships.

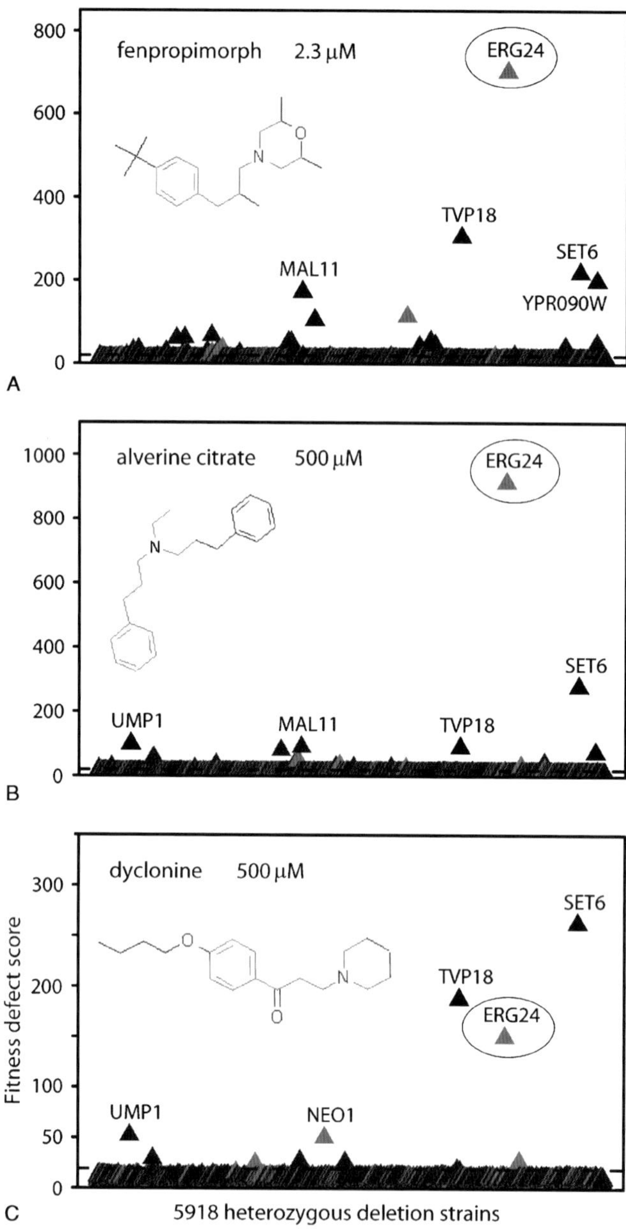

Figure 5. Comparison of the profiles of three compounds and their structures. Axes are the same as described in Figure 3. All three profiles look very similar and identify *ERG24* as the likely drug target. Interestingly, all three compounds share structural similarity, demonstrating that compounds with structural similarity can produce similar HIP profiles. Taken from Giaever *et al.* (2004).

G. HIP: Caveats

There are several caveats to be aware of when using the HIP assay to characterize new compounds or uncover novel targets. First, the compound must inhibit growth or else it is unlikely that any heterozygotes will display increased sensitivity. Second, for any human therapeutics, the target must have a homolog in yeast. Third, the compound must be able to penetrate yeast and remain at inhibitory intracellular concentrations throughout the course of the assay. Finally, if the compound must be metabolized in order to be activated, these metabolic pathways must exist in yeast.

With respect to analyzing the data from HIP assays, false negatives will arise from strains whose barcodes do not hybridize well. At the time of writing the chapter approximately 200 strains (~3.3%) do not hybridize well enough to be included in the analysis ($<3 \times$ background). The next generation of the YKO collection version 2.0 is expected to correct many of these errors and will be available in the summer of 2006 (A. Chu, personal communication).

◆◆◆◆◆◆ III. THE HOP ASSAY: BACKGROUND AND METHOD

Methodologically, Homozygous profiling (or HOP) is essentially the same assay as the HIP assay but the homozygous deletion collection is screened instead of the heterozygous deletion collection. Biologically, however, the types of questions that can be addressed by HOP are conceptually quite distinct and the interpretation of the data very different. While in the case of HIP, the essential genes are included and drug targets can be identified, in the case of the HOP assay, none of the essential genes are included and because the genes are complete homozygous deletions, nothing can be learned about the drug target from a single experiment. In this sense, HOP is much more similar to classical genetics, where many mutants are often complete loss-of-function mutants. The sensitive strains in the HOP assay are informative in that they reveal information about strains that may interact with the drug target pathway and are important for survival in the presence of compound.

Another difference in the HOP assay is that a significant fraction (~15%: Deutschbauer *et al.*, 2005) of the homozygous diploid deletion strains exhibit a slow-growth phenotype in the absence of perturbation. This can be problematic because these strains become too depleted in the pool for accurate analysis. There are at least three approaches to account for the slow-growers: (1) test slow-growers in a separate pool of slow-growing strains (2) expose cells to compound immediately following thawing from the frozen state (surprisingly, omitting the recovery step does not cause nonspecific growth effects (Lee *et al.*, 2005) and (3) constructing the initial pool (Protocol 1) with 3–5 fold greater number of each slow-growing strain.

Homozygous profiling (or fitness profiling) has been performed on the complete deletion collections after exposure to environmental perturbation (e.g. high osmolarity, high salt, high pH, different carbon sources: Birrell *et al.*, 2001; Steinmetz *et al.*, 2002; Winzeler *et al.*, 1999) as well as with several anticancer compounds (Birrell *et al.*, 2001; Giaever *et al.*, 2004; Lum *et al.*, 2004; Wu *et al.*, 2004). These studies have yielded valuable information by identifying genes required for survival in diverse conditions.

A. HOP: Uncovering Drug Mechanism When no Protein Target Exists

The results of HIP profiling with cisplatin underscore an important principle of these competitive assays. When yeast cells are treated with compounds that lack *bona fide* protein targets, the HIP profile will be uninformative as no significantly sensitive strains will be identified. In these cases, a HOP assay can illuminate a cell's response to compound. An illustration of this phenomenon is seen in the case of cisplatin (Figure 6). The HIP assay reveals no sensitive strains, whereas in the HOP assay several strains deleted for genes involved in DNA repair are significantly sensitive. It becomes clear then, that in this case, the target of the compound is the DNA itself. On the basis of this observation, Lee *et al.* (2005) interrogated the homozygous deletion pool for sensitivity to 12 agents known to damage DNA. Strains sensitive to these compounds were deleted for genes known to be involved in DNA metabolism and uncovered genes not previously known to be related to the DNA damage response. In this study, the HOP profiles were clustered to classify the mechanism of drug action and to understand gene function. This analysis allowed

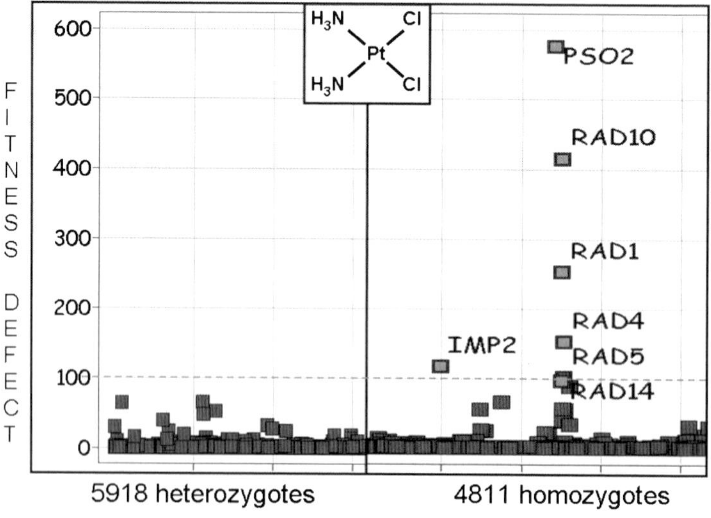

Figure 6. Comparison of the HIP and HOP assay results. Axes are the same as in Figure 3. On the left the profile of cisplatin in the HIP assay is shown, with no strains appearing significantly sensitive. On the right the profile of cisplatin in the HOP assay is shown. In this case, several sensitive strains are detected with many in this group involved in DNA repair.

for a clear discrimination of the genome-wide response of a cell to agents that damage its DNA by forming interstrand cross-links from those that do not, and also uncovered several surprises where compounds that are thought to have identical mechanisms of action in fact do not (see Figure 7, e.g. cisplatin and carboplatin).

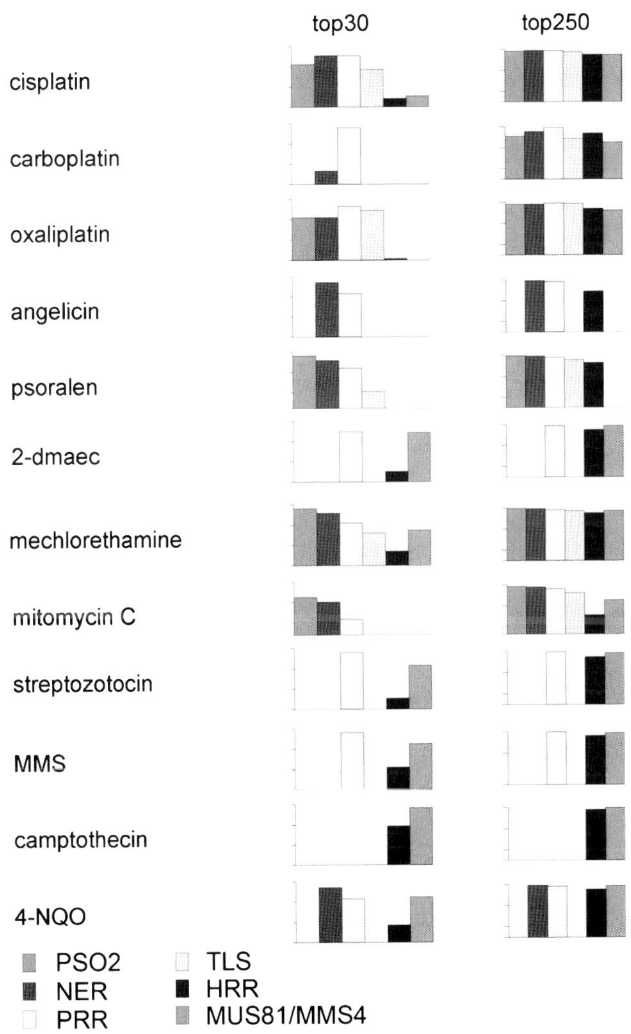

Figure 7. Relative importance of the DNA damage response modules for resistance to DNA-damaging agents. Each bar graph represents strains that were found to be among the top 30 (or 250) most sensitive strains in that compound and are known to be members of a well-characterized DNA-damage-response pathway. The bars represent the median rank for genes in each of the gene groups listed in the visual key. The gene groups were defined in the following way: x-linking genes (*PSO2*); NER (*RAD2*, *RAD4*, *RAD10*, *RAD1*, and *RAD14*); PRR (*RAD6*, *RAD18*, and *RAD5*); error-prone TLS (*REV1*, *REV3*, and *REV7*); HRR (*RAD57*, *RAD55*, *RAD51*, *RAD52*, *RAD54*, and *RAD59*); stalled replication-fork repair (*MUS81* and *MMS4*). Those compounds that form interstrand cross-links are labeled with an asterisk. Taken from Lee *et al.* (2005). (See color plate section).

♦♦♦♦♦♦ IV. THE ADVANTAGES OF COMBINING THE HIP AND HOP ASSAYS

A. HIP-HOP: Complementary Approaches

When used in combination with HIP, the HOP profile becomes particularly informative. For example, in the case of methotrexate, the homozygous profiling reveals several strains deleted for genes required for repair of DNA damage (data not shown). This is likely due to the fact that inhibiting dihydrofolate reductase with methotrexate eventually blocks nucleotide formation which in turn inhibits DNA replication. To resolve the resulting stalled replication forks, DNA repair genes are required. In the absence of the HIP assay it would be extremely difficult to discern the target from the HOP assay alone. Rather, for the purposes of target identification and characterization, the HOP assay is a powerful complement to the HIP assay as it allows the identification of genes and pathways required for recovery from inhibition of a particular target. These genes are important in that they allow the identification and potential deconvolution of pathways functionally related to the drug target.

B. Validating the HIP and HOP Microarray Results by Individual Strain Confirmation

As for any microarray experiment, it is important to confirm the results using an independent measure. The HIP and HOP assays make this task relatively straightforward because all sensitive strains identified in the genome-wide assays can be confirmed in individual liquid culture before the results are considered complete (in a manner analogous to that described in protocol 2). Furthermore, these confirmation assays can define how well the data from the microarray experiments agree the growth experiments. In our DNA damage experiments, we confirmed the array results by exposing the individual homozygous diploid deletion strains to the same concentration of compound in a microculture assay (Lee *et al.*, 2005: sample growth curves are shown in Figure 8A; see Colour

Figure 8. Confirmation of microarray results by individual strain growth. (A) Growth curves of 16 strains grown in the presence of solvent (DMSO, dashed black line) and 62.5 μM mechlorethamine (solid black line). Growth was monitored by optical density ($O.D._{600}$) of cultures every 15 min for 30 h. Fitness of each strain was defined by the difference between the average doubling times in mechlorethamine and in DMSO. (B) Correlation between growth rates of individual strains and microarray-based fitness estimates. The ratio of growth rates of 186 individual homozygous deletion strains divided by the average wildtype growth rate are plotted on the x-axis against the average fitness defect scores from three pool experiments on the y-axis. The correlation ($R^2 = 0.7462$) is highly significant ($p = 5.4 \times 10^{-57}$). Taken from Lee *et al.* (2005).

Plate section). Growth values were normalized to wildtype and plotted against their corresponding fitness-defect scores as measured from the microarray (Figure 8B), yielding a highly significant correlation ($R^2 = 0.7462$; $p = 5.4 \times 10^{-57}$). In addition, 203 of 233 strains exhibited significant strain sensitivity, suggesting our

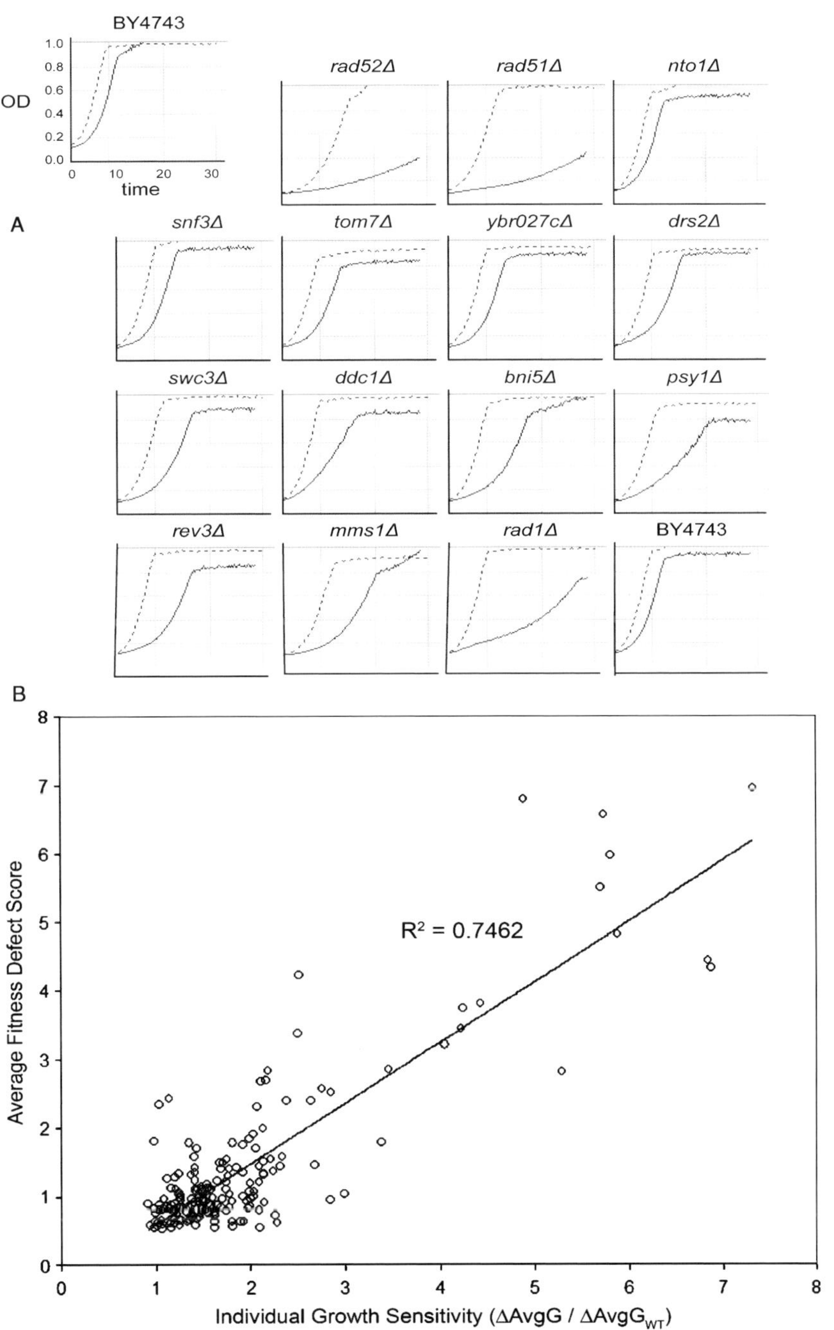

false positive rate is very low. It is important to keep in mind that, at present, the confirmation assays only address the false positive rate and not the false negative rate. With respect to false negatives (i.e. strains that erroneously score as unaffected by treatment, we know that we will have false negatives for any sensitive strains whose barcodes do not hybridize well (see above).

C. Caveats of HOP

The presence of slow-growing strains that nonspecifically drop out of the pool is probably the biggest limitation of the HOP assays. However, this issue can be addressed as discussed above. An additional complication is the potential presence of secondary site mutations. Although the homozygote diploid collection is less likely to be affected by second-site mutations compared to haploid *MATa* or *MATα* collections (because the diploids were derived from mating two independently constructed haploids) such lesions will occur at some frequency (Giaever *et al.*, 2002; Pan *et al.*, 2004). Five years of user feedback to the Yeast Deletion Database (http://www.yeastdeletion.stanford.edu) has been collected and these strains, in addition to newly identified ORFs are currently being deleted with additional barcodes as part of the YKO Collection Version 2.0 (A. Chu personal communication). There will remain a small number of genes that will be difficult to include into the YKO collections because of the unique biology of their mutant phenotypes. For example, nonessential genes that confer a sterile phenotype will only be included in the YKO if they were specifically requested.

D. Comparison to Expression Array Technology

Because both expression profiling and fitness profiling interrogate the whole genome simultaneously, the question arises: what is the correlation between the change in a gene's expression level and the requirement for that gene for growth in the same condition? Comparing expression changes to fitness defects in several conditions (Birrell *et al.*, 2001) showed little correlation (Figure 9). In the most highly correlated test condition (Birrell *et al.*, 2001), i.e. media containing galactose as the sole carbon source, only 7% of the genes that exhibited a significant increase in mRNA expression also exhibited a significant decrease in fitness, and other conditions (1 M salt, 1 M sorbitol) showed correlations of less than 1%. This lack of correlation between fitness requirement and expression change observation suggests that many gene products required for rapid adaptation to environmental changes or compound treatment are already present within the cell, and that the expression changes observed may have other adaptive roles. This observation highlights that HIP and HOP fitness profiling can identify gene products required for a particular condition even if there is no change in their expression.

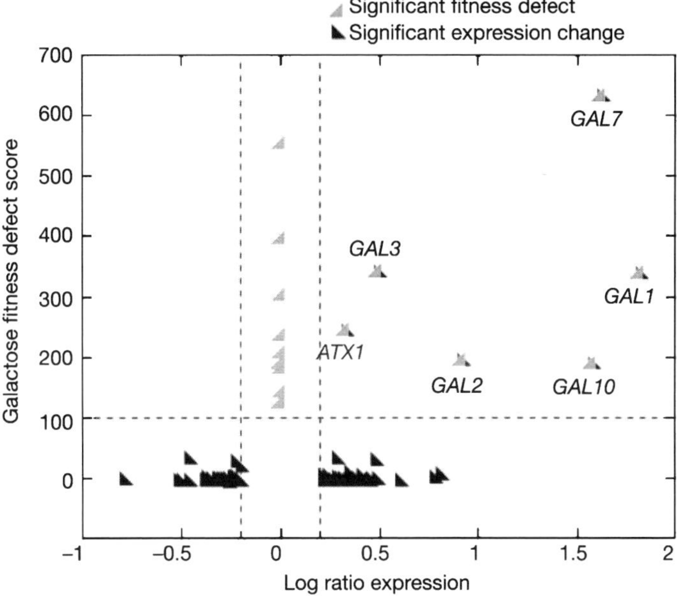

Figure 9. Comparison of HOP fitness data to expression data. For clarity, only those genes designated as sensitive in galactose by the fitness defect score are plotted. Light gray triangles represent genes with significant fitness defect scores (above the dashed line) plotted as a function of their corresponding values for log ratio expression: log(expression in galactose/reference expression). Black triangles represent genes with significant log ratio expression (outside the two vertical dashed lines) plotted with their corresponding fitness defect scores. The values of the fitness defects plotted are the minimum score from two experiments. Taken from Giaever *et al.* (2002).

♦♦♦♦♦♦ V. COMPARISON TO OTHER TECHNOLOGIES

A. Array Technologies

In this chapter we focus on performing the HIP and HOP assays using Affymetrix arrays and scanners. There are of course several DNA microarray platforms that one could use; including other photolithographic technologies (e.g. Nimblegen), commercially available ink-jet arrays (e.g. Agilent: Lum *et al.*, 2004; Yuan *et al.*, 2005) as well as spotted oligonucleotide arrays prepared either by contract with a company or home made in one's own laboratory. In principle, the HIP/HOP assay should perform on any of these platforms (Lum *et al.*, 2004).

B. Pinning Technologies

Despite the power of the molecular barcodes in parallel assays to increase the speed and sensitivity of genome-wide functional analysis, many investigators continue to use the yeast deletion collection by pinning individual colonies onto agarose plates (see, for example, Bennett *et al.*, 2001; Tong *et al.*, 2001). While extremely powerful, the results of pin assays are less quantitative than the competitive pool assays. On the basis of a direct comparison of microarray data to growth rate, arrays can detect <5% growth difference, whereas pinning is not likely to detect differences less than 25% unless sophisticated imaging and analysis tools are used. Moreover, the protocols that rely on pinning of individual strains face the challenge of plate and strain tracking, and of the odd colony that fails to replicate. To minimize these issues barcodes can be applied to plates and vials, and commercial pinning robots used.

Pinning assays have been widely used on a small-scale in almost every yeast laboratory. The Boone laboratory introduced their **S**ynthetic **G**enetic **A**rray (**SGA**) technology enabling new, large-scale, high throughput interrogations of the yeast genome and the YKO collections (Tong *et al.*, 2001, 2004). The SGA method is covered in another chapter in this volume, and we will focus solely on the use of SGA and related techniques as they have been used in drug studies on the YKO collections. A study by Parsons *et al.* examined the phenotypic effects of drug–gene deletion combinations on the haploid deletion collection with 12 compounds and compared these results to SGA by clustering the data with other SGA screens (Parsons *et al.*, 2004). They found that they could infer the target from the profile it clustered next to (e.g. a screen using a temperature-sensitive *ERG11* mutant clustered near the azole compounds). The power of this technique for drug/compound screening is limited, however, due to the large amount of compound required for plate assays and the fact that many drug targets are essential and therefore do not exist in the haploid collection. Xie *et al.* recently offered a solution to the compound consumption drawback of plate assays (Xie *et al.*, 2005). Using a commercially available DNA microarrayer they printed cells onto agar medium at a density of >10 000 strains/plate, which effectively consolidates one of the four deletion collections into a single 50 ml agar dish. Although this format still requires an order of magnitude more compound vs. the HIP or HOP assays, this proof-of-principle study with rapamycin suggests that this technique holds promise for future studies.

C. Individual Microculture Assays

An alternative way to assay each strain individually is to measure the growth rate of each haploid deletion strain in microculture in much the same manner as we describe for dose-ranging and confirmation assays in protocol boxes 2 and 4. These data are disseminated via the PROPHECY Project (Warringer and Blomberg, 2003). PROPHECY quantifies growth aberrations by estimating the rate of growth, the efficiency of growth and the adaptation time. In their proof-of-principle study, the authors present a methodology for functional prediction based on extraction of growth variables from all viable haploid YKO mutants. They dub this approach "quantitative phenomics" and have applied the technology to assay growth in high salt media. The number of phenotypes assayed at Prophecy continues to grow and can be accessed in several visualizations from the project's website (http://www.prophecy.lundberg.gu.se/). The resolution of this approach is quite high, although the throughput is modest and the compound consumption similar in scale to a plate assay.

The Synthetic Lethality Analyzed by Microarray (SLAM) (Ooi *et al.*, 2003) and Diploid Synthetic Lethality Analyzed by Microarray (dSLAM) (Pan *et al.*, 2004) assays are methods for construction of double deletions in a high-throughput, parallel manner using barcode microarrays. These studies are compelling in that they have the potential to identify all synthetic lethal strains from a total of ~12.5E6 double deletion strains. At present, however, while the final readout of these assays is a quantitative intensity value, it is difficult to achieve saturation of all strains because of the requirement for extremely efficient transformation of deletion pools with specific query constructs. Theoretical simulations and empirical analyses suggest that 5E5 to 5E6 independent transformants are necessary to prevent stochastic loss of individual YKO mutants (Pan *et al.*, 2004). Such technological hurdles once surmounted, should provide an extremely powerful technology.

◆◆◆◆◆◆ VI. PERSPECTIVES AND FUTURE DIRECTIONS

Whole genome modifications of the deletion collection have been made by transforming the deletion pools "en masse" with linear DNA or circular plasmids. In one case this method was used to screen for components of the nonhomologous end-joining (NHEJ) pathway (Ooi *et al.*, 2001). Known components of the pathway were identified, as well as one gene not previously known to be involved in NHEJ. The paper demonstrates that pools of mutants generated by DNA transformation can be analyzed in parallel. This approach is likely to have many important applications in any genetic screen requiring a plasmid (e.g. Giorgini *et al.*, 2005) as well as in screens for synthetic dosage lethality.

To export the power of the yeast barcodes to other strains and collections, we have embarked on an effort to design a "universal collection" in which 6000 wildtype *MAT*α strains are barcoded by homologous recombination of unique barcode cassettes at the inactive HO locus. Such a collection would be unlinked to the gene of interest, but it could serve as a mating partner for any existing mutant collection in a *MATa* strain background to enable rapid generation of a barcoded diploid colletion. Variants of this schema can be imagined by incorporating the methods developed for SGA (Tong *et al.*, 2001, 2004) to generate new barcoded haploid collections.

A limitation of the HIP and HOP protocols is the difficulty in determining which gene products, if any, confer resistance to a particular treatment. This difficulty is due in part from the experimental design, and in part from the analysis methods currently used. One way to assess if a gene product does confer resistance to a particular treatment would be to have a complete collection of strains that contain barcoded episomal versions of every ORF with expression driven by its own promoter. Pooling this collection and assaying it as described for HIP would serve to validate any targets identified in HIP as well as to uncover any *bona fide* resistant deletion mutants. Such mutants should comprise important members of protein complexes and pathways. Collections of overexpressing strains have been reported (Zhu *et al.*, 2001), but these collections are difficult to treat in parallel as expression, typically driven by a galactose-inducible promoter, is difficult to control.

The power of the deletion collection is limited to the range of phenotypes that its genetic background (SC288C) can manifest. For example, experiments might benefit from the presence of additional auxotrophic markers whereas others might benefit from having no markers at all in a prototrophic strain. Additionally it would be extremely valuable to improve the sporulation efficiency of the collection and confer invasive growth capability in SK-1 or sigma backgrounds, respectively. A recent study shows that this may be quite feasible (Deutschbauer and Davis, 2005). In practice the creation of such new collections should be faster than the original deletion project because of the lessons learned regarding the behavior of each particular tag cassette (based on sequencing and hybridization analysis) and a greater reliance on automation.

A common complaint about competitive genome-wide assays described here has been the up front costs for the technology. Though it is true that the expense of the technology has, in the past, provided an entry barrier for some, the amount of data generated greatly offsets this concern. Furthermore, as the number of published studies continues to grow and standardized protocols are employed, the array-based studies described in this chapter should become increasingly accessible.

The tools that have been developed using the YKO collections are continually growing and have been a uniquely useful resource for the scientific community (see Table 1). Further, the chapters in this volume attest to the creativity that will drive the next generation of genome-wide studies. A full understanding of cellular physiology will require "systems" level experiments and analysis on several fronts including genomics, proteomics, and metabolomics. Furthermore, we will need to understand all of those gene products that interact with small molecules and other perturbations. The HIP and HOP assays and future variations will aid the drug discovery process by identifying new compounds that define those proteins in the cell that can be targeted by small molecules. A dividend of this effort will be a suite of probes for chemical biology.

Table 1. Important landmarks in large scale analyses of the YKO collections

Sequence completed (Goffeau *et al.*, 1997)	1996
Proof of concept parallel deletion analysis (Shoemaker *et al.*, 1996)	1996
Proof of concept HIP assay (Giaever *et al.*, 1999)	1999
Synthetic Genetic Arrays (Tong *et al.*, 2001)	
Deletion collection completed (Giaever *et al.*, 2002)	2002
Prophecy (Warringer and Blomberg, 2003)	2003
dSLAM (Pan *et al.*, 2004)	
HIP reduction to practice (Giaever *et al.*, 2004; Lum *et al.*, 2004)	2004
Routinue use of HIP (Baetz *et al.*, 2004a, 2004b; Dorer *et al.*, 2005)	2004–2005
YKO 2.0 projected completion date (A. Chu personal communication)	2006

Acknowledgements

We thank the entire Yeast Deletion Consortium, in particular Mark Johnston for his vision in undertaking the YKO effort and Ronald W. Davis for his inspiration to barcode each strain. We also thank all members of the HIP-HOP lab for critical readings of the manuscript.

Research in the Stanford Chemogenomics Laboratory has been supported by grants from The National Institutes of Health divisions of the NCI, NIBIB and the NHGRI.

Chemical Genomic Tools

References

(Given space constraints and the pace at which genome-wide yeast assays are appearing, the authors apologize for the omission of many excellent papers that both led up to and have extended upon the YKO collections).

Baetz, K., McHardy, L., Gable, K., Tarling, T., Reberioux, D., Bryan, J., Andersen, R. J., Dunn, T., Hieter, P. and Roberge, M. (2004a). Yeast genome-wide drug-induced haploinsufficiency screen to determine drug mode of action. *Proc. Natl. Acad. Sci. USA* **101**, 4525–4530.

Baetz, K. K., Krogan, N. J., Emili, A., Greenblatt, J. and Hieter, P. (2004b). The ctf13-30/CTF13 genomic haploinsufficiency modifier screen identifies the yeast chromatin remodeling complex RSC, which is required for the establishment of sister chromatid cohesion. *Mol. Cell Biol.* **24**, 1232–1244.

Benjamini, Y. and Hochberg, Y. (1995). Controlling the False Discovery Rate – a Practical and Powerful Approach to Multiple Testing. *J. R. Stat. Soc. Ser. B-Methodol.* **57**, 289–300.

Bennett, C. B., Lewis, L. K., Karthikeyan, G., Lobachev, K. S., Jin, Y. H., Sterling, J. F., Snipe, J. R. and Resnick, M. A. (2001). Genes required for ionizing radiation resistance in yeast. *Nat. Genet.* **29**, 426–434.

Birrell, G. W., Giaever, G., Chu, A. M., Davis, R. W. and Brown, J. M. (2001). A genome-wide screen in *Saccharomyces cerevisiae* for genes affecting UV radiation sensitivity. *Proc. Natl. Acad. Sci. USA* **98**, 12608–12613.

Brenner, C. (2004). Chemical genomics in yeast. *Genome Biol.* **5**, 240.
Chabner, B. A., Ryan, D. P., Paz-Ares, L., Garcia-Carbonero, R. and Calabresi, P. (2001). *Chemotherapy of Neoplastic Diseases*. McGraw-Hill, New York.
Coelho, A. M., Jacob, L., Fioramonti, J. and Bueno, L. (2001). Rectal antinociceptive properties of alverine citrate are linked to antagonism at the 5-HT1A receptor subtype. *J. Pharm. Pharmacol.* **53**, 1419–1426.
Deutschbauer, A. M. and Davis, R. W. (2005). Quantitative trait loci mapped to single-nucleotide resolution in yeast. *Nat. Genet.* **37**, 1333–1340.
Deutschbauer, A. M., Jaramillo, D. F., Proctor, M., Kumm, J., Hillenmeyer, M. E., Davis, R. W., Nislow, C. and Giaever, G. (2005). Mechanisms of haploinsufficiency revealed by genome-wide profiling in yeast. *Genetics* **169**, 1915–1925.
Dorer, R. K., Zhong, S., Tallarico, J. A., Wong, W. H., Mitchison, T. J. and Murray, A. W. (2005). A small-molecule inhibitor of Mps1 blocks the spindle-checkpoint response to a lack of tension on mitotic chromosomes. *Curr. Biol.* **15**, 1070–1076.
Efron, B. (2004). Large-scale simultaneous hypothesis testing: the choice of a null hypothesis. *J. Am. Stat. Assoc.* **99**, 96–104.
Engelbrecht, C., Ljungquist, I., Lewan, L. and Yngner, T. (1984). Modulation of 5-fluorouracil metabolism by thymidine. *In vivo* and *in vitro* studies on RNA-directed effects in rat liver and hepatoma. *Biochem. Pharmacol.* **33**, 745–750.
Giaever, G. (2003). A chemical genomics approach to understanding drug action. *Trends Pharmacol. Sci.* **24**, 444–446.
Giaever, G., Chu, A. M., Ni, L., Connelly, C., Riles, L., Veronneau, S., Dow, S., Lucau-Danila, A., Anderson, K., Andre, B. *et al.* (2002). Functional profiling of the *Saccharomyces cerevisiae* genome. *Nature* **418**, 387–391.
Giaever, G., Flaherty, P., Kumm, J., Proctor, M., Nislow, C., Jaramillo, D. F., Chu, A. M., Jordan, M. I., Arkin, A. P. and Davis, R. W. (2004). Chemogenomic profiling: identifying the functional interactions of small molecules in yeast. *Proc. Natl. Acad. Sci. USA* **101**, 793–798.
Giaever, G., Shoemaker, D. D., Jones, T. W., Liang, H., Winzeler, E. A., Astromoff, A. and Davis, R. W. (1999). Genomic profiling of drug sensitivities via induced haploinsufficiency. *Nat. Genet.* **21**, 278–283.
Giorgini, F., Guidetti, P., Nguyen, Q., Bennett, S. C. and Muchowski, P. J. (2005). A genomic screen in yeast implicates kynurenine 3-monooxygenase as a therapeutic target for Huntington disease. *Nat. Genet.* **37**, 526–531.
Goffeau, A., Aert, R., Agostini-Carbone, M. L., Ahmed, A., Aigle, M., Alberghina, L., Albermann, K., Albers, M., Aidea, M., Alexandraki, D. *et al.* (1997). The yeast genome directory. *Nature* **387**, 5.
Grunenfelder, B. and Winzeler, E. A. (2002). Treasures and traps in genome-wide data sets: case examples from yeast. *Nat. Rev. Genet.* **3**, 653–661.
Haggarty, S. J., Mayer, T. U., Miyamoto, D. T., Fathi, R., King, R. W., Mitchison, T. J. and Schreiber, S. L. (2000). Dissecting cellular processes using small molecules: identification of colchicine-like, taxol-like and other small molecules that perturb mitosis. *Chem. Biol.* **7**, 275–286.
Hausknecht, R. U. (1995). Methotrexate and misoprostol to terminate early pregnancy. *N. Engl. J. Med.* **333**, 537–540.
Kapoor, T. M., Mayer, T. U., Coughlin, M. L. and Mitchison, T. J. (2000). Probing spindle assembly mechanisms with monastrol, a small molecule inhibitor of the mitotic kinesin, Eg5. *J. Cell. Biol.* **150**, 975–988.

Kazlow Stern, D., Tripp, J. M., Ho, V. C. and Lebwohl, M. (2005). The use of systemic immune moderators in dermatology: an update. *Dermatol. Clin.* **23**, 259–300.

Lai, M. H., Bard, M., Pierson, C. A., Alexander, J. F., Goebl, M., Carter, G. T. and Kirsch, D. R. (1994). The identification of a gene family in the *Saccharomyces cerevisiae* ergosterol biosynthesis pathway. *Gene* **140**, 41–49.

Lee, W., St Onge, R. P., Proctor, M., Flaherty, P., Jordan, M. I., Arkin, A. P., Davis, R. W., Nislow, C. and Giaever, G. (2005). Genome-wide requirements for resistance to functionally distinct DNA-damaging agents. *PLoS Genet.* **1**, e24.

Linke, S. P., Clarkin, K. C., Di Leonardo, A., Tsou, A. and Wahl, G. M. (1996). A reversible, p53-dependent G0/G1 cell cycle arrest induced by ribonucleotide depletion in the absence of detectable DNA damage. *Genes Dev.* **10**, 934–947.

Longley, D. B., Harkin, D. P. and Johnston, P. G. (2003). 5-Fluorouracil: mechanisms of action and clinical strategies. *Nat. Rev. Cancer* **3**, 330–338.

Lum, P. Y., Armour, C. D., Stepaniants, S. B., Cavet, G., Wolf, M. K., Butler, J. S., Hinshaw, J. C., Garnier, P., Prestwich, G. D., Leonardson, A. *et al.* (2004). Discovering modes of action for therapeutic compounds using a genome-wide screen of yeast heterozygotes. *Cell* **116**, 121–137.

Marcireau, C., Guilloton, M. and Karst, F. (1990). *In vivo* effects of fenpropimorph on the yeast *Saccharomyces cerevisiae* and determination of the molecular basis of the antifungal property. *Antimicrob. Agents Chemother.* **34**, 989–993.

Ooi, S. L., Shoemaker, D. D. and Boeke, J. D. (2001). A DNA microarray-based genetic screen for nonhomologous end-joining mutants in *Saccharomyces cerevisiae*. *Science* **294**, 2552–2556.

Ooi, S. L., Shoemaker, D. D. and Boeke, J. D. (2003). DNA helicase gene interaction network defined using synthetic lethality analyzed by microarray. *Nat. Genet.* **35**, 277–286.

Pan, X., Yuan, D. S., Xiang, D., Wang, X., Sookhai-Mahadeo, S., Bader, J. S., Hieter, P., Spencer, F. and Boeke, J. D. (2004). A robust toolkit for functional profiling of the yeast genome. *Mol. Cell.* **16**, 487–496.

Parker, W. B. and Cheng, Y. C. (1990). Metabolism and mechanism of action of 5-fluorouracil. *Pharmacol. Ther.* **48**, 381–395.

Parsons, A. B., Brost, R. L., Ding, H., Li, Z., Zhang, C., Sheikh, B., Brown, G. W., Kane, P. M., Hughes, T. R. and Boone, C. (2004). Integration of chemical-genetic and genetic interaction data links bioactive compounds to cellular target pathways. *Nat. Biotechnol.* **22**, 62–69.

Peyser, B. D., Irizarry, R. A., Tiffany, C. W., Chen, O., Yuan, D. S., Boeke, J. D. and Spencer, F. A. (2005). Improved statistical analysis of budding yeast TAG microarrays revealed by defined spike-in pools. *Nucl. Acids Res.* **33**, e140.

Pritchard, D. M., Watson, A. J., Potten, C. S., Jackman, A. L. and Hickman, J. A. (1997). Inhibition by uridine but not thymidine of p53-dependent intestinal apoptosis initiated by 5-fluorouracil: evidence for the involvement of RNA perturbation. *Proc. Natl. Acad. Sci. USA* **94**, 1795–1799.

Scherens, B. and Goffeau, A. (2004). The uses of genome-wide yeast mutant collections. *Genome Biol.* **5**, 229.

Shoemaker, D. D., Lashkari, D. A., Morris, D., Mittmann, M. and Davis, R. W. (1996). Quantitative phenotypic analysis of yeast deletion mutants using a highly parallel molecular bar-coding strategy [see comments]. *Nat. Genet.* **14**, 450–456.

Specht, K. M. and Shokat, K. M. (2002). The emerging power of chemical genetics. *Curr. Opin Cell Biol.* **14**, 155–159.

Steinmetz, L. M., Scharfe, C., Deutschbauer, A. M., Mokranjac, D., Herman, Z. S., Jones, T., Chu, A. M., Giaever, G., Prokisch, H., Oefner, P. J. and Davis, R. W. (2002). Systematic screen for human disease genes in yeast. *Nat. Genet.* **31**, 400–404.

Straight, A. F., Cheung, A., Limouze, J., Chen, I., Westwood, N. J., Sellers, J. R. and Mitchison, T. J. (2003). Dissecting temporal and spatial control of cytokinesis with a myosin II Inhibitor. *Science* **299**, 1743–1747.

Tong, A. H., Evangelista, M., Parsons, A. B., Xu, H., Bader, G. D., Page, N., Robinson, M., Raghibizadeh, S., Hogue, C. W., Bussey, H. *et al.* (2001). Systematic genetic analysis with ordered arrays of yeast deletion mutants. *Science* **294**, 2364–2368.

Tong, A. H., Lesage, G., Bader, G. D., Ding, H., Xu, H., Xin, X., Young, J., Berriz, G. F., Brost, R. L., Chang, M. *et al.* (2004). Global mapping of the yeast genetic interaction network. *Science* **303**, 808–813.

Warringer, J. and Blomberg, A. (2003). Automated screening in environmental arrays allows analysis of quantitative phenotypic profiles in *Saccharomyces cerevisiae. Yeast* **20**, 53–67.

Weiss, E. and Winey, M. (1996). The *Saccharomyces cerevisiae* spindle pole body duplication gene *MPS1* is part of a mitotic checkpoint. *J. Cell Biol.* **132**, 111–123.

Winzeler, E. A., Shoemaker, D. D., Astromoff, A., Liang, H., Anderson, K., Andre, B., Bangham, R., Benito, R., Boeke, J. D., Bussey, H. *et al.* (1999). Functional characterization of the *S. cerevisiae* genome by gene deletion and parallel analysis. *Science* **285**, 901–906.

Wood, K. W., Cornwell, W. D. and Jackson, J. R. (2001). Past and future of the mitotic spindle as an oncology target. *Curr. Opin Pharmacol.* **1**, 370–377.

Wu, H. I., Brown, J. A., Dorie, M. J., Lazzeroni, L. and Brown, J. M. (2004). Genome-wide identification of genes conferring resistance to the anticancer agents cisplatin, oxaliplatin, and mitomycin C. *Cancer Res.* **64**, 3940–3948.

Xie, M. W., Jin, F., Hwang, H., Hwang, S., Anand, V., Duncan, M. C. and Huang, J. (2005). Insights into TOR function and rapamycin response: chemical genomic profiling by using a high-density cell array method. *Proc. Natl. Acad. Sci. USA* **102**, 7215–7220.

Yuan, D. S., Pan, X., Ooi, S. L., Peyser, B. D., Spencer, F. A., Irizarry, R. A. and Boeke, J. D. (2005). Improved microarray methods for profiling the Yeast Knockout strain collection. *Nucl. Acids Res.* **33**, e103.

Zhu, H., Bilgin, M., Bangham, R., Hall, D., Casamayor, A., Bertone, P., Lan, N., Jansen, R., Bidlingmaier, S., Houfek, T. *et al.* (2001). Global analysis of protein activities using proteome chips. *Science* **293**, 2101–2105.

♦♦♦♦♦♦ NOTE ADDED IN PROOF

A recent paper showcases an improved affymetrix TA6 array and describes a suite of updated protocols with performance statistics.

Pieice, S. E., Fung, E. L., Jaramillo, D. F., Chu, A. M., Davis, R. W., Nislow, C. and Giaever, G. (2006). A unique and universal molecules barcode array. *Nat. Methods* **3**, 601–603.

18 RNA Gene Analysis

Cosmin Saveanu, Micheline Fromont-Racine and Alain Jacquier
Unité de Génétique des Interactions Macromoléculaires, Institut Pasteur (CNRS-URA2171), 25–28 rue du Dr. Roux 75724, Paris cedex 15, France

♦♦♦

CONTENTS

Introduction
Identification of ncRNAs genes
Transcripts characterization
RNP characterization, isolation and identification
Conclusion

♦♦♦♦♦♦ I. INTRODUCTION

Most genes encode proteins. Yet, a fraction of the genes do not. They are usually called non-coding RNA genes (ncRNA genes), albeit it would probably be preferable to call them non-protein-coding RNA genes. The most commonly known ncRNA genes are, of course, those of the nuclear or mitochondrial rRNA and tRNA, but this class also includes snRNAs and snoRNAs, the RNA components of the nucleases RNase P and MRP, of the telomerase and of the signal recognition particle. While these ncRNA genes only represent a relatively small fraction of the genome, they can hardly be considered as a minor class of genes because their products generally carry fundamental cellular functions. Moreover, their transcripts can represent up to 95% of the total mass of RNAs in the cells. Analysis of the products of RNA genes can involve a number of very specific techniques that will not be treated here because they are beyond the scope of this article. These include techniques such as those used to study nucleotide modifications within transcripts, the secondary and tertiary structures of RNAs or their localization within cells. In this chapter, we will start with a rapid survey of some of the approaches used to identify these peculiar genes and then we will describe some classical or more recent techniques used to study the products of these genes. Finally, we will have a look at some approaches used to analyze the RNAs in the context of ribonucleoprotein (RNP) complexes.

METHODS IN MICROBIOLOGY, VOLUME 36
0580-9517 DOI:10.1016/S0580-9517(06)36018-7

♦♦♦♦♦♦ II. IDENTIFICATION OF ncRNAs genes

In yeast, the identification of protein coding genes is relatively straightforward. In its most unsophisticated form, it relies on the identification of open reading frames (ORFs) longer than a statistically defined threshold, refined with the search of features such as putative splice sites. More recently, protein coding gene determination was made considerably more accurate by comparative genomic studies in which protein sequence conservations allowed a reliable annotation of the genome for protein coding gene (Dujon, 2005). Non-coding RNA genes lack features like ORFs, are not polyadenylated and are thus not present in EST databases, resulting in many of them, aside from the abundant rRNAs and tRNAs, being mostly overlooked for a long time. For example, until 1986, whether spliceosomal small nuclear RNAs (snRNAs) existed in yeast, as it had been shown in humans, remained a mystery. Over time, the combination of genetic, biochemical and functional studies eventually identified many of the ncRNAs that we now known in yeast. Surprisingly for a model organism such as *Saccharomyces cerevisiae*, very few systematic searches of ncRNAs have been ever performed.

Aside from the pioneering work of the groups of Christine Guthrie, in the 1980s (for review, see Guthrie, 1986) or, later, of Roy Parker (Olivas *et al.*, 1997), systematic searches of ncRNAs have essentially been focused on identifying additional members of already known RNA families such as tRNAs (Fichant and Burks, 1991; Lowe and Eddy, 1997), box C/D small nucleolar RNAs (snoRNAs) (Lowe and Eddy, 1999) or, recently, H/ACA snoRNAs (Schattner *et al.*, 2004; Torchet *et al.*, 2005). The only recent systematic and general search for ncRNAs that we are aware of was performed by McCutcheon and Eddy (2003) (see below).

Essentially, there are two entirely different, yet complementary, strategies to search for ncRNAs: experimental and computational. Of course, global computational searches were impossible before the completion of the whole genome sequencing and the first fishing for ncRNAs were thus experimental.

A. Experimental Strategies to Identify ncRNAs

The group of Christine Guthrie performed pioneering works in the 1980s to identify spliceosomal RNAs (for review, see Guthrie, 1986). The strategy used an enrichment step by immunoprecipitation with antibodies directed against the trimethyl guanosine cap because this structure had been shown in higher eukaryotes to be a specific feature of snRNAs. The enriched-RNA fraction was used for either direct characterization of *in vivo* or *in vitro* labeled RNAs or for cDNA cloning. These early studies identified more than a dozen RNAs including most of the snRNAs together with some RNAs that we now know as snoRNAs.

Affinity enrichment probably remains the most powerful approach to experimentally identify non-coding RNAs characterized by their association with specific proteins. Today's approaches take advantage of the availability of DNA microarrays that cover the whole genome of *S. cerevisiae* (and not only protein coding genes) to readily identify the position of the enriched sequences over the genome. A typical example of such an approach can be found in Torchet *et al.* (2005), where all, but one, known and previously unknown yeast H/ACA snoRNAs were identified at once by affinity purification of H/ACA snRNPs using tagged Nhp2 and Gar1 proteins, components of these RNPs. A detailed protocol for affinity purification of RNPs is in Section IV (''RNP characterization''). Several key points are worth mentioning for the success of the experiment when used in our laboratory for the identification of non-coding RNAs. First, it is important to note that an unambiguous distinction between the stably bound over unbound RNAs was possible only by the use of tandem affinity purification (Rigaut *et al.*, 1999). When only one step was used, the populations of bound and unbound RNAs could not be completely resolved. Second, following suggestions made in Peng *et al.* (2003), RNAs were chemically labeled with fluorescent dyes and directly used as probes on the arrays. This avoided an experimental bias introduced when using cDNAs as probes because both random oligonucleotide hybridization and reverse transcriptase processivity are affected by stable RNA structures that often characterize non-coding RNAs (note that direct RNA labeling precludes the use of sense strand oligonucleotides arrays commonly used for analyses of ORFs because it requires anti-sense features for detection). The sensitivity and specificity of this approach should be greatly enhanced by the use of tiling arrays (Mockler *et al.*, 2005), which are becoming available for the yeast genome.

One of the major drawbacks of the affinity purification approach is that it is restricted to the subclass of RNAs interacting with the tagged protein used as bait. Therefore, alternative strategies must be used for unbiased search of non-coding RNAs. Pioneering experimental works have been performed by Northern-blot analyses. For example, systematic Northern-blot hybridization for large (>2 kb) intergenic regions devoid of annotated features revealed a dozen new transcripts, coding or non-coding (Olivas *et al.*, 1997). This strategy was limited by the fact that only large intergenic regions were tested (89 altogether). Alternatively, systematic Northern analyses over complete chromosomes have been performed (for example, for the analysis of chromosome XI, see Richard *et al.* (1997) and also revealed a few transcripts that could not be assigned to known features. But, since this work was intended to characterize transcripts of known coding regions, the unassigned RNAs were not analyzed further.

Another straightforward systematic ncRNA search strategy consists of systematic sequencing of a cDNA library generated from total RNAs selected by their size (usually 50–500 nucleotide long) on polyacrylamide denaturing gels (for review, see Huttenhofer *et al.*, 2005). This approach has been very successful with a number of model organisms, but, to our knowledge, was not yet applied to *S. cerevisiae*. One of the major drawbacks of this approach is that it is strongly biased against low-abundance RNAs.

The SAGE approach (serial analysis of gene expression: Velculescu *et al.*, 1995, 1997) allows a more quantitative analysis of a transcriptome. In this strategy, short 14 nucleotide long DNA tag sequences are generated from the 3′ ends of polyadenylated transcritps, concatenated and sequenced. Since the tags are concatenated, a number of them (26 in average in Velculescu *et al.*, 1997) can be determined per sequencing reaction. Hence, with today's capillary sequencing machines, the determination of 20 000 tag sequences represents a moderate effort (with 20 000 tags, it was estimated that a transcript present at one copy per yeast cell has a probability of 72% to be detected). Analysis of 60 000 SAGE tags from yeast allowed the identification of transcripts present at levels as low as 0.3 transcripts per cell. Surprisingly, up to 10% of the SAGE tags, an important proportion, came from intergenic regions (Velculescu *et al.*, 1997). We have recently shown that this intergenic RNA class was enriched for transcripts stabilized by mutations affecting the nuclear exosome or an associated nuclear polyadenylation complex (TRAMP: LaCava *et al.*, 2005; Vanacova *et al.*, 2005) thought to specifically target aberrant transcripts (Wyers *et al.*, 2005). These intergenic unstable transcripts have been called CUTs for ''cryptic unstable transcript''. It is thought that the majority of them are devoid of genetic information. They would originate from spurious transcription and are normally discarded by the combined action of the TRAMP and exosome complexes (Wyers *et al.*, 2005). Nevertheless, it is likely that some of the intergenic SAGEs originate from genuine stable transcripts that remain to be characterized.

One important drawback of the initial SAGE technique when looking for ncRNAs is that it is based on the presence of a poly(A) tail. A more recent version of the SAGE technique, called Long-SAGE, allows the identification of longer tags (hence giving rise to much fewer ambiguities in the identification of the corresponding genomic sequence) derived either from the 5′ or 3′ polyadenylated ends of transcripts (Wei *et al.*, 2004). The 5′ Long-SAGE technique should allow the identification of non-polyadenylated ncRNAs, but, to our knowledge, this approach has not yet been used in yeast.

In conclusion, it appears that systematic, unbiased experimental search for ncRNAs is still incomplete in yeast although the required tools are available.

B. Computational Strategies to Identify ncRNAs

As for experimental strategies, computational approaches to search ncRNAs can be either generic or directed for given classes of RNAs, such as tRNAs or snoRNAs (Schattner *et al.*, 2005). Although these dedicated programs are very powerful and efficient at performing the task for which they have been designed, they will, by their nature, miss objects that present atypical features for their class. For example, the search for H/ACA snoRNA was particularly difficult because these ncRNAs are widely divergent, carry very little conserved nucleotides and, although the snoGPS program (Schattner *et al.*, 2004) proved amazingly efficient in identifying H/ACA snoRNAs in yeast, it still missed a few atypical guiding sequences (compare with the affinity purification approach by Torchet *et al.*, 2005). Most non-specialized ncRNA search programs take advantage of the fact that all known ncRNAs carry some conserved secondary structure (that can be restricted to their precursor forms in the case of the small interfering RNAs found in higher eukaryotes).

By far the most efficient way to identify RNA structures is to look for their conservation during evolution despite changes in sequence. The principle of these algorithms will thus be to look, among aligned conserved sequences found in non-coding regions, for nucleotide changes that will conserve potential base pairings (compensating or conservative changes). Such an approach should be particularly efficient in yeast because quite a large number of yeast genomes are available (12 at present see for review Dujon, 2005), with various degrees of divergence, and this number is likely to grow. One of these programs, QRNA, already identified quite a few ncRNAs when genomic sequences of only six yeast related species were compared (McCutcheon and Eddy, 2003). It is likely that applying this approach on more sequences will be even more sensitive. There are other recent programs to search conserved structures, such as RNAz (Washietl *et al.*, 2005), which has the advantage over QRNA to be able to compare more than two sequences at a time, but to our knowledge, it has not yet been used to search whole yeast genomes.

♦♦♦♦♦♦ III. TRANSCRIPTS CHARACTERIZATION

A. Characterizing the Size of RNAs

Probably the most widely used technique so far to characterize transcripts is Northern-blot hybridization. New technologies such as those based on DNA microarrays are now available and are powerful tools for genome-wide transcript analysis, but, aside from the new tiling-array technology that is emerging or specialized arrays (for a tiling array specialized for ncRNA maturation analysis, see Hiley *et al.* (2005), it is essentially useful for relative quantifications but will not be able to detect any structural variations or heterogeneities. Therefore, not only will Northern blots remain useful to confirm quantification data obtained from microarrays, but they will remain invaluable as a first approach for primary structure characterization, essentially size analysis, in particular because even tiling arrays do not allow the simple identification of multiple transcripts arising from a single sequence.

Depending on their size, isolated transcripts are separated either on agarose gels (above 500 nucleotides) or acrylamide gels (below 800 nucleotides). Of course, agarose and acrylamide gels can be used outside these limits, but the separation and resolution of the bands will not be optimal.

Hybridization is performed in the same way whether the RNAs were separated on either gel types. For low-abundance transcripts, we use random primed probes (Sambrook and Russell, 2001) with the ULTRAhyb buffer from Ambion, which we found in our hands to give an excellent signal over noise ratio. For abundant RNAs, such as pre-rRNAs or snoRNAs, 5′-end [^{32}P]-labeled oligonucleotides, used either with the ULTRAhyb-oligo from Ambion or the Rapid-hyb buffer from Amersham, give rapid and reliable results.

1. Northern blots from agarose gels

Since RNA molecules form secondary and tertiary structures, it is necessary to denature the RNA molecules before loading and keep them denatured during migration. A variety of methods are available to prepare denaturant agarose gels for Northern blotting: methylmercury, formaldehyde or glyoxal/DMSO gels (detailed protocols are published in Sambrook and Russell (2001). Methylmercury is certainly the most efficient denaturating agent, however, this method was largely abandoned because of the very high toxicity and volatility of the product. Formaldehyde is also toxic, and formaldehyde gels must be handled under a hood. Finally, in our hands, glyoxal/DMSO gels did not allow an optimal resolution of the large RNA molecules. We thus prefer the formaldehyde protocol when we need to precisely determine the size of transcripts on an agarose gel. The RNA samples are denatured prior loading by heating at 65°C for 2 min in deionized formamide (we use Formazol, a purified and stabilized formamide from Molecular Research Center, Inc., OH), containing 10 mM EDTA (pH 8.0), 0.05% xylene cyanol and 0.05% bromophenol blue.

Precise size determination is generally not required (except when characterizing a transcript for the first time) and we found that non-denaturing gels are adequate for most applications provided the RNA samples are denatured prior to loading (as above). This technique only works well with short runs and it is thus most effective on minigels. Ethidium bromide (0.3 μg/ml) can be included in the gel and the running buffer (1 × TBE composition). Although some secondary structures most likely form when the RNAs enter the gel, we nevertheless found that the method adequately separates transcripts to correct approximate sizes. The RNAs are then transferred for 2 h on a positively charged nylon membrane using a vacuum blotter in the presence of 50 mM NaOH that contributes to RNA denaturation during blotting.

2. Northern blots from polyacrylamide gels

Whereas 1500 bases is the mean length of transcripts coding for proteins in yeast, many non-coding RNAs are less than a few hundred nucleotides long. To separate these small RNA molecules, polyacrylamide gels are most useful. The gels are identical in composition to DNA sequencing gels, using 8.3 M (50% w/v) urea as a denaturing agent with 1 × TBE as running buffer (detailed protocols are published in Sambrook and Russell, 2001). Smaller gels, about 20 cm long, are most often sufficient since one nucleotide resolution is rarely necessary. The samples are denatured as for agarose gels. A semi-dry electro-blotting apparatus ("Trans-blot SD" from Bio-Rad, for example) is used to transfer the RNA molecules onto a membrane by electrophoresis (see detailed protocol 1).

Protocol 1. Electro-blotting of a polyacrylamide gel.

1. Prepare a positively-charged nylon membrane (we use Hybond-N+ from Amersham) by cutting it at the desired size and dipping it on 0.5 × TBE to get it wet.
   Prepare six 3-M Whatman paper sheets of the same size as the membrane and soak them into 0.5 × TBE.
2. Once the run is finished, stack on the anode plate three-soaked Whatman papers and the membrane on top, carefully removing trapped air bubbles by rolling a pipet on the membrane. Add some 0.5 × TBE buffer on top of the membrane such that it will be kept well soaked. Remove one glass plate from the gel and stick a dry 3-M Whatman sheet on it. Carefully peel off the gel from the remaining glass plate with the Whatman paper.
3. Cut it to the size of the membrane and lay it, the gel face down, on the wet membrane (note that when the acrylamide concentration exceeds 6%, the gel will not efficiently stick to the paper and it might be easier to directly peel the gel by hand off the plate and lay it directly on the wet membrane). Add 0.5 × TBE to the dry Whatman paper on the gel, so that it will be completely wet. Add three soaked 3-M Whatman paper and the cathode plate. Electroblotting is performed for 45 min at 10 V. Although this is probably not necessary with positively charged nylon membranes, we cross-link the wet membrane after transfer by UV irradiation (we use a "Stratalinker" from Stratagen, on the auto-cross-link position, which corresponds to 120 mJ).

B. Characterizing the Ends of Transcripts

Characterizing the extremities of transcripts implies two things; first the determination of the start and end sequences and second the determination of the modifications of these extremities. 5′-end and

3′-end processing and modification are essential for the localization, the stability and the functionality of transcripts and their determination is one of the first steps in describing RNAs.

1. Reverse transcription

There are many ways to estimate, with various degrees of precision, at which nucleotide a transcript starts and ends. For the 5′ end, the most straightforward method is primer extension. Nevertheless one must be aware of some common artifacts that can be observed with this technique. First, premature termination is common because reverse transcriptase is sensitive to secondary structure. In some instances, blocks to reverse transcription can be so strong to become misleading for the determination of the 5′ end of a transcript (some base modifications can also induce strong stops, but this problem is essentially restricted to rRNA). To minimize this problem, it is thus advisable to perform reverse transcription at the highest possible temperature. We generally use the RNase H^- SuperScript II reverse transcriptase from Invitrogen and, although its standard incubation temperature is 42°C, we often use it at 45°C for structured RNAs. Moreover, a new form of the enzyme, SuperScript III, is given to work between 45°C and 55°C. Invitrogen also propose another reverse transcriptase, Thermoscript that should work at temperatures up to 70°C. While we have not tested these new enzymes in our laboratory, we think that they should be particularly useful with highly structured transcripts. Note that one should make sure that the Tm of the primer is adapted to these increased temperatures. The use of up to 10% DMSO has also been reported to help the generation of full-length cDNAs from G+C rich structured RNAs. A detailed standard protocol is given below (Protocols 2 and 3).

Protocol 2. Reverse transcription with SuperScript II and labeled oligonucleotides.

1. 1–5 μg total RNA and 0.5 μl 5′-end-labeled oligonucleotide (prepared as described below; ~0.3 pmol) were mixed in a volume of 2.5 μl completed with H_2O.
2. After denaturation for 2 min at 85°C, the labeled oligonucleotide was annealed to specific RNA by 5 min at 42–45°C. This is best performed in a thermocycler with a heated lid. 5 μl of RT reaction mix containing 1.5 μl 5 × first strand buffer (5 ×: 250 mM Tris-HCl pH 8.3 room temperature, 375 mM KCl, 15 mM $MgCl_2$); 1.5 μl 100 mM dithiothreitol (DTT); 0.2 μl 25 mM dNTP (25 mM each dNTP at neutral pH); 0.5 μl Actinomycin D (1 mg/ml; optional); 0.5 μl Superscript II from Invitrogen (200 units/μl); 0.8 μl H_2O (to 5 μl) were added to each tube.
3. After incubation for 30–60 min at 42–45°C, the reaction was stopped by adding 6-μl loading buffer (deionized formamide

containing 10 mM EDTA (pH 8.0), 0.05% xylene cyanol and 0.05% bromophenol blue). Samples were denatured for 3 min at 85°C prior to loading on a urea-containing denaturing sequencing gel.

Protocol 3. [^{32}P] 5′-end labeling of oligonucleotides.

1. 1 µl of oligonucleotide at 10 µM (10 pmol); 1 µl 10 × kinase buffer (10 × : 0.7 M Tris-HCl (pH 7.6), 0.1 M $MgCl_2$, 50 mM dithiothreitol); 3 µl [γ-^{32}P] ATP (≥3000 Ci/mmol; 10 mCi/ml (370 MBq/ml), i.e. ~9 pmol) and 1 µl T4-polynucleotide kinase at 10 units/µl were mixed in a final volume of 10 µl completed with H_2O.
2. After incubation at 37°C for 30 min the enzyme was inactivated at 85°C for 5 min.
3. Use directly or, preferentially, after purification on a spin column. We use MicroSpin G-25 columns from Amersham (note that we apply the 10 µl reaction mix without dilution, even though the recommended minimum volume is 25 µl; the elution volume is then around 15–20 µl).

We add Actinomycin D to the reaction because it avoids second strand synthesis, which occurs when the reverse transcriptase copies the cDNA it just synthesized. Longer cDNA artifacts can be misleading when interpreting the reverse transcription results.

The efficiency of the reaction varies depending on the accessibility of the oligonucleotide on the RNA. When performing primer extension for the first time on a transcript, it is thus a good practice to perform at least two reactions, each with a different oligonucleotide. It is also important to make sure that the observed products are specific. Two primers hybridizing to two different sites located at a distance of *N* nucleotides should give rise to cDNAs differing in length by N nucleotides.

2. RNase protection

RNase protection is a versatile approach to determine the primary structure of a transcript as it can characterize both the 5′ and the 3′ ends and even spliced products. The principle of this method is to use anti-sense transcripts that can be labeled either internally or at their 5′ or 3′ ends. These anti-sense transcripts are annealed with the cellular RNAs and digested with RNases specific for single-stranded RNAs. If a complementary transcript is present in the cell, the corresponding portion of the labeled anti-sense RNA will be protected from RNase digestion (Figure 1A).

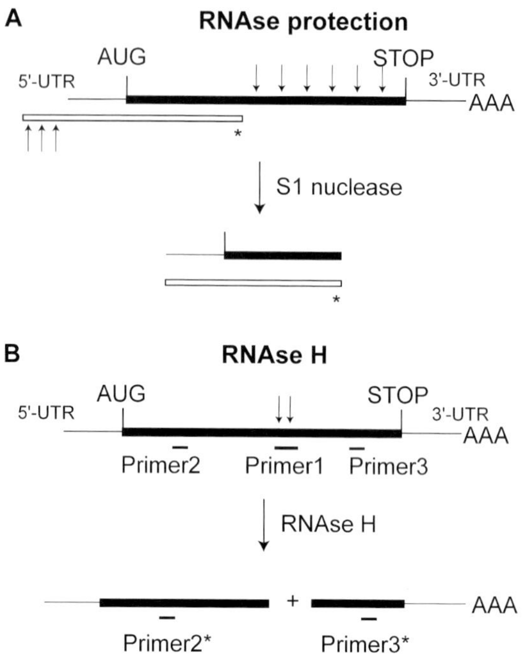

Figure 1. (A) RNAse protection. An RNA sample is hybridized with a labeled anti-sense template. The radiolabeled anti-sense probe can be synthesized using Ambion's MAXIscript™ Kit with [^{32}P] UTP (800 Ci/mmol, 10 Ci/ml). After hybridization, all remaining single stranded RNAs are digested by single-strand-specific RNases (many RNases are single strand specific; a mixture of RNase A and T1 is commonly used). The radiolabeled digested products are then separated on a denaturing polyacrylamide gel and visualized by autoradiography or phosphorimaging. (B) RNAse H treatment. Primer 1 targets the cleavage site for RNAse H. After RNAse H treatment, the digested products are separated on a denaturating polyacrylamide gel, transferred on a membrane and the size of the molecules is determined by hybridization with a radiolabeled primer (as described above) specific for either the 5′ (Primer 2) or the 3′ end (Primer 3) of the molecule. The asterisks indicate radiolabeled primers.

It is essential to have a portion of the anti-sense RNA that is not complementary to any sequence in yeast, to serve as internal control for digestion (the protected product must always be smaller than the starting labeled RNA). Leader sequences on T7, T3 or SP6 PCR fragments or plasmid are useful for this. Conversely, if an end label is used, one must be sure that it will correspond to a portion of the anti-sense RNA that is fully complementary to the cellular RNA. Internally labeled RNAs are thus more versatile and easier to use. For example of use of this procedure and a protocol, see Chanfreau *et al.* (1998).

3. RNAse H digestion

RNAse H digestion is an alternative to the nuclease protection assay that permits, depending of the position of the oligonucleotide complementary to the RNA, the determination of either the 5′ end or the 3′ end of an RNA molecule (Sambrook and Russell, 2001). RNAse H is an endonuclease that degrades specifically the RNA in RNA/DNA duplexes, producing 3′-OH and 5′-P ends. RNase H digestion is performed after annealing of the RNA with the appropriate oligonucleotide, cleaving the RNA molecule in a 5′ and a 3′ part (Figure 1B). The size of each part of the digested RNA is then determined by Northern-blot analysis (as described above). Knowing the size of each digested product and the site of cleavage, it is simple to deduce the approximate 5′ and 3′ positions of the transcript.

RNAse H treatment can also be used to determine the poly(A) tail size of a given RNA by using oligo(dT) to target cleavage of the poly(A) tail, in addition to a specific oligonucleotide close to the 3′ end of the mRNA. By comparing the sizes of the RNase H digestion products generated with or without oligo(dT), the size of the poly(A) can be deduced (for an example, see Badis *et al.*, 2004).

4. RACE

Rapid amplification of cDNA ends (RACE) is a sensitive technique to determine either the 5′ end or the 3′ end of mRNAs having low levels of expression (Frohman *et al.*, 1988). In many cases, mRNAs are not expressed to a sufficient level to be detected by a simple primer extension procedure and a PCR step is used to amplify the signal. Using this procedure, the isolation of poly(A) RNA is usually no longer necessary.

To determine the 3′ end of an mRNA, cDNAs are synthetized by reverse transcriptase with an oligonucleotide containing about 17-dT residues and an adapter sequence. The 3′ ends from the entire population of mRNAs can be reverse transcribed in one experiment. A particular 3′ end is then amplified using the adapter linker and an oligonucleotide that is specific of the mRNA of interest. The specificity and efficiency of the experiment can be increased by performing an additional PCR step using a second specific primer internal to the initially amplified region (nested PCR).

The initial 5′ RACE procedure consisted of the addition of either a linker with a ligase or a poly(A) tail with the terminal deoxytransferase to the first strand of the cDNAs synthesized from the mRNAs. Then, an oligonucleotide complementary to the added sequence, in conjunction with a primer specific for the mRNA, was used to amplify the sequence between the linker and the specific primer.

Different variants of this initial RACE method appeared over time. One of the reasons the original RACE procedure was not always successful was that every cDNA can be used as a template for the addition of the linker, even those corresponding to premature terminations of the reverse transcriptase or generated from partially degraded RNAs. To circumvent this problem, an improvement of this technique, called RL-PCR (Fromont-Racine *et al.*, 1993) or RLM-RACE (Liu and Gorovsky, 1993) has been described (Figure 2).

The RNA sample is first treated with calf intestine phosphatase (CIP) or bovine alkaline phosphatase (BAP) to remove the 5′-PO_4 end from degraded RNAs and DNAs. This step is important to avoid background due to the degradation of the RNAs. Second, the cap structure of the mRNA is removed by using the tobacco acid pyrophosphatase (TAP), generating a 5 monophosphate end. The following step consists in the addition of an RNA linker with T4 RNA ligase. Because of the two preceding treatments, only the genuine mRNA 5′ ends receive an adapter. Then, a cDNA is synthesized by primer extension using an oligonucleotide (Primer 1) specific for the RNA of interest. Finally, the cDNA is amplified by PCR with a RNA specific reverse primer (Primer 2) and a forward primer (Primer 4) complementary to the linker added to 5′ ends (see Figure 2). Sometimes, the use of an additional nested PCR step, using a specific oligonucleotide that hybridized just upstream to Primer 2, is required to increase the sensitivity and specificity of the procedure. In that case, it is this primer that is labeled. Finally, the

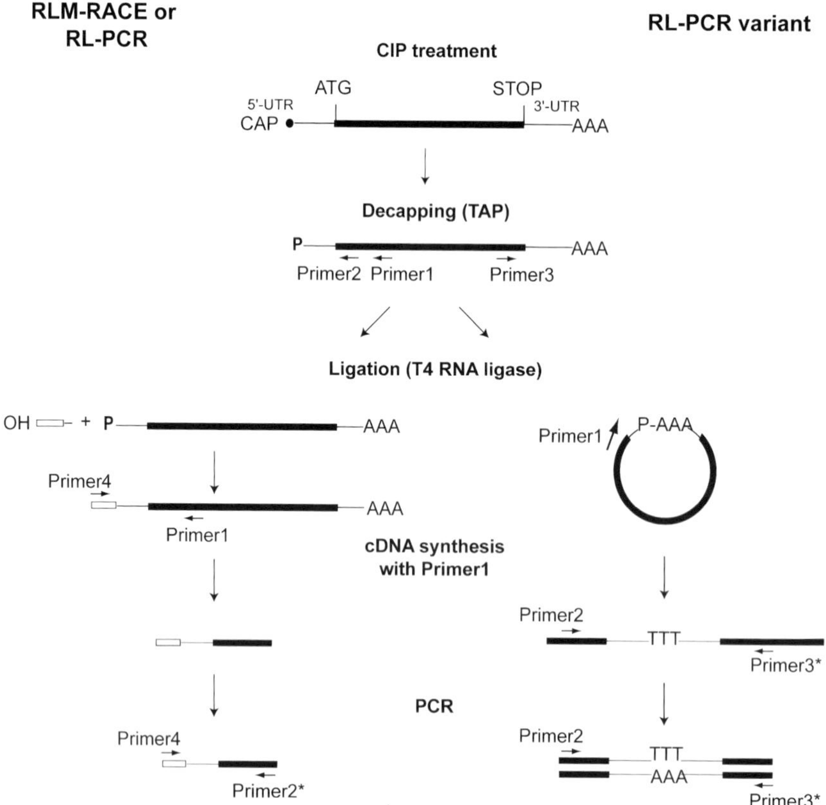

Figure 2. Principle of the RL-PCR procedure and its variants (see text for explanations).

PCR products are separated on a denaturating polyacrylamide gel followed by autoradiography. The modified RACE procedure has a number of advantages compared to the initial 5′ RACE. Only cDNAs from capped mRNA are generated, leading to an increase of the specificity and sensitivity. Restriction enzyme sites can also be included in the primers to enable the cloning and sequencing of the PCR products.

An adaptation of the RL-PCR has been described that allows the determination, in a single experiment, of both the 5′ and 3′ ends as well as the size of the poly(A) tail of an mRNA (Couttet *et al.*, 1997). The first two steps are unchanged (Figure 2), but the ligation of the RNA adapter to the 5′ end of the mRNA is replaced by a self-ligation of the RNA molecules under low RNA concentration, which promotes self-ligation (circularization). A cDNA is then synthesized with a primer specific for the RNA of interest. The poly(A) tail and the 5′ and 3′ ends surrounding it are then amplified by PCR using nested divergent primers that flank the junction between the 5′ and the 3′ ends (see Figure 2). The reverse oligonucleotide (Primer 3) is radiolabeled. An aliquot of the RNA population, subjected to oligo dT/RNAse H treatment to remove the poly(A) tail prior to circularization, is treated in parallel. Finally, the PCR products are separated on a denaturating polyacrylamide gel. By comparison with the RNA subjected to an oligo dT/RNAse H treatment, the poly(A) tail size can be deduced. The PCR products can also be sequenced, either directly or after cloning.

♦♦♦♦♦♦ IV. RNP CHARACTERIZATION, ISOLATION AND IDENTIFICATION

In living cells, RNAs are not free but associate with proteins in complexes known as ribonucleoproteins or RNPs. For ncRNAs, the associated proteins often contribute to catalytic activities or specificities. The proteins are also essential determinants of the stability of the complexes. For example, the stability of mRNAs, and implicitly the expression of the corresponding proteins, is often conditioned by their association with specific proteins. As hypothesized by Keene and Tenenbaum (2002), the protein composition of mRNPs appears to be a key element in post-transcriptional regulation of classes of functionally related mRNAs. For example, in yeast, the isolation of mRNAs associated with Puf1, 2, 3, 4 or Puf5 and their analysis using microarrays revealed a strong correlation between the functions of mRNAs and their association to specific Puf proteins. Changes in the levels of only one Puf protein potentially affect the stability and thus the expression of classes of functionally related mRNAs (Gerber *et al.*, 2004).

Large and complex RNP particles such as the ribosomal subunits require an unexpectedly high number of proteins, over 200, for correct ribosomal RNA maturation, RNP assembly and intracellular transport (for a review see Fromont-Racine *et al.*, 2003). On many occasions a function for these pre-ribosomal factors was hypothesized only on the basis of their identification as components of large RNPs, precursors to the ribosomal subunits. The discovery of RNA association with proteins of known function may also provide the necessary hints as for the potential role and mechanisms of RNAs action. For example, a transcriptional regulatory role for the abundant bacterial 6S RNA was established in *E. coli* on the basis of its physical association with the RNA polymerase holoenzyme (Wassarman and Storz, 2000). The first hint to this physical association was RNA polymerase and 6S RNA co-sedimentation on glycerol gradients.

The study of RNPs makes use of all known biochemical methods used for protein–protein complexes. In most cases, first estimates about the formation and size of an RNP are obtained by sedimentation analysis, a method that is easy to perform, robust and highly informative especially for very large complexes that are difficult to separate by other means. Affinity purifications of RNPs are the method of choice to isolate and identify the components of these complexes. However, other biochemical methods were successfully used when a specific catalytic activity of a particular fraction was followed during several purification steps. Two forms of an RNA polymerase II carboxy-terminal domain (CTD) kinase complex, separated on glycerol gradients and immunoprecipitation allowed the identification, in the larger complex, of the 7SK human RNA, an abundant RNA of previously undescribed function (Nguyen *et al.*, 2001). The so-called genetic methods derived from the two-hybrid system allow very sensitive confirmation or even screening for novel RNA–protein interactions by the three-hybrid approach (SenGupta *et al.*, 1996; Hook *et al.*, 2005).

In the following section we will describe some of the methods used to isolate and characterize RNPs with examples of successful use of these different methods in yeast, when such examples are available.

A. Sedimentation to Estimate RNP Size

As stated in the previous section, the sedimentation behavior of many if not all RNPs studied to date has been tested. Separation of molecules is obtained by centrifugal acceleration in solution and sedimentation through increasing concentrations of sucrose or glycerol. Larger molecules have higher sedimentation rates and since most RNPs are globular, sedimentation rates correlate well with molecular weight. The approximate sedimentation rate or coefficient expressed in Svedberg units (S – 10^{-13} s) is currently used for large particles like the 40S and the 60S ribosomal subunits. Even larger macromolecules are easily distinguished by ultracentrifugation. For example, visualization of ribosomes actively translating mRNAs on sucrose gradients, first described in 1963 when the term polyribosomes was coined, is still a method of choice to study mRNAs translation (for an historical account of polysome discovery see Warner and Knopf, 2002) (Figure 3).

With DNA microarrays, the distribution of mRNAs extracted from various fractions of the polysome gradients can be examined and the translational status of every mRNA can be described under different growth conditions. For example, when glucose is withdrawn from the yeast culture medium, a drastic reduction in global translation is witnessed by a decrease in the level of polysomes (Ashe *et al.*, 2000). The redistribution of yeast mRNA to lower or higher translation rate populations could be assessed by using sucrose gradients followed by mRNA level measurements using DNA microarrays (Kuhn *et al.*, 2001). More elaborate procedures have been recently described to test not only the number of ribosomes bound to a specific mRNA but also their density along the RNA (ribosome density mapping, Arava *et al.*, 2005). A general protocol for obtaining cell extracts and running sucrose gradients for polysome analysis is provided below (Protocol 4). An example of RNA detection after sedimentation on sucrose gradients is shown in Figure 4.

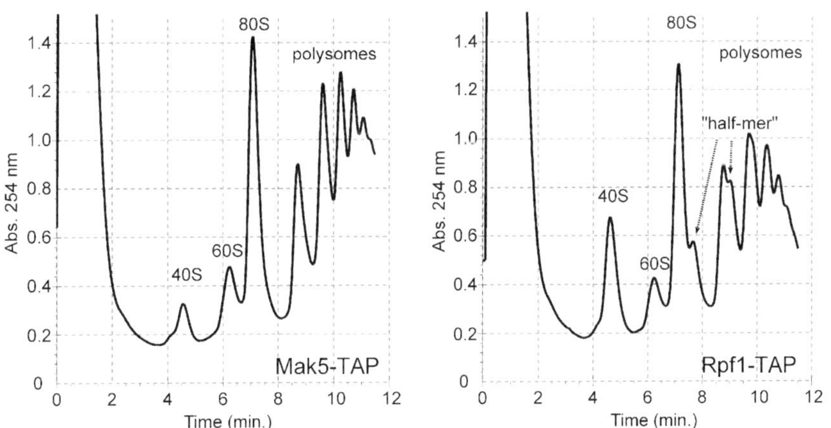

Figure 3. Example of polysome profile analysis and tag effects. Two strains are compared in which the pre-ribosomal factors Mak5 or Rpf1 are expressed as fusion proteins with the TAP tag (Rigaut *et al.*, 1999). Both proteins are involved in 60S ribosomal subunit biogenesis. The C-terminal fusion with the TAP is only deleterious for Rpf1 function as seen for the disequilibrium between free 40S and free 60S ribosomal subunits. Polysome levels are also slightly lower in the Rpf1-TAP strain, which is also slow growing. A shouldering of the 80S and disome peaks is visible and corresponds to "half-mers", initiation complexes that are blocked with 40S loaded on the mRNAs but without joining with 60S particles because of the relative 60S ribosomal subunits deficit.

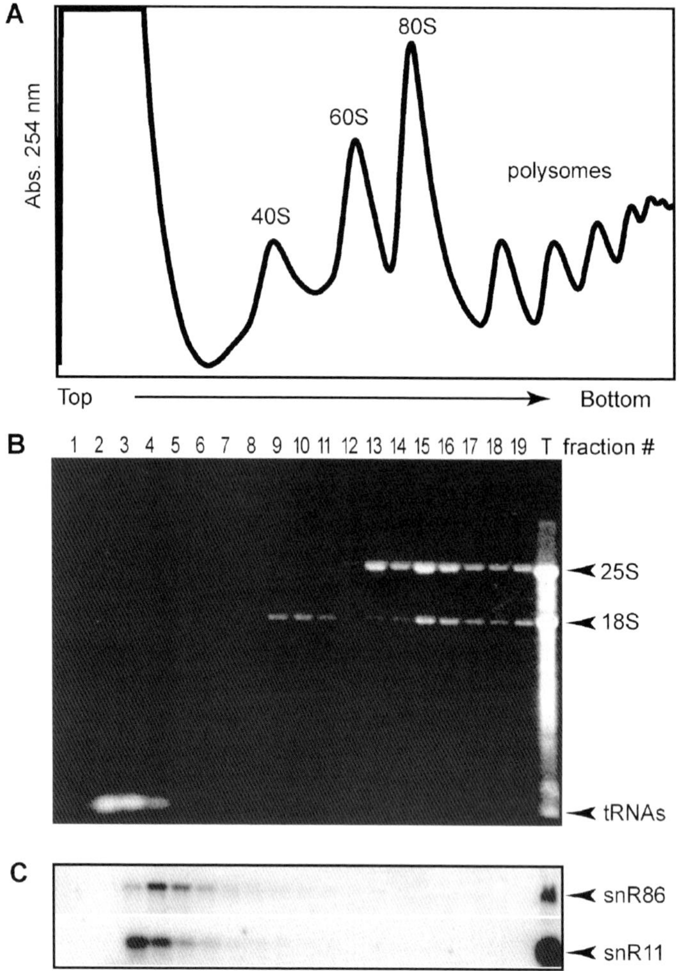

Figure 4. Detection of RNAs in sucrose gradient fractions. (A) UV absorbance profile allows the designation of the major peaks, free ribosomal subunits 40S and 60S, 80S ribosomes and polyribosomes. (B) RNA was extracted from fractions collected from the top of the gradient with phenol–chloroform, precipitated and solubilized in water. After addition of deionized formamide and brief denaturation at 85°C, RNAs were separated on a 1% agarose gel. (C) After transfer to a charged nylon membrane, specific snoRNA were detected by hybridization with [^{32}P]-labeled specific primers (see above).

Protocol 4. Polysome profile analysis in yeast.

1. *Gradient preparation*: To obtain many identical sucrose gradients for polysome profiles we use a convenient method originally described by Luthe (1983). Equal volumes of sucrose solutions (2.3 ml) are added to the ultracentrifuge tube (Beckman SW41-for other tubes adjust depending on their volume) beginning with a 50% sucrose solution. After quick

freezing at −80°C for 30 min, a 40% sucrose solution layer is added and frozen. The procedure is repeated with 30, 20 and 10% solutions. A uniform sucrose gradient forms during thawing (overnight at 4°C or on the bench in 2 h).

2. *Cell breaking buffer*: 10 mM Tris-HCl pH 7.4, 100 mM NaCl, 30 mM $MgCl_2$, 50 μg/ml cycloheximide and 1 tablet of protease inhibitor for 50 ml buffer.
3. *50% sucrose solution*: 50% sucrose (w/v), 50 mM Tris-HCl pH 7.4, 12 mM $MgCl_2$, 50 mM NH_4Cl and 1 mM DTT. The other concentrations of sucrose are easily obtained by diluting the 50% solution with buffer containing no sucrose to the appropriate concentration.
4. *Extracts*: Exponentially growing cells are disrupted with glass beads. A single experiment requires 150 ml of a culture at 0.4–0.6 $OD_{600\,nm}$. Cycloheximide from a 1000 × stock in ethanol (50 mg/ml) is added to the culture prior to cell collection to block translation and preserve ribosome association with the mRNAs. The cultures are cooled to 4°C in an ice bath and cells are collected by centrifugation at 4°C. The pellet is resuspended into 10 ml of pre-cooled breaking buffer into pre-cooled 15 ml tubes. After a second centrifugation step, the pellet is resuspended into 200 μl of pre-cooled breaking buffer. The buffer contains salts and a high concentration of $MgCl_2$ required for polysome integrity. The cell suspension, in a final volume of about 600 μl, is transferred into a 1.5 ml tube containing 500 μl of pre-cooled glass beads (425–600 μm, acid washed, Sigma). The cells are disrupted by 10 vigorous vortexing cycles of 30 s, separated by cooling on ice for 30 s. To eliminate the glass beads and cell debris, the tubes are centrifuged at maximum speed (16 000*g*) for 10 min at 4°C. The cellular extract is transferred to another tube and centrifuged again at 16 000*g* for 10 min. The absorbance of the extract is measured from a 1/500 dilution. An absorbance of 50–100 at 260 nm should be obtained. The cellular extract is stored at −80°C as aliquots corresponding to 10 $OD_{260\,nm}$. The aliquots are sensitive to freeze thawing, thaw each sample just before layering on the gradient.
5. *Ultracentrifugation*: 10 units of the extract are carefully loaded onto the top of the 10–50% gradient and sedimented by centrifugation at 39 000 rpm (188 000*g* average) for 2.45 h at 4°C in an SW41Ti rotor (Beckman). Lower centrifugation speeds can be used when the polysome distribution of mRNA is tested (35 000 rpm, 151 000*g* average). Twenty-four fractions of 0.5 ml are collected using an ISCO/Brandel (Gaithersburg, MD) fractionator and an in-line UV detector (254 or 260 nm). We do not favor the method of gradient recovery by piercing the bottom of the tube and found preferable to inject the dense liquid Fluorinert FC40 (3*M*) through a thin

stainless-steel tube (supplied as an accessory) inserted from the top to the bottom of the tube to displace the gradient upward.

Equally effective is the collection of the gradient from the top to the bottom by using a gradient fractionator originally from Haake Buchler instruments and now from Labconco (Kansas City, MO). However the passage of fractions through the peristaltic pumps tends to lower the resolution of gradient separation. If the gradient profile is not to be followed, simple manual collection of 0.4–0.5 ml from top to the bottom of the tube will nicely separate the different fractions.

Estimates of the sedimentation rate of a given complex may be obtained by calculation methods based on the isokinetic properties of 5–20% sucrose gradients where an equilibrium between increasing acceleration and increasing sucrose concentration allows movement of the molecules along the gradient at constant speed (for details and theory refer to Steensgaard *et al.*, 1992). In most cases calibration by the use of total cell extracts and identification of 40S, 60S and 80S peaks or the use of known proteins allows raw estimates of the sedimentation rates to be obtained. The kit of high molecular weight protein calibration for gel filtration experiments from Amersham Biosciences is excellent to perform calibration of sucrose or glycerol gradients up to a sedimentation coefficient of 19S (thyroglobulin).

Gradients also offer a global view of the different complexes to which a cellular RNA might be associated. For example, the small nucleolar U3 RNA, required for ribosome biogenesis, can be found to sediment into at least two different peaks, one at about 10–15S and another, broader, around 80S (see for example Figure 7 in Billy *et al.*, 2000). This sedimentation pattern is due to U3 association with a few "core" proteins in a small RNP particle. The association of this U3 particle to pre-ribosomes is most probably responsible for the presence of the second peak at higher sedimentation coefficients.

For a known RNA, the sedimentation analysis of a total cellular extract, with detection of the RNA in the different fractions with or without proteinase K treatment of the extract should indicate if the RNA is present in a complex with proteins or other RNAs. In rare cases, specific RNP complexes are exceptionally resistant to proteinase K treatment, for example the bacterial 50S ribosomal subunits (Noller *et al.*, 1992).

Sometimes, a cellular RNA is difficult to detect or the gradient distribution does not show clear peaks. One alternative is to use *in vitro* transcribed, radiolabeled RNA incubated with the cellular extract and analyze it on a gradient. Gradients run in parallel with RNA alone should be used as a control of the sedimentation pattern without cellular extract. An even better control is point-mutated RNA, known to have lost its function. Rpl28/Cyh2 pre-mRNA, used as a model RNA to study splicing *in vitro*, associates with the spliceosome and sediments with a rate of about 40S, while the same RNA with a point mutation that removes the 5′ splicing acceptor site is no longer found in the 40S peak (Brody and Abelson, 1985). These simple sedimentation experiments could show the association of the RNA with a large 40S complex and confirmed the specificity of 5′ splice site recognition by the spliceosome.

B. Affinity Purification of RNP Complexes

Both classes of components of the RNPs, the RNA or the associated proteins, can be used for affinity purification of the complexes. When available, methods of enrichment prior to the affinity purification step are recommended. One example of highly successful RNP isolation by affinity methods made use of antibodies against the 2,2,7-trimethylguanosine cap structure of the uridine rich, U snRNAs. The pre-requisite to successful purification of the complexes was the isolation of nuclear content from mammalian cells as a first enrichment step (Bringmann *et al.*, 1983). While nuclear isolation in yeast is more difficult, methods have been described and used to uncover, for example, the intermediates in ribosomal RNA maturation (Kruiswijk *et al.*, 1978). The method was recently applied to determine the kinetics of synthesis, maturation and export of ribosomal RNAs in the absence of the proteins of the small ribosomal subunit (Ferreira-Cerca *et al.*, 2005). Pulse-chase experiments combined with nuclear-cytoplasmic fractionation allowed the authors to distinguish between mutants that blocked ribosomal RNA maturation in the nucleus and mutants that also affected or only affected the export of pre-40S particles to the cytoplasm.

1. Protein tagging and TAP

Since its introduction in 1999, the tandem affinity purification method (TAP) has become a technique of choice in isolating protein complexes in yeast (Protocol 5). On many occasions, once the protein composition of a complex had been established, an RNA component could be also recovered and analyzed. Protein association is thus a good method to identify novel RNAs and has been used with success, for example to identify novel bacterial regulatory RNAs that all link the Lsm-like protein Hfq (Wassarman *et al.*, 2001). If a protein is common to many RNPs that only differ by their RNA component, microarray analysis of the pulled-down RNAs can be used to identify novel RNPs of the same family. This method was successfully used in yeast for the H/ACA box small nucleolar RNAs (snoRNAs) (Torchet *et al.*, 2005). Protein tagging does not preclude the use of specific antibodies, if available.

Protocol 5. TAP for RNPs.

1. A typical purification starts with 4 l of yeast culture in YPD medium under exponential growth. Cells harvested by centrifugation and washed once in cold water should weight 5–10 g and may be kept frozen at −80°C. If nitrogen grinding is envisaged for cell disruption, cells can be suspended directly in breaking buffer (1:1) and frozen as beads by dropping the suspension directly to a tube containing liquid nitrogen.
2. Breaking buffer: 10% glycerol, 0.1*M* NaCl, 0.1*M* TrisHCl pH 8, complete protease inhibitors (Roche)-2 tablets/50 ml, ribonucleoside vanadyl complexes (New England Biolabs, ref. S1402S, 200 mM; keep in single-use aliquots at −20°C, final concentration 10–20 mM heat at 65°C for 5 min and add to suspension in breaking buffer). A comparison of different methods of cell disruption has shown that grinding a liquid nitrogen frozen yeast pellet in the presence of RNAse inhibitors best preserves the integrity of mRNAs (Lopez de Heredia and Jansen, 2004). For the study of stable RNAs, the French Press proved to be very convenient, especially because handling of large volumes of cell suspension is straightforward. Liquid nitrogen grinding of cell suspensions is rather laborious to perform manually on large volumes. Different grinding devices may be used with frozen cell suspensions: Mortar Grinder RM100 (Retsch, Germany), 6750 Freeze/Mill (SPEX SamplePrep LLC, Metuchen, NJ) or domestic grinders. Note that Vanadyl complexes are dissociated and completely inactivated by EDTA at stoichiometric concentration. RNAsin (Promega) may also be added to the initial extract to a concentration of 200 U/ml (Promega suggests 1000 U/ml).
3. Rapidly thaw the cells, add an equal volume of breaking buffer and homogenize. Pass once through the French Press at high pressure (20 000 psi). A second passage improves the amount of recovered extract by 30–40% but may be deleterious for the RNAs since heating of the sample occurs at each passage. Transfer to centrifuge tubes that support high speeds (Nalgene in Beckman JS13–1 11 300 rpm or JA25–50 at 130 000 rpm); centrifuge for 40 min at 22 000*g*, 4°C. Decant the supernatant in 50 ml Falcon tubes. The extract should be yellow opalescent (unless vanadyl has been added; in that case it should be brown; if it is green the vanadyl has been oxidized and is ineffective as an RNAse inhibitor). When decanting, most of the lipid phase (seen at the top of the extract) should stick to the wall of the centrifugation tube. Finally, around 8–12 ml of total extract are obtained with a total protein concentration superior to 10 mg/ml. Freeze the extract in liquid nitrogen and keep it at −80°C or continue with the purification. A high extract concentration improves the stability of the complexes.

4. Optional step – to eliminate large and abundant complexes like the ribosomes, it is possible to perform an ultracentrifugation step as described in (Krogan *et al.*, 2004). However, we had difficulties in reproducing the exact conditions described by the authors. As a starting point, use 38 000 rpm in a Beckman SW41 Ti rotor (average rcf of 178 000*g*) for 40 min.
5. Optional step – it may be of interest to tightly control the composition of the extract by dialysis against a buffer with the desired characteristics. Salt concentration as well as pH might be critical for the complex integrity (an example may be found in Krogan *et al.*, 2002); compare the Spt16-TAP complex in 150 mM NaCl and in 125 mM NaCl).
6. *IgG Sepharose binding* is done in a binding buffer (IPP100) containing 20 mM Tris-HCl pH 7.4, 0.1 M NaCl, 0.1% NP40 (IGEPAL CA-630, Sigma). The same buffer with variations will be used for TEV protease digestion and calmodulin binding.
   Equilibrate 0.2 ml of IgG Sepharose beads (Amersham Biosciences) for each purification in the IPP100 binding buffer. Recover beads by centrifugation at 150*g* for 1 min. Do not exceed the centrifugation speed since Sepharose beads are fragile and collapse at higher centrifugal forces.
   After thawing the extract, *adjust its pH to neutrality* (tested with pH paper) by adding Tris-HCl pH 8, 1*M* (generally 0.3–0.5 ml). Add NP40 to a final concentration of 0.1%. Add the IgG Sepharose beads suspension to the cell extract. Allow binding for 2 h at 4°C on a rotating wheel.
7. Beads washing may be performed in many ways, one that is very convenient is by using small Mobicol columns, provided with filters and Luer adapters. This method allows quick washing of the beads with large volumes of buffer without beads loss and also complete recovery of the eluate. Washing of the IgG Sepharose beads is performed using the TEV buffer (IPP100 supplemented with 1 mM DTT), which is required for the activity of the TEV enzyme, a cysteine protease. Recover the beads by low-speed (150*g*) centrifugation for 1 min. Resuspend the beads in 1-ml TEV buffer in Eppendorf tubes, repeat centrifugation. Decant the supernatant and add 0.5 ml TEV digestion buffer and transfer to a MobiCol (0.8 ml-MoBiTec M1002 with M2135 filters) microcolumn. Amersham Microspin G25 columns are convenient too, after removal of the Sephadex. Let run dry or push air with a syringe.
8. After washing with 15 ml of TEV digestion buffer, add 400 μl of TEV digestion buffer supplemented with 100 units (10 μl) of TEV (AcTEV, Invitrogen) protease and 400 U of RNAsin and incubate for 1.5 h at 16°C on a rotating wheel. The TEV eluate is recovered by pushing air with a syringe. You may keep 10–30 μl of the eluate as a control. On several occasions, the second purification step might not be efficient.

9. *Binding to calmodulin beads and final elution.* A buffer containing calcium ions is required for the binding of the calmodulin binding peptide to the calmodulin. Equilibrate 150 μl (300 μl suspension) of calmodulin Sepharose beads (Amersham Biosciences) with the calmodulin-binding buffer (IPP100+2 mM $CaCl_2$). Add 1.5 μl $CaCl_2$ 0.5 M to the TEV eluate, transfer to a MobiCol and mix with 150 μl pre-equilibrated calmodulin beads resuspended in 200 μl calmodulin-binding buffer. Allow binding on a rotating wheel at 4°C for 1 h. The calmodulin-binding buffer is the IgG Sepharose binding buffer to which $CaCl_2$ is added to 2 mM. Salt may be added to the elution buffer if, for example, an enzymatic activity of the TAP eluate will be tested.
10. Wash with 10 ml of calmodulin binding buffer. Push air to remove the last droplets of buffer. Seal the MobiCol. Add 450 μl calmodulin elution buffer (20 mM Tris HCl pH 8 with 5 mM calcium chelator EGTA), mix well and leave for 5 min at room temperature. RNAs and proteins are next extracted from the TAP eluate by usual methods. We recommend the use of the methanol–chloroform method for protein precipitation, especially when the concentration of the TAP eluate is low (Wessel and Flugge, 1984).

C. RNA Tagging *In Vivo* and *In Vitro*

If an RNA of interest is fused to an RNA motif specifically recognized by a protein, the isolation of RNPs with the use of a tagged RNA becomes very similar to their recovery when using a tagged RNP protein or specific antibodies. A successful example of *in vivo* RNA tagging and RNP purification is the isolation of U3 snoRNP complexes from a yeast strain expressing a U3 RNA in which a fragment of the human U1A pre-mRNA was inserted (Watkins *et al.*, 2000). The high-affinity (K_d of about 0.1 nM) interaction of the tag with the *N*-terminal domain, first 101 amino acids of the hU1A spliceosomal protein (van Gelder *et al.*, 1993) was used to isolate the RNPs. The hU1A protein was itself expressed in fusion with a repetition of HA epitopes that allowed purification with monoclonal anti-HA antibodies. An example of the efficiency of the U1A system when we tested its use to detect a specific RNA–protein interaction is shown in Figure 5A–C. Other couples of RNA–RNA binding proteins are potentially useful for the same purpose and the interaction of the bacteriophage MS2 coat protein and an MS2 recognition RNA motif (a hairpin) is currently used to detect RNA–protein interactions in the three-hybrid system (see the next section).

On many occasions, it is easier to obtain the tagged RNA *in vitro* and to test its assembly with proteins from a cell extract. An early example of this strategy was the use of a known RNA element,

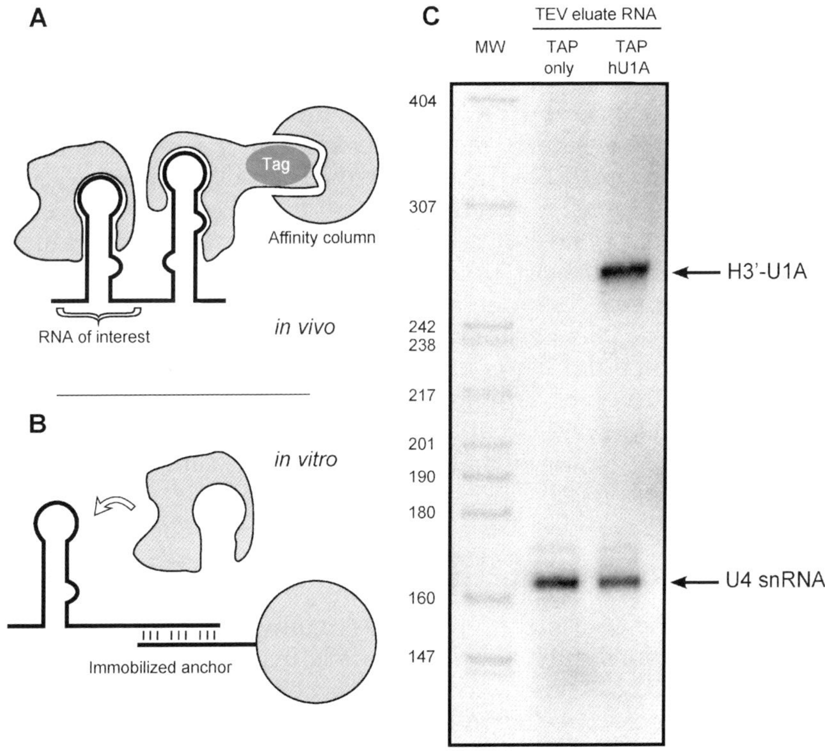

Figure 5. Methods allowing the identification of RNPs components, *in vitro* or *in vivo*. (A) An RNA tag, specifically recognized by an RNA-binding protein may be fused with the RNA sequence of interest. Co-expression of the fusion RNA and the RNA-binding protein allows the purification of the RNA-binding protein and the isolation of the attached RNA, which in turn could be bound to other proteins. (B) The assembly of an RNP *in vitro* may also be tested by producing an "anchor" RNA, fixed to a support to which the RNA of interest is attached. A protein extract is next incubated with the RNA and the proteins specifically associated may be eluted. RNAs of interest may also be directly linked to a support (example Allerson *et al.*, 2003) or biotinylated (von Ahsen and Noller, 1995). (C) Efficient recovery of an RNA (hairpin of the 3′ end of Rps28B transcript) expressed in yeast in fusion with a fragment of the human U1A pre-messenger RNA with a fragment of hU1A spliceosomal protein fused to the TAP tag. Both the RNA and protein were expressed from high-copy vectors (pIIIA derivative for the hybrid RNA (Good and Engelke, 1994; SenGupta *et al.*, 1996; Zhang *et al.*, 1997) and pCM190-TAP derivative for the fusion protein (Gari *et al.*, 1997; Rigaut *et al.*, 1999). Total extracts, obtained in the presence of vanadyl nucleoside complexes as RNAse inhibitor, were incubated with IgG Sepharose. After washing, the RNP complexes were eluted by TEV protease cleavage of the protein tag. RNA was extracted with phenol–chloroform directly from the eluate. Isolated RNA was separated on a urea–polyacrylamide gel and transferred to a positively charged nylon membrane. Yeast cells expressing the RNA and the TAP tag alone were used as a control. To verify similar levels of gel loading, residual U4 snRNA was detected with a specific oligonucleotide probe. Real-time PCR on reverse transcribed total RNA confirmed similar amounts of expressed RNA in both extracts. Real-time PCR quantitation of the co-purified RNA showed a ratio of 1000 to 1 for the pulled-down RNA in the strain expressing TAP-hU1A as compared with the strain expressing the TAP tag alone.

required for translational regulation in response to iron (IRE, iron-responsive element), to pull down a specific IRE binding protein. The RNA of interest was biotinylated and a streptavidin column was used during the purification protocol (Rouault *et al.*, 1989). We used a similar method to identify the interaction of a regulatory bacterial RNA with RNAse III (Huntzinger *et al.*, 2005). Studies of the splicing reactions and the spliceosome made use of *in vitro* transcribed substrates or blocked substrates either immobilized on a solid support or recovered after their assembly with the spliceosomal components. Different RNA tags were used like MS2 (Das *et al.*, 2000) or tobramycin binding aptamers (Hartmuth *et al.*, 2002). While the list is not exhaustive other aptamers have been used as RNA tags: streptotag selected for streptomycin binding (Bachler *et al.*, 1999), an aptamer that binds to streptavidin and another one selected for Sephadex binding (Srisawat and Engelke, 2002).

D. Protein–RNA Interactions by Three-Hybrid

One step further from the two-hybrid method, RNA–protein interactions may also be detected by a similar assay, the three-hybrid method, described in the Wickens laboratory (SenGupta *et al.*, 1996; Hook *et al.*, 2005). This method may be used to identify novel RNA-binding proteins (see, for example, Long *et al.*, 2000) and conversely, is potentially useful to identify RNAs that interact with a given protein as demonstrated for the Snp1 protein and the identification of the U1 snRNA as a binder in a genomic screen (Sengupta *et al.*, 1999).

As for the two-hybrid method, a specific RNA–protein interaction leads to transcriptional activation of chromosomal integrated reporter genes such as *lacZ* and *HIS3* downstream bacterial LexA or other transcription factor binding sites (see schematics in Figure 6). Two fusion proteins are produced in yeast cells: one containing a LexA DNA-binding domain fused to an RNA-binding protein

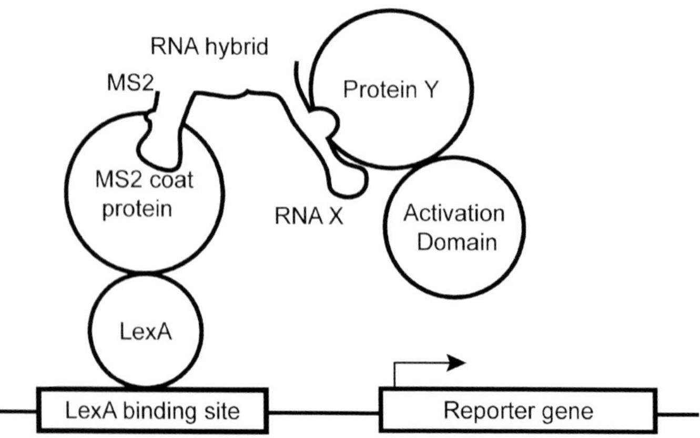

Figure 6. Principle of the three-hybrid system.

(bacteriophage MS2 coat protein in the original method) and another one containing a protein, "Y" fused to a transcription activation domain (AD). An RNA hybrid formed of a short MS2 RNA hairpin bound by the MS2 coat protein and an RNA sequence of interest ("X"), which binds to polypeptide "Y" is transcribed from a RNA polymerase III promoter. If an interaction between the RNA X and the protein "Y" occurs, the expression of reporter genes is activated. The specific interaction between the mammalian iron responsive element (IRE) RNA and the iron regulatory protein 1 (IRP1) has been used to validate the system and can also be used as a positive control (SenGupta *et al.*, 1996).

The major interest of the three-hybrid is its potential ability to be used in genetic screens for novel protein partners of known RNA sequences. A successful example of such a screen with *Caenorhabditis elegans* RNA and proteins was described by the authors of the original three-hybrid method (Zhang *et al.*, 1997); however the number of successful screens remained very limited, mainly because of a high level of false–positive interactions that obscure the interactions of interest. The main source of spurious results was the direct interaction between the two different fusion proteins independently of the RNA hybrid. Tinkering with the MS2 coat fusion protein allowed an important reduction in the number of recovered RNA-independent interactions with a concomitant increase in the sensitivity of the assay (Hook *et al.*, 2005). In the improved version of the three-hybrid screens, the MS2 coat protein domain was duplicated, since dimers of MS2 protein are required to bind the MS2 RNA hairpin. A point mutation that increases the affinity of the protein for the RNA was also introduced in the novel strain. With this new version of the three-hybrid more library screen results are to be expected. Meanwhile, three-hybrid remains a versatile method to test RNA–protein interactions and to dissect the sequence requirements for specific interactions.

Owing to its high sensitivity, three-hybrid may be successful when direct biochemical methods would fail or when additional, unknown, cellular factors are required for the interaction. An example of three-hybrid sensitivity from our laboratory is the description of the interaction between the yeast Rps28 protein and *RPS28B* mRNA 3′UTR regulatory hairpin (Badis *et al.*, 2004). Mutations of nucleotides in positions that were predicted to be critical for the RNA/protein interaction were sufficient to abolish the activation of the reporter gene, confirming the specificity of the interaction and validating the proposed structure.

♦♦♦♦♦♦ V. CONCLUSION

Part of this chapter non-exhaustively described basic techniques to characterize the primary structure of RNA gene products as well as the proteins that associate with these RNA to form ribonucleoprotein particles. One must be aware, however, that technology is evolving more and more rapidly and some new essential approaches that might drastically change the way we will address some of the problems discussed here are likely to develop in the near future. This was true already for the impact that mass spectrometry had on the determination of the composition of complexes or for DNA microarrays on the genomic quantification of transcripts. Likewise, some rapidly developing techniques, which remain at this time not widely accessible yet, might rapidly become essential for transcripts analyses. Tiling microarrays are certainly among those approaches. Another one, that emerged very recently, is the massively multiplexed sequencing that allow the determination of sequences tags of millions of tags at a time (Margulies *et al.*, 2005; Shendure *et al.*, 2005). Although these new techniques have thus far mainly been discussed in term of genome sequencing, they are likely to become also invaluable to characterize transcriptomes, not only quantitatively, but also qualitatively by cDNA and/or SAGE sequencing.

Acknowledgements

We thank Claire Torchet for providing the Northern-blot experiment shown in Figure 4C and Gwenael Badis for performing the Northern blot presented in Figure 5C.

References

Allerson, C. R., Martinez, A., Yikilmaz, E. and Rouault, T. A. (2003). A high-capacity RNA affinity column for the purification of human IRP1 and IRP2 overexpressed in *Pichia pastoris*. *RNA* **9**, 364–374.

Arava, Y., Boas, F. E., Brown, P. O. and Herschlag, D. (2005). Dissecting eukaryotic translation and its control by ribosome density mapping. *Nucleic Acids Res.* **33**, 2421–2432.

Ashe, M. P., De Long, S. K. and Sachs, A. B. (2000). Glucose depletion rapidly inhibits translation initiation in yeast. *Mol. Biol. Cell* **11**, 833–848.

Bachler, M., Schroeder, R. and von Ahsen, U. (1999). StreptoTag: a novel method for the isolation of RNA-binding proteins. *RNA* **5**, 1509–1516.

Badis, G., Saveanu, C., Fromont-Racine, M. and Jacquier, A. (2004). Targeted mRNA degradation by deadenylation-independent decapping. *Mol. Cell* **15**, 5–15.

Billy, E., Wegierski, T., Nasr, F. and Filipowicz, W. (2000). Rcl1p, the yeast protein similar to the RNA 3′-phosphate cyclase, associates with U3 snoRNP and is required for 18S rRNA biogenesis. *EMBO J.* **19**, 2115–2126.

Bringmann, P., Rinke, J., Appel, B., Reuter, R. and Luhrmann, R. (1983). Purification of snRNPs U1, U2, U4, U5 and U6 with 2,2,7-trimethylguanosine-specific antibody and definition of their constituent proteins reacting with anti-Sm and anti-(U1)RNP antisera. *EMBO J.* **2**, 1129–1135.

Brody, E. and Abelson, J. (1985). The "spliceosome": yeast pre-messenger RNA associates with a 40S complex in a splicing-dependent reaction. *Science* **228**, 963–967.

Chanfreau, G., Rotondo, G., Legrain, P. and Jacquier, A. (1998). Processing of a dicistronic small nucleolar RNA precursor by the RNA endonuclease Rnt1. *EMBO J.* **17**, 3726–3737.

Couttet, P., Fromont-Racine, M., Steel, D., Pictet, R. and Grange, T. (1997). Messenger RNA deadenylylation precedes decapping in mammalian cells. *Proc. Natl. Acad. Sci. USA* **94**, 5628–5633.

Das, R., Zhou, Z. and Reed, R. (2000). Functional association of U2 snRNP with the ATP-independent spliceosomal complex E. *Mol. Cell* **5**, 779–787.

Dujon, B. (2005). Hemiascomycetous yeasts at the forefront of comparative genomics. *Curr. Opin. Genet. Dev.* **15**, 614–620.

Ferreira-Cerca, S., Poll, G., Gleizes, P. E., Tschochner, H. and Milkereit, P. (2005). Roles of eukaryotic ribosomal proteins in maturation and transport of pre-18S rRNA and ribosome function. *Mol. Cell* **20**, 263–275.

Fichant, G. A. and Burks, C. (1991). Identifying potential tRNA genes in genomic DNA sequences. *J. Mol. Biol.* **220**, 659–671.

Frohman, M. A., Dush, M. K. and Martin, G. R. (1988). Rapid production of full-length cDNAs from rare transcripts: amplification using a single gene-specific oligonucleotide primer. *Proc. Natl. Acad. Sci. USA* **85**, 8998–9002.

Fromont-Racine, M., Bertrand, E., Pictet, R. and Grange, T. (1993). A highly sensitive method for mapping the 5′ termini of mRNAs. *Nucleic Acids Res.* **21**, 1683–1684.

Fromont-Racine, M., Senger, B., Saveanu, C. and Fasiolo, F. (2003). Ribosome assembly in eukaryotes. *Gene* **313**, 17–42.

Gari, E., Piedrafita, L., Aldea, M. and Herrero, E. (1997). A set of vectors with a tetracycline-regulatable promoter system for modulated gene expression in *Saccharomyces cerevisiae*. *Yeast* **13**, 837–848.

Gerber, A. P., Herschlag, D. and Brown, P. O. (2004). Extensive association of functionally and cytotopically related mRNAs with Puf family RNA-binding proteins in yeast. *PLoS Biol.* **2**, E79.

Good, P. D. and Engelke, D. R. (1994). Yeast expression vectors using RNA polymerase III promoters. *Gene* **151**, 209–214.

Guthrie, C. (1986). Finding functions for small nuclear RNAs in yeast. *TIBS* **11**, 430–434.

Hartmuth, K., Urlaub, H., Vornlocher, H. P., Will, C. L., Gentzel, M., Wilm, M. and Luhrmann, R. (2002). Protein composition of human prespliceosomes isolated by a tobramycin affinity-selection method. *Proc. Natl. Acad. Sci. USA* **99**, 16719–16724.

Hiley, S. L., Babak, T. and Hughes, T. R. (2005). Global analysis of yeast RNA processing identifies new targets of RNase III and uncovers a link between tRNA 5′ end processing and tRNA splicing. *Nucleic Acids Res.* **33**, 3048–3056.

Hook, B., Bernstein, D., Zhang, B. and Wickens, M. (2005). RNA–protein interactions in the yeast three-hybrid system: affinity, sensitivity, and enhanced library screening. *RNA* **11**, 227–233.

Huntzinger, E., Boisset, S., Saveanu, C., Benito, Y., Geissmann, T., Namane, A., Lina, G., Etienne, J., Ehresmann, B., Ehresmann, C., Jacquier, A., Vandenesch, F. and Romby, P. (2005). Staphylococcus aureus RNAIII and the endoribonuclease III coordinately regulate spa gene expression. *EMBO J.* **24**, 824–835.

Huttenhofer, A., Schattner, P. and Polacek, N. (2005). Non-coding RNAs: hope or hype?. *Trends Genet.* **21**, 289–297.

Keene, J. D. and Tenenbaum, S. A. (2002). Eukaryotic mRNPs may represent posttranscriptional operons. *Mol. Cell* **9**, 1161–1167.

Krogan, N. J., Kim, M., Ahn, S. H., Zhong, G., Kobor, M. S., Cagney, G., Emili, A., Shilatifard, A., Buratowski, S. and Greenblatt, J. F. (2002). RNA polymerase II elongation factors of *Saccharomyces cerevisiae*: a targeted proteomics approach. *Mol. Cell. Biol.* **22**, 6979–6992.

Krogan, N. J., Peng, W. T., Cagney, G., Robinson, M. D., Haw, R., Zhong, G., Guo, X., Zhang, X., Canadien, V., Richards, D. P., Beattie, B. K., Lalev, A., Zhang, W., Davierwala, A. P., Mnaimneh, S., Starostine, A., Tikuisis, A. P., Grigull, J., Datta, N., Bray, J. E., Hughes, T. R., Emili, A. and Greenblatt, J. F. (2004). High-definition macromolecular composition of yeast RNA-processing complexes. *Mol. Cell* **13**, 225–239.

Kruiswijk, T., Planta, R. J. and Krop, J. M. (1978). The course of the assembly of ribosomal subunits in yeast. *Biochim. Biophys. Acta* **517**, 378–389.

Kuhn, K. M., DeRisi, J. L., Brown, P. O. and Sarnow, P. (2001). Global and specific translational regulation in the genomic response of *Saccharomyces cerevisiae* to a rapid transfer from a fermentable to a nonfermentable carbon source. *Mol. Cell. Biol.* **21**, 916–927.

LaCava, J., Houseley, J., Saveanu, C., Petfalski, E., Thompson, E., Jacquier, A. and Tollervey, D. (2005). RNA degradation by the exosome is promoted by a nuclear polyadenylation complex. *Cell* **121**, 713–724.

Liu, X. and Gorovsky, M. A. (1993). Mapping the 5′ and 3′ ends of *Tetrahymena thermophila* mRNAs using RNA ligase mediated amplification of cDNA ends (RLM-RACE). *Nucleic Acids Res.* **21**, 4954–4960.

Long, R. M., Gu, W., Lorimer, E., Singer, R. H. and Chartrand, P. (2000). She2p is a novel RNA-binding protein that recruits the Myo4p-She3p complex to ASH1 mRNA. *EMBO J.* **19**, 6592–6601.

Lopez de Heredia, M. and Jansen, R. P. (2004). RNA integrity as a quality indicator during the first steps of RNP purifications: a comparison of yeast lysis methods. *BMC Biochem.* **5**, 14.

Lowe, T. M. and Eddy, S. R. (1997). tRNAscan-SE: a program for improved detection of transfer RNA genes in genomic sequence. *Nucleic Acids Res.* **25**, 955–964.

Lowe, T. M. and Eddy, S. R. (1999). A computational screen for methylation guide snoRNAs in yeast. *Science* **283**, 1168–1171.

Luthe, D. S. (1983). A simple technique for the preparation and storage of sucrose gradients. *Anal. Biochem.* **135**, 230–232.

Margulies, M., Egholm, M., Altman, W. E., Attiya, S., Bader, J. S., Bemben, L. A., Berka, J., Braverman, M. S., Chen, Y. J., Chen, Z., Dewell, S. B., Du, L., Fierro, J. M., Gomes, X. V., Godwin, B. C., He, W., Helgesen, S., Ho, C. H., Irzyk, G. P., Jando, S. C., Alenquer, M. L., Jarvie, T. P., Jirage, K. B., Kim, J. B., Knight, J. R., Lanza, J. R., Leamon, J. H., Lefkowitz, S. M., Lei, M., Li, J., Lohman, K. L., Lu, H., Makhijani, V. B., McDade, K. E., McKenna, M. P., Myers, E. W., Nickerson, E., Nobile, J. R., Plant, R., Puc, B. P., Ronan, M. T., Roth, G. T., Sarkis, G. J., Simons, J. F., Simpson, J. W., Srinivasan, M., Tartaro, K. R., Tomasz, A., Vogt, K. A., Volkmer, G. A., Wang, S. H., Wang, Y., Weiner, M. P., Yu, P., Begley, R. F. and Rothberg, J. M. (2005). Genome sequencing in microfabricated high-density picolitre reactors. *Nature* **437**, 376–380.

McCutcheon, J. P. and Eddy, S. R. (2003). Computational identification of non-coding RNAs in *Saccharomyces cerevisiae* by comparative genomics. *Nucleic Acids Res.* **31**, 4119–4128.

Mockler, T. C., Chan, S., Sundaresan, A., Chen, H., Jacobsen, S. E. and Ecker, J. R. (2005). Applications of DNA tiling arrays for whole-genome analysis. *Genomics* **85**, 1–15.

Nguyen, V. T., Kiss, T., Michels, A. A. and Bensaude, O. (2001). 7SK small nuclear RNA binds to and inhibits the activity of CDK9/cyclin T complexes. *Nature* **414**, 322–325.

Noller, H. F., Hoffarth, V. and Zimniak, L. (1992). Unusual resistance of peptidyl transferase to protein extraction procedures. *Science* **256**, 1416–1419.

Olivas, W. M., Muhlrad, D. and Parker, R. (1997). Analysis of the yeast genome: identification of new non-coding and small ORF-containing RNAs. *Nucleic Acids Res.* **25**, 4619–4625.

Peng, W. T., Robinson, M. D., Mnaimneh, S., Krogan, N. J., Cagney, G., Morris, Q., Davierwala, A. P., Grigull, J., Yang, X., Zhang, W., Mitsakakis, N., Ryan, O. W., Datta, N., Jojic, V., Pal, C., Canadien, V., Richards, D., Beattie, B., Wu, L. F., Altschuler, S. J., Roweis, S., Frey, B. J., Emili, A., Greenblatt, J. F. and Hughes, T. R. (2003). A panoramic view of yeast noncoding RNA processing. *Cell* **113**, 919–933.

Richard, G. F., Fairhead, C. and Dujon, B. (1997). Complete transcriptional map of yeast chromosome XI in different life conditions. *J. Mol. Biol.* **268**, 303–321.

Rigaut, G., Shevchenko, A., Rutz, B., Wilm, M., Mann, M. and Seraphin, B. (1999). A generic protein purification method for protein complex characterization and proteome exploration. *Nat. Biotechnol.* **17**, 1030–1032.

Rouault, T. A., Hentze, M. W., Haile, D. J., Harford, J. B. and Klausner, R. D. (1989). The iron-responsive element binding protein: a method for the affinity purification of a regulatory RNA-binding protein. *Proc. Natl. Acad. Sci. USA* **86**, 5768–5772.

Sambrook, J. and Russell, D. W. (eds) (2001). *Molecular Cloning. A Laboratory Manual.* Cold Spring Harbor Laboratory Press, New York.

Schattner, P., Brooks, A. N. and Lowe, T. M. (2005). The tRNAscan-SE, snoscan and snoGPS web servers for the detection of tRNAs and snoRNAs. *Nucleic Acids Res.* **33**, W686–W689.

Schattner, P., Decatur, W. A., Davis, C. A., Ares, M., Jr., Fournier, M. J. and Lowe, T. M. (2004). Genome-wide searching for pseudouridylation guide snoRNAs: analysis of the *Saccharomyces cerevisiae* genome. *Nucleic Acids Res.* **32**, 4281–4296.

Sengupta, D. J., Wickens, M. and Fields, S. (1999). Identification of RNAs that bind to a specific protein using the yeast three-hybrid system. *RNA* **5**, 596–601.

SenGupta, D. J., Zhang, B., Kraemer, B., Pochart, P., Fields, S. and Wickens, M. (1996). A three-hybrid system to detect RNA–protein interactions *in vivo*. *Proc. Natl. Acad. Sci. USA* **93**, 8496–8501.

Shendure, J., Porreca, G. J., Reppas, N. B., Lin, X., McCutcheon, J. P., Rosenbaum, A. M., Wang, M. D., Zhang, K., Mitra, R. D. and Church, G. M. (2005). Accurate multiplex polony sequencing of an evolved bacterial genome. *Science* **309**, 1728–1732.

Srisawat, C. and Engelke, D. R. (2002). RNA affinity tags for purification of RNAs and ribonucleoprotein complexes. *Methods* **26**, 156–161.

Steensgaard, J., Humphries, S. and Spragg, P. (1992). Measurements of sedimentation coefficients. In: *Preparative Centrifugation. A Practical Approach* (D. Rickwood, ed.), pp. 187–232. IRL Press at Oxford University Press, Oxford.

Torchet, C., Badis, G., Devaux, F., Costanzo, G., Werner, M. and Jacquier, A. (2005). The complete set of H/ACA snoRNAs that guide rRNA pseudouridylations in *Saccharomyces cerevisiae*. *RNA* **11**, 928–938.

van Gelder, C. W., Gunderson, S. I., Jansen, E. J., Boelens, W. C., Polycarpou-Schwarz, M., Mattaj, I. W. and van Venrooij, W. J. (1993). A complex secondary structure in U1A pre-mRNA that binds two molecules of U1A protein is required for regulation of polyadenylation. *EMBO J.* **12**, 5191–5200.

Vanacova, S., Wolf, J., Martin, G., Blank, D., Dettwiler, S., Friedlein, A., Langen, H., Keith, G. and Keller, W. (2005). A new yeast poly(A) polymerase complex involved in RNA quality control. *PLoS Biol.* **3**, e189.

Velculescu, V. E., Zhang, L., Vogelstein, B. and Kinzler, K. W. (1995). Serial analysis of gene expression. *Science* **270**, 484–487.

Velculescu, V. E., Zhang, L., Zhou, W., Vogelstein, J., Basrai, M. A., Bassett, D. E., Jr., Hieter, P., Vogelstein, B. and Kinzler, K. W. (1997). Characterization of the yeast transcriptome. *Cell* **88**, 243–251.

von Ahsen, U. and Noller, H. F. (1995). Identification of bases in 16S rRNA essential for tRNA binding at the 30S ribosomal *P* site. *Science* **267**, 234–237.

Warner, J. R. and Knopf, P. M. (2002). The discovery of polyribosomes. *Trends Biochem. Sci.* **27**, 376–380.

Washietl, S., Hofacker, I. L. and Stadler, P. F. (2005). Fast and reliable prediction of noncoding RNAs. *Proc. Natl. Acad. Sci. USA* **102**, 2454–2459.

Wassarman, K. M., Repoila, F., Rosenow, C., Storz, G. and Gottesman, S. (2001). Identification of novel small RNAs using comparative genomics and microarrays. *Genes Dev.* **15**, 1637–1651.

Wassarman, K. M. and Storz, G. (2000). 6S RNA regulates *E. coli* RNA polymerase activity. *Cell* **101**, 613–623.

Watkins, N. J., Segault, V., Charpentier, B., Nottrott, S., Fabrizio, P., Bachi, A., Wilm, M., Rosbash, M., Branlant, C. and Luhrmann, R. (2000). A common core RNP structure shared between the small nucleoar box C/D RNPs and the spliceosomal U4 snRNP. *Cell* **103**, 457–466.

Wei, C. L., Ng, P., Chiu, K. P., Wong, C. H., Ang, C. C., Lipovich, L., Liu, E. T. and Ruan, Y. (2004). 5′ Long serial analysis of gene expression (LongSAGE) and 3′ LongSAGE for transcriptome characterization and genome annotation. *Proc. Natl. Acad. Sci. USA* **101**, 11701–11706.

Wessel, D. and Flugge, U. I. (1984). A method for the quantitative recovery of protein in dilute solution in the presence of detergents and lipids. *Anal. Biochem.* **138**, 141–143.

Wyers, F., Rougemaille, M., Badis, G., Rousselle, J. C., Dufour, M. E., Boulay, J., Regnault, B., Devaux, F., Namane, A., Seraphin, B., Libri, D. and Jacquier, A. (2005). Cryptic pol II transcripts are degraded by a nuclear quality control pathway involving a new poly(A) polymerase. *Cell* **121**, 725–737.

Zhang, B., Gallegos, M., Puoti, A., Durkin, E., Fields, S., Kimble, J. and Wickens, M. P. (1997). A conserved RNA-binding protein that regulates sexual fates in the *C. elegans* hermaphrodite germ line. *Nature* **390**, 477–484.

19 Analysis of Gene Function of Mitochondria

Stéphane Duvezin-Caubet, Andreas S Reichert and Walter Neupert
Adolf-Butenandt-Institut für Physiologische Chemie, Ludwig-Maximilians-Universität München, Butenandtstr. 5, 81377 München, Germany

♦♦

CONTENTS

Introduction
Identification of genes and proteins relevant for the function of mitochondria
Identification of the function of mitochondrial genes and proteins
Conclusion

Abbreviations

APAF	apoptotic protease activating factor
BrdU	bromodeoxyuridine
DASPMI	dimethylaminostyrylpyridiniummethyl iodine
DAPI	4′,6-diamidino-2-phenylindole
DsRed	Discosoma red fluorescent protein
GFP	green fluorescent protein
MPP	matrix processing peptidase
mtDNA	mitochondrial DNA
MTS	mitochondrial targeting sequence
NADH	nicotinamide adenine dinucleotide, reduced
OTC	ornithine transcarbamylase
OXPHOS	oxidative phosphorylation
PMSF	phenylmethanesulfonyl fluoride
ROS	reactive oxygen species

♦♦♦♦♦♦ I. INTRODUCTION

Mitochondria are ubiquitous organelles of eukaryotic cells. Their structural organisation is rather complex (see Figure 1). They consist of two different membranes, the outer membrane and the inner

0580-9517 DOI:10.1016/S0580-9517(06)36019-9

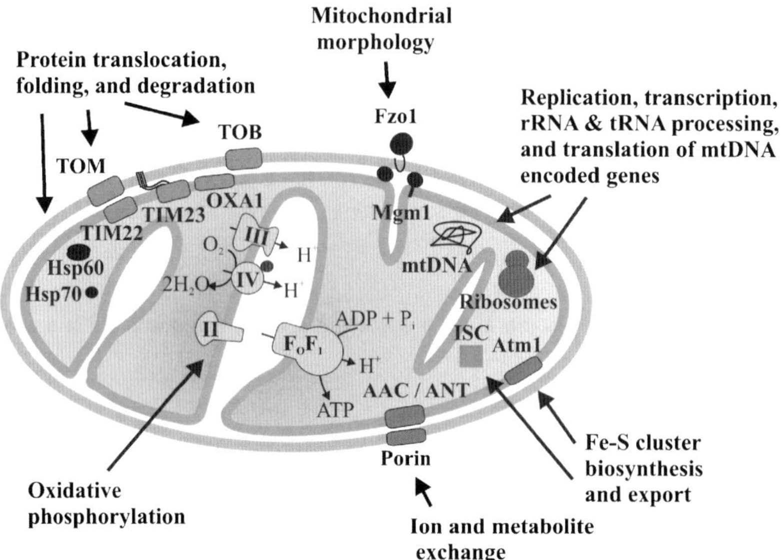

Figure 1. Mitochondrial functions in *S. cerevisiae*: Protein translocation (TOM, translocase complex of the outer membrane; TOB, complex mediating topogenesis of outer membrane beta-barrel proteins; TIM22, TIM23 and OXA1, the three translocase complexes of the inner membrane). Protein folding (Hsp60, Hsp70) and degradation. Mitochondrial morphology (Fzo1, Mgm1). Replication, transcription of mitochondrial DNA (mtDNA), rRNA and tRNA processing, translation (ribosomes) of mtDNA-encoded genes. Fe/S cluster biosynthesis (ISC, iron–sulfur cluster assembly complex) and export (Atm1). Ion and metabolite exchange (AAC/ANT, ADP/ATP carrier or adenine nucleotide translocase; porin/VDAC, channel in outer membrane). Oxidative phosphorylation (complexes II, III, IV and the F_1F_O-ATP synthase; complex I is not present in *S. cerevisiae*). ATP synthesis in the mitochondrial matrix using the proton gradient across the inner membrane is indicated. Basic metabolic processes such as the Krebs cycle, the metabolism of certain amino acids and lipids, and heme biosynthesis are omitted (adapted from Reichert and Neupert, 2004).

membrane. These serve as boundaries for the intermembrane space located between the two membranes, and the matrix space that is enclosed by the inner membrane. The inner membrane can be further subdivided into the inner boundary membrane which is juxtaposed to the outer membrane, and the cristae which are formed by invagination of the inner membrane. Mitochondria are indispensable organelles, as they fulfil a multitude of important functions for the cell (see Figure 1).

A. Mitochondrial Functions

One prominent function is energy transduction by oxidative phosphorylation (Saraste, 1999). Substrates derived from foodstuffs are metabolised in the Krebs cycle which is located in the mitochondrial matrix. The resulting reduction equivalents, mostly NADH, are funnelled into the respiratory chain. The respiratory chain complexes transduce the redox energy into an electrochemical proton gradient as electrons are passed to the final acceptor, oxygen. The energy of the proton gradient is transduced into the chemical energy stored in ATP by the F_1F_O-ATP synthase (complex V). The process of oxidative phosphorylation is uniform in almost all eukaryotic cells. The structures and functions of the components involved have been largely conserved during evolution. However, there are also variations. For instance, the yeast *Saccharomyces cerevisiae* does not contain an energy transducing complex I. It is replaced by two NADH-dehydrogenases pointing to opposite sides of the inner membrane (Marres *et al.*, 1991; De Vries *et al.*, 1992).

A further important function of mitochondria is the biosynthesis of iron sulfur clusters for Fe–S proteins. This is essential for the cell since the mitochondrial system is producing Fe–S clusters also for proteins in the cytosol (Lill and Kispal, 2000; Lill and Muhlenhoff, 2005). In addition, mitochondria are involved in the formation of lipids, amino acids and heme (Scheffler, 2001). With several of these biosynthetic pathways, certain steps take place in the mitochondria, whereas other steps are located in the cytosol or in other organelles (Abadjieva *et al.*, 2001).

B. Nuclear and Mitochondrial Genes for Mitochondrial Proteins

Currently, there are about 700 proteins predicted to be present in yeast mitochondria, about 500 of them are established mitochondrial constituents (see Reichert and Neupert, 2004 for review, and included references). The genes coding for mitochondrial proteins are present either in the nucleus or within the mitochondria. The vast majority of proteins are synthesised from nuclear genes and subsequently imported into mitochondria.

Most subunits of oxidative phosphorylation complexes and numerous assembly factors involved in the biogenesis of these complexes are encoded by nuclear genes, including proteins for the synthesis and modification of cofactors. This group comprises at least some 100 proteins. The various enzymes involved in the anabolic and catabolic pathways, in particular enzymes of the Krebs cycle and biosynthesis of certain amino acids, account for a minimum of 130 proteins. The biogenesis of Fe–S clusters requires about 15 known proteins and the system for detoxification and protection against oxidative stress at least an equal number. Numerous proteins have functions in the transport of substrates, ions and coenzymes across the membranes of mitochondria, among them about 35 carriers or transporters for solutes that are located in the inner membrane.

A particularly large number of proteins are involved in the biogenesis of mitochondria (see Figure 1). Mitochondrial DNA (mtDNA) replication, transcription and translation comprise at least 100 different proteins. More than 50 proteins are involved in the import and sorting of proteins into the various subcompartments of the mitochondria. A large set of molecular chaperones play a role in the folding of proteins and in the prevention or repair of damage due to cellular stress, such as heat and oxidative stress. In addition, quite a number of proteases exist that are known to degrade misfolded or unassembled proteins in the various subcompartments.

The mtDNA encodes a few subunits of complexes catalysing oxidative phosphorylation and constituents of the mitochondrial protein synthesis machinery. Most of these proteins represent key components of the respiratory chain complexes, in particular cytochrome *b* of complex III, subunits Cox I, II and III of complex IV, and subunits Atp 6, 8 and 9 which belong to the proton channel forming F_O subcomplex of the F_1F_O-ATP synthase. These proteins are all integral membrane proteins with very high hydrophobicity, which may contribute to the retention of their genes in the mitochondrial genome. In yeast, Var1, a subunit component of the mitochondrial ribosome, is the only hydrophilic protein encoded by the mitochondrial genome. The mtDNA encodes, in addition to proteins, ribosomal RNAs (21S and 16S RNAs) and a complete set of transfer RNAs, whose genes are distributed all over the genome (Foury *et al.*, 1998). Another feature of yeast mtDNA is that genes located within introns of *CYTb*, *COXI* and 21S rRNA genes encode endonucleases that mediate splicing and the mobility of introns.

C. Mitochondrial Fusion, Fission and Morphology

A new group of proteins was discovered more recently which are involved in mitochondrial structural dynamics and morphology (reviewed in Okamoto and Shaw, 2005). In yeast, like in most eukaryotes, mitochondria are organised as a network of interconnected tubules (see Figure 2a). They form a highly dynamic network resulting from the balance between frequent events of fusion and division (Bereiter-Hahn and Voth, 1994; Yaffe, 1999). A mutant phenotype of fragmented mitochondria can be caused by impairment of fusion as shown for a number of examples. On the other hand, impairment of fission results in an extensive interconnected "fishnet-like" tubular arrangement of mitochondria. Determination of mitochondrial morphology involves at least some 20 proteins (Dimmer *et al.*, 2002). Moreover, a number of proteins play a critical role in the positioning of mtDNA in mitochondria and of mitochondria in the cell (Yaffe, 1999). A number of these proteins are involved, directly or indirectly, in the inheritance of mtDNA (Berger and Yaffe, 2000).

A further structural complexity is the organisation of the inner membrane, in particular, the folding of the cristae membrane (see Figure 2b). Cristae generate a large surface of inner membrane to

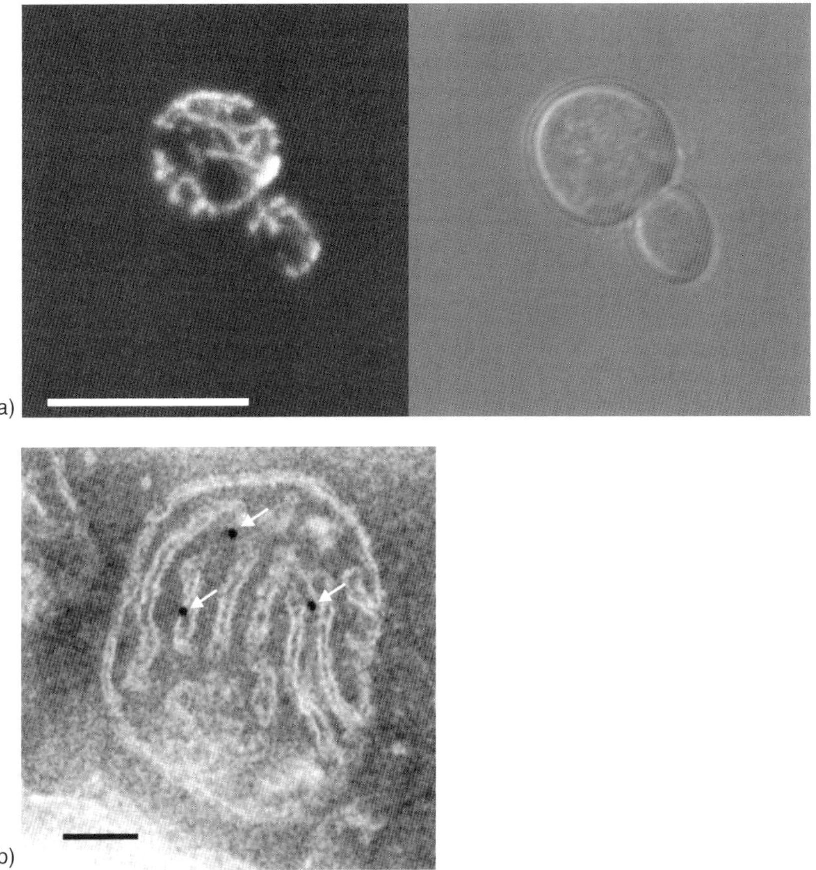

Figure 2. Mitochondrial network and ultrastructure of mitochondria in *S. cerevisiae.* (a) Right panel, Nomarski view. Left panel, the tubular network of mitochondria is visualised by mitochondrially targeted GFP (courtesy of Dr Ravi Jagasia). Scale bar 5 μm. (b) Ultrastructure of mitochondria by electron microscopy. Fixed and cryo-sectioned cells were immunodecorated with antibodies raised against the mitochondrial ribosomal protein MRPL36 and secondary antibody-gold conjugate (courtesy of Dr Frank Vogel). Arrows indicate gold particles. Scale bar 100 nm.

provide space for the complexes of the oxidative phosphorylation machinery and associated complexes for the exchange of substrates and products across the inner membrane. It was believed for a long time that the infolding of cristae is essentially a passive process. However, it appears now that the oligomerisation of the ATP synthase is required for the proper folding of the inner membrane. Mutants in which ATP synthase assembly does not occur as an oligomeric supracomplex have apparently functional mitochondria but show an abnormal onion like structure (Paumard *et al.*, 2002).

For many proteins detected in mitochondria the molecular function is still unknown or insufficiently understood. Moreover, the mitochondrial proteome of yeast (and other organisms) is still not complete, largely due to technical limitations to detect proteins of low abundance (Reichert and Neupert, 2004). Genomic information is extremely useful, but the power of predictive methods for mitochondrial localisation of proteins is also limited.

D. *S. cerevisiae*, an Excellent Model Organism for Functional Analysis of Mitochondrial Proteins by Genetics

Yeast offers a number of advantages for the analysis of mitochondrial functions. A primary reason is that the methods of forward and reverse genetics are extremely powerful and efficient means to study mitochondria. In addition, yeast energy metabolism has several peculiarities which are of particular advantage for studying mitochondria. Owing to the exceptional ability of yeast for fermentative growth, oxidative phosphorylation and mtDNA are dispensable. Moreover, the presence of glucose strongly represses the formation of mitochondria and thereby respiration. This shift towards fermentation leads to the complete consumption of glucose and formation of ethanol and glycerol as end products in a first phase (anaerobic). Only when all glucose is used up, respiration is induced leading over to a second growth phase (aerobic) in which ethanol and glycerol are metabolised. Other fermentable carbon sources, such as galactose, do not repress mitochondrial biogenesis and both fermentation and respiration can take place at the same time. In addition, non-fermentable carbon sources like glycerol, ethanol or lactate can only be metabolised through the aerobic pathway that requires mitochondria.

For these reasons, it is possible to isolate and study mutants which inactivate oxidative phosphorylation directly or indirectly. Originally, mutants in mtDNA were detected by the "*cytoplasmic petite*" phenotype, a trait that gives rise to small colonies and is segregated in a non-Mendelian fashion (Ephrussi *et al.*, 1949a, b; Ephrussi and Slonimski, 1955). These small colonies are due to the lack of a respiratory growth phase after the initial fermentation phase. They were later shown to be due to mutations in the mitochondrial genome, initially called *rho* factor. Wild-type cells were named rho^+ and the petites rho^-/rho°. Cells with no mtDNA (rho°) or mtDNA including large deletions (50% or more of the mitochondrial genome) and reiteration of the remaining sequence (rho^-) display the same petite phenotype because mitochondrial protein synthesis and therefore respiratory capacity are lost (Mounolou *et al.*, 1966; Piskur, 1994). Indeed, due to the organisation of the mitochondrial genome any large deletion removes at least one tRNA gene and consequently inactivates completely mitochondrial protein synthesis. *S. cerevisiae* can survive the loss of functional mtDNA and is referred to that respect as a "petite positive" organism. It should be noted that rho^- and rho° mutations are quite frequent and represent about 1–10% of cells in cultures of wild-type strains, depending on genetic backgrounds and conditions of culture. Some mutants exhibit a great tendency to lose mtDNA, increasing the proportion of *rho*$^-$/*rho*$^\circ$ up to 100% (reviewed in Contamine and Picard, 2000).

Later on, large collections of point mutations or limited lesions were obtained in the different mitochondrial genes inactivating respiratory function. They were referred to as *mit*$^-$ mutants. These mutations helped defining the mitochondrial genes and the rules governing their transmission (Slonimski and Tzagoloff, 1976). The genetic transformation of yeast mitochondria using the biolistic gun allowed researchers to create directed mutations in mitochondrial genes and insert new genes into mtDNA (Johnston *et al.*, 1988; Steele *et al.*, 1996; Bonnefoy and Fox, 2001). In particular, a much better understanding of the mode of expression of mitochondrial genes was achieved using this tool (Fox, 1996; Green-Willms *et al.*, 1998; Sanchirico *et al.*, 1998; Bonnefoy *et al.*, 2001).

Nuclear mutations were also shown to yield a respiratory deficiency phenotype that is segregated in a Mendelian fashion. These mutants affect nuclear-encoded components and are called "*pet*" mutants. They are particularly useful in identifying new proteins involved in mitochondrial biogenesis and function (Tzagoloff *et al.*, 1975; Grivell, 1989; Tzagoloff and Dieckmann, 1990; Bolotin-Fukuhara and Grivell, 1992; Grivell *et al.*, 1999; Steinmetz *et al.*, 2002). Not only proteins located inside mitochondria but also proteins acting outside or temporarily on the surface of mitochondria can be revealed by this approach, in contrast to proteomic approaches. This latter class of non-mitochondrial proteins comprises regulatory and signalling proteins as well as proteins that are involved in transporting components to mitochondria, including, e.g. chaperones and metal chelators.

This chapter concentrates on the analysis of nuclear-encoded proteins, as the functions of the proteins encoded by the organellar DNA are essentially known. Functional analysis of nuclear genes for mitochondrial proteins rely on established genetic and biochemical methodology used for any kind of nuclear gene. The emphasis of this chapter is therefore to point out the specific aspects related to mitochondrial functions. Many of the techniques described in the other chapters of this volume are relevant here as well.

◆◆◆◆◆◆ II. IDENTIFICATION OF GENES AND PROTEINS RELEVANT FOR THE FUNCTION OF MITOCHONDRIA

A. Common Phenotypes of Mutants Impaired in Mitochondrial Functionality

The first step in the traditional procedure of identifying nuclear genes with a role in mitochondrial functions and their products consists in isolating mutants with a mitochondrial phenotype.

Therefore, it is important to define precisely what a mitochondrial phenotype is. Because of the numerous processes depending on mitochondria, mutants affecting mitochondrial function lead to very diverse and pleiotropic phenotypes. We focus here on the most intensely studied mitochondrial phenotype, namely respiratory deficiency, but we describe also examples of other phenotypes which proved to be of interest in screening for mutants affected in mitochondrial function.

1. Respiratory growth deficiency

The classical mitochondrial phenotype is a respiratory growth defect. Mutations in about 500 different genes lead to this phenotype (http://www.yeastgenome.org). Complete growth arrest or slow growth on respiratory carbon sources such as glycerol or lactate is indicative of a deficiency in oxidative phosphorylation. This can be due to loss of function of an enzyme or subunit of complexes of oxidative phosphorylation, or the lack of assembly of one of these complexes (Grivell, 1989, 1995; Tzagoloff and Dieckmann, 1990; Grivell *et al*., 1999). It should be noted that this respiratory growth deficiency is sometimes only revealed under stress conditions, such as high or low temperature. For example, Fmc1p is a mitochondrial protein encoded by a nuclear gene, the deletion of which leads to a growth deficiency on non-fermentable carbon sources only at 37°C. Further analyses confirmed the mitochondrial location of this protein. The function of the F_1F_O-ATP synthase was specifically affected due to depletion of the F_1 sector. This protein was not found in the assembled complex. Moreover, F_1F_O-ATP synthase assembly was fully restored upon overexpression of Atp12p, a well-known chaperone of the F_1 component of the ATP synthase. Fmc1p, like several other proteins, is involved in the assembly of alpha and beta subunits to the heterohexameric complex in the F_1, an initial step in the formation of the ATP synthase (Lefebvre-Legendre *et al*., 2001).

Another example is the nuclear mutation ts2858 isolated in a screen for temperature-sensitive *pet* mutants that can grow on lactate at 23°C but not at 36°C. It leads to the accumulation of a precursor intermediate of Cox2 and cytochrome *b2* and turned out to affect the inner membrane protease Imp1p responsible for the final maturation of these substrates (Pratje *et al*., 1983; Pratje and Guiard, 1986).

A genetic defect leading to a respiratory growth defect may not only be of nuclear origin but also of mitochondrial origin, as crucial components of the OXPHOS are encoded by mtDNA. Moreover, loss of maintenance or expression of mtDNA also leads to a deficiency of both the respiratory chain and ATP synthase. Therefore, any mutation of a mitochondrial gene or of a nuclear-encoded component in charge of the maintenance, inheritance or expression of the mtDNA induces a respiratory-deficient phenotype as a secondary effect (Grivell, 1995; Grivell *et al.*, 1999; Contamine and Picard, 2000). This illustrates nicely the importance of nucleo–mitochondrial interactions to sustain the function of mitochondria. Thus, the integrity of the mitochondrial genome and its expression must be checked initially in respiratory-deficient mutants in order to localise the gene and mitochondrial function affected.

2. Lethality or drastic alterations of growth

There is a class of nuclear mutants that are characterised by a general deficiency of mitochondrial functionality leading to drastic growth defects. A large number of essential biochemical steps take place in mitochondria, critical not only for respiration and energy metabolism. Therefore, defects for instance, in pathways of import and maturation of proteins or in the biosynthesis of amino acids and lipids are generally lethal. However, the number of essential genes encoding mitochondrial proteins is rather limited. About 40 gene products were predicted to be located in mitochondria by *in silico* approaches and some 100 genes were found which show severe defects of mitochondrial structure when downregulated (Altmann and Westermann, 2005).

Strategies have been developed to screen for a phenotype which results from a deficiency in the biogenesis of essential mitochondrial proteins. For example, Pollock *et al.* generated a library of temperature-sensitive (ts) lethal mutants. For this they used a yeast strain which expressed in an inducible manner ornithine transcarbamylase (OTC), a nuclear-encoded mitochondrial enzyme whose activity depends on correct import and oligomerisation. Cells of the library were shifted to 37°C and simultaneously OTC was induced. Cells were harvested and the OTC enzyme activity was determined. Further analysis focused on cells lacking this activity. This approach led to the characterisation of the gene encoding the chaperonin Hsp60 protein that is required for the folding and assembly of the functional OTC trimer (Pollock *et al.*, 1988; Cheng *et al.*, 1989). Another strategy consists in screening temperature-sensitive lethal mutants for accumulation of the precursor to the beta subunit of the F_1-ATPase. A mutation affecting the matrix processing peptidase (MPP) was isolated by this approach (Yaffe and Schatz, 1984; Yaffe *et al.*, 1985).

Another example illustrates the pleiotropic effects which are sometimes obtained with mitochondrial mutants. Thorsness and colleagues used an original approach to screen for mutants of mitochondrial function based on the observation that DNA fragments can be transferred, at very low frequency, from mtDNA to the nuclear genome. They isolated nuclear mutants in which the frequency of mtDNA escape to the nucleus was increased. One of these mutations, *yme1*, in addition, showed a heat-sensitive respiratory-deficient phenotype at 37°C, a cold-sensitive growth defect

their growth. Non-fermentable carbon sources (ethanol, glycerol, lactate) vs. fermentable carbon sources (glucose, galactose, raffinose) can be used either in rich or in minimal media. Mutants affected in mitochondrial function may not grow on one or more non-fermentable carbon sources. 0.1% glucose may be added to non-fermentable carbon sources. Respiration-incompetent cells yield only small colonies (*petite*) as compared with respiratory-competent cells. Such media notably are useful to score the proportion of petite cells in a population. The growth should always be compared with the corresponding wild type, as some laboratory strains do not grow on minimal media containing non-fermentable carbon sources (Tzagoloff and Dieckmann, 1990; Yaffe, 1991). In addition, growth rate defects may indicate a quantitative loss of mitochondrial function while lower biomass production is rather indicating inefficient use of nutrients to produce biomass, e.g. when mitochondria are partially uncoupled due to leakiness of inner membrane to protons.

2. Mitochondrial protein import in intact cells

Many precursor proteins are processed during or after their import into mitochondria. This can be followed by the shift in molecular size of precursor and mature forms. Therefore, a defect in import of protein precursors can be detected by accumulation of the precursor form of a protein from yeast total protein extracts by western blotting or after immunoprecipitation of such a marker protein after radioactive labelling *in vivo*. This method has been successfully used in numerous studies with marker proteins such as the beta subunit of F_1F_O-ATP synthase and Hsp60 (Nelson and Schatz, 1979; Kang *et al.*, 1990; Yaffe, 1991; Kozany *et al.*, 2004).

3. Visualisation of mitochondria

Yeast mitochondria are organised as a dynamic network of tubular structures distributed all over the cell cortex. Mitochondria can be stained and visualised by membrane potential-dependent vital dyes such as DASPMI (Sigma, Germany) or MitoTracker CMXRos (Molecular probes, The Netherlands), or mitochondria-targeted forms of GFP variants (see Figure 2a: Matz *et al.*, 1999; Westermann and Neupert, 2000; Duvezin-Caubet *et al.*, 2003). The first method is fast and does not require any genetic manipulation, but is dependent on a sufficiently high membrane potential, making it less suitable for use in cells lacking functional mtDNA. Efficient targeting of GFP variants does not require high membrane potential. Furthermore, the different compartments of mitochondria can be visualised by using known targeting sequences and components fused to GFP variants. For example, matrix-targeting presequences of Subunit 9 of F_1F_O-ATP synthase from *N. crassa* (amino acids 1–69) or CoxIV from *S. cerevisiae* (amino acids 1–21) fused to GFP are very commonly used (Sesaki and Jensen, 1999; Westermann and Neupert, 2000). The fusion of GFP or DsRed to outer membrane components (Tom70, Dnm1) (Bleazard *et al.*, 1999; Westermann and Neupert, 2000), inner membrane components (Yta10) (Okamoto *et al.*, 1998), inter membrane space components (Yme1) (Campbell *et al.*, 1994), or even nucleoid proteins (Abf2, Mgm101) (Okamoto *et al.*, 1998; Meeusen *et al.*, 1999) allows the visualisation of these different compartments (see Nunnari *et al.*, 2002; Okamoto *et al.*, 2001 for reviews). Mitochondria can also be visualised by indirect immunofluorescence using primary antibodies raised against a mitochondrial protein and secondary antibodies conjugated to fluorescent dyes. Using these various techniques, mitochondrial behaviour can be assessed in live cells and after fixation of cells (Westermann and Neupert, 2000; Nunnari *et al.*, 2002).

These approaches allow the study of the structure and motility of mitochondria. Defects in morphology of mitochondria are often associated with growth defects. For example the absence of Mgm1, a protein of the fusion machinery of mitochondria, leads to a complete loss of mtDNA and respiratory function in addition to fragmentation of mitochondria (Guan *et al.*, 1993; Wong *et al.*, 2000a). However, this correlation does not hold in all the cases. The morphology of the organelle can be highly aberrant without any other detectable phenotype. A detailed analysis of deletion mutants of non-essential genes leading to various types of altered mitochondrial morphology has been described (Dimmer *et al.*, 2002). Defective morphology can, for example, result in fragmented or aggregated mitochondria, or in the presence of a single net structure of mitochondria (Okamoto and Shaw, 2005). A double deletion mutant of the fusion component Mgm1 and the fission component Dnm1 shows quite normal tubular mitochondrial morphology (Wong *et al.*, 2000a). Fusion competence can be assessed by mating yeast cells with differently labelled mitochondria and looking for mixing of dyes under the microscope (Nunnari *et al.*, 1997). A lack of mixing of dyes in the zygote indicates a lack of fusion of mitochondria. This assay has been used and developed further to identify the specific function of proteins involved in mitochondrial fusion and fission (Hermann *et al.*, 1998; Bleazard *et al.*, 1999; Wong *et al.*, 2000b; Nunnari *et al.*, 2002). Very recently, an *in vitro* assay was described to assess fusion of isolated mitochondria. The principle of this approach is to look for the mixing of two types of mitochondria, harbouring a matrix-targeted GFP or DsRed, respectively (Meeusen *et al.*, 2004; Meeusen and Nunnari, 2005). The development of novel techniques of time-resolved high-resolution fluorescence microscopy has added a new dimension to the observation of the dynamic behaviour of mitochondria (Hell *et al.*, 2004).

The ultrastructure of mitochondria including inner membrane morphology can only be studied using electron microscopic techniques (see Figure 2b). Tomographic techniques enable three-dimensional reconstruction of the network and of the inner structure of mitochondria (Perkins *et al.*, 1997; Nicastro *et al.*, 2000; Medalia *et al.*, 2002). For example, in the case of deletions of subunits *e* or *g* of ATP synthase, no clear phenotype was observed initially. Cells were growing slightly slower and the mitochondrial genome was unstable. The ATP synthase was present at normal levels and fully functional. The main difference observed in those mutants was the absence of oligomers of ATP synthase. Electron microscopic examination revealed a striking phenotype. The mitochondrial inner membrane was disorganised and instead of cristae, "onion like" structures were observed (Paumard *et al.*, 2002).

4. Visualisation of mitochondrial nucleoids

The mitochondrial genome, present in 50–100 copies per yeast cell, is packaged into nucleoid structures. Nucleoids comprise several copies of mtDNA, a number of different proteins, and are anchored to the matrix face of the inner membrane. These structures play a crucial role in the inheritance of mtDNA in daughter cells. Their major protein constituents were identified during the past years (Berger and Yaffe, 2000; Chen and Butow, 2005). There are two classical ways to visualise nucleoids. In a first approach, mtDNA itself is visualised, either by staining with 4′,6-diamidino-2-phenylindole (DAPI: Okamoto *et al.*, 1998), a fluorescent dye which binds selectively to mtDNA *in vivo*, or alternatively after incorporation of the thymidine analogue bromodeoxyuridine (BrdU) into mtDNA (Meeusen *et al.*, 1999). In the latter case, indirect immunofluorescence using anti-BrdU antibodies after fixation of cells is used. The second approach relies on the expression of fusion proteins consisting of constituents of nucleoids and GFP variants, or by detection of nucleoid proteins by indirect immunofluorescence (Okamoto *et al.*, 1998; Meeusen *et al.*, 1999). The use of DAPI or detection of mtDNA binding proteins allows the rapid study of nucleoid formation, distribution and segregation. On the other hand, the incorporation of BrdU can be employed in addition to assess mtDNA replication (Nunnari *et al.*, 2002).

C. *In Vitro* Approaches

1. Isolation of mitochondria

Standard methods have been developed and optimised to isolate mitochondria of different purity and that maintain their functional integrity (Daum *et al.*, 1982; Diekert *et al.*, 2001; Meisinger *et al.*, 2006). It is possible to obtain a crude mitochondrial fraction by a fast method but this is not suitable for functional studies (Lange *et al.*, 1999; Diekert *et al.*, 2001). The most classical method to isolate mitochondria from yeast requires spheroplast preparation and isolation of mitochondria by differential centrifugation after breaking the cells by a subtle osmotic shock (Daum *et al.*, 1982, see Protocol 1). Mitochondria isolated by this method maintain their integrity and are very suitable for functional studies on import of proteins or measurement of coupled respiration rates. Nevertheless, it is possible to further purify those mitochondria by gradient centrifugation (Lewin *et al.*, 1990; Diekert *et al.*, 2001). This last step allows one to eliminate or reduce contamination by other organelles, in particular endoplasmic reticulum and plasma membrane. The purity of mitochondrial fractions can be tested by western blotting using antibodies against proteins of various cellular compartments. Such purified mitochondria can be used to investigate the cellular location of proteins and to reduce the background in studies *in vitro*. The location of a protein can be studied by using antibodies against the protein of interest and marker proteins for the various cellular fractions (see Figure 3a). This is particularly important in view of the increasing number of mitochondrial proteins with dual or multiple locations in the cells. Mitochondria can share proteins with the cytosol, the peroxisomes, and with the nucleus.

2. Submitochondrial fractionation of isolated mitochondria

The integrity of isolated mitochondria can be assessed by the so-called protease protection assay with or without osmotic swelling of mitochondria under hypotonic conditions (Glick *et al.*, 1992; Diekert *et al.*, 2001). The rationale of these experiments is to treat the organelles with protease, usually proteinase K or trypsin, and analyse by immunodecoration the degree of protection of marker proteins in the intermembrane space and the matrix (see Protocol 2A and Figure 3b, left panel). When the organelles are intact, both kinds of markers are protected from degradation. In a second step, mitochondria are first subjected to hypoosmotic treatment in order to swell the matrix and achieve selective rupturing of the outer membrane. In mitoplasts generated in this way the intermembrane space, but not the matrix, is accessible for the added protease. Therefore, these assays also allow determination of the submitochondrial location of proteins. Proteins that are degraded in non-swollen mitochondria are outer membrane proteins, whereas after swelling also intermembrane space and inner membrane proteins are degraded. Matrix proteins will remain protected even after swelling.

Another assay consists in incubating mitochondria with increasing concentrations of digitonin together with proteinase K. The outer membrane only will be solubilised by low digitonin concentrations while the inner membrane requires higher concentrations (Hartl *et al.*, 1986). Using a range of digitonin concentrations will result in the progressive accessibility of proteins in the different compartments to the added protease. Immunodecoration of the protein of interest compared to appropriate markers will reveal its submitochondrial location (Ryan *et al.*, 2001).

Protocol 1. Isolation of Mitochondria.

Materials and Solutions:

1. *YPEG agar plates*: For preparing growth medium, dissolve 10 g/l yeast extract, 20 g/l Bacto-peptone, and 20 g/l agar in water. Adjust pH to 5.5 with concentrated HCl. Autoclave 20 min at 120°C. Add 100 ml/l of sterile 30% (w/v) glycerol and 20 ml/l of ethanol to autoclaved growth medium. Then pour agar plates. Store the solid plates at 4°C.
2. *Lactate medium*: Dissolve 3 g/l yeast extract, 1 g/l KH_2PO_4, 1 g/l NH_4Cl, 0.5 g/l $CaCl_2 \cdot 2H_2O$, 0.5 g/l NaCl, 0.6 g/l $MgSO_4 \cdot 7H_2O$ and 0.3 ml/l of 1% (w/v) $FeCl_3$ solution in water. Add 22 ml/l of 90% (w/v) lactic acid. Adjust the pH to 5.5 with 10 M KOH solution and bring to final volume with water. Autoclave 20 min at 120°C. Store at room temperature.
3. *YPGal*: Dissolve 10 g/l yeast extract and 20 g/l Bacto-Peptone in water. Adjust pH to 5.5 with concentrated HCl. Autoclave 20 min at 120°C. Add 67 ml/l of sterile 30% (w/v) galactose solution and store at room temperature.
4. *100 mM Tris-*SO_4: Solubilise 12.11 g/l of Tris in distilled water. Adjust pH to 9.4 with H_2SO_4. Store at 4°C.
5. *1 M DTT*: Dissolve 154 mg dithiothreitol in 1 ml distilled water. Prepare freshly.
6. *KPi-buffer*: First make 100 ml of a 1 M K_2HPO_4 solution and 100 ml of a 1 M KH_2PO_4 solution. Add KH_2PO_4 solution to the K_2HPO_4 until pH 7.4 is achieved. Store at 4°C.
7. *1.2 M sorbitol buffer*: 1.2 M sorbitol, 20 mM *KPi buffer*, pH 7.4.
8. *HS-buffer*: 0.6 M sorbitol, 20 mM HEPES, pH 7.4.

Steps:

A. Growth and Harvesting of Cells

1. Streak out the yeast strain onto a *YPEG agar plate* and grow for 2–3 days at 30°C.
2. Inoculate 20 ml of *lactate medium* or *YPGal* in a 100-ml Erlenmeyer flask with a loop full of the culture. Grow overnight at 30°C under shaking at 120 rpm.
3. Use the overnight culture to inoculate fresh *lactate medium* or *YPGal* (100 ml in an 500-ml Erlenmeyer flask). The initial OD_{600} should be 0.05–0.1. Grow the culture overnight as described in step 2 and evaluate the growth rate. In order to determine more precisely the growth rate, such a preculture may be repeated once or twice.
4. For the main culture inoculate 2 l of *lactate medium* in a 5-l Erlenmeyer flask with the preculture. Grow the culture overnight at 30°C under shaking at 120 rpm to an OD_{600} of 1–2.
5. Collect cells by centrifugation at 3000 rpm (Beckman JA10 rotor) for 5 min at room temperature and discard supernatant.

6. Resuspend the cell pellet in 200 ml of distilled water and spin again as described in step A.5.
7. Determine the weight of the cell pellet (wet weight). It is usually 5–10 g of yeast cells with a wild-type strain.

B. Preparation of Spheroplasts

1. Resuspend the cells at 3 ml/g of cells wet weight (from step A.7) in *100 mM Tris-SO_4*, pH 9.4. Add *1 M DTT* to a final concentration of 10 mM.
2. Incubate the cells for 10 min at 30°C under shaking.
3. Spin down the cells at 3000 rpm (Beckman rotor JA10) for 5 min at room temperature and discard supernatant.
4. Resuspend cell pellet in *1.2 M sorbitol buffer* at 4 ml/g of cells wet weight (from step A.7).
5. Repeat steps B.3 and B.4.
6. Dissolve 3 mg Zymolyase 20 T (Seikagaku) per gram of cells wet weight (from step A.7) in 1 ml *1.2 M sorbitol buffer*. Remove a small aliquot of about 50 µl of the cell suspension to use as a control for the efficiency of spheroplasting (see step B.8). Add Zymolyase to the cell suspension.
7. Incubate the cells for 20–40 min at 30°C in a shaking water bath.
8. To check for the degree of spheroplast formation, withdraw a 20 µl sample, add to 1 ml water and measure OD_{600}. Incubation should be continued until the OD_{600} is in the range of 10–20% of the control measured prior to the addition of Zymolyase (see step B.6). Alternatively, add 20 µl of water or *1.2 M sorbitol buffer* to 5 µl aliquots of cells, mix thoroughly and inspect under the microscope. Incubation should be continued until more than 90% of cells have burst when water was added.
9. Transfer spheroplast suspension into a 50 ml-centrifuge tube. Spin at 4000 rpm (Beckman rotor JA20) for 5 min at 4°C and discard supernatant.
10. Resuspend the spheroplast pellet in 40 ml *1.2 M sorbitol buffer* and spin again at 4000 rpm (JA20 rotor) for 5 min at 4°C. Discard supernatant.

C. Isolation of Mitochondria

1. Resuspend the spheroplasts at 3 ml/g of cells wet weight (from step A.7) in *HS-buffer* supplemented with 0.5 mM PMSF, 5 mM EDTA and 0.1% fatty acid free BSA (For this and all further steps the samples should be kept at 0–4°C).
2. Transfer the spheroplast suspension to a glass–glass homogeniser (tight fitting pestle) and perform 15 strokes, avoiding foaming of the sample. Transfer into four clean centrifuge tubes and fill up with same buffer (see step C.1).

3. Spin at 4000 rpm (JA20 rotor) for 5 min at 4°C. Pour the supernatant into fresh centrifuge tubes.
4. *Optional*: repeat step C.1 with the pellets. This will increase the yield of mitochondria but may affect their quality in functional studies.
5. Spin the supernatants at 12 000 rpm (JA20 rotor) for 12 min at 4°C. Supernatants may be used as a post-mitochondrial fraction.
6. Resuspend carefully and pool the mitochondrial pellets in 1 ml of *HS-buffer* using a pipette tip with a wide opening (e.g. a tip with cut end).
7. Transfer resuspended mitochondria to a small-scale glass–glass homogeniser and perform 5 strokes, avoiding foaming of the sample. Transfer to one fresh centrifuge tube, fill up with *HS-buffer*, and repeat steps C.3 and C.5.
8. Resuspend carefully the mitochondrial pellet in 300 μl of *HS-buffer* and determine the protein concentration using the Bradford method. It is usually between 10 and 30 mg protein/ml with a wild-type strain.
9. Freeze the mitochondrial suspension as 50 μl aliquots in liquid nitrogen and store at −70°C.

To distinguish soluble or peripherally attached intermembrane space proteins from outer or inner membrane proteins, salt and carbonate extractions are performed (Fujiki *et al.*, 1982; Diekert *et al.*, 2001). Release of the protein into the supernatant upon centrifugation is determined by treatment of the mitochondria with high ionic strength buffers or carbonate buffer at alkaline pH (see Protocol 2B and C). Soluble proteins of the intermembrane space or the matrix are released upon salt or carbonate extraction while integral membrane proteins will be recovered in the mitochondrial pellet (see Figure 3b, right panel). It should be noted that some proteins having a single transmembrane segment are often extracted at least partially upon carbonate extraction. Marker proteins of the different compartments are used as controls.

3. Oxidative phosphorylation and other metabolic activities

A number of individual enzyme activities can be measured in isolated mitochondria. In particular, established assays exist for determination of ATP synthesis/hydrolysis activities, cytochrome c oxidase activity, succinate dehydrogenase activity and a number of metabolic enzyme activities (Somlo, 1968; Lanzetta *et al.*, 1979; Velours *et al.*, 2001). Furthermore, spectrophotometric analysis of the mitochondrial cytochromes often helps in the determination of molecular defects in the respiratory chain.

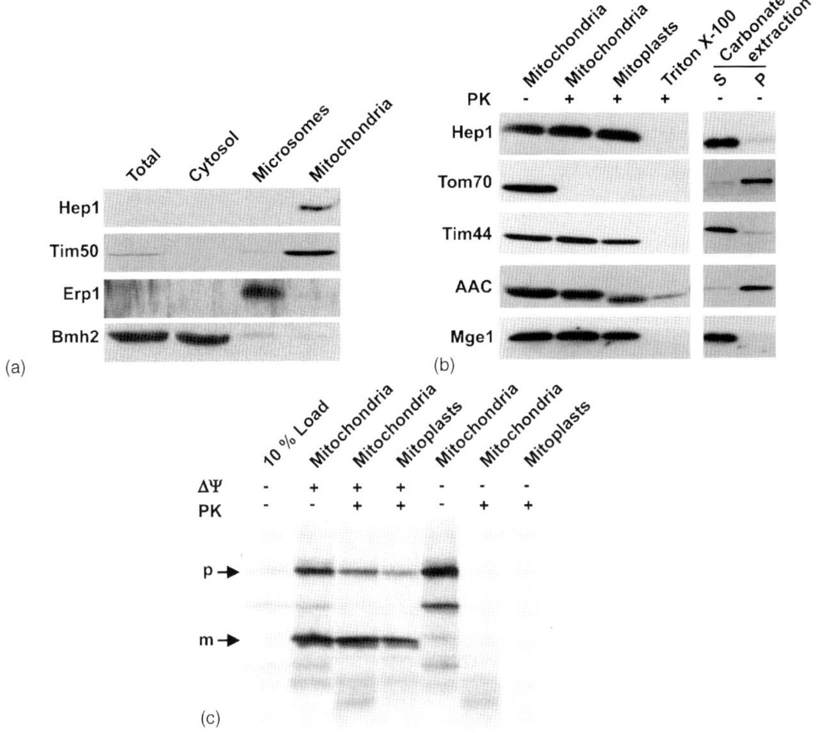

Figure 3. Subcellular and submitochondrial localisation of the mitochondrial matrix protein Hep1. Hep1 (mtHsp70 escort protein 1) has a role in preventing mitochondrial Hsp70 (mtHsp70) from aggregation. (a), Subcellular fractionation of yeast. Equal amounts of protein of subcellular fractions were subjected to SDS-PAGE and immunoblotting with antibodies against Hep1 and marker proteins of mitochondria (Tim50), microsomes (Erp1) and cytosol (Bmh2). Hep1 colocalises with the mitochondrial protein Tim50. (b) Submitochondrial localisation of Hep1. Mitochondria and mitoplasts were prepared and treated with proteinase K (PK); Hep1 was not degraded. After disruption of the inner membrane with detergent (Triton X-100), the protein was degraded by the added protease; upon alkaline extraction with carbonate, Hep1 was found in the supernatant fraction (S). This demonstrates that Hep1 is not an integral membrane protein but rather a soluble protein of the mitochondrial matrix. Samples were subjected to SDS-PAGE and immunoblotting with antibodies against Hep1 and various marker proteins of mitochondria: Tom70, integral outer membrane protein; Tim44, matrix protein attached to the matrix side of the inner membrane; AAC (ADP/ATP carrier), integral inner membrane protein; Mge1, soluble matrix protein. (S) supernatant, (P) pellet fraction. (c) The Hep1 precursor (p) is imported into isolated mitochondria in a membrane potential-dependent manner and processed to the mature species (m) of *ca.* 17 kDa. Reticulocyte lysate containing ^{35}S-labelled Hep1 was incubated with mitochondria in the presence or absence of membrane potential ($\Delta\Psi$). Mitochondria were collected by centrifugation; aliquots were converted to mitoplasts and treated with proteinase K. Samples were subjected to SDS-PAGE and autoradiography. 10% load (left lane) represents 10% of reticulocyte lysate used per import reaction (taken from Sichting *et al.*, 2005).

Other assays assess the function of complete pathways like electron transport through the respiratory chain or the entire oxidative phosphorylation pathway. In this case, one cannot only estimate the presence of all components or steps necessary for respiration, but

Protocol 2. Submitochondrial Fractionation.

Solutions:

1. *HS-buffer*: 0.6 M sorbitol, 20 mM HEPES-KOH, pH 7.4.
2. *H-buffer*: 20 mM HEPES-KOH, pH 7.4.
3. *HNaCl-buffer*: 20 mM HEPES-KOH, pH 7.4, 2 M NaCl.
4. *Carbonate solution*: 200 mM Na_2CO_3 (make fresh).

Steps:

A. Protease Protection Assay

1. Thaw quickly a frozen sample of mitochondria and transfer three aliquots each corresponding to 100 μg protein to microcentrifuge tubes.
2. Spin immediately at 13 400*g* for 10 min at 4°C and discard supernatants.
3. Resuspend two pellets (1 and 2) in 500 μl *HS-buffer* and one (3) in 500 μl *H-buffer*.
4. Add 5 μl distilled water to pellet 1 and 5 μl 10 mg/ml proteinase K to pellets 2 and 3.
5. Incubate 25 min on ice.
6. Add 5 μl 0.2 M PMSF to each sample and incubate 5 min on ice.
7. Spin at 13 400*g* for 10 min at 4°C and discard supernatants.
8. Resuspend the mitochondria and mitoplast pellets in 500 μl *HS-buffer*.
9. Precipitate proteins of the pellet fractions with 12% (w/v) trichloroacetic acid. Vortex and incubate 10 min on ice.
10. Spin at 25 000*g* for 10 min at 4°C and discard supernatants.
11. Add 1 ml cold (−20°C) acetone to the pellets and vortex.
12. Spin at 25 000*g* for 10 min at 4°C and discard supernatants.
13. Let the pellets dry at 55°C for 5 min and resuspend them in 50 μl Laemmli buffer.
14. Analyse proteins of the different samples by SDS-PAGE and immunodecoration.

B. NaCl Extraction

1. Thaw quickly a frozen sample of mitochondria and transfer an aliquot corresponding to 100 μg protein to a microcentrifuge tube.
2. Spin immediately at 13 400*g* for 10 min at 4°C and discard supernatant.
3. Resuspend mitochondrial pellet in 500 μl *H-buffer*.
4. Vortex.
5. Add 500 μl *HNaCl-buffer*.
6. Incubate 30 min on ice.
7. Spin at 90 000*g* for 30 min at 4°C to pellet membranes (P). Save supernatant (S).

8. Rinse once the pellet with 100 μl ice-cold distilled water and resuspend in 200 μl *H-buffer*.
9. Precipitate proteins of the supernatant and pellet fractions as described in A. steps 9–13.
10. Analyse proteins of membrane pellet (P) and supernatant fraction (S) by SDS-PAGE and immunodecoration.

C. Carbonate Extraction

1. Thaw quickly a frozen sample of mitochondria and transfer an aliquot corresponding to 100 μg proteins to a microcentrifuge tube.
2. Spin immediately at 13 400*g* for 10 min at 4°C and remove supernatant.
3. Resuspend mitochondrial pellet in 100 μl *H-buffer*. Add 100 μl *carbonate solution*.
4. Vortex and incubate 30 min on ice.
5. Spin at 90 000*g* for 30 min at 4°C to pellet membranes (P). Save supernatant (S).
6. Rinse once the pellet with 100 μl ice-cold distilled water and resuspend in 200 μl *H-buffer*.
7. Precipitate proteins of the supernatant (S) and pellet (P) fractions as described in A. steps 9–13.
8. Analyse proteins of membrane pellet (P) and supernatant fraction (S) by SDS-PAGE and immunodecoration.

also the intactness and efficiency of such multienzymatic pathways. For example, oxidative phosphorylation involves a large number of enzymes in mitochondria that catalyse the different steps of the Krebs cycle, the electron transfer to and within the respiratory chain, coupled or not to proton transfer across the inner membrane. The electrochemical gradient of protons thereby generated is used by a number of ion and solute transporters as well as by the F_1F_O-ATP synthase to generate ATP from ADP and inorganic phosphate. Therefore, dysfunction or uncoupling of any of those enzymes can lead to inefficient oxidative phosphorylation.

One of the most widely used assays to investigate oxidative phosphorylation relies on the determination of the oxygen consumption rate by intact mitochondria (Guerin *et al.*, 1979; Rigoulet and Guerin, 1979; Yaffe, 1991). A chamber equipped with a Clark electrode is used for this assay. In such a coupled assay, mitochondria are incubated with substrates of the respiratory chain, and other substrates such as ADP or drugs (inhibitors or protonophores) may be added. In this way, the response of the respiratory chain activity to changes in the ATP synthase activity or to a collapse of the membrane potential can be studied. The behaviour of mitochondria isolated from a mutant and the respective wild-type parental strain are then compared. The measurement of ATP produced per oxygen consumed (ATP/O or P/O ratio) allows a direct estimation of the efficiency of oxidative phosphorylation. As a complementary assay, the membrane potential produced by the respiratory chain can be evaluated by recording changes in the fluorescence of lipophilic cations accumulating in the mitochondrial matrix such as rhodamine 123 using a spectrofluorometer (Emaus *et al.*, 1986; Yaffe, 1991). Mitochondrial uncoupling, defective respiratory chain or defective ATP synthase can all be detected with such assays (Duvezin-Caubet *et al.*, 2003; Meier *et al.*, 2005).

4. Study of the biogenesis of mitochondria by labelling the mitochondrial translation products

Only eight major proteins are encoded by the mitochondrial genome of yeast. *In vivo* and *in organello* labelling of mitochondrial translation products are well suitable techniques for analysing the biosynthesis of these proteins. These approaches allow the investigation not only of mutants defective in mitochondrial translation but also mutants affected in other steps of mitochondrial gene expression such as transcription of a gene or maturation and stabilisation of messenger RNAs (Douglas and Butow, 1976; McKee and Poyton, 1984; McKee *et al.*, 1984; Yaffe, 1991; Herrmann *et al.*, 1994; Westermann *et al.*, 2001).

The biogenesis of mitochondrially encoded proteins, like that of other cellular proteins, requires prevention of aggregation of newly synthesised polypeptides, targeting to their final location, maturation by processing peptidases, folding into native conformation, integration into a membrane and assembly into higher order molecular protein complexes. In addition, misfolded or unassembled polypeptides are often degraded by proteases. The specific radioactive labelling of mitochondrial translation products provides a highly useful tool to study these processes (Arlt *et al.*, 1996; Westermann *et al.*, 1996, 2001; van Dyck *et al.*, 1998; Herrmann *et al.*, 2001; Tzagoloff *et al.*, 2004).

5. Protein import into isolated mitochondria

An important aspect of mitochondrial biogenesis is protein import into the mitochondria, as the vast majority of mitochondrial proteins are nuclear-encoded and synthesised in the cytosol. The import machinery involves more than 50 different proteins, all nuclear-encoded, and is organised in at least six complex molecular machines (Koehler *et al.*, 2000; Paschen and Neupert, 2001; Endo *et al.*, 2003; Herrmann and Neupert, 2003; Pfanner *et al.*, 2004; Rehling *et al.*, 2004; Mokranjac and Neupert, 2005). It notably comprises the TOM (translocase of outer membrane) and TIM (translocase of inner membrane) complexes. A preprotein is directed to the mitochondria by virtue of its targeting signal, in most cases situated at the *N*-terminus. The TOM complex includes receptors for mitochondrial preproteins and allows their translocation to the intermembrane space. The major TIM complex, TIM23, forms a channel in the inner membrane that is connected to the TOM complex and to the import motor in the matrix. Precursors targeted to the matrix or the inner membrane are imported by the TIM23 machinery using the membrane potential and matrix ATP as driving forces. Mainly precursors directed to the matrix require additionally the help of the import motor. For inner membrane proteins, there are two pathways of import. Either they are arrested during import and laterally inserted into the membrane (stop-transfer), or they are completely imported into the matrix and then integrated into the membrane (conservative sorting). In addition, most precursors are proteolytically processed at their *N*-terminus to their mature form during or after import.

In order to study such complex pathways, various assays of *in vitro* import of proteins in isolated mitochondria have been developed. They can be used to dissect the various import complexes and to identify molecular defects in mutants and thereby identify new components and functions. Isolated mitochondria, mitoplasts, or mitochondrial subfractions are incubated with *in vitro* translated radioactively labelled precursor proteins. Import into the correct compartment, status of maturation of the precursor, and energy dependence of the import can be assessed (see Figure 3c). Therefore, a subsequent protease protection assay is usually performed (see Section III.C and Protocol 2) before analysing protein extracts on SDS-PAGE. This is a quite robust assay and established protocols are available (see Protocol 3). A number of variations and subsequent treatments of imported proteins can be performed as detailed elsewhere (Yaffe, 1991; Herrmann *et al.*, 2001; Reichert *et al.*, 2005; Ryan *et al.*, 2001; Wiedemann *et al.*, 2006). Import into mitochondria isolated from mutants blocked in different steps of mitochondrial biogenesis will provide information on the sorting pathway of the protein (Rissler *et al.*, 2005; Sichting *et al.*, 2005). A potential targeting sequence may be fused to a passenger protein to check its functionality in sorting proteins to mitochondria. As an additional tool, the *C*-terminal part of a protein can be replaced by mouse dihydrofolate reductase (DHFR). DHFR can be stably folded by addition of the substrate analogue methotrexate, and thereby an arrest of translocation intermediates is achieved (Wienhues *et al.*, 1991; Chacinska *et al.*, 2005). This tool proved to be useful in studying the interactions of precursors with the mitochondrial translocation machineries (Bomer *et al.*, 1997; Dekker *et al.*, 1997).

6. Analysis of protein–protein interactions in isolated mitochondria

Many mitochondrial proteins are present as subunits of larger molecular complexes. In addition, a number of these complexes are

Protocol 3. Protein Import into Isolated Mitochondria.

Solutions and Material:

1. *HS-buffer*: 0.6 M sorbitol, 20 mM HEPES-KOH, pH 7.4.
2. *H-buffer*: 20 mM HEPES-KOH, pH 7.4.
3. 2 × *IP-buffer*: 0.1 M HEPES-KOH, pH 7.2, 1.2 M sorbitol, 160 mM KCl, 20 mM $MgAc_2$, 5 mM EDTA, 4 mM KH_2PO_4, 2 mg/ml fatty acid free BSA.
4. *Radiolabelled lysate*: Precursor proteins labelled with $[^{35}S]$methionine are obtained by *in vitro* coupled transcription/translation of plasmid DNA using the TnT Coupled Reticulocyte Lysate System (Promega, USA).

Steps:

1. Thaw quickly a frozen mitochondrial sample and take an aliquot corresponding to 200 μg proteins (see protocol "Isolation of mitochondria" for details) in a microcentrifuge tube.
2. Spin immediately at 13 400*g* for 10 min at 4°C and discard supernatant.
3. Resuspend mitochondrial pellet in 20 μl *HS-buffer* (10 μg/μl final) and keep on ice.
4. Prepare the import mix: 100 μl 2 × *IP-buffer,* 5 μl 0.2 M NADH, 2.5 μl 0.2 M ATP, 2 μl malate/succinate 0.25 M each and 2 μl radiolabelled lysate. Bring to a final volume of 180 μl with distilled water.
5. Add the mitochondria (20 μl) and incubate for desired time periods (usually 10–20 min) at room temperature.
6. Take three aliquots of 50 μl each. Add 450 μl of *HS-buffer* to samples 1 and 2 and 450 μl of *H-buffer* to sample 3.
7. Add 5 μl of 10 mg/ml proteinase K to samples 2 and 3 and incubate all for 25 min on ice.
8. Add 5 μl 0.2 M PMSF to all samples and incubate for further 5 min on ice.
9. Spin at 13 400*g* for 10 min at 4°C and discard supernatant.
10. Rinse gently the tube without resuspending the pellet with 500 μl *HS-buffer* supplemented with 2.5 μl 0.2 M PMSF.
11. Spin at 13 400*g* for 10 min at 4°C and discard supernatant.
12. Analyse imported proteins by SDS-PAGE, BN-PAGE, immunoprecipitation or other suitable method.

membrane embedded. This is illustrated by the respiratory chain complexes and the F_1F_O-ATP synthase, as well as the protein import and fusion/fission machineries. Therefore, the characterisation of protein–protein interactions in mitochondria is of particular importance for the determination of mitochondrial function and regulation. A broad spectrum of biochemical techniques is available to determine as to whether a mitochondrial protein is present in a supramolecular complex.

In a first approach, various size-exclusion and ion-exchange chromatography methods allow the determination of the size of complexes containing a given protein (Cruciat *et al.*, 1999, 2000). As a number of complexes comprise at least some hydrophobic membrane integral proteins, the analysis of protein–protein interactions is usually conducted after extraction of proteins from mitochondria with non-ionic detergents such as digitonin or TritonX100. Thus, special attention should be given to possible artifactual interactions generated after lysis of mitochondria, and interactions should be confirmed by different methods. Native electrophoresis methods were developed during the last decade, like blue native or colourless native gels (BN- and CN-PAGE) to separate large complexes such as the F_1F_O-ATP synthase (about 550 kDa: Schagger and von Jagow, 1991; Schagger *et al.*, 1994; Schagger, 2001b; Wittig and Schagger, 2005). Moreover, in order to identify such complexes, it is possible to specifically stain different complexes of the respiratory chain and the ATP synthase by their enzymatic activities directly inside the gel (Grandier-Vazeille and Guerin, 1996). A second dimension analysis (SDS-PAGE) of these complexes allows one to identify the different components of a given complex. Such electrophoresis studies led to the characterisation of supramolecular complexes such as complexIII/complexIV dimers or dimers and oligomers of F_1F_O-ATP synthase (Arnold *et al.*, 1998; Schagger and Pfeiffer, 2000; Schagger, 2001a).

A direct way to investigate interactions between proteins is to use immunoprecipitation. When the extraction of mitochondrial proteins is performed at low concentrations of non-ionic detergents, many protein–protein interactions but not all, are preserved and interacting partners can be copurified (Herrmann *et al.*, 2001). In most multisubunit complexes, a given protein has several interacting partners which can be copurified. The co-immunoprecipitation of two different polypeptides using antibodies against one or the other will provide additional evidence about the specificity of their interaction. Using these different techniques, significant insights have been achieved in the characterisation of the different complexes of mitochondria and the analysis of their assembly state and composition under different conditions or in mutants (Cruciat *et al.*, 1999; Boldogh *et al.*, 2003; Duvezin-Caubet *et al.*, 2003; Wong *et al.*, 2003; Chacinska *et al.*, 2005; Waizenegger *et al.*, 2005; Adam *et al.*, 2006).

It should be noted that protein–protein interactions are often rather weak and transient. A way to capture such interactions is to perform chemical cross-linking in intact mitochondria (see Protocol 4). A large choice of bifunctional cross-linkers is available with different reactive groups and lengths of spacers. Among them are some which can penetrate mitochondrial membranes. A limitation of the procedure is that in most cases the efficiency of

Protocol 4. Chemical Cross-Linking of Proteins in Intact Mitochondria.

Solutions:

1. *20 mM MBS*: Dissolve 3.2 mg MBS (*m*-maleimidobenzoyl-*N*-hydroxysuccinimide ester; pierce biotechnology) in 500 μl dimethyl sulfoxide (DMSO). Make fresh immediately before use.
2. *1 M cysteine, pH 8.0*: Dissolve 12.12 g cysteine in 90 ml distilled water; adjust to pH 8.0 with 10 M KOH and fill up to a total volume of 100 ml with distilled water. Store at −20°C.

Steps:

1. Thaw quickly a frozen sample of mitochondria. Take two aliquots corresponding to 50 μg protein in microcentrifuge tubes and dilute to a final concentration of 0.5 mg protein/ml using *HS-buffer* (recipe in Protocol 1).
2. Add 1 μl DMSO to one sample for mock treatment and 1 μl *20 mM MBS* to other sample for cross-linking reaction.
3. Incubate the samples for 30 min on ice.
4. Stop the cross-linking reaction by addition of *1 M cysteine, pH 8.0*, to a final concentration of 0.1 M and incubate 10 min on ice.
5. Spin samples at 13 400*g* for 10 min at 4°C and discard supernatants.
6. Resuspend the mitochondrial pellets in 25 μl Laemmli buffer.
7. Resolve mitochondrial proteins and cross-linked products by SDS-PAGE. The cross-linked products are detected by western blotting.

Tips: MBS is a heterobifunctional cross-linking agent reacting with cysteines on one side and with primary amines on the other side with a spacer region of 10 Å. Numerous other agents are available with spacers of different sizes and different reactive groups (see Pierce Biotechnology catalogue). For example, 1,5-difluoro-2,4-dinitrobenzene (DFDNB), disuccinimidyl glutarate (DSG), and disuccinimidyl suberate (DSS) react on both sides with primary amines and have spacer regions of 3, 7.7 and 11.4 Å, respectively. The same procedure as described above can be used for other cross-linking agents. In addition, a final concentration of 5–1000 μM final of the different cross-linking agents can be applied. The cysteine solution used in this protocol allows blocking of excess amine and cysteine-reactive groups of cross-linking agents. However, quenching can be achieved by addition of 1 M glycine when cross-linkers are used that react only with amine groups.

cross-linking is rather low. Still, this approach can be very successfully applied for instance to determine the dynamic interactions between precursor protein and protein translocases, to establish the topology of large complexes, or to identify novel interacting partners (Davis *et al.*, 2000; Velours and Arselin, 2000; Herrmann *et al.*, 2001; Mokranjac *et al.*, 2003; Reichert *et al.*, 2005).

♦♦♦♦♦♦ IV. CONCLUSION

The genetic basis of the numerous functions of mitochondria has been analysed in impressive detail during the past decades. In this way, many biochemical pathways were characterised and their protein components were identified. Still, a plethora of questions remain open. In particular, the role of mitochondrial proteins in processes such as apoptosis, fusion and fission, morphology, positioning and movement in the cell, is far from clear. In recent years new pathways, for instance, Ca^{2+}-dependent cellular signalling and iron–sulfur cluster biosynthesis, have been discovered. It is likely that new important and unexpected pathways and functions will be revealed in the next future. Yeast genetics and molecular biology will certainly play a major role in detecting and dissecting these new functions.

Acknowledgements

We are very grateful to Michael Zick, Dr Vincent Soubannier, Dr Carsten Bornhövd, Dr Dejana Mokranjac and Dr Soledad Funes for helpful discussions and critical reading of the manuscript, and to Dr Ravi Jagasia and Dr Frank Vogel for kindly providing photographs of yeast cells.

References

Abadjieva, A., Pauwels, K., Hilven, P. and Crabeel, M. (2001). A new yeast metabolon involving at least the two first enzymes of arginine biosynthesis: acetylglutamate synthase activity requires complex formation with acetylglutamate kinase. *J. Biol. Chem.* **276**, 42869–42880.

Adam, A. C., Bornhovd, C., Prokisch, H., Neupert, W. and Hell, K. (2006). The Nfs1 interacting protein Isd11 has an essential role in Fe/S cluster biogenesis in mitochondria. *EMBO J.* **25**, 174–183.

Altmann, K. and Westermann, B. (2005). Role of essential genes in mitochondrial morphogenesis in *Saccharomyces cerevisiae*. *Mol. Biol. Cell* **16**, 5410–5417.

Andersson, S. G., Zomorodipour, A., Andersson, J. O., Sicheritz-Ponten, T., Alsmark, U. C., Podowski, R. M., Naslund, A. K., Eriksson, A. S., Winkler, H. H. and Kurland, C. G. (1998). The genome sequence of *Rickettsia prowazekii* and the origin of mitochondria. *Nature* **396**, 133–140.

Andreoli, C., Prokisch, H., Hortnagel, K., Mueller, J. C., Munsterkotter, M., Scharfe, C. and Meitinger, T. (2004). MitoP2, an integrated database on mitochondrial proteins in yeast and man. *Nucl. Acids Res.* **32**(Database issue), D459–D462.

Arlt, H., Tauer, R., Feldmann, H., Neupert, W. and Langer, T. (1996). The YTA10-12 complex, an AAA protease with chaperone-like activity in the inner membrane of mitochondria. *Cell* **85**, 875–885.

Arnold, I., Pfeiffer, K., Neupert, W., Stuart, R. A. and Schagger, H. (1998). Yeast mitochondrial F1F0-ATP synthase exists as a dimer: identification of three dimer-specific subunits. *EMBO J.* **17**, 7170–7178.

Arselin, G., Vaillier, J., Salin, B., Schaeffer, J., Giraud, M. F., Dautant, A., Brethes, D. and Velours, J. (2004). The modulation in subunits *e* and *g* amounts of yeast ATP synthase modifies mitochondrial cristae morphology. *J. Biol. Chem.* **279**, 40392–40399.

Bereiter-Hahn, J. and Voth, M. (1994). Dynamics of mitochondria in living cells: shape changes, dislocations, fusion, and fission of mitochondria. *Microsc. Res. Tech.* **27**, 198–219.

Berger, K. H. and Yaffe, M. P. (2000). Mitochondrial DNA inheritance in *Saccharomyces cerevisiae. Trends Microbiol.* **8**, 508–513.

Bleazard, W., McCaffery, J. M., King, E. J., Bale, S., Mozdy, A., Tieu, Q., Nunnari, J. and Shaw, J. M. (1999). The dynamin-related GTPase Dnm1 regulates mitochondrial fission in yeast. *Nat. Cell Biol.* **1**, 298–304.

Boldogh, I. R., Nowakowski, D. W., Yang, H. C., Chung, H., Karmon, S., Royes, P. and Pon, L. A. (2003). A protein complex containing Mdm10p, Mdm12p, and Mmm1p links mitochondrial membranes and DNA to the cytoskeleton-based segregation machinery. *Mol. Biol. Cell* **14**, 4618–4627.

Bolotin-Fukuhara, M. and Grivell, L. A. (1992). Genetic approaches to the study of mitochondrial biogenesis in yeast. *Antonie Van Leeuwenhoek* **62**, 131–153.

Bomer, U., Meijer, M., Guiard, B., Dietmeier, K., Pfanner, N. and Rassow, J. (1997). The sorting route of cytochrome b2 branches from the general mitochondrial import pathway at the preprotein translocase of the inner membrane. *J. Biol. Chem.* **272**, 30439–30446.

Bonnefoy, N., Bsat, N. and Fox, T. D. (2001). Mitochondrial translation of *Saccharomyces cerevisiae* COX2 mRNA is controlled by the nucleotide sequence specifying the pre-Cox2p leader peptide. *Mol. Cell Biol.* **21**, 2359–2372.

Bonnefoy, N. and Fox, T. D. (2001). Genetic transformation of *Saccharomyces cerevisiae* mitochondria. *Methods Cell Biol.* **65**, 381–396.

Bonnefoy, N. and Fox, T. D. (2002). Genetic transformation of *Saccharomyces cerevisiae* mitochondria. *Methods Enzymol.* **350**, 97–111.

Butow, R. A., Henke, R. M., Moran, J. V., Belcher, S. M. and Perlman, P. S. (1996). Transformation of *Saccharomyces cerevisiae* mitochondria using the biolistic gun. *Methods Enzymol.* **264**, 265–278.

Cabiscol, E., Belli, G., Tamarit, J., Echave, P., Herrero, E. and Ros, J. (2002). Mitochondrial Hsp60, resistance to oxidative stress, and the labile iron pool are closely connected in *Saccharomyces cerevisiae. J. Biol. Chem.* **277**, 44531–44538.

Campbell, C. L., Tanaka, N., White, K. H. and Thorsness, P. E. (1994). Mitochondrial morphological and functional defects in yeast caused by yme1 are suppressed by mutation of a 26S protease subunit homologue. *Mol. Biol. Cell* **5**, 899–905.

Chacinska, A., Lind, M., Frazier, A. E., Dudek, J., Meisinger, C., Geissler, A., Sickmann, A., Meyer, H. E., Truscott, K. N., Guiard, B., Pfanner, N. and Rehling, P. (2005). Mitochondrial presequence translocase: switching between TOM tethering and motor recruitment involves Tim21 and Tim17. *Cell* **120**, 817–829.

Chen, X. J. and Butow, R. A. (2005). The organization and inheritance of the mitochondrial genome. *Nat. Rev. Genet.* **6**, 815–825.

Cheng, M. Y., Hartl, F. U., Martin, J., Pollock, R. A., Kalousek, F., Neupert, W., Hallberg, E. M., Hallberg, R. L. and Horwich, A. L. (1989). Mitochondrial heat-shock protein hsp60 is essential for assembly of proteins imported into yeast mitochondria. *Nature* **337**, 620–625.

Claros, M. G. (1995). MitoProt, a Macintosh application for studying mitochondrial proteins. *Comput. Appl. Biosci.* **11**, 441–447.

Claros, M. G. and Vincens, P. (1996). Computational method to predict mitochondrially imported proteins and their targeting sequences. *Eur. J. Biochem.* **241**, 779–786.

Colson, A. M. (1993). Random mutant generation and its utility in uncovering structural and functional features of cytochrome *b* in *Saccharomyces cerevisiae*. *J. Bioenerg. Biomembr.* **25**, 211–220.

Contamine, V. and Picard, M. (2000). Maintenance and integrity of the mitochondrial genome: a plethora of nuclear genes in the budding yeast. *Microbiol. Mol. Biol. Rev.* **64**, 281–315.

Cruciat, C. M., Brunner, S., Baumann, F., Neupert, W. and Stuart, R. A. (2000). The cytochrome *bc1* and cytochrome *c* oxidase complexes associate to form a single supracomplex in yeast mitochondria. *J. Biol. Chem.* **275**, 18093–18098.

Cruciat, C. M., Hell, K., Folsch, H., Neupert, W. and Stuart, R. A. (1999). Bcs1p, an AAA-family member, is a chaperone for the assembly of the cytochrome *bc*(1) complex. *EMBO J.* **18**, 5226–5233.

Daum, G., Bohni, P. C. and Schatz, G. (1982). Import of proteins into mitochondria. Cytochrome *b2* and cytochrome *c* peroxidase are located in the intermembrane space of yeast mitochondria. *J. Biol. Chem.* **257**, 13028–13033.

Davis, A. J., Sepuri, N. B., Holder, J., Johnson, A. E. and Jensen, R. E. (2000). Two intermembrane space TIM complexes interact with different domains of Tim23p during its import into mitochondria. *J. Cell Biol.* **150**, 1271–1282.

De Vries, S., Van Witzenburg, R., Grivell, L. A. and Marres, C. A. (1992). Primary structure and import pathway of the rotenone-insensitive NADH-ubiquinone oxidoreductase of mitochondria from *Saccharomyces cerevisiae*. *Eur. J. Biochem.* **203**, 587–592.

Dekker, P. J., Martin, F., Maarse, A. C., Bomer, U., Muller, H., Guiard, B., Meijer, M., Rassow, J. and Pfanner, N. (1997). The Tim core complex defines the number of mitochondrial translocation contact sites and can hold arrested preproteins in the absence of matrix Hsp70-Tim44. *EMBO J.* **16**, 5408–5419.

DeRisi, J. L., Iyer, V. R. and Brown, P. O. (1997). Exploring the metabolic and genetic control of gene expression on a genomic scale. *Science* **278**, 680–686.

di Rago, J. P. and Colson, A. M. (1988). Molecular basis for resistance to antimycin and diuron, Q-cycle inhibitors acting at the Qi site in the mitochondrial ubiquinol-cytochrome *c* reductase in *Saccharomyces cerevisiae*. *J. Biol. Chem.* **263**, 12564–12570.

Diekert, K., de Kroon, A. I., Kispal, G. and Lill, R. (2001). Isolation and subfractionation of mitochondria from the yeast *Saccharomyces cerevisiae*. *Methods Cell Biol.* **65**, 37–51.

Dimmer, K. S., Fritz, S., Fuchs, F., Messerschmitt, M., Weinbach, N., Neupert, W. and Westermann, B. (2002). Genetic basis of mitochondrial function and morphology in *Saccharomyces cerevisiae*. *Mol. Biol. Cell* **13**, 847–853.

Doudican, N. A., Song, B., Shadel, G. S. and Doetsch, P. W. (2005). Oxidative DNA damage causes mitochondrial genomic instability in *Saccharomyces cerevisiae*. *Mol. Cell Biol.* **25**, 5196–5204.

Douglas, M. G. and Butow, R. A. (1976). Variant forms of mitochondrial translation products in yeast: evidence for location of determinants on mitochondrial DNA. *Proc. Natl. Acad. Sci. USA* **73**, 1083–1086.

Drawid, A. and Gerstein, M. (2000). A Bayesian system integrating expression data with sequence patterns for localizing proteins: comprehensive application to the yeast genome. *J. Mol. Biol.* **301**, 1059–1075.

Duvezin-Caubet, S., Caron, M., Giraud, M. F., Velours, J. and di Rago, J. P. (2003). The two rotor components of yeast mitochondrial ATP synthase are mechanically coupled by subunit delta. *Proc. Natl. Acad. Sci. USA* **100**, 13235–13240.

Emanuelsson, O., Nielsen, H., Brunak, S. and von Heijne, G. (2000). Predicting subcellular localization of proteins based on their *N*-terminal amino acid sequence. *J. Mol. Biol.* **300**, 1005–1016.

Emaus, R. K., Grunwald, R. and Lemasters, J. J. (1986). Rhodamine 123 as a probe of transmembrane potential in isolated rat-liver mitochondria: spectral and metabolic properties. *Biochim. Biophys. Acta* **850**, 436–448.

Endo, T., Yamamoto, H. and Esaki, M. (2003). Functional cooperation and separation of translocators in protein import into mitochondria, the double-membrane bounded organelles. *J. Cell Sci.* **116**, 3259–3267.

Ephrussi, B., Hottinguer, H. and Chimenes, Y. (1949a). Action de l'acriflavine sur les levures. I. La mutation "petite colonie". *Ann. Inst. Pasteur* **76**, 351–367.

Ephrussi, B., Hottinguer, H. and Tavlitzki, J. (1949b). Action de l'acriflavine sur les levures. II. Etude génétique du mutant "petite colonie". *Ann. Inst. Pasteur* **76**, 419–442.

Ephrussi, B. and Slonimski, P. P. (1955). Yeast mitochondria. Subcellular units involved in the synthesis of respiratory enzymes in yeast. *Nature* **176**, 1207–1208.

Foury, F., Roganti, T., Lecrenier, N. and Purnelle, B. (1998). The complete sequence of the mitochondrial genome of *Saccharomyces cerevisiae*. *FEBS Lett.* **440**, 325–331.

Foury, F. and Tzagoloff, A. (1976). Localization on mitochondrial DNA of mutations leading to a loss of rutamycin-sensitive adenosine triphosphatase. *Eur. J. Biochem.* **68**, 113–119.

Fox, T. D. (1996). Translational control of endogenous and recoded nuclear genes in yeast mitochondria: regulation and membrane targeting. *Experientia* **52**, 1130–1135.

Fox, T. D., Folley, L. S., Mulero, J. J., McMullin, T. W., Thorsness, P. E., Hedin, L. O. and Costanzo, M. C. (1991). Analysis and manipulation of yeast mitochondrial genes. *Methods Enzymol.* **194**, 149–165.

Fujiki, Y., Hubbard, A. L., Fowler, S. and Lazarow, P. B. (1982). Isolation of intracellular membranes by means of sodium carbonate treatment: application to endoplasmic reticulum. *J. Cell Biol.* **93**, 97–102.

Gari, E., Piedrafita, L., Aldea, M. and Herrero, E. (1997). A set of vectors with a tetracycline-regulatable promoter system for modulated gene expression in *Saccharomyces cerevisiae*. *Yeast* **13**, 837–848.

Gavin, A. C., Bosche, M., Krause, R., Grandi, P., Marzioch, M., Bauer, A., Schultz, J., Rick, J. M., Michon, A. M., Cruciat, C. M., Remor, M., Hofert, C., Schelder, M., Brajenovic, M., Ruffner, H., Merino, A., Klein, K., Hudak, M., Dickson, D., Rudi, T., Gnau, V., Bauch, A., Bastuck, S., Huhse, B., Leutwein, C., Heurtier, M. A., Copley, R. R., Edelmann, A., Querfurth, E., Rybin, V., Drewes, G., Raida, M., Bouwmeester, T., Bork, P., Seraphin, B., Kuster, B., Neubauer, G. and Superti-Furga, G. (2002). Functional organization of the yeast proteome by systematic analysis of protein complexes. *Nature* **415**, 141–147.

Glick, B. S., Brandt, A., Cunningham, K., Muller, S., Hallberg, R. L. and Schatz, G. (1992). Cytochromes c1 and b2 are sorted to the intermembrane space of yeast mitochondria by a stop-transfer mechanism. *Cell* **69**, 809–822.

Grandier-Vazeille, X. and Guerin, M. (1996). Separation by blue native and colorless native polyacrylamide gel electrophoresis of the oxidative phosphorylation complexes of yeast mitochondria solubilized by different detergents: specific staining of the different complexes. *Anal. Biochem.* **242**, 248–254.

Green-Willms, N. S., Fox, T. D. and Costanzo, M. C. (1998). Functional interactions between yeast mitochondrial ribosomes and mRNA 5′ untranslated leaders. *Mol. Cell Biol.* **18**, 1826–1834.

Grivell, L. A. (1989). Nucleo–mitochondrial interactions in yeast mitochondrial biogenesis. *Eur. J. Biochem.* **182**, 477–493.

Grivell, L. A. (1995). Nucleo–mitochondrial interactions in mitochondrial gene expression. *Crit. Rev. Biochem. Mol. Biol.* **30**, 121–164.

Grivell, L. A., Artal-Sanz, M., Hakkaart, G., de Jong, L., Nijtmans, L. G., van Oosterum, K., Siep, M. and van der Spek, H. (1999). Mitochondrial assembly in yeast. *FEBS Lett.* **452**, 57–60.

Gu, Z., Valianpour, F., Chen, S., Vaz, F. M., Hakkaart, G. A., Wanders, R. J. and Greenberg, M. L. (2004). Aberrant cardiolipin metabolism in the yeast taz1 mutant: a model for Barth syndrome. *Mol. Microbiol.* **51**, 149–158.

Guan, K., Farh, L., Marshall, T. K. and Deschenes, R. J. (1993). Normal mitochondrial structure and genome maintenance in yeast requires the dynamin-like product of the MGM1 gene. *Curr. Genet.* **24**, 141–148.

Guda, C., Fahy, E. and Subramaniam, S. (2004a). MITOPRED: a genome-scale method for prediction of nucleus-encoded mitochondrial proteins. *Bioinformatics* **20**, 1785–1794.

Guda, C., Guda, P., Fahy, E. and Subramaniam, S. (2004b). MITOPRED: a web server for the prediction of mitochondrial proteins. *Nucl. Acids Res.* **32**(Web Server issue), W372–W374.

Guerin, B., Labbe, P. and Somlo, M. (1979). Preparation of yeast mitochondria (*Saccharomyces cerevisiae*) with good P/O and respiratory control ratios. *Methods Enzymol.* **55**, 149–159.

Habib, S. J., Neupert, W. and Rapaport, D. (2007). Analysis and prediction of mitochondrial targeting signals. *Methods Cell Biol.* (Mitochondria 2nd edn), Chapter 35, Vol. 80. Academic Press, Elsevier, Amsterdam. ISBN: 978-0-12-544173-5 Book/Hardback.

Hampsey, M. (1997). A review of phenotypes in *Saccharomyces cerevisiae*. *Yeast* **13**, 1099–1133.

Hartl, F. U., Schmidt, B., Wachter, E., Weiss, H. and Neupert, W. (1986). Transport into mitochondria and intramitochondrial sorting of the Fe/S protein of ubiquinol-cytochrome c reductase. *Cell* **47**, 939–951.

Hell, K., Herrmann, J., Pratje, E., Neupert, W. and Stuart, R. A. (1997). Oxa1p mediates the export of the *N*- and *C*-termini of pCoxII from the mitochondrial matrix to the intermembrane space. *FEBS Lett.* **418**, 367–370.

Hell, S. W., Dyba, M. and Jakobs, S. (2004). Concepts for nanoscale resolution in fluorescence microscopy. *Curr. Opin. Neurobiol.* **14**, 599–609.

Hermann, G. J., Thatcher, J. W., Mills, J. P., Hales, K. G., Fuller, M. T., Nunnari, J. and Shaw, J. M. (1998). Mitochondrial fusion in yeast requires the transmembrane GTPase Fzo1p. *J. Cell Biol.* **143**, 359–373.

Herrmann, J. M., Fölsch, H., Neupert, W. and Stuart, R. A. (1994). Isolation of yeast mitochondria and study of mitochondrial protein translation. In: *Cell Biology: A Laboratory Handbook*, Vol. 1 (J. E. Celis, ed.), pp. 538–544. Academic Press, San Diego.

Herrmann, J. M. and Neupert, W. (2003). Protein insertion into the inner membrane of mitochondria. *IUBMB Life* **55**, 219–225.

Herrmann, J. M., Westermann, B. and Neupert, W. (2001). Analysis of protein–protein interactions in mitochondria by coimmunoprecipitation and chemical cross-linking. *Methods Cell Biol.* **65**, 217–230.

Ho, Y., Gruhler, A., Heilbut, A., Bader, G. D., Moore, L., Adams, S. L., Millar, A., Taylor, P., Bennett, K., Boutilier, K., Yang, L., Wolting, C., Donaldson, I., Schandorff, S., Shewnarane, J., Vo, M., Taggart, J., Goudreault, M., Muskat, B., Alfarano, C., Dewar, D., Lin, Z., Michalickova, K., Willems, A. R., Sassi, H., Nielsen, P. A., Rasmussen, K. J., Andersen, J. R., Johansen, L. E., Hansen, L. H., Jespersen, H., Podtelejnikov, A., Nielsen, E., Crawford, J., Poulsen, V., Sorensen, B. D., Matthiesen, J., Hendrickson, R. C., Gleeson, F., Pawson, T., Moran, M. F., Durocher, D., Mann, M., Hogue, C. W., Figeys, D. and Tyers, M. (2002). Systematic identification of protein complexes in *Saccharomyces cerevisiae* by mass spectrometry. *Nature* **415**, 180–183.

Hughes, T. R., Marton, M. J., Jones, A. R., Roberts, C. J., Stoughton, R., Armour, C. D., Bennett, H. A., Coffey, E., Dai, H., He, Y. D., Kidd, M. J., King, A. M., Meyer, M. R., Slade, D., Lum, P. Y., Stepaniants, S. B., Shoemaker, D. D., Gachotte, D., Chakraburtty, K., Simon, J., Bard, M. and Friend, S. H. (2000). Functional discovery via a compendium of expression profiles. *Cell* **102**, 109–126.

Huh, W. K., Falvo, J. V., Gerke, L. C., Carroll, A. S., Howson, R. W., Weissman, J. S. and O'Shea, E. K. (2003). Global analysis of protein localization in budding yeast. *Nature* **425**, 686–691.

Ito, T., Chiba, T., Ozawa, R., Yoshida, M., Hattori, M. and Sakaki, Y. (2001). A comprehensive two-hybrid analysis to explore the yeast protein interactome. *Proc. Natl. Acad. Sci. USA* **98**, 4569–4574.

Jazwinski, S. M. (2005). Yeast longevity and aging – the mitochondrial connection. *Mech. Ageing Dev.* **126**, 243–248.

Jiang, F., Ryan, M. T., Schlame, M., Zhao, M., Gu, Z., Klingenberg, M., Pfanner, N. and Greenberg, M. L. (2000). Absence of cardiolipin in the crd1 null mutant results in decreased mitochondrial membrane potential and reduced mitochondrial function. *J. Biol. Chem.* **275**, 22387–22394.

Johnston, S. A., Anziano, P. Q., Shark, K., Sanford, J. C. and Butow, R. A. (1988). Mitochondrial transformation in yeast by bombardment with microprojectiles. *Science* **240**, 1538–1541.

Kang, P. J., Ostermann, J., Shilling, J., Neupert, W., Craig, E. A. and Pfanner, N. (1990). Requirement for hsp70 in the mitochondrial matrix for translocation and folding of precursor proteins. *Nature* **348**, 137–143.

Koehler, C. M., Murphy, M. P., Bally, N. A., Leuenberger, D., Oppliger, W., Dolfini, L., Junne, T., Schatz, G. and Or, E. (2000). Tim18p, a new subunit of the TIM22 complex that mediates insertion of imported proteins into the yeast mitochondrial inner membrane. *Mol. Cell Biol.* **20**, 1187–1193.
Kozany, C., Mokranjac, D., Sichting, M., Neupert, W. and Hell, K. (2004). The J domain-related cochaperone Tim16 is a constituent of the mitochondrial TIM23 preprotein translocase. *Nat. Struct. Mol. Biol.* **11**, 234–241.
Kumar, A., Agarwal, S., Heyman, J. A., Matson, S., Heidtman, M., Piccirillo, S., Umansky, L., Drawid, A., Jansen, R., Liu, Y., Cheung, K. H., Miller, P., Gerstein, M., Roeder, G. S. and Snyder, M. (2002). Subcellular localization of the yeast proteome. *Genes Dev.* **16**, 707–719.
Lange, H., Kispal, G. and Lill, R. (1999). Mechanism of iron transport to the site of heme synthesis inside yeast mitochondria. *J. Biol. Chem.* **274**, 18989–18996.
Lanzetta, P. A., Alvarez, L. J., Reinach, P. S. and Candia, O. A. (1979). An improved assay for nanomole amounts of inorganic phosphate. *Anal. Biochem.* **100**, 95–97.
Lefebvre-Legendre, L., Vaillier, J., Benabdelhak, H., Velours, J., Slonimski, P. P. and di Rago, J. P. (2001). Identification of a nuclear gene (FMC1) required for the assembly/stability of yeast mitochondrial F(1)-ATPase in heat stress conditions. *J. Biol. Chem.* **276**, 6789–6796.
Leonhard, K., Herrmann, J. M., Stuart, R. A., Mannhaupt, G., Neupert, W. and Langer, T. (1996). AAA proteases with catalytic sites on opposite membrane surfaces comprise a proteolytic system for the ATP-dependent degradation of inner membrane proteins in mitochondria. *EMBO J.* **15**, 4218–4229.
Lewin, A. S., Hines, V. and Small, G. M. (1990). Citrate synthase encoded by the CIT2 gene of *Saccharomyces cerevisiae* is peroxisomal. *Mol. Cell Biol.* **10**, 1399–1405.
Lill, R. and Kispal, G. (2000). Maturation of cellular Fe–S proteins: an essential function of mitochondria. *Trends Biochem. Sci.* **25**, 352–356.
Lill, R. and Muhlenhoff, U. (2005). Iron–sulfur–protein biogenesis in eukaryotes. *Trends Biochem. Sci.* **30**, 133–141.
Madeo, F., Herker, E., Wissing, S., Jungwirth, H., Eisenberg, T. and Frohlich, K. U. (2004). Apoptosis in yeast. *Curr. Opin. Microbiol.* **7**, 655–660.
Marres, C. A., de Vries, S. and Grivell, L. A. (1991). Isolation and inactivation of the nuclear gene encoding the rotenone-insensitive internal NADH: ubiquinone oxidoreductase of mitochondria from *Saccharomyces cerevisiae*. *Eur. J. Biochem.* **195**, 857–862.
Matz, M. V., Fradkov, A. F., Labas, Y. A., Savitsky, A. P., Zaraisky, A. G., Markelov, M. L. and Lukyanov, S. A. (1999). Fluorescent proteins from nonbioluminescent Anthozoa species. *Nat. Biotechnol.* **17**, 969–973.
McConnell, S. J., Stewart, L. C., Talin, A. and Yaffe, M. P. (1990). Temperature-sensitive yeast mutants defective in mitochondrial inheritance. *J. Cell Biol.* **111**, 967–976.
McKee, E. E., McEwen, J. E. and Poyton, R. O. (1984). Mitochondrial gene expression in *Saccharomyces cerevisiae*. II. Fidelity of translation in isolated mitochondria from wild type and respiratory-deficient mutant cells. *J. Biol. Chem.* **259**, 9332–9338.
McKee, E. E. and Poyton, R. O. (1984). Mitochondrial gene expression in *Saccharomyces cerevisiae*. I. Optimal conditions for protein synthesis in isolated mitochondria. *J. Biol. Chem.* **259**, 9320–9331.

Medalia, O., Weber, I., Frangakis, A. S., Nicastro, D., Gerisch, G. and Baumeister, W. (2002). Macromolecular architecture in eukaryotic cells visualized by cryoelectron tomography. *Science* **298**, 1209–1213.

Meeusen, S., McCaffery, J. M. and Nunnari, J. (2004). Mitochondrial fusion intermediates revealed *in vitro*. *Science* **305**, 1747–1752.

Meeusen, S., Tieu, Q., Wong, E., Weiss, E., Schieltz, D., Yates, J. R. and Nunnari, J. (1999). Mgm101p is a novel component of the mitochondrial nucleoid that binds DNA and is required for the repair of oxidatively damaged mitochondrial DNA. *J. Cell Biol.* **145**, 291–304.

Meeusen, S. L. and Nunnari, J. (2005). How mitochondria fuse. *Curr. Opin. Cell Biol.* **17**, 389–394.

Meier, S., Neupert, W. and Herrmann, J. M. (2005). Conserved *N*-terminal negative charges in the Tim17 subunit of the TIM23 translocase play a critical role in the import of preproteins into mitochondria. *J. Biol. Chem.* **280**, 7777–7785.

Meisinger, C., Pfanner, N. and Truscott, K. N. (2006). Isolation of yeast mitochondria. *Methods Mol. Biol.* **313**, 33–39.

Miranda-Vizuete, A., Damdimopoulos, A. E. and Spyrou, G. (2000). The mitochondrial thioredoxin system. *Antioxid Redox Signal* **2**, 801–810.

Mokranjac, D. and Neupert, W. (2005). Protein import into mitochondria. *Biochem. Soc. Trans.* **33**, 1019–1023.

Mokranjac, D., Sichting, M., Neupert, W. and Hell, K. (2003). Tim14, a novel key component of the import motor of the TIM23 protein translocase of mitochondria. *EMBO J.* **22**, 4945–4956.

Mounolou, J. C., Jakob, H. and Slonimski, P. P. (1966). Mitochondrial DNA from yeast "petite" mutants: specific changes in buoyant density corresponding to different cytoplasmic mutations. *Biochem. Biophys. Res. Commun.* **24**, 218–224.

Mueller, D. M. (2000). Partial assembly of the yeast mitochondrial ATP synthase. *J. Bioenerg. Biomembr.* **32**, 391–400.

Myers, A. M., Pape, L. K. and Tzagoloff, A. (1985). Mitochondrial protein synthesis is required for maintenance of intact mitochondrial genomes in *Saccharomyces cerevisiae*. *EMBO J.* **4**, 2087–2092.

Nakai, K. and Horton, P. (1999). PSORT: a program for detecting sorting signals in proteins and predicting their subcellular localization. *Trends Biochem. Sci.* **24**, 34–36.

Nedeva, T. S., Petrova, V. Y., Zamfirova, D. R., Stephanova, E. V. and Kujumdzieva, A. V. (2004). Cu/Zn superoxide dismutase in yeast mitochondria-a general phenomenon. *FEMS Microbiol. Lett.* **230**, 19–25.

Nelson, N. and Schatz, G. (1979). Energy-dependent processing of cytoplasmically made precursors to mitochondrial proteins. *Proc. Natl. Acad. Sci. USA* **76**, 4365–4369.

Neupert, W. (1997). Protein import into mitochondria. *Annu. Rev. Biochem.* **66**, 863–917.

Nicastro, D., Frangakis, A. S., Typke, D. and Baumeister, W. (2000). Cryo-electron tomography of neurospora mitochondria. *J. Struct. Biol.* **129**, 48–56.

Nohl, H., Kozlov, A. V., Gille, L. and Staniek, K. (2003). Cell respiration and formation of reactive oxygen species: facts and artefacts. *Biochem. Soc. Trans.* **31**, 1308–1311.

Nunnari, J., Marshall, W. F., Straight, A., Murray, A., Sedat, J. W. and Walter, P. (1997). Mitochondrial transmission during mating in *Saccharomyces cerevisiae* is determined by mitochondrial fusion and fission and the intramitochondrial segregation of mitochondrial DNA. *Mol. Biol. Cell* **8**, 1233–1242.

Nunnari, J., Wong, E. D., Meeusen, S. and Wagner, J. A. (2002). Studying the behavior of mitochondria. *Methods Enzymol.* **351**, 381–393.

O'Brien, E. A., Badidi, E., Barbasiewicz, A., deSousa, C., Lang, B. F. and Burger, G. (2003). GOBASE – a database of mitochondrial and chloroplast information. *Nucl. Acids Res.* **31**, 176–178.

Okamoto, K., Perlman, P. S. and Butow, R. A. (1998). The sorting of mitochondrial DNA and mitochondrial proteins in zygotes: preferential transmission of mitochondrial DNA to the medial bud. *J. Cell Biol.* **142**, 613–623.

Okamoto, K., Perlman, P. S. and Butow, R. A. (2001). Targeting of green fluorescent protein to mitochondria. *Methods Cell Biol.* **65**, 277–283.

Okamoto, K. and Shaw, J. M. (2005). Mitochondrial morphology and dynamics in yeast and multicellular eukaryotes. *Annu. Rev. Genet.* **39**, 503–536.

Ozawa, T., Sako, Y., Sato, M., Kitamura, T. and Umezawa, Y. (2003). A genetic approach to identifying mitochondrial proteins. *Nat. Biotechnol.* **21**, 287–293.

Paschen, S. A. and Neupert, W. (2001). Protein import into mitochondria. *IUBMB Life* **52**, 101–112.

Paumard, P., Vaillier, J., Coulary, B., Schaeffer, J., Soubannier, V., Mueller, D. M., Brethes, D., di Rago, J. P. and Velours, J. (2002). The ATP synthase is involved in generating mitochondrial cristae morphology. *EMBO J.* **21**, 221–230.

Perkins, G., Renken, C., Martone, M. E., Young, S. J., Ellisman, M. and Frey, T. (1997). Electron tomography of neuronal mitochondria: three-dimensional structure and organization of cristae and membrane contacts. *J. Struct. Biol.* **119**, 260–272.

Pfanner, N., Wiedemann, N., Meisinger, C. and Lithgow, T. (2004). Assembling the mitochondrial outer membrane. *Nat. Struct. Mol. Biol.* **11**, 1044–1048.

Piskur, J. (1994). Inheritance of the yeast mitochondrial genome. *Plasmid* **31**, 229–241.

Pollock, R. A., Hartl, F. U., Cheng, M. Y., Ostermann, J., Horwich, A. and Neupert, W. (1988). The processing peptidase of yeast mitochondria: the two co-operating components MPP and PEP are structurally related. *EMBO J.* **7**, 3493–3500.

Pratje, E. and Guiard, B. (1986). One nuclear gene controls the removal of transient pre-sequences from two yeast proteins: one encoded by the nuclear the other by the mitochondrial genome. *EMBO J.* **5**, 1313–1317.

Pratje, E., Mannhaupt, G., Michaelis, G. and Beyreuther, K. (1983). A nuclear mutation prevents processing of a mitochondrially encoded membrane protein in *Saccharomyces cerevisiae*. *EMBO J.* **2**, 1049–1054.

Prokisch, H., Scharfe, C., Camp, D. G., Jr., Xiao, W., David, L., Andreoli, C., Monroe, M. E., Moore, R. J., Gritsenko, M. A., Kozany, C., Hixson, K. K., Mottaz, H. M., Zischka, H., Ueffing, M., Herman, Z. S., Davis, R. W., Meitinger, T., Oefner, P. J., Smith, R. D. and Steinmetz, L. M. (2004). Integrative analysis of the mitochondrial proteome in yeast. *PLoS Biol.* **2**, e160.

Rapaport, D., Brunner, M., Neupert, W. and Westermann, B. (1998a). Fzo1p is a mitochondrial outer membrane protein essential for the

biogenesis of functional mitochondria in *Saccharomyces cerevisiae*. *J. Biol. Chem.* **273**, 20150–20155.
Rapaport, D., Brunner, M., Neupert, W. and Westermann, B. (1998b). Fzo1p is a mitochondrial outer membrane protein essential for the biogenesis of functional mitochondria in *Saccharomyces cerevisiae*. *J. Biol. Chem.* **273**, 20150–20155.
Rehling, P., Brandner, K. and Pfanner, N. (2004). Mitochondrial import and the twin-pore translocase. *Nat. Rev. Mol. Cell Biol.* **5**, 519–530.
Reichert, A. S., Mokranjac, D., Neupert, W. and Hell, K. (2005). Analysis of protein–protein interactions by chemical crosslinking. In: *Cell Biology: Laboratory Manual* (J. Celis, ed.). CSHL Press, New York 3rd edn..
Reichert, A. S. and Neupert, W. (2004). Mitochondriomics or what makes us breathe. *Trends Genet.* **20**, 555–562.
Reinhardt, A. and Hubbard, T. (1998). Using neural networks for prediction of the subcellular location of proteins. *Nucl. Acids Res.* **26**, 2230–2236.
Rigoulet, M. and Guerin, B. (1979). Phosphate transport and ATP synthesis in yeast mitochondria: effect of a new inhibitor: the tribenzylphosphate. *FEBS Lett.* **102**, 18–22.
Rissler, M., Wiedemann, N., Pfannschmidt, S., Gabriel, K., Guiard, B., Pfanner, N. and Chacinska, A. (2005). The essential mitochondrial protein Erv1 cooperates with Mia40 in biogenesis of intermembrane space proteins. *J. Mol. Biol.* **353**, 485–492.
Ryan, M. T., Voos, W. and Pfanner, N. (2001). Assaying protein import into mitochondria. *Methods Cell Biol.* **65**, 189–215.
Sanchirico, M. E., Fox, T. D. and Mason, T. L. (1998). Accumulation of mitochondrially synthesized *Saccharomyces cerevisiae* Cox2p and Cox3p depends on targeting information in untranslated portions of their mRNAs. *EMBO J.* **17**, 5796–5804.
Saraste, M. (1999). Oxidative phosphorylation at the fin de siecle. *Science* **283**, 1488–1493.
Schagger, H. (2001a). Respiratory chain supercomplexes. *IUBMB Life* **52**, 119–128.
Schagger, H. (2001b). Blue-native gels to isolate protein complexes from mitochondria. *Methods Cell Biol.* **65**, 231–244.
Schagger, H., Cramer, W. A. and von Jagow, G. (1994). Analysis of molecular masses and oligomeric states of protein complexes by blue native electrophoresis and isolation of membrane protein complexes by two-dimensional native electrophoresis. *Anal. Biochem* **2**, 220–230.
Schagger, H. and Pfeiffer, K. (2000). Supercomplexes in the respiratory chains of yeast and mammalian mitochondria. *EMBO J.* **19**, 1777–1783.
Schagger, H. and von Jagow, G. (1991). Blue native electrophoresis for isolation of membrane protein complexes in enzymatically active form. *Anal. Biochem.* **199**, 223–231.
Scheffler, I. E. (2001). A century of mitochondrial research: achievements and perspectives. *Mitochondrion* **1**, 3–31.
Schwikowski, B., Uetz, P. and Fields, S. (2000). A network of protein–protein interactions in yeast. *Nat. Biotechnol.* **18**, 1257–1261.
Sesaki, H. and Jensen, R. E. (1999). Division versus fusion: Dnm1p and Fzo1p antagonistically regulate mitochondrial shape. *J. Cell Biol.* **147**, 699–706.
Sesaki, H. and Jensen, R. E. (2001). *UGO1* encodes an outer membrane protein required for mitochondrial fusion. *J. Cell Biol.* **152**, 1123–1134.

Sichting, M., Mokranjac, D., Azem, A., Neupert, W. and Hell, K. (2005). Maintenance of structure and function of mitochondrial Hsp70 chaperones requires the chaperone Hep1. *EMBO J.* **24**, 1046–1056.

Sickmann, A., Reinders, J., Wagner, Y., Joppich, C., Zahedi, R., Meyer, H. E., Schonfisch, B., Perschil, I., Chacinska, A., Guiard, B., Rehling, P., Pfanner, N. and Meisinger, C. (2003). The proteome of *Saccharomyces cerevisiae* mitochondria. *Proc. Natl. Acad. Sci. USA* **100**, 13207–13212.

Slonimski, P. P. and Tzagoloff, A. (1976). Localization in yeast mitochondrial DNA of mutations expressed in a deficiency of cytochrome oxidase and/or coenzyme QH2-cytochrome *c* reductase. *Eur. J. Biochem.* **61**, 27–41.

Small, I., Peeters, N., Legeai, F. and Lurin, C. (2004). Predotar: a tool for rapidly screening proteomes for *N*-terminal targeting sequences. *Proteomics* **4**, 1581–1590.

Somlo, M. (1968). Induction and repression of mitochondrial ATPase in yeast. *Eur. J. Biochem.* **5**, 276–284.

Steele, D. F., Butler, C. A. and Fox, T. D. (1996). Expression of a recoded nuclear gene inserted into yeast mitochondrial DNA is limited by mRNA-specific translational activation. *Proc. Natl. Acad. Sci. USA* **93**, 5253–5257.

Steinmetz, L. M., Scharfe, C., Deutschbauer, A. M., Mokranjac, D., Herman, Z. S., Jones, T., Chu, A. M., Giaever, G., Prokisch, H., Oefner, P. J. and Davis, R. W. (2002). Systematic screen for human disease genes in yeast. *Nat. Genet.* **31**, 400–404.

Stewart, L. C. and Yaffe, M. P. (1991). A role for unsaturated fatty acids in mitochondrial movement and inheritance. *J. Cell Biol.* **115**, 1249–1257.

Su, X. and Dowhan, W. (2006). Translational regulation of nuclear gene COX4 expression by mitochondrial content of phosphatidylglycerol and cardiolipin in *Saccharomyces cerevisiae*. *Mol. Cell Biol.* **26**, 743–753.

Thorsness, P. E. and Fox, T. D. (1993). Nuclear mutations in *Saccharomyces cerevisiae* that affect the escape of DNA from mitochondria to the nucleus. *Genetics* **134**, 21–28.

Thorsness, P. E., White, K. H. and Fox, T. D. (1993). Inactivation of *YME1*, a member of the ftsH-SEC18-PAS1-CDC48 family of putative ATPase-encoding genes, causes increased escape of DNA from mitochondria in *Saccharomyces cerevisiae*. *Mol. Cell Biol.* **13**, 5418–5426.

Tzagoloff, A., Akai, A. and Needleman, R. B. (1975). Assembly of the mitochondrial membrane system. Characterization of nuclear mutants of *Saccharomyces cerevisiae* with defects in mitochondrial ATPase and respiratory enzymes. *J. Biol. Chem.* **250**, 8228–8235.

Tzagoloff, A., Barrientos, A., Neupert, W. and Herrmann, J. M. (2004). Atp10p assists assembly of Atp6p into the F0 unit of the yeast mitochondrial ATPase. *J. Biol. Chem.* **279**, 19775–19780.

Tzagoloff, A. and Dieckmann, C. L. (1990). PET genes of *Saccharomyces cerevisiae*. *Microbiol. Rev.* **54**, 211–225.

Uetz, P., Giot, L., Cagney, G., Mansfield, T. A., Judson, R. S., Knight, J. R., Lockshon, D., Narayan, V., Srinivasan, M., Pochart, P., Qureshi-Emili, A., Li, Y., Godwin, B., Conover, D., Kalbfleisch, T., Vijayadamodar, G., Yang, M., Johnston, M., Fields, S. and Rothberg, J. M. (2000). A comprehensive analysis of protein–protein interactions in *Saccharomyces cerevisiae*. *Nature* **403**, 623–627.

van Dyck, L., Neupert, W. and Langer, T. (1998). The ATP-dependent PIM1 protease is required for the expression of intron-containing genes in mitochondria. *Genes Dev.* **12**, 1515–1524.
van Loon, A. P., Pesold-Hurt, B. and Schatz, G. (1986). A yeast mutant lacking mitochondrial manganese-superoxide dismutase is hypersensitive to oxygen. *Proc. Natl. Acad. Sci. USA* **83**, 3820–3824.
Velours, J. and Arselin, G. (2000). The *Saccharomyces cerevisiae* ATP synthase. *J. Bioenerg. Biomembr.* **32**, 383–390.
Velours, J., Vaillier, J., Paumard, P., Soubannier, V., Lai-Zhang, J. and Mueller, D. M. (2001). Bovine coupling factor 6, with just 14.5% shared identity, replaces subunit *h* in the yeast ATP synthase. *J. Biol. Chem.* **276**, 8602–8607.
von Mering, C., Krause, R., Snel, B., Cornell, M., Oliver, S. G., Fields, S. and Bork, P. (2002). Comparative assessment of large-scale data sets of protein–protein interactions. *Nature* **417**, 399–403.
Waizenegger, T., Schmitt, S., Zivkovic, J., Neupert, W. and Rapaport, D. (2005). Mim1, a protein required for the assembly of the TOM complex of mitochondria. *EMBO Rep.* **6**, 57–62.
Weber, E. R., Hanekamp, T. and Thorsness, P. E. (1996). Biochemical and functional analysis of the *YME1* gene product, an ATP and zinc-dependent mitochondrial protease from *S. cerevisiae*. *Mol. Biol. Cell* **7**, 307–317.
Westermann, B., Gaume, B., Herrmann, J. M., Neupert, W. and Schwarz, E. (1996). Role of the mitochondrial DnaJ homolog Mdj1p as a chaperone for mitochondrially synthesized and imported proteins. *Mol. Cell Biol.* **16**, 7063–7071.
Westermann, B., Herrmann, J. M. and Neupert, W. (2001). Analysis of mitochondrial translation products *in vivo* and in organello in yeast. *Methods Cell Biol.* **65**, 429–438.
Westermann, B. and Neupert, W. (2000). Mitochondria-targeted green fluorescent proteins: convenient tools for the study of organelle biogenesis in *Saccharomyces cerevisiae*. *Yeast* **16**, 1421–1427.
Wiedemann, N., Pfanner, N. and Rehling, P. (2006). Import of precursor proteins into isolated yeast mitochondria. *Methods Mol. Biol.* **313**, 373–383.
Wienhues, U., Becker, K., Schleyer, M., Guiard, B., Tropschug, M., Horwich, A. L., Pfanner, N. and Neupert, W. (1991). Protein folding causes an arrest of preprotein translocation into mitochondria *in vivo*. *J. Cell Biol.* **115**, 1601–1609.
Wittig, I. and Schagger, H. (2005). Advantages and limitations of clear-native PAGE. *Proteomics* **5**, 4338–4346.
Wong, E. D., Wagner, J. A., Gorsich, S. W., McCaffery, J. M., Shaw, J. M. and Nunnari, J. (2000a). The dynamin-related GTPase, Mgm1p, is an intermembrane space protein required for maintenance of fusion competent mitochondria. *J. Cell Biol.* **151**, 341–352.
Wong, E. D., Wagner, J. A., Gorsich, S. W., McCaffery, J. M., Shaw, J. M. and Nunnari, J. (2000b). The dynamin-related GTPase, Mgm1p, is an intermembrane space protein required for maintenance of fusion competent mitochondria. *J. Cell Biol.* **151**, 341–352.
Wong, E. D., Wagner, J. A., Scott, S. V., Okreglak, V., Holewinske, T. J., Cassidy-Stone, A. and Nunnari, J. (2003). The intramitochondrial dynamin-related GTPase, Mgm1p, is a component of a protein complex that mediates mitochondrial fusion. *J. Cell Biol.* **160**, 303–311.

Yaffe, M. P. (1991). Analysis of mitochondrial function and assembly. *Methods Enzymol.* **194**, 627–643.

Yaffe, M. P. (1995). Isolation and analysis of mitochondrial inheritance mutants from *Saccharomyces cerevisiae*. *Methods Enzymol.* **260**, 447–453.

Yaffe, M. P. (1999). The machinery of mitochondrial inheritance and behavior. *Science* **283**, 1493–1497.

Yaffe, M. P., Ohta, S. and Schatz, G. (1985). A yeast mutant temperature-sensitive for mitochondrial assembly is deficient in a mitochondrial protease activity that cleaves imported precursor polypeptides. *EMBO J.* **4**, 2069–2074.

Yaffe, M. P. and Schatz, G. (1984). Two nuclear mutations that block mitochondrial protein import in yeast. *Proc. Natl. Acad. Sci. USA* **81**, 4819–4823.

Zhang, M., Mileykovskaya, E. and Dowhan, W. (2002). Gluing the respiratory chain together. Cardiolipin is required for supercomplex formation in the inner mitochondrial membrane. *J. Biol. Chem.* **277**, 43553–43556.

Zhong, Q., Gohil, V. M., Ma, L. and Greenberg, M. L. (2004). Absence of cardiolipin results in temperature sensitivity, respiratory defects, and mitochondrial DNA instability independent of pet56. *J. Biol. Chem.* **279**, 32294–32300.

20 Yeast Prions and Their Analysis *In Vivo*

Mick F Tuite, Lee J Byrne, Lyne Jossé, Frederique Ness, Nadejda Koloteva-Levine and Brian Cox
Department of Biosciences, University of Kent, Canterbury, Kent CT2 7NJ, UK

CONTENTS

Introduction
Yeast prions: a primer
Analysis of prion-associated phenotypes
Genetic analysis of yeast prions
Analysis of prion protein aggregates formed *in vivo*
Eliminating yeast prions
Propagon counting
Studying prion protein polymerisation *in vitro*
How to recognise a new yeast prion

Abbreviations

GdnHCl	Guanidine hydrochloride
GFP	Green fluorescent protein
PrD	Prion-forming domain
TSE	Transmissible spongiform encephalopathies
USA	Ureidosuccinic acid

I. INTRODUCTION

The existence of protein-only infectious agents ('prions') was first established in animals and humans through their association with the fatal neurodegenerative diseases classified as the transmissible spongiform encephalopathies (TSE). The infectious entity in the TSEs is associated with a protease-resistant and conformationally distinct version of the PrP protein (Prusiner *et al.*, 1998). The

METHODS IN MICROBIOLOGY, VOLUME 36
0580-9517 DOI:10.1016/S0580-9517(06)36020-5

infectious form of PrP (called PrP^{Sc}) has an identical amino acid sequence to the non-infectious, membrane-associated form (PrP^{c}), but is found largely as high-molecular-weight deposits in the brain. These aggregates have the biophysical characteristics of an amyloid: ordered protein polymers in the form of non-branching fibrils that are rich in β-sheet and, which when stained with Congo Red, exhibit red-green birefringence under polarised light. Prions however differ from most other disease-associated amyloids in that they are transmissible, i.e. the prion form of PrP can be propagated within the host and passed on to other individuals of the same, or closely related, species (Soto *et al.*, 2006, review).

While a great deal of attention has been – and continues to be – paid to the disease mechanisms associated with the mammalian TSEs, prions are not unique to mammals. There is now irrefutable evidence that the yeast *Saccharomyces cerevisiae* has at least three different proteins – Sup35p, Ure2p and Rnq1p – that can generate extrachromosomally inherited phenotypes as a direct consequence of an inherited change in their conformation via a prion-like mechanism (Uptain and Lindquist, 2002; Wickner *et al.*, 2004; Table 1). Het-s, an unrelated prion protein, has also been described in the filamentous fungus *Podospora anserina* where it controls vegetative incompatibility (Coustou *et al.*, 1997; Maddelein *et al.*, 2002).

In contrast to the life-threatening consequences associated with the appearance of, or infection by, PrP^{Sc} in humans and animals, yeast cells show no overt 'disease' phenotypes if they contain the prion form of any one of these three proteins. While the prion form of two of the proteins – Sup35p and Ure2p – have yet to be found in natural isolates of *S. cerevisiae* (Jensen *et al.*, 2001; Resende *et al.*, 2003; Nakayashiki *et al.*, 2005), strains carrying the prion form of the Rnq1p protein are frequently found (Resende *et al.*, 2003). This suggests that the presence of the prion form of Sup35p or Ure2p might have a negative impact on the fitness of yeast cells in the wild. However, several studies have shown that yeast cells carrying the prion form of Sup35p are more resistant to various physical and chemical stresses (Eaglestone *et al.*, 1999; True & Lindquist, 2000). Nevertheless, there are characteristic phenotypes associated with the presence of a yeast prion (i.e. [$PRION^{+}$]) that do not reflect any underlying change in the sequence of the cell's genome. Yeast prions can therefore be considered as protein-based epigenetic determinants. In addition to phenotypic differences, one can also define biochemical differences between [$PRION^{+}$] cells and their [$prion^{-}$] (but otherwise isogenic) counterparts, especially with regards the degree of solubility of the underlying prion protein.

Yeast prions are transmitted efficiently from cell-to-cell during mitosis and meiosis, indicating that a very effective mechanism must exist that ensures their continued propagation even in rapidly dividing cells (Tuite and Koloteva-Levin, 2004, review). Two different but not mutually exclusive models have been put forward to explain the self-propagation of prions: template-directed refolding

Table 1. Native yeast prions and their associated phenotypes

Prion	Protein	Prion-associated phenotype(s)	Reference
[*URE3*]	Ure2p	1. Utilisation of poor N_2 sources in *ura2* cells	Lacroute (1971)
		2. Excretion of uracil in wild-type cells in the presence of excess ureidosuccinic acid	Chernoff *et al.* (2002)
[PSI^+]	Sup35p	1. Suppression of *ade1-14* (red/white colonies)	Chernoff *et al.* (1995)
		2. Suppression of *ade2-1* in an *SUQ5* strain (red/white colonies)	Cox (1965)
		3. Resistance to certain physical and chemical stresses	Eaglestone *et al.* (1999) and True and Lindquist (2000)
		4. Growth inhibition when Sup35p is over expressed	Chernoff *et al.* (1993)
[RNQ^+]	Rnq1p	No phenotype known	Sondheimer and Lindquist (2000)
[PIN^+]	Rnq1p[a]	High frequency *de novo* conversion to [PSI^+]	Derkatch *et al.* (2001) and Osherovich and Weissman (2001)

[a]Although the Rnq1p is most commonly associated with the [PIN^+] prion, at least eight other proteins, including Ure2p, can also give rise to a prion with this property (Derkatch *et al.*, 2001; Osherovich and Weissman, 2001).

(Prusiner, 1991) and seeded polymerisation (Jarrett and Lansbury, 1993). The more widely accepted seeded polymerisation model is based on the prion protein existing in an altered conformational state which is in a reversible dynamic equilibrium with a soluble form of that protein. The seeding of prion protein polymerisation would be triggered by smaller, perhaps transient, oligomeric form(s) of the protein that in turn arise from the association of conformationally altered prion protein molecules. These forms of the protein have historically been referred to as seeds although we have recently coined the term 'propagon' which we define as a self-replicating hereditary particle that is required to maintain the [$PRION^+$] state (Cox *et al.*, 2003). Thus the propagon, by providing a nucleating activity, drives the polymerisation of both existing and newly synthesised prion protein molecules into the characteristic prion aggregates.

Stable propagation of the yeast [*PRION*$^+$] state also requires propagons to be efficiently distributed during mitosis and meiosis, although whether this is achieved by an active or a passive mechanism has yet to be established. The generation and transmission of propagons in yeast appears to be dependent upon a number of different molecular chaperones with one particular chaperone, the heat-shock-inducible protein Hsp104p, being essential for the propagation of all three native prions (Chernoff *et al.*, 1995; Moriyama *et al.*, 2000; Sondheimer and Lindquist, 2000).

In this chapter, we first provide a brief overview of the three yeast (*S. cerevisiae*) prions in terms of their associated phenotypes and their cell biological and biochemical properties and then go on to review the experimental approaches that can be taken to study the three native prions *in vivo*. Finally, we consider how one establishes whether or not a new or novel phenotype is associated with the prion-like behaviour of a cellular protein.

◆◆◆◆◆◆ II. YEAST PRIONS: A PRIMER

Saccharomyces cerevisiae has at least three proteins that meet the necessary genetic and biochemical criteria to be defined as a prion-forming protein: Sup35p which gives rise to the [*PSI*$^+$] prion, Ure2p which gives rise to the [*URE3*] prion and Rnq1p which gives rise to the [*RNQ*$^+$] or, as it now more routinely referred to, the [*PIN*$^+$] prion (Uptain and Lindquist, 2002, a review). None of these proteins share any amino acid sequence identity with the mammalian PrP protein or with each other although they do share some sequence features, as described below.

The three prions have impacts on very different biological processes: Sup35p is a translation termination factor, Ure2p regulates nitrogen metabolism via transcriptional modulation, while no specific cellular role has yet been attributed to Rnq1p. Nevertheless, they all share a number of properties in common as originally defined by Wickner (1994) whose pioneering studies on the yeast Ure2p/[*URE3*] prion provided the first evidence that prions exist in yeast. The key properties are:

- The [*PRION*$^+$] state shows a non-Mendelian pattern of inheritance in [*PRION*$^+$] × [*prion*$^-$] genetic crosses consistent with a cytoplasmically located 'genetic' determinant.
- The [*PRION*$^+$] state can only be established and propagated if the nuclear gene that encodes the prion protein is present, i.e. the [*PRION*$^+$] state is not maintained if the prion protein gene has been deleted. This can only be verified for Ure2p/[*URE3*] and Rnq1p/[*PIN*$^+$] since the *SUP35* gene is essential for viability.
- Elimination of the [*PRION*$^+$] state by certain non-mutagenic agents results in a viable [*prion*$^-$] cell that retains the ability to re-establish the [*PRION*$^+$] state *de novo*.

- Overproduction of the underlying protein in a [*prion*$^-$] cell results in a significant elevation in the rate of *de novo* appearance of the [*PRION*$^+$] form of that protein. The proviso here is that the [*prion*$^-$] must be [*PIN*$^+$] (see Section II.C below).

A. The [*PSI*$^+$] Prion

In its [*PRION*$^+$] state, the translation termination factor Sup35p (also known as eRF3 – eukaryotic release factor 3) gives rise to [*PSI*$^+$] cells, which show a defect in translation termination that can be readily detected by a simple nonsense suppression-based assay (Cox, 1965; Figure 1, see Colour Plate Section). Sup35p physically interacts with at least one other protein, eRF1 (Sup45p), to form the functional release factor needed for polypeptide chain release (Stansfield *et al.*, 1995). The inactivation of Sup35p via prion-mediated aggregation in a [*PSI*$^+$] strain would be expected to result in a reduction in levels of the Sup35p:Sup45p functional complex required for translation termination. This in turn would lead to an increase in the frequency with which ribosomes can read through a defined nonsense codon thus giving rise to nonsense suppression.

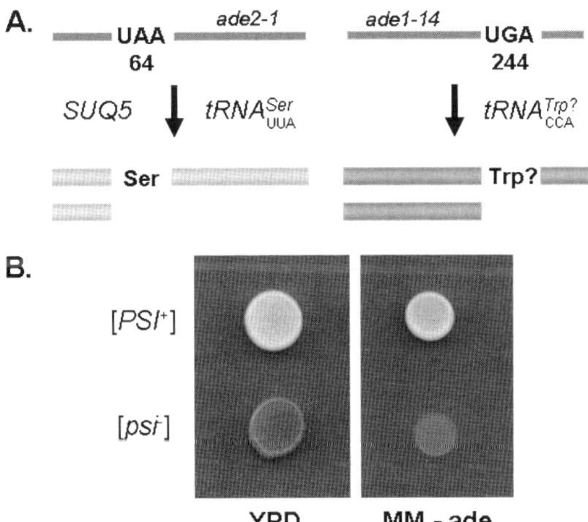

Figure 1. A simple colony colour assay for the presence of the [*PSI*$^+$] prion in *Saccharomyces cerevisiae*. Either of two different suppressible alleles can be used, the *ade2-1* allele and the *ade1-14* allele. In both the cases, when the mutation is expressed, i.e. in a [*psi*$^-$] strain, the cells form red colonies that signal an adenine auxotrophic phenotype. In [*PSI*$^+$] cells, suppression of either allele leads to white colonies that are prototrophic and can grow without the provision of exogenous adenine. Note that strains carrying the *ade2-1* allele must also carry the weak ochre suppressor $tRNA^{Ser}$ encoded by the *SUQ5* (*SUP16*) gene (Cox, 1965), whereas the *ade1-14* allele can be suppressed directly by [*PSI*$^+$] in the absence of a suppressor tRNA. The identity of the amino acid inserted when the UGA codon in the *ade1-14* allele is suppressed is unknown but is likely to be tryptophan (encoded by the UGG codon) (See color plate section).

In [PSI^+] cells, a significant proportion of the Sup35p in the cell is present in the form of one or more high-molecular-weight aggregates that can be readily sedimented from cell lysates by ultracentrifugation. In prion-free [psi^-] cells Sup35p is largely soluble (Patino *et al.*, 1996; Paushkin *et al.*, 1996). The [PSI^+]-associated Sup35p aggregates contain both protease resistant (Paushkin *et al.*, 1996) and SDS-resistant forms of Sup35p (Kryndushkin *et al.*, 2003). While it is usually assumed that these aggregates are most likely to be amyloid in nature, this has not been formally demonstrated *in vivo*, although *in vitro* polymerisation studies with Sup35p show that it can form amyloid-like fibres *in vitro* (Glover *et al.*, 1997; King *et al.*, 1997).

Two basic [PSI^+] variants have been described which show no differences in the primary amino acid sequence of Sup35p, but which do show phenotypic and biochemical differences. In cells carrying the 'strong' [PSI^+] variant, 90% or more of the Sup35p in the cell is usually present in the form of high-molecular-weight aggregates leading to efficient nonsense suppression. In the 'weak' [PSI^+] variants, a much greater proportion of the Sup35p is present in the soluble fraction (Uptain *et al.*, 2001) and consequently the nonsense suppression phenotype is weaker (Figure 2). The differences in phenotype most likely arise due to subtle differences in Sup35p conformation, which in turn alter the rate at which new propagons are formed in growing cells.

Critical for the *de novo* formation and propagation of the [PSI^+] prion is the Gln/Asn-rich prion-forming region (PrD) located at the N-terminus of Sup35p, between residues 1 and 97 (Figure 3; Ross *et al.*, 2005, review). This largely unstructured region of the protein consists of two functionally distinct sub-regions (Osherovich *et al.*, 2004):

(a) The QN-rich (QNR) region that spans residues 1–40 and is particularly rich in Asn and Gln residues. This region is necessary

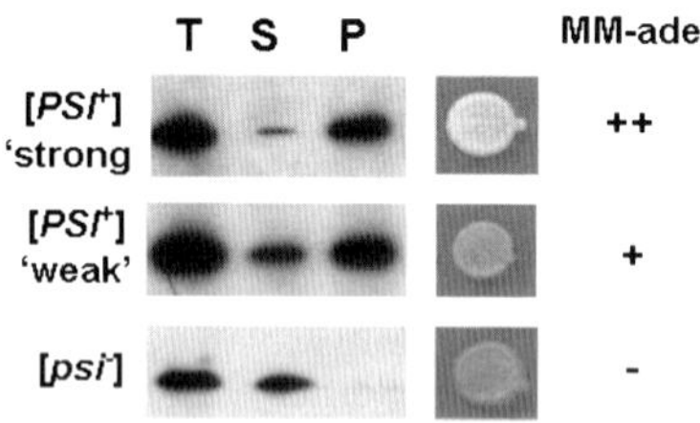

Figure 2. Sub-cellular fractionation analysis can be used to distinguish between both [PSI^+] and [psi^-] strains and between 'weak' and 'strong' variants of [PSI^+]. Three different samples, analysed by SDS-PAGE and Western blotting using an anti-Sup35p antibody, are shown: T, total un-fractionated extract; S, soluble fraction after centrifugation at 100 000*g* and P, the pellet fraction remaining after the ultracentrifugation step. Note that the 'weak' [PSI^+] variant has more soluble Sup35p and has a weaker nonsense suppression phenotype compared to the 'strong' [PSI^+] variant as judged by both colony colour (where dark tones represent red colony pigmentation) and relative growth on an adenine-deficient medium.

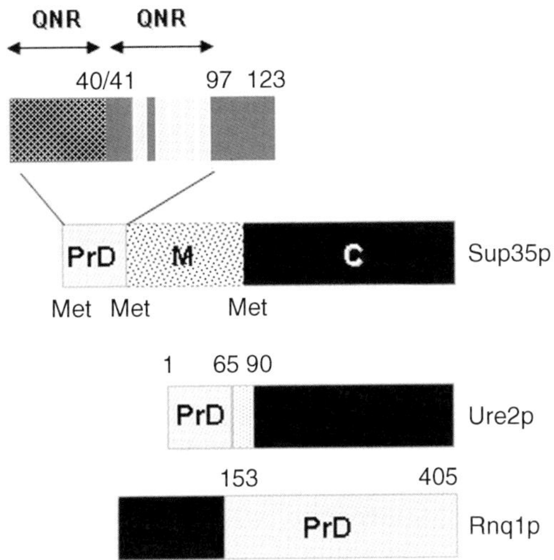

Figure 3. Basic organisation of the three native prion proteins of *Saccharomyces cerevisiae*. For each protein the location of the prion-forming domain (PrD) is indicated. For the Sup35p protein, a more detailed description of the PrD is provided in which the location of the Gln/Asn-rich region (QNR) and the region containing the five copies (and one part copy) of an oligopeptide repeat (OPR) are shown, as defined by Osherovich *et al.* (2004). The numbers indicate the residue number with the initiator Met being taken as residue 1. The positions of the Met residues that are used to demark the three functionally distinct regions of Sup35p (i.e. PrD, M and C) are also shown.

for prion aggregate formation and contains a highly amyloidogenic region GNNQQNY between residues 7 and 13 (Diaz-Avalos *et al.*, 2003).

(b) The oligopeptide repeat-containing region (OPR), which is required for propagation of the prion form of the protein and contains five imperfect copies of an oligopeptide repeat.

The Sup35p-PrD is separated from the functional C-terminal region of the protein molecule by a highly charged M region which may also contribute to the prion-like properties of the Sup35p protein (Figure 3; Liu *et al.*, 2002). Over expression of the Sup35p-PrD or the Sup35p-PrD+M leads to an increase in the rate of *de novo* induction in a [*psi*$^-$] cell provided the [*PIN*$^+$] prion is present in the cell (Wickner *et al.*, 2001, review).

B. The [*URE3*$^+$] Prion

The main cellular role of the Ure2p prion protein is to regulate nitrogen catabolic gene expression at the level of transcription in response to nitrogen levels. In the presence of excess nitrogen, Ure2p forms a complex with the transcription factor Gln3p and this in turn leads to the sequestration of Gln3p in the cytoplasm and a concomitant reduction in the transcription of genes regulated by Gln3p (Kulkarni *et al.*, 2001). In [*URE3*] cells, the aggregation of

Ure2p results in the activation of Gln3p and its downstream targets, e.g. the *DAL5* gene, which encodes a permease for the uptake of poor nitrogen sources such as allantoate and the structurally related metabolic intermediate ureidosuccinic acid (USA).

The Ure2p protein has a C-terminal region (residues 65–354) that shows sequence and structural identity to glutathione-*s*-transferases and contains the nitrogen regulatory function(s). However, Ure2p does not have glutathione transferase activity, but does appear to have an associated glutathione peroxidase activity even when in its amyloid-like, fibrillar form (Bai *et al.*, 2004). The amino-terminal region of Ure2p (residues 1–65) is required for both its prion-like behaviour *in vivo* and its seeded polymerisation *in vitro*, and is particularly rich in Asn residues. The propagation of the [*URE3*] prion is dependent upon the Hsp104p chaperone although other chaperones, particularly the Ssa1/2p and the Hsp40 Ydj1p have also been implicated in the mechanism which propagates [*URE3*] (Moriyama *et al.*, 2000).

C. The [RNQ^+]/[PIN^+] Prion and *De Novo* Conversion

The ability of the Rnq1p protein to form a prion-like determinant was established by Sondheimer and Lindquist (2000) who showed that this protein could take up an insoluble, aggregated form that was propagated by an Hsp104-dependent mechanism. Although the cellular function of soluble Rnq1p remains to be established, when in its prion form, it can facilitate the *de novo* formation of other prions. Both [PSI^+] and [*URE3*] prions can also arise *de novo*, either spontaneously (at a frequency of $\sim 10^{-5}$) or by over expression of the corresponding protein or its PrD (which elevates the rate of *de novo* conversion some 100–1000 fold) (Chernoff *et al.*, 1993; Wickner, 1994).

Both spontaneous and induced *de novo* conversion of cells to [PSI^+] require the presence of a second prion originally called [PIN^+] (for [**P**SI]-**in**ducing prion). [PIN^+] is the prion form of the Rnq1p protein in most [PIN^+] laboratory strains studied although other proteins can form [PIN^+] (Derkatch *et al.*, 2001; Osherovich and Weissman, 2001). How [PIN^+] mediates *de novo* conversion of a sequence unrelated protein is unknown, but presumably it either sequesters an anti-aggregation factor from cells, which leads to an increase in the rate of spontaneous aggregation of the Sup35p (Osherovich and Weissman, 2001) or the [PIN^+] prion nucleates the polymerisation of soluble forms of Sup35p or Ure2p leading to the formation of the seeds necessary for the propagation of the [$PRION^+$] state (Derkatch *et al.*, 2001). Recent *in vitro* studies have provided strong support for the latter model (Derkatch *et al.*, 2004).

Rnq1p is a non-essential 405 residue protein rich in Gln and Asn residues (hence RNQ, **r**ich in **N** and **Q**) and, like Ure2p and Sup35p, is also able to form amyloid-like fibrils *in vitro* (Sondheimer and

Lindquist, 2000). Although the precise location of the Rnq1p PrD has yet to be mapped, this function resides between residues 130 and 405 in a C-terminal region rich in Asn and Gln residues, i.e. it is a C-terminally located PrD (Sondheimer and Lindquist, 2000). Maintenance of the prion form of Rnq1p requires the Hsp104p chaperone (Sondheimer and Lindquist, 2000), but in addition requires another member of the Hsp40 family, Sis1p (Sondheimer *et al.*, 2001).

◆◆◆◆◆◆ III. ANALYSIS OF PRION-ASSOCIATED PHENOTYPES

The [PSI^+] and [*URE*3] prions were uncovered in classical genetic screens: a rare [PSI^+] mutant emerged from a genetic screen for nonsense suppressor mutants using an *ade2-1*-based assay (Cox, 1965), while the [*URE3*] mutant emerged from a screen for yeast mutants that allowed a *ura2* mutant to grow on a minimal medium supplemented with ureidosuccinic and glutamic acids (Lacroute, 1971). In both the cases, it was the unusual genetic behaviour of these mutants that lead their discoverers to conclude that neither mutant could be simply explained by a nuclear gene mutation. It was not until 1994 however that these properties were linked to the existence of prions in yeast (Wickner, 1994). In contrast, the [*RNQ*/PIN^+] prion was identified by a rational approach: following the demonstration that the amino-terminal Sup35p-PrD was rich in Gln and Asn residues and a feature important for its prion-like behaviour, sequence-led searches were undertaken for other yeast proteins with similar QNR regions. Over 100 such proteins were identified (Michelitsch and Weissman, 2000; Sondheimer and Lindquist, 2000) but only one of these – to date – has proven to be a prion, i.e. Rnq1p. At least one other protein – New1p – has an QNR region that can functionally replace the equivalent QNR region in the Sup35p-PrD (Osherovich *et al.*, 2004), but whether full-length New1p forms a prion in the cell remains to be established. Although Rnq1p satisfies all the criteria for to be classified as a prion protein, no cell-level phenotype was originally detected that could readily identify [RNQ^+] strains. With the subsequent discovery that the [RNQ^+] prion can act as a [*PSI*]-inducing prion (see above), a phenotypic assay has now become available that allows for detection of its presence in yeast strains.

Yeast Prions and Their Analysis *In Vivo*

A. The [PSI^+] Prion Phenotype

[PSI^+] was identified as a mutation that enhanced the ability of the weak suppressor $tRNA^{Ser}$ encoded by the *SUQ5* (also called *SUP16*) gene to suppress the *ade2-1* allele. This allele of *ade2* contains a premature UAA codon at codon position 64 (Prokopi, M.,

Koloteva-Levin, N. and Tuite, M. F., unpublished) and when the mutant phenotype is expressed, this leads to red-coloured colonies. There was also some evidence early on that [*PSI*$^+$] could weakly suppress nonsense mutations in the absence of the *SUQ5* tRNA, e.g. *cyc1-72* (Liebman and Sherman, 1979). However, the identification of an *ade1* allele (*ade1-14*) that was relatively efficiently suppressed by [*PSI*$^+$] in the absence of a cognate suppressor tRNA yet also had the red/white colour selection, and has lead to it being widely used to assay for [*PSI*$^+$]. The *ade1-14* allele has a UGA codon in place of UGG codon at position 244 (Nakayashiki *et al.*, 2001; see Figure 1).

The availability of the *ade1-14* and *ade2-1* nonsense alleles therefore provides two simple colony-level assays for the presence of the [*PSI*$^+$] prion, namely colony colour and growth on medium lacking adenine. Furthermore, the strength of the phenotype, as defined by varying shades of white and pink and the degree of adenine auxotrophy, allows the ready differentiation between weak and strong [*PSI*$^+$] variants: the strong variants are usually white/pale pink and show a strong Ade$^+$ phenotype, while the weak [*PSI*$^+$] variants usually give rise to pink/dark pink colonies that only grow relatively weakly on adenine-deficient medium. In our hands, the medium that gives the best distinction in red/white colouration when plating for single colonies is 1/4 YEPD (1% peptone, 0.25% yeast extract, 4% glucose plus 2% agar). Suppression of the auxotrophy associated with the *ade1-14* and *ade2-1* alleles can be relatively inefficient (especially for weak [*PSI*$^+$] variants). Consequently, the addition of a small amount of adenine (usually 1% w/v of normally added levels) or of 2.5 ml YEPD per 100 ml of the standard YNB-based minimal medium lacking adenine, can give greater distinction between the [*PSI*$^+$] and [*psi*$^-$] cells after only 2–3 days growth.

Suppression of the *ade1-14/ade2-1* markers gives only a qualitative estimate of the efficiency of suppression. For a more quantitative estimate of the relative efficiencies of nonsense suppression, a plasmid-based stop codon read-through system originally described by Firoozan *et al.* (1991), can be used. This involves the expression of a plasmid-borne *PGK-lacZ* gene fusion where the two reading frames are separated by one or other of the three stop codons UAA, UAG or UGA. A fourth construct has a sense codon at this position and is used to determine the control (100%) value. Subsequently, a number of variations of this type of bi-cistronic assay have been developed; for example, the plasmid pAC99 which involves two functional cistrons whose products can be independently assayed for in cells, i.e. *lacZ* encoding β-galactosidase and *luc* encoding luciferase, separated by a stop codon (Namy *et al.*, 2002). Using extracts prepared from pAC99 transformed cells both the β-galactosidase and luciferase activities are quantified and the ratio of luciferase activity to β-galactosidase activity calculated. These levels are then compared to the equivalent activities in cells expressing a *bone fide* β-galactosidase–luciferase fusion protein.

The relative efficiency of nonsense suppression for a given strain/stop codon can then be calculated by dividing the luciferase/β-galactosidase ratio obtained by the same ratio obtained with the in-frame protein fusion control.

B. The [*URE3*] Prion Phenotype

The change in nitrogen metabolism that occurs in [*URE3*] cells can be used to detect the presence of this prion in *ura2* cells. The *URA2* gene encodes the enzyme aspartate transcarbamylase that catalyses the synthesis of USA. In *ura2* cells that are deficient in this enzyme, if the [*URE3*] prion is present, such cells can utilise sodium ureidosuccinate in the absence of uracil because the Dal5p permease is present. [*URE3*] can also be detected in cells carrying the wild-type $URA2^+$ gene because [*URE3*] cells excrete uracil in the presence of excess USA and this can be detected by haloes formed by cross-feeding on a lawn of *ura2* cells (Chernoff *et al.*, 2002).

An alternative way of detecting [*URE3*] using a colony colour-based assay has also been established, which avoids the need to select for growth on a USA-containing medium. Schlumpberger *et al.* (2001) developed a novel *ADE2*-based reporter for detecting the [*URE3*] prion in which the wild-type *ADE2* gene is placed under the control of the *DAL5* promoter (P_{DAL5}). This promoter is regulated by Ure2p and the *DAL5* gene product, the Dal5p transporter, is necessary for the uptake of USA. Consequently, when the $P_{DAL5}ADE2$ reporter is expressed in an *ade2* mutant in the presence of ammonium ions, Ure2p binds to Gln3p thereby preventing transcription of the $P_{DAL5}ADE2$ gene and thus the colony is red. If expressed in a [*URE3*] *ade2* mutant, the release of the Gln3p block results in transcription of the $P_{DAL5}ADE2$ gene and the colonies are white, adenine prototrophs. Such a colour-based screen for [*URE3*] has been used to search for compounds that eliminate [*URE3*] from cells (Bach *et al.*, 2003) and to identify natural variants of the [*URE3*] prion (Schlumpberger *et al.*, 2001). Because of the intrinsic mitotic instability of many [*URE3*] isolates it is important to maintain selection for the [*URE3*] prion in such strains by growth on minimal medium lacking adenine (Kyprianidou, C., Byre, L. J. and Tuite, M. F., unpublished).

C. The [PIN^+] Prion Phenotype

The prion-based inactivation of Rnq1p function does not lead to a detectable change in phenotype other than allowing for a high rate of *de novo* formation of the [PSI^+] prion in cells. The constitutive or transient overexpression of either full-length Sup35p or just the Sup35p-PrD in a [psi^-] [PIN^+] strain usually leads to a 10^2–10^3-fold increase in the frequency of appearance of [PSI^+] cells and in some genetic backgrounds this can be as high as 30% of the cells. To carry out such an assay, a single copy or multicopy plasmid carrying the *SUP35* construct under the transcriptional regulation of the *GAL1* promoter is used (Chernoff *et al.*, 2002). This plasmid and a suitable control plasmid (e.g. the expression vector without the *SUP35* sequence) are independently introduced into an otherwise isogenic pair of *ade1-14* [*pin*$^-$] or [*PIN*$^+$] strains and the cells patched onto an agar plate-containing medium that retains selection for the plasmid and contains galactose to induce expression of the *SUP35* gene construct. After 2–3 days growth the cells are replica plated onto a glucose-based adenine-deficient medium to score the Ade$^+$ [*PSI*$^+$] cells while switching off the overexpression of the *SUP35* construct to avoid any associated toxicity; overexpression of *SUP35* is lethal in most [*PSI*$^+$] strains (Chernoff *et al.*, 1993). While a relatively straightforward assay, there are a number of complications in assessing the outcome of the assay which need to be considered (Chernoff *et al.*, 2002); for example, nuclear *SUP* gene mutations can also lead to suppression of the *ade1-14* allele.

◆◆◆◆◆◆ IV. GENETIC ANALYSIS OF YEAST PRIONS

A. Meiotic Transmission

The unusual pattern of non-Mendelian inheritance for both the [*PSI*$^+$] (Cox, 1965) and [*URE3*] (Lacroute, 1971) traits implicates a genetic determinant located in the cytoplasm. For example, when a [*PSI*$^+$] haploid strain is crossed with a [*psi*$^-$] haploid strain of the opposite mating type, the resulting diploid invariably shows the [*PSI*$^+$]-associated nonsense suppression phenotype, i.e. [*PSI*$^+$] is genetically dominant. Analysis of the meiotic spores from such a cross usually reveals a clear 4[*PSI*$^+$]:0[*psi*$^-$] segregation pattern for the spores in a single tetrad although some 'weak' [*PSI*$^+$] variants will often give rise to tetrads containing 3[*PSI*$^+$]:1[*psi*$^-$] spores or even 2[*PSI*$^+$]:2[*psi*$^-$] spores (Derkatch *et al.*, 1996). There are nuclear gene mutations that will also give rise to a [*PSI*$^+$]-like phenotype; for example, the ochre tRNA^{Tyr} suppressor mutants such as *SUP4* which are able to suppress *ade2-1* in a [*psi*$^-$] genetic background, but these give rise to two Ade$^+$ (white) to two Ade$^-$ (red) spores per tetrad.

Inheritance of [*URE3*] shows similar genetic behaviour to [*PSI*$^+$], but while the majority of asci give 4[*URE3*]:0[*ure3*] spores, most crosses also produce asci showing 3:1 and 2:2 [*URE3*]:[*ure3*] patterns of inheritance indicating that the [*URE3*] prion shows the high level of meiotic instability as is seen with the weak [*PSI*$^+$] variants (see above). The prion form of Rnq1p, in the form of the [*PIN*$^+$] prion, is also genetically dominant and when crossed to a [*pin*$^-$] strain, the majority of asci carry 4[*PIN*$^+$]:0[*pin*$^-$] spores (Derkatch *et al.*, 1996).

When carrying out genetic crosses to confirm the presence of a yeast prion ideally the mating partner should be an otherwise isogenic haploid strain in which the mating type has been switched. There is significant variation in the genetic backgrounds of many of the laboratory strains used by yeast researchers and the presence of a number of undefined genetic modifiers can affect the strength of the prion phenotype. A suitable strain can be obtained by taking the prion-free form of the original strain and introducing a plasmid expressing the *HO* gene. The majority of laboratory strains are defective in the *HO* gene (i.e. are ho^-) and so do not switch their mating type. Expression of the *HO*-encoded endonuclease triggers the mating type switch and cells can then be readily identified that have lost the plasmid-borne copy of the *HO* and who had their mating type switched.

B. Transmission by Cytoduction

One classical method for demonstrating transmission of a genetic determinant through the cytoplasm in fungi is cytoduction. Although not a normal part of the yeast life cycle, nevertheless cytoduction can be used due to the availability of the karyogamy defective *kar1* mutant, which blocks the fusion of two haploid nuclei during mating but does not affect plasmogamy, i.e. cell fusion and mixing of the cytoplasm from the two parent strains (Conde and Fink, 1976). In a *kar1* × $KAR1^+$ cross, after the cells have fused to form a cell containing two haploid nuclei, and thus equivalent to a dikaryon, new haploid daughter cells arise from the dikaryon which contain one or other of the parental nuclei but a mixture of cytoplasm arising from both parents. Figure 4 outlines the strategy that can be used to demonstrate the transfer of the [PSI^+] prion by cytoduction, but the other two yeast prions can also be efficiently transferred by cytoduction.

♦♦♦♦♦♦ V. ANALYSIS OF PRION PROTEIN AGGREGATES FORMED *IN VIVO*

As with the mammalian prion protein PrP, the aggregated prion forms of Ure2p (Masison and Wickner, 1995) and Sup35p (Paushkin *et al.*, 1996) show an increased resistance to digestion by proteinase K compared to the soluble forms of these proteins, which in turn are readily digested by this proteinase. However, unlike PrP^{Sc} digestion with proteinase K, diagnostic and consistent proteinase-resistant protein fragments are not produced for either Sup35p or Ure2p. Proteinase K resistance is therefore usually monitored simply by the relative rate of loss of the full-length prion protein with time of incubation in the presence of the proteinase (Chernoff *et al.*, 2002).

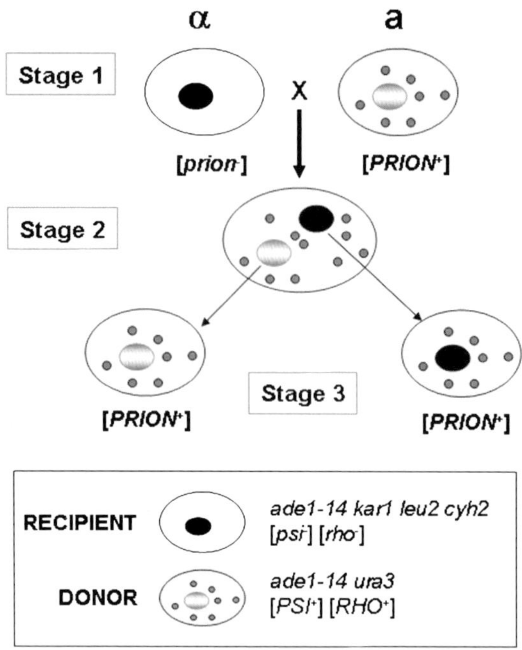

Figure 4. The use of cytoduction to demonstrate the cytoplasmic transmission of yeast prions. The experiment requires a [*prion*$^-$] recipient strain of the opposite mating type to the strain that is [*PRION*$^+$]. The recipient strain carries the *kar1* mutation (Conde and Fink, 1976) together with one or more nuclear genetic markers that are different to the markers in the donor strain. In addition, the recipient strain is made a [*rho*^{0}] *petite* by growth in the presence of ethidium bromide. Following mating of the donor and recipient strain (Stage 1), the mating mixture will contain a large number of heterokaryons (Stage 2). The mix is then plated onto a selective medium that does not allow either cells carrying the donor nucleus or any rare diploid cells that may form to grow, but does allow haploid cells (cytoductants) carrying the recipient nucleus to grow. The use of a [*rho*^{0}] *petite* recipient strain also allows the experimenter to confirm that cytoplasmic transfer has occurred since this would result in cytoductants carrying the recipient nucleus becoming [*RHO*$^+$] *grande* and thus being able to utilise a non-fermentable carbon source such as glycerol. The example shown is for studying the [*PSI*$^+$] prion, but can equally well be applied to other prions.

That the Ure2p and Sup35p prion aggregates show increased resistance to proteinase K digestion has been taken as one line of evidence that the yeast prion aggregates are amyloids. Certainly, the recombinant forms of these proteins form proteinase K resistant, amyloid-like structures *in vitro* (see Section VIII below), but the evidence that they form amyloid-like structures in the cell is less convincing. [*URE3*] cells engineered to over express Ure2p certainly contain distinctive, filamentous networks of Ure2p in the cytoplasm (Speransky *et al.*, 2001) consistent with an amyloid-like structure, although Ripaud *et al.* (2003) have suggested, based on proteinase K digestion patterns, that aggregated Ure2p in [*URE3*] yeast cells is conformationally distinct from the amyloid form of this protein generated *in vitro*. Another approach is to use amyloid-specific stains on [$PRION^+$] cells and Kimura *et al.* (2003) have shown that aggregates made by over expressing either the Rnq1p-PrD or the Sup35p-PrD can be stained in yeast cells by thioflavin-S, an amyloid-binding dye. However, staining [PSI^+] cells with another amyloid-specific dye, 2-(4′-methylaminophenyl) benzothiazole (BTA-1), an uncharged derivative of thioflavin-T (Mathis *et al.*, 2002) also stains a number of aggregates or structures in the [PSI^+] cell but because these structures are also seen in [psi^-] cells in which the *HSP104* gene deleted, they are unlikely to represent amyloid-like aggregates associated with the prion form of Sup35p or Rnq1p (Byrne, L. J. and Tuite, M. F., unpublished data).

In [PSI^+] cells, a certain proportion of the Sup35p protein remains soluble and functional and although the amount is sufficient to ensure cell viability, it is not enough for efficient termination. The relative proportion of soluble:aggregated Sup35p also varies depending on the [PSI^+] variant being studied (Figure 2; Uptain *et al.*, 2001) and this is reflected in the termination phenotype as 'strong' variants have a proportionally more severe termination defect than 'weak' variants (see Section II.A).

A. Sub-Cellular Fractionation of Prion Aggregates

The standard way of establishing whether or not a specific prion protein is present as a high-molecular-weight aggregate is by the use of differential centrifugation of total cell extracts usually prepared from exponentially growing cells. This results in the generation of soluble and pellet fractions and the presence of the respective prion protein in each fraction can be assessed by SDS-PAGE and Western blotting (Figure 2). To prepare total cell extracts for such an analysis, cells are usually disrupted using glass bead lysis and the extract then subjected to centrifugation at 100 000*g* for between 15 and 30 min at 4°C. No preliminary slow speed spin is carried out to remove cell debris as this can result in loss of a significant proportion of the high-molecular-weight forms of the prion protein prior centrifugation. Consequently, it is important to solubilise the pellet fraction after the high-speed spin and prior to electrophoresis by boiling in an SDS-based sample buffer (Ness *et al.*, 2002).

The standard sub-cellular fractionation protocol does however need to be optimised for each of the three different yeast prion proteins and/or for different yeast strains under test. For example, although good separation of Sup35p between the soluble and pellet fractions in [PSI^+] and otherwise isogenic [psi^-] strains can be achieved by centrifugation for 15 min (Figure 2), good fractionation of Rnq1p in [PIN^+]/[pin^-] strains requires a longer spin (30 min) with a higher centrifugal force (Figure 5). Soluble Rnq1p is also a very unstable protein in the presence of a range of protease inhibitors and, following differential centrifugation, the samples need to be analysed immediately by Western blot analysis, and not stored for

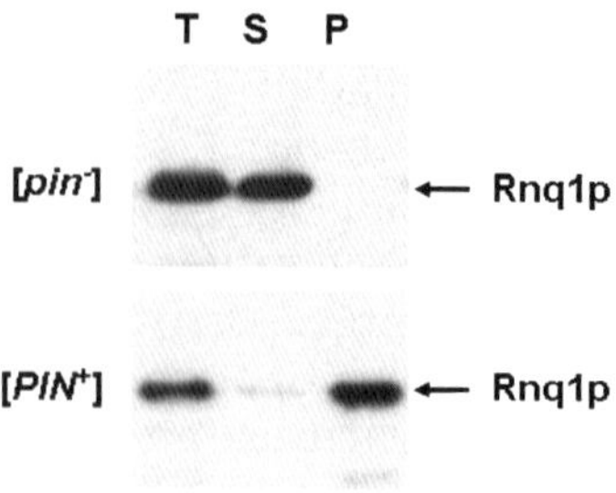

Figure 5. Sub-cellular fractionation analysis of the Rnq1p protein can be used to distinguish between [*PIN*$^+$] and [*pin*$^-$] strains. Three different samples prepared from the strain 74D-694 were analysed by SDS-PAGE and Western blotting using an anti-Rnq1p antibody: T, total un-fractionated extract; S, soluble fraction after centrifugation at 100 000*g* and P, the pellet fraction remaining after the ultracentrifugation step.

any length of time, in order to avoid degradation of soluble Rnq1p (Koloteva-Levin, N. and Tuite, M. F., unpublished). Fractionation by differential centrifugation does not result in good fractionation of Ure2p into the expected soluble and pellet fractions in some [*URE3*]/[*ure3*] strain pairs. This raises the question of whether Ure2p actually forms high-molecular-weight aggregates in all [*URE3*] strains expressing wild-type levels of Ure2p (Fernandez-Bellot *et al.*, 2002). What is also clear is that the nuclear genetic background of the strain can influence the relative distribution of protein between the soluble and pellet fractions in [*PRION*$^+$] yeast strains irrespective of the prion type or variant present.

B. Separation of Prion Protein Oligomers Using Agarose Gel Electrophoresis

Sedimentation analysis provides a relatively quantitative method for assessing the fraction of the respective prion protein that forms high-molecular-weight aggregates in a [*PRION*$^+$] strain compared to a control [*prion*$^-$] strain. What it does not do is provide information on the nature of the aggregates that are formed, nor on any oligomeric sub-particles that may be present. Such information can be obtained by using a novel electrophoretic method that uses agarose rather than acrylamide as the separation matrix, a method termed semi-denaturing detergent agarose gel electrophoresis (SDD-AGE: Kryndushkin *et al.*, 2003).

SDD-AGE was originally developed to study Sup35p aggregates in [*PSI*$^+$] strains. In preparing samples for such analysis, Kryndushkin *et al.* (2003) discovered that treating total yeast cell extracts with SDS and incubating at various temperatures between room temperature and 65°C resulted in the disassembly of the high-molecular-weight Sup35p aggregates into more discrete SDS-stable oligomeric forms of Sup35p ranging in size between 10 and 50 monomers of Sup35p. The relationship between these oligomers and the forms of Sup35p that are important for Sup35p oligomerisation and [PSI^+] propagation *in vivo* remain to be established. Nevertheless, they show different size distributions in different [PSI^+] variants and also in [PSI^+] cells where the activity of Hsp104p has been inhibited by guanidine hydrochloride (GdnHCl) (Kryndushkin *et al.*, 2003). The SDS-resistant Sup35p-containing oligomers detected in [PSI^+] strains are not discrete entities, but represent a range of sizes between 1.5 and 3.0 MDa (Figure 6). The challenge of finding suitable molecular weight markers for SDD-AGE analysis can be met by using chicken pectoralis extract which contains myosin heavy chain (205 kDa), nebulin (740 kDa) and titin (approximately 4 MDa: Kryndushkin *et al.*, 2003).

SDD-AGE can also be used to study SDS-stable oligomers of Rnq1p that form in [PIN^+] strains and these oligomers contain between 20 and 100 Rnq1p monomers (Bagriantsev and Liebman, 2004). Differential thermal stability of the Rnq1p oligomers can be used to distinguish between two different variants of the [PIN^+] prion.

An alternative approach to fractionating the high-molecular-weight aggregates formed by yeast prion proteins is to use sucrose gradient centrifugation (Paushkin *et al.*, 1996). Centrifugation of total cell extracts through a 15–40% sucrose gradient at 170 000*g* can separate out aggregates from soluble forms of Sup35p although, as with SDD-AGE, the aggregates are relatively dispersed throughout the gradient. The need to remove cell debris prior to loading on the sucrose gradient can however lead to loss of prion aggregates.

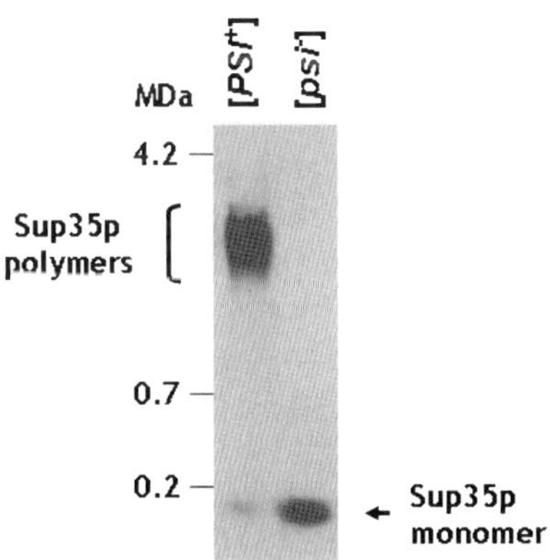

Figure 6. The use of semi-denaturing detergent agarose gel electrophoresis (SDD-AGE) to study prion aggregates in a [PSI^+] and [psi^-] pair of strains. Samples were prepared and fractionated on an agarose gel in the presence of SDS as described by Kryndushkin *et al.* (2003). Following transfer of the proteins, the membrane was blotted using an anti-Sup35p antibody. The location of the disperse Sup35p-containing polymers in the [PSI^+] strain and the monomeric form of Sup35p present in the [psi^-] strain, are indicated. Molecular weight markers used were myosin heavy chain (205 kDa), nebulin (740 kDa) and titin (approximately 4.2 MDa).

C. GFP Fusion Technology

One direct means of assessing whether or not a given strain contains prion-like aggregates of a specific protein is to express in those cells, a fusion protein between the prion-forming domain (PrD) of the protein in question and the widely exploited reporter gene encoding green fluorescent protein (GFP). [A detailed account of the use of GFP as a reporter gene in yeast can be found in Chapter 8, this volume, von der Haar *et al.*] If cells of a given strain contain the prion in question, the fusion protein is seeded to form discrete fluorescent foci, whereas in the [*prion*$^-$] cell the fusion protein is usually detected as diffuse fluorescence in the cell's cytoplasm.

The standard strategy is to introduce the gene encoding the prion protein PrD-GFP fusion into the cell on a single copy plasmid with the gene under the control of either the galactose-inducible *GAL* promoter or the copper-inducible *CUP1* promoter (Patino *et al.*, 1996). Using the weaker *SUP35* gene promoter results in lower levels of expression of the fusion protein and this can make visualisation of the prion-related foci relatively difficult. The problem with using an efficient promoter to express the fusion protein is that this will also induce the *de novo* formation of the prion state if the cell is [*PIN*$^+$] (see above).

In [*PSI*$^+$] cells the number and nature of the Sup35pPrD-GFP aggregates (or foci as they are usually referred to as) can show significant variation in both numbers and morphology both within and between different strains (Figure 7, see Colour Plate Section). In some strains a few large discrete foci can be seen while in others, numerous small foci are detected. Different [*PSI*$^+$] variants in the same nuclear genetic background also give very different types of Sup35pPrD-GFP foci; 'weak' variants have a few large foci while 'strong' variants have many small foci (Fernandez-Bellot, E. and Tuite, M. F., unpublished). In cells undergoing *de novo* conversion as a consequence of over expression of the Sup35p-PrD, one can also detect striking elongated structures (Figure 7, Panel e) which disappear after the [*PSI*$^+$] state has stabilised (Zhou *et al.*, 2001).

Both Rnq1pPrD-GFP (Sondheimer and Lindquist, 2000) and Ure2pPrD-GFP (Edskes *et al.*, 1999) fusions have also been used. In the case of Rnq1pPrD-GFP, invariably only one of the two types of fluorescence pattern have been reported for [*PIN*$^+$] cells: either 'single dot' or multidot' foci and this may diagnose the existence of two different variants of this prion (Bradley and Liebman, 2003). With Ure2pPrD-GFP the situation is slightly more complex. For example, not all [*URE3*] cells produce distinct foci following the expression of the Ure2pPrD-GFP fusion and problematically, the expression of this fusion in a [*URE3*] cell results in loss of the [*URE3*] prion (Edskes *et al.*, 1999). Furthermore, Fernandez-Bellot *et al.* (2002) showed that while expression of Ure2-GFP leads to the formation of aggregates, this aggregation is not apparently

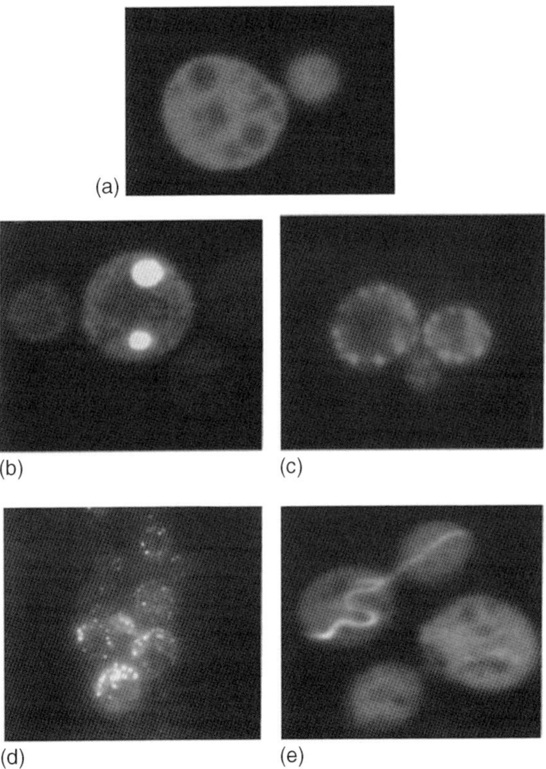

Figure 7. The use of green fluorescent protein (GFP) fusions to visualise Sup35p-based aggregates in [*PSI*$^+$] strains. All cells shown contain a plasmid expressing an identical Sup35pPrD+M-GFP fusion protein whose synthesis was induced using the copper-inducible *CUP1* promoter (Patino *et al.*, 1996); (a) a [*psi*$^-$] strain; (b)–(d) different [*PSI*$^+$] variants; (e) a [*PIN*$^+$][*psi*$^-$] strain undergoing *de novo* conversion to [*PSI*$^+$] as a consequence of the over expression of the Sup35pPrD+M-GFP fusion protein (see Zhou *et al.*, 2001) (See color plate section).

associated with [*URE3*] formation *per se*. The use of Ure2pPrD-GFP to study the [*URE3*] prion is therefore of questionable value.

There are also a number of other issues that raise concerns about the value of using GFP fusions to study yeast prions *in vivo*:

(a) The variability seen within a single population of cells expressing the fusion where not all cells in the populations show the same behaviour.
(b) The way in which cells are prepared for microscopic analysis can modify the appearance of the GFP foci (Chernoff *et al.*, 2002).
(c) The change in the number and/or morphology of the foci as cells move from exponential growth to stationary phase (Zhou *et al.*, 2001).
(d) A significant proportion of cells showing clear distinct foci are usually dead particularly in the case of [*PSI*$^+$] cells (Byrne, L. J. and Tuite, M. F., unpublished).

The relationship between the GFP foci and the oligomeric forms of the native prion protein necessary for propagation of the prion state is also not established and most likely the observable fluorescent foci are the dead-end products of the aggregation process and play no direct role in prion propagation.

◆◆◆◆◆◆ VI. ELIMINATING YEAST PRIONS

Native yeast prions can be readily and rapidly eliminated – in some cases with almost 100% efficiency – from yeast cells either by growing in the presence of various chemical agents or by rational manipulation of the levels of the molecular chaperone Hsp104p (Table 2). None of these agents cause a change in the DNA sequence of the prion gene, i.e. are 'non-mutagenic', but rather give rise to a defect in the process by which prions are propagated. Chemical agents that are known to cause gene mutation via DNA sequence damage, e.g. ethyl methane sulphonate (EMS) and ultraviolet light (UV), can also induce prion loss (Cox *et al.*, 1980), but the rate at which this occurs is consistent with the underlying event being a mutation in either the prion gene, that impairs its ability to take up and/or maintain the prion state, or in the *HSP104* gene.

A. Elimination by Guanidine Hydrochloride

A number of chemical agents have now been described that can eliminate one or more of the native prions (Table 2). The most

Table 2. Chemical agents that can eliminate yeast prions

Chemical agent	Concentration	Mode of action	Reference
Guanidine hydrochloride (GdnHCl)	1–5 mM	Inhibits ATPase activity of Hsp104p	Tuite *et al.* (1981)
Methanol	5–10% v/v	Unknown	Tuite *et al.* (1981)
Latrunculin A	200–500 μM	Unknown (inhibits actin cytoskeleton)	Bailleul-Winslett *et al.* (2000)
KCl	2 M	Unknown	Singh *et al.* (1979)
6-Aminophenanthridine	0.2 mM	Unknown (requires 200 μM GdnHCl)	Bach *et al.* (2003)
Kastellpaolitines (various)	nd	Unknown (requires 200 μM GdnHCl)	Bach *et al.* (2003)

nd, not determined.

effective and widely used means of eliminating [PSI^+], [PIN^+] or [*URE3*] is by growing cells in the presence of 3–5 mM GdnHCl, a chaotropic protein denaturant (Tuite *et al.*, 1981). Although such treatment also generates a high frequency of mitochondrial [rho^-] *petite* mutants (Juliani *et al.*, 1975), provided the cells are allowed to continually grow for at least 10 generations in the presence of the GdnHCl, then approaching 100% of the cells remaining at the end of the experiment will be [$prion^-$] (Eaglestone *et al.*, 2000; Figure 8, see Colour Plate Section). The reason that GdnHCl has this dramatic effect is not because it is a protein denaturant *per se*; the normal concentrations used for protein denaturation are normally in the 1–5 M range. Rather, the GdnHCl appears to act as a potent and specific inhibitor of the ATPase activity of the Hsp104p chaperone that is required for propagation of all three prions (Ferreira *et al.*, 2001; Jung *et al.*, 2002; Grimminger *et al.*, 2004). Therefore, in the presence of GdnHCl, Hsp104 chaperone function is impaired and this in turns leads to a failure to produce the new propagons required for the continued propagation of the prion form. Consequently, those propagons that were present at the point in time at which the GdnHCl is added simply dilute out of the population through cell division leading to the emergence of [$prion^-$] cells after only a few generations of growth. Other guanidinium salts, e.g. guanidine dihydrogen sulphate, can also effectively eliminate [PSI^+] from growing cells (Lawrence, C. W., Eaglestone, S. S. and Tuite, M. F., unpublished).

In practice, all that is needed to generate [$prion^-$] cells is to patch cells onto an agar plate with a rich medium (e.g. YEPD) containing

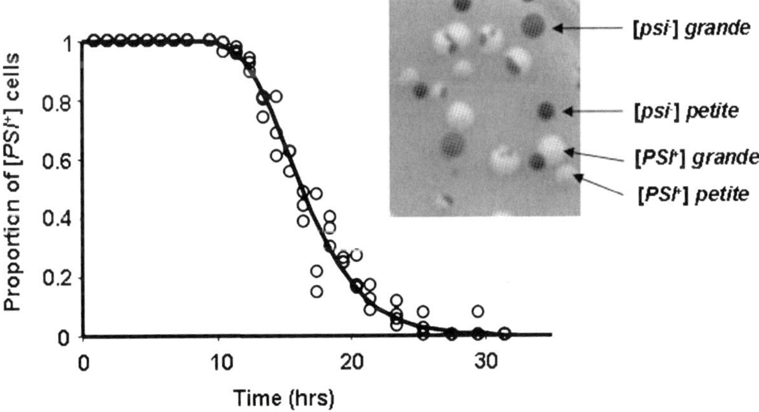

Figure 8. The elimination of the [PSI^+] prion from cells grown in the presence of 3 mM guanidine hydrochloride (GdnHCl) over a 30-h period. The experiment was carried out as described in the Protocol 1 and the percent [PSI^+] with time was plotted as shown. The data are combined from three independent experiments. The inset shows the types of colonies that one observes in such an experiment noting in particular that (a) GdnHCl induces a high frequency of mitochondrial *petites*, and (b) the *petite* mutation does cause a change in colouration when compared to *grande* strains with functional mitochondria. Colonies sectored red and white are counted as [PSI^+]. Further details can be found in the text (See color plate section).

3–5 mM GdnHCl. After 48–72 h growth, the cells should be transferred to a fresh YEPD+GdnHCl plate and to ensure efficient elimination of the prion from all cells, this should then be repeated one or two more times. However, because this treatment does not destroy the propagons that were present in the cells prior to the addition of GdnHCl, there will always be a few [$PRION^+$] cells remaining in the population (Cox *et al.*, 2003). This phenomenon can be exploited as a means of estimating the number of propagons in a cell (see Section VII.B below). The efficiency of GdnHCl-mediated prion elimination does depend on the medium used with lower concentrations of GdnHCl being required in rich (YEPD) medium than those required to achieve the same effect in defined (YNB-based) medium, being 3 and 5 mM, respectively for efficient [PSI^+] elimination.

B. Elimination by Other Chemical Agents

A number of other chemical agents have also been described that cause the loss of [PSI^+] from growing cells (Table 2) but they are generally less efficient than GdnHCl and generally the underlying mechanism is unknown. These include methanol and DMSO (Tuite *et al.*, 1981) and the toxin latrunculin A (Bailleul-Winslett *et al.*, 2000). For latrunculin A, continuing cell growth is not required for the loss of [PSI^+] which suggests that it may act to dissociate key oligomeric forms of Sup35p directly in non-dividing cells. That latrunculin A is known to disrupt the polymeric actin cytoskeleton (Coue *et al.*, 1987) is consistent with this.

The kastellpaolitines, a new group of compounds, have recently been reported to eliminate both [PSI^+] and [*URE3*] and were identified using a cell-based screen for anti-prion compounds (Bach *et al.*, 2003). In this assay, 200 μM GdnHCl was added to the assay medium and although this concentration does not lead to prion elimination *per se*, it was found to be necessary for detecting the prion-eliminating properties of the kastellpaolitines. Several other prion-eliminating compounds were also identified by Bach *et al.* (2003) including phenanthridine and 6-aminophenanthridine. Since none of these compounds are effective in the absence of the 200 μM GdnHCl, they most probably work synergistically with the GdnHCl to inhibit Hsp104p, although it is also possible that they are simply membrane-active agents that increase uptake of GdnHCl from the medium leading to a higher than normal intracellular level of GdnHCl than would normally be achieved when cells are grown in 200 μM GdnHCl. Importantly, several of the compounds identified by Bach *et al.* (2003) also inhibit the formation of PrP^{Sc} in cultured animal cells and since mammalian cells do not appear to have an orthologue of Hsp104p, this would suggest that these compounds actually affect prion conversion directly.

C. Elimination of Yeast Prions by Manipulating the Levels of Chaperone Proteins

The molecular chaperone Hsp104p is essential for the propagation of all three yeast prions and, as originally reported by Chernoff *et al.* (1995), either ablation or overexpression of the *HSP104* gene under non-stress conditions, results in the immediate loss of [PSI^+] from growing cells. Overexpression of a mutant form of Hsp104p that is ATPase negative (i.e. carries a double mutation: K218T and K620T) also results in rapid elimination of [PSI^+] even in the presence of the wild-type *HSP104* gene, i.e. it has a dominant negative effect (Chernoff *et al.*, 1995).

The kinetics of elimination of [PSI^+] by overexpression of wild-type Hsp104p (Figure 9a) are different to what is seen if Hsp104p function is inhibited by GdnHCl (Figure 8) or by overexpression of the ATPase negative (K218T/K620T) allele (Figure 9b) indicating that it may cause prion loss by a different mechanism. The most plausible explanation is that the elevated levels of Hsp104p fully disaggregate prion oligomers including those required for continued propagation of the prion form, although recent biochemical studies provide conflicting evidence for and against this model (Inoue *et al.*, 2004; Shorter and Lindquist, 2004; Krzewska and Melki, 2006). In contrast to the strict dependence of all native yeast prions on Hsp104p for their continued propagation (Sondheimer and Lindquist, 2000; Moriyama *et al.*, 2000) only [PSI^+] is eliminated when Hsp104p is overexpressed (Chernoff *et al.*, 1995; Moriyama *et al.*, 2000), so this does not represent a generic method for yeast prion elimination. However, [PSI^+] elimination by Hsp104p over expression does allow for the easy construction of [PIN^+][psi^-] strains.

To establish whether or not Hsp104p is required for prion propagation, the most straightforward approach is to use GdnHCl, since the generation of an *HSP104* gene knockout can be relatively time consuming. It should be noted however that several artificial prions have been created in yeast that do not apparently require Hsp104p for their continued propagation (e.g. Crist *et al.*, 2003), while some of these 'non-native' prions are also susceptible to overexpression of three other chaperone proteins, the Hsp70's Ssa1p and Ssb1p and the Hsp40 Ydj1p (Kushnirov *et al.*, 2000). [*URE3*] is also eliminated by overexpression of Ydj1p (Moriyama *et al.*, 2000).

♦♦♦♦♦♦ VII. PROPAGON COUNTING

As described in Section VI.A, the addition of GdnHCl to growing yeast [PSI^+] cells results in the failure to generate any new propagons and subsequently, what propagons were present in the cells at the point at which the GdnHCl was added, are diluted out by cell division (Eaglestone *et al.*, 2000). At the end point of this experiment,

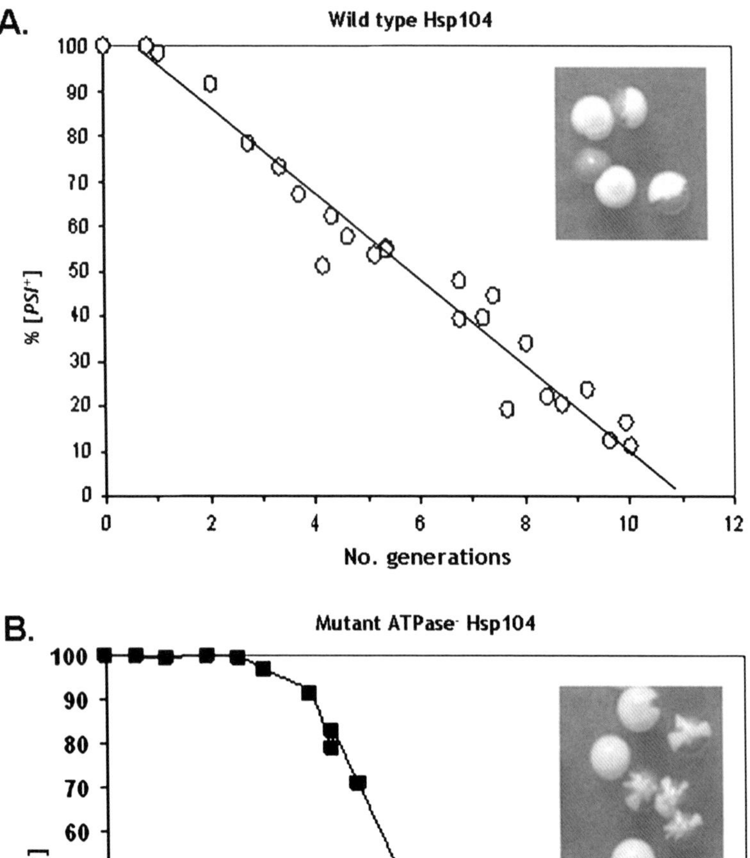

Figure 9. The kinetics of elimination of $[PSI^+]$ from cells overexpressing either (A) wild-type Hsp104p or (B) a mutant of Hsp104p which is unable to hydrolyse ATP as a consequence of a double K218T and K620T mutation. Following induction of expression of the respective *HSP104* genes by use of a galactose-inducible *GAL1* promoter, the percent $[PSI^+]$ cells arising when cells were plated onto 1/4 YEPD were scored over a period of 10–12 generations of growth. The inset shows colonies from this study. Note the difference both in terms of the kinetics of loss and in the types of sectored colonies that arise in the two different experiments.

there will be a number of cells in the population that contain one of the original propagons and the number of these cells gives a readout of the average number of propagons in cells in the population at time = 0. This observation provides a novel means for estimating the number of propagons (n_0) in a given strain (Eaglestone *et al.*, 2000).

A. The Kinetic Method

In this method cells are allowed to grow continuously for up to 36 h after the addition of GdnHCl by serial dilution into fresh GdnHCl-containing medium (Protocol 1). At various time points throughout the 36 h, cell samples are taken and plated onto YEPD to count the number of [PSI^+] and [psi^-] cells in the culture, and the percent [PSI^+] vs. time data are plotted. A significant proportion of the colonies that arise are sectored (Figure 8) and, for the purposes of this analysis, any colony that has a [PSI^+] component is counted as a [PSI^+] cell because at the point at which the cells were plated they must have contained at least one prion seed.

Initially, Eaglestone *et al.* (2000) estimated n_0 by a simple binomial method using the kinetics of [PSI^+] loss data, i.e. Bin (n_0, 2^{-g}), where g is the number of generations of growth and n_0 the number of propagons at time = 0. This gives an estimate for n_0 of around 60 per cell (Eaglestone *et al.*, 2000). Subsequently, this model has been replaced by a stochastic model that takes into account a number of key parameters that must be considered, in particular the growth rate of both mother and daughter cells and the relative distribution of propagons between mother and daughter cells (Cole *et al.*, 2004; Cole *et al.*, submitted for publication). Using this model together with the experimentally determined parameters, the estimation for n_0 is closer to 600.

B. The Colony Method

The number of propagons can also be estimated by an alternative and technically simpler protocol originally described by Cox *et al.* (2003). A number of single cells (usually between 10 and 20) from a given [PSI^+] strain are individually micromanipulated onto solid YEPD medium containing 3–5 mM GdnHCl. These cells are allowed to go through at least 10 generations of growth to form a visible microcolony. In the resulting colony there will be a certain number of cells that contain one of the original [PSI^+] propagons. When these cells are returned to a GdnHCl-free medium they become [PSI^+] because there is nothing to prevent the generation of new propagons, which occurs with the numbers doubling every approximately 20 min (Ness *et al.*, 2002). The number of these cells in a given colony therefore gives a direct readout of the number of propagons in the cell at the time the cell is exposed to GdnHCl. The remaining [PSI^+] cells are detected by plating the entire microcolony onto minimal medium selecting for Ade^+/[PSI^+] cells. Although some of the Ade^+ cells that arise will be due to nuclear tRNA

Protocol 1. Estimating the number of prion seeds in a [*PSI*$^+$] cell using GdnHCl-induced curing.

1. Inoculate a single [*PSI*$^+$] colony into 50 ml liquid YEPD medium and grow overnight at 30°C in order for the culture to reach mid-to-late exponential phase (OD_{600} ~0.5–0.8) by the following morning.
2. Prepare two flasks each containing 50 ml YEPD. To Flask 1 add 3 mM GdnHCl. Flask 2 is the no GdnHCl control.
3. Inoculate both Flasks 1 and 2 with 100 µl of the overnight culture (step 1) and immediately take 3 × 100 µl aliquots from Flask 1, dilute and plate onto 1/4 YEPD plates so that between 100 and 200 colonies appear on each plate. This is time point $t = 0$. Continue to incubate both flasks at 30°C.
4. At 2 h intervals repeat step 3, taking 3 × 100 µl aliquots from Flask 1 and making allowances in the dilutions needed, for the increasing density of the culture. Continue taking samples until $t = 8$ h.
5. Prepare two fresh flasks each containing 50 ml YEPD both containing 3 mM GdnHCl and label Flasks 3 and 4.
6. At $t \sim 14$ h inoculate Flask 3 with 100 µl from Flask 1 and Flask 4 with the same volume from Flask 2. Allow these cultures to grow at 30°C but no sample needs to be taken until the following morning. This staggered inoculation allows the experimenter to get a good night's sleep.
7. At $t \sim 24$ h take a fresh set of aliquots from Flasks 1 and 2. The cultures in Flasks 3 and 4 should be diluted (if necessary) into 50 ml fresh YEPD containing 3 mM GdnHCl and 3 × 100 µl aliquots taken from each flask and process as in step 3. This will be $t = 24$ for Flask 1 and $t = 10$ for Flasks 3 and 4. Continue to take two hourly samples for the next 8 h, i.e. till $t \sim 32$, ensuring the culture remains in exponential phase and diluting into fresh YEPD if necessary.
8. All 1/4 YEPD plates containing cells plated from the different time points in the experiment should be incubated at 30°C for a minimum of 4–5 days but no longer than 7 days to allow the full colour of the resulting colonies to develop. Transferring the grown plates to a 4°C cold room for 24 h will often enhance colour development allowing best distinction between [*PSI*$^+$] and [*psi*$^-$] colonies.
9. From the resulting colony counts:
   (a) Determine the percent [*PSI*$^+$] at each time point counting any colony that has some white sectors as [*PSI*$^+$]. Note that the mitochondrial *petite* mutation can change the colours of colonies (see Figure 8).
   (b) Calculate the number of viable cell/ml for both [*PSI*$^+$]/[*psi*$^-$] remembering to allow for the different dilutions that were made during the experiment.

10. These data can then be used to determine n_0, the average number of prion seeds at $t = 0$ using either a simple binomial distribution model (Eaglestone *et al.*, 2000) or the advanced stochastic model (Cole *et al.*, 2004).

Solutions and Media Needed

YEPD liquid medium: 1% (w/v) yeast extract, 1% (w/v) bactopeptone, 2% (w/v) glucose).
1/4 YEPD solid medium: 0.25% (w/v) yeast extract, 1% (w/v) bactopeptone, 2% (w/v) glucose, 2% (w/v) granulated agar. (NB: This medium gives a better red/white distinction for *petites* in many strains.)
3 M GdnHCl stock, sterilised by autoclaving.
Phosphate buffered saline (PBS): per litre add 8 g NaCl, 0.2 g KCl, 1.44 g Na_2HPO_4, 0.24 g KH_2PO_4. Adjust pH 7.4 and sterilise by autoclaving. Use for diluting cell aliquots prior to plating onto YEPD/1/4 YEPD.

suppressor mutations, these can be easily identified by replica plating onto fresh YEPD+GdnHCl; only true [*PSI*$^+$] turns red. The number of propagons estimated by this method are in good agreement with those obtained using the kinetic method (Cox *et al.*, 2003; Byrne, L. J. *et al.*, unpublished).

◆◆◆◆◆◆ VIII. STUDYING PRION PROTEIN POLYMERISATION *IN VITRO*

The study of yeast prions is not restricted to the *in vivo* approaches, which are the focus of this chapter. Much has been learnt about the conformational rearrangements and subsequent protein aggregation for two of the yeast prion proteins (Ure2p and Sup35p) from *in vitro* studies. In the test tube, both proteins are able to spontaneously undergo conformational rearrangement in the absence of any other proteins or nucleic acids, to generate highly stable fibrils which have the biophysical characteristics of amyloid fibrils (Glover *et al.*, 1997; King *et al.*, 1997; Taylor *et al.*, 1999; Thual *et al.*, 1999). The bulk of these studies have been carried out using the amino-terminal fragments of the protein that have been defined *in vivo* as being required for prion formation, i.e. the PrDs, although some studies have used full-length proteins (e.g. Krzewska and Melki, 2006).

The standard assay for *in vitro* polymerisation of either full-length or isolated PrDs uses the amyloid-specific dyes Congo Red or Thioflavin-T to monitor the formation of amyloid *in vitro*. Starting with soluble protein, without any seeding, amyloid fibres usually form within 30 90 h. This time can be reduced to under 15 h by seeding the assay with either preformed fibres of the protein under investigation (Glover *et al.*, 1997) or by using whole cell extracts prepared from the relevant [*PRION*$^+$] strain (Uptain *et al.*, 2001). Importantly, sonication of the preformed fibres prior to their addition to the assay, or gentle agitation of the assay by use of a roller drum can significantly reduce the length of the assay to less than 2 h. At the end point of such assays a significant fraction of the starting protein sample is now found as distinct amyloid-like fibres that can be readily visualised by atomic force microscopy (AFM), transmission electron microscopy (TEM) or scanning transmission electron microscopy (STEM). These high-resolution microscopic techniques can also be used to monitor the aggregation process, for example, by identifying key intermediate oligomeric forms (e.g. Serio *et al.*, 2000) and measuring the rate of amyloid fibre growth (e.g. DePace and Weissman, 2002).

To prepare sufficient quantities of the desired PrD for *in vitro* polymerisation studies requires the use of an *E. coli*-based high-efficiency expression system and engineered versions of the respective PrD with suitable purification tags such as hexa-histidine placed in-frame at either the N- or the C-terminus of the protein. The preferred version of Sup35p used for many *in vitro* studies contains both the N (PrD) and M domain encompassing residues 1–254, while for Ure2p fragments encompassing either residues 1–65 or residues 1–89 (Taylor *et al.*, 1999; Thual *et al.*, 1999; Baxa *et al.*, 2002; Jiang *et al.*, 2004) are routinely used.

The major challenge in preparing suitable protein for *in vitro* polymerisation studies is the need to have soluble protein, but both the Ure2p and Sup35p-PrDs are highly prone to aggregation when expressed in *E. coli*, which is perhaps not surprising given their amyloidogenic properties. Consequently, the proteins are usually purified under denaturing conditions using, for example, buffers containing 8 M urea or 4 M GdnHCl. A detailed description of how recombinant Sup35p fragments can be prepared can be found in Chernoff *et al.* (2002). Encouragingly, some researchers have however been able to produce native, soluble protein from *E. coli*; for example, Krzewska and Melki (2006) were able to produce soluble full-length Sup35p in *E. coli* for their *in vitro* studies.

One of the most important questions arising from the use of *in vitro* polymerisation assays is inevitably 'What is the relevance of the observed *in vitro* behaviour to the *in vivo* behaviour of a yeast prion protein'? The most direct way of answering is to show that the amyloid-like aggregates formed *in vitro* are able to seed the formation of the prion state of the protein in the living cell (King and Diaz-Avalos, 2004; Tanaka *et al.*, 2004). That this has now been achieved using aggregated forms of the Sup35pN or NM regions provides perhaps the most definitive proof of the protein-only hypothesis of prion replication. In the most straightforward of these assays, Tanaka *et al.* (2004) generated amyloid-like fibres of Sup35pNM *in vitro* and then introduced them into [*psi*$^-$] *ade1-14* yeast cells using a 'protein transformation' protocol. This protocol is essentially the same as that originally developed for plasmid transformation (see Chapter 3, this volume) and involves the generation of sphaeroplasts using lyticase and then, after addition of the protein sample together with a selectable *URA3* plasmid, the sphaeroplasts are regenerated and transformed cells selected. The use of a plasmid enables the researcher to first select for Ura^+ transformed cells and then, by re-streaking these cells onto a medium that selects for Ura^+ Ade^+ cells, identify the cells transformed by the protein and have become [*PSI*$^+$].

♦♦♦♦♦♦ IX. HOW TO RECOGNISE A NEW YEAST PRION

The unusual nature of the phenotypic traits associated with the [*PSI*$^+$] and [*URE3*] prions first came to our attention because of their non-Mendelian mode of inheritance (see Section IV.A). Over the years a large number of other mutants have also emerged from yeast genetic screens, which have an underlying genetic determinant that is inherited in a non-Mendelian fashion. While most of these mutations map to the mitochondrial genome, i.e. are *petite* mutants, several do not and their basis remains unexplained. For example, Kunz and Ball (1977) reported on a glucosamine-resistant mutant whose determinant was cytoplasmically inherited but was not eliminated by agents (e.g. ethidium bromide) that effectively eliminate mitochondrial DNA from yeast. Now that we recognise that prions exist in yeast and that they can give rise to distinct phenotypes that are not necessarily detrimental to the host cell, there are a number of steps a researcher can take to establish whether or not they have stumbled upon a new prion-based determinant.

Based on what we know about the three native prions so far described in *S. cerevisiae*, in addition to the failure to show Mendelian inheritance patterns for the underlying determinant, four questions need to be answered using relatively straightforward experiments in order to establish the nature of the underlying genetic determinant:

(a) Is the genetic determinant lost when cells are cultured in the presence of GdnHCl (see Section VI.A)?
(b) Can the 'mutation' spontaneously reappear in such cured cells?
(c) Is the genetic determinant not maintained in cells carrying a knockout of the non-essential *HSP104* gene (see Section VI.C)?
(d) Can the determinant be transmitted to other cells in the absence of karyogamy, i.e. cytoduced (see Section IV.B)?

While positive answers emerging from these experiments would give one confidence to explore the nature of the genetic determinant, they are not sufficient to unambiguously conclude that a prion-based determinant is involved in inheritance of the trait under examination. Critical is identification of the underlying prion protein and the demonstration that

(a) the maintenance of the [$PRION^+$] state depends on the presence of the gene encoding that protein;
(b) the protein forms transmissible Hsp104-dependent high–molecular-weight aggregates in [$PRION^+$] cells but not in [$prion^-$] cells;
(c) overexpression of the gene leads to an increase in the *de novo* appearance of [$PRION^+$] cells in a [PIN^+] but not in a [pin^-] strain; and
(d) the protein polymerises *in vitro* to form self-seeding amyloid-like fibres that, when transformed into a [$prion^-$] cell gives rise at high frequency to the [$PRION^+$] state (see Section VIII).

Should the protein identified satisfy all of these criteria, then one can be confident that one was dealing with a prion-based phenomenon.

How does one identify the underlying prion protein? For both [PSI^+] and [*URE3*], a short list of candidate genes could – and indeed was – quickly drawn up based on what was known about the molecular basis of the associated phenotype. On the other hand, the Rnq1p prion protein was discovered through an attempt to predict what a yeast prion might look like and behave like, i.e. contains a Gln/Asn-rich region at either its N- or C-terminus (Sondheimer and Lindquist, 2000). However, in excess of 100 different yeast proteins contain such regions (Michelitsch and Weissman, 2000) but with the exception of Rnq1p, none of the other candidates has yet emerged as a true prion. Caution must therefore be taken using the rational approach; it is not sufficient to conclude that the protein is a prion just because a Gln/Asn-rich region exists in the protein sequence. That said, there must be a strong possibility that further prions remain to be discovered in *S. cerevisiae* and perhaps in other fungi?

Acknowledgements

The work on yeast prions carried out in the authors' laboratory is funded by the Biotechnology and Biological Sciences Research Council, the Wellcome Trust, the European Union (APOPIS: LSHM-CT-2003-503330) and by the award of a Leverhulme Trust Emeritus Fellowship to BSC.

References

Bach, S., Talarek, N., Andrieu, T., Vierfond, J. M., Mettey, Y., Galons, H., Dormont, D., Meijer, L., Cullin, C. and Blondel, M. (2003). Isolation of drugs active against mammalian prions using a yeast-based screening assay. *Nat. Biotech.* **21**, 1075–1081.

Bagriantsev, S. and Liebman, S. W. (2004). Specificity of prion assembly *in vivo*. [*PSI*$^+$] and [*PIN*$^+$] form separate structures in yeast. *J. Biol. Chem.* **279**, 51042–51048.

Bai, M., Zhou, J. M. and Perrett, S. (2004). The yeast prion protein Ure2 shows glutathione peroxidase activity in both native and fibrillar forms. *J. Biol. Chem.* **279**, 50025–50030.

Bailleul-Winslett, P. A., Newnam, G. P., Wegrzyn, R. D. and Chernoff, Y. O. (2000). An antiprion effect of the anticytoskeletal drug latrunculin A in yeast. *Gene Expr.* **9**, 145–156.

Baxa, U., Speransky, V., Steven, A. C. and Wickner, R. B. (2002). Mechanism of inactivation on prion conversion of the *Saccharomyces cerevisiae* Ure2 protein. *Proc. Natl. Acad. Sci. USA* **99**, 5253–5260.

Bradley, M. E. and Liebman, S. W. (2003). Destabilizing interactions among [*PSI*$^+$] and [*PIN*$^+$] yeast prion variants. *Genetics* **165**, 1675–1685.

Chernoff, Y. O., Derkach, I. L. and Inge-Vechtomov, S. G. (1993). Multicopy *SUP35* gene induces *de-novo* appearance of PSI-like factors in the yeast *Saccharomyces cerevisiae. Curr. Genet.* **24**, 268–270.

Chernoff, Y. O., Lindquist, S. L., Ono, B., Ingevechtomov, S. G. and Liebman, S. W. (1995). Role of the chaperone protein Hsp104 in propagation of the yeast prion-like factor Psi$^+$. *Science* **268**, 880–884.

Chernoff, Y. O., Uptain, S. M. and Lindquist, S. L. (2002). Analysis of prion factors in yeast. *Methods Enzymol.* **351**, 499–538.

Cole, D. J., Morgan, B. J. T., Ridout, M. S., Byrne, L. J. and Tuite, M. F. (2004). Estimating the number of prions in yeast cells. *Math. Med. Biol.* **21**, 369–395.

Conde, J. and Fink, G. R. (1976). A mutant of *Saccharomyces cerevisiae* defective for nuclear fusion. *Proc. Natl. Acad. Sci. USA* **73**, 3651–3655.

Coue, M., Brenner, S. L., Spector, I. and Korn, E. D. (1987). Inhibition of actin polymerization by latrunculin A. *FEBS Lett.* **213**, 316–318.

Coustou, V., Deleu, C., Saupe, S. and Begueret, J. (1997). The protein product of the het-s heterokaryon incompatibility gene of the fungus *Podospora anserina* behaves as a prion analog. *Proc. Natl. Acad. Sci. USA* **94**, 9773–9778.

Cox, B. (1965). *PSI*, a cytoplasmic suppressor of super-suppressor in yeast. *Heredity* **20**, 505–521.

Cox, B. S., Ness, F. and Tuite, M. F. (2003). Analysis of the generation and segregation of propagons: entities that propagate the [*PSI*$^+$] prion in yeast. *Genetics* **165**, 23–33.

Cox, B. S., Tuite, M. F. and Mundy, C. R. (1980). Reversion from suppression to non-suppression in *SUQ5* [psi$^+$] strains of yeast: the classification of mutations. *Genetics* **95**, 589–609.

Crist, C. G., Nakayashiki, T., Kurahashi, H. and Nakamura, Y. (2003). [*PHI*$^+$], a novel Sup35-prion variant propagated with non-Gln/Asn oligopeptide repeats in the absence of the chaperone protein Hsp104. *Genes Cells* **8**, 603–618.

DePace, A. H. and Weissman, J. S. (2002). Origins and kinetic consequences of diversity in Sup35 yeast prion fibers. *Nat. Struct. Biol.* **9**, 389–396.

Derkatch, I. L., Bradley, M. E., Hong, J. Y. and Liebman, S. W. (2001). Prions affect the appearance of other prions: the story of [*PIN*$^+$]. *Cell* **106**, 171–182.

Derkatch, I. L., Chernoff, Y. O., Kushnirov, V. V., Inge-Vechtomov, S. G. and Liebman, S. W. (1996). Genesis and variability of [*PSI*] prion factors in *Saccharomyces cerevisiae*. *Genetics* **144**, 1375–1386.

Derkatch, I. L., Uptain, S. M., Outeiro, T. F., Krishnan, R., Lindquist, S. L. and Liebman, S. W. (2004). Effects of Q/N-rich, polyQ, and non-polyQ amyloids on the *de novo* formation of the [PSI^+] prion in yeast and aggregation of Sup35 *in vitro*. *Proc. Natl. Acad. Sci. USA* **101**, 12934–12939.

Diaz-Avalos, R., Long, C., Fontano, E., Balbirnie, M., Grothe, R., Eisenberg, D. and Caspar, D. (2003). Cross-beta order and diversity in nanocrystals of an amyloid-forming peptide. *J. Mol. Biol.* **330**, 1165–1175.

Eaglestone, S. S., Cox, B. S. and Tuite, M. F. (1999). Translation termination efficiency can be regulated in *Saccharomyces cerevisiae* by environmental stress through a prion-mediated mechanism. *EMBO J.* **18**, 1974–1981.

Eaglestone, S. S., Ruddock, L. W., Cox, B. S. and Tuite, M. F. (2000). Guanidine hydrochloride blocks a critical step in the propagation of the prion-like determinant [PSI^+] of *Saccharomyces cerevisiae*. *Proc. Natl. Acad. Sci. USA* **97**, 240–244.

Edskes, H., Gray, V. and Wickner, R. (1999). The [*URE3*] prion is an aggregate form of Ure2 that can be cured by overexpression of Ure2P fragments. *Proc. Natl. Acad. Sci. USA* **96**, 1498–1503.

Fernandez-Bellot, E., Guillemet, E., Ness, F., Baudin-Baillieu, A., Ripaud, L., Tuite, M. and Cullin, C. (2002). The [*URE3*] phenotype: evidence for a soluble prion in yeast. *EMBO Rep.* **3**, 76–81.

Ferreira, P. C., Ness, F., Edwards, S. R., Cox, B. S. and Tuite, M. F. (2001). The elimination of the yeast prion [PSI^+] by guanidine hydrochloride is the result of Hsp104 inactivation. *Mol. Microbiol.* **40**, 1357–1369.

Firoozan, M., Grant, C. M., Duarte, J. and Tuite, M. F. (1991). Quantitation of readthrough of termination codons in yeast using a novel gene fusion assay. *Yeast* **7**, 173–183.

Glover, J. R., Kowal, A. S., Schirmer, E. C., Patino, M. M., Liu, J. J. and Lindquist, S. (1997). Self-seeded fibers formed by Sup35, the protein determinant of [PSI^+], a heritable prion-like factor of *S. cerevisiae*. *Cell* **89**, 811–819.

Grimminger, V., Richter, K., Imhof, A., Buchner, J. and Walter, S. (2004). The prion curing agent guanidinium chloride specifically inhibits ATP hydrolysis by Hsp104. *J. Biol. Chem.* **279**, 7378–7383.

Inoue, Y., Taguchi, H., Kishimoto, A. and Yoshida, M. (2004). Hsp104 binds to yeast Sup35 prion fiber but needs other factor(s) to sever it. *J. Biol. Chem.* **279**, 52319–52323.

Jarrett, J. T. and Lansbury, P. T. (1993). Seeding "one-dimensional crystallization" of amyloid: a pathogenic mechanism in Alzheimer's disease and scrapie?. *Cell* **73**, 1055–1058.

Jensen, M. A., True, H. L., Chernoff, Y. O. and Lindquist, S. (2001). Molecular population genetics and evolution of a prion-like protein in *Saccharomyces cerevisiae*. *Genetics* **159**, 527–535.

Jiang, Y., Li, H., Zhu, L., Zhou, J. M. and Perrett, S. (2004). Amyloid nucleation and hierarchical assembly of Ure2p fibrils. Role of asparagine/glutamine repeat and nonrepeat regions of the prion domains. *J. Biol. Chem.* **279**, 3361–3369.

Juliani, M. H., Gambarini, A. G. and Costa, S. O. (1975). Induction of rhominus mutants in *Saccharomyces cerevisiae* by guanidine hydrochloride. I. Genetic analysis. *Mutat. Res.* **29**, 67–75.

Jung, G. M., Jones, G. and Masison, D. C. (2002). Amino acid residue 184 of yeast Hsp104 chaperone is critical for prion-curing by guanidine, prion propagation, and thermotolerance. *Proc. Natl. Acad. Sci. USA* **99**, 9936–9941.

Kimura, Y., Koitabashi, S. and Fujita, T. (2003). Analysis of yeast prion aggregates with amyloid-staining compound *in vivo*. *Cell Struct. Funct.* **28**, 187–193.

King, C. Y. and Diaz-Avalos, R. (2004). Protein-only transmission of three yeast prion strains. *Nature* **428**, 319–323.

King, C. Y., Tittmann, P., Gross, H., Gebert, R., Aebi, M. and Wuthrich, K. (1997). Prion-inducing domain 2-114 of yeast Sup35 protein transforms *in vitro* into amyloid-like filaments. *Proc. Natl. Acad. Sci. USA* **94**, 6618–6622.

Kryndushkin, D. S., Alexandrov, I. M., Teravanesyan, M. D. and Kushnirov, V. V. (2003). Yeast [*PSI*$^+$] prion aggregates are formed by small Sup35 polymers fragmented by Hsp104. *J. Biol. Chem.* **278**, 49636–49643.

Krzewska, J. and Melki, R. (2006). Molecular chaperones and the assembly of the prion Sup35p, an *in vitro* study. *EMBO J.* **25**, 822–833.

Kulkarni, A. A., Abul-Hamd, A. T., Rai, R., El Berry, H. and Cooper, T. G (2001). Gln3p nuclear localization and interaction with Ure2p in *Saccharomyces cerevisiae*. *J. Biol. Chem.* **276**, 32136–32144.

Kunz, B. A. and Ball, A. J. (1977). Glucosamine resistance in yeast. II. Cytoplasmic determinants conferring resistance. *Mol. Gen. Genet.* **153**, 169–177.

Kushnirov, V. V., Kryndushkin, D. S., Boguta, M., Smirnov, V. N. and Ter-Avanesyan, M. D. (2000). Chaperones that cure yeast artificial [*PSI*$^+$] and their prion-specific effects. *Curr. Biol.* **10**, 1443–1446.

Lacroute, F. (1971). Non-Mendelian mutation allowing ureidosuccinic acid uptake in yeast. *J. Bact.* **106**, 519–522.

Liebman, S. W. and Sherman, F. (1979). Extrachromosomal psi+ determinant suppresses nonsense mutations in yeast. *J. Bact.* **139**, 1068–1071.

Liu, J. J., Sondheimer, N. and Lindquist, S. L. (2002). Changes in the middle region of Sup35 profoundly alter the nature of epigenetic inheritance for the yeast prion [*PSI*$^+$]. *Proc. Natl. Acad. Sci. USA* **99**(Suppl. 4), 16446–16453.

Maddelein, M. L., Dos Reis, S., Duvezin-Caubet, S., Coulary-Salin, B. and Saupe, S. J. (2002). Amyloid aggregates of the HET-s prion protein are infectious. *Proc. Natl. Acad. Sci. USA* **99**, 7402–7407.

Masison, D. C. and Wickner, R. B. (1995). Prion-inducing domain of yeast Ure2p and protease resistance of Ure2p in prion-containing cells. *Science* **270**, 93–95.

Mathis, C. A., Bacskai, B. J., Kajdasz, S. T., Mclellan, M. E., Frosch, M. P., Hyman, B. T., Holt, D. P., Wang, Y., Huang, G-F., Debnath, M. L. and Klunk, W. E. (2002). A lipophilic thioflavin-T derivative for positron emission tomography (PET) imaging of amyloid in brain. *Bioorg. Med. Chem. Lett.* **12**, 295–298.

Michelitsch, M. D. and Weissman, J. S. (2000). A census of glutamine/asparagine-rich regions: implications for their conserved function and the prediction of novel prions. *Proc. Natl. Acad. Sci. USA* **97**, 11910–11915.

Moriyama, H., Edskes, H. K. and Wickner, R. B. (2000). [*URE3*] prion propagation in *Saccharomyces cerevisiae*: requirement for chaperone Hsp104 and curing by overexpressed chaperone Ydj1p. *Mol. Cell. Biol.* **20**, 8916–8922.

Nakayashiki, T., Ebihara, K., Bannai, H. and Nakamura, Y. (2001). Yeast [*PSI*$^+$] "prions" that are cross-transmissible and susceptible beyond a species barrier through a quasi-prion state. *Mol. Cell* **7**, 1121–1130.

Nakayashiki, T., Kurtzman, C. P., Edskes, H. K. and Wickner, R. B. (2005). Yeast prions [*URE3*] and [*PSI*$^+$] are diseases. *Proc. Natl. Acad. Sci. USA* **102**, 10575–10580.

Namy, O., Duchateau-Nguyen, G. and Rousset, J. P. (2002). Translational readthrough of the *PDE2* stop codon modulates cAMP levels in *Saccharomyces cerevisiae*. *Mol. Microbiol.* **43**, 641–652.

Ness, F., Ferreira, P., Cox, B. S. and Tuite, M. F. (2002). Guanidine hydrochloride inhibits the generation of prion "seeds" but not prion protein aggregation in yeast. *Mol. Cell. Biol.* **22**, 5593–5605.

Osherovich, L. Z., Cox, B. S., Tuite, M. F. and Weissman, J. S. (2004). Dissection and design of yeast prions. *PLoS Biol.* **2**, 442–451.

Osherovich, L. Z. and Weissman, J. S. (2001). Multiple Gln/Asn-rich prion domains confer susceptibility to induction of the yeast [*PSI*$^+$] prion. *Cell* **106**, 183–194.

Patino, M. M., Liu, J. J., Glover, J. R. and Lindquist, S. (1996). Support for the prion hypothesis for inheritance of a phenotypic trait in yeast. *Science* **273**, 622–626.

Paushkin, S. V., Kushnirov, V. V., Smirnov, V. N. and Ter-Avanesyan, M. D. (1996). Propagation of the yeast prion-like [*PSI*$^+$] determinant is mediated by oligomerization of the SUP35-encoded polypeptide chain release factor. *EMBO J.* **15**, 3127–3134.

Prusiner, S. B. (1991). Molecular biology of prion diseases. *Science* **252**, 1515–1522.

Prusiner, S. B., Scott, M. R., DeArmond, S. J. and Cohen, F. E. (1998). Prion protein biology. *Cell* **93**, 337–348.

Resende, C. G., Sands, L., Outerio, T. F., Cox, B. S., Lindquist, S. and Tuite, M. F. (2003). Prion protein gene polymorphisms in *Saccharomyces cerevisiae*. *Mol. Microbiol.* **49**, 1005–1017.

Ripaud, L., Maillet, L. and Cullin, C. (2003). The mechanisms of [*URE3*] prion elimination demonstrate that large aggregates of Ure2p are dead-end products. *EMBO J.* **22**, 5251–5259.

Ross, E. D., Minton, A. and Wickner, R. B. (2005). Prion domains: sequences, structures and interactions. *Nat. Cell Biol.* **7**, 1039–1044.

Schlumpberger, M., Prusiner, S. B. and Herskowitz, I. (2001). Induction of distinct [*URE3*] yeast prion strains. *Mol. Cell. Biol.* **21**, 7035–7046.

Serio, T. R., Cashikar, A. G., Kowal, A. S., Sawicki, G. J., Moslehi, J. J., Serpell, L., Arnsdorf, M. F. and Lindquist, S. L. (2000). Nucleated conformational conversion and the replication of conformational information by a prion determinant. *Science* **289**, 1317–1321.

Shorter, J. and Lindquist, S. (2004). Hsp104 catalyzes formation and elimination of self-replicating Sup35 prion conformers. *Science* **304**, 1793–1797.

Singh, A., Helms, C. and Sherman, F. (1979). Mutation of the non-mendelian suppressor, psi, in yeast by hypertonic media. *Proc. Natl. Acad. Sci. USA* **76**, 1952–1956.

Sondheimer, N. and Lindquist, S. L. (2000). Rnq1: an epigenetic modifier of protein function in yeast. *Mol. Cell* **5**, 163–172.

Sondheimer, N., Lopez, N., Craig, E. A. and Lindquist, S. (2001). The role of Sis1 in the maintenance of the [*RNQ*$^+$] prion. *EMBO J.* **20**, 2435–2442.

Soto, C., Estrada, L. and Castilla, J. (2006). Amyloids, prions and the inherent infectious nature of misfolded protein aggregates. *Trends Biochem. Sci.* **31**, 150–155.

Speransky, V. V., Taylor, K. L., Edskes, H. K., Wickner, R. B. and Steven, A. C. (2001). Prion filament networks in [*URE3*] cells of *Saccharomyces cerevisiae*. *J. Cell. Biol.* **153**, 1327–1336.

Stansfield, I., Jones, K. M., Kushnirov, V. V., Dagkesamanskaya, A. R., Poznyakovski, A. I., Paushkin, S. V., Nierras, C. R., Cox, B. S., Ter-Avanesyan, M. D. and Tuite, M. F. (1995). The products of the *SUP45* (eRF1) and *SUP35* genes interact to mediate translation termination in *Saccharomyces cerevisiae*. *EMBO J.* **14**, 4365–4373.

Tanaka, M., Chien, P., Naber, N., Cooke, R. and Weissman, J. S. (2004). Conformational variations in an infectious protein determine prion strain differences. *Nature* **428**, 323–328.

Taylor, K. L., Cheng, N., Williams, R. W., Steven, A. C. and Wickner, R. B. (1999). Prion domain initiation of amyloid formation *in vitro* from native Ure2p. *Science* **283**, 1339–1343.

Thual, C., Komar, A. A., Bousset, L., Fernandez-Bellot, E., Cullin, C. and Melki, R. (1999). Structural characterization of *Saccharomyces cerevisiae* prion-like protein Ure2. *J. Biol. Chem.* **274**, 13666–13674.

True, H. L. and Lindquist, S. L. (2000). A yeast prion provides a mechanism for genetic variation and phenotypic diversity. *Nature* **407**, 477–483.

Tuite, M. F. and Koloteva-Levin, N. (2004). Propagating prions in fungi and mammals. *Mol. Cell* **14**, 541–552.

Tuite, M. F., Mundy, C. R. and Cox, B. S. (1981). Agents that cause a high frequency of genetic change from [psi$^+$] to [psi$^-$] in *Saccharomyces cerevisiae*. *Genetics* **98**, 691–711.

Uptain, S. M. and Lindquist, S. (2002). Prions as protein-based genetic elements. *Annu. Rev. Microbiol.* **56**, 703–741.

Uptain, S. M., Sawicki, G. J., Caughey, B. and Lindquist, S. (2001). Strains of [*PSI*$^+$] are distinguished by their efficiencies of prion-mediated conformational conversion. *EMBO J.* **20**, 6236–6245.

Wickner, R. B. (1994). [*URE3*] as an altered URE2 protein: evidence for a prion analog in *Saccharomyces cerevisiae*. *Science* **264**, 566–569.

Wickner, R. B., Edskes, H. K., Roberts, B. T., Pierce, M. M., Baxa, U. and Ross, E. (2001). Prions beget prions: the [*PIN*$^+$] mystery!. *Trends Biochem. Sci.* **26**, 697–699.

Wickner, R. B., Edskes, H. K., Ross, E. D., Pierce, M. M., Baxa, U., Brachmann, A. and Shewmaker, F. (2004). Prion genetics: new rules for a new kind of gene. *Annu. Rev. Genet.* **38**, 681–707.

Zhou, P., Derkatch, I. L. and Liebman, S. W. (2001). The relationship between visible intracellular aggregates that appear after overexpression of Sup35 and the yeast prion-like elements [*PSI*$^+$] and [*PIN*$^+$]. *Mol. Microbiol.* **39**, 37–46.

21 Metabolic Control in the Eukaryotic Cell, a Systems Biology Perspective

Juan I Castrillo and Stephen G Oliver
Centre for the Analysis of Biological Complexity (CABC), Faculty of Life Sciences, Michael Smith Building, The University of Manchester, Oxford Road, Manchester M13 9PT, UK

♦♦♦

CONTENTS

Introduction
Metabolic control analysis (MCA) as a tool for metabolism and functional analysis
Metabolic control in the eukaryotic cell
Metabolic control at the gene expression and metabolome-level integrative strategies
Metabolic control. A systems biology perspective

List of Abbreviations

ANOVA	ANalysis Of VAriance between groups
BST	biochemical systems theory
FANCY	Functional ANalysis by Co-responses in Yeast
MCA	metabolic control analysis
MFA	metabolic flux analysis
MIRIAM	minimum information requested in the annotation of biochemical models
SBML	systems biology mark-up language

♦♦♦♦♦♦ I. INTRODUCTION

The regulation of metabolic fluxes constitutes a central subject in Biology. How are metabolic fluxes controlled physiologically? To what extent are metabolic fluxes regulated at the level of gene expression and to what extent by modulating enzyme activity? What mechanisms are mainly responsible for a particular metabolic response (i.e. distribution of metabolic fluxes) under specific

METHODS IN MICROBIOLOGY, VOLUME 36
0580-9517 DOI:10.1016/S0580-9517(06)36021-7

conditions? This chapter focuses on integrative strategies for studies of *in vivo* patterns of metabolic control in yeast.

The first part of the chapter outlines the main concepts and methodological frameworks for the study of metabolic control, focusing mainly on metabolic control analysis (MCA) and the need for extended descriptions that reflect the exquisitely complex, modular architecture of the eukaryotic cell. In the central part of the chapter, we present a demonstration that illustrates the possibilities that integrative genomic studies and the incorporation of advanced high-throughput technologies are opening for the understanding of *in vivo* physiological patterns of metabolic control, particularly the relevance of control at the level of gene expression. Hence, as a reference example, we present the application of high-throughput transcriptional and proteomic studies coupled to chemostat culture, which allows the elucidation of *in vivo* patterns of metabolic control at the proteome (enzyme) level and the relevance of multienzyme regulation during cell growth. We also present a perspective on advanced studies of metabolic control at the metabolome level, and raise the need for more integrative studies combining post-genomics strategies and techniques at different functional genomic levels.

In the last section, we present some of the latest advances in metabolic control frameworks and conceptual strategies towards an integrative theory of metabolic control, the first applications of these conceptual frameworks and formal rules in yeast, and a perspective of what integrative advanced studies at the gene expression and metabolome level are beginning to reveal about the *in vivo* control of metabolic fluxes in the eukaryotic cell. From this perspective, we propose the integration of kinetic data from quantitative metabolomics and proteomics, the two levels most directly involved in metabolic control, as the most comprehensive way to integrate metabolomics into genome-scale, System Biology models that should permit the design of comprehensive metabolic engineering strategies.

♦♦♦♦♦♦ II. METABOLIC CONTROL ANALYSIS (MCA) AS A TOOL FOR METABOLISM AND FUNCTIONAL ANALYSIS

Two main global theories of metabolic control have been reported. The biochemical systems theory (BST: Savageau, 1976) encompasses the large number of interacting components of biochemical systems and the non-linear character of these interactions but it is mathematically complex, which makes its application difficult. In contrast, the 'MCA' theory (Kacser and Burns, 1973; Fell, 1997), focused mainly on metabolic networks and their regulation, is much more simple and has become the main systems approach in the study of metabolic control (Thomas and Fell, 1998). Thus, MCA and metabolic flux analysis (MFA: Varma and Palsson, 1994) are commonly used for quantitative descriptions of metabolism. The main principles of the formal theory of MCA are illustrated in Figure 1, and can be summarized as follows (Kacser and Burns, 1973; Fell, 1997; Wagner, 2005):

(1) In a metabolic pathway the control of the flux is 'shared', distributed, between the different enzymatic steps. Thus, under essentially constant physiological conditions in which metabolic systems arrive at a steady state (with metabolite concentrations not changing significantly and flux, J, constant), for a simple linear metabolic pathway the flux can be expressed (Kacser and Burns, 1973) as:

$$J = A/(1/E_1 + 1/E_2 + \ldots + 1/E_n) \tag{1}$$

where J equals flux through the metabolic pathway, E enzyme activity, and A is a constant dependent on the concentration of the first substrate and product. This equation shows that steady-state flux depends on all enzyme activities and the existence of a hyperbolic relationship between flux and each enzymatic activity (Kacser and Burns, 1973; Wagner, 2005; see Figure 1).

(2) The control exerted by each individual enzyme over the total flux (J) is determined by the 'flux control coefficient', C^J_E, which can be defined as the relative change in flux caused by an infinitesimal change of the activity of the specific enzyme (E) at steady state.

$$C^J_{E_i} = (\partial J/J)/(\partial E_i/E_i) = (\partial J/\partial E_i)(E_i/J) = \partial \ln J/\partial \ln E_i \tag{2}$$

Based on equation (1), and applying differentiation, it follows that each control coefficient can be expressed as:

$$C^J_{E_i} = (\partial J/\partial E_i)(E_i/J) = (1/E_i)/(1/E_1 + 1/E_2 + \ldots + 1/E_n) \tag{3}$$

From here, the average contribution (flux control coefficient) of an enzyme in a pathway of n equally active enzymes equals $(1/n)$.

(3) The distribution of control through a metabolic pathway can be different under different conditions (Wagner, 2005).

(4) For an un-branched pathway, the Summation Theorem states that the sum of the flux control coefficients of all steps is equal to unity.

The main contribution of MCA is the concept of distributed control (as opposed to the old concept of a 'rate-limiting step'). The theory can successfully predict that flux control is distributed among the different steps. MCA has been successfully applied in descriptions of metabolism and metabolic control studies in many

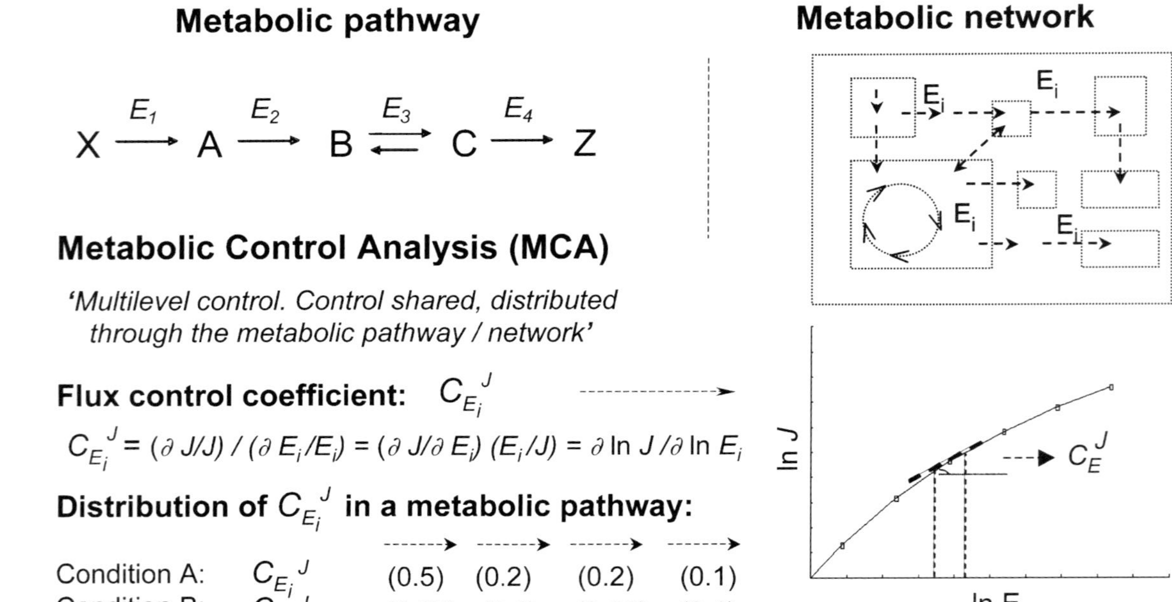

Figure 1. Metabolic control analysis (MCA). Schematic representation of metabolic pathways, networks and main MCA principles: flux control coefficient; distribution of control through a metabolic pathway; summation and connectivity theorems.

different systems (e.g. Westerhoff *et al.*, 1991; Krauss and Quant, 1996; Fell, 1997; Ainscow and Brand, 1999; Hornberg *et al.*, 2005) as well as in functional analysis, with MCA together with metabolic co-responses studies (FANCY method) being used as a valid strategy for the elucidation of the phenotype of silent mutations (Teusink *et al.*, 1998; Raamsdonk *et al.*, 2001). For a more extensive review on the main principles and examples of applications of metabolic control analysis in biological systems, the reader can refer to Kell and Westerhoff (1990), Westerhoff *et al.*, (1991), Fell (1997), the MCA web (http://dbkgroup.org/mca_home.htm), the MCA frequently asked questions website (http://bip.cnrs-mrs.fr/bip10/mcafaq.htm), and references therein.

However, MCA theory cannot predict the distributions of enzyme activities and flux coefficients *in vivo*; the dynamics of *in vivo* regulatory patterns in metabolic pathways and networks during non-infinitesimal perturbations; or how the control distribution will change from one experimental condition to another (Thomas and Fell, 1998; Wagner, 2005). These constitute specific challenges for which limited kinetic information is available, and such data will be essential for the implementation of an extended theory of metabolic control (see below). Recent studies based on a quantitative genetic model of flux control (Bost *et al.*, 1999; Wagner, 2005) predict that flux control coefficients should show a skewed distribution, with most enzymes in a pathway exhibiting very small flux control coefficients and only a few having large flux control. Experimental evidence in bacterial and eukaryotic cells is accumulating supporting these predictions (Hornberg *et al.*, 2005; Wagner, 2005).

♦♦♦♦♦♦ III. METABOLIC CONTROL IN THE EUKARYOTIC CELL

A. The Eukaryotic Cell. Multilevel Control

The eukaryotic cell is exquisitely complex, performing coordinate reactions between biological entities at different genomic levels (genome, transcriptome, proteome and metabolome), in different sub-cellular compartments, with the involvement of regulatory mechanisms at the (epi)genomic, transcriptional, post-transcriptional, post-translational and metabolic levels (Choudhuri, 2004; Castrillo and Oliver, 2006). A schematic representation of the eukaryotic cell as a global system is presented in Figure 2. The eukaryotic cell, as displayed, is characterized by coordinate integration of four interconnected regulatory modules, in direct relation with the environment: (i) environmental sensing; (ii) signal transduction regulatory networks; (iii) gene expression at the transcriptional and translational level and (iv) metabolic networks. Each has its specific role and relevance in the global cellular network. Thus,

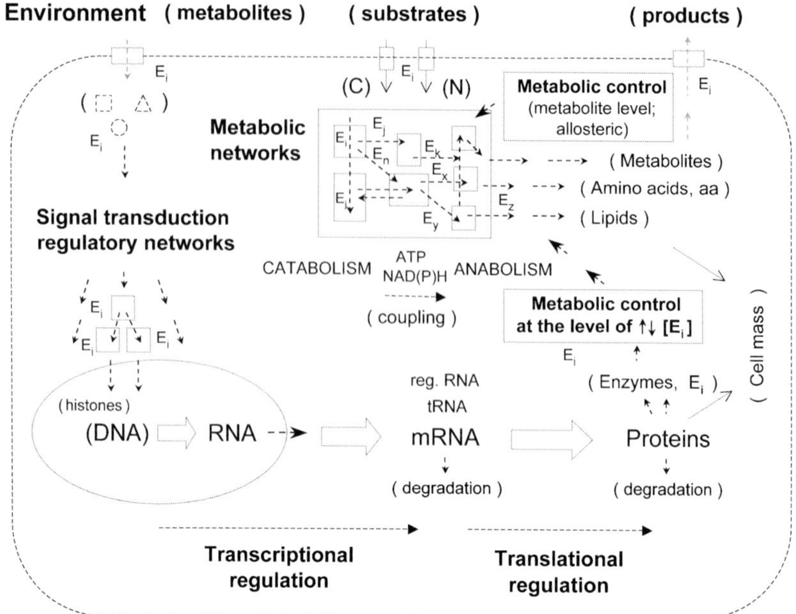

Figure 2. Eukaryotic cell. Genomic levels and relationships between main regulatory levels. Main regulatory modules: environmental sensing; signal transduction regulatory networks; gene expression at the transcriptomic and proteomic levels; metabolic networks, including metabolic control at the metabolome and protein (enzyme) level.

from this viewpoint, we can see global cellular metabolic activity (e.g. cell growth) to be a consequence of coordinate, multilevel control of the different regulatory modules. Moreover, this representation based on regulatory modules shows that metabolic networks are regulated at both the metabolite and enzyme production (net protein expression) levels.

The complexity of the cell makes it difficult to obtain information on the contribution of regulatory modules to global fitness (i.e. growth rate). Nevertheless, the existence of multilevel, distributed control in sub-cellular systems has been confirmed in different studies, not only on metabolic networks but also on signalling regulatory networks, with a small number of reactions exhibiting a large degree of flux control (Fell, 1997; Thomas and Fell, 1998; Hornberg *et al.*, 2005). Together with this, approaches to extend the MCA theory to incorporate regulatory modules, complex regulatory networks and sub-cellular compartmentation have been progressively developed (Khan and Westerhoff, 1991; Bruggeman *et al.*, 2002; Peletier *et al.*, 2003). Among these, modular MCA (Khan and Westerhoff, 1991) allows the study of signal transduction, enzyme activation and other factors influencing metabolic control under the aegis of MCA theory. Briefly, this is achieved by dividing the complex network into modules. Within a module, intermediates are

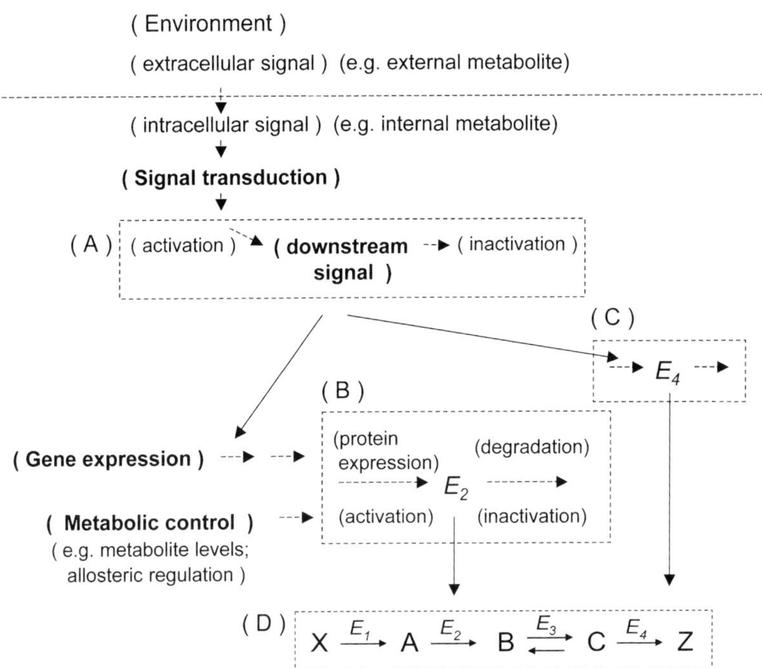

Figure 3. Control of metabolic fluxes at the gene expression and metabolic level. Modular MCA. Structural organization of modules affecting enzyme activities at the gene expression (hierarchical) and metabolic levels in a specific pathway. Environmental signals (e.g. variations in concentrations of external nutrients) can be sensed and transmitted intracellularly through signal transduction networks resulting in activation of intracellular signals (e.g. intracellular metabolites), which can modify enzymatic levels (via synthesis of degradation) or the activation state of specific enzymes (Khan and Westerhoff, 1991; Thomas and Fell, 1998).

inter-converted by mass action, but between modules information is transmitted by signals, with negligible mass transfer (Khan and Westerhoff, 1991; Thomas and Fell, 1998). This is illustrated, in Figure 3, for the linear metabolic pathway displayed in Figure 1. In this example, the enzymatic activity of two enzymes (modules B and C) of the metabolic pathway (module D) can be regulated at the gene expression or metabolite level in response to changes in the level of an intracellular signalling molecule (module A), in response to changes in the environment.

Although incorporating regulatory modules, the modular MCA theory (just like classical MCA) is not able to predict the dynamics of *in vivo* regulatory patterns in metabolic networks during non-infinitesimal perturbations, nor the relative contributions of gene expression and metabolic regulation (Figure 2). Extended conceptual frameworks incorporating data from *in vivo* kinetic studies are necessary, which will require the implementation of comprehensive experimental designs and the application of advanced high-throughput integrative strategies.

B. *In vivo* Studies on Physiological Control of Metabolic Fluxes – Challenges

Some of the main challenges in advanced studies of *in vivo* control of metabolic fluxes can be summarized as follows:

(i) *Relative contributions of control at the gene expression and metabolic level.* In a metabolic pathway, to what extent are metabolic fluxes regulated by gene expression or by metabolites themselves? Are specific enzymes predominantly regulated at the gene expression or the metabolic level?
(ii) *Control of metabolic fluxes under different culture conditions.* Do the relative contributions to the global control (i.e. gene expression and metabolic regulation) vary under different culture conditions?
(iii) *Relative changes in patterns of control.* In a transition from an initial to a new steady state characterized by a net change in metabolic flux, can these relative contributions vary, or be readjusted?
(iv) *An extended theory of metabolic control.* Can the concept of modular MCA, or some equivalent, be extended towards a simple global framework, which could be applied to the studies of integral metabolic control in the eukaryotic cell?

To address these challenges, we suggest, will require studies at the gene expression (i.e. transcriptional and protein expression) and metabolic regulation levels, using advanced high-throughput technologies, incorporating kinetic data from carefully programmed experiments (steady-state and time-course dynamic studies; see below).

♦♦♦♦♦♦ IV. METABOLIC CONTROL AT THE GENE EXPRESSION AND METABOLOME-LEVEL INTEGRATIVE STRATEGIES

A. Integration of Protein and Transcriptional Expression Studies – Translational Control

Global changes in gene expression at the transcriptional and proteome level, coupled to intracellular protein turnover, lead to relative changes in enzyme concentrations. This constitutes a central mechanism of control of metabolic flux for which limited *in vivo* information is available. Increases in enzyme levels result in proportionally higher enzymatic rates (v_i) *in vivo*, and determine the maximum attainable rate ($v_{max} = k_{cat}$ [E]; Fersht, 1999).

Using high-throughput transcriptional and proteomic studies coupled to chemostat culture, we conducted a study to reveal *in vivo* patterns of metabolic control at the level of gene expression in

Saccharomyces cerevisiae. Using hybridization-array technology (Hayes *et al.*, 2002) and isotope tags for quantitative proteomics analysis (iTRAQ: Ross *et al.*, 2004), we analysed patterns of gene expression at the transcriptional and proteomic levels for steady-state chemostat cultures at two growth rates (0.1 and $0.2\,h^{-1}$; equivalent to a population doubling time (T_d) of 7 and 3.5 h, respectively), under carbon, nitrogen, phosphate and sulphate-limiting conditions. With this approach we were able to detect and quantify a significant proportion of the yeast proteome (*ca.* 600 proteins per nutrient-limiting condition). This allowed a comprehensive study of *in vivo* patterns of regulation at the level of gene expression during cell growth (transition from 0.1 to $0.2\,h^{-1}$). In an initial step, we analysed the relative changes in protein versus transcript levels from growth rate 0.1 to $0.2\,h^{-1}$ (proteome–transcriptome correlations) and the presence of specific outliers, which may be the subjects of considerable translational or post-translational control. Thus an example case (Figure 4) shows the results obtained for chemostat series under phosphate-limiting conditions. The fact that mRNA

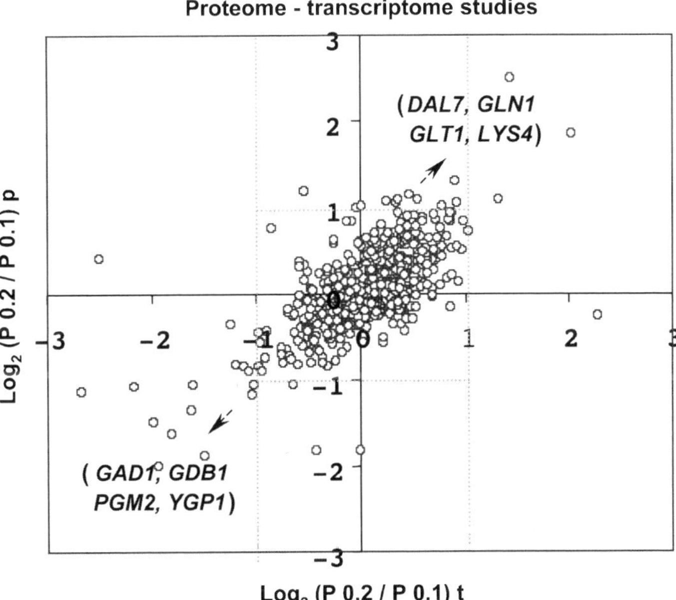

Figure 4. Proteome–transcriptome correlations – translational control of gene expression. Relative changes in protein levels vs. mRNA changes for phosphate-limited cultures of *Saccharomyces cerevisiae* from growth rate (μ) 0.1 to $0.2\,h^{-1}$ (doubling time, $T_d = 7$–$3.5\,h$; 597 identified proteins), including relevant outliers: *DAL7*, malate synthase; *GLN1*, glutamine synthetase; *GLT1*, NAD(+)-dependent glutamate synthase; *LYS4*, homoaconitase; *GAD1*; glutamate decarboxylase; *GDB1*; glycogen debranching enzyme, containing glucanotranferase and alpha-1,6-amyloglucosidase activities; *PGM2*, phosphoglucomutase; *YGP1*, cell wall-related secretory glycoprotein. High-throughput techniques applied: hybridization array technology and isotope tags for multiplexed relative and absolute quantification (iTRAQ) coupled to chemostat culture.

Metabolic Control in the Eukaryotic Cell, A Systems Biology Perspective changes do not globally correlate with protein changes (linear correlation coefficient obtained, $r = 0.66$) suggests a relevant role for post-transcriptional mechanisms. In this case, a number of biosynthetic enzymes appear as significant outliers, pointing to the existence of specific enzymatic steps particularly regulated at the translational level.

B. Metabolic Control at the Gene Expression (Protein) Level – Multiple Enzyme Regulation

Mechanisms of regulation of gene expression at the transcriptional and translational level, coupled to intracellular protein turnover, regulate net global changes in intracellular protein (enzyme) levels (Figure 2). The application of high-throughput isotope tags for multiplexed relative and absolute quantification (iTRAQ: Ross *et al.*, 2004) permits a wide-ranging quantitative analysis of quantitative proteomic changes in protein levels (proteomic signatures) between two (steady) states. Thus, in our example case, the global pattern of relative increases in protein levels for a shift from growth rate 0.1 to $0.2\,h^{-1}$ (growth rate doubling) of *S. cerevisiae* under phosphate limitation is presented in Figure 5. Comprehensive analyses of these global patterns are of primary importance to reveal groups of growth-regulated proteins and processes (e.g. ribosomal proteins). However, in this case, we decided to focus our study on how the expression patterns/levels of specific enzymes are physiologically regulated within a specific pathway.

The patterns of *in vivo* relative changes in enzyme levels (net protein expression) of enzymes involved in lysine biosynthesis, corresponding to a shift in growth rate (μ) from 0.1 to $0.2\,h^{-1}$ (i.e. a doubling of the growth rate) are presented in Figure 6, for carbon- and phosphate-limited conditions. These show that, within a particular metabolic pathway, specific enzymes appear more selectively regulated at the protein expression level (e.g. Aco1p, Aco2p, Lys4p), whereas others exhibit negligible protein expression regulation (e.g. Lys12p). This means that net changes in the *in vivo* internal fluxes through a metabolic pathway can be achieved via multiple enzyme regulation of different enzymes at the gene expression and/or metabolome level. Some enzymes are controlled predominantly at the gene expression level, whereas others are regulated predominately at the metabolic level. Moreover, this selective multienzyme regulation appears specific for different nutrient-limited conditions (i.e. different patterns are seen for relative changes in protein expression under carbon and phosphate limitation, Figure 6).

The possibility of multienzyme activation in response to some long-term metabolic stimulus has been suggested by some authors, based on studies of comparison of enzyme rates assayed under conditions of maximal activity (enzymatic extracts: Fell and Thomas, 1995; Thomas and Fell, 1998), but has not been rigorously

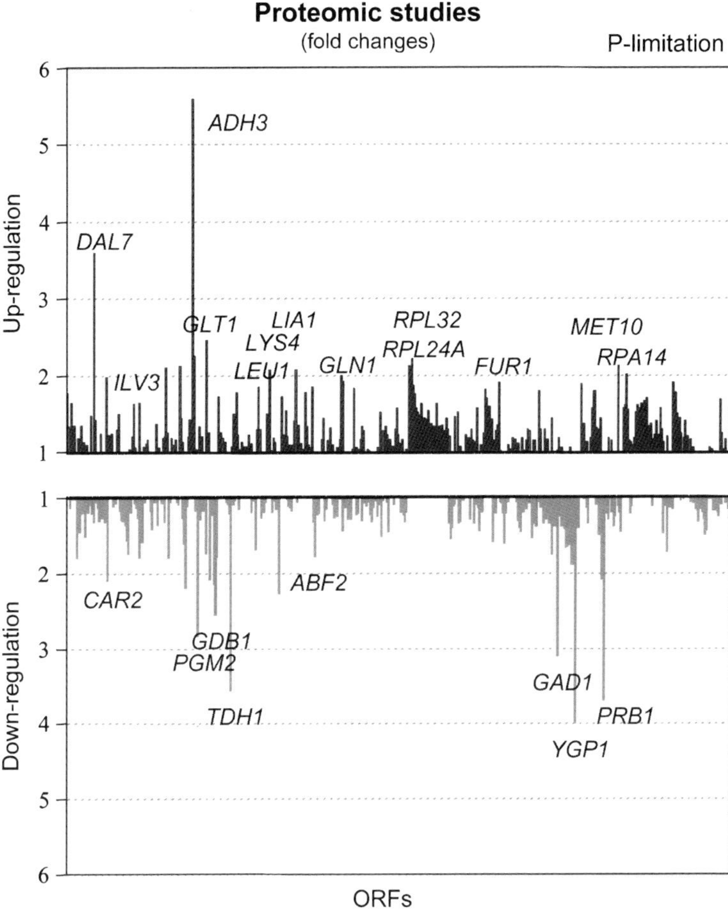

Figure 5. Proteomic studies. Metabolic control at the level of protein (enzyme) expression. Relative changes in protein levels of 597 identified proteins in phosphate-limited cultures of *Saccharomyces cerevisiae*, from growth rate (μ) 0.1 to 0.2 h^{-1} (doubling time, $T_d = 7$–3.5 h), including relevant outliers. ORFs/gene names sorted by biological process (Nomenclature: *Saccharomyces Genome* Database, SGD; http://www.yeastgenome.org). High-throughput method: isotope tags for multiplexed relative and absolute quantification (iTRAQ) coupled to chemostat culture.

demonstrated *in vivo*. Our studies point to the possibility of using high-throughput quantitative proteomics combined with metabolomics (see below) to advance our knowledge of the *in vivo* patterns of metabolic control.

C. Control of Metabolic Fluxes at the Metabolome Level

Gene expression studies provide relevant information at the level of protein abundance (e.g. net increases/decreases in enzyme levels). However, these do not provide information on the influence of variations of metabolite concentrations on enzymatic activities and

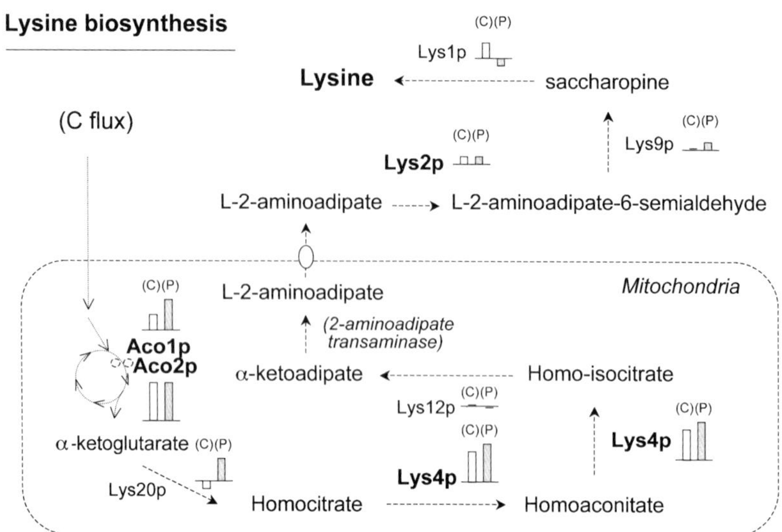

Figure 6. Metabolic control at the level of enzyme synthesis. Multiple enzyme regulation. *In vivo* relative changes in enzyme levels within the lysine biosynthetic pathway from growth rate (μ) 0.1 to 0.2 h^{-1} (doubling time, $T_d = 7$–3.5 h), under carbon- (C) and phosphate- (P) limiting conditions. Lysine biosynthesis diagram (*Saccharomyces Genome* Database, SGD; http://www.yeastgenome.org). Enzymes consistently up-regulated under both conditions are marked in bold. High-throughput method: isotope tags for multiplexed relative and absolute quantification (iTRAQ) coupled to chemostat culture. Increasing/decreasing bars show relative increases/decreases in levels of the specific enzymes between growth rate 0.1 and 0.2 h^{-1}, under carbon- (C) and phosphate- (P) limiting conditions (shown by open and shaded bars, respectively).

metabolic fluxes. Regulation at the metabolic level constitutes the second element contributing to the actual flux through a metabolic pathway (e.g. the second term in Michaelis–Menten kinetic formulation: $v_i = (k_{cat}\,[E]) \times ([S]/(K_M + [S]))$; Fersht, 1999).

The complete pool of internal and external cellular metabolites (the metabolome) plays a relevant role in metabolic control. Intracellular metabolite concentrations not only influence actual reaction rates, but also intermediary metabolites (e.g. ATP, ADP, citrate) can exert rapid short-term regulation upon central metabolic fluxes. In addition to this, external metabolites (i.e. substrates, products) may act as environmental signals whose information is transmitted intracellularly, leading to variations in internal levels of proteins or metabolites and subsequent changes in gene expression. The role of metabolites as regulatory molecules in signal transduction pathways and regulatory networks constitutes a field of increasingly relevant interest (Castrillo and Oliver, 2005, 2006).

We expect comprehensive high-throughput studies directed to the *in vivo* study of dynamics of internal and external metabolites under well defined conditions, whose information can be integrated with studies at other ‘omic ’ levels, will come to be regarded as of primary importance (Castrillo and Oliver, 2005; Nielsen and Oliver,

2005; Villas-Boas *et al.*, 2005). Among the most relevant studies and strategies for comprehensive analyses at the metabolome level, using yeast as a model are: functional studies of co-responses in yeast (Raamsdonk *et al.*, 2001), experimental approaches for studies at the level of the exometabolome (Allen *et al.*, 2003; Kell *et al.*, 2005), and studies on short-term and long-term metabolome dynamic responses (Wu *et al.*, 2005a, 2006a, 2006b). For an updated perspective on advanced methods in metabolome analyses the reader can refer to Vaidyanathan *et al.* (2005) and references therein.

D. Strategies for Integrative Studies on Metabolic Control

From the previous sections, a global conclusion arises, which is the convenience of application of integrative approaches that combine studies at the gene expression and metabolic levels for *in vivo* analysis of metabolic control. Although this approach can be basically correct a word of caution is appropriate. These two levels constitute important approaches to the study of metabolic control. However, they have their limitations. Thus, for example, high-throughput proteomic studies on enzyme abundances do not provide information on the proportions of the enzyme present as active or inactive forms (e.g. phosphorylated–dephosphorylated; ubiquitinated–de-ubiquitinated). Similarly, high-throughput metabolome studies on groups of metabolites may need to evaluate their actual concentrations in different sub-cellular organelles, or the proportion of protein-associated metabolites in the cell. These facts just highlight the need for progressive incorporation of more information on regulation at different levels (gene expression at transcriptional and translational levels, mRNA and protein turnover and post-translational mechanisms affecting the activity of enzymes), by incorporation of new methods and comprehensive strategies. A global perspective of advanced methods and strategies at different levels, which can be considered in future integrative studies on metabolic control is presented in Table 1, together with most relevant conceptual approaches towards an extended theory on metabolic control (see next section).

♦♦♦♦♦♦ V. METABOLIC CONTROL. A SYSTEMS BIOLOGY PERSPECTIVE

A. Towards a Global Integrative Theory of Metabolic Control

Modular MCA (Khan and Westerhoff, 1991) describes the main regulatory levels (gene expression and metabolic control) involved in control of metabolic fluxes (Figure 3). It would be desirable to incorporate the contributions of these levels into an extended theory

Table 1. Post-genomic strategies for integral studies of metabolic control. Relevant techniques and strategies for integrative studies of dynamics of metabolic control at the gene expression (transcriptional, proteomic) and metabolome levels

Genomic level	Techniques/strategies[a]	References
(Epi)genome and transcriptional gene expression	Epigenome	Wilson *et al.* (2006)
	Sequence studies and molecular genetics techniques	Chapters 2–5, this volume
	Mutant collections	Chapter 25, this volume
(protein–DNA binding; chromatin studies)	ChIP analysis	Ezhkova and Tansey (2006)
Regulatory networks	*Protein–protein interaction networks*	von Mering *et al.* (2002)
	Modular control	Khan and Westerhoff (1991)
	Dynamic control studies	Hornberg *et al.* (2005)
Transcriptional gene expression (relative and absolute quantification)	Hybridization array technology	Hayes *et al.* (2002) and Chapter 9, this volume
	SAGE, DD, QRT-PCR	
	Transcriptional regulatory networks	Balaji *et al.* (2006)
Proteome expression (relative and absolute quantification)	Multiplexed quantification	
	2D-DIGE	Lilley and Friedman (2004)
	iTRAQ	Ross *et al.* (2004)
	Absolute quantification methods	Beynon *et al.* (2005)
	Protein array technology	Chapter 14, this volume
Noise (stochasticity) in signal transduction and gene expression	Cell-to-cell variation. Noise in signal transduction pathways and gene expression	Ramanathan and Swain (2005), Colman-Lerner *et al*, (2005), Raser and O'Shea (2004, 2005), Newman *et al.* (2006)
Metabolome studies (internal and external metabolites pools)	Metabolome analysis and strategies	Vaidyanathan *et al.* (2005), Castrillo and Oliver (2005), Nielsen and Oliver (2005), Kell *et al.* (2005)
	FANCY	Raamsdonk *et al.* (2001)

Table 1. *(Continued)*

Genomic level	Techniques/strategies[a]	References
	Metabolic coupling/ connectivity (GC-TOF-MS, NMR, isotope dilution LC-MS; FT-ICR studies)	Becker *et al.* (2006), Allen *et al.* (2003), Villas-Boas *et al.* (2005), Wu *et al.* (2005a, 2006a, 2006b), O'Hagan *et al.* (2005)
Metabolic control	(MCA, BST, MFA, modular MCA)	Kacser and Burns (1973), Savageau (1976), Varma and Palsson (1994), Wagner (2005), Khan and Westerhoff (1991)
	Metabolome conceptual strategy	Wu *et al.* (2005b)
	Regulation analysis	Ter Kuile and Westerhoff (2001), Rossell *et al.* (2005, 2006)

[a]Abbreviations: ChIP, chromatin immunoprecipitation; SAGE, serial snalysis of gene expression; DD, differential display; QRT-PCR, quantitative real time PCR; 2D-DIGE, 2-dimensional fluorescence difference gel electrophoresis; iTRAQ, isotope tags for multiplexed relative and absolute quantification; GFP, green fluorescent protein; GC-TOF-MS, gas chromatography time-of-flight mass spectrometry; NMR, nuclear magnetic resonance; isotope dilution LC-MS, isotope dilution liquid chromatography/mass spectrometry; FT-ICR, fourier-transform ion cyclotron resonance mass spectrometry; FANCY, functional analysis by co-responses in yeast; MCA, metabolic control analysis; BST, biochemical systems theory; MFA, metabolic flux analysis.

of metabolic control, while keeping the simple mathematical description that is one of the strengths of MCA. In this respect, two recent approaches can be highlighted: a metabolome conceptual strategy (Wu *et al.*, 2005b), and the regulation analysis theory (ter Kuile and Westerhoff, 2001; Rossell *et al.*, 2005, 2006). Both approaches exploit enzyme kinetics (Fersht, 1999), which defines the flux (v_i) through an enzyme i as a function of enzyme activity (e_i) and metabolite concentrations ([S]):

$$v_i = f_i(e_i, [S])$$

In the metabolome strategy, provided that the kinetic parameters of f_i are known, *in vivo* enzyme activities can be calculated from metabolite concentrations and fluxes obtained *in vivo*. This approach can calculate changes of *in vivo* enzyme activities relative to a reference (e.g. mutant versus wild type), which can provide information on the site(s) of action of an altered gene (Wu *et al.*, 2005b).

In the case of the regulation analysis approach, this defines the flux through an enzymatic step as a function of regulation at the gene expression and metabolite levels.

$$v_i = f_i(E_i, M)$$

where E_i = enzyme expression level and M = metabolite regulation.

The basic concepts of regulation analysis are illustrated in Figure 7. Regulation analysis exploits the fact that, in many kinetic models, these two contributions can be easily dissected. For example, in a simple Michaelis–Menten formulation:

$$v_i = (K_{cat}[E]) \times ([S]/(K_M + [S])) \text{ i.e.}$$
$$v_i = f(\text{gene expression}) \times g(\text{metabolic regulation})$$

The theory shows that the total contribution to global flux through each enzymatic step can be expressed as the sum of two main contributions at the gene expression and metabolic levels. From here, after applying simple mathematical transformations, this can be formulated as the "summation theorem for the regulation of flux," which states that, for each enzyme, the sum of the hierarchical regulation coefficient (ρ_h) and the metabolic regulation coefficient (ρ_m) equals unity (Figure 7).

$$\rho_h + \rho_m = 1 \qquad \text{Rossell } et\ al. \text{ (2005)}$$

The use of the term 'hierarchical' to indicate regulation at the level of gene expression may cause confusion. Nonetheless, regulation analysis is compatible with, and complementary to MCA, and modular MCA, and can be applied not only to small changes but also to large changes in flux (Rossell *et al.*, 2005).

Application of regulation analysis to studies on the *S. cerevisiae* glycolytic pathway has revealed the existence of individual steps predominantly regulated at the gene expression (hierarchical) level, whereas others are mainly metabolically regulated, and that the pathway's regulation profile differs radically between two different starvation conditions (Rossel *et al.*, 2005, 2006). These results are in clear agreement with the results presented in this chapter for the case of the lysine biosynthetic pathway (Figure 6), showing the existence of enzymes fundamentally regulated at the level of gene (protein) expression (e.g. Aco1p, Aco2p, Lys4p), whereas other enzymes in the same pathway (e.g. Lys12p) are mainly regulated metabolically.

From these results, metabolic flux regulation is beginning to emerge as a dynamic process in which subtle regulation, at the gene expression and metabolite levels, of selective enzymes may be the origin of the intrinsic adaptability and distributed robustness of the

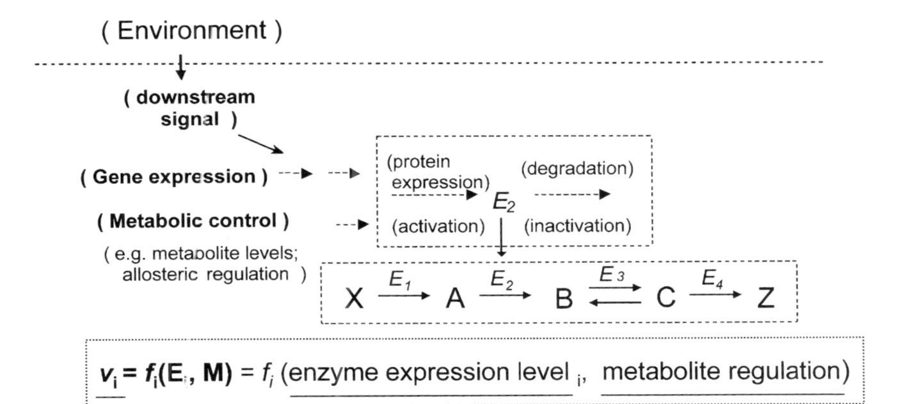

Regulation analysis

v_i = f(gene expression) • g(metabolic control) --> (e.g. $v_2 = \overbrace{(\,Kcat\,[E_2]\,)}^{v_{max2}} \cdot (\,[A]/(K_m + [A])\,)$)

$\ln v_i = \ln f\,(\text{g.exp}) + \ln g\,(\text{metabolic control})$. Since in steady state, $v_i = J$ (flux through the pathway):

$$1 = \Delta\ln v_i / \Delta\ln J = [\,\Delta\,(\ln f\,(\text{g, exp.}) + \ln g\,(\text{metabolic control})\,]\,/\,\Delta\ln J$$

$$= [\,\Delta\ln f\,(\text{g, exp.}) / \Delta\ln J\,] + [\,\Delta\ln g\,(\text{metabolic control}) / \Delta\ln J\,]$$

$$= \quad \rho_h \quad + \quad \rho_m$$

(ρ_h = hierarchical, gene expression regulation coefficient)
(ρ_m = metabolic regulation coefficient)

Summation theorem for the regulation of flux: $\rho_h + \rho_m = 1$ (e.g. provided J and v_{max} (i.e., ρ_h) can be measured, ρ_m can be derived)

Figure 7. Conceptual strategies for studies of control of metabolic fluxes at the gene expression and metabolic level. Case example: Regulation analysis. The flux through an enzymatic step (v_i) can be expressed as a function of gene expression and metabolic regulation, with kinetic models (Fersht, 1999) defining the specific formulation for each particular enzyme. Strategies for *in vivo* estimation of two of these magnitudes (e.g. steady-state fluxes and quantitative changes in metabolites) open the way to the determination of the rest (e.g. gene expression contribution; Wu *et al.*, 2005b). In the regulation analysis theory (ter Kuile and Westerhoff, 2001; Rossell *et al.*, 2005, 2006) the hierarchical gene expression (ρ_h) and metabolic regulation (ρ_m) contributions can be dissected, which results in simple determination of *in vivo* gene expression and metabolic regulation contributions (Rossell *et al.*, 2005, 2006).

eukaryotic cell, allowing it to accommodate both short- and long-term environmental changes (Wagner, 2005). All this emphasizes the need for more dynamic studies at the different levels (Table 1), under defined and well-controlled conditions, to unravel *in vivo* patterns of control of metabolic fluxes, which can be integrated into genome-scale Systems Biology models of the eukaryotic cell.

B. Integration of High-Throughput Metabolic and Protein Expression Studies into Genome-Scale, Systems Biology Models – Applications

The generation of molecular and genome-wide data at different genomic levels is opening the way to new strategies for integration of 'omics' data sets (Joyce and Palsson, 2006). Thus, in a recent study, an integrated metabolic and regulatory network model allowed the *in silico* prediction of growth phenotypes of transcription factor knockout yeast strains (Herrgard *et al.*, 2006). In metabolic control studies, however, the ultimate objective is the integration of experimental data under an appropriate control framework (e.g. regulation analysis), which allows the investigation of enzymatic steps and interactions relevant to flux control, and their hierarchical and metabolic contribution to the *in vivo* metabolic fluxes under different conditions. From this perspective, metabolic network models (e.g. Förster *et al.*, 2003; Duarte *et al.*, 2004) need to evolve from basic metabolic descriptions to incorporate progressively the contribution of cellular regulatory modules such as signal transduction and the regulation of gene expression at the (epi)genomic, transcriptomic and proteomic levels (i.e. hierarchical regulation) and true kinetic metabolic models (Fersht, 1999), towards a more realistic description of the eukaryotic cell (Figure 2).

Among some of the most relevant modelling approaches and initiatives incorporating modularity (regulatory modules) representative of the intracellular complexity are: approaches on modelling signal transduction pathways and gene expression at the transcriptional level (Khan and Westerhoff, 1991; Kofahl and Klipp, 2004; Hornberg *et al.*, 2005; Klipp *et al.*, 2005; Nordlander *et al.*, 2005) and the Silicon Cell initiative, which is a modular approach towards a detailed description at the cellular level (Snoep, 2005; Snoep *et al.*, 2006). As an earnest of their importance, these and future systems biology models are due to comply with the standard for defining biochemical models: Minimum Information Requested In the Annotation of biochemical Models (MIRIAM: Le Novere *et al.*, 2005). They should also be encoded in SBML – the Systems Biology Markup Language (Hucka *et al.*, 2003; http://www.sbml.org).

In this chapter, we have analysed the potential of coupling high-throughput transcriptional and proteomic studies to chemostat culture in order to reveal *in vivo* patterns of multiple enzyme regulation during cell growth. We have also examined the currently available high-throughput metabolomic strategies, which can be used

Table 2. Amino acid biosynthetic enzymes with protein expression levels consistently up- and down-regulated with growth rate under nutrient-limiting conditions (C-, N-, P- and S-limiting conditions; synthetic medium with glucose as carbon source and ammonium as nitrogen source)

Amino acid biosynthetic pathway	Enzymes[a]	
	Up-regulated	Down-regulated
Cysteine, homocysteine, methionine and sulphur compounds	Ecm17p, Met13p, Sam2p, Met6p, Ado1p	Sam1p
Glutamine	Gln1p	
Lysine	Aco1p, Aco2p, Lys2p, Lys4p	

[a]Nomenclature: Aco1p, aconitase; Aco2p, putative aconitase isozyme; Ado1p, adenosine kinase; Ecm17p, sulphite reductase beta subunit; Gln1p, glutamine synthetase; Lys2p, alpha aminoadipate reductase; Lys4p, homoaconitase; Met6p, methionine synthase; Met13p, methylenetetrahydrofolate reductase isozyme; Sam1p, *S*-adenosylmethionine synthetase isozyme; Sam2p, *S*-adenosylmethionine synthetase isozyme.

together with them for exhaustive studies on metabolic control patterns *in vivo*. From this perspective, and on the basis of the modular description of the eukaryotic cell presented in Figure 2, we submit the integration of kinetic data from quantitative metabolomic and proteomic analyses, the 'omic' levels most closely associated with function, as the most comprehensive way of incorporating metabolome studies into genome-scale, system biology models.

The strategies and approaches presented here open the way to advanced studies on metabolic control as well as to the design of comprehensive metabolic engineering strategies (Stephanopoulos, 1999). Thus, as a relevant example, groups of amino acid biosynthetic enzymes significantly up- and down-regulated with growth rate irrespective of the nutrient-limiting condition, extracted from proteomics ANOVA analysis, are shown in Table 2. These are enzymes that appear consistently up- or down-regulated with cell growth rate at the protein expression (i.e. hierarchical) level. Such enzymatic steps may constitute attractive targets for metabolic engineering towards increased fluxes of amino acid synthesis.

Acknowledgments

This work was supported by grants from BBSRC's Investigating Gene Function Initiative for COGEME (Consortium for the Functional Genomics of Microbial Eukaryotes; http://www.cogeme.manchester.ac.uk/), BBSRC Project Grants, and a grant from the Wellcome Trust (all to SGO). We thank Dr. A. Hayes, Dr. L. Zeef and Mrs. L. Wardleworth (Microarray Facility; Faculty of Life Sciences, University of Manchester) for their help on microarray analyses and the processing of transcriptional data; Dr. K. S. Lilley and the Cambridge Centre for Proteomics (University of Cambridge, UK), for the proteomics work; Prof. D. Kell, Dr. W. B. Dunn and Dr. M. Brown (Manchester Centre for Integrative Systems Biology, University of Manchester) for analyses of the metabolome data, and Dr. D. Hoyle (Northwest Institute for Bio-Health Informatics, NIBHI) for his guidance on statistical analysis of 'omic' data. This is a contribution from the Centre for the Analysis of Biological Complexity (CABC; Faculty of Life Sciences, University of Manchester) and the Manchester Centre for Integrative Systems Biology (http://www.mcisb.org/).

References

Ainscow, E. K. and Brand, M. D. (1999). Top-down control analysis of ATP turnover, glycolysis and oxidative phosphorylation in rat hepatocytes. *Eur. J. Biochem.* **263**, 671–685.

Allen, J., Davey, H. M., Broadhurst, D., Heald, J. K., Rowland, J. J., Oliver, S. G. and Kell, D. B. (2003). High-throughput classification of yeast mutants for functional genomics using metabolic footprinting. *Nat. Biotechnol.* **21**, 692–696.

Becker, S. A., Price, N. D. and Palsson, B. O. (2006). Metabolite coupling in genome-scale metabolic networks. *BMC Bioinform.* **7**, 111.

Balaji, S., Babu, M. M., Iyer, L. M., Luscombe, N. M. and Aravind, L. (2006). Comprehensive analysis of combinatorial regulation using the transcriptional regulatory network of yeast. *J. Mol. Biol.* **360**, 213–227.

Beynon, R. J., Doherty, M. K., Pratt, J. M. and Gaskell, S. J. (2005). Multiplexed absolute quantification in proteomics using artificial QCAT proteins of concatenated signature peptides. *Nat. Methods* **2**, 587–589.

Bost, B., Dillmann, C. and de Vienne, D. (1999). Fluxes and metabolic pools as model traits for quantitative genetics. I: The L-shaped distribution of gene effects. *Genetics* **153**, 2001–2012.

Bruggeman, F. J., Westerhoff, H. V., Hoek, J. B. and Kholodenko, B. N. (2002). Modular response analysis of cellular regulatory networks. *J. Theor. Biol.* **218**, 507–520.

Castrillo, J. I. and Oliver, S. G. (2005). Towards integrative functional genomics using yeast as a reference model. In: *Metabolome Analyses: Strategies for Systems Biology* (S. Vaidyanathan, G. G. Harrigan and R. Goodacre, eds), pp. 9–29. Springer, New York.

Castrillo, J. I. and Oliver, S. G. (2006). Metabolomics and Systems Biology in *Saccharomyces cerevisia*e. In: *The Mycota. Vol. XIII. Fungal Genomics-* (K. Esser and A. Brown, eds), pp. 1–18. Springer, New York.

Choudhuri, S. (2004). The nature of gene regulation. *Int. Arch. Biosci.*, 1001–1015.

Colman-Lerner, A., Gordon, A., Serra, E., Chin, T., Resnekov, O., Endy, D., Pesce, C. G. and Brent, R. (2005). Regulated cell-to-cell variation in a cell-fate decision system. *Nature* **437**, 699–706.

Duarte, N. C., Herrgard, M. J. and Palsson, B. O. (2004). Reconstruction and validation of *Saccharomyces cerevisiae* iND750, a fully compartmentalized genome-scale metabolic model. *Genome Res.* **14**, 1298–1309.

Ezhkova, E. and Tansey, W. P. (2006). Chromatin immunoprecipitation to study protein–DNA interactions in budding yeast. *Methods Mol. Biol.* **313**, 225–244.

Fell, D. A. (1997). *Understanding the Control of Metabolism*. Portland Press Ltd, London.

Fell, D. A. and Thomas, S. (1995). Physiological control of metabolic flux: the requirement for multisite modulation. *Biochem. J.* **311**, 35–39.

Fersht, A. (1999). *Structure and Mechanism in Protein Sciensb: A Guide to Enzyme Catalysis and Protein Folding*. Freeman, New York.

Förster, J., Famili, I., Fu, P., Palsson, B. O. and Nielsen, J. (2003). Genome-scale reconstruction of the *Saccharomyces cerevisiae* metabolic network. *Genome Res.* **13**, 244–253.

Hayes, A., Zhang, N., Wu, J., Butler, P. R., Hauser, N. C., Hoheisel, J. D., Lim, F. L., Sharrocks, A. D. and Oliver, S. G. (2002). Hybridization array technology coupled with chemostat culture: tools to interrogate gene expression in *Saccharomyces cerevisiae*. *Methods* **26**, 281–290.

Herrgard, M. J., Lee, B. S., Portnoy, V. and Palsson, B. O. (2006). Integrated analysis of regulatory and metabolic networks reveals novel regulatory mechanisms in *Saccharomyces cerevisiae*. *Genome Res.* **16**, 627–635.

Hornberg, J. J., Binder, B., Bruggeman, F. J., Schoeberl, B., Heinrich, R. and Westerhoff, H. V. (2005). Control of MAPK signalling: from complexity to what really matters. *Oncogene* **24**, 5533–5542.

Hucka, M., Finney, A., Sauro, H. M., Bolouri, H., Doyle, J. C., Kitano, H., Arkin, A. P., Bornstein, B. J., Bray, D., Cornish-Bowden, A., Cuellar, A. A., Dronov, S., Gilles, E. D., Ginkel, M., Gor, V., Goryanin, I. I., Hedley, W. J., Hodgman, T. C., Hofmeyr, J. H., Hunter, P. J., Juty, N. S., Kasberger, J. L., Kremling, A., Kummer, V., Le Novere, N., Loew, L. W., Lucio, D., Mendes, P., Minch, E., Mjolsness, E. D., Nakayama, Y., Nelson, M. R., Nielsen, P. F., Sakurada, T., Schaff, J. C., Shapiro, B. E., Shimizu, T. S., Spence, H. D., Stelling, J., Takahashi, K., Tomita, M., Wagner, J. and Wang, J. (2003). The systems biology markup language (SBML): a medium for representation and exchange of biochemical network models. *Bioinformatics* **19**, 524–531.

Joyce, A. R. and Palsson, B. O. (2006). The model organism as a system: integrating 'omics' data sets. *Nat. Rev. Mol. Cell Biol.* **7**, 198–210.

Kacser, H. and Burns, J. A. (1973). The control of flux. *Symp. Soc. Exp. Biol.* **27**, 65–104.

Kahn, D. and Westerhoff, H. V. (1991). Control theory of regulatory cascades. *J. Theor. Biol.* **153**, 255–285.

Kell, D. B. and Westerhoff, H. V. (1990). Metabolic control analysis: theory and practice. In: *Mixed and Multiple Substrates and Feedstocks* (G. Hamer, T. Egli and M. Snozzi, eds), pp. 107–119. Hartung-Gorre, Konstanz.

Kell, D. B., Brown, M., Davey, H. M., Dunn, W. B., Spasic, I. and Oliver, S. G. (2005). Metabolic footprinting and systems biology: the medium is the message. *Nat. Rev. Microbiol.* **3**, 557–565.

Klipp, E., Herwig, R., Kowald, A., Wierling, C. and Lehrach, H. (2005). *Systems Biology in Practisb: Concepts, Implementation and Application*. Wiley-VCH, Weinheim.

Kofahl, B. and Klipp, E. (2004). Modelling the dynamics of the yeast pheromone pathway. *Yeast* **21**, 831–850.

Krauss, S. and Quant, P. A. (1996). Regulation and control in complex, dynamic metabolic systems: experimental application of the top-down approaches of metabolic control analysis to fatty acid oxidation and ketogenesis. *J Theor. Biol.* **182**, 381–388.

Le Novere, N., Finney, A., Hucka, M., Bhalla, U. S., Campagne, F., Collado-Vides, J., Crampin, E. J., Halstead, M., Klipp, E., Mendes, P. *et al.*

(2005). Minimum information requested in the annotation of biochemical models (MIRIAM). *Nat. Biotechnol.* **23**, 1509–1515.

Lilley, K. S. and Friedman, D. B. (2004). All about DIGE: quantification technology for differential-display 2D-gel proteomics. *Expert Rev. Proteomics* **1**, 401–409.

Newman, J. R., Ghaemmaghami, S., Ihmels, J., Breslow, D. K., Noble, M., Derisi, J. L. and Weissman, J. S. (2006). Single-cell proteomic analysis of *S. cerevisiae* reveals the architecture of biological noise. *Nature* **441**, 840–846.

Nielsen, J. and Oliver, S. (2005). The next wave in metabolome analysis. *Trends Biotechnol.* **23**, 544–546.

Nordlander, B., Klipp, E., Kofahl, B. and Hohmann, S. (2005). Modelling signalling pathways – a yeast approach. In: *Systems Biology. Definitions and Perspectives*, (L. Alberghina, H. V. Westerhoff, eds), In Series *Topics in Current Genetics* (S. Hohmann, series editor), pp. 277–302. Springer, Heidelberg.

O'Hagan, S., Dunn, W. B., Brown, M., Knowles, J. D. and Kell, D. B. (2005). Closed-loop, multiobjective optimization of analytical instrumentation: gas chromatography/time-of-flight mass spectrometry of the metabolomes of human serum and of yeast fermentations. *Anal. Chem.* **77**, 290–303.

Peletier, M. A., Westerhoff, H. V. and Kholodenko, B. N. (2003). Control of spatially heterogeneous and time-varying cellular reaction networks: a new summation law. *J. Theor. Biol.* **225**, 477–487.

Raamsdonk, L. M., Teusink, B., Broadhurst, D., Zhang, N., Hayes, A., Walsh, M. C., Berden, J. A., Brindle, K. M., Kell, D. B., Rowland, J. J., Westerhoff, H. V., van Dam, K. and Oliver, S. G. (2001). A functional genomics strategy that uses metabolome data to reveal the phenotype of silent mutations. *Nat. Biotechnol.* **19**, 45–50.

Ramanathan, S. and Swain, P. S. (2005). Tracing the sources of cellular variation. *Dev. Cell.* **9**, 576–578.

Raser, J. M. and O'Shea, E. K. (2004). Control of stochasticity in eukaryotic gene expression. *Science* **304**, 1811–1814.

Raser, J. M. and O'Shea, E. K. (2005). Noise in gene expression: origins, consequences, and control. *Science* **309**, 2010–2013.

Ross, P. L., Huang, Y. N., Marchese, J. N., Williamson, B., Parker, K., Hattan, S., Khainovski, N., Pillai, S., Dey, S., Daniels, S., Purkayastha, S., Juhasz, P., Martin, S., Bartlet-Jones, M., He, F., Jacobson, A. and Pappin, D. J. (2004). Multiplexed protein quantitation in *Saccharomyces cerevisiae* using amine-reactive isobaric tagging reagents. *Mol. Cell. Proteomics* **3**, 1154–1169.

Rossell, S., van der Weijden, C. C., Kruckeberg, A. L., Bakker, B. M. and Westerhoff, H. V. (2005). Hierarchical and metabolic regulation of glucose influx in starved *Saccharomyces cerevisiae*. *FEMS Yeast Res.* **5**, 611–619.

Rossell, S., van der Weijden, C. C., Lindenbergh, A., van Tuijl, A., Francke, C., Bakker, B. M. and Westerhoff, H. V. (2006). Unraveling the complexity of flux regulation: a new method demonstrated for nutrient starvation in *Saccharomyces cerevisiae*. *Proc. Natl. Acad. Sci. USA* **103**, 2166–2171.

Savageau, M. A. (1976). *Biochemical Systems Analysis: A Study of Function and Design in Molecular Biology.* Addison-Wesley, Reading, MA.

Snoep, J. L. (2005). The silicon cell initiative: working towards a detailed kinetic description at the cellular level. *Curr. Opin. Biotechnol.* **16**, 336–343.

Snoep, J. L., Bruggeman, F., Olivier, B. G. and Westerhoff, H. V. (2006). Towards building the silicon cell: a modular approach. *Biosystems* **83**, 207–216.

Stephanopoulos, G. (1999). Metabolic fluxes and metabolic engineering. *Metab. Eng.* **1**, 1–11.

ter Kuile, B. H. and Westerhoff, H. V. (2001). Transcriptome meets metabolome: hierarchical and metabolic regulation of the glycolytic pathway. *FEBS Lett.* **500**, 169–171.

Teusink, B., Baganz, F., Westerhoff, H. V. and Oliver, S. G. (1998). Metabolic control analysis as a tool in the elucidation of the function of novel genes. In: *Methods in Microbiology Yeast Gene Analysis*, Vol. 26 (A. J. P. Brown and M. F. Tuite, eds), pp. 297–336. Academic Press, London.

Thomas, S. and Fell, D. A. (1998). The role of multiple enzyme activation in metabolic flux control. *Adv. Enzyme Reg.* **38**, 65–85.

Vaidyanathan, S., Harrigan, G. G. and Goodacre, R. (2005). *Metabolome Analyses: Strategies for Systems Biology.* Springer, New York.

Varma, A. and Palsson, B. O. (1994). Metabolic flux balancing: Basic concepts, scientific and practical use. *Bio/Technology* **12**, 994–998.

von Mering, C., Krause, R., Snel, B., Cornell, M., Oliver, S. G., Fields, S. and Bork, P. (2002). Comparative assessment of large-scale data sets of protein–protein interactions. *Nature* **417**, 399–403.

Villas-Boas, S. G., Moxley, J. F., Akesson, M., Stephanopoulos, G. and Nielsen, J. (2005). High-throughput metabolic state analysis: the missing link in integrated functional genomics of yeasts. *Biochem. J.* **388**, 669–677.

Wagner, A. (2005). *Robustness and Evolvability in Living Systems. Princeton Studies in Complexity.* Princeton University Press, Princeton, NJ.

Westerhoff, H. V., van Heeswijk, W., Kahn, D. and Kell, D. B. (1991). Quantitative approaches to the analysis of the control and regulation of microbial metabolism. *Antonie van Leeuwenhoek* **60**, 193–207.

Wilson, I. M., Davies, J. J., Weber, M., Brown, C. J., Alvarez, C. E., MacAulay, C., Schubeler, D. and Lam, W. L. (2006). Epigenomics: mapping the methylome. *Cell Cycle* **5**, 155–158.

Wu, L., Mashego, M. R., van Dam, J. C., Proell, A. M., Vinke, J. L., Ras, C., van Winden, W. A., van Gulik, W. M. and Heijnen, J. J. (2005a). Quantitative analysis of the microbial metabolome by isotope dilution mass spectrometry using uniformly ^{13}C-labeled cell extracts as internal standards. *Anal. Biochem.* **336**, 164–171.

Wu, L., van Winden, W. A., van Gulik, W. M. and Heijnen, J. J. (2005b). Application of metabolome data in functional genomics: a conceptual strategy. *Metab. Eng.* **7**, 302–310.

Wu, L., van Dam, J., Schipper, D., Kresnowati, M. T., Proell, A. M., Ras, C., van Winden, W. A., van Gulik, W. M. and Heijnen, J. J. (2006a). Short-term metabolome dynamics and carbon, electron, and ATP balances in chemostat-grown *Saccharomyces cerevisiae* CEN PK 113-7D following a glucose pulse. *Appl. Environ. Microbiol.* **72**, 3566–3577.

Wu, L., Mashego, M. R., Proell, A. M., Vinke, J. L., Ras, C., van Dam, J., van Winden, W. A., van Gulik, W. M. and Heijnen, J. J. (2006b). *In vivo* kinetics of primary metabolism in *Saccharomyces cerevisiae* studied through prolonged chemostat cultivation. *Metab. Eng.* **8**, 160–171.

22 Phylogenetic Footprinting

Paul F Cliften
Department of Biology, Utah State University, 5305 Old Main Hill, Logan, UT 84322, USA

◆◆◆

CONTENTS

Introduction
Genomic resources
Algorithms for phylogenetic footprinting
Footprinting yeast genes

Abbreviations

BLAST	basic local alignment sequence tool
$NADP^+$	nicotinamide adenine dinucleotide phosphate, oxidized form
NAD^+	nicotinamide adenine dinucleotide, oxidized form
OOPS	one occurrence per sequence
ZOOPS	zero or one occurrence per sequence
MSA	multiple sequence alignment

◆◆◆◆◆◆ I. INTRODUCTION

Phylogenetic footprinting is a method for identifying functional elements in a genome by their conservation across evolutionary lineages. This approach facilitates identification of any functional sequence, but is most often used to identify sequence elements that regulate gene expression. Its name originated from the small regions of conservation that were detected in early comparisons of orthologous mammalian promoters (Tagle *et al.*, 1988; Gumucio *et al.*, 1992, 1993). Like the footprints made *in vitro* by DNA-binding proteins, phylogenetic footprints often correspond to sequences that are bound by transcription factors.

In this chapter, we will focus on the use of phylogenetic footprinting to identify gene regulatory elements in *Saccharomyces cerevisiae*. My objective is to make phylogenetic footprinting more accessible and more universally employed in the common yeast lab that does not have expertise in bioinformatics. This is important

METHODS IN MICROBIOLOGY, VOLUME 36
0580-9517 DOI:10.1016/S0580-9517(06)36022-9

because there is a tremendous amount of sequence data available from other *Saccharomyces* species that could greatly inform our knowledge of yeast gene regulation. However, our knowledge of yeast gene regulatory elements is insufficient to train computational methods to differentiate between real and false binding sites, and computational methods cannot distinguish between regulatory elements and conserved sequence elements that have other functions. It is important to engage bench scientists who understand the biology of the organism in this kind of sequence analysis, because experimental approaches will be necessary for computational comparisons such as phylogenetic footprinting to contribute to our understanding of yeast gene regulation.

Our current understanding of yeast gene regulation is somewhat rudimentary. On one hand, we know a large amount about a handful of specific gene promoters and transcription factors. On the other, our knowledge is biased toward transcription factors that regulate many genes or that heavily modify the expression of the genes they control. For instance, we know a great deal about Mbp1, the MluI-box binding protein that helps to control the cell cycle and thus regulates many genes (Verma *et al.*, 1991). We also know much about Gal4, which strongly activates expression of about 10 genes that are required for galactose metabolism (Ren *et al.*, 2000). However, our experimental and computational approaches have likely overlooked many regulatory factors that subtly affect the transcription of a handful of genes. Many different experimental and computational techniques will be required to decipher the regulation of each yeast gene. Phylogenetic footprinting is one powerful tool that can be applied in every yeast lab with a computer and access to the Internet.

The use of phylogenetic footprinting to identify gene regulatory sequences is still in its infancy. Investigators have been using sequence comparisons for many years to identify genes and the functional sequences of proteins. The idea of using phylogenetic comparisons to identify sequences regulating gene expression has been around for about 20 years, since the term 'phylogenetic footprint' was coined in the late 1980s (Tagle *et al.*, 1988). However, the data required for developing the technique have only recently become available, and it has not been fully exploited.

There are two major considerations for phylogenetic footprinting. First, it is critical to compare genomes of species that are appropriately diverged. The greater the evolutionary distance between the compared sequences, the more the non-functional sequences will have diverged, increasing the definition of the conserved sequences. However, as the sequences to be compared become more diverged, it becomes increasingly difficult to align them, making it difficult to identify orthologous conserved sequences. Second, orthologous genes from the species being compared must be regulated similarly for phylogenetic footprinting to be successful. We do not know how widely gene regulation is conserved through evolution, but the more related two species are the more likely their orthologous genes will be regulated similarly. One approach has been to compare genome sequences of multiple closely related species to provide enough cumulative genetic distance for non-functional sequences to diverge. But the chance that an orthologous gene is regulated differently in one of the species increases with the number of species that are used for the comparison. Addressing these basic questions will require more experience with phylogenetic footprinting and, ultimately, experimental results.

The potential and the limitations of phylogenetic footprinting have not been fully explored, but the following examples provide much optimism for the potential to apply this method productively. Here we provide two examples that help to illustrate the diverse potential of phylogenetic footprinting. The first is of the promoter of a weakly expressed gene, *GAL4*, which is repressed approximately five-fold by the Mig1 repressor. A thorough genetic analysis of this promoter identified its functional regions (Griggs and Johnston, 1993). Nearly 10 years after that analysis, we determined the phylogenetic footprints of the *GAL4* promoter using several closely related *Saccharomyces* species (Cliften *et al.*, 2001). The conserved sequences we identified were strikingly coincident to the functional regions identified from the genetic analysis (see Figure 1). The amount of time and effort that went into the phylogenetic analysis was much less than was required for the genetic analysis. Had we done the phylogenetic footprinting first, the genetic analysis would have been much easier (and cheaper). Thus, phylogenetic footprinting has great potential for guiding design of experiments to reveal functional sequences. Currently, there are ample sequence data available to be able to shed light on the regulation of *every* yeast gene.

Another illustrative example involves the promoter of *YDR374C*, a gene of unknown function. A phylogenetic footprint of this promoter revealed three obvious conserved sequence motifs (see Figure 2) that are canonical binding sites for Reb1, Ume6, and Ndt80. Since Ume6 and Ndt80 are transcription factors that regulate expression of genes involved in sporulation, it seems likely that *YDR374C* functions in this process. Thus, phylogenetic footprinting can provide clues to the function of genes. These two examples aptly demonstrate the ability of the phylogenetic footprinting to inform our knowledge of gene regulation and gene function.

◆◆◆◆◆◆ II. GENOMIC RESOURCES

Genome sequence data of many yeast and other fungi is available for phylogenetic comparisons with *S. cerevisiae*. Low coverage genome survey sequences are available from 13 different hemiascomycete species that span the entire clade (Souciet *et al.*, 2000). The complete genome sequences of four of these species – *Candida glabrata*,

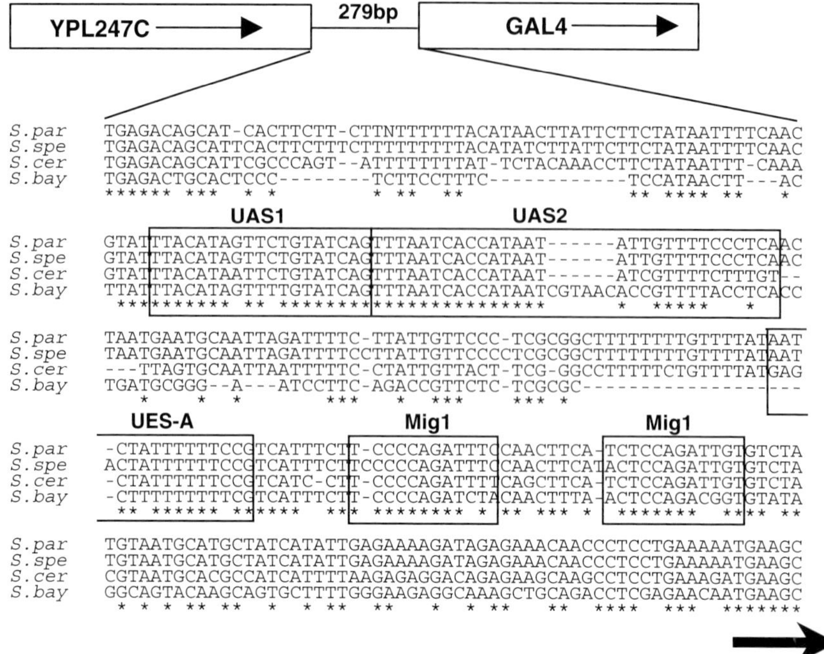

Figure 1. A multispecies alignment of the *GAL4* promoter region. Functional promoter elements identified experimentally by Griggs and Johnston (1993) are boxed and labeled. UAS and UES refer to upstream activator and upstream essential sequence respectively. The alignment begins with the stop codon (TGA) of *YPL247C* and includes the first 7 nucleotides of *GAL4*. The arrow indicates the *GAL4* coding region. The species within the alignment are abbreviated; they include *S. paradoxus, S. cerevisiae, S. bayanus* and an unknown *Saccharomyces* species.

```
                                                Reb1
S.kud   CCAAA-GCATCTAGGATAAATAAGATGTGAATGTATTACCCGTTTT-GTATTCAAGATCACCTC
S.mik   TCTGA-GCAACCAAAAATAAACAGTTCAAGTGTTGCTACCCGTTTTTGCAGTTAAGATCACTTA
S.cer   CTTG--GTGACCGAAAATAGACACG----AAATCGCTACCCGTTTC--CCCAGAATATCACTCC
S.bay   CCTAAAGTAAACAAGAATAAATATACTGCATGGGGCTACCCGTT-C--CATATGATATCATCGG

                Ume6                        Ndt80
S.kud   TCACGGAGGGGTTTCGGCGGCTAATCGTTATTAG-CGCCTTTTGTGATATGCGTATAAATAAAG
S.mik   CCACGGATAAGTATCGGCGGCTAATCCTCATGGGACGCCTTTTGTGATATATAAATACATGCAT
S.cer   TCACG-ATGTACCTCGGCGGCTAATCTTTTTGGTA-GCCTTTTGTGATATATATATAAATAAAT
S.bay   TCACG-AAGTG--TCGGCGGCTAAT--TTAGAGTACGCCTTTTGTGATATATATATA-------

S.kud   T---------------GACT-ACTTCTAGCTTCAAAAAATTGC----TTACTGCTATACCCCTC-
S.mik   C---------------TAGTGAAACCTTTTCTTCAAAATTCAC----TCGCTG----ACTATAA-
S.cer   AAGTATACATACATATATATATATATATATTTATACAGCTACATTGTTTTCCTCCAAAATTTTC
S.bay   ---------------TATATATATATACATAGAATGAACTAC--CGCTATTTTAAAACTCTTTT

S.kud   -----GCTCTAA---GCGC--GAAGTTTCAAAATTGTCTGTTCTACCATTCCTTGGTTAAGAAA
S.mik   -----GCCCCAA---ACA---GAAGCTTTAAAACTACGTATTCTACTACTAATTGATTAGAAAA
S.cer   TGTTGGTTATGAATCGCAAAAGAAGTTTTCAGATTGTGTCCTCTGTTACTATTTCGTTAAGAAA
S.bay   TGGTGGCTATGA-TTGCAGAAAAAGTGTCTA-ATAATAAG-TGTGTT-CTGTCACTTTGAGAAA

S.kud   ------------ATACTGCTAGG-GTGGTGTGAACATTGTCTTGT--GCTTGAGAAAATG
S.mik   T-ATCACTTCATACACGGTTGAA-GTGGCTTAAGCATTGT-TTGT--GCTTGAAAAAATG
S.cer   GGAAGATATCGTCTACGGCTGGT-GTGACGTAAGTATTGCGTTGT--GCTCTAAAA-ATG
S.bay   G-AATATTGCATATACGGTAAACAGTGGTGTGAGCTTTCTATTTTTTATTTTAAGAAATG
```

Figure 2. A multispecies alignment of the *YDR374C* promoter region. Known sequence motifs are boxed and labeled with the transcription factor that recognizes the motif. The start of *YPL247C* is indicated by an arrow. The aligned sequences are from *S. kudriavzevii, S. mikatae, S. cerevisiae* and *S. bayanus*.

Kluyveromyces lactis, Debaryomyces hansenii, and *Yarrowia lipolitica* – are available (Dujon *et al.*, 2004), as is sequence of the filamentous fungus *Ashbya gossypii* (Brachat *et al.*, 2003; Dietrich *et al.*, 2004). It should be noted that *Candida glabrata* is misclassified: it diverged from *S. cerevisiae* after the whole genome duplication that occurred in this lineage (Dujon *et al.*, 2004), and is thus more closely related to *S. cerevisiae* than is *S. kluyveri,* which clearly diverged from other *Saccharomyces* species before the whole genome duplication (Cliften *et al.*, 2006). Partial genome sequences are available for several species more closely related to *S. cerevisiae,* including the *sensu stricto* species *S. paradoxus, S. mikatae, S. kudriavzevii,* and *S. bayanus,* and the more distantly related *sensu lato* species *S. castellii* and the petite negative species *S. kluyveri*. The sequences of these six *Saccharomyces* species are draft sequences with 6–8-fold coverage (produced at the Broad Institute, Kellis *et al.*) and from 2–4-fold coverage combined with some directed sequence "finishing" (produced at Washington University; Cliften *et al.*, 2003, 2006).

While *Schizosaccharomyces pombe* and many hemiascomycete fungi are too distantly related to *S. cerevisiae* to be useful for promoter comparisons, the genome sequences of the *Saccharomyces* species provide very informative comparisons. The most useful comparisons for *S. cerevisiae* promoters are usually with the *Saccharomyces sensu stricto* species that are relatively closely related to *Saccharomyces cerevisiae.*

All of the sequence data from these projects are available at Genbank. BLAST searches are usually sufficient for identifying and obtaining the relevant sequences. An assembly of the *S. bayanus* genome sequence that includes data from two sequencing centers (The Broad Institute and Washington University) is available; a joint assembly of the *S. mikatae* sequence data was hindered by genetic differences in the strains sequenced by the two groups. Advanced options on the NCBI BLAST server allow the user to limit the search to the fungal kingdom, thus restricting the number of homologous sequences that are returned from the search. There are other sources of the genome sequences of fungal species that often provide enhanced access to specific portions of the data. A fungal genome resource at Duke University "http://fungal.genome.duke.edu/" allows BLAST searches limited to ascomycete species (see Table 1). Several fungal genome sequences are available at the *Saccharomyces* Genome Database (SGD). The "Fungal BLAST" page "http://seq.yeastgenome.org/cgi-bin/blast-fungal.pl" allows the user to limit the search to specific species. All pertinent DNA sequence can be identified through BLAST searches at SGD and then downloaded for more specific comparisons.

Although the sequence data is freely available, it is not centralized. For instance, there is currently no convenient list of all orthologous gene sequences from each species. Where orthologous sequences have been tabulated, these are usually limited to one or a few species that are the subject of specific studies or databases.

Table 1. List of private Internet databases containing genomic data for *Saccharomyces* and closely related species

Institution	Web location (URL)	Species
Ashbya Genome Database	http://agd.unibas.ch	*Ashbya gossypii*
Broad Institute	http://www.broad.mit.edu/annotation/fungi/fgi	Many fungal species
Duke University	http://fungal.genome.duke.edu	Various fungal species
Genolevures	http://cbi.labri.fr/Genolevures/index.php	*Hemiascomycete* species
SGD (Saccharomyces Genome Database)	http://www.yeastgenome.org	Various fungal species
Washington University	http://genomeold.wustl.edu/projects/yeast/	*Saccharomyces* species

Much of the problem lies in the difficulty in inferring orthologs between species with incomplete genome sequences, due to gene duplications and large gene families. The difficulty of this task is exacerbated in the *Saccharomyces* lineage, which contains duplicated genes. Many of these duplicated genes do not have a one-to-one orthologous relationship to genes in other species. Once orthologous genes are identified, it would be useful to have their upstream sequences available. Some intergenic sequence is available from the Broad Institute and from Washington University (see Table 1) but the data are limited to that produced at each sequencing center, and the data is provided *en masse*. In other words, one must download the entire data set to obtain the sequence upstream of a gene of interest. Even when orthologous sequences are readily available, one should be somewhat skeptical of the orthology assignments provided by a third party because they can be misleading for a number of reasons, including incomplete data or incorrect assumptions. Automated methods for assigning orthology usually involve several general assumptions that do not necessarily hold for any particular gene of interest. Worse, the assumptions are usually unavailable or are difficult to find.

♦♦♦♦♦♦ III. ALGORITHMS FOR PHYLOGENETIC FOOTPRINTING

Algorithms for phylogenetic footprinting are of two different classes: alignment methods and motif detection methods.

Alignment methods such as BLAST (Altschul *et al.*, 1990: for two sequences) or ClustalW (Thompson *et al.*, 1994: for two or more sequences) typically require sequences that are at least 60% identical. In these cases the sequences are clearly related and the alignment reveals the sequences that are most highly conserved within the related sequences. Motif-finding algorithms, on the other hand, were not designed for phylogenetic footprinting, but were optimized to find sequence motifs that are present in multiple unrelated sequences. They are most often used to search for motifs upstream of co-regulated genes. In this case, the sequences being compared are all unrelated (except for paralogs); the algorithm assumes an average identity of about 25% (which would be present between random sequences). The motif-finding algorithms are effective in finding sequence motifs conserved in related sequences, but they overestimate the significance of the identified motifs, since they assume that the input sequences are unrelated. Several new, second generation motif-finding tools account for the relatedness of the sequences in reporting the statistical significance for finding the motifs (Blanchette *et al.*, 2002; Blanchette and Tompa, 2003; Wang and Stormo, 2003). However, it is likely that differences in the discovered sequence motifs are more attributable to the underlying algorithms than to the methods that they use to determine statistical significance.

A. Pair-wise Alignment Methods

The Smith–Waterman algorithm (Smith and Waterman, 1981) is the method of choice for aligning two sequences for phylogenetic footprinting, because it is guaranteed to give the best possible alignment. Other well-known alignment programs such as BLAST and FASTA (Pearson, 1990) rapidly produce an alignment, but it is not necessarily the best alignment. These programs were designed for iterative searches involving large datasets where speed is more important than the accuracy of the alignment. BLAST or FASTA are sufficient for most comparisons, but if a Smith–Waterman alignment tool is available it is worth using to ensure that the alignments are optimal. The Needleman–Wunsch algorithm (Needleman and Wunsch, 1970), which is closely related to the Smith–Waterman algorithm, produces a global alignment, whereas Smith–Waterman produces the best local alignment. If the two sequences are similar over their entire lengths, the two algorithms will produce similar if not identical alignments. However, if the similarity of the two sequences is confined to a specific portion of the sequences, the Smith–Waterman algorithm will produce an alignment of just the similar region. Implementations of the two algorithms are available in the Wisconsin Package from the Accelrys GCG Corporation. BESTFIT uses the Smith–Waterman algorithm; GAP uses the Needleman–Wunsch algorithm.

B. Multiple Sequence Alignments

Although optimal pair-wise alignments are straightforward to produce, there usually is not enough information in the alignment to precisely locate conserved motifs. This is because alignment tools cannot align DNA sequences that are much less than 60% identical. Thus, sequences must be fairly closely related to be accurately aligned. On the other hand, closely related sequences are often too similar to provide definition of functional sequence motifs in pair-wise alignments. Multiple sequence alignments (MSAs) are, therefore, usually required to provide the cumulative sequence divergence necessary to reveal the conserved sequences. However, optimal MSAs are much more difficult to achieve than pair-wise alignments. MSA tools therefore use heuristics to produce very good alignments in a much shorter time than it would take to guarantee the optimal sequence alignment. The key here is to remember that the alignment is not necessarily the best alignment. Additionally, several equally valid alignments may exist in addition to the alignment that is generated, but only one of several equally scoring alignments will be displayed. It is often advisable to view alterations of the alignment and alternative alignments created with different software (e.g., see Figure 3). In this example, the well-defined motif bound by *RPN4* (Mannhaupt *et al.*, 1999) is split by a gap. However, a slightly higher scoring alignment occurs when the gap is shifted several bases to the left.

Many MSA tools are available on the Web, each based on different assumptions and algorithms. Some of the more popular tools include: ClustalW (Thompson *et al.*, 1994), Dialign (Morgenstern, 1999; Morgenstern *et al.*, 1998), Pileup (Wisconsin package), and T-Coffee (Notredame *et al.*, 2000), but many additional packages are also available. Most programs such as ClustalW provide global sequence alignments, but Dialign can provide both global and local alignments. No alignment program has been shown to work best in all

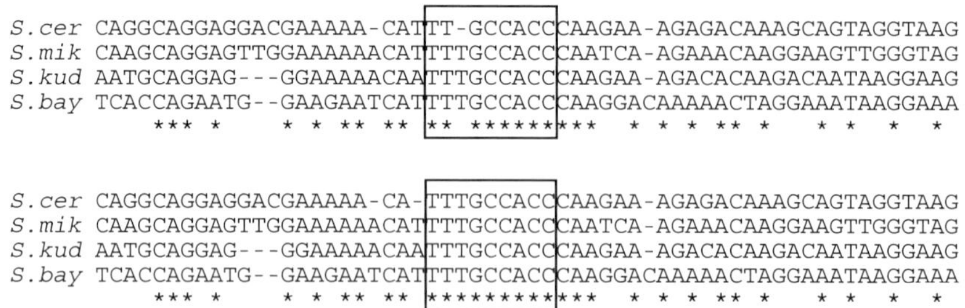

Figure 3. Alternative alignments of a putative Rpn4 binding site within the *RPN3* promoter. The top alignment was created by ClustalW. The bottom alignment was manually edited to move the gap within the motif three places to the left. The nonamer binding sequence of Rpn4 is boxed. Note that the modified alignment produces an extra column of identical residues meaning that it would score slightly higher than the first alignment. Equally scoring alignment modifications should also be considered.

conditions. The speed of the alignment algorithm is important only if hundreds or thousands of promoter sequences need to be aligned. Most of the MSA software is optimized to work with protein rather DNA sequence alignments, and many of the new advances being made to the field are specific to protein alignments. Even with these shortcomings, most MSA programs are fairly accurate and they provide a visual and intuitive view of the conserved promoter sequences.

C. Motif-Finding Algorithms

As mentioned previously, motif-finding algorithms were not designed for phylogenetic footprinting. However, used correctly, they can add a great deal to a promoter analysis. There are several compelling reasons to use these methods in conjunction with MSA methods. First, protein-binding motifs are often degenerate: the same protein may recognize several similar sequences. Since this is the case, the sequence motifs in the different species may have changes (or silent mutations) that have little effect on their function. These sequence motifs may not be identical within an MSA and may easily be overlooked. Second, most MSA methods assume that sequence motifs will be in the same order and with the same relative location in the orthologous promoters of different species. This assumption is not always true, so motif-finding algorithms that do not rely on the linear order of sequence motifs will likely help to identify features that are missed in MSAs. Third, because most sequence motifs are relatively short, new motifs can likely arise in a promoter by random mutations. Promoters containing several binding sites for the same protein could easily gain and lose sites in evolution with little effect on the regulation of the gene. This process would tend to rearrange the sequence motifs within the promoter, thereby hindering their alignment in multiple sequence comparisons. Finally, motif-finding algorithms allow more distantly related species (having less than 50% identity) to be included in the analysis, thus enabling one to take advantage of the increased sequence divergence. In our analysis of the yeast genome, we observed that 11% of yeast promoters are 50% or greater in identity in four-way alignments of *sensu stricto Saccharomyces* species. Although much of this conservation likely represents functional sequence, the high degree of sequence conservation of these promoters makes it difficult to identify individual transcription factor binding sites in their sequence alignments. More diverged organisms, such as the *sensu lato Saccharomyces* species, are often useful for identifying the protein-binding sites in these cases of high conservation.

D. Motif-Finding Software

Several motif-finding programs are available and are based on different (often complementary) algorithms. One of the first algorithms developed to find motifs in unrelated sequences was the Gibbs sampler (Lawrence *et al.*, 1993; Thompson *et al.*, 2003). There are several different programs based on this algorithm, including AlignAce (Roth *et al.*, 1998). Another example is Consensus, which uses a greedy algorithm (Hertz *et al.*, 1990; Hertz and Stormo, 1999). MEME uses a third algorithm, called expectation maximization (Bailey and Elkan, 1994, 1995), and has become quite popular due to its ease of use and the ability to change several assumptions about the search. FootPrinter is another algorithm worth considering, since it was the first program to take into account the idea that the input sequences may be related (Blanchette *et al.*, 2002; Blanchette and Tompa, 2003). This software is now in its second release and has several improvements over the original version. Phylocon (Wang and Stormo, 2003), a significant modification to the Consensus algorithm, also takes into account the relatedness of the sequences being compared.

E. Software Comparisons

To our knowledge, the different algorithms have not been rigorously compared to benchmark their performance in the identification of known DNA motifs. This is due in large part to the reality that the data set of experimentally characterized promoters is not large enough to adequately assess their performance in motif finding. In comparisons of yeast promoters for which no regulatory information is known, we observed a significant number of unique sequence motifs predicted by the different algorithms. In the absence of experimental data on the function of these sequence motifs, we are left to assume that the programs are somewhat complimentary in their ability to find conserved motifs.

F. Noteworthy Features of Motif-Finding Algorithms

The MEME software has several options that allow the user to choose between three different assumptions about the regulatory motif. One option is that the motif will have only one occurrence per sequence (OOPS). Or, the motif may be assumed to have zero or one occurrence per sequence (ZOOPS), a good assumption for a group of similarly expressed genes that might be regulated by two or more transcription factors. A third option searches for sequence motifs using a model in which the motif may have any number of occurrences in each sequence. This option tends to be superior for motif finding in promoters that contain several similar or identical motifs.

Most of the motif-finding algorithms have difficulty in identifying sequence motifs where there are large gaps between conserved critical nucleotides, as for the Gal4 DNA-binding sequence (as a spaced dyad: CGG-N_{11}-CCG). AlignAce attempts to account for gaps within a motif by allowing for a large number of non-critical residues to be present within the alignment. By default, the motif can extend three times the length of the conserved columns. For example, in searching for a motif six nucleotides long, the program will look over a stretch of 18 nucleotides. Despite this feature, AlignAce also appears to miss some spaced dyad motifs that are likely to be real. Therefore, gapped motifs appear to be problematic for most motif-finding algorithms. This is of some concern, since approximately one-third of the transcription factors in *S. cerevisiae* contain Gal4-type DNA-binding domains that recognize spaced dyad sequences. However, the sequences recognized by most of these transcription factors will likely have a gap smaller than the 11 base pairs that separate the CGG palindrome recognized by Gal4. One side effect of the liberal spacing requirement of AlignAce is that many of the motifs it identifies are not typical of known transcription factor-binding sites. Most of these unconventional motifs are likely spurious, but some of them could be binding sites of uncharacterized classes of DNA-binding proteins.

G. Online Bioinformatics Resources

Many bioinformatics tools are generously hosted on the Web and can be accessed by the public. Table 2 contains a current list of Web servers that host useful software programs that are commonly used in phylogenetic footprinting. The list is divided into sites hosting software for pair-wise sequence alignment, MSA, or motif identification. The list also contains several sites that host a wide variety of bioinformatics tools. These sites are generally hosted by larger institutions or centers for use by the general public and are meant to be "one stop" bioinformatics centers. The list is not exhaustive, but should serve as a suitable starting point for identifying Web-based comparative tools.

♦♦♦♦♦♦ IV. FOOTPRINTING YEAST GENES

In this section, we will walk through the process of footprinting a typical yeast promoter using sequence and analysis tools that are available on the Web. In the descriptions, we aim to assist readers who have no experience with Unix/Linux command-line bioinformatics tools. Experience and use of bioinformatics tools would make the exercise easier to complete, but it is well beyond the scope of this chapter to immerse the readers in the field of bioinformatics. We assume only that the reader is familiar with the *Saccharomyces* Genome Database and the various features that are maintained there.

For this exercise, we will use a typical yeast gene, *ALD6*, encoding a cytosolic aldehyde dehydrogenase that is required for conversion of acetaldehyde to acetate. The enzyme is different than the other cytosolic aldehyde dehydrogenases (Ald2 and Ald3) in that it is

Table 2. List of Web servers hosting useful sequence analysis tools

Pairwise Sequence Alignment

Smith Waterman
http://www-hto.usc.edu/software/seqaln/seqaln-query.html
http://clavius.bc.edu/~clotelab/TeachingTools/SmithWaterman/index.html
http://www.ebi.ac.uk/emboss/align/

Needleman Wunsch
http://clavius.bc.edu/~clotelab/TeachingTools/NeedlemanWunschPathMatrix/index.html
http://www.ebi.ac.uk/emboss/align/
http://bioweb.pasteur.fr/seqanal/interfaces/needle.html

BLAST
http://www.ncbi.nlm.nih.gov/blast/bl2seq/wblast2.cgi

Multiple Sequence Alignment

ClustalW
http://www.ebi.ac.uk/clustalw/
http://www.ch.embnet.org/software/ClustalW.html
http://align.genome.jp/
http://npsa-pbil.ibcp.fr/cgi-bin/npsa_automat.pl?page=/NPSA/npsa_clustalw.html

Dialign
http://dialign.gobics.de/chaos-dialign-submission
http://bibiserv.techfak.uni-bielefeld.de/dialign/submission.html

T-Coffee
http://www.ch.embnet.org/software/TCoffee.html
http://igs-server.cnrs-mrs.fr/Tcoffee/tcoffee_cgi/index.cgi

Motif Identification

MEME
http://meme.sdsc.edu/meme/meme.html
http://bioweb.pasteur.fr/seqanal/motif/meme/

AlignACE
http://atlas.med.harvard.edu/cgi-bin/alignace.pl

Consensus
http://adric.wustl.edu/oldconsensus/html/Html/main.html
http://bioweb.pasteur.fr/seqanal/interfaces/consensus-simple.html

FootprinterII
http://wingless.cs.washington.edu/htbin-post/unrestricted/FootPrinterWeb/FootPrinterInput2.pl

Gibbs Sampler
http://bayesweb.wadsworth.org/cgi-bin/gibbs.4.pl?data_type=DNA

Servers Hosting Many Different Sequence Analysis Tools

http://searchlauncher.bcm.tmc.edu/multi-align/multi-align.html
http://ca.expasy.org/tools/
http://www.ebi.ac.uk/Tools/sequence.html
http://bioweb.pasteur.fr/seqanal/alignment/intro-uk.html

activated by Mg^{2+} and it uses $NADP^+$ rather than NAD^+ as a coenzyme.

A. Identifying Homologs

To begin, we will look up *ALD6* in SGD and retrieve the amino acid sequence of the encoded protein. Use the drop-down menu under "Retrieve Sequences" (see Figure 4) to select the "ORF Translation" and to obtain the amino acid sequence of the gene. The sequence is then copied into the "Query Sequence" box of the Fungal BLAST search form (http://seq.yeastgenome.org/cgi-bin/blast-fungal.pl). The Fungal BLAST page can also be accessed by selecting the "BLAST" link (see Figure 4) near the top of the SGD summary page and then selecting the "Fungal BLAST search form" link at the top of the BLAST page.

TBLASTN is used to identify protein-coding homologs in DNA sequences of other fungal species. A protein sequence rather than nucleotide sequence search is a better choice for identifying the homologs outside of the *Saccharomyces sensu stricto* species since the genes often have little similarity at the nucleotide level. Protein searches will also aid in identifying the ends of the genes, which tend to be more diverged at the nucleotide level. Next, select the species that are most appropriate for the comparison to the *S. cerevisiae* protein. For this example, we will choose all of the

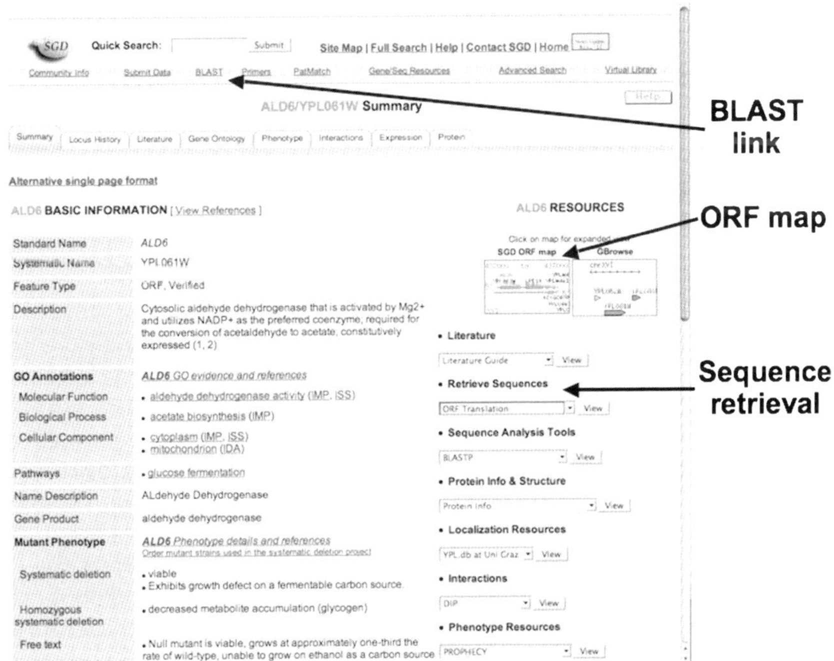

Figure 4 The *ALD6* summary page at SGD with the location of various useful features for retrieving sequences marked.

Saccharomyces species. *S. cerevisiae* could be left out of the BLAST comparison, but it is often useful to identify how many homologs of the gene are recognized in *S. cerevisiae*. The default BLAST parameters should be sufficient to identify most if not all of the relevant orthologous *Saccharomyces* sequences. However, one may want to limit the number of alignments shown to 25 or 50 to reduce the amount of output.

B. Identifying Orthologs

The next task is to identify the orthologs among the homologs that were identified by the BLAST search. Theoretically, each *S. cerevisiae* gene should have a similar gene in the other species that performs the exact function. However, *S. cerevisiae* may also have related genes that perform similar functions. In that case, the other species will likely have similar genes that do not perform the exact function of the gene of interest. These 'paralogs' will usually be identified by the TBLASTN search. In most cases the ortholog will be easy to identify based on its close similarity to the *S. cerevisiae* protein. However, there are several scenarios in which the orthology assignment is more difficult, such as incomplete sequence data for the ortholog or a gene duplication that is specific to one or more species. It should be kept in mind that since the assembled genomes are incomplete, the highest scoring alignment or the alignment with the highest statistical probability is not necessarily the ortholog for that species. Therefore, lower scoring alignments where only a portion of the orthologous sequence is present in the assembly (and in the alignment) should be considered. In ambiguous orthology cases, it is often helpful to retain both sequences for MSAs since non-orthologous sequences stand out in the alignment. Table 3 shows the abbreviated output of the TBLASTN comparison and indicates likely orthologs of *ALD6*.

C. Retrieving Intergenic Sequence

Once the target orthologs are identified their nucleotide sequences need to be obtained for the comparisons. It is a good idea to start with *S. cerevisiae* since its genome is the most complete and best annotated one. Because the transcription start sites of most yeast genes are not known, it is customary to start promoter analyses at the ATG start codon and work upstream. This seems to be a good idea, since many yeast genes contain conserved sequences adjacent to the start of transcription that are likely part of the transcript and could regulate gene expression (Cliften *et al.*, 2003). The amount of upstream sequence to consider is another consideration. A common distance is 1000 bp, but most intergenic regions are less than 500 bp, so including 1000 bp in the sequence alignment would include coding sequence of the upstream gene. Since coding sequences are

Table 3. Summary of TBLASTN output from a Fungal Genomes Search at SGD showing the top 25 TBLASTN hits to *ALD6**

Sequences producing significant alignments:	Score (bits)	E-value
gb|U39205.1|SCU39205 Saccharomyces cerevisiae chromosome XVI, left arm, cosmid...	2579	7.1e-267
gb|U56604.1|SCU56604 Saccharomyces cerevisiae cytosolic aldehyde dehydrogenase...	2566	8.5e-267
gb|AAFW01000148.1| Saccharomyces cerevisiae YJM789 chromosome 16	2579	9.6e-267
gb|AAEG01000033.1| Saccharomyces cerevisiae RM11–1a	2573	4.2e-266
gb|AACH01000598.1| Saccharomyces mikatae IFO 1815	2543	1.1e-264
gb|AABZ01000171.1| Saccharomyces mikatae IFO 1815	2546	1.4e-263
gb|AABY01000069.1| Saccharomyces paradoxus NRRL Y-17217	2541	8.3e-263
gb|AACA01000129.1| Saccharomyces bayanus MCYC 623	2534	3.4e-262
gb|AACI01000066.1| Saccharomyces kudriavzevii IFO 1802	1838	3.4e-256
gb|AACF01000055.1| Saccharomyces castellii NRRL Y-12630	2119	4.7e-218
gb|AACE01000118.1| Saccharomyces kluyveri NRRL Y-12651	2006	2.4e-206

(*Continued*)

Table 3. *(Continued)*

Sequences producing significant alignments:	Score (bits)	E-value
gb|AACE01000018.1| Saccharomyces kluyveri NRRL Y-12651	1477	4.6e-150
emb|Z75282.1|SCYOR374W S. cerevisiae chromosome XV reading frame ORF YOR374w	1464	6.1e-150
emb|AX536772.1| Sequence 373 from Patient WO02064766	1464	6.8e-150
gb|AACG01000544.1| Saccharomyces bayanus 623–6C	1466	8.9e-150
gb|AACH01000121.1| Saccharomyces mikatae IFO 1815	1467	3.4e-149
gb|AABZ01000073.1| Saccharomyces mikatae IFO 1815	1467	5.5e-149
gb|AAFW01000135.1| Saccharomyces cerevisiae YJM789 chromosome 15	1464	1.3e-148
gb|AACF01000056.1| Saccharomyces castellii NRRL Y-12630	1462	2.1e-148
gb|AABY01000016.1| Saccharomyces paradoxus NRRL Y-17217	1462	2.2e-148
gb|AACI01000023.1| Saccharomyces kudriavzevii IFO 1802	1459	3.8e-148
gb|AACG01000001.1| Saccharomyces bayanus 623–6C	1434	4.2e-146

gb|AACA01000002.1| Saccharomyces bayanus MCYC 623	1434	4.6e-146
emb|AL407336.1|CNS06NWY T7 end of clone XAU0AA002G03 of library XAU0AA from strai...	1425	1.8e-145
gb|AAEG01000067.1| Saccharomyces cerevisiae RM11–1a	1367	2.3e-144

*The amino acid sequence of Ald6 was compared by TBLASTN to assembled sequences of *Saccharomyces* species. The summary is ordered according to the E-value rather than the score. Potential orthologs of the *S. cerevisiae* sequence are underlined. The most similar *S. cerevisiae* paralog, *ALD4 YOR374W*, is double underlined. Note that two *S. kluyveri* homologs are more similar to *ALD6* than is the *S. cerevisiae ALD4*. The top scoring homolog (GenBank accession number AACE01000118.1) is likely the ortholog to *ALD6*, but both homologs could be retained for further analysis. Another ortholog of *ALD6* is found in a different assembly of *S. bayanus* (dashed underline). Note that only part of the gene is present in the assembly, so the alignment score is lower than that of *S. cerevisiae ALD4*. Since the 5′ end of the gene is available, you could use this assembly to obtain the orthologous promoter sequence, if it was not available in the other *S. bayanus* assembly. Two different assemblies of *S. mikatae* are also available so only the top scoring sequence was underlined.

also conserved through evolution, current comparative approaches to identify regulatory elements must be limited to non-coding sequences. If the intergenic region of interest is obviously longer than average, the additional sequence should be analyzed. While most regulatory sequences will be within 500 bp of the translational start codon, regulatory elements have been identified over 1300 bp upstream of the ATG codon of the *HO* gene (Brazas and Stillman, 1993; Sidorova and Breeden, 1993).

On each gene page, SGD provides an ORF map (see Figure 4) that shows surrounding genes. Clicking on the map gives a larger view of the area and brings up a display called a "Chromosomal Features Map" (Figure 5, see Colour Plate section) that can be used to identify the next gene or other known sequence element upstream of the gene of interest. Next use the "Chromosomal coordinates" (listed at the bottom of each Gene or Feature page) to calculate the distance between the start of the gene and the next feature. Keep in mind that the coordinates are listed 5′ to 3′ so genes or features on the Crick (bottom) strand actually start at the 3′ end or at the larger coordinate. Use the "Custom Retrieval" feature (in the pull-down menu under "Retrieve Sequences") to obtain the intergenic region for *S. cerevisiae*. One can either (1) indicate how many nucleotides of upstream sequence that you wish to be displayed (this method returns the gene sequence) or (2) indicate the exact coordinates of the sequence of interest. The sequences must be oriented in the same direction for most MSA programs, and both options allow retrieval of the reverse complementary sequences if needed. Since the other *Saccharomcyes* genomes are not well annotated, it takes a considerable amount of

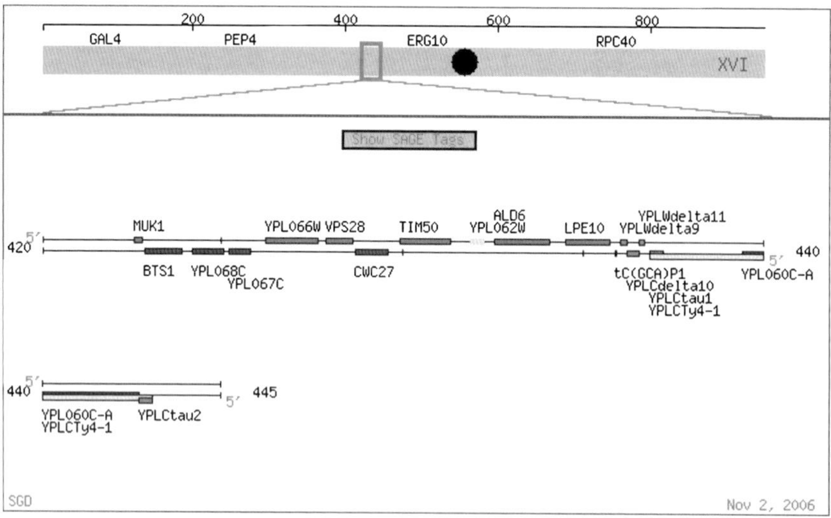

Figure 5. An SGD "Chromosomal Features Map" of the *Saccharomyces* genome near the *ALD6* gene. This expanded view is obtained by selecting the ORF map on the gene (*ALD6* in this case) summary page. The map provides a quick overview of genes or other chromosomal features adjacent to the gene of interest. (See color plate section)

effort to identify the upstream gene or feature. A quicker alternative is to use a similar length of sequence obtained for the *S. cerevisiae* promoter. It may help to add an additional 10 or 15% to the length of promoter sequence that you retrieve to help ensure that important elements of the sequence were not truncated.

The Fungal BLAST report page contains GenBank accession numbers for each alignment that are linked to the Entrez Nucleotide sequence retrieval system at NCBI. One click brings up the NCBI website; a second click on the accession number opens the GenBank file containing the sequence. The nucleotide range of the desired sequence can be indicated, and the reverse complement of the sequence can be obtained if needed. Additionally, the sequences can be displayed in FASTA rather than the GenBank format and sent to a file or clipboard as an alternative to cutting and pasting the sequence. For the *ALD6* data we will obtain 1000 bp of sequence upstream of the initiation codon. The SGD ORF map shows a dubious ORF, YPL061W, upstream of *ALD6*, which is completely contained within the 1000 bp region. The ORF does not appear to code for a gene based on comparisons with other species (see the MSA below in Figure 6).

D. Multiple Sequence Alignment

With the sequences in hand the comparison can begin. An MSA is the best place to start because it will allow investigation of any questionable orthology assignments and determination of what steps should be taken next. Many of the MSA programs are available via web servers and the sequences of interest can be pasted into the appropriate query box. Only the *sensu stricto* species should initially be aligned, since more diverged sequences will hinder accurate alignments or complicate the interpretation of the alignment.

Most of the time some conserved features will be apparent in a 4-way alignment of *sensu stricto* species. In cases where the promoter is highly conserved, it may be useful to employ motif searches including the more diverged species to search for smaller functional regions. Alternatively, the MSA alignment could be used directly to inform experimental approaches. In a small number of 4 or 5-way alignments of *sensu stricto* species there is too much sequence divergence for any motifs to be discovered. In these cases, individual sequences can easily be removed from the alignment until a more suitable level of divergence is found. Figure 7 shows an alignment for *ANP1*, which encodes a subunit of the alpha-1,6 mannosyl transferase. Little similarity is seen in the alignment, but removing the *S. kudriavzevii* or *S. bayanus* sequence greatly improves the percentage of conserved residues in the alignment. The low level of similarity within an MSA could be due to several factors including miscalled orthologs, high rates of sequence divergence of the specific region of the respective genomes, or species-specific differences in the regulation of the gene. In a ClustalW alignment the two

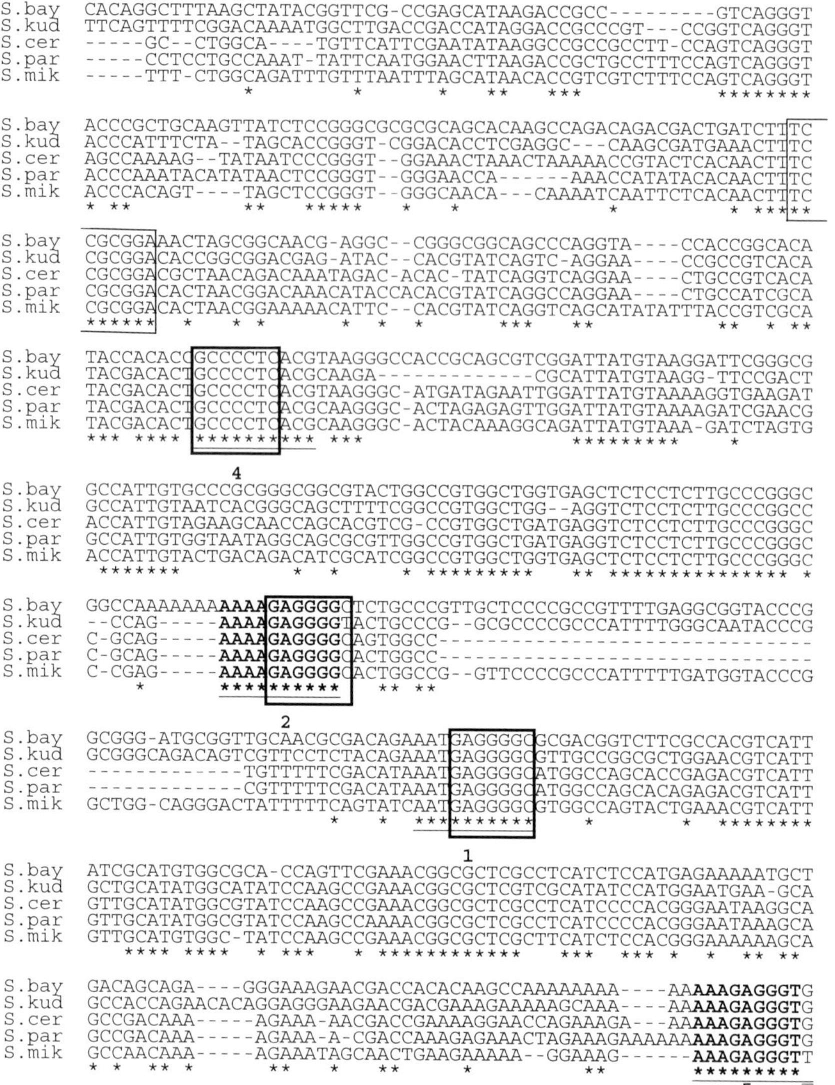

Figure 6. A multiple sequence alignment of the *ALD6* promoter region. The start of the protein-coding region is depicted with an arrow. The two regions of the alignment that were manually edited to produce longer motifs are shown in bold. A putative TATA box is underlined. The top five motifs identified by MEME (in a search of sequences of seven *Saccharomyces* species) are underlined and number in order of decreasing significance. Notice that the two modified regions of the alignment correspond to motifs predicted by MEME. Two motifs identified by MEME in the "any number of repetition per sequence" mode are boxed. The top motif (shown with a thicker line) is present four times within the alignment whereas the second motif (shown with a thinner line) is present twice in the alignment but is not conserved in *S. castellii* or *S. kluyveri*. Species included in the MEME analysis included *S. bayanus, S. kudriavzevii, S. cerevisiae, S. paradoxus, S. mikatae, S. castellii* and *S. kluyveri*. The last two sequences were not included in the ClustalW alignment because they are too diverged from the other species to inform the alignment.

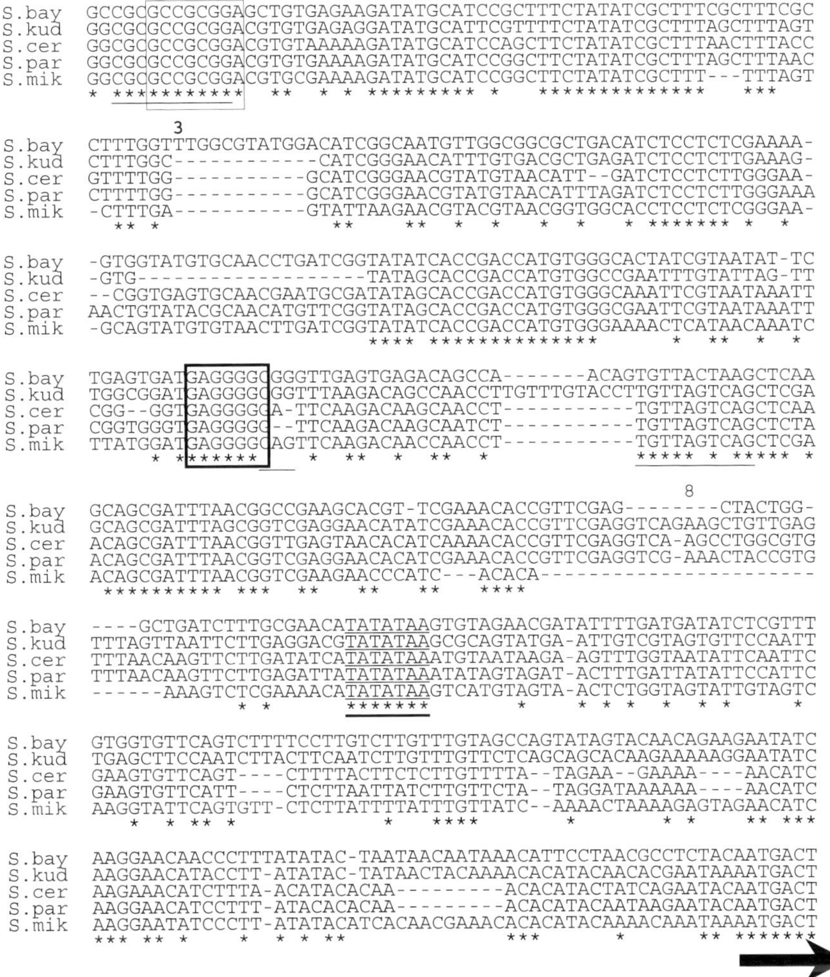

Figure 6. (*Continued*)

most similar sequences are placed at the top of the alignment and then the next most similar sequence is added until the alignment is complete. This ordering makes it simple to identify the most diverged sequence and to remove it from the alignment. If the most diverged sequence in the alignment is from *S. cerevisiae*, it is probably necessary to reassess the orthology assignments, since it will be of little value to identify conserved elements that are only present in the other species. It appears that a very small fraction of *S. cerevisiae* genes do not share much if any regulatory sequence conservation with their orthologs in the *sensu stricto* species.

E. Evaluation of the Multiple Sequence Alignment

After creating an MSA, the alignment is inspected for possible mistakes in the annotation (i.e., make sure that there are no genes in

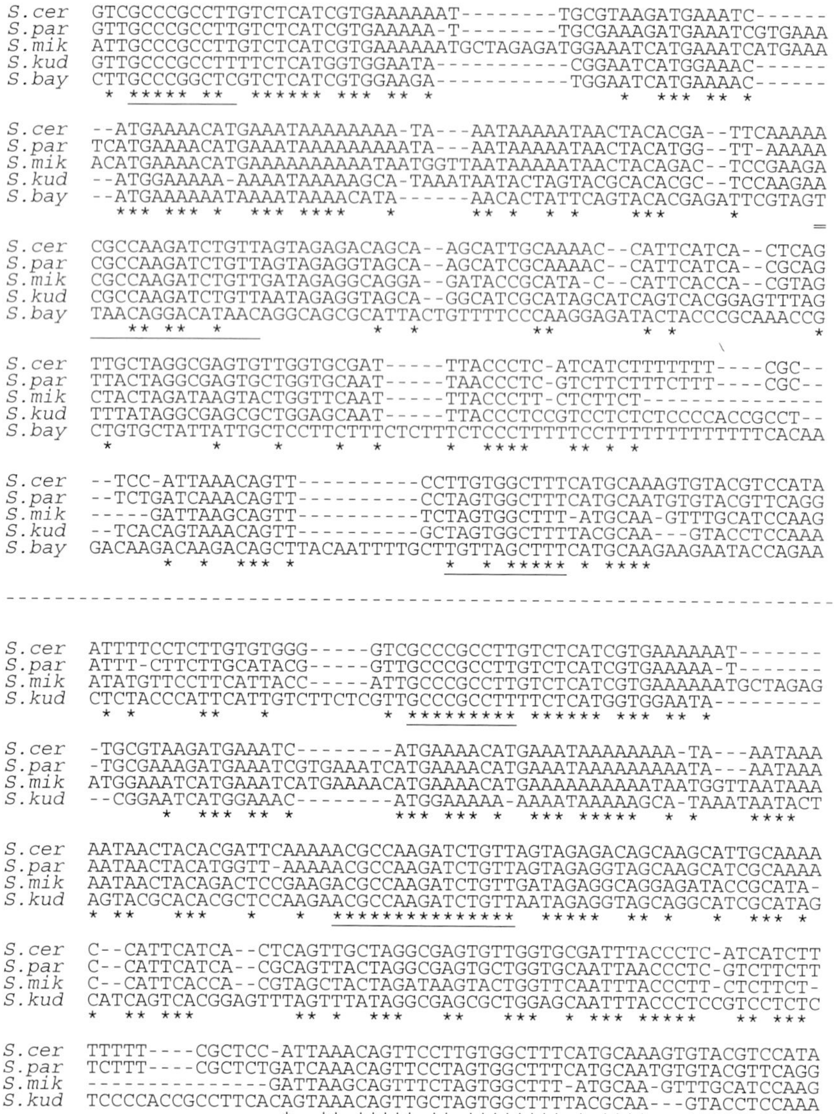

Figure 7. A comparison of a 4- and 5-way alignment of the *ANP1* promoter region. The upper 5-way alignment includes sequences of *S. cerevisiae, S. paradoxus, S. mikatae, S. kudriavzevii* and *S. bayanus*. A careful visual inspection of the 5-way alignment reveals that the *S. bayanus* sequence is quite diverged from the other sequences especially in the last three rows of the alignment. The *S. bayanus* sequence was omitted from the lower alignment. The 5-way alignment contains 23.7% identical nucleotides compared to 40.4% in the 4-way alignment. Three motifs (of at least 6 consecutive nucleotides) that are only identical in the 4-way alignment are underlined. The position of the motifs in the 5-way alignment is also depicted.

the "intergenic region" of interest). Many annotation errors were discovered through phylogenetic comparisons (Brachat *et al.*, 2003; Cliften *et al.*, 2003; Kellis *et al.*, 2003), but apparently some remain to be found because experimental approaches continue to uncover errors (Zhang and Dietrich, 2005). Annotation errors could include: extensions to the annotated genes, small ORFs, unrecognized introns, or unrecognized RNA coding genes. Alignments containing protein-coding sequence are often characterized by higher levels of sequence conservation (but sometimes only slightly higher), with non-conserved nucleotides tending to appear three or multiples of three nucleotides apart (because of the degenerate nature of the triplet code). The "TACTAAC" box, a conserved sequence near the branch point of the splice site, is usually very well conserved in MSAs containing introns. However, there are several examples in which the TACTAAC box is not easily aligned within an MSA. RNA coding genes could potentially be problematic since they tend to show conservation of secondary structure rather than primary sequence. However, a systematic search for unidentified genes encoding only RNA suggests that such cases are rare (or very difficult to identify) within the *Saccharomyces* lineage (McCutcheon and Eddy, 2003).

The *ALD6* MSA does not contain any hallmarks of protein-coding sequences, but does contain a remarkable number of conserved sequences. The original ClustalW alignment contains 14 conserved blocks of at least eight nucleotides in length. Two additional conserved blocks of sequence are obtained by slightly adjusting gap positions in the alignment. The modified alignment contains at least 16 potential regulatory motifs and a potential TATA box with 7 nucleotides of conservation (see Figure 6).

F. Motif Detection Using Unalignable Sequences

Conserved sequence motifs in the *ALD6* promoter that may have been missed in the MSA can be searched for using any of the motif-finding algorithms. It would be wise to use several algorithms to search for the motifs, but for the sake of brevity we will show only the results of analysis of the sequences with MEME. On the MEME website (http://rocks50.sdsc.edu/meme/meme.html), the FASTA formatted sequence was pasted into the sequence box. The maximum number of motifs to find was set at five, and their length was set to between six and ten nucleotides. Using these parameters, MEME was run three times using the three different options for motif distribution: one occurrence per sequence (OOPS), zero or one occurrence per sequence (ZOOPS), or any number of repetitions per sequence. The first two options produced nearly identical motifs (these are highlighted in the MSA in Figure 7). The third option produced two interesting motifs present multiple times in each sequence but the first motif was not very well conserved near the edges. MEME was therefore run three more times using the "any number of repetitions per sequence" mode restricting the length of the motif to six, seven or eight nucleotides to "clean up" the motifs. The two motifs that were revealed using the motif length of seven are depicted in Figure 7. There are four copies of the top scoring motif in each sequence of the alignment, but only two copies in the sequences of the more distantly related species *S. castellii* and *S. kluyveri* (data not shown). The second motif is present only in the aligned species, but is found twice per sequence.

As with many scientific endeavors, phylogenetic analysis of the *ALD6* provides more questions than answers. Such questions as "why are there so many conserved motifs in the MSA?" or "why don't many of these motifs appear to be present in the more distantly related species?" will likely only be answered with carefully designed experiments. On the other hand, the MSA reveals that many regions of the *ALD6* promoter are conserved through evolution and likely have some function. The MSA and the MEME motif analysis also strongly suggest that *ALD6* is regulated by a factor that binds to the motif GAGGGGC or GAGGGGT. These insights should be sufficient to stimulate experiments aimed at discovering the mechanisms of regulation of *ALD6* expression.

Acknowledgement

I express great appreciation to Mark Johnston, a terrific mentor who edited the manuscript and provided strategic ideas for the chapter.

References

Altschul, S. F., Gish, W., Miller, W., Myers, E. W. and Lipman, D. J. (1990). Basic local alignment search tool. *J. Mol. Biol.* **215**, 403–410.

Bailey, T. L. and Elkan, C. (1994). Fitting a mixture model by expectation maximization to discover motifs in biopolymers. *Proc. Int. Conf. Intell. Syst. Mol. Biol.* **2**, 28–36.

Bailey, T. L. and Elkan, C. (1995). The value of prior knowledge in discovering motifs with MEME. *Proc. Int. Conf. Intell. Syst. Mol. Biol.* **3**, 21–29.

Blanchette, M., Schwikowski, B. and Tompa, M. (2002). Algorithms for phylogenetic footprinting. *J. Comput. Biol.* **9**, 211–223.

Blanchette, M. and Tompa, M. (2003). FootPrinter: A program designed for phylogenetic footprinting. *Nucleic Acids Res.* **31**, 3840–3842.

Brachat, S., Dietrich, F. S., Voegeli, S., Zhang, Z., Stuart, L. *et al.* (2003). Reinvestigation of the *Saccharomyces cerevisiae* genome annotation by comparison to the genome of a related fungus: *Ashbya gossypii*. *Genome Biol.* **4**, R45.

Brazas, R. M. and Stillman, D. J. (1993). Identification and purification of a protein that binds DNA cooperatively with the yeast SWI5 protein. *Mol. Cell Biol.* **13**, 5524–5537.

Cliften, P. F., Fulton, R. S., Wilson, R. K. and Johnston, M. (2006). After the duplication: gene loss and adaptation in *Saccharomyces* genomes. *Genetics* **172**, 863–872. Epub 2005 Dec 1.

Cliften, P., Sudarsanam, P., Desikan, A., Fulton, L., Fulton, B. *et al.* (2003). Finding functional features in Saccharomyces genomes by phylogenetic footprinting. *Science* **301**, 71–76.

Cliften, P. F., Hillier, L. W., Fulton, L., Graves, T., Miner, T. *et al.* (2001). Surveying Saccharomyces genomes to identify functional elements by comparative DNA sequence analysis. *Genome Res.* **11**, 1175–1186.

Dietrich, F. S., Voegeli, S., Brachat, S., Lerch, A., Gates, K. *et al.* (2004). The Ashbya gossypii genome as a tool for mapping the ancient *Saccharomyces cerevisiae* genome. *Science* **304**, 304–307.

Dujon, B., Sherman, D., Fischer, G., Durrens, P., Casaregola, S. *et al.* (2004). Genome evolution in yeasts. *Nature* **430**, 35–44.

Griggs, D. W. and Johnston, M. (1993). Promoter elements determining weak expression of the *GAL4* regulatory gene of *Saccharomyces cerevisiae. Mol. Cell Biol.* **13**, 4999–5009.

Gumucio, D. L., Heilstedt-Williamson, H., Gray, T. A., Tarle, S. A., Shelton, D. A. *et al.* (1992). Phylogenetic footprinting reveals a nuclear protein which binds to silencer sequences in the human gamma and epsilon globin genes. *Mol. Cell Biol.* **12**, 4919–4929.

Gumucio, D. L., Shelton, D. A., Bailey, W. J., Slightom, J. L. and Goodman, M. (1993). Phylogenetic footprinting reveals unexpected complexity in trans factor binding upstream from the epsilon-globin gene. *Proc. Natl. Acad. Sci. USA* **90**, 6018–6022.

Hertz, G. Z., Hartzell 3rd, G. W. and Stormo, G. D. (1990). Identification of consensus patterns in unaligned DNA sequences known to be functionally related. *Comput. Appl. Biosci.* **6**, 81–92.

Hertz, G. Z. and Stormo, G. D. (1999). Identifying DNA and protein patterns with statistically significant alignments of multiple sequences. *Bioinformatics* **15**, 563–577.

Kellis, M., Patterson, N., Endrizzi, M., Birren, B. and Lander, E. S. (2003). Sequencing and comparison of yeast species to identify genes and regulatory elements. *Nature* **423**, 241–254.

Lawrence, C. E., Altschul, S. F., Boguski, M. S., Liu, J. S., Neuwald, A. F. *et al.* (1993). Detecting subtle sequence signals: A Gibbs sampling strategy for multiple alignment. *Science* **262**, 208–214.

Mannhaupt, G., Schnall, R., Karpov, V., Vetter, I. and Feldmann, H. (1999). Rpn4p acts as a transcription factor by binding to PACE, a nonamer box found upstream of 26S proteasomal and other genes in yeast. *FEBS Lett.* **450**, 27–34.

McCutcheon, J. P. and Eddy, S. R. (2003). Computational identification of non-coding RNAs in Saccharomyces cerevisiae by comparative genomics. *Nucleic Acids Res.* **31**, 4119–4128.

Morgenstern, B. (1999). DIALIGN 2: Improvement of the segment-to-segment approach to multiple sequence alignment. *Bioinformatics* **15**, 211–218.

Morgenstern, B., Frech, K., Dress, A. and Werner, T. (1998). DIALIGN: Finding local similarities by multiple sequence alignment. *Bioinformatics* **14**, 290–294.

Needleman, S. B. and Wunsch, C. D. (1970). A general method applicable to the search for similarities in the amino acid sequence of two proteins. *J. Mol. Biol.* **48**, 443–453.

Notredame, C., Higgins, D. G. and Heringa, J. (2000). T-Coffee: A novel method for fast and accurate multiple sequence alignment. *J. Mol. Biol.* **302**, 205–217.

Pearson, W. R. (1990). Rapid and sensitive sequence comparison with FASTP and FASTA. *Methods Enzymol.* **183**, 63–98.

Ren, B., Robert, F., Wyrick, J. J., Aparicio, O., Jennings, E. G. *et al.* (2000). Genome-wide location and function of DNA binding proteins. *Science* **290**, 2306–2309.

Roth, F. P., Hughes, J. D., Estep, P. W. and Church, G. M. (1998). Finding DNA regulatory motifs within unaligned noncoding sequences clustered by whole-genome mRNA quantitation. *Nat. Biotechnol.* **16**, 939–945.

Sidorova, J. and Breeden, L. (1993). Analysis of the SWI4/SWI6 protein complex, which directs G1/S-specific transcription in Saccharomyces cerevisiae. *Mol. Cell Biol.* **13**, 1069–1077.

Smith, T. F. and Waterman, M. S. (1981). Identification of common molecular subsequences. *J. Mol. Biol.* **147**, 195–197.

Souciet, J., Aigle, M., Artiguenave, F., Blandin, G., Bolotin-Fukuhara, M. *et al.* (2000). Genomic exploration of the hemiascomycetous yeasts: 1. A set of yeast species for molecular evolution studies. *FEBS Lett.* **487**, 3–12.

Tagle, D. A., Koop, B. F., Goodman, M., Slightom, J. L., Hess, D. L. *et al.* (1988). Embryonic epsilon and gamma globin genes of a prosimian primate (Galago crassicaudatus). Nucleotide and amino acid sequences, developmental regulation and phylogenetic footprints. *J. Mol. Biol.* **203**, 439–455.

Thompson, J. D., Higgins, D. G. and Gibson, T. J. (1994). CLUSTAL W: Improving the sensitivity of progressive multiple sequence alignment through sequence weighting, position-specific gap penalties and weight matrix choice. *Nucleic Acids Res.* **22**, 4673–4680.

Thompson, W., Rouchka, E. C. and Lawrence, C. E. (2003). Gibbs Recursive Sampler: Finding transcription factor binding sites. *Nucleic Acids Res.* **31**, 3580–3585.

Verma, R., Patapoutian, A., Gordon, C. B. and Campbell, J. L. (1991). Identification and purification of a factor that binds to the Mlu I cell cycle box of yeast DNA replication genes. *Proc. Natl. Acad. Sci. USA* **88**, 7155–7159.

Wang, T. and Stormo, G. D. (2003). Combining phylogenetic data with co-regulated genes to identify regulatory motifs. *Bioinformatics* **19**, 2369–2380.

Zhang, Z. and Dietrich, F. S. (2005). Mapping of transcription start sites in Saccharomyces cerevisiae using 5′ SAGE. *Nucleic Acids Res.* **33**, 2838–2851.

23 *Saccharomyces cerevisiae* as a Tool for Human Gene Function Discovery

Hans R Waterham and Ronald JA Wanders
Laboratory Genetic Metabolic Diseases, Departments of Pediatrics and Clinical Chemistry, Academic Medical Centre, Amsterdam, The Netherlands

♦♦

CONTENTS

Introduction
Expressing human genes in yeast
Identification of functional homologues
Functional testing of human genes

List of Abbreviations

cDNA	complementary DNA
GPCRs	G-protein coupled receptors
HGI	human gene of interest
HPI	human protein of interest
mRNA	messenger RNA
PCR	polymerase chain reaction

♦♦♦♦♦♦ I. INTRODUCTION

Saccharomyces cerevisiae is an attractive organism for functional studies on human gene products. The main features that make this yeast so suitable for studies on processes and proteins from higher eukaryotic organisms, including human, are the well-developed and easily accessible genetic tools, rapid growth, the simple and inexpensive culturing conditions and the fact that many of the basic cellular and metabolic processes found in higher eukaryotes are conserved in this unicellular eukaryote. These features and the

METHODS IN MICROBIOLOGY, VOLUME 36
0580-9517 DOI:10.1016/S0580-9517(06)36023-0

availability of the entire yeast genome sequence, including the annotation of the genes and the functions of their encoded products (see: http://www.yeastgenome.org), provides the researcher with straightforward, relatively inexpensive and versatile opportunities for genetic and physiological manipulation of *S. cerevisiae* to allow the expression, identification and functional studies of human proteins. In this chapter we describe and discuss the use, advantages and limitations of a variety of approaches that can be used for this purpose. Because most of the techniques used for these studies are straightforward and commonly employed, we focus on the applicability rather than on the experimental details. For this latter we refer to other chapters in this volume and sources such as the Volume 194 of the Methods in Enzymology series (Guthrie and Fink, 1991).

♦♦♦♦♦♦ II. EXPRESSING HUMAN GENES IN YEAST

The most suitable approach to express a certain human protein of interest (HPI) very much depends on the aim of the study and, among others, on whether problems in the expression are expected, such as toxicity for the host. Owing to the well-developed and versatile molecular genetic tools available for *S. cerevisiae* there is ample choice from a large variety of promoters, expression vectors (with various selectable markers), yeast strains and growth conditions that enables the researcher to control the timing and levels of expression of the HPI and to tackle many of the potential problems herein. In this chapter we describe some strategies and considerations that may aid in choosing a suitable approach for the expression of the HPI.

A. Introducing Human Genes in Yeast Expression Plasmids

The architecture of human genes differs greatly from the vast majority of yeast genes in that human genes are commonly composed of multiple exons, harboring the protein-encoding regions as well as translation regulatory elements, alternated with intronic sequences, which are non-coding and may vary widely in size. After transcription of a gene, the primary RNA transcript will undergo extensive splicing, an ordered process during which the intronic sequences are removed and the exons interconnected, eventually resulting in messenger RNAs (mRNAs) composed of a continuous protein-encoding region flanked by 5′ and 3′ non-coding regions. In contrast, the yeast genome primarily consists of relatively small genes composed of only the protein-encoding regions and essentially lacking introns.

Since the mammalian splicing machinery is not well conserved in yeast, one should only express the coding region of the human gene of interest (*HGI*) in yeast. Usually this coding region is amplified by the polymerase chain reaction (PCR) from human complementary DNA (cDNA) using specific oligonucleotide primers that are complementary to 5′ and 3′ sequences of the cDNA of the *HGI*. These cDNAs can be obtained by reverse transcription, using the enzyme reverse transcriptase, from mRNAs isolated from human cells or tissues. Alternatively, one can use as template commercially available annotated human cDNAs.

To allow easy unidirectional subcloning into a yeast expression vector, it is recommended to extend the oligonucleotide primers used for the PCR amplification of the coding regions of the human cDNAs with nucleotides containing recognition sites for restriction endonucleases identical to those located downstream of the yeast promoter (e.g. in the multiple cloning site). Obviously, the choice of these recognition sites depends on the absence of such sites in the cDNA sequence itself.

Whereas for mammalian expression by means of mammalian expression vectors the length of the 5′ non-coding region of the cDNA (i.e. the sequence preceding the translation initiation ATG codon) is less important and in fact may improve transcription and mRNA translation/stability, it is recommended for optimal transcription in yeast to keep this sequence in yeast expression vectors relatively short (<10 nucleotides). Furthermore, since translation initiation in yeast occurs via a ribosome scanning mechanism (Yoon and Donahue, 1992) it is important to assure that upstream of the authentic translation initiation ATG codon of the human cDNA no other ATG codon is introduced, since this severely reduces or even prevents translation initiation at the authentic ATG codon in yeast (Yun *et al.*, 1996).

Finally, before embarking in the actual yeast expression studies with the yeast expression vector containing the cDNA of the *HGI* it is strongly advised to sequence the entire cDNA to assure that no sequence errors have been introduced by the PCR amplification.

B. Which Promoter to Choose?

The levels and timing of expression of your HPI can be controlled by the use of different promoters. For this purpose a variety of constitutive and inducible, often culture condition-dependent, promoters of variable strength are available, the properties and applicability of some of which are briefly discussed below.

1. Constitutive promoters

Constitutive promoters that are commonly used for yeast expression studies are the promoters of the yeast genes encoding phosphoglycerate kinase-1 (*PGK1*; Hitzeman *et al.*, 1982), glyceraldehyde-3-phosphate dehydrogenase (*GPD* or *TDH3*; Holland and

Holland, 1980), and alcohol dehydrogenase-1 (*ADH1* or *ADC1*; Ammerer, 1983). These promoters are all similar in strength and rather strong, because they are derived from endogenous yeast genes encoding glycolytic enzymes that each are relatively abundant in yeast. Although considered constitutive, the levels of expression may vary somewhat dependent on the specific culture conditions (e.g. carbon source used for growth). For example, dependent on the length of the *ADC1* promoter fragment used a decrease in expression during growth on ethanol or in high density culture may occur (Ammerer, 1983).

2. Inducible promoters

The *GAL1/GAL10* promoter, of the yeast galactokinase gene, is regulated by the carbon source used for growth and can be used to drive expression of proteins that may affect yeast growth or viability but also in general for proteins the expression of which is not required or detrimental for growth of the yeast. The transcriptional activity of the promoter is strongly repressed during growth on glucose, becomes de-repressed during growth on lactose and is strongly induced during growth on galactose (1000-fold higher than on glucose; Johnston and Davis, 1984).

Transcriptional activity of the *CTA1* promoter, of the yeast peroxisomal catalase gene, is also under regulation of the carbon source and varies from moderate on glucose and glycerol to strong during growth on oleic acid. This promoter is especially suitable for studies on peroxisomal processes because growth on oleic acid (and other fatty acids) leads to a strong induction of the beta oxidation pathway, which in yeast is exclusively localized in peroxisomes (Filipits *et al.*, 1993).

Transcription by the *MET3* promoter, which is much weaker than the *GAL1* promoter, is independent of the carbon source and occurs in the absence of methionine in the medium but is inhibited when 2 mM methionine is added (Zhang *et al.*, 1999). The promoter is a good alternative for the *GAL1* promoter in those cases where expression needs to be inducible but less abundant and when carbon sources other than galactose are required.

The *tetO* promoter is an artificial promoter (Belli et al., 1998a, b) the transcriptional activity of which is also not dependent on the carbon source. The promoter strength is comparable to the *GAL1* promoter, and transcription occurs in the absence of doxycycline (or other tetracycline analogues), whereas addition of >10 μg/ml doxycycline to the medium prevents transcription.

C. Which Expression Vector to Choose?

Expression of heterologous/exogenous genes in *S. cerevisiae* can be readily achieved through site-directed integration of the expression unit in yeast chromosomal DNA. In general, however, autonomous replicating vectors are used for this purpose. These vectors can be selected and propagated in both *Escherichia coli* and yeast and usually carry as selectable marker in yeast a gene for which the host strain contains a non-reversible mutant allele and the presence of which can thus be selected for in the yeast after transformation. These mostly are genes involved in amino acid metabolism, including *LEU2*, *HIS3*, and *TRP1*, or nucleotide metabolism, such as *URA3*, but one can also use vectors with dominant selectable markers that allow transformed strains to grow in media supplemented with antibiotics such as Zeocin, blasticidin, and geneticin (G-418) or others.

In addition to the choice of promoter, the levels of expression of the HPI can also be modulated by the choice of the yeast expression vector. Yeast expression vectors that contain the origin of replication of the yeast endogenous 2-μm plasmid are maintained in high-copy number in most host strains (from 10 up to 200 copies per cell; Broach, 1983) and thus usually result in high expression levels. In contrast, yeast expression vectors containing a yeast chromosomal centromeric sequence (*CEN*) are maintained as single copy per cell and thus are better suited for low or moderate expression.

Yeast expression vectors that harbor the different features discussed above can be obtained from commercial suppliers, such as Invitrogen (e.g. the pYES vector series including 2-μm and *CEN* vectors, various selectable markers and the *GAL1* promoter, and the pTEF1/ZEO and pTEF1Bsd vectors with dominant selectable markers and the *TEF1* promoter), Stratagene (yeast vectors with divergent *GAL1/GAL10* promoters), and New England Biolabs (vectors based on the original pRS series reported by Sikorski and Hieter, 1989). From the European *Saccharomyces cerevisiae* Archives for Functional Analysis (EUROSCARF), many additional yeast expression vectors for various purposes are available for a nominal price. These include vectors with tetracycline-inducible promoters, various dominant selectable markers and epitope cassettes.

D. Which Yeast Strain to Use?

Some consideration has to be made with respect to the choice of the yeast strain for your study. In particular, there is a variety of 'wild-type' strains which each differ considerably in properties, which may influence the expression of your phenotype. For example, to study functional aspects of the yeast mitochondrial FAD transporter Flx1p different host strains have been used, which led to different interpretations and conclusions. Deletion of *FLX1* in the W303-1A strain resulted in a respiratory deficient phenotype, a 7-fold decrease in the mitochondrial FAD/FMN ratio and a marked decrease in activity of mitochondrial flavoproteins from which it was concluded that Flx1p is responsible for the import of FAD into mitochondria (Tzagoloff *et al.*, 1996). When *FLX1* is deleted in the EBY157A strain, however, no change in mitochondrial FAD content was observed, although the activity of mitochondrial flavoproteins was reduced 2-fold from which it was concluded that the deletion results in a defective mitochondrial FAD export, which affects nuclear gene expression leading to lowered expression of the flavoproteins (Bafunno *et al.*, 2004). Functional complementation by the candidate human orthologue was determined by regained growth on plates and appeared possible in the original *flx1* mutant in the W303-1A host strain but not in a similar mutant with another host strain background (Spaan *et al.*, 2005).

Furthermore, a particular yeast strain that is considered a wild-type strain in some studies, may turn out to have a specific genetic defect in other studies. For example, a commonly used laboratory 'wild-type' strain YP102 turned out to be defective in growth on oleic acid due to a mutation in the yeast *PAS8* gene preventing the correct assembly of peroxisomes, which are required for these specific culture conditions (Voorn-Brouwer *et al.*, 1993). Thus, for functional expression studies it can be important to test different *S. cerevisiae* 'wild-type' strains to determine which one is most suitable for your purpose.

For virtually every annotated yeast gene, commercially available knock out strains have been generated by replacement of (part of) the coding regions of each yeast gene through homologous recombination with the *kanMX* module flanked by gene-specific sequences using the PCR-based gene disruption approach (Wach *et al.*, 1994, and this volume, Chapter 4). Knockout strains for non-essential genes are available as haploids and homozygous or heterozygous diploids, while knockout strains for essential genes are available only as heterozygous diploids. Different wild-type host strains have been used for the generation of these knockouts, which, as mentioned above, may not all be suitable for the specific studies with a certain HPI and thus it is advised to first test this with the various host strains. Commercial suppliers for these strains are the American Type Culture Collection (ATCC), EUROSCARF and Invitrogen/Research Genetics (see Chapter 27 for a discussion of genetic strain collections).

E. Which Growth Conditions to Use

S. cerevisiae can use different carbon sources for growth, the choice of which depends on the aim of your study. In particular in the case of (functional complementation) studies with human proteins or enzymes that function in certain metabolic pathways, it is important to first confirm that a certain host strain is well capable of growing on the specific culture conditions needed to perform the functional studies because this can be strain-specific in particular for more uncommon conditions such as growth on oleic acid (for peroxisomal studies) or acetate (for mitochondrial studies).

Most functional studies use either rich media including 2% (w/v) peptone and 1% (w/v) yeast extract or regular synthetic dropout media (i.e. yeast nitrogen base) with amino acid supplements required for growth, that vary only in the amount and type of carbon source, the latter needed for the induction of certain metabolic pathways and/or a regulatable promoter used for expression.

F. Verification of Expression

To perform functional studies with an HPI expressed in yeast, one needs to verify that the protein is expressed correctly and at appreciable levels. Such verification of functional expression may be straightforward in those cases in which a functionally expressed HPI is capable of reversing a mutant phenotype into normal when expressed in a yeast strain lacking the yeast orthologue of this protein (see below under Section IIIA). When no functional complementation occurs or is attempted, however, one needs to confirm correct expression by other means, such as detection of the HPI by specific antibodies or, as for example, in the case of enzymes, by activity assays. The first approach only confirms expression *per se* whereas the second approach in addition confirms that the expressed protein is functional. If no specific activity assay or antibodies are available for the HPI, one may attempt to express the HPI fused to green fluorescent protein or an epitope-tag, such as HA (from hemagglutinin), *myc*, PROTEIN A, or 6 × His, which will allow the verification of its expression by using antibodies that recognize these tags. A potential drawback of this approach is, however, that such tags may interfere with the function of the protein.

G. If Expression Fails

Even though there are many examples in literature of successful functional expression of human proteins in *S. cerevisiae*, there can be several reasons why a particular protein is not correctly expressed or is correctly expressed but not functionally active in *S. cerevisiae*. Below are listed some possibilities and recommendations to address these.

1. No protein expressed

As mentioned above, correct expression of the HPI can be verified by functional complementation, an activity assay and/or immunodetection. If the protein is not expressed or only at very low levels, one could consider the following approaches to determine at which level the problem occurs:

(a) Transformed yeast strain does not contain (correct) plasmid

To be sure that the transformant contains the correct plasmid, confirm its presence by PCR on DNA extracted from the transformant using specific oligonucleotide primers or by rescuing the plasmids by (electro)transformation of yeast DNA to *E. coli* (e.g. Hoffman and Winston, 1987), but also possible with a commercially available plasmid yeast mini kit (Invitrogen).

(b) Coding region of HGI is not transcribed in yeast

The coding region of the cDNA of the HGI may not be transcribed efficiently in yeast or result in an unstable mRNA for several reasons (see also below). Check for correct expression of the mRNA of the HGI by RT-PCR using total RNA isolated from the transformants.

(c) Sequence errors in the expression plasmid

Owing to the infidelity of the PCR reaction, sequence errors may have been introduced in the coding region amplified from the cDNA of the *HGI*. In addition, errors may be present in the oligonucleotide primers used for the amplification of the coding region. Both may abrogate transcription or lead to the production of a non-functional protein. Sequence verification of the coding region and junctions in the expression plasmid should be carried out.

(d) Variable expression in transformants

The levels of protein expression may vary considerably among different yeast transformants, even when the same host strain and plasmids are used. It is therefore recommended that at least 10 independent transformants are analyzed for the level of expression of the HPI by activity measurements or immunodetection.

(e) No correct induction of expression

If the HPI is expressed under transcriptional control of a regulatable promoter, it is important to assure that the culture conditions are optimal for induction of the promoter. For example, for optimal induction of the *GAL1* promoter by galactose, there should not be any glucose in the culture medium. This can be achieved by preculturing yeast transformants in lactose prior to induction with galactose or by including a low-speed centrifugation step to remove the culture medium prior to switching transformants from a glucose to a galactose culture (e.g. Ijlst *et al.*, 1998).

(f) Protein is proteolytically degraded

Some proteins are highly sensitive to proteolysis and thus may not be expressed efficiently in normal host strains. Expression can be attempted in a protease-deficient yeast strain that contains a knockout of the vacuolar endopeptidase Pep4p, which controls the activation of other vacuolar hydrolases (Woolford *et al.*, 1986).

(g) *Host strain not suitable*

As discussed above, the growth and physiological properties of the various wild-type host strains may vary considerably. It may be worth analyzing different host strains for the expression of the HPI to find the most suitable host.

(h) *Promoter too strong or too weak*

The promoter chosen for the expression of the HPI may be either too strong or too weak. When the promoter is too strong it may lead to aggregation of the HPI, which often triggers proteolysis or may be cytotoxic to the yeast strain leading to slow growth or a lethal phenotype. If the gene product is potentially toxic to the yeast, then it would be better to express this under transcriptional control of a regulatable promoter, such as the *GAL1/10* or *MET3* promoters, which can be switched on after the yeast has reached a certain growth phase. If the promoter is too weak, it may not lead to detectable levels of the protein. For these reasons it may be worth testing different promoters. Variations in the expression levels can also be modulated by the choice of multi- or single-copy plasmids.

(i) *Problems with codon usage*

The preferred codon usages in *S. cerevisiae* and humans are rather different as indicated also by the overall GC contents of both organisms: 38% for yeast and 46% for human. Accordingly, human cDNAs can be rather GC-rich and thus contain many unfavorable codons for expression in yeast, which may lead to low or no expression in yeast. One way to overcome this is to use synthetic cDNAs in which all human codons are replaced with preferred yeast codons (e.g. Kotula and Curtis, 1991; Yadava and Ockenhouse, 2003). Alternatively, one may first try to express a partial synthetic cDNA in which only the human codons in the 5′ coding region are replaced with preferred yeast codons (e.g. Krynetski *et al.*, 1995; Flis *et al.*, 2005).

2. Correct expression but not functionally active

(a) *Use different functional tests*

Failure to restore function or growth of a transformant in functional complementation studies does not necessarily mean that the HPI is not functionally expressed. To demonstrate functional expression one should consider the use of more subtle functional tests such as measuring the specific catalytic activity of an enzyme or flux through a metabolic pathway. For example, functional expression of human desmosterol reductase in yeast could be demonstrated by measurement of efficient conversion of desmosterol into cholesterol in homogenates of transformants. The same activity measurement could also be used to determine the effect of mutations on the function of this enzyme (Waterham *et al.*, 2001). Furthermore, functional complementation of a *cact/cit2* deletion strain by the wild-type human mitochondrial carnitine-acylcarnitine translocase restored growth on oleic acid. However, the residual activity of a mutant protein identified in a patient with a mild clinical presentation was not sufficient to restore growth but nevertheless could be demonstrated by measurement of a partial restored fatty acid oxidation activity in *cact/cit2* cells expressing the mutant protein (Ijlst *et al.*, 2001).

(b) Tags introduced that interfere with the function

Expressing the HPI with an epitope tag allows one to determine that the protein is correctly expressed in those cases where no other detection method is available. Unfortunately, the presence of the tag can also affect correct folding, targeting, multimerization, and/or catalytic function of the protein. When a tag is preferred, one should consider comparing the relative effects of a C-terminal tag to that of an N-terminal tag.

(c) Protein not targeted to the correct subcellular location

Human proteins that normally function in certain organelles or cellular membranes, may not end up in these locations when expressed in yeast, e.g. because the human targeting sequences are not correctly recognized in yeast (Waterham, H. and Wanders, R.J., unpublished). As a consequence, these proteins may not become active, become active at the wrong subcellular location or are degraded. When antibodies or activity assays are available for the protein, the subcellular localization can be studied through cell fractionation studies or immunomicroscopy. Replacing the human targeting sequence with a yeast targeting sequence may solve the problem.

(d) Protein requires certain modifications that do not occur in yeast

Expression of secretory proteins in yeast may be problematic because protein glycosylation in yeast is very different from that in mammalian cells. Although yeast is capable of both *O*- and *N*-glycosylation, *O*-linked oligosaccharides in yeast are composed solely of mannose residues, whereas in mammals the *O*-linked oligosaccharides are composed of a variety of sugars including *N*-acetylgalactosamine, galactose and sialic acid (Spiro, 2002). *N*-glycosylation in yeast diverges after the trimming of the oligosaccharide core $Glc_3Man_9GlcNAc_2$ unit to $Man_8GlcNAc_2$. In yeast, these latter *N*-linked core oligosaccharide units are elongated through the addition of mannose outer chains, which often leads to hyperglycosylation (Knauer and Lehle, 1999), whereas in mammals more complex oligosaccharides are generated composed of a mixture of different sugars. This may present a problem in cases where glycosylation is important for the function of the heterologous protein. In addition, exogenous secretory proteins may not pass the biosynthetic quality control machinery in *S. cerevisiae* and be retained in the endoplasmic reticulum where they ultimately are degraded via the unfolded protein response (Zhang and Kaufman, 2006). Also other posttranslational protein modifications required for the function of an HPI may not occur correctly in yeast, such as methylation, phosphorylation, acetylation, and processing.

(e) Protein not correctly folded in yeast or requires additional partners

Human proteins may not become correctly folded in yeast leading to aggregation, which often triggers proteolysis, or may be cytotoxic to the yeast strain leading to slow growth or a lethal phenotype (Zhang and Kaufman, 2006). In addition, some proteins are only functionally active when in a heteroduplex with a natural partner protein. Expressing only one of the partners may not result in a functional active protein. A nice example for this is the calcitonin receptor-like receptor, the ligand specificity of which depends on heteromer formation with different members of the receptor activity modifying protein (RAMP) family. Co-expression of the calcitonin receptor-like receptor with the RAMP1 protein results in the formation of a calcitonin gene-related peptide receptor, whereas co-expression with RAMP2 or RAMP3 results in the formation of an adrenomedullin receptor (Miret *et al.*, 2002).

♦♦♦♦♦♦ III. IDENTIFICATION OF FUNCTIONAL HOMOLOGUES

The availability of the genomic sequence information of an increasing number of organisms, including yeast and humans, has led to the development of comparative genomic approaches to elucidate and study the functions of human genes and their encoded products. These approaches elaborate on formerly established methods and variations thereof, such as functional complementation testing and functional interactions screens, but also on bioinformatics approaches, including selective database searches for putative (functional) homologues based on sequence similarity, domain conservation and regulatory aspects, for example, coordinated expression patterns.

A. Functional Complementation

Ever since the first successful functional complementation of a *cdc2* mutant of the yeast *Schizosaccharomyces pombe* by the human *CDC2* orthologue (Lee and Nurse, 1987), yeast has been successfully used to identify human genes that encode proteins functionally similar to yeast proteins. Because such functional homologues often only share limited sequence similarity, they are commonly referred to as orthologues.

In most functional complementation studies, a human cDNA, which is predicted or suspected to encode a protein with a certain function, is expressed in a yeast mutant in which the gene encoding the corresponding yeast orthologue has been mutated or deleted, followed by selection for restored function. This selection may be robust, e.g. regained growth of the transformant on defined medium, or subtle and only partial, requiring more refined functional tests such as measuring the specific catalytic activity of an enzyme or flux through a metabolic pathway (Ijlst *et al.*, 1998; Zhang *et al.*, 2003). The latter approach also has been used to demonstrate that of the 12 different in-frame splice variants of the human *TAZ* gene, implicated in Barth syndrome, only one was capable of restoring growth when expressed in a yeast strain that lacked the yeast orthologue of this gene (Vaz *et al.*, 2003).

Another approach is screening human cDNA expression libraries for cDNAs that can functionally complement defined yeast mutants to identify any human orthologue of a yeast protein with known function, for which the gene has been mutated or deleted in those mutants (Zhang *et al.*, 2003).

If the yeast gene for which a human orthologue is sought is essential, the promoter of the essential gene can be replaced by homologous recombination by a conditional promoter including the carbon-source dependent *GAL1* or the repressible *MET3* and *tetO* promoters (see Zhang *et al.*, 2003, and references therein). The resulting strain will be viable in medium promoting transcription but will no longer grow under promoter repressing conditions unless functional complementation occurs by the human orthologue. This strategy can be used to screen on a gene-to-gene basis as well as for screening of human cDNA expression libraries (Zhang *et al.*, 2003).

Functional complementation studies have revealed or confirmed the function of many human genes. It should be noted, however, that the approach may not always be successful even in cases in which there exists a significant conservation in a certain cellular process and the components involved. For example, the various human *PEX* genes that encode proteins essential for peroxisome biogenesis share significant structural and sequence similarity with their yeast orthologues, but are not capable of functionally complementing the corresponding yeast *pex* mutants (Weller *et al.*, 2003). On the other hand, there are also examples of human and yeast orthologues that share no or only limited sequence similarity for which functional complementation was feasible. For example, the human lysosomal cystine transporter cystinosin, which is defective in the lysosomal storage disease cystinosis, shares only limited sequence homology with the yeast vacuolar Ers1 protein, but was capable of functionally complementing a yeast strain with a deletion of the *ERS1* gene (Gao *et al.*, 2005).

B. *In Silico* Identification

The availability of the genomic sequence information of an increasing number of organisms in conjunction with the increased knowledge on functional domains within amino acid sequences has led to the development of a variety of bioinformatics tools that nowadays often are used as a first step toward the identification of candidate human orthologues of yeast proteins. Bioinformatics approaches may include searches of human databases for proteins with overall sequence similarity to yeast proteins, for example by using the BLAST algorithm (e.g. Foury, 1997), but may also focus on conservation of functional domains within protein sequences, e.g. catalytic sites, interaction domains, targeting signals, etc. Examples of this *in silico* strategy are the identification of most of the currently 12 different *PEX* genes required for human peroxisome biogenesis and defective in the human disease Zellweger syndrome by virtue of their similarity in sequence and/or domain conservation with their yeast orthologues (Weller *et al.*, 2003). Because this disease is very rare and genetically heterogeneous, these genes could not have been detected by classical human genetic approaches such as linkage analysis.

Bioinformatics can be very helpful in identifying candidate human orthologues of yeast proteins or vice versa and, through this, in elucidating the function of these proteins, but definite proof for this still requires experimentation. This may be accomplished in different ways, including functional complementation and/or functional testing studies. Because, as mentioned above, functional complementation may not always be successful or an option, one can also perform functional tests, such as expression of the human orthologue in yeast followed by analysis of its subcellular localization, activity, interactions, etc. (see also below). Alternatively, if the human orthologue identified *in silico* is a candidate for a human disease, one can analyze the corresponding human gene in patients for mutations, as exemplified above for Zellweger syndrome (Weller *et al.*, 2003).

C. Screening for Interacting Proteins

The development of the yeast two-hybrid system (Fields and Song, 1989) and variations and improvements thereof has provided an easy accessible, straightforward tool to study functional interactions of human proteins (see Chapters 6 and 7, this volume). The methodology can be used to investigate the interaction between known proteins as well as for screening for potentially interacting proteins using human cDNA expression libraries and the HPI as "bait". The latter approach may also provide important clues about a possible function of the HPI, as exemplified for several proteins involved in the rare genetic defect Fanconi anemia, which is associated with bone marrow failure, chromosome instability, and predisposition to cancer (Reuter *et al.*, 2003). Using three of the human Fanconi anemia proteins as "baits" in a two-hybrid screen, 69 interacting proteins were identified, most of which are associated with four functional classes including transcription regulation (21 proteins), signaling (13 proteins), oxidative metabolism (10 proteins), and intracellular transport (11 proteins), suggesting that these proteins are functionally involved in several complex cellular pathways (Reuter *et al.*, 2003).

It is important to note that not all interactions revealed with yeast two-hybrid screens are physiologically significant, because often proteins that show interaction in the yeast screen appear physically separated in the cell. Details on this approach and potential drawbacks are discussed in chapters 6 and 7 of this volume.

◆◆◆◆◆◆ IV. FUNCTIONAL TESTING OF HUMAN GENES

A. Functional Characterization of Human Proteins

Expression in yeast can also be used to study functional aspects of an HPI other than by functional complementation. This includes studies on the function of the wild-type proteins as well as on the function of certain structural domains of these proteins using site-directed mutagenesis to introduce specific mutations. An example for this latter approach is the yeast-based screening method for trans-activation activity of the human estrogen receptor alpha. In the assay, the trans-activation activity of this ligand-activated transcription factor is coupled to the growth rate of yeast cells, through the use of an integrated *HIS3* reporter gene with a minimal promoter preceded by three tandem copies of the estrogen response element, allowing screening of libraries of randomly mutated estrogen receptors (Chen and Zhao, 2003).

Such heterologous protein expression can be readily done when yeast does not have a functional orthologue of the HPI. However, if there are known yeast proteins with comparable function or activity, it may be better to first delete the yeast genes encoding these proteins to obtain a clean background in which the functional aspects of the HPI can be determined more accurately.

B. Studying Disease Alleles

An important application of *S. cerevisiae* for studies on human diseases is the possibility to test the effect of mutations on the activity and/or function of a human protein. This can be assessed by the ability of the mutant allele to complement functionally a yeast strain lacking the yeast orthologue or, by evaluating the effect on the catalytic activity of a human enzyme expressed in yeast. Examples of the former approach are the analysis of the effect on growth of mutations found in the gene encoding carnitine acylcarnitine translocase (CACT) of several patients when expressed in a yeast *cact*-deletion strain (Ijlst *et al.*, 2001) and the analysis of the effect of mutations in the gene encoding cystinosin on hygromycin B sensitive growth of an *ers1*-deletion strain (Gao *et al.*, 2005). An example of the latter approach is the confirmation that mutations found in the gene encoding desmosterol reductase of some patients with a specific defect in cholesterol biosynthesis are indeed affecting the catalytic activity of the encoded protein (Waterham *et al.*, 2001).

C. Subcellular Localization Studies

The advantage of using of *S. cerevisiae* as expression system for human proteins when compared to bacterial expression systems is the fact that *S. cerevisiae* is a eukaryotic organism sharing the same subcellular make up with higher eukaryotic cells. This allows the study of aspects of human proteins that are required for targeting the correct subcellular compartments. Furthermore, this is in particular advantageous for the expression of membrane proteins that function in organellar membranes and thus can be functionally expressed in *S. cerevisiae* and further analyzed. Examples for this are the already mentioned CACT (Ijlst *et al.*, 2001) and desmosterol reductase (Waterham *et al.*, 2001) that function in the mitochondrial and endoplasmic reticulum membrane, respectively.

D. Ligand and Drugs Screening

The ability to functionally complement yeast cells or to functionally express human proteins in yeast is also of considerable pharmaceutical interest, because it allows the development of sensitive cell-based high-throughput screens for natural and synthetic ligands and drugs. In particular for the screening of the functions of G-protein coupled receptors (GPCRs), expression in yeast has turned out to be very successful (Beukers and Ijzerman, 2005; Minic *et al.*, 2005). In humans, GPCRs constitute a large family of cell surface receptors (~650) that couple interaction with certain extracellular ligands to the activation of intracellular G-proteins and downstream signal transduction pathways. Activating ligands are very diverse and include light, peptides, lipids, organic odorants, nucleotides, and taste molecules. GPCRs are one of the major targets in medicine (Beukers and Ijzerman, 2005).

The yeast cell-based screening method is based on the high homology between the yeast pheromone signaling pathway and that of human GPCRs. Upon expression in yeast, human GPCRs can couple functionally to the endogenous yeast G protein G-alpha (Gpa1p), to co-expressed mammalian G-alpha proteins or chimeric yeast-human G-alpha proteins, leading to induction of the yeast MAPK pathway. As read-out for GPCR activity, different reporter genes can be used that are expressed under control of an MAPK pathway-induced promoter, including *lacZ* (screening for blue colonies with X-gal), *HIS3* (screening for growth), and luciferase (screening for bioluminescence) (Minic *et al.*, 2005).

Yeast has also been used as a model system for protein conformational diseases (Coughlan and Brodsky, 2005; Li and Harris, 2005) and for screening of prion-inhibiting drugs, using two unconventional phenotypes, [*PSI*$^+$] and [*URE3*], caused by autocatalytic aggregation of the Sup35p and Ure2p proteins, respectively, with mechanistic similarities to prion formation in humans (Bach *et al.*, 2003). The screen is based on the changed protein synthesis fidelity when cells show the [*PSI*$^+$] phenotype (i.e. Sup35 in its aggregated prion conformation), because in this condition, ribosomes have the tendency to read through translation opal stop codons allowing growth in minimal synthetic medium of *ade1*-mutated cells that express an *ADE1* reporter gene with an opal stop codon in its open reading frame. When Sup35 is normally soluble, translation will terminate at the stop codon and cells will not grow in minimal synthetic medium lacking adenine, while on YPD plates cells will form red colonies. Alternatively, one can use the [*URE3*] phenotype for screening for growth dependent on the expression of the *ADE1*/*ADE2* gene under transcriptional control of the promoter of the yeast *DAL5* gene. Expression of the *ADE1* gene only occurs when Ure2p is in its aggregated prion conformation, but not when Ure2p is in its normal conformation (Schlumpberger *et al.*, 2001 see Chapter 20).

References

Ammerer, G. (1983). Expression of genes in yeast using the *ADCI* promoter. *Methods Enzymol.* **101**, 192–201.

Bach, S., Talarek, N., Andrieu, T., Vierfond, J. M., Mettey, Y., Galons, H., Dormont, D., Meijer, L., Cullin, C. and Blondel, M. (2003). Isolation of drugs active against mammalian prions using a yeast-based screening assay. *Nat. Biotechnol.* **21**, 1075–1081.

Bafunno, V., Giancaspero, T. A., Brizio, C., Bufano, D., Passarella, S., Boles, E. and Barile, M. (2004). Riboflavin uptake and FAD synthesis in *Saccharomyces cerevisiae* mitochondria: Involvement of the Flx1p carrier in FAD export. *J. Biol. Chem.* **279**, 95–102.

Belli, G., Gari, E., Aldea, M. and Herrero, E. (1998a). Functional analysis of yeast essential genes using a promoter-substitution cassette and the tetracycline-regulatable dual expression system. *Yeast* **14**, 1127–1138.

Belli, G., Gari, E., Piedrafita, L., Aldea, M. and Herrero, E. (1998b). An activator/repressor dual system allows tight tetracycline-regulated gene expression in budding yeast. *Nucleic Acids Res.* **26**, 942–947.

Beukers, M. W. and Ijzerman, A. P. (2005). Techniques: How to boost GPCR mutagenesis studies using yeast. *Trends Pharmacol. Sci.* **26**, 533–539.

Broach, J. R. (1983). Construction of high copy yeast vectors using 2-microns circle sequences. *Methods Enzymol.* **101**, 307–325.

Chen, Z. and Zhao, H. (2003). A highly efficient and sensitive screening method for trans-activation activity of estrogen receptors. *Gene* **306**, 127–134.

Coughlan, C. M. and Brodsky, J. L. (2005). Use of yeast as a model system to investigate protein conformational diseases. *Mol. Biotechnol.* **30**, 171–180.

Fields, S. and Song, O. (1989). A novel genetic system to detect protein–protein interactions. *Nature* **340**, 245–246.

Filipits, M., Simon, M. M., Rapatz, W., Hamilton, B. and Ruis, H. (1993). A *Saccharomyces cerevisiae* upstream activating sequence mediates induction of peroxisome proliferation by fatty acids. *Gene* **132**, 49–55.

Flis, K., Hinzpeter, A., Edelman, A. and Kurlandzka, A. (2005). The functioning of mammalian ClC-2 chloride channel in *Saccharomyces cerevisiae* cells requires an increased level of Kha1p. *Biochem. J.* **390**, 655–664.

Foury, F. (1997). Human genetic diseases: A cross-talk between man and yeast. *Gene* **195**, 1–10.

Gao, X. D., Wang, J., Keppler-Ross, S. and Dean, N. (2005). ERS1 encodes a functional homologue of the human lysosomal cystine transporter. *FEBS J.* **272**, 2497–2511.

Guthrie, C. and Fink, G. R. (eds) (1991). Guide to yeast genetics and molecular biology. *Methods in Enzymology,* vol. 194. Published by Academic Press, Inc., San Diego.

Hitzeman, R. A., Hagie, F. E., Hayflick, J. S., Chen, C. Y., Seeburg, P. H. and Derynck, R. (1982). The primary structure of the *Saccharomyces cerevisiae* gene for 3-phosphoglycerate kinase. *Nucleic Acids Res.* **10**, 7791–7808.

Hoffman, C. S. and Winston, F. (1987). A ten-minute DNA preparation from yeast efficiently releases autonomous plasmids for transformation of *Escherichia coli*. *Gene* **57**, 267–272.

Holland, J. P. and Holland, M. J. (1980). Structural comparison of two nontandemly repeated yeast glyceraldehyde-3-phosphate dehydrogenase genes. *J. Biol. Chem.* **255**, 2596–2605.

Ijlst, L., van Roermund, C. W., Iacobazzi, V., Oostheim, W., Ruiter, J. P., Williams, J. C., Palmieri, F. and Wanders, R. J. A. (2001). Functional analysis of mutant human carnitine acylcarnitine translocases in yeast. *Biochem. Biophys. Res. Commun.* **280**, 700–706.

Ijlst, L., Mandel, H., Oostheim, W., Ruiter, J. P., Gutman, A. and Wanders, R. J. A. (1998). Molecular basis of hepatic carnitine palmitoyltransferase I deficiency. *J. Clin. Invest.* **102**, 527–531.

Johnston, M. and Davis, R. W. (1984). Sequences that regulate the divergent GAL1-GAL10 promoter in *Saccharomyces cerevisiae*. *Mol. Cell. Biol.* **4**, 1440–1448.

Knauer, R. and Lehle, L. (1999). The oligosaccharyltransferase complex from yeast. *Biochim. Biophys. Acta* **1426**, 259–273.

Kotula, L. and Curtis, P. J. (1991). Evaluation of foreign gene codon optimization in yeast: Expression of a mouse Ig kappa chain. *Biotechnology* **9**, 1386–1389.

Krynetski, E. Y., Drutsa, V. L., Kovaleva, I. E. and Luzikov, V. N. (1995). High yield expression of functionally active human liver CYP2D6 in yeast cells. *Pharmacogenetics* **5**, 103–109.

Lee, M. G. and Nurse, P. (1987). Complementation used to clone a human homologue of the fission yeast cell cycle control gene *cdc2*. *Nature* **327**, 31–35.

Li, A. and Harris, D. A. (2005). Mammalian prion protein suppresses Bax-induced cell death in yeast. *J. Biol. Chem.* **280**, 17430–17434.

Minic, J., Sautel, M., Salesse, R. and Pajot-Augy, E. (2005). Yeast system as a screening tool for pharmacological assessment of G protein coupled receptors. *Curr. Med. Chem.* **12**, 961–969.

Miret, J. J., Rakhilina, L., Silverman, L. and Oehlen, B. (2002). Functional expression of heteromeric calcitonin gene-related peptide and adrenomedullin receptors in yeast. *J. Biol. Chem.* **277**, 6881–6887.

Reuter, T. Y., Medhurst, A. L., Waisfisz, Q., Zhi, Y., Herterich, S., Hoehn, H., Gross, H. J., Joenje, H., Hoatlin, M. E., Mathew, C. G. and Huber, P. A. (2003). Yeast two-hybrid screens imply involvement of Fanconi anemia proteins in transcription regulation, cell signaling, oxidative metabolism, and cellular transport. *Exp. Cell. Res.* **289**, 211–221.

Schlumpberger, M., Prusiner, S. B. and Herskowitz, I. (2001). Induction of distinct [URE3] yeast prion strains. *Mol. Cell. Biol.* **21**, 7035–7046.

Sikorski, R. S. and Hieter, P. (1989). A system of shuttle vectors and yeast host strains designed for efficient manipulation of DNA in *Saccharomyces cerevisiae*. *Genetics* **122**, 19–27.

Spaan, A. N., Ijlst, L., van Roermund, C. W., Wijburg, F. A., Wanders, R. J. A. and Waterham, H. R. (2005). Identification of the human mitochondrial FAD transporter and its potential role in multiple acyl-CoA dehydrogenase deficiency. *Mol. Genet. Metab.* **86**, 441–447.

Spiro, R. G. (2002). Protein glycosylation: Nature, distribution, enzymatic formation, and disease implications of glycopeptide bonds. *Glycobiology* **12**, 43R–56R.

Tzagoloff, A., Jang, J., Glerum, D. M. and Wu, M. (1996). *FLX1* codes for a carrier protein involved in maintaining a proper balance of flavin nucleotides in yeast mitochondria. *J. Biol. Chem.* **271**, 7392–7397.

Vaz, F. M., Houtkooper, R. H., Valianpour, F., Barth, P. G. and Wanders, R. J. A. (2003). Only one splice variant of the human TAZ gene encodes a functional protein with a role in cardiolipin metabolism. *J. Biol. Chem.* **278**, 43089–43094.

Voorn-Brouwer, T., van der Leij, I., Hemrika, W., Distel, B. and Tabak, H. F. (1993). Sequence of the *PAS8* gene, the product of which is essential for biogenesis of peroxisomes in Saccharomyces cerevisiae. *Biochim. Biophys. Acta.* **1216**, 325–328.

Wach, A., Brachat, A., Pohlmann, R. and Philippsen, P. (1994). New heterologous modules for classical or PCR-based gene disruptions in *Saccharomyces cerevisiae*. *Yeast* **10**, 1793–1808.

Waterham, H. R., Koster, J., Romeijn, G. J., Hennekam, R. C. M., Vreken, P., Andersson, H. C., FitzPatrick, D. R., Kelley, R. I. and Wanders, R. J. A. (2001). Mutations in the 3beta-hydroxysterol Delta24-reductase gene cause desmosterolosis, an autosomal recessive disorder of cholesterol biosynthesis. *Am. J. Hum. Genet.* **69**, 685–694.

Weller, S., Gould, S. J. and Valle, D. (2003). Peroxisome biogenesis disorders. *Annu. Rev. Genomics Hum. Genet.* **4**, 165–211.

Woolford, C. A., Daniels, L. B., Park, F. J., Jones, E. W., Van Arsdell, J. N. and Innis, M. A. (1986). The *PEP4* gene encodes an aspartyl protease implicated in the posttranslational regulation of *Saccharomyces cerevisiae* vacuolar hydrolases. *Mol. Cell. Biol.* **6**, 2500–2510.

Yadava, A. and Ockenhouse, C. F. (2003). Effect of codon optimization on expression levels of a functionally folded malaria vaccine candidate in prokaryotic and eukaryotic expression systems. *Infect. Immun.* **71**, 4961–4969.

Yoon, H. and Donahue, T. F. (1992). Control of translation initiation in *Saccharomyces cerevisiae*. *Mol. Microbiol.* **6**, 1413–1419.

Yun, D. F., Laz, T. M., Clements, J. M. and Sherman, F. (1996). mRNA sequences influencing translation and the selection of AUG initiator codons in the yeast *Saccharomyces cerevisiae*. *Mol. Microbiol.* **19**, 1225–1239.

Zhang, N., Osborn, M., Gitsham, P., Yen, K., Miller, J. R. and Oliver, S. G. (2003). Using yeast to place human genes in functional categories. *Gene* **303**, 121–129.

Zhang, N., Gardner, D. C., Oliver, S. G. and Stateva, L. I. (1999). Genetically controlled cell lysis in the yeast *Saccharomyces cerevisiae*. *Biotechnol. Bioeng.* **64**, 607–615.

Zhang, K. and Kaufman, R. J. (2006). The unfolded protein response: A stress signaling pathway critical for health and disease. *Neurology* **66**, S102–S109.

24 Bioinformatic Prediction of Yeast Gene Function

Insuk Lee, Rammohan Narayanaswamy and Edward M Marcotte
Center for Systems and Synthetic Biology, Institute for Cellular & Molecular Biology, University of Texas at Austin, Austin, TX 78712, USA

♦♦

CONTENTS

Introduction
Predicting function through guilt-by-association
Recognizing and assessing error in functional genomics data
A quantitative error model for yeast two-hybrid, mass spectrometry, and other interactions
Stronger inferences via data integration
Methods and protocols for employing pre-calculated functional predictions
An example application to the partially characterized gene *PRP43*

♦♦♦♦♦♦ I. INTRODUCTION

The bioinformatic prediction of gene function is, although young, already an extensive field, and with the high quality of the yeast genome sequence and the already large and rapidly growing volume of yeast functional genomics data, the prediction of yeast gene function is a substantial subfield in itself. A wide variety of approaches have been developed to predict gene function, ranging from sequence analyses to assign genes into functional families (Bork and Koonin, 1998; Ponting, 2001; Bateman *et al.*, 2004), to structural analyses to assign protein folds (Honig, 1999; Schonbrun *et al.*, 2002; Godzik, 2003) and active sites (Fetrow and Skolnick, 1998; Madabushi *et al.*, 2002), to phylogenetic analyses for subdividing gene families into functional subgroups (Eisen, 1998a, b; Abhiman and Sonnhammer, 2005; Engelhardt *et al.*, 2005) or predicting interacting partners (Pazos and Valencia, 2002). As 'gene function' takes such a wide variety of forms, from the corresponding protein's biochemical activity to its physical interaction partners to membership in a given pathway, we focus here only on the latter 'network' aspects of gene function: a protein's interaction and pathway partners, and the inferences of function that derive from these.

METHODS IN MICROBIOLOGY, VOLUME 36
0580-9517 DOI:10.1016/S0580-9517(06)36024-2

One of the most effective strategies for inferring pathway-type functional information has turned out to be the general strategy of 'guilt by association' (e.g., as in Eisen *et al.*, 1998; Marcotte *et al.*, 1999b; Aravind, 2000; Eisenberg *et al.*, 2000; Oliver, 2000; Wu *et al.*, 2002; Huynen *et al.*, 2003; Xia *et al.*, 2004; Jiang and Keating, 2005; Wolfe *et al.*, 2005, to name but a few). This chapter will discuss the inference of yeast gene function via guilt-by-association approaches, discussing a variety of relevant functional and comparative genomics approaches, and their integration to predict gene function more accurately. We focus in particular on how these approaches can be made quantitative by estimating the error rates in these data and in the predicted gene functions.

♦♦♦♦♦♦ II. PREDICTING FUNCTION THROUGH GUILT-BY-ASSOCIATION

A. A General Principle for Finding Yeast Gene Function

The general strategy of guilt by association involves implicating genes in the same biological processes. Linking an uncharacterized gene to genes known to function in ribosome biogenesis carries an implication that the uncharacterized gene functions in this general area as well. The specific linkages may imply more specific function. This strategy can be employed with many different classes of functional and comparative genomics data, some of which allow stronger inferences than others. The strength of inferences vary depending not only on the immediate links, the type of data, but also the larger dataset beyond the immediate genes of interest (i.e., a dataset might, for example be strong for *certain* classes of genes but weak for others), as well as the *prior* chances of such inferences being correct, an aspect that is frequently overlooked in these analyses.

In this section, we will first introduce the various classes of data useful for guilt-by-association inferences, discussing the forms of inferences that are commonly made from them. As we will see in Section III, all of these approaches can be made quantitative without explicit development of statistic models through supervised methods of benchmarking and measuring error. Many of the experimental techniques are treated in more detail in other chapters, including yeast two-hybrid assays (Chapters 6 and 7), expression analysis (Chapter 9), protein localization (Chapter 13), and synthetic genetic arrays (Chapter 16).

B. Guilt-by-Association via Functional Genomics

1. Protein interaction mapping by yeast two-hybrid and mass spectrometry

Yeast protein–protein interaction data are primarily derived from two approaches: (1) genome-wide, high-throughput yeast two-hybrid experiments, by which over 4000 unique protein interactions were observed between yeast proteins in three large-scale experiments (Ito *et al.*, 2000; Uetz *et al.*, 2000; Ito *et al.*, 2001), and (2) affinity purification of complexes of yeast proteins, followed by identification of the proteins by mass spectrometry (Gavin *et al.*, 2002; Ho *et al.*, 2002), identifying thousands more interactions among yeast proteins.

In addition to the large-scale experimental approaches, a number of groups have collected previously measured protein–protein interactions from the biological literature (Blaschke *et al.*, 1999; Humphreys *et al.*, 2000; Proux *et al.*, 2000; Thomas *et al.*, 2000; Marcotte *et al.*, 2001). This systematic collection of known protein interaction data provides necessary checks on the quality of the large-scale interaction data; large-scale protein interaction data have varied widely in accuracy (Mrowka *et al.*, 2001; Deane *et al.*, 2002; von Mering *et al.*, 2002).

Protein interaction databases combine the interactions from large-scale screens with interactions extracted from the literature, and include the biomolecular interaction network database (BIND) (Bader *et al.*, 2003) and the general repository for interaction datasets (GRID) (Breitkreutz *et al.*, 2003) databases and the database of interacting proteins (DIP). As of this writing, the DIP (http://dip.doe-mbi.ucla.edu/; Salwinski *et al.*, 2004) currently contains >18 000 protein–protein interactions among >4900 yeast proteins. The GRID database (http://www.thebiogrid.org) includes >20 000 yeast protein–protein interactions. The BIND database (http://www.unleashedinformatics.com/) includes >71 000 yeast molecular interactions, although these include non-protein–protein interactions in the count. For example, protein–DNA interaction data are also accumulating rapidly, primarily due to the scaling of chromatin immunoprecipitation methods to genome scale using DNA microarrays (Ren *et al.*, 2000; Bulyk *et al.*, 2001; Iyer *et al.*, 2001; Mukherjee *et al.*, 2004), allowing large-scale assays of ~200 yeast transcription factor binding specificities (Lee *et al.*, 2002; Harbison *et al.*, 2004).

For the purposes of inferring function from these interaction data, it is important to consider the model under which inferences are drawn. In particular, in direct measurements of protein interactions, such as the two-hybrid and mass spectrometry data, the experiments are typically performed by measuring interactions between a 'bait' protein and whatever 'prey' proteins it may interact with. If one protein (the 'bait') is observed to interact with multiple 'prey' proteins, there is no guarantee that the 'prey' will also interact with each other, although this may be likely in the case when they are members of the same protein complex. As shown in Figure 1A, there is a distinction made (Bader and Hogue, 2002) between a 'spoke' interpretation, in which only directly observed interactions between 'bait' and 'prey' are considered, and a 'matrix' interpretation, in which 'prey' bound by the same 'bait' protein are also inferred to interact with each other. Intuitively, the spoke model may seem too restrictive at times and the matrix model too permissive. As we

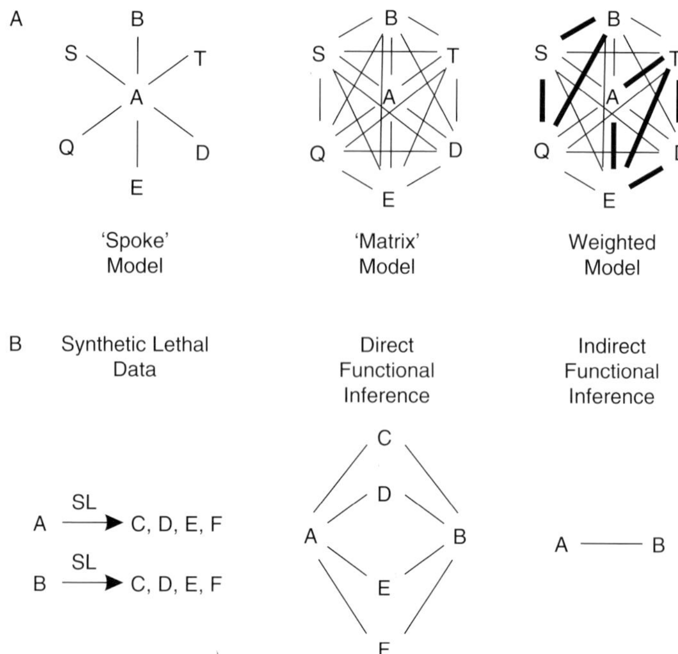

Figure 1. Alternate models for determining associations from functional genomics data. (A) Direct measurements of interactions, such as by yeast two-hybrid or mass spectrometry, can be interpreted as only providing evidence of 'bait'—'prey' associations (the 'spoke' model), as providing evidence for 'bait'—'prey' and 'prey'—'prey' associations (the 'matrix' model), or can be assigned weighted confidence scores based on interactions from the rest of the screen, as described in Section IV. (B) Genetic interaction data can provide evidence for associations between the synthetic lethal partners, or, less obviously, can provide evidence for linkages between genes synthetic lethal to the same set of other proteins.

will demonstrate in Section IV, there is an alternative model to these approaches, the weighted interaction model, which outperforms both of these strategies.

2. Genetic interactions and synthetic genetic arrays

Functional associations, far from being limited to physical interactions, can be drawn from more general associations between genes, such as those provided by genetic interactions. In yeast, the bulk of these data are from synthetic genetic array experiments, in which two mutant strains are robotically mated, sporulated, and the double mutant progeny examined for synthetic phenotypes, such as lethality (Tong *et al.*, 2001; Tong *et al.*, 2004). Unlike physical interactions, synthetic lethal relationships are not necessarily simple to interpret. They clearly represent legitimate constraints on the cell to grow properly, and it is generally perceived that the experiments have low false-positive rates (although this is hard to measure) that generally stem from technical errors, such as occasional defects in the original yeast deletion strain collection (Giaever *et al.*, 2002), rather than biological artifacts in the screens.

Nonetheless, it has been shown (Wong *et al.*, 2004; Kelley and Ideker, 2005) that only a fraction (perhaps half) of synthetic lethal interaction partners belong to the same biological pathway. Therefore, synthetic lethal interactions give two alternate interpretations for the purposes of inferring gene function, as illustrated in Figure 1B. Given a synthetic lethal interaction between two genes, one can interpret this as partial evidence that they belong to the same pathway. However, the same inference can often be drawn in the case where two genes are not themselves synthetic lethal to each other, but have synthetic interactions *with the same set of other proteins.*

3. Co-expression and co-localization

Owing to the prevalence of publicly available large-scale mRNA expression datasets, strong functional inferences can be drawn through analyses of genes' expression patterns. These data are primarily in the form of thousands of DNA microarray experiments stored in the Stanford Microarray Database (Gollub *et al.*, 2003) and the GEO database (Barrett *et al.*, 2005). These data have proved powerful in the guilt-by-association style transfer of function, with diverse algorithms developed to mine the data, ranging from simple calculations of correlations between genes expression profiles across a bank of microarray experiments to a rich variety of clustering, classification, and deconvolution algorithms for more sophisticated grouping of genes into functional groups (e.g., as reviewed in Slonim, 2002).

Complementing the mRNA expression data is yeast protein localization data, primarily from large-scale analyses of fusion protein localizations (Habeler *et al.*, 2002; Kumar *et al.*, 2002; Huh *et al.*, 2003). These data provide an important source of functional associations that vary from extremely specific (e.g., both proteins of interest localize to the spindle pole body) to very general (e.g., both are cytosolic). These data have proved most useful for functional inference when combined with other datasets (Jansen *et al.*, 2003).

C. Guilt-by-Association via Comparative Genomics

A number of comparative genomics methods has been employed to identify yeast gene function. Here, we summarize three of these approaches, in particular:

(1) The discovery of functional associations via the observation that bacterial orthologs of the genes occur in the same operons.
(2) The discovery of functional associations based upon co-inheritance of genes across many organisms.
(3) The discovery of functional associations by observation of gene fusion events.

1. Deriving yeast gene function from bacterial genome organization

This approach relies upon the trend for bacterial genes of related function to be organized into operons. Therefore, yeast orthologs of these genes are also likely to function together. Although many operons are known for some organisms (e.g., see the RegulonDB database for known operons of *E. coli*, Salgado *et al.*, 2004), many more are uncharacterized. Two computational methods, illustrated in Figure 2A and B, have proven effective for predicting functional relationships between genes by their orthologs' tendencies to co-occur

A Bacterial orthologs form operon, judged by intergenic distances

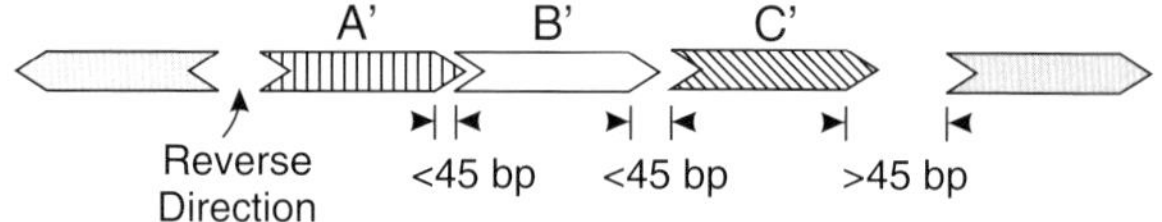

B Bacterial orthologs conserved as neighboring genes

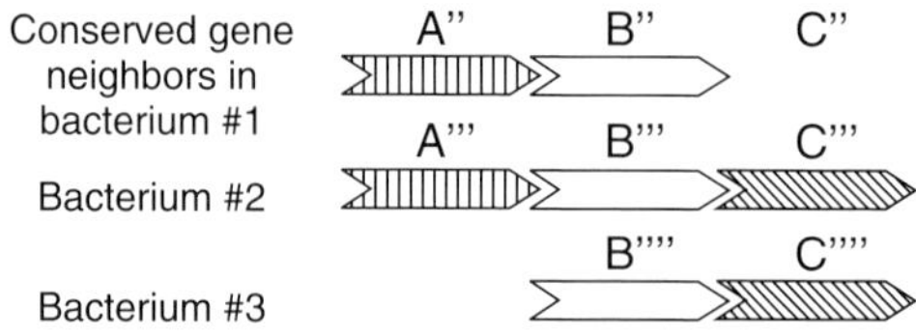

C Yeast genes co-inherited across sequenced genomes, judged by phylogenetic profiles

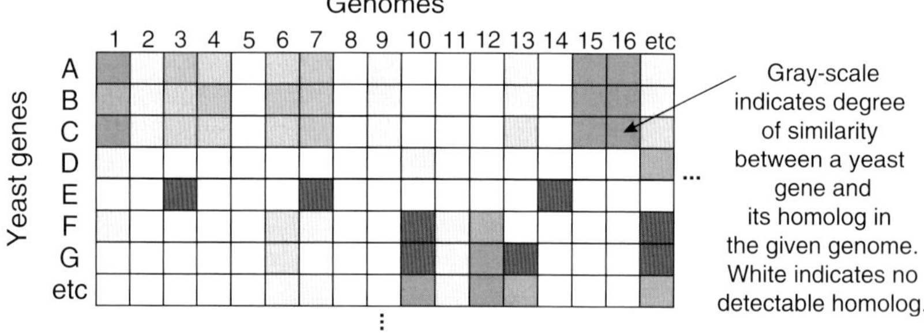

D Homologs in other organism fused into composite 'Rosetta Stone' genes

A' B' or B' C'

Figure 2. Functional associations derived from comparative genomics, in which the genomic organization of homologs or orthologs of the yeast genes may imply associations between the yeast genes. The genomic organization of yeast genes' bacterial orthologs into operons may imply functional associations between yeast genes, judged by (A) intergenic distances and (B) conservation of gene neighbors. Functional associations may also be inferred based upon (C) co-inheritance of genes, as measured by calculating and comparing phylogenetic profiles, or upon (D) fusions of homologous genes into composite genes.

in bacterial operons. Both of these methods ignore the identification of promoters and regulatory sequences, and instead exploit other properties to identify the operons. One approach exploits the tendency for adjacent genes in the same operon to be separated by short intergenic distances, while adjacent genes in different operons are separated by longer intergenic distances (Salgado *et al.*, 2000). The second approach, often referred to as the 'conserved gene neighbor' approach, exploits the tendency for genes in operons to be conserved as adjacent, neighboring genes in multiple bacterial genomes (Tamames *et al.*, 1997; Dandekar *et al.*, 1998; Overbeek *et al.*, 1999; Snel *et al.*, 2002; Yanai *et al.*, 2002). These approaches have proved remarkably powerful in yeast, as at least 1302 yeast proteins have bacterial orthologs with conserved gene-order information (Huynen *et al.*, 2003).

2. Phylogenetic profiles and the principle of co-inheritance

The second approach for identifying pathway and interaction partners exploits the tendency for proteins in the same pathway to be co-inherited across organisms (Pellegrini *et al.*, 1999; Huynen *et al.*, 2000), as illustrated in Figure 2C. A protein sequence is compared to sequences of all known proteins from the available fully sequenced genomes. From this analysis, a phylogenetic profile is calculated, describing which genomes contain homologs of the query gene. This phylogenetic profile can then be compared to those of all other genes in a genome to suggest functional partners of the query gene, with variations on the approach using orthologs rather than homologs (Eisen and Wu, 2002), employing probabilistic measures of profile similarity (Wu *et al.*, 2003) or measuring mutual information between the phylogenetic profiles (Huynen *et al.*, 2000; Date and Marcotte, 2003), and taking into account the phylogenetic relationships among the genomes (Vert, 2002; Sun *et al.*, 2005).

3. The Rosetta Stone/gene fusion approach

In the third approach, illustrated in Figure 2D, two genes in one organism can be inferred to be functionally linked due to the discovery of a third gene, which is the fusion of the two separate genes (Enright *et al.*, 1999; Marcotte *et al.*, 1999a). Empirically, it appears that gene fusions typically occur only between genes of related function; consequently, this method can be used to rapidly suggest functional partners for a gene. As with the previous examples, this approach is rarely used without enforcing some statistical criteria (Enright *et al.*, 1999; Yanai *et al.*, 2001; Verjovsky *et al.*, 2002; Bowers *et al.*, 2004) to rule out spurious associations that arise from the tendency of certain 'promiscuous domains' to be found in multidomain proteins with many other protein domains (Marcotte *et al.*, 1999a). This approach has also recently been combined with the gene neighbor approach by looking for conservation of protein domains as neighbors (i.e., ignoring whether the domains belong to the same protein or different proteins, merely requiring that they be adjacent on the chromosome, Pasek *et al.*, 2005).

♦♦♦♦♦♦ III. RECOGNIZING AND ASSESSING ERROR IN FUNCTIONAL GENOMICS DATA

It is a fact that all biological data, experimental or computational, low-throughput or high-, contain errors. An error-free experiment has yet to be conducted. It is clearly critical, then, to determine the extent of errors in both experiments and predictions. Are these large or small, and can they be measured precisely? Many of the original protein interaction and genetic interaction mapping experiments were presented only as experimental results, some of these experimental *tours de force*, but rarely (at first) accompanied by estimates of the error in these techniques (e.g., Chien *et al.*, 1991; Rigaut *et al.*, 1999). Much of our appreciation of the errors in these approaches (e.g., appreciation of the classes of false positive interactions identified in typical yeast two-hybrid assays) comes from empirical observations of many individual investigators (e.g., Bartel *et al.*, 1993; Estojak *et al.*, 1995) and from computational analyses performed after the original screens (e.g., Mrowka *et al.*, 2001; Deane *et al.*, 2002; von Mering *et al.*, 2002; Patil and Nakamura, 2005). Unfortunately, data for the failure rates of these techniques are rarely published, making it difficult to systematically measure error rates from small-scale assays (Jansen and Gerstein, 2004). For several of these approaches, specialized statistics have been developed to better measure the associated errors in specific large-scale screens, such as the prediction of true and false positive yeast two-hybrid interactions based upon internal properties of a large scale screen (Bader *et al.*, 2004).

It is clear that without any appreciation of the errors involved, one can be strongly misled as to the likely functions and interactions of a gene of interest. Functional and comparative genomics have the advantage that their errors can be measured and the data interpreted accordingly. In this section, we introduce the primary approaches for measuring error in these data, in this manner estimating how correct functional inferences drawn from these data are likely to be.

A. Reference Sets for Evaluating Functional Association

In predicting gene function by guilt by association, false predictions arise when associations are identified between functionally irrelevant or uncoupled genes. These false associations are generally products of non-biological variance in the data, even with extensive filtering of the data. Therefore, one of the more successful strategies has been to rely upon external, independent datasets to act as a form of 'gold-standard' reference for assessing the primary data quality. With high-quality reference sets, we can then evaluate different datasets using a unified quality criterion.

As with any 'supervised' approach, the quality of the reference set is critical and directly determines the quality of evaluations performed with the set. A faulty reference set can lead to faulty evaluation and, consequently, to faulty biological inferences. We consider three aspects of the reference set quality: First, the size (or coverage) of the reference set should be large enough to offer a statistically reliable evaluation of the data. Second, the set should be of high accuracy, containing a minimum number of false examples. Although the evaluation methods we will discuss are noise tolerant, a significant amount of noise can interfere with the analysis and lead to erroneous conclusions. Third, the resolution (or specificity) of information must be sufficiently high for the types of associations to be drawn. Even with all true examples in the reference set, highly generalized associations will be of less utility. For example, the fact that two proteins both localize to the cytosol only weakly indicates that they might be functionally associated, as the association is too general in nature to draw a strong association.

Diverse associations between genes, indicating different functional associations, can be used as reference sets. For example, reference sets might consist of the following:

1. protein pairs sharing functional annotation(s),
2. protein pairs sharing pathway annotation(s),
3. protein pairs found in the same complex(es), and
4. protein pairs sharing cellular localization annotation(s).

Box 1 lists a variety of yeast gene annotation sets of these types that are useful as reference sets. (Whenever possible, the precise locations of data files are given in the box, rather than the project web site gateways.) These annotations vary in both specificity and coverage. Moreover, several of these are hierarchically organized, and choosing different levels of the annotation hierarchy may generate quite different evaluations for the same dataset. Generally, top-level annotations provide extensive coverage but low-specificity (resolution), while low-level annotations decrease coverage but increase specificity. Therefore, the choice of an appropriate level(s) of hierarchical annotation must be considered carefully to achieve the optimal trade-off between coverage and specificity. Once chosen, the reference set should be consistent throughout the entire data analysis.

It is striking that the current yeast annotation and reference sets are quite non-overlapping. For example, fewer than half of the Kyoto Encyclopedia of Genes and Genomes (KEGG) pathway database associations are also contained in the gene ontology (GO) annotation set (Bork *et al.*, 2004). The low overlap is primarily due to different data mining methods and to inclusion bias among the annotation sets. This, however, provides the opportunity to generate

Box 1. Datasets recommended for benchmarking functional predictions.

Functional annotation

GO (Gene ontology) biological process
http://www.geneontology.org/ontology/process.ontology
GO is hierarchically organized, with the top level annotation most general and the bottom level most specific. Generally, the middle range of annotation provides a reasonable reference set. GO is notable for only annotating genes with the lowest annotation available and not listing the (implied) higher level terms, requiring the user to reconstruct these implied terms for a complete reference set.

Eukaryotic clusters of orthologous groups (KOGs)
ftp://ftp.ncbi.nih.gov/pub/COG/KOG/kog
KOGs is a comprehensive but very general annotation set, with <30 different functional terms. Due to its availability for many organisms, it is valuable for assessing poorly studied organisms.

CYGD (the comprehensive yeast genome database) functional category, hosted by MIPS (Munich Information Center for Protein Sequences)
ftp://ftpmips.gsf.de/yeast/catalogues/funcat/
CYGD is a comprehensive and detailed annotation set specific for yeast.

Pathway annotation

Kyoto Encyclopedia of Genes and Genomes (KEGG) pathway
ftp://ftp.genome.jp/pub/kegg/pathways/sce/sce_gene_map.tab
KEGG offers a three-level hierarchical annotation of pathways.

Complex annotation

CYGD (the comprehensive yeast genome database) complex, hosted by MIPS (Munich Information Center for Protein Sequences)
ftp://ftpmips.gsf.de/yeast/catalogues/complexcat/

GO (Gene ontology) cellular components
http://www.geneontology.org/ontology/component.ontology
Note: includes both complexes and subcellular locations, with similar organization as GO biological process.

Cellular localization annotation

Yeast GFP fusion localization database (hosted by UCSF)
http://yeastgfp.ucsf.edu/
The most comprehensive experimental annotation for yeast protein sub-cellular localization. It has, however, <30 different location terms, ranging widely in specificity.

GO (Gene ontology) cellular components
http://www.geneontology.org/ontology/component.ontology
Note: includes both complexes and subcellular locations, with similar organization as GO biological process.

TRIPLES (TRansposon-insertion phenotypes, localization, and expression in *Saccharomyces*)
ftp://ygac.med.yale.edu/ygac_pub_ftp/localization_pub_data_9_4_01.tab
Another localization annotation set, generated by random transposon-tagging of yeast genes.

YPL.db (Yeast Protein Localization Database)
http://ypl.uni-graz.at/pages/home.html

more comprehensive reference sets by combination of different annotation sets. In practice, both GO and KEGG have proven to be generally reliable for benchmarking datasets.

B. Evaluating Function Predictions Using a Reference Set

Having selected a reference, the next step in functional prediction is to select a method for evaluating functional predictions against the reference set. There are multiple methods to choose from, differing primarily in the parameters for measuring data quality. Two key parameters expressing data quality are *coverage* and *accuracy*. For example, we can measure coverage of the proteome (the complete set of yeast proteins) by the proteins in the dataset, or we can measure coverage of the interactome itself. As we do not currently know any complete interaction map, or even the size of a complete interaction map, the latter measure of coverage cannot be expressed as a percentage of the whole, but only as a total number of interactions. The coverage also can be represented as a fraction of the reference interactions. The accuracy can be measured by as the percentage of interaction data confirmed by the reference set (von Mering *et al.*, 2002), or alternately, as the likelihood of being true associations (Jansen *et al.*, 2003; Lee *et al.*, 2004). In the latter approach, likelihood is usually calculated as a log of the odds ratio that the data are correct to incorrect. (For example, a 2:1 odds ratio of being correct:incorrect corresponds to a 67% chance of being correct.)

This likelihood approach is often based upon Bayesian statistics, which allow relatively straightforward calculation of such likelihood ratios. To evaluate a dataset, a log likelihood ratio (LLR) can be calculated as

Bioinformatic Prediction of Yeast Gene Function

$$LLR = \ln\left(\frac{P(D|I)}{P(D|\sim I)}\right)$$

where $P(D|I)$ and $P(D|{\sim}I)$ are the probability of observing the functional genomics data (D) conditioned on the genes being associated in the reference set (I) or not being associated in the reference set (${\sim}I$). By Bayes' theorem, this equation can be rewritten as:

$$LLR = \ln\left(\frac{P(I|D)/P(\sim I|D)}{P(I)/P(\sim I)}\right)$$

where $P(I|D)$ and $P({\sim}I|D)$ are the frequencies of functional associations observed in the given dataset (D) between annotated genes that are associated in the reference set (I) or not associated in the reference set (${\sim}I$), respectively. $P(I)$ and $P({\sim}I)$ represent the prior expectations (the total frequencies of all reference set genes being associated in the reference set or not, respectively). This latter version of the equation is simpler to compute. A score of zero indicates interaction partners in the dataset being tested are no more likely than random to belong to the same pathway or to interact; higher scores indicate a more accurate dataset.

Another variation to consider for benchmarking functional inferences depends on whether we desire a single measured value for characterizing an entire dataset or instead use a continuous measurement to differentiate among different quality subsets within the data. A small dataset with a fairly uniform probability distribution of error can be described by a single evaluation score for the entire dataset. However, if we have a parameter associated with the dataset that correlates with data accuracy, it is often better to sort the data by this parameter and measure the accuracy as a function of a certain parameter range. An example of this is given in Section IV.

One formal way to evaluate data by coverage and accuracy is to plot a recall–precision curve. *Recall* (defined as the percentage of positives in the reference set correctly predicted as positives in the dataset) provides a measure of coverage and *precision* (defined as the percentage of predicted positives in the dataset confirmed as true positives by the reference set) provides a measure of accuracy (Figure 3). In the case of using a log-likelihood scoring scheme, the log-likelihood score would be substituted for the accuracy or precision parameter, plotted in its place on the y-axis of the plot.

Another formal assessment method is the receiver operating characteristic (ROC) curve. The ROC curve is named from its popularity in communications research, where it is used to detect the 'hit rate' of true positives at a given cost of false alarms over a noisy communications channel. The ROC approach can be applied as well to biological data, but has several shortcomings. We consider the true positive rate (defined as 100 times the number of true positives divided by the number of positives in the reference set) as the true hit rate and the false positive rate (defined as 100 times the number of false positive divided by the number of negatives in the reference set) as the false hit rate. All curves in the ROC plot (each

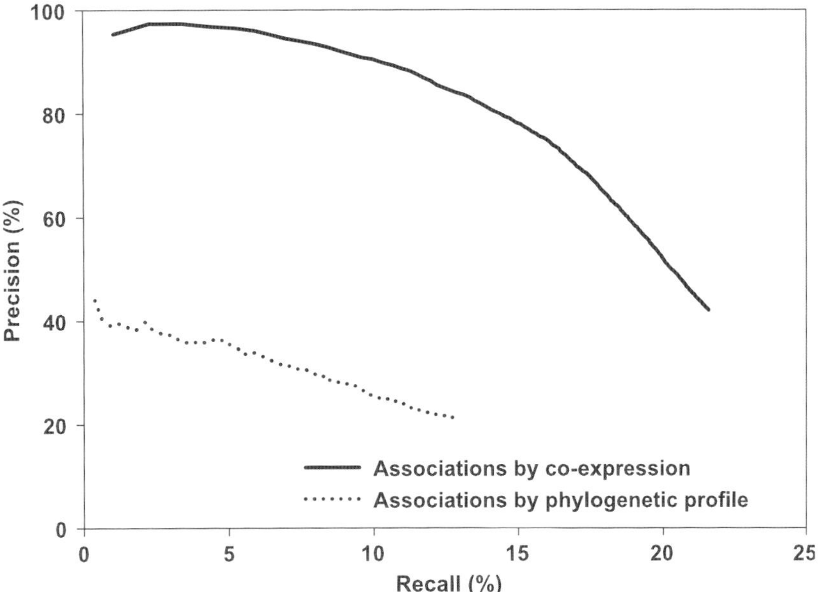

Figure 3. An example of a recall–precision curve, displaying the performance of mRNA co-expression and phylogenetic profiling at predicting functional associations. Recall (percentage of reference set positives correctly predicted as positive) and precision (percentage of predicted positives confirmed as positive by reference set) are based on a cumulative confusion matrix for the given recall level. For good predictions, we expect the evaluation curve to tend toward to the right upper corner. The given recall–precision curves indicate that these particular functional associations (Lee *et al.*, 2004) inferred from mRNA co-expression are better than those inferred from phylogenetic profiling, generally showing twice the precision at comparable levels of recall. In both Figures 3 and 4, KEGG pathways are used as the reference set (Kanehisa *et al.*, 2004).

corresponding to one dataset) start at 0% in both the true- and false-positive rates and ultimately arrive at 100% in each, as illustrated in Figure 4. For a dataset of randomly chosen associations, we would expect to observe the same rates of true and false positives, giving rise to a diagonal line. For a real dataset, however, we hope to observe a higher rate of gain of true positives. The area under each ROC curve is therefore proportional to the quality of the datasets. The ROC analysis has a notable shortcoming – to be strictly comparable, each dataset must be evaluated over the same reference set. So, for comparisons of multiple datasets, we have to define a subset of the original reference set common to all datasets in the plot, which can substantially diminish the size of the reference set used and which may even introduce a bias in the evaluation favoring specific datasets – it is possible that the *common subset* of reference associations has a significantly higher or lower proportion of matches for specific datasets than do the remaining set of associations.

A final benchmarking strategy that has proven useful is the modeling of a dataset as a mixture of true and false positives

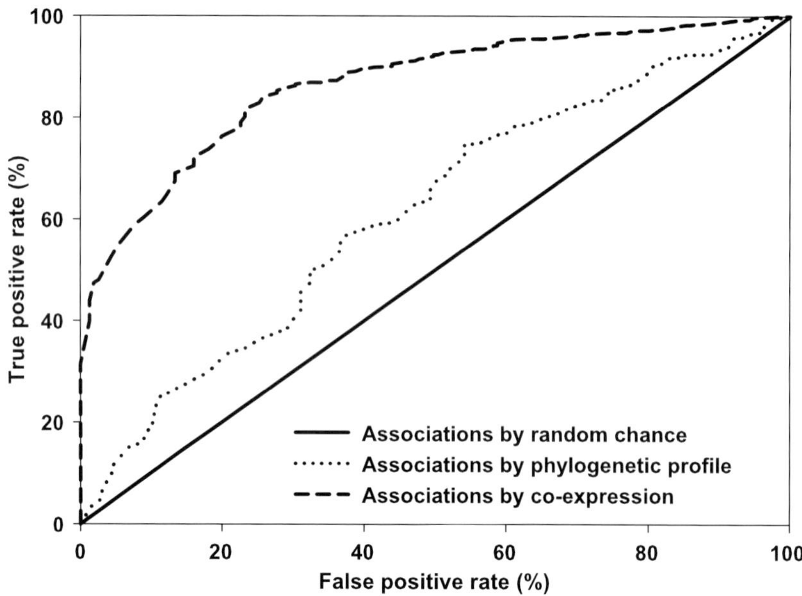

Figure 4. An example of a ROC curve, plotting the performance of mRNA co-expression and phylogenetic profiling at predicting functional associations. True positive rate (or true hit rate = percentage of reference set positives correctly predicted as positive) and false positive rate (or false hit rate = percentage of reference set negatives incorrectly predicted as positives) are plotted by ranking all predictions by their associated scores, then calculating the hit rates as a function of increasing rank. Random guesses generally show equal rates of true and false hits, generating a diagonal line in the ROC curve. Any prediction that is better than random guessing shows a curve above the diagonal. Overall performance can be estimated by the area under the ROC curve. For these particular data (the same data as in Figure 3), co-expression-based functional inferences out-perform phylogenetic profiling-based inferences.

(Mrowka *et al.*, 2001; Deane *et al.*, 2002). Therefore, the behavior of the dataset on some external benchmark (such as testing the extent of correlation of the gene pairs' mRNA expression patterns) can be mathematically fit as a linear combination of the behaviors of two reference sets (a positive and a negative sets), with the fit percentage of false positives providing the false positive rate of the data. An example of this strategy is calculating error rates for yeast two-hybrid data based upon the co-expression of the genes across a bank of DNA microarray experiments. Random gene pairs show one distribution of co-expression, true positives show another, skewed distribution, and the two-hybrid set being tested shows a mixture of the two distributions (Deane *et al.*, 2002; Kemmeren *et al.*, 2002).

Figure 5 illustrates the benchmarking of a number of the functional genomics datasets introduced earlier, using the recall–precision analysis described above. The relative scores indicate the utility of these datasets for inferring functional associations between genes linked by the particular methods. The reference set

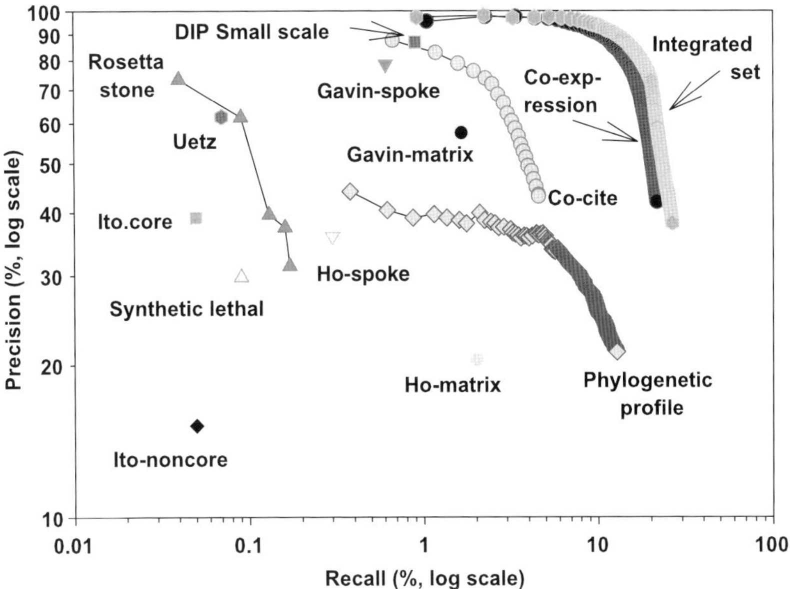

Figure 5. Benchmarking functional interactions from various experimental and computational approaches against the KEGG pathway reference set. The data plotted (analyzed as described in Lee *et al.*, 2004) include a collection of small scale experimental interaction data from the DIP (Small scale; Xenarios *et al.*, 2002), yeast proteins co-cited in Medline abstracts (co-cite), yeast genes whose mRNAs are co-expressed across 491 microarrays from the Stanford Microarray Database (co-expression; Gollub *et al.*, 2003), co-inherited yeast genes (phylogenetic profile; Date and Marcotte, 2003), yeast genes linked by gene fusion analysis (Rosetta stone), two spoke models and two matrix models of mass spectrometry data (Gavin-spoke, Gavin-matrix from Gavin *et al.*, 2002; and Ho-spoke, Ho-matrix from Ho *et al.* 2002), three yeast two-hybrid datasets (Uetz *et al.*, 2000; Ito-core, Ito-noncore from Ito *et al.*, 2001), and synthetic lethal interactions (synthetic lethal, Tong *et al.*, 2004). Datasets with associated continuous scores were evaluated in binned groups of 1000 gene pairs each. The reference set consists of ~49 000 positive examples, made by pairing genes belonging to the same KEGG pathway (Kanehisa *et al.*, 2004), and ~629 000 negative examples, made by pairing genes that do not belong to the same KEGG pathways. The integrated dataset was generated by the weighted sum log-likelihood method of (Lee *et al.*, 2004).

was generated by pairs of genes belonging to the same KEGG pathways (~49 000 positive examples). Each dataset is thus evaluated for its utility in inferring two genes belong to the same pathway. Better inferences tend toward the upper right corner of the plot. As can be seen, there is a wide spectrum of accuracy and coverage across the different datasets. For example, two proteins found to interact in the 'Uetz' yeast two-hybrid assay (Uetz *et al.*, 2000) are ~60% likely to belong to the same KEGG pathway, giving an indication of their general functional relatedness, while two genes that are synthetic lethal to each other in the Tong *et al.* (2004) dataset are ~30% likely to belong to the same KEGG pathway. The best functional inference is accomplished by integrating the individual datasets.

◆◆◆◆◆◆ IV. A QUANTITATIVE ERROR MODEL FOR YEAST TWO-HYBRID, MASS SPECTROMETRY, AND OTHER INTERACTIONS

Most interaction datasets, such as yeast two hybrid (Uetz *et al.*, 2000; Ito *et al.*, 2001) and protein complex affinity purification followed by mass spectrometry (Gavin *et al.*, 2002; Ho *et al.*, 2002), are accompanied by the simple observation of 'interacting' or 'not interacting'. However, for the purposes of predicting gene function, we would prefer to have some more fine-grained measure of confidence in the interactions and in the value of the transferred function (as in the 'weighted' interaction model of Figure 1A). One particularly simple and general theoretical framework for error is based upon the hypergeometric distribution. Variants of this approach appear to work well for many linkage and interaction types (Verjovsky *et al.*, 2002; Samanta and Liang, 2003; Schlitt *et al.*, 2003). In this section, we present an approach for calculating the hypergeometric error model and demonstrate its effectiveness.

The essential notion is that when data is assembled from many interaction experiments, such as deriving from a large-scale protein interaction assay, proteins take on varied numbers of interaction partners, with some proteins interacting promiscuously with many partners. This may be due to errors in the interaction screen or may be a legitimate reflection of the multi-functionality for the given proteins. In either case, assignment of specific function by guilt-by-association is non-trivial, with our intuition that we probably want to assign less confident scores for these particular interactions. Following this line of reason, we can apply a statistical re-interpretation to the interaction data to assign different confidence scores for different interactions depending on their tendency to participate in many or few interactions. With this approach, even the matrix model of mass spectrometry data becomes highly informative, with subsets showing extensive coverage with reliable quality.

Rather than use these coarse-grained models, we suggest a simple hypergeometric error model for calculating the probability of two proteins interacting by random chance given their behavior in the large-scale screen, assigning a probability (p-value) to the pair as:

$$p(\#\text{interactions} \geq k|n, m, N) = \sum_{i=k}^{\min(n,m)} p(i|n, m, N)$$

where :

$$p(i|n, m, N) = \frac{\binom{n}{i}\binom{N-n}{m-i}}{\binom{N}{m}} = \frac{n!(N-n)!m!(N-m)!}{(n-i)!i!(m-i)!(N-n-m+i)!N!},$$

and where k is the number of experiments in which an interaction is observed between proteins A and B (e.g., the number of yeast two-hybrid interactions or co-purifications involving both A and B), n the total number of experiments in which protein A is observed, m the total number of experiments in which protein B is observed, and N the total number of experiments with ≥ 1 interaction measured (e.g., the overall number of yeast two-hybrid interactions observed in a large-scale two-hybrid experiment, or the overall number of pull-down experiments with at least one interaction partner observed in a large-scale affinity purification mass spectrometry experiment). The resulting probability score indicates how likely the specific interaction between proteins A and B was to be observed by random chance given the other interactions observed in the screen, in effect taking into account how promiscuous A and B are in their interactions.

Figure 6 demonstrates the dramatic improvement in the confidence in yeast protein–protein interactions under this hypergeometric error model, using as examples a genome scale yeast two-hybrid experiment dataset (Ito *et al.*, 2001) and interacting protein data from affinity purifications followed by mass spectrometry (Gavin *et al.*, 2002). The cumulative prediction accuracy with the hypergeometric error model shows equivalent power to the whole dataset under the matrix model. However, the model achieves a much higher resolution of information, separating out the high- and low-quality interactions in the set, leading to more powerful data integration and interpretation. As discussed in the earlier sections, the 'spoke' model provides more accurate but less extensive interactions, while the 'matrix' model increases coverage at the expense of accuracy.

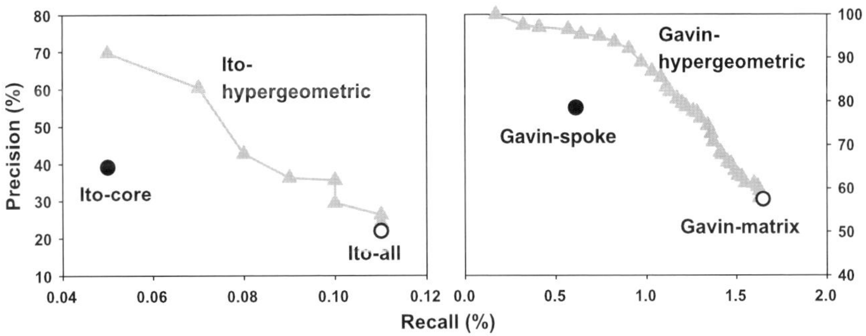

Figure 6. Re-interpreting interaction datasets using the hypergeometric error model. Large-scale yeast two-hybrid interaction sets and mass spectrometry interaction sets generally provide binary 'scores' (observed or not-observed). The hypergeometric error model will generate continuous accuracy scores for these data. These scores are illustrated for yeast two hybrid data (Ito *et al.*, 2001) and mass spectrometry interaction data (Gavin *et al.*, 2002). Recall–precision curves for the error models are plotted using bins of 500 gene pairs. For the two-hybrid data, the core dataset ('Ito-core') is more accurate than the entire dataset ('Ito-all'). The hypergeometric model shows substantially higher accuracies for the same level of coverage. For the mass spectrometry data, the 'spoke' model ('Gavin-spoke') is more accurate, but has lower coverage, than the 'matrix' model ('Gavin-matrix'). The hypergeometric model is considerably more accurate than the spoke model at the same level of coverage, and eventually converges on the matrix model.

However, the hypergeometric error model can differentiate the reliable interactions from the others in the original experimental data, resulting in generally more extensive coverage at accuracy equal to the spoke model. Note that this approach is quite generic, and can be applied to many different forms of linkages, interactions, and predicted linkages.

♦♦♦♦♦♦ V. STRONGER INFERENCES VIA DATA INTEGRATION

Data integration for predicting gene function has recently drawn considerable attention for a couple of major reasons. First, no single experimental or computational approach has proven able to analyze the yeast gene set to completion, due either to biases in the system or to increasing technical difficulties in taking screens to 100% of the genes/proteins. Second, different approaches show bias in the sets of genes and classes of functional inference that they are more useful for. Integrating inferences from multiple data types is therefore mandated. Agreement between inferences drawn from distinct datasets also increases our confidence in their correctness. In general, we expect better functional inferences when integrating multiple datasets than from even the best single dataset. There are many different approaches for data integration, ranging from quick-and-dirty methods, such as the voting method, to more statistically empowered approaches, such as Bayesian networks.

A. Simple Voting Methods

There are many variations to the voting approach. If we have only binary (interact or not interact) inferences from the given dataset, an intersection or a union method can be adopted (e.g., as in Marcotte *et al.*, 1999b; Mellor *et al.*, 2002). While the former takes a subset of the data with support from multiple datasets after applying suitable thresholds for each analysis, the latter takes a combined set with support from at least one dataset. The 'intersection' approach works best with redundant datasets and gives the better integration of the two while the 'union' approach performs well when relatively independent datasets are involved and tends to give improved integration without significant loss in accuracy. If the datasets have associated confidence scores, the method could be extended by taking the maximum or the averaged scores of multiple lines of evidence to generate an integrated dataset.

B. Bayesian Statistical Methods

The foremost goal of a formal statistical approach to data integration entails the normalization of scores from different datasets with different features. Normalizing all the datasets for integration can be performed using the reference sets discussed earlier. The final integrated data will be biased and characterized by the types of information implemented in it. For example, if we want to learn protein–protein physical interactions using an integrated dataset, our evaluation of the pre-integration datasets should be based on a reference set of protein–protein physical interaction.

One of the common approaches to data integration is using a Bayesian network (e.g., as in Jansen *et al.*, 2003; Troyanskaya *et al.*, 2003). In this approach, a graph is created with each node (vertex) representing a specific condition (a combination of dataset and parameter) and probability of true instance. If two different conditions are not completely independent (i.e., if A happens, B tends to happen as well), they are connected with some probability. In this case, the construction of the graph is not entirely straightforward. Finding the correct degree of dependence between conditions is critical for choosing an appropriate model for integration.

There are two distinct styles of Bayesian network approach. In the fully connected Bayesian network, we assume that all conditions are connected, and find a model for the best integration by evaluating all possible combinations of conditions. For a simple model, this approach is feasible, but for a complex one, the problem of exponentially increasing combinations precludes this approach. Biological data, being highly complex, often face this problem. In an attempt to simplify the data integration, a naïve Bayesian network approach has been adopted in some instances. This simpler approach, however, assumes that the different datasets are completely independent. Interestingly, such simple naïve Bayesian network approaches have performed well in functional genomics data integration (Jansen *et al.*, 2003; Troyanskaya *et al.*, 2003; Patil and Nakamura, 2005), which may indicate that existing genomics data are relatively sparse, with only minimal overlap between different datasets.

Into the future, however, we expect genomics data will continue to increase and the problem of measuring correlation between different datasets will have to be addressed. Therefore, it is imperative to identify an efficient way to handle information redundancy for integration of large datasets. For example, one simple but incomplete approach to this is the weighted sum method (Lee *et al.*, 2004), which has the merit of simplicity but at the cost of only treating a subset of correlations in the datasets. An example of integration by this approach is shown in Figure 5.

C. Other Methods

A number of more complex probabilistic and kernel-based supervised approaches have been proposed (e.g., see Yamanishi *et al.*,

2004 for kernel canonical correlation analysis, Jiang and Keating (2005) for a combined approach using probabilistic assignment of function combined with decision trees, and Tanay *et al.* (2004) for graph partitioning based approaches) that reliably assign biological function from combinations of high-throughput data.

Note that we have taken a fairly narrow view of data integration for function prediction here, focusing primarily on the guilt-by-association method as determined from network models of gene associations. However, other methods worth noting include models of data integration that are completely independent of network structure, such as those based on learning association rules by training on reference sets (Clare and King, 2003), as well as approaches for predicting genetic interactions (Wong *et al.*, 2004) and diverse approaches for learning specific ontology descriptions or gene annotations for uncharacterized genes, such as by the effective propagation of information through gene networks (see, e.g., Vazquez *et al.*, 2003; Deng *et al.*, 2004a, b; Karaoz *et al.*, 2004; Nabieva *et al.*, 2005) or through algorithmic classifiers (Pavlidis *et al.* 2002).

◆◆◆◆◆◆ VI. METHODS AND PROTOCOLS FOR EMPLOYING PRE-CALCULATED FUNCTIONAL PREDICTIONS

In this section, we highlight several of the main functional- and comparative-genomics based integrated models of yeast gene function. Each is available either as binary functional linkages between yeast genes or in the form of an internet server for interactively searching for functional linkages (Box 2).

A. AVID (Jiang and Keating, 2005)

URL: http://web.mit.edu/biology/keating/AVID/

(1) The user can enter the gene name or the GO ID in the query box.
(2) AVID is a computational function prediction framework that integrates data from high-throughput experiments, such as yeast two-hybrid, mass spectrometry, DNA microarrays, protein localization data, and protein sequence similarity.
(3) AVID initially assumes that the query gene interacts with all other genes, then successively filters low-confidence partners from the list based on varied criteria, resulting in a list of functional interactions with candidate genes at the level of three GO categories – molecular function, biological process, and cellular component.

Box 2. Predictions of yeast gene function via integrating multiple datasets.

AVID (Jiang and Keating, 2005)
http://web.mit.edu/biology/keating/AVID/

FinalNet (Lee *et al.*, 2004)
Online supplemental data with paper, also available for download from SGD database: ftp://genome-ftp.stanford.edu/pub/yeast/data_download/systematic_results/published_computational_predictions/lee_pmid_15567862/

LIANG (Samanta and Liang, 2003)
http://www.systemix.org/PP/partners/index.php

MAGIC (Troyanskaya *et al.*, 2003)
http://genome-www.stanford.edu/magic/

PIT (Jansen *et al.*, 2003)
http://networks.gersteinlab.org/intint/index-2.html

PLEX (Date and Marcotte, 2005)
http://bioinformatics.icmb.utexas.edu/plex/

Prediction of CCPs (Zhang *et al.*, 2004)
Online supplemental data with paper

PREDICTOME (Mellor *et al.*, 2002)
http://predictome.bu.edu/index.php

PROLINKS (Bowers *et al.*, 2004)
http://dip.doe-mbi.ucla.edu/pronav

STRING (von Mering *et al.*, 2005)
http://string.embl.de/

B. FinalNet (Lee *et al.*, 2004)

Available as supplemental data from the on-line publication, or from the following URL:

ftp://genome-ftp.stanford.edu/pub/yeast/data_download/systematic_results/published_computational_predictions/lee_pmid_15567862/

(1) This dataset awaits an interactive web feature. It can be readily accessed, however, through the above *Saccharomyces* genome database (SGD) FTP site.
(2) The user can download the yeast_FinalNet.txt.gz and unzip the file as a text (.txt) file.
(3) This file has comprehensive listings of the interacting partners of all yeast genes computed by an integrated Bayesian formalism. The probabilistic functional prediction is calculated by computing

a LLS for each functional genomics dataset to arrive at an integrated LLS score. The datasets include mRNA co-expression, Rosetta Stone gene fusions, phylogenetic profiles, literature mining (co-citation), and genetic and protein interaction experiments. This initial integrated network is further rescored by identifying linkages that are dependent on the network context. A final integrated network is arrived at, on which is applied a threshold based upon 'gold-standard' small-scale protein interaction assays. The resulting 'ConfidentNet' constitutes 34 000 linkages between 4681 genes. Further hierarchical clustering yields a network with 627 modules of functionally linked genes spanning 3285 genes, giving an estimate of protein systems in a yeast cell.

C. LIANG (Samanta and Liang, 2003)

URL: http://www.systemix.org/PP/partners/index.php

(1) The user can enter the name of the query gene in the query box. The server will return candidate interacting partners.
(2) This function prediction works on the hypothesis that two proteins sharing common interacting partners more than just by random chance are likely to interact with each other with higher probability. The known interacting proteins from DIP (Salwinski *et al.*, 2004) are taken to illustrate this hypothesis. Further analyses of the interacting partners of these proteins are done to generate the complete dataset.

D. MAGIC (Troyanskaya *et al.*, 2003)

URL: http://genome-www.stanford.edu/magic/

Currently, the Multisource Association of Genes by Integration of Clusters (MAGIC) dataset does not have an interactive web interface. However the user can access the final output at the above internet site (also available from the FTP download site of the SGD database). The predicted interactions for a given gene can be seen as different GO category IDs at different stringencies or cut-off scores (cut-off 0.9 and cut-off 0.75 available at the FTP site)

(1) MAGIC defines gene pairs having 'functional' relationships (i.e., involved in the same biological processes, defining processes as in the GO database).
(2) The Bayesian network that is implemented in MAGIC draws from known protein–protein interactions in the GRID database and protein–DNA interactions in the promoter database of *Saccharomyces cerevisiae* (Zhu and Zhang, 1999). Expression data is also incorporated. A score for each data source is calculated that represents the strength of each method's confidence in the

existence of a relationship between a gene pair. A combined probability score is then calculated.

E. PIT (Jansen *et al.*, 2003)

URL: http://networks.gersteinlab.org/intint/index-2.html

(1) A web interface exists to search these physical interaction predictions, in which a user can paste or type in the systematic name of the query gene and be given predicted interaction partners.
(2) Briefly, a probabilistic combination of multiple datasets is used to predict physical protein–protein interactions. The protein complexes represented in the MIPS database are used as the positive 'gold-standard dataset' whereas a negative gold-standard dataset consisting of protein pairs in separate sub-cellular compartments are used to train the Bayesian network. Two separate probabilistic interactomes (PIs) are constructed, one called the PIP (PI predicted) uses genomic context, mRNA co-expression, GO process, MIPS function and gene essentiality as datasets and the other called PIE (PI experimental) that uses data from co-immunoprecipitation/mass spectrometry and yeast two-hybrid experiments. These two interactomes along with the gold-standards are further integrated to a total PI (PIT) that represents a comprehensive view of known and predicted protein complexes in yeast.
(3) The user can choose a likelihood ratio cut-off (Lcut) as well as choose the PI version to be browsed for a particular analysis. An Lcut of 600 defines a threshold where a given protein pair is predicted to exist in the same complex with a better than 50% chance and therefore serves as a useful default.
(4) The output lists the proteins interacting with the query in an easily interpretable tabular form.

F. PLEX (Date and Marcotte, 2005)

URL: http://bioinformatics.icmb.utexas.edu/plex/

(1) PLEX returns functional associations predicted from phylogenetic profiles, operon neighbors, and gene fusions. Click on 'Submit a new job' and paste a protein sequence into the Protein Link Explorer (PLEX) query box or its Genbank GI number from the NCBI non-redundant database.
(2) Once the sequence is obtained, PLEX proceeds to compare the query sequence against ~350 000 proteins from 89 fully sequenced genomes using BLAST (Altschul *et al.*, 1997) sequence alignments. A phylogenetic profile is constructed from BLAST scores of the top-matching homologs in each genome. The

profile is displayed as a series of colored boxes, with blue indicating absence of the query in a given genome as opposed to red, which indicated varying degrees of confidence with which the query is present in different genomes.

(3) After a profile is created, the user can compare it to the profiles of all known proteins in any of the 89 genomes by entering a mutual information score cut-off for the comparison of phylogenetic profiles.

(4) PLEX searches are iterative: the first search associates the query gene (which may be from any source) with genes from the database genomes; successive iterations return pre-calculated associations (including operon neighbors and gene fusions) among database genes.

G. Predictome (Mellor *et al.*, 2002)

URL: http://predictome.bu.edu/index.php

(1) Predictome contains pre-calculated associations among genes from yeast and other organisms. The associations are derived from a variety of functional and comparative genomics approaches. It can be searched via a web interface, in which the user can paste the gene name (derived from SGD) into the query box to get the proteins predicted to be associated with the query protein.

(2) The output yields putative protein links with the query protein by integrating both experimental (yeast two-hybrid, co-immunoprecipitation and co-expression) and computational (phylogenetic profiles, gene fusion, and gene neighbor) datasets.

(3) The user can further view the interaction results in a network format using the VISANT applet.

H. PROLINKS (Bowers *et al.*, 2004)

URL: http://128.97.39.94/cgi-bin/functionator/pronav

(1) Enter the GI number of the query protein or the gene name and the genome to query.

(2) The user can then choose the sequence ID corresponding to the query gene name.

(3) The annotation page that is displayed lists general information such as sequence, chromosome, and existing annotation. Additional tabs in the page allow the user to display the homologs, the functional linkages 'PROLINKS' or to a graph depicting the protein interaction network involving the query protein.

(4) PROLINKS arrives at the output by computing the phylogenetic profile, Rosetta Stone, gene-neighbor and gene-cluster scores of a query protein. The user can further set a minimum confidence threshold to test for the strength of interactions.

I. STRING (von Mering *et al.*, 2005)

URL: http://string.embl.de/

(1) Enter the SWISS-PROT identifier of the query protein or its amino acid sequence in the query box.
(2) A search can be performed in the 'COG' mode which links with the COG database to classify the query protein to a particular COG category in an all-or-none fashion. A 'protein' mode search can also be done that relies on a matrix of pre-computed all-against-all protein similarity scores consisting of ~750 000 proteins.
(3) Once the query protein is associated with a COG category or with partner proteins as the case may be, search tool for the retrieval of interacting genes/proteins (STRING) computes a 'combined' score from seven individual sub-scores using a Bayesian prediction scheme. These include gene neighbors, gene fusions, phylogenetic profiles, mRNA co-expression, large-scale experiments (e.g., yeast two-hybrid), database imports that constitute previous knowledge, and literature mining (co-citation). This 'joint' probability score is often of higher confidence as compared to individual sub-scores.
(4) A final result is displayed as a network depiction with the individual sub-scores and the combined scores for every functional prediction. These scores serve to give a quick insight into the possible nature of interactions of a query protein, especially for proteins that are un-annotated.

◆◆◆◆◆◆ VII. AN EXAMPLE APPLICATION TO THE PARTIALLY CHARACTERIZED GENE *PRP43*

As an example, we will now take *PRP43* (*YGL120C*), a yeast RNA helicase, through several of the function prediction databases discussed above. *PRP43* is an essential yeast DEAH box protein, one of the family of proteins thought to possess RNA helicase activity and that function extensively in the coordination and catalysis of the pre-mRNA splicing reaction. *PRP43* itself, although not shown to exhibit an RNA helicase activity, was implicated to function in the disassembly of the U2/U5.U6 spliceosomal complex post-catalysis and subsequent release of the lariat RNA (Staley and Guthrie, 1998; Martin *et al.*, 2002).

Figure 7 shows the functional associations of *PRP43* predicted from the different methods introduced in Section VI. The genes linked to *PRP43* are labeled according to their broad functional categories (based upon their GO annotations and literature) – 'mRNA processing', including mRNA transactions, such as pre-mRNA splicing; 'rRNA processing and biogenesis', constituting the

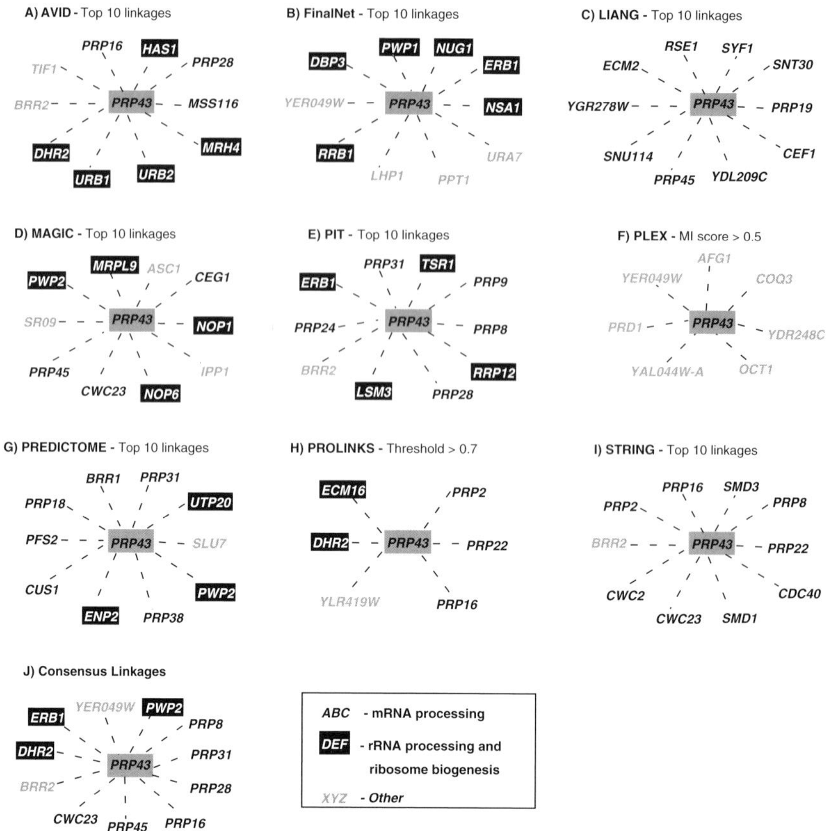

Figure 7. Functional associations of *PRP43* predicted by seven algorithms. Default parameters were used for querying the gene in each of seven functional prediction internet servers, and in each case, only the top 10 functional associations are shown (or fewer, if less than 10 are returned passing the default confidence threshold). Genes involved in mRNA processing are labeled in black text with white background, gene involved in rRNA processing and ribosome biogenesis are labeled in white fonts on black background, and gene of other functions are labeled in gray text on a white background. (A) Depicts the top 10 linkages as predicted by AVID, (B) depicts the top 10 linkages as predicted by FinalNet, (C) depicts the top 10 linkages as predicted by Liang *et al.*, (D) depicts the top 10 linkages as predicted by MAGIC, (E) depicts the top 10 linkages as predicted by PIT , (F) depicts the top 10 linkages as predicted by PLEX, (G) depicts the top 10 linkages predicted by PREDICTOME, (H) depicts all the linkages predicted by PROLINKS at a threshold greater than 0.7, (I) depicts the top 10 linkages predicted by STRING, and (J) depicts the consensus linkages that are predicted by more than one server, considering only their top 10 linkages.

synthesis, processing, and assembly of rRNAs into ribosomes; and 'other', referring to functions besides these two categories.

Although it may appear at first glance in Figure 7 that the algorithms are returning different predictions, most of the apparent disagreement is simply a function of including only the top 10 associations from each – as the methods include different data types and have different scoring functions, they tend to exhibit trivial differences in the ranking of associated genes, and we have omitted the full set of predictions from many methods for reasons of space, concentrating only on the top 10 predictions per method. Nonetheless, although the specific predictions vary, certain linkages are identified by multiple algorithms, such as the linkages of *PRP43* to *ERB1, YER049W, CWC23, PWP2, PRP2, PRP28, DHR2, PRP8, BRR2, PRP31,* and *PRP16* (see Figure 7J). An examination of the broad functional categories of these predicted genes shows that the majority lie in mRNA processing such as pre-mRNA splicing, the known function of *PRP43* (Martin *et al.*, 2002). However, a number of associations are inferred with genes implicated in rRNA processing and ribosome biogenesis. This observation turns out to be in agreement with very recent data (Lebaron *et al.*, 2005; Combs *et al.*, 2006; Leeds *et al.*, 2006) that indicate that *PRP43* serves an essential role in the biogenesis of both ribosome subunits while having a non-essential role in pre-mRNA processing. Thus, the two primary functions of *PRP43* are correctly inferred by the algorithms. Although further experiments would be needed to verify the exact candidates involved in this process with *PRP43* and their manner of involvement, it is clear that integrated function prediction databases can be immensely valuable at generating new and testable hypotheses.

Acknowledgment

This work was supported by grants from the N.S.F. (IIS-0325116, EIA-0219061, 0241180), N.I.H. (GM06779-01), Welch (F1515) and a Packard Fellowship (E.M.M.).

References

Abhiman, S. and Sonnhammer, E. L. (2005). Large-scale prediction of function shift in protein families with a focus on enzymatic function. *Proteins* **60**, 758–768.

Altschul, S. F. *et al.* (1997). Gapped BLAST and PSI-BLAST: a new generation of protein database search programs. *Nucleic Acids Res.* **25**, 3389–3402.

Aravind, L. (2000). Guilt by association: contextual information in genome analysis. *Genome Res.* **10**, 1074–1077.

Bader, G. D., Betel, D. and Hogue, C. W. (2003). BIND: the biomolecular interaction network database. *Nucleic Acids Res.* **31**, 248–250.

Bader, G. D. and Hogue, C. W. (2002). Analyzing yeast protein–protein interaction data obtained from different sources. *Nat. Biotechnol.* **20**, 991–997.

Bader, J. S., Chaudhuri, A., Rothberg, J. M. and Chant, J. (2004). Gaining confidence in high-throughput protein interaction networks. *Nat. Biotechnol.* **22**, 78–85.

Barrett, T., Suzek, T. O., Troup, D. B., Wilhite, S. E., Nagu, W. C., Ledoux, P., Rudnev, D., Lash, A. E., Fujibuchi, W. and Edgar, R. (2005). NCBI GEO: mining millions of expression profiles – database and tools. *Nucleic Acids Res.* **33**, D562–D566.

Bartel, P., Chien, C. T., Sternglanz, R. and Fields, S. (1993). Elimination of false positives that arise in using the two-hybrid system. *Biotechniques* **14**, 920–924.

Bateman, A. *et al.* (2004). The Pfam protein families database. *Nucleic Acids Res* **32**, D138–D141.

Blaschke, C., Andrade, M. A., Ouzounis, C. and Valencia, A. (1999). Automatic extraction of biological information from scientific text: protein–protein interactions. *Proc. Int. Conf Intell. Syst. Mol. Biol.* 60–67.

Bork, P. *et al.* (2004). Protein interaction networks from yeast to human. *Curr. Opin. Struct. Biol.* **14**, 292–299.

Bork, P. and Koonin, E. V. (1998). Predicting functions from protein sequences – where are the bottlenecks?. *Nat. Genet.* **18**, 313–318.

Bowers, P. M. *et al.* (2004). Prolinks: a database of protein functional linkages derived from coevolution. *Genome Biol.* **5**, R35.

Breitkreutz, B. J., Stark, C. and Tyers, M. (2003). The GRID: the general repository for interaction datasets. *Genome Biol.* **4**, R23.

Bulyk, M. L., Huang, X., Choo, Y. and Church, G. M. (2001). Exploring the DNA-binding specificities of zinc fingers with DNA microarrays. *Proc. Natl. Acad. Sci. USA* **98**, 7158–7163.

Chien, C. T., Bartel, P. L., Sternglanz, R. and Fields, S. (1991). The two-hybrid system: a method to identify and clone genes for proteins that interact with a protein of interest. *Proc. Natl. Acad. Sci. USA* **88**, 9578–9582.

Clare, A. and King, R. D. (2003). Predicting gene function in *Saccharomyces cerevisiae*. *Bioinformatics* **19**(Suppl 2), II42–II49.

Combs, D. J., Nagel, R. J., Ares, M. J. and Stevens, S. W. (2006). Prp43p is a DEAH-box spliceosome disassembly factor essential for ribosome biogenesis. *Mol. Cell. Biol.* **26**, 523–534.

Dandekar, T., Snel, B., Huynen, M. and Bork, P. (1998). Conservation of gene order: a fingerprint of proteins that physically interact. *Trends Biochem. Sci.* **23**, 324–328.

Date, S. V. and Marcotte, E. M. (2003). Discovery of uncharacterized cellular systems by genome-wide analysis of functional linkages. *Nat. Biotechnol.* **21**, 1055–1062.

Date, S. V. and Marcotte, E. M. (2005). Protein function prediction using the Protein Link EXplorer (PLEX). *Bioinformatics* **21**, 2558–2559.

Deane, C. M., Salwinski, L., Xenarios, I. and Eisenberg, D. (2002). Protein interactions: two methods for assessment of the reliability of high-throughput observations. *Mol. Cell. Proteomics* **1**, 349–356.

Deng, M., Chen, T. and Sun, F. (2004a). An integrated probabilistic model for functional prediction of proteins. *J. Comput. Biol.* **11**, 463–475.

Deng, M., Tu, Z., Sun, F. and Chen, T. (2004b). Mapping gene ontology to proteins based on protein–protein interaction data. *Bioinformatics* **20**, 895–902.

Eisen, J. A. (1998a). A phylogenomic study of the MutS family of proteins. *Nucleic Acids Res.* **26**, 4291–4300.

Eisen, J. A. (1998b). Phylogenomics: improving functional predictions for uncharacterized genes by evolutionary analysis. *Genome Res.* **8**, 163–167.

Eisen, J. A. and Wu, M. (2002). Phylogenetic analysis and gene functional predictions: phylogenomics in action. *Theor. Popul. Biol.* **61**, 481–487.

Eisen, M. B., Spellman, P. T., Brown, P. O. and Botstein, D. (1998). Cluster analysis and display of genome-wide expression patterns. *Proc. Natl. Acad. Sci. USA* **95**, 14863–14868.

Eisenberg, D., Marcotte, E. M., Xenarios, I. and Yeates, T. O. (2000). Protein function in the post-genomic era. *Nature* **405**, 823–826.
Engelhardt, B. E., Jordan, M. I., Muratore, K. E. and Brenner, S. E. (2005). Protein molecular function prediction by Bayesian phylogenomics. *PLoS Comput. Biol.* **1**, e45.
Enright, A. J., Iliopoulos, I., Kyrpides, N. C. and Ouzounis, C. A. (1999). Protein interaction maps for complete genomes based on gene fusion events. *Nature* **402**, 86–90.
Estojak, J., Brent, R. and Golemis, E. A. (1995). Correlation of two-hybrid affinity data with *in vitro* measurements. *Mol. Cell. Biol.* **15**, 5820–5829.
Fetrow, J. S. and Skolnick, J. (1998). Method for prediction of protein function from sequence using the sequence-to-structure-to-function paradigm with application to glutaredoxins/thioredoxins and T1 ribonucleases. *J. Mol. Biol.* **281**, 949–968.
Gavin, A. C. *et al.* (2002). Functional organization of the yeast proteome by systematic analysis of protein complexes. *Nature* **415**, 141–147.
Giaever, G. *et al.* (2002). Functional profiling of the *Saccharomyces cerevisiae* genome. *Nature* **418**, 387–391.
Godzik, A. (2003). Fold recognition methods. *Methods Biochem. Anal.* **44**, 525–546.
Gollub, J. *et al.* (2003). The Stanford Microarray Database: data access and quality assessment tools. *Nucleic Acids Res.* **31**, 94–96.
Habeler, G. *et al.* (2002). YPL.db: the yeast protein localization database. *Nucleic Acids Res.* **30**, 80–83.
Harbison, C. T. *et al.* (2004). Transcriptional regulatory code of a eukaryotic genome. *Nature* **431**, 99–104.
Ho, Y. *et al.* (2002). Systematic identification of protein complexes in *Saccharomyces cerevisiae* by mass spectrometry. *Nature* **415**, 180–183.
Honig, B. (1999). Protein folding: from the levinthal paradox to structure prediction. *J. Mol. Biol.* **293**, 283–293.
Huh, W. K. *et al.* (2003). Global analysis of protein localization in budding yeast. *Nature* **425**, 686–691.
Humphreys, K., Demetriou, G. and Gaizauskas, R. (2000). Two applications of information extraction to biological science journal articles: enzyme interactions and protein structures. *Pac. Symp. Biocomput.* 505–516.
Huynen, M., Snel, B., Lathe, W., 3rd and Bork, P. (2000). Predicting protein function by genomic context: quantitative evaluation and qualitative inferences. *Genome Res.* **10**, 1204–1210.
Huynen, M. A., Snel, B., von Mering, C. and Bork, P. (2003). Function prediction and protein networks. *Curr. Opin. Cell Biol.* **15**, 191–198.
Ito, T. *et al.* (2000). Toward a protein–protein interaction map of the budding yeast: a comprehensive system to examine two-hybrid interactions in all possible combinations between the yeast proteins. *Proc. Natl. Acad. Sci. USA* **97**, 1143–1147.
Ito, T. *et al.* (2001). A comprehensive two-hybrid analysis to explore the yeast protein interactome. *Proc. Natl. Acad. Sci. USA* **98**, 4569–4574.
Iyer, V. R. *et al.* (2001). Genomic binding sites of the yeast cell-cycle transcription factors SBF and MBF. *Nature* **409**, 533–538.
Jansen, R. *et al.* (2003). A Bayesian networks approach for predicting protein–protein interactions from genomic data. *Science* **302**, 449–453.
Jansen, R. and Gerstein, M. (2004). Analyzing protein function on a genomic scale: the importance of gold-standard positives and negatives for network prediction. *Curr. Opin. Microbiol.* **7**, 535–545.

Jiang, T. and Keating, A. E. (2005). AVID: an integrative framework for discovering functional relationships among proteins. *BMC Bioinform.* **6**, 136.

Kanehisa, M. *et al.* (2004). The KEGG resource for deciphering the genome. *Nucleic Acids Res.* **32**(Database issue), D277–D280.

Karaoz, U. *et al.* (2004). Whole-genome annotation by using evidence integration in functional-linkage networks. *Proc. Natl. Acad. Sci. USA* **101**, 2888–2893.

Kelley, R. and Ideker, T. (2005). Systematic interpretation of genetic interactions using protein networks. *Nat. Biotechnol.* **23**, 561–566.

Kemmeren, P. *et al.* (2002). Protein interaction verification and functional annotation by integrated analysis of genome-scale data. *Mol. Cell.* **9**, 1133–1143.

Kumar, A. *et al.* (2002). The TRIPLES database: a community resource for yeast molecular biology. *Nucleic Acids Res.* **30**, 73–75.

Lebaron, S., Froment, C., Fromont-Racine, M., Rain, J. C., Monsarrat, B., Caizergues-Ferrer, M. and Henry, Y. (2005). The splicing ATPase prp43p is a component of multiple preribosomal particles. *Mol. Cell. Biol.* **25**, 9269–9282.

Lee, I., Date, S. V., Adai, A. T. and Marcotte, E. M. (2004). A probabilistic functional network of yeast genes. *Science* **306**, 1555–1558.

Lee, T. I. *et al.* (2002). Transcriptional regulatory networks in *Saccharomyces cerevisiae*. *Science* **298**, 799–804.

Leeds, N. B. *et al.* (2006). The splicing factor Prp43p, a DEAH box ATPase, functions in ribosome biogenesis. *Mol. Cell. Biol.* **26**, 513–522.

Madabushi, S. *et al.* (2002). Structural clusters of evolutionary trace residues are statistically significant and common in proteins. *J. Mol. Biol.* **316**, 139–154.

Marcotte, E. M. *et al.* (1999a). Detecting protein function and protein–protein interactions from genome sequences. *Science* **285**, 751–753.

Marcotte, E. M. *et al.* (1999b). A combined algorithm for genome-wide prediction of protein function. *Nature* **402**, 83–86.

Marcotte, E. M., Xenarios, I. and Eisenberg, D. (2001). Mining literature for protein–protein interactions. *Bioinformatics* **17**, 359–363.

Martin, A., Schneider, S. and Schwer, B. (2002). Prp43 is an essential RNA-dependent ATPase required for release of lariat-intron from the spliceosome. *J. Biol. Chem.* **277**, 17743–17750.

Mellor, J. C. *et al.* (2002). Predictome: a database of putative functional links between proteins. *Nucleic Acids Res.* **30**, 306–309.

Mrowka, R., Patzak, A. and Herzel, H. (2001). Is there a bias in proteome research?. *Genome Res.* **11**, 1971–1973.

Mukherjee, S. *et al.* (2004). Rapid analysis of the DNA-binding specificities of transcription factors with DNA microarrays. *Nat. Genet.* **36**, 1331–1339.

Nabieva, E. *et al.* (2005). Whole-proteome prediction of protein function via graph-theoretic analysis of interaction maps. *Bioinformatics* **21**(Suppl 1), i302–i310.

Oliver, S. (2000). Guilt-by-association goes global. *Nature* **403**, 601–603.

Overbeek, R. *et al.* (1999). The use of gene clusters to infer functional coupling. *Proc. Natl. Acad. Sci. USA* **96**, 2896–2901.

Pasek, S. *et al.* (2005). Identification of genomic features using microsynthesis of domains: domain teams. *Genome Res.* **15**, 867–874.

Patil, A. and Nakamura, H. (2005). Filtering high-throughput protein–protein interaction data using a combination of genomic features. *BMC Bioinform.* **6**, 100.

Pavlidis, P., Weston, J., Cai, J. and Noble, W. S. (2002). Learning gene functional classifications from multiple data types. *J. Comput. Biol.* **9**, 401–411.

Pazos, F. and Valencia, A. (2002). *In silico* two-hybrid system for the selection of physically interacting protein pairs. *Proteins* **47**, 219–227.

Pellegrini, M. *et al.* (1999). Assigning protein functions by comparative genome analysis: protein phylogenetic profiles. *Proc. Natl. Acad. Sci. USA* **96**, 4285–4288.

Ponting, C. P. (2001). Issues in predicting protein function from sequence. *Brief. Bioinform.* **2**, 19–29.

Proux, D., Rechenmann, F. and Julliard, L. (2000). A pragmatic information extraction strategy for gathering data on genetic interactions. *Proc. Int. Conf. Intell Syst. Mol. Biol.* **8**, 279–285.

Ren, B. *et al.* (2000). Genome-wide location and function of DNA binding proteins. *Science* **290**, 2306–2309.

Rigaut, G. *et al.* (1999). A generic protein purification method for protein complex characterization and proteome exploration. *Nat. Biotechnol.* **17**, 1030–1032.

Salgado, H. *et al.* (2004). RegulonDB (version 4.0): transcriptional regulation, operon organization and growth conditions in *Escherichia coli* K-12. *Nucleic Acids Res.* **32**, D303–D306.

Salgado, H., Moreno-Hagelsieb, G., Smith, T. F. and Collado-Vides, J. (2000). Operons in *Escherichia coli*: genomic analyses and predictions. *Proc. Natl. Acad. Sci. USA* **97**, 6652–6657.

Salwinski, L. *et al.* (2004). The database of interacting proteins: 2004 update. *Nucleic Acids Res.* **32**(Database issue), D449–D451.

Samanta, M. P. and Liang, S. (2003). Predicting protein functions from redundancies in large-scale protein interaction networks. *Proc. Natl. Acad. Sci. USA* **100**, 12579–12583.

Schlitt, T. *et al.* (2003). From gene networks to gene function. *Genome Res.* **13**, 2568–2576.

Schonbrun, J., Wedemeyer, W. J. and Baker, D. (2002). Protein structure prediction in 2002. *Curr. Opin. Struct. Biol.* **12**, 348–354.

Slonim, D. K. (2002). From patterns to pathways: gene expression data analysis comes of age. *Nat. Genet.* **32**(Suppl), 502–508.

Snel, B., Bork, P. and Huynen, M. A. (2002). The identification of functional modules from the genomic association of genes. *Proc. Natl. Acad. Sci. USA* **99**, 5890–5895.

Staley, J. P. and Guthrie, C. (1998). Mechanical devices of the spliceosome: motors, clocks, springs, and things. *Cell* **92**, 315–326.

Sun, J. *et al.* (2005). Refined phylogenetic profiles method for predicting protein–protein interactions. *Bioinformatics* **21**, 3409–3415.

Tamames, J., Casari, G., Ouzounis, C. and Valencia, A. (1997). Conserved clusters of functionally related genes in two bacterial genomes. *J. Mol. Evol.* **44**, 66–73.

Tanay, A., Sharan, R., Kupiec, M. and Shamir, R. (2004). Revealing modularity and organization in the yeast molecular network by integrated analysis of highly heterogeneous genomewide data. *Proc. Natl. Acad. Sci. USA* **101**, 2981–2986.

Thomas, J., *et al.* (2000). Automatic extraction of protein interactions from scientific abstracts. *Pac. Symp. Biocomput.* 541–552.
Tong, A. H. *et al.* (2001). Systematic genetic analysis with ordered arrays of yeast deletion mutants. *Science* **294**, 2364–2368.
Tong, A. H. *et al.* (2004). Global mapping of the yeast genetic interaction network. *Science* **303**, 808–813.
Troyanskaya, O. G. *et al.* (2003). A Bayesian framework for combining heterogeneous data sources for gene function prediction (in *Saccharomyces cerevisiae*). *Proc. Natl. Acad. Sci. USA* **100**, 8348–8353.
Uetz, P. *et al.* (2000). A comprehensive analysis of protein–protein interactions in *Saccharomyces cerevisiae. Nature* **403**, 623–627.
Vazquez, A., Flammini, A., Maritan, A. and Vespignani, A. (2003). Global protein function prediction from protein–protein interaction networks. *Nat. Biotechnol.* **21**, 697–700.
Verjovsky Marcotte, C. J. and Marcotte, E. M. (2002). Finding functionally linked proteins from gene fusions with confidence. *Appl. Bioinform.* **2**, 93–100.
Vert, J. P. (2002). A tree kernel to analyse phylogenetic profiles. *Bioinformatics* **18**(Suppl 1), S276–S284.
von Mering, C. *et al.* (2002). Comparative assessment of large-scale datasets of protein–protein interactions. *Nature* **417**, 399–403.
von Mering, C. *et al.* (2005). STRING: known and predicted protein–protein associations, integrated and transferred across organisms. *Nucleic Acids Res.* **33**(Database issue), D433–D437.
Wolfe, C. J., Kohane, I. S. and Butte, A. J. (2005). Systematic survey reveals general applicability of "guilt-by-association" within gene coexpression networks. *BMC Bioinform.* **6**, 227.
Wong, S. L. *et al.* (2004). Combining biological networks to predict genetic interactions. *Proc. Natl. Acad. Sci. USA* **101**, 15682–15687.
Wu, J., Kasif, S. and DeLisi, C. (2003). Identification of functional links between genes using phylogenetic profiles. *Bioinformatics* **19**, 1524–1530.
Wu, L. F. *et al.* (2002). Large-scale prediction of *Saccharomyces cerevisiae* gene function using overlapping transcriptional clusters. *Nat. Genet.* **31**, 255–265.
Xenarios, I. *et al.* (2002). DIP, the database of interacting proteins: a research tool for studying cellular networks of protein interactions. *Nucleic Acids Res.* **30**, 303–305.
Xia, Y. *et al.* (2004). Analyzing cellular biochemistry in terms of molecular networks. *Ann. Rev. Biochem.* **73**, 1051–1087.
Yamanishi, Y., Vert, J. P. and Kanehisa, M. (2004). Protein network inference from multiple genomic data: a supervised approach. *Bioinformatics* **20**(Suppl 1), I363–I370.
Yanai, I., Derti, A. and DeLisi, C. (2001). Genes linked by fusion events are generally of the same functional category: a systematic analysis of 30 microbial genomes. *Proc. Natl. Acad. Sci. USA* **98**, 7940–7945.
Yanai, I., Mellor, J. C. and DeLisi, C. (2002). Identifying functional links between genes using conserved chromosomal proximity. *Trends Genet.* **18**, 176–179.
Zhang, L. V., Wong, S. L., King, O. D. and Roth, F. P. (2004). Predicting co-complexed protein pairs using genomic and proteomic data integration. *BMC Bioinform.* **5**, 38.
Zhu, J. and Zhang, M. Q. (1999). SCPD: a promoter database of the yeast *Saccharomyces cerevisiae. Bioinformatics* **15**, 607–611.

25 Yeast Genetic Strain and Plasmid Collections

Karl-Dieter Entian[1] and Peter Kötter[2]
[1] *Center of Exellence: Macromolecular Complexes and Institute for Molecular Biosciences, Johann Wolfgang Goethe University, Max-von-Laue Str. 9, 60438 Frankfurt/Main, Germany;*
[2] *Institute for Molecular Biosciences, Johann Wolfgang Goethe University, Max-von-Laue Str. 9, 60438 Frankfurt/Main, Germany*

♦♦♦

CONTENTS

Introduction
Yeast gene deletion strain collections
Genetic tools, strains and plasmids
Strain stability and services
Access to the strain and plasmid collections
Conclusions and impact of the deletion collections on eukaryotic research

♦♦♦♦♦♦ I. INTRODUCTION

A. General Introduction

The yeast *Saccharomyces cerevisiae* is a very suitable organism for genetic analysis (Lindegren and Lindegren, 1943) and now its genome has been completely sequenced, deletion mutants for each gene can be easily established by reverse genetics. Using the very precise recombination apparatus of *S. cerevisiae,* each gene locus can be replaced by selection markers such as amino acid auxotrophies, nucleoside auxotrophies or dominant resistance markers. This allowed the generation of a collection of mutants for each of the approximately 6000[1] open reading frames (ORFs) within the *S. cerevisiae* genome (Kowalczuk *et al.*, 1999; Wood *et al.*, 2001; Mackiewicz *et al.*, 2002; Cliften *et al.*, 2003). Even deletions within essential genes could be collected in the form of heterozygous

Yeast Genetic Strain and Plasmid Collections

[1]Depending on the bioinformatic tools the number of annotated yeast genes varies from 5800 to 6400 genes.

METHODS IN MICROBIOLOGY, VOLUME 36
0580-9517 DOI:10.1016/S0580-9517(06)36025-4

diploid strains. Additionally, the deletion cassettes can be amplified by PCR from the respective deletion mutant, so that the deletion can be easily introduced into any *S. cerevisiae* strain of interest. At present, deletion mutants for about 5900 genes are available from the various *S. cerevisiae* strain collections. Furthermore, an increasing number of strains and plasmids can be received where genes are under regulated expression, have affinity tags for easy purification of the respective protein, are fused to GFP (green fluorescence protein) for their easy cellular localization, etc. Additionally, a large number of genetic tools for gene functional analysis became available which are also accessible.

The first collection of *S. cerevisiae* deletion mutants, the EUROSCARF collection,[2] was established in 1994 and originated from the German functional analysis network[3] (about 325 mutants) and the European EUROFAN I network[4] (about 825 mutants), both of which aimed to study *S. cerevisiae* genes of unknown function.

In 1999, a consortium of US (3000 deletions), Canadian (600 deletions) and European (EUROFAN II, 2400 deletions) groups could generate bar-coded deletions in nearly all 6,000 yeast genes (Winzeler *et al.*, 1999). Such mutants are now available at EUROSCARF/SRD (web.uni-frankfurt.de/fb15/mikro/euroscarf), ATCC (www.atcc.org), Open Biosystems (www.openbiosystems.com) and Invitrogen (www.invitrogen.com). Furthermore, the strain collections were also extended by various tools for yeast molecular genetics. This chapter describes the respective biological materials, their use in yeast research and how to access these materials.

B. Yeast Genetic Analysis

The yeast *S. cerevisiae* provides a simple model for the organization of eukaryotic cells. As a model microorganism it displays several advantages compared with multicellular eukaryotic organisms, such as its fast growth (with generation times less than 2 h for haploid cells), its ease of handling and storage as well as its accessibility to genetic manipulation, allowing easy acquisition of genetic linkage information. From such mapping data, a genetic map of *S. cerevisiae* was established by Mortimer and co-workers (Mortimer and Schild, 1980, 1985; Mortimer *et al.*, 1989). From this map it was concluded that *S. cerevisiae* contains 16 chromosomes, a fact that was later confirmed by pulsed-field gel electrophoresis (Carle and Olson, 1984).

The well-established genetic map of *S. cerevisiae* and its compact genomic organization provided the basis for the complete DNA

[2]EUROSCARF stands for **EURO**pean *Saccharomyces cerevisiae* **Ar**chive for **F**unctional Analysis.

[3]Supported by the German Ministry for Science (BMBF, Bundesministerium für Bildung und Forschung).

[4]Supported by the European Community (EUROFAN standing for **EURO**pean **F**unctional **A**nalysis **N**etwork).

sequencing of the yeast genome. This worldwide effort was successfully completed in April 1996 (Goffeau *et al.*, 1996). DNA sequencing revealed a genome size of 12.5 Mb, encoding approximately 6000 genes that are now accessible for systematic functional analysis. The precise homologous recombination apparatus of *S. cerevisiae* allows one to perform specific gene deletions by selective integration of prototrophic gene markers or dominant resistance genes. In addition to the isolation of mutants by classical genetics, this allows the creation of gene-specific mutations by reverse genetics. Previously, the genetic map of *S. cerevisiae* was mainly derived from mapping data (Mortimer *et al.*, 1989), but genome sequencing has now established a physical map for gene localization. In general, the DNA sequencing confirmed the high quality of the earlier genetic map (Mortimer *et al.*, 1989) and we now know that, on average, 1 cM corresponds to approximately 3 kb. However, this varies because of the differing recombination frequencies at different regions of the chromosome.

The ability to create well-defined mutations and deletions for each gene (Rothstein, 1983, 1991) allows one to investigate the biological function of yeast genes with unknown function by searching for phenotypic abnormalities of the respective mutants. Such mutants can be collected, a genetic archive of gene deletion mutants can be established and these mutants can be made available to the scientific community. Here we report such efforts which now cover nearly all genes of the *S. cerevisiae* genome. The major advantage of the present deletion collections compared with previous mutant collections is that all deletions have been created in the isogenic backgrounds of not more than three strains. This was a major breakthrough for the functional analysis of genes because the genetic background strongly interferes with the mutant phenotype in many cases.

At least two types of null mutant can be created by reverse genetics. In gene disruption mutants the ORF of a particular gene is interrupted by the insertion of a genetic marker. In general this results in a non-functional protein. However, depending on the position of the disruption within the ORF, the residual proteins may display partial function. Furthermore, if there is significant selective pressure against the mutation, revertants will probably arise through selection for outlooping of the disruption marker to restore the original protein. Alternatively, gene function can be destroyed via a gene deletion. In this approach, the major part of the target ORF is replaced with a selection marker thereby deleting part of the ORF. In general, such gene deletions abolish the gene function irreversibly, even if there are selective pressures against such mutations. However, in many cases mutations in a second gene (epistatic suppressor mutations) can compensate for such deletions. Such epistatic mutations can easily be detected in tetrad analysis because in a cross with a wild-type haploid they result in a 3:1 segregation of the wild-type phenotype in tetratype asci.

Table 1. Yeast genetic nomenclature

Gene/Protein	Abbreviation	Description	Protein
Wild-type gene name	*HXK1*	*Italic*: three upper case letters + number	Hexokinase PI
Mutant gene name (recessive)	*hxk1*	*Italic*: three lower case letters + number	Mutated hexokinase PI
Mutant allele designation	*nep1-1*	*Italic: mutant gene name + allele number*	Nucleolar essential protein 1 mutation
Gene deletion (recessive)	*Δhxk1*	*Italic*: mutant gene name + Δ as a pre- or suffix	Deleted hexokinase PI gene
Gene deletion (recessive)		*Δhxk1::LEU2*	*Italic*: gene deletion name followed by two double dots and the deletion marker
Hexokinase PI deletion			
Mutant gene name (dominant)	*CAT1-2*^{d}	*Italic*: three upper case letters + allele number + superior d	Dominant mutation within Cat1-kinase
ts mutant allele	*nep1-1*^{ts}	*Italic*: mutant gene name + allele number + superior lower case letters	Temperature-sensitive Nep1 protein
Wild-type protein name	Nep1p	Three letter gene name with first letter in upper case + suffix p	Nep1 protein
Mutant protein name	nep1p	Three lower case letters	nep1 mutant protein

C. Gene Nomenclature

In here we use the following genetic nomenclature for genes and proteins (Table 1). Three upper case italic letters followed by a number are characteristic of a wild-type gene, e.g. *HXK1* for the gene encoding the glycolytic enzyme hexokinase PI. Lower case italic letters are used for the corresponding recessive[5] mutant gene,

[5]In diploid cells which have the mutant gene together with the wild-type gene (heterozygous diploids) the phenotype of a recessive mutation is lost, whereas the phenotype of a dominant mutation is still expressed.

Table 2. Two examples for ORFs named according to the sequencing nomenclature

Systematic gene name *YAL001c* Symbol	Meaning	Systematic gene name *YBR003w* Symbol	Meaning
Y	Yeast	Y	Yeast
A	Chromosome I	B	Chromosome II
L	Left arm	R	Right arm
C	Encoded by the Crick stand	W	Encoded by the Watson strand
001c	First gene to the left side of the centromere	003w	Third gene to the right side of the centromere

e.g. *hxk1*. A certain allele of a gene (in general a certain mutation) is characterized by an additional number separated by a hyphen form the gene or mutant name, e.g. *nep1-1*. Temperature-sensitive alleles are symbolized by a superior ts suffix, e.g. *nep1-1^{ts}*. A dominant mutation is symbolized by the wild-type gene name and the allele number followed by a superior d, e.g. *CAT1-2^{d}*. The encoded proteins are also characterized by the gene letters and numbers, however, written in Roman, starting with a capital letter and the suffix p for protein after the gene number, e.g. Hxk1p. The suffix is omitted if the term protein is used in addition to the protein name.

Gene deletion mutations are often symbolized by a Greek Δ in front of the mutant gene name, e.g. *Δhkx1*. Additionally, the deletion marker should also be provided after two double dots, e.g. *Δhkx1::LEU2*. In those cases where the exact positions of deleted nucleotides are known these are provided in brackets, e.g. *Δhkx1(4,850)::LEU2*.

During the *S. cerevisiae* genome sequencing project an alternative nomenclature was developed (Table 2). The sequencing nomenclature also uses three upper case letters, where the first letter Y stands for the organism yeast. The second letter symbolizes the respective chromosome (A–P, where from the 16 *S. cerevisiae* chromosomes A corresponds to chromosome 1 and P to chromosome 16). The third letter shows if the gene is located on the left (L) or right (R) arm of the chromosome with respect to the centromere. The orientation of the chromosome was assigned based on the genetic map of Mortimer *et al.* (1989). The genes are numbered starting from the centromere either to the right or to the left side. As ORFs are located on both nucleotide strands the orientation is given either by the suffix c (for Crick strand) or w (for Watson strand), e.g. *YBR003w* or *YAL001c*.

Today, for most of the genes physiological functions have been described and names according to the genetic nomenclature have been provided. The sequencing nomenclature is mainly used for annotated genes with unknown function and after physiological characterization a name following the genetic nomenclature is preferred.

For many of the genes which were genetically identified synonyms exist. At the EUROSCARF homepage the gene names according to the sequencing nomenclature and the genetic nomenclature as well as synonymous gene designations (data from SGD) are easily accessible.

D. Strategies for Yeast Gene Deletions

Deletions are being created either by replacement of the target gene with suitable prototrophic markers such as *LEU2, HIS3, URA3* and *TRPl* (Rothstein, 1983, 1991), or with a dominant resistance marker such as kanamycin (*kanMX*), nourseothricin (*natMX*), hygromycin B (*hphMX*), phosphinotricin (*patMX*) and a D-serine deaminase from *E. coli* which causes resistance against D-serine (see Section III.B).

In the early times of yeast molecular genetics cassettes for making gene deletions and disruptions[6] were mainly constructed by cloning using suitable restriction sites. For gene replacements with prototrophic markers the respective deletion cassettes were usually obtained after replacing approximately 6–90% of the target's coding region by the prototrophic marker. This deletion strategy, however, is time-consuming and its success generally depends on suitable restriction sites within the target gene.

Today, based on a PCR deletion strategy (Baudin *et al.*, 1993), an efficient deletion method is available that uses heterologous modules such as the kanamycin resistance gene (*kanMX*) as a selection marker (Wach *et al.*, 1994). In contrast to the prototrophic markers that usually share homologous sequences within the *S. cerevisiae* genome, no sequences homologous to the *kanMX* deletion cassette are present in the *S. cerevisiae* genome. Therefore, short sequences of about 40 bases can be used for recombination which target the *kanMX* module to the gene of interest. This allows one to construct a deletion cassette with short flanking recombination sequences by PCR. Even with strains that are not strongly isogenic to the S288C sequencing strain, this deletion strategy works with a more than 90% rate of success.

The PCR-mediated deletion strategy (Baudin *et al.*, 1993) was further improved by the introduction of directed repeats of *loxP* recombination sites flanking the *kanMX* deletion cassette (Güldener *et al.*, 1996). These 34 bases lie at each end of the kanamycin resistance gene and this deletion cassette also generates stable deletions. However, the kanamycin resistance gene can subsequently be removed from the genome after transformation of the deletion mutant with the Cre-recombinase of bacteriophage PI, which catalyzes a

[6]Disruption is used when the ORF is destroyed after the insertion of new nucleotides within the ORF. In general, disruptions also destroy the function of the ORF. Deletion is used when part of the ORF is removed during transformation.

site-specific recombination event between the *loxP* sites. Using this *loxP-kanMX-loxP* deletion strategy allows one to remove the kanamycin resistance gene from the genome, thereby making it possible to perform a second deletion using the same resistance marker. This is a remarkable advantage, which allows the sequential replacement of isogenes, and it is also suitable for the generation of deletions in industrial strains in which removal of the resistance marker is recommended after transformation.

E. Isogenic Gene Deletion Strains

At present, gene deletions are available in the three *S. cerevisiae* strains BY (derived from FY1679), CEN.PK2 and W303, each of which have different genetic backgrounds. They are all available with various combinations of the auxotrophic mutations, which are often used as selectable markers for yeast transformation or for different genetic test systems (Table 3).

1. BY (FY) strains

BY strains were derived from strain FY1679, which was constructed by Bernard Dujon after mating strains FY23 and FY73 (note that $23 \times 73 = 1679$). The FY strains are isogenic to strain S288C which was used as the source of DNA for *S. cerevisiae* genome sequencing (Winston *et al.*, 1995). The vast majority of bar-coded deletions are being made in strain BY4743, so that mutants are isogenic to the sequenced strain (Brachmann *et al.*, 1998). The BY strains have no *trp*-auxotrophy as it is said that such auxotrophy makes these strains more cold sensitive (Singh and Manney, 1974; Hampsey, 1997). Unfortunately, this strain has an extremely poor sporulation frequency, which makes its genetic analysis more difficult. Furthermore, there is an instability of heterozygous diploids at the *MET15/met15-Δ0* locus, which converts the heterozygous locus either into *MET15/MET15* or *met15-Δ0/met15-Δ0* by gene conversion.

2. CEN.PK strains

The CEN.PK strains were constructed in an interdisciplinary project of geneticists, biochemists and biotechnological engineers to study metabolic fluxes, which was supported by the Volkswagen Stiftung, Germany in 1993/1994 (see Figure 1 for genotypes and strain history). From several laboratory strains tested only ENY.WA strains (a series of isogenic strains obtained after 10-fold inbreeding by Karl-Dieter Entian and Werner Albig, Frankfurt, in 1988) provided good growth during chemostat cultivation. However, ENY.WA strains were slightly flocculent. To abolish this disadvantage strain ENY.WA-1A was mated with strain MC996A (a strain from the lab of Michael Ciriacy, Düsseldorf). The progeny of these strains was

Table 3. Genotypes of diploid strains used for gene deletions

Strain name	Genotype						
CEN.PK2	*MAT***a** / *MATα*	*ura3-52* / *ura3-52*	*trp1-289* / *trp1-289*	*leu2-3,112*[a] / *leu2-3,112*	*his3-Δ1* / *his3-Δ1*	*MAL2-8*[cb] / *MAL2-8*[c]	*SUC2*[c] / *SUC2*
W303	*MAT***a** / *MATα*	*ura3-1* / *ura3-1*	*trp1-Δ1* / *trp1-Δ1*	*leu2-3,112* / *leu2-3,11,*	*his3-11* / *his3-11*	*ade2-1*[d] / *ade2-1*	*can1-100*[e] / *can1-100*
FY1679	*MAT***a** / *MATα*	*ura3-52* / *ura3-52*	*trp1-Δ63* / *TRP1*	*leu2-Δ1* / *LEU2*	*his3-Δ200* / *HIS3*	*GAL2* / *GAL2*	
BY4743	*MAT***a** / *MATα*	*ura3-Δ0*[f] / *ura3-Δ0*	*met15-Δ0* / *MET15*	*leu2-Δ0* / *leu2-Δ0*	*his3-Δ1* / *his3-Δ1*	*lys2-Δ0* / *LYS2*	

[a]The two numbers symbolize two independent point mutations within the *leu2* gene which prevent reversion and ensure a stable auxotrophy.
[b]*MAL2-8*[c] is a dominant mutant allele that causes a partially constitutive (not requiring maltose as an inducer) but still glucose-repressible *MAL2* gene expression (Zimmermann & Eaton, 1974).
[c]SUC2 encodes the extracellular enzyme invertase and enables sucrose fermentation.
[d]Deletion within the *ADE2* gene (coding for phosphoribosylamino imidazol carboxylase) causes red colonies due to the accumulation of phosphoribosylamino imidazole (AIR), which oxidises to a red-coloured polymer.
[e]Mutation within the arginine permease gene causes resistance to canavanine which is an analogon to arginine.
[f]The *Δ0* symbolizes that the ORF and part of the 5′ and 3′ flanking regions have been removed from the genome.

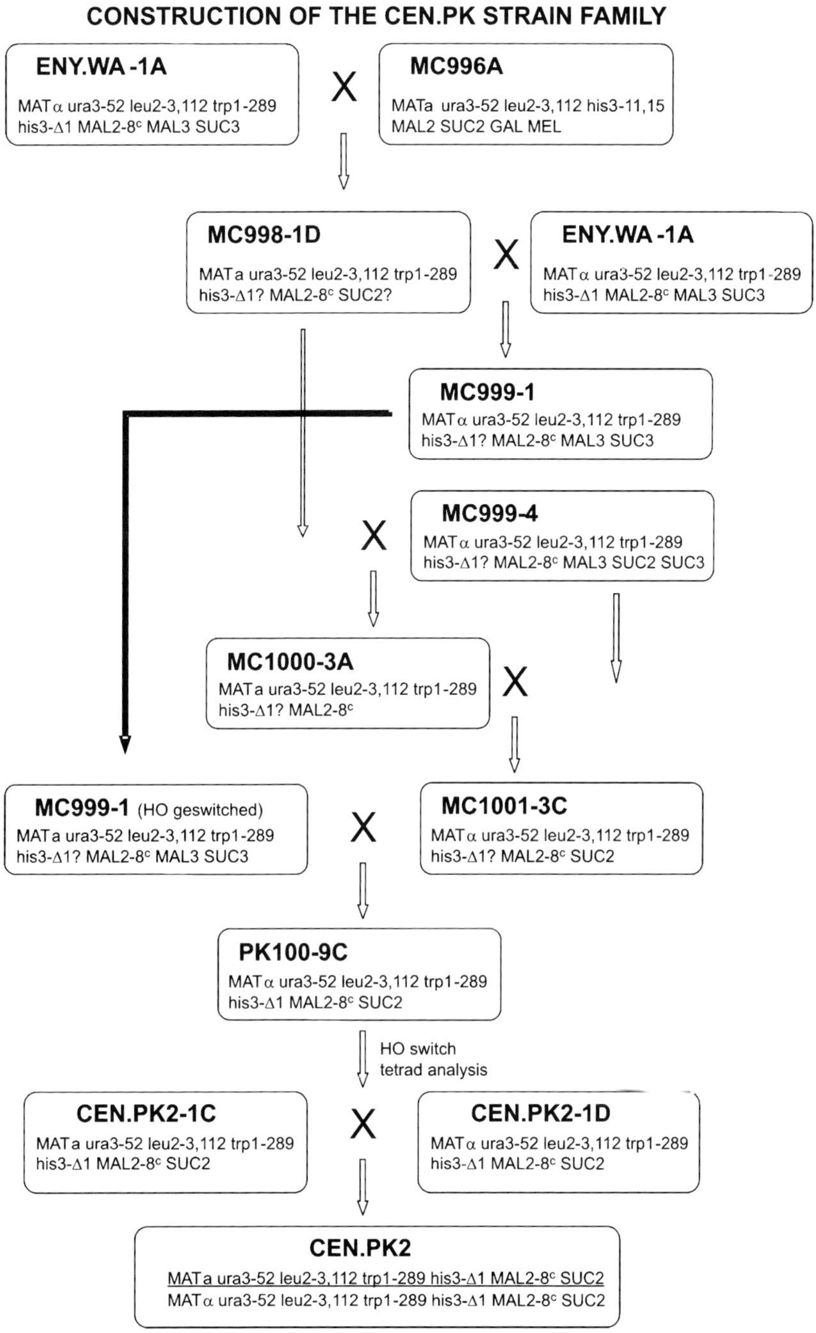

Figure 1. Construction of the CEN.PK2 strain family.

tested for the following properties: (a) high growth rates in complete and defined media under aerobic and anaerobic conditions in batch, fed-batch and continuous culture; (b) high mating and sporulation efficiency; (c) high spore viability; (d) high transformation efficiency;

and (e) good single-cell formation (no flakiness). According to these requests, the most suitable segregant, PK100-9C was selected. To obtain a set of isogenic strains Peter Kötter diploidized strain PK100-9C after transformation with plasmid YCp50-HO (kindly provided by Rob Jensen and Ira Hershkowitz) and after tetrad analysis the resulting segregants were named according to the construction labs CEN.PK2-1C and CEN.PK2-1D (for **C**iriacy **EN**tian.**P**eter **K**ötter). According to the history of the strain, such strains are also sometimes referred to as VW strains. All CEN.PK strains contain the *SUC2* gene for sucrose utilization and the *MAL2-8*^{c} gene for maltose utilization. *MAL2-8*^{c} is a dominant mutant allele that causes a partially constitutive (not requiring maltose as an inducer) but still glucose-repressible *MAL2* gene expression (Zimmermann and Eaton, 1974). CEN.PK strains have various combinations of auxotrophies *ura3-52 his3-Δ1 leu2-3,112* and *trp1-289*. For industrial application, the four auxotrophies of the CEN.PK2 strains were replaced by the corresponding wild-type genes. CEN.PK strains are available from EUROSCARF (see Section II.A).

Owing to their excellent growth characteristics CEN.PK strains are now used by many groups as a laboratory strain in chemostat cultures (Ostergaard *et al.*, 2000; van Dijken *et al.*, 2000; Stückrath *et al.*, 2002; Saldanha *et al.*, 2004) and also some strains isogenic to CEN.PK2 are commercially used (Gothia's ALCOFREE^{TM} Yeast01, see www.gothiayeast.com).

In addition to its use in chemostats CEN.PK strains show high transformation and sporulation efficiencies. A major advantage is the fast growth rate, with doubling times of about 80 min for haploid strains. The available CEN.PK2 derivatives are isogenic and mutations that reduce germination or growth can be detected easily by a clear 2:2 segregation in tetrads in which wild-type segregants become visible within 24 h after tetrad dissection. Therefore the strain was used in various gene function analysis projects (Entian *et al.*, 1999) and about 2000 genes have been successfully deleted in CEN.PK strains (P. Kötter, unpublished) from which 800 are freely available from EUROSCARF (see Section II.A).

3. W303 strains

Strain W303[7] was used in a European pilot project for functional analysis which mainly involved the screening of the respective mutants for altered sensitivities to drugs and metabolic inhibitors (Rieger *et al.*, 1997, 1999) and also as a reference strain for EUROFAN I deletions. W303 has a much better sporulation efficiency than FY1679, which makes the former strain more suitable for genetic analysis. About 300 W303 deletion mutants are available from

[7]Although the strains in the deletion collection are isogenic, some supposed W303 strains are possibly not strongly isogenic.

EUROSCARF and most of them were generated during the EUROFAN I (EUROpean Functional Analysis Network) project.

♦♦♦♦♦♦ II. YEAST GENE DELETION STRAIN COLLECTIONS

A. The EUROSCARF (SRD) Collection

The EUROSCARF strain collection (web.uni-frankfurt.de/fb15/mikro/EUROSCARF) was established in 1994 after the sequencing of the *S. cerevisiae* genome had revealed a considerable number of genes with unknown function. Funded by the German government,[8] a consortium of 16 research groups established a scientific network, the aim of which was to delete approximately 325 genes of unknown function from the yeast genome (BMBF Functional Analysis Network). The resulting 325 deletion mutants provided the basis for a systematic screen for possible phenotypes of the respective mutations (Entian *et al.*, 1999). Using the strain CEN.PK2, approximately one-third of the deletion mutants first revealed a phenotype during tetrad analysis according to the slower growth or germination of the mutant segregants. In about 20% of cases the deleted genes were essential. For approximately two-thirds of the mutants, phenotypes were discovered, some of which were indicative of the function of the corresponding gene (Entian *et al.*, 1999).

The deletion mutants were stored centrally and made available to the participating research groups. In 1996, the European Community also decided to support the systematic study of yeast genes of unknown function. The resulting EUROFAN projects (EUROFAN I and II) were a collaboration in which more than 100 European research groups were involved. Within the EUROFAN I project approximately 1000 genes of unknown function were studied (Oliver, 1996). These genes were deleted from the yeast genome by reverse genetics and the resulting deletion mutants were made accessible to the participating laboratories. The selection of genes to be targeted was made in collaboration with the BMBF Functional Analysis Network, resulting in a EUROSCARF collection of approximately 1200 deletions (1000 FY strains, 800 CEN.PK2 strains and 300 W303 strains).

In the second phase of the EUROFAN project (EUROFAN II), which started in October 1997, 2400 *S. cerevisiae* genes were deleted as part of a worldwide effort with US and Canadian research groups, finally resulting in a collection of deletion mutants for all 6200 *S. cerevisiae* genes. In addition to the earlier used deletion strategies, the deletions were generated in a two-step PCR-based

[8]German Ministry for Science (BMBF, Bundesministerium für Bildung und Forschung).

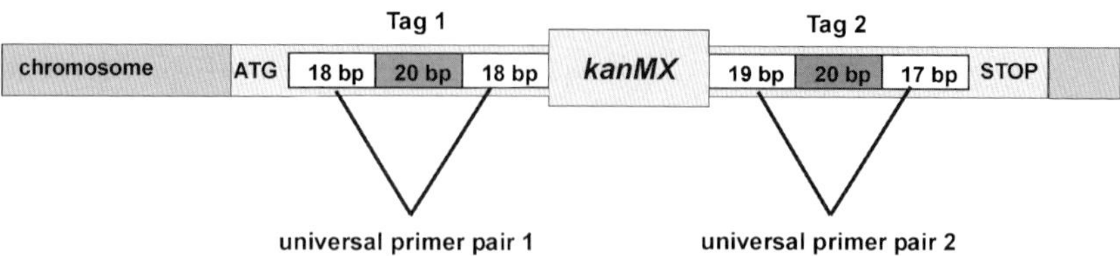

Figure 2. Common genomic structure of bar-coded yeast deletion mutants. Tag 1 and Tag 2 correspond to the gene-specific bar-coding sequences. The universal primer pair 1 is present in all bar-coded deletions and corresponds to two PCR-priming sequences (both 18 bp) for the easy determination of the Tag 1 sequence. The universal primer pair 2 is also present in all bar-coded deletions and allows the easy determination of the Tag 2 sequence.

gene replacement where each of the deletion mutants were additionally tagged by two specific and unique DNA sequences that are not present in the yeast genome (see Section III.A and Figure 2).

When the collection was started efficient quality control systems were established by Peter Kötter. Upon receipt of the strains at EUROSCARF, each strain was streaked out for single colonies on YEPD plates (1% yeast extract, 2% bacto peptone and 4% glucose), replica-plated onto a routine set of synthetic drop-out plates (each lacking adenine, uracil, leucine, histidine or tryptophan) and tested for kanamycin resistance. The colony growth was estimated and particular care was taken to monitor any form of papillary growth because such growth is indicative of a heterogeneous population or genetic instability of the deletion mutant. If colonies of uniform size were obtained on plates, a single colony was inoculated into 5 ml YEPD liquid medium and grown to stationary phase. For storage, glycerol was added to a final concentration of 17.5% (w/w) and strains were stored at −80°C. The strains were cultivated in three-fold redundancy. If stored cultures have to be replicated, the number of generation times is kept to a minimum to avoid the accumulation of second-site mutations.

Upon request, the strains are streaked out onto YEDP plates and tested for uniform growth. For mailing, strains are either soaked onto sterile filter papers wrapped in sterile aluminum foil, or alternatively, streaked out onto a YEPD agar slope in a screw-capped vial.

Strains carrying mutations that affect essential genes or genes that strongly reduce growth on glucose are collected as heterozygous diploids to minimize any selective pressure against the respective mutation. Mutants with strongly reduced growth on glucose need special attention and a tetrad analysis is recommended upon receipt to confirm the 2:2 segregation of the mutation. This reduces the risk that second-site mutations have accumulated in the diploids which, in most cases, will result in a non-2:2 segregation of the growth phenotype. Upon request, such analyses can be performed by EUROSCARF and haploid segregants obtained directly from the tetrad analysis will be mailed.

After the termination of the EUROFAN projects, Scientific Research and Development GmbH,[9] which was also involved in establishing the bar-coded mutant collection, together with Entian's group at the Goethe University, Frankfurt took over the preservation and distribution of EUROSCARF and extended the collection further.

Currently, the EUROSCARF(SRD) strain collection contains a large number of gene deletions in various strain backgrounds, the bar-coded gene deletion strains generated by the US/European/Canadian consortium and various tools for yeast molecular analysis (see also below).

B. The ATCC Collection

The American Type Culture Collection (ATCC) (www.atcc.org) was established in 1925 as a central collection of microorganisms. Today, the collection is extended to all kinds of biological materials including bacteria, bacteriophages, cell lines and hybridomas, filamentous fungi, yeasts, plant seeds, protozoa, algae, viruses, animal and plant cells.

The ATCC also provides several plasmids useful for molecular biological research with *S. cerevisiae*. Furthermore, vectors used for *S. cerevisiae* transformations and *S. cerevisiae* recipient strains are available. ATCC also distributes the set of bar-coded gene deletion strains generated by the US/European/Canadian consortium (see also above and below) and offers a set of clones constructed by Olson and co-workers (Olson *et al.*, 1986; Link and Olson, 1991; Riles *et al.*, 1993) for the physical mapping of *S. cerevisiae* genes, and some of the cosmids that were used for the sequencing of the yeast genome.

ATCC also has a large collection of strains with *S. cerevisiae* point mutations and in 1999 took over the mutant collection of the Yeast Genetic Stock Center (YGSC), which was started by Bob Mortimer as early as 1960 at the University of California at Berkeley. Approximately, 450 genes are covered and the 1200 strains also include strains for gene mapping, teaching of yeast genetics, transformation, allelism tests, estimation of recombination frequencies and tests for aneuploidy. Most of the collected mutants result from point mutations, and only in a few cases have the DNA mutations been identified within the respective genes. Furthermore, the mutations were made in different genetic backgrounds, which makes their use difficult. Today, with the excellent tools of reverse genetics, and the large number of isogenic gene deletions available, the use of the genetically less defined mutants is limited to some special applications such as the use of temperature-sensitive strains and strains suitable

[9]Scientific Research and Development GmbH Oberursel/Frankfurt, Köhlerweg 20, D-61440 Oberursel, see also www.srd-biotec.de.

for the analysis of genetic instabilities such as gene conversion and recombination events.

C. The Open Biosystems Collection

Open Biosystems started in 2002 and provides genetic tools for molecular biological research. Open Biosystems (www.openbiosystems.com) also distribute the bar-coded gene deletion strains generated by the US/European/Canadian consortium and various tools for yeast molecular analysis (see also below).

D. The Collection from Invitrogen

In 2000, Research Genetics Inc. also stored the bar-coded deletion collection of the US/European/Canadian consortium. After its merger with Invitrogen (www.invitrogen.com) the deletion collection and other tools for yeast molecular analysis are available from Invitrogen (see also below).

♦♦♦♦♦♦ III. GENETIC TOOLS, STRAINS AND PLASMIDS

A. The Bar-Coded Mutant Collection

Starting in 1997, a worldwide collaboration of yeast laboratories from the US, Europe (EUROFAN II) and Canada deleted 5900 genes of the yeast genome using the bar-coding system (Giaever *et al.*, 2002). Such tagged mutants are very useful for functional analysis in selective screening procedures involving rapid identification of the enriched mutations (see below). Recently, 200 small ORFs were additionally deleted with this strategy and deposited at the strain collections (M. Snyder, unpublished) resulting in a current number of 6000 gene deletions and another 300 deletions are in progress (M. Snyder, personal communication).

Each gene in the bar-coded mutant collection was precisely deleted from the start to stop codon (non-inclusive) and replaced by the *kanMX* deletion cassette (see Section III.B). The methods for the identification of deletion mutants were greatly improved by the introduction of a unique 20-base sequence tag for each yeast deletion mutant (Shoemaker *et al.*, 1996). Approximately, 12 000 tags were designed that had similar melting temperatures for nucleic acid hybridization, displayed no secondary structure, showed no similarity to each other and were not present in the yeast genome. Hence, two particular tags (molecular bar-coding) were assigned to each of the yeast deletions. Adjacent to the tags common PCR-priming sites were also introduced together with the deletion tags (Figure 2).

In addition to its advantages for phenotypic analysis, the bar-coding of deletion mutants also improved the quality of the collection of deletion mutants. Because the genetic markers are very similar for all deletion mutants, no simple tests are available to distinguish a particular deletion. Therefore, in contrast to previous mutant collections, there is a much higher risk that mix-ups will occur. The bar-coding of the deletions allows the easy identification of each mutant either by hybridization or by DNA sequencing of the respective PCR fragments (see below).

At EUROSCARF, ATCC, Open Biosystems and Invitrogen (see Section II.A–II.D) about 4900 haploid deletions within non-essential genes are available in both mating types and as homozygous diploids.[10] Furthermore, 6100 heterozygous diploids where only one allele is deleted are deposited, which also include about 1100 essential gene deletions.

The availability of gene deletion mutants for all *S. cerevisiae* ORFs enables easy access for the scientific community to each *S. cerevisiae* mutant of interest, and such deletions can be easily introduced into any *S. cerevisiae* strain background of interest (see Section III.B). This avoids redundant efforts in the study of genome functions. Furthermore, the easy access to deletion mutants also improves the use of *S. cerevisiae* as a model organism because they can be used by scientists studying other eukaryotic organisms if functional similarities become obvious.

1. Phenotype analysis of haploid and homozygous mutants (non-essential genes)

After 95–96% of *S. cerevisiae* genes are accessible as bar-coded deletion mutants new tools for phenotype analysis using the entire collection of haploid or homozygous diploid deletions became available. These methods use the different fitness of the respective deletion mutants with respect to various growth conditions. Owing to the easy identification of the deletion mutants by oligonucleotide hybridization or PCR sequencing the entire population of deletion mutants can be assayed in population dynamic experiments. For example, the fitness of the entire population of haploid deletion mutants can be tested after exposure to various drugs and growth under well-conditioned media (vitamins, amino acids, high and low salt concentration, temperature, etc.: Giaever *et al.*, 2002). In the case with galactose as carbon source for example, all known *GAL* genes and also some so far unknown gene were identified in a single experiment as being important for the utilization of galactose as carbon source (Giaever *et al.*, 2002; see also this volume, Chapter 17).

[10] The promoters and terminators of the *kanMX* cassettes are patent restricted (EP536192B1, US5,650294, JP 3059214, CA 2080482-3). No patent fees are requested for non-commercial fundamental research.

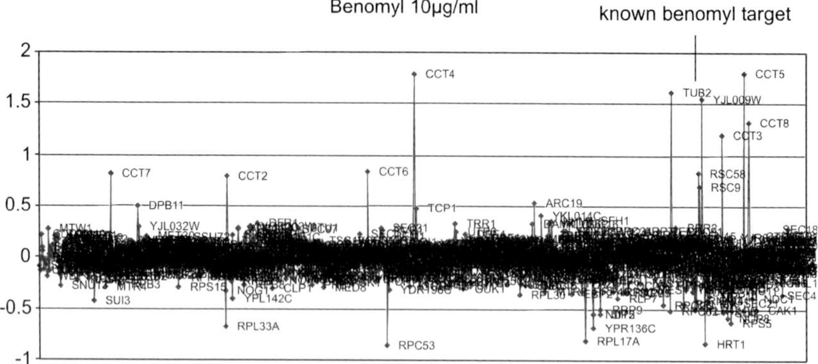

Figure 3. Haplotoxicity assay with benomyl. In total, 1100 diploid mutants with heterozygous deletions of all essential *S. cerevisiae* genes were grown in the presence of 10 µg/ml benomyl in microtiter plates for 24 h. Growth was estimated by optical density. Positive RGR numbers (relative growth rate numbers, *Y*-axis) indicate diploids with decreased growth and negative RGR numbers indicate diploids with improved growth as compared to wild type.

2. Phenotype analysis of heterozygous mutants (essential genes)

On the first view, population dynamic experiments seem to be restricted to non-essential genes, as the respective deletions are viable as haploids. However, it has been known for long time by yeast geneticists that heterozygous diploids where only one allele of the essential gene is deleted in many cases also reveal a phenotype, an observation which is now referred to as haploinsufficiency (Giaever *et al.*, 1999). As many essential genes of *S. cerevisiae* have functional homologs in human and plant pathogens, e.g. *Candida* species and *Aspergillus fumigatus*, or they share functional similarities with genes involved in the manifestation of human diseases, such genes are of special interest to functional analysis.

A set of 1100 heterozygous deletions within all known essential genes is available either from EUROSCARF, Open Biosystems or Invitrogen. An example for haplotoxicity screening[11] with benomyl is shown in Figure 3 and Table 4. Interestingly, not only diploids with the known heterozygous tubulin genes, such as *TUB2*, were more sensitive to benomyl, but also gene *YJL009w* with a so far unknown function showed an increased sensitivity to benomyl. Additionally, some heterozygous diploids showed a slightly improved growth (relative growth values −0.71 to −0.88). However, data with relative growth values lower than 1 or higher than −1 need further confirmation by specific experiments (data from Recktenwald, Eschrich, Kötter, Sättler, Hauf and Entian, joint

[11]If the set of diploids with heterozygous essential genes is used for growth competitive assays, we refer to this as haplotoxicity screening.

Table 4. Heterozygous diploids with decreased (positive RGR numbers) or increased growth (negative RGR numbers) upon exposure to benomyl. RGR values lower than 1 or higher than −1 are less significant and need further confirmation by specific experiments

No	Gene	Relative growth value	Function	Remarks
Decreased growth				
1	*CCT5*	1.79	Subunit of the Cct-ring complex	Tubulin-associated
2	*CCT4*	1.78	Subunit of the Cct-ring complex	Tubulin-associated
3	*TUB2*	1.61	β-Tubulin	β-Tubulin
4	*YJL009w*	1.54	Unknown function	
5	*CCT8*	1.32	Subunit of the Cct-ring complex	Tubulin-associated
6	*CCT3*	1.19	Subunit of the Cct-ring complex	Tubulin-associated
7	*CCT6*	0.83	Subunit of the Cct-ring complex	Tubulin-associated
8	*CCT7*	0.82	Subunit of the Cct-ring complex	Tubulin-associated
9	*RCS58*	0.82	Component of yeast chromatin remodelling complex	
10	*CCT2*	0.79	Subunit of the Cct-ring complex	Tubulin-associated
11	*RCS9*	0.71	Component of yeast chromatin remodelling complex	
Improved growth				
12	*RPC53*	−0.88	Subunit RNA polymerase III	Transcribing tRNAs and 5S rRNA
13	*HRT1(RBX1)*	−0.87	Subunit E3 ubiquitin ligase	
14	*RPL17A*	−0.84	Ribosomal protein L17	60S subunit
15	*RPL33A*	−0.72	Ribosomal protein L37	60S subunit
16	*YPR136c*	−0.72	Unknown function	

project with University of Frankfurt, Phenion GmbH & CoKG and Scientific Research and Development GmbH). Such heterozygous deletion analysis is available from Scientific Research and Development GmbH, www.srd-biotec.de.

B. Deletion Cassettes

In order to prove that the deletion cassettes that were generated within the EUROFAN I project were suitable for constructing deletions in genetic backgrounds other than strain FY1679, all deletions were also being introduced into a second strain, which was either CEN.PK2 or W303. Although the deletion strategy using short flanking homology regions works for most of the genes for all three strain backgrounds described here, problems may arise if the strains are more diverged. To overcome this, deletion cassettes with long flanking sequences of 0.5–1 kb were constructed in the EUROFAN I project. In general, the large flanking sequences were obtained by PCR reactions and most of them were checked for PCR-induced errors by sequencing. At present about 1200 deletion cassettes with long flanking sequences are available in the EUROSCARF collection.

With today's PCR-based deletion techniques the collection of deletion cassettes lost its importance. To introduce a specific deletion into any *S. cerevisiae* strain of interest we recommend to amplify the deletion cassette from the respective bar-coded deletion mutant with appropriate PCR-primers. By using the *kanMX* marker for selection (see below), the resulting PCR fragment can be easily used to delete the respective gene in any *S. cerevisiae* strain of interest.

1. Dominant resistance markers

The development of dominant drug resistance markers was a breakthrough for the generation of *S. cerevisiae* deletion mutants. The first MX cassette for gene deletions was based on the observation that the *E. coli* kanamycin resistance gene makes *S. cerevisiae* resistant against the aminoglycoside drug G418. Peter Philippsen and his co-workers used promoter and terminator sequences of the translational elongation factor 1α from the fungus *Ashbia gossipii* to express the G418 resistance gene as a resistance marker (*kanMX*) in *S. cerevisiae* (Wach *et al.*, 1994; Wach, 1996).

The G418 drug resistance marker has several advantages:

- It provides strong G418 resistance even in single copy expression.
- The cassette can be easily amplified by PCR.
- There is no homology to the yeast genome.
- Its application does not need auxotrophic mutations and is independent from the strain background.

The *kanMX* cassette was used in most yeast function analysis projects (see above) and is available from EUROSCARF.

In recent years, new dominant drug resistance cassettes such as *natMX, hphMX, patMX* and *dsdAMX* were developed (see also Table 5, Goldstein and McCusker, 1999; Vorachek-Warren and McCusker, 2004). The *natMX* cassette contains the *nat1* gene from *Streptomyces noursei* encoding a nourseothricin *N*-acetyltransferase

Table 5. Deletion cassettes and selection markers available from EUROSCARF-SRD

MX4[a] (CEN/ARS)[d]	MX4[a] (pFA6)[e]	MX3[b] (CEN/ARS)[d]	MX3[b] (pFA6)[e]	*loxP*-MX4-*loxP*[c]
	kan		*Kan*	*kan*
nat	*nat*		*Nat*	
nph	*hph*		*Hph*	
	Pat		*Pat*	
CaURA3	*CaURA3*	*CaURA3*	*CaURA3*	
LYS5	*LYS5*	*LYS5*	*LYS5*	*LYS5*
CaLYS5	*CaLYS5*	*CaLYS5*	*CaLYS5*	*CaLYS5*
	DsdA			*dsdA*
				Sphis5
				ble
				KlURA3[f]
				KlLEU2[f]

[a]MX4: The corresponding selection marker is flanked by the *AgTEF1* promoter and terminator sequences.
[b]The MX3 cassettes are additionally flanked by 470 bp direct repeats of the 3′-regions of the *AgLEU2* gene.
[c]The MX4 cassettes are flanked by *loxP* sites.
[d]CEN, plasmids with a centromer sequence, ARS plasmids with an autonomous replication sequence.
[e]pFA6: Basic cloning **p**lasmid for **f**unctional **a**nalysis (Wach *et al.*, 1994).
[f]*KlURA3* and *KlLEU2* are expressed under the control of their own promoter and terminator sequences.

and confers resistance against the aminoglycoside nourseothricin. The *hph* gene from *Klebsiella pneumoniae* encoding a hygromycin B phosphotransferase is part of the *hphMX* cassette which provides resistance to the aminoglycoside hygromycin B produced by *Streptomyces hygroscopicus*. The resistance to phosphinothricin (a glutamate analog which is also named bialaphos) produced by *Streptomyces hygrocopicus* is conferred by the *pat* gene from *Streptomyces viridochromogenes* Tü94 encoding a phosphinothricin *N*-acetyltransferase which is part of the *patMX* cassette.

Recently, a new dominant resistant marker based on the *dsdA* gene of *E. coli* which encodes a D-serine deaminase was developed (Vorachek-Warren and McCusker, 2004). For *S. cerevisiae* the non-proteinogenic amino acid D-serine is toxic and expression of the *dsdA* gene confers resistance to D-serine and additionally D-serine can be used as the sole nitrogen source. This enforces selection in media without ammonium, as a twofold selection pressure favors the transformants: First, resistance to D-serine and, second, D-serine as the sole nitrogen source.

2. Heterologous prototrophic markers

Prototrophic markers from other organisms which can heterologously complement the respective *S. cerevisiae* auxotrophies were also useful tools for targeted gene deletion. The *his5* gene of *Schizosaccharomyces pombe* gene (the homolog to *S. cerevisiae HIS3*), the *Candida albicans* genes *URA3* (Goldstein *et al.*, 1999) and *LYS5*

(Ito-Harashima and McCusker, 2004) were used to develop the respective MX cassettes (see also Table 5).

3. The Cre-*loxP* marker rescue system

The rescue of selection markers is very important for successive deletions in one strain. This is especially important for the construction of industrial yeast strains and for the functional analysis of large gene families. The rescue of genetic markers was strongly improved when Hans Hegemann and his co-workers adapted the bacterial Cre-*loxP* system for its application in yeast (Güldener *et al.*, 1996, 2002). In these deletion cassettes the 34 bp *loxP* sites are placed on both sites of the *kanMX4* module. The expression of bacteriophage P1 Cre-recombinase mediates an efficient recombination between the *loxP* sites resulting in excision of the marker gene.[12] One prominent example for the multiple use of the *loxP-kanMX-loxP* deletion cassette is the analysis of the hexose transporter family: concurrent deletions of at least 20 transporter genes were required to block the uptake of hexose completely (Wieczorke *et al.*, 1999).

Alternatively, the *URA3MX* cassette (see above and Goldstein *et al.*, 1999) can also be used for marker rescue, if 5-FOA[13] is used to select against the *URA3* prototrophic marker (counterselection). By using α-aminoadipate the *LYS5MX* and *CaLYS5MX* cassettes can also be counterselected as cells become α-aminoadipate-resistant upon loss of the *LYS5* prototrophic marker (Ito-Harashima and McCusker, 2004). The *LYS5MX* and *CaLYS5MX* cassettes were constructed as MX3 and MX4 modules as well as *loxP*-flanked module. These modules are also available as yeast centromeric vectors. An overview on the currently available MX and *loxP*-MX cassettes from EUROSCARF is provided in Table 5.

C. Cognate Clones

Most of the genes that were deleted are also available on centromeric plasmids as complete genes with their native promoter and terminator sequences. These can be used for *trans*-complementation. This is especially important after a phenotype had been observed for the corresponding deletion mutant, and the plasmid is used to confirm that the phenotype is due to this deletion rather than some

[12]Yeast *CEN*/ARS plasmids expressing the Cre-recombinase under the control of the *GAL1* promoter are available with the prototrophic *URA3*, *HIS3* and *TRP1* markers and the dominant marker ble^R from the bacterial transposon Tn5 which confers resistance to the antibiotic phleomycin (Güldener *et al.*, 2002).

[13]5-FOA stands for 5-fluoroorotic acid, which is used as an antimetabolite. 5-FOA is converted to the toxic product, 5-fluorouracil, by the action of the *URA3* gene product. *URA3* prototrophic cells are killed by 5-FOA, *ura3* auxotrophic mutants are resistant against 5-FOA.

second-site mutation. About 600 genes with native promoter and terminator sequences are available from EUROSCARF.

D. Regulated Gene Expression

Regulated promoter expression is important for the study of gene functions and makes it possible to conditionally switch off the expression of the gene under investigation. Regulated expression is especially important for essential genes thereby making it possible to investigate the manner of cell death upon switching off the essential function. The *GAL1* promoter has often been used for regulated gene expression (Schneider and Guarente, 1991), but, unfortunately is not really tight and protein depletion effects often only occur after extended growth on glucose. A strongly improved system is based on the bacterial *tetO* promoter where gene expression depends on the absence or presence of doxycycline (Gossen and Bujard, 1992; Gari *et al.*, 1997; Belli *et al.*, 1998a,b).

1. Plasmids for tetracycline-regulated gene expression

In recent years, Enrique Herrero and his co-workers adapted the tetracycline-regulatable promoter system which was originally developed for the use in mammalian cells (Gossen and Bujard, 1992) for regulated gene expression in *S. cerevisiae* (Gari *et al.*, 1997; Belli *et al.*, 1998a,b). The basic element of this regulatory system is a hybrid *tetO-CYC1* promoter, which is expressed upon binding of the tTA transactivator. The tTA transactivator is a fusion protein between the VP16 activator domain[14] and the tetracycline-inducible repressor tetR.[15] Addition of doxycycline (a tetracycline derivative) removes the tTA transactivator from the *tetO-CYC1* promoter and rapidly prevents its transcription. Three sets of vectors for doxycycline-regulated expression are available from EUROSCARF.

The first set of cloning vectors (TET-SET1) are centromeric or episomal plasmids with *URA3* as selection marker where the tTA transactivator and the *tetO-CYC1* promoter are followed by multicloning sites (MCS) with either two (*tetO*$_2$) or seven (*tetO*$_7$) sites.

In a second set of cloning vectors (TET-SET2) the tet reverse system was adapted for yeast cells and for a tighter regulation of gene expression a dual tet regulatory system was additionally developed. In the reverse system the tTA transactivator is modified and consists of a mutated tetR domain (tetR′) which binds in the presence of doxycycline to the *tetO* site(s) and therefore allows the rapid induction of *tetO*-driven gene expression. For stringent control, a dual system consists where a tetR-VP16 activator and a tetR′-Tup1

Yeast Genetic Strain and Plasmid Collections

[14]The VP16 protein of the herpes simplex virus.
[15]tetR from the Tn10-encoded tetracycline-resistance operon of *E. coli*.

repressor[16] are both expressed in the same cell (Barnett and Entian, 2005). Therefore, in the presence of doxycycline the expression of the *tetO-CYC1* promoter is prevented in two ways, first, upon binding of the tetR′-Tup1 repressor and, second, upon removal of the tetR-VP16 activator. The *vice versa* combination has also been developed where doxycycline induces the *tetO-CYC1* promoter upon binding of a tetR′-VP16 activator and removal of a tetR-Tup1 repressor.

In a third set of vectors (TET-SET3) promoter-substitution cassettes have been constructed containing the tetR and tetR′ transactivator genes, with either $tetO_2$ or $tetO_7$ promoters, and the *kanMX4* module as selection marker. These cassettes are suitable for one-step replacement of the chromosomal target gene promoters by the entire tetracycline-regulatable *tetO* promoter system in yeast cells.

2. Strains for tetracycline-regulated gene expression

With some modification the *tetO* promoter-substitution system was used within the EUROFAN 2 B3 node (essential genes and gene families). To facilitate the promoter-substitution, yeast strain YUG37 which is isogenic to the FY strains (see Section I.E.1), was designed in which the tTA transactivator has been integrated at the genomic *LEU2* locus (Genotype: *MAT***a** *ura3-52 trp1-63 LEU2::tTA GAL2*). For PCR-mediated promoter-substitution plasmid pAH3 was constructed which carries the $tetO_7$-*CYC1* hybrid promoter linked to the *loxP-KanMX4-loxP* module. $TetO_7$ promoter strains for about 100 essential genes were constructed and are available from EUROSCARF.

About 600 ORFs of essential genes under shut-off control of the $tetO_7$ promoter were constructed in strain R1158[17] (Mnaimneh *et al.*, 2004). Strain R1158 was constructed by one-step integration of the tTA transactivator at the *ura3-Δ0* locus (Genotype: *MAT***a** *his3-Δ1 leu2-Δ0 met15-Δ0 URA3::CMVp-tTA*[18]). $TetO_7$ promoter alleles were constructed by replacing the 100 bp upstream of the start codon with a kan^R-$tetO_7$-$TATA_{CYC1}$ cassette from plasmid RP188. The yeast *tet* promoter shut-off strains (Hughes strains) are available from Open Biosystems.

3. Zero background reporter plasmids

For the measurement of the activity of promoter elements a series of yeast reporter plasmids were constructed by Karsten Melcher and

[16]The Tup1 repressor prevents the expression of a large number of *S. cerevisiae* genes (for review see Barnett and Entian, 2005).

[17]R1158 is a derivative of BY4741 which is the haploid *MAT***a** wild-type strain used in the deletion consortium (see Section III.D.1).

[18]CMVp-tTA: the human cytomegalovirus promoter IE which directs expression of the tTA activator gene.

his co-workers (Melcher *et al.*, 2000) based on the *MEL1* gene encoding α-galactosidase, a secreted but largely cell-wall-associated enzyme (Lazo *et al.*, 1977). Therefore α-galactosidase activities can be followed in whole cells which makes the determination of promoter activities highly convenient. After insertion into the *MEL1* zero background reporter plasmids the activity of promoter elements can be easily followed. UAS (upstream activating sequences) can be directly followed and URS (upstream repressing sequences) can be followed after previous insertion of a UAS element. Such plasmids are available from EUROSCARF.

E. Regulated Protein Degradation

Functional analysis of essential genes is still a challenge and methods are needed which rapidly inhibit the protein function. Therefore, temperature-sensitive mutations in which the respective protein becomes inactive upon shift either to higher or to lower temperatures (ts- and cs-mutants[19]) provide an important analytical tool to fulfill such requirements. However, the generation of temperature-sensitive mutations is difficult and labor-intensive. Therefore, for the analysis of a large number of essential *S. cerevisiae* genes, replacing the promoter by a regulated promoter became the method of choice[20] (see Section III.D). The major disadvantage of protein depletion upon shift to repressing conditions is that – although the mRNA is rapidly removed – the final phenotype depends either on the decay of the essential protein or on its dilution during further growth. So regulated gene expression mainly provide long-term phenotypes and secondary responses may falsify the phenotype.

A direct approach to inactivate the function of proteins more efficiently is their fusion with the so-called degrons, e.g. sequences of the Cup1 protein, which conditionally trigger the rapid proteolytic degradation of the fusion protein (this volume, Chapter 5). Such degrons were first described by Dohmen and co-workers (Dohmen *et al.*, 1994) and it is supposed that they expose lysine residues which are the sites of ubiquitinylation upon unfolding at higher temperature so that the fusion protein is labeled for rapid proteasomal degradation. This depends on the Ubr1 protein which is associated with a ubiquitin-conjugating enzyme (Labib *et al.*, 2000). For large-scale analysis essential *S. cerevisiae* proteins, most of them having functional human homologs, were N-terminally tagged with the Cup1 degron sequence (Kanemaki *et al.*, 2003). Currently, 130

[19]The term ts-mutation is used when proteins are inactivated at higher temperatures (in general 34–37°C), whereas cs-mutations refer to proteins which are inactivated at lower temperatures (in general 25°C).

[20]Destabilizing mRNAs via RNA interference is an alternative method in higher eukaryotes, but this cannot be applied in *S. cerevisiae* as the enzymes needed for RNA interference are not present in *S. cerevisiae*.

such strains and constructs as well as the plasmids for making such constructs are available from EUROSCARF.

F. TAP-Tagged Strains and Plasmids

The TAP-tagging (tandem affinity purification) of proteins in combination with MALDI-TOF (**M**atrix **A**ssociated **L**aser **D**esorption **I**onisation – **T**ime-**O**f-**F**light) mass spectrometry has become a powerful tool for the analysis of the composition of heteromeric protein complexes (Rigaut *et al.*, 1999; Puig *et al.*, 2001). The protein of interest is C-terminally fused with two affinity tags, the calmodulin binding protein and the two IgG binding domains of protein A. Between both affinity tags a TEV (**T**obacco **e**tch **v**irus) cleavage site is inserted (Figure 4). After mild cell disruption protein complexes are bound to IgG beads which have an extremely high affinity to the protein A tags. After washing steps the tagged protein complex is resolved from the IgG column after incubation with TEV protease.

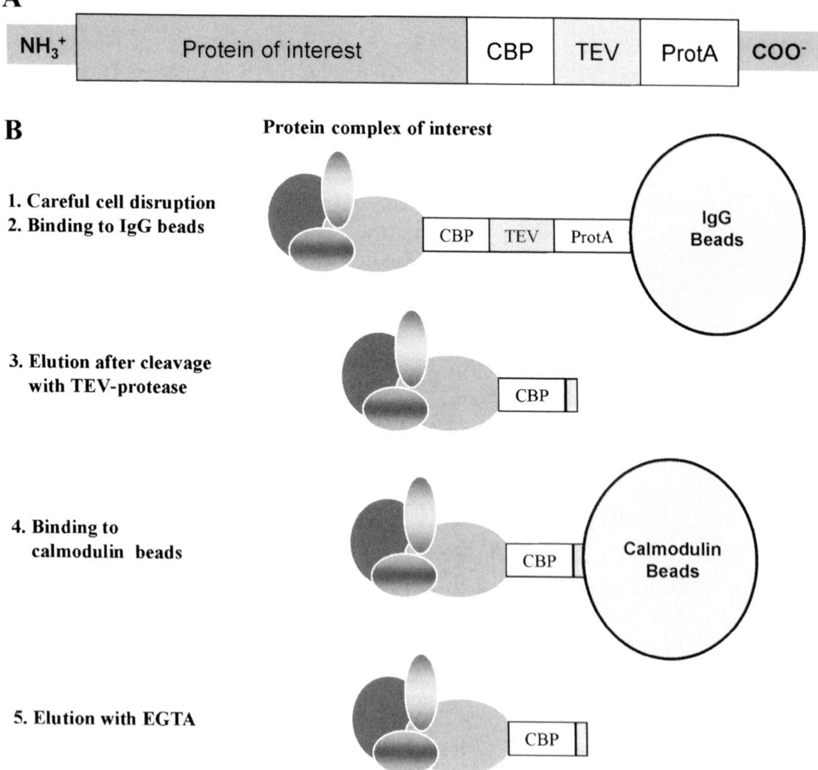

Figure 4. Affinity purification of protein complexes. By using protein A as affinity tag the protein complex is bound to IgG beads. Owing to the strong binding the complex is eluted after incubation with TEV protease and thereafter bound to calmodulin beads using the CBP domain. Finally, the protein complex is eluted with increasing concentrations of EGTA.

This cleaves at the TEV cleavage site that is adjacent to the C-terminus of the calmodulin tag. In a second chromatography step the released protein complex is bound to a calmodulin affinity column. The binding of the calmodulin binding protein is Ca^{2+} dependent, so that the protein complex can be eluted with increasing EGTA concentrations (Figure 4). The purified protein complex is thereafter dissociated with SDS and the heteromeric subunits are separated via SDS polyacrylamide gel electrophoreses. A more detailed technical description of this method can be found in this volume, Chapter 18.

For further analysis the gel is cut into slices which are incubated with proteases (in general trypsin) and the mass of the resulting tryptic fragments is analyzed via MALDI-TOF mass spectrometry (Figure 5). Owing to the known *S. cerevisiae* genome sequence the obtained masses can be compared to the expected tryptic masses of each *S. cerevisiae* protein and the bioinformatics analysis results in a probability score for the respective peptides.

The TAP-tagged/MALDI MS method was applied in a high-throughput analysis with 800 different proteins (Gavin *et al.*, 2002) and later extended to all 6000 *S. cerevisiae* proteins (Gavin *et al.*,

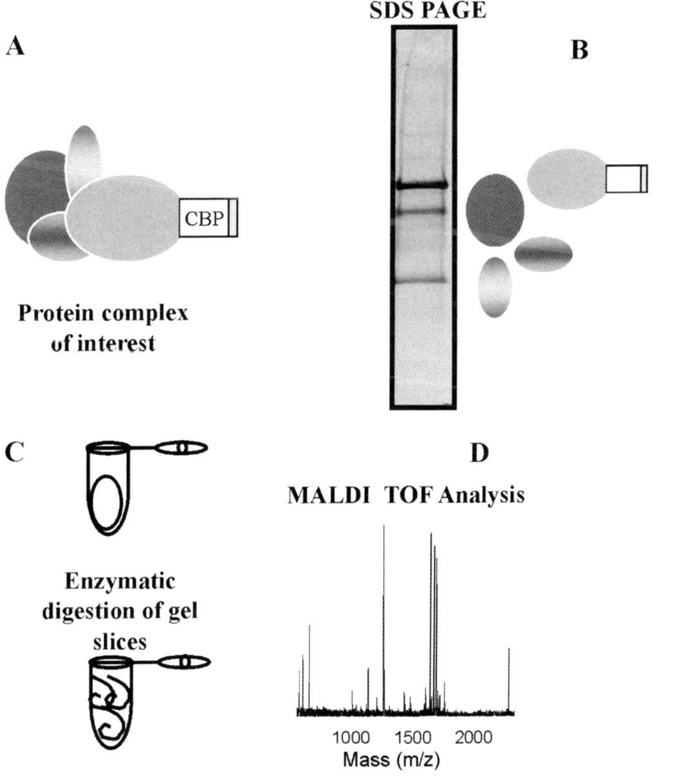

Figure 5. Analysis of TAP-purified protein complexes. (A) Enriched protein complex, (B) separation by SDS-PAGE, (C) enzymatic protein digestion and (D) MALDI-TOF analysis.

2006). About 5400 TAP-tagged strains are available from EUROSCARF. These strains are probably W303 derivatives and were inserted into the genome with the *URA3* gene of *Kluyveromyces lactis* (*KlURA3*).

About 5000 strains also suitable for TAP-affinity purification are available from Open Biosystems in a BY-strain background. The TAP-tags are the same as in the TAP-tagged collection of EUROSCARF (see Section III.B.2) but the tags are inserted with the HIS3MX6 module which uses the *his5* gene of *S. pombe* (Ghaemmaghami *et al.*, 2003).

Additionally, plasmids for N- and C-terminal TAP-tagging using either the *KlTRP1* or the *KlURA3* gene as selectable marker (Puig *et al.*, 2001) can be received from EUROSCARF.

G. Epitope-Tagged Strains and Plasmids

Using a 93 bp in frame insertion via transposon insertion generated a collection of strains with HA-tagged proteins (Ross-Macdonald *et al.*, 1999; Kumar *et al.*, 2002). Such strains were generated after random transposon insertion followed by the removal of the transposon using Cre-recombination sites. About 25 000 strains with HA-expression were sequenced at the transposon insertion site and the insertion sequence was determined. Finally, 3600 strains with in-frame HA-tags within their ORFs were obtained which are available from Open Biosystems.

Several plasmids for PCR-based tagging are available from EUROSCARF. These include plasmids for the conditional expression of N-terminally epitope-tagged genes, where the *GAL10* promoter was fused to epitope tags 2 × ProtA, 3 × *c-myc* and 8 × His, respectively (Lafontaine and Tollervey, 1996). Recently, a toolbox for PCR-based tagging of *S. cerevisiae* genes with the yeast-optimized version of the red-fluorescent RedStar2 (Patterson and Lippincott-Schwartz, 2002) protein, other fluorescent proteins and tags became available (Janke *et al.*, 2004). Proteins can be tagged either at the N-terminus with various heterologous promoters or at the C-terminus. This toolbox is highly economic as only four primers are needed for the various taggings and is available from EUROSCARF.

In many cases, epitopes are inserted together with heterologous promoters which consequently change the strength of gene expression as compared to the natural promoter. This can artificially interfere with the protein function. Therefore, a set of plasmids was recently developed for the N-terminal tagging of proteins which leave the endogenous promoter intact and therefore retain the endogenous expression level of the respective tagged protein (Gauss *et al.*, 2005). The tags include 6 × HA, 9 × *c-myc*, yEGFP, TEV-GST-6 × HIS, ProtA, TEV-ProtA and TEV-ProtA-7 × HIS in combination with different heterologous selection markers and can be obtained from EUROSCARF.

H. GFP-Tagged Strains and Plasmids

In a genome-wide approach a collection of yeast strains expressing full-length, chromosomally tagged GFP fusion proteins were constructed and analyzed (Huh *et al.*, 2003). The *Aequorea victoria* coding sequence of GFP (S65T) and the *HIS6MX6* module as selectable marker were inserted at the C-terminus of each yeast ORF through homologous recombination with a success rate of 97% for all ORFs and 87% for the essential ORFs. About 70% of all tagged ORFs showed GFP signals above the background level and were classified in 22 distinct subcellular localization categories. The GFP-tagged yeast strains are available from Invitrogen either as a single yeast strain or as the whole collection in MTP format.

Plasmids for N- and C-terminal tagging with the yeast-optimized red-fluorescent RedStar2 protein (Janke *et al.*, 2004) can be received from EUROSCARF. Plasmids for PCR-based N-terminal GFP-tagging (yEGFP) which leaves the endogenous promoters unchanged have recently been described and are available from EUROSCARF (Gauss *et al.*, 2005).

I. Expression of Recombinant Proteins

A modular set of prokaryotic and eukaryotic expression vectors was developed by Karsten Melcher (Melcher, 2000) and is available from EUROSCARF. These expression vectors allow improved affinity purification of recombinant fusion proteins after expression either in *E. coli* or *S. cerevisiae*. Features of these vectors include serial affinity tags (hexahistidine-GST) followed by a TEV cleavage site.

A yeast ORF collection is available from Open Biosystems which was designed for regulated protein overexpression and purification where 5500 OFRs were cloned in plasmid BG1805 (unpublished, for further information see Open Biosystems webpage). All ORFs are expressed under the control of the *GAL1* promoter and contain a C-terminal tag for affinity purification. Plasmid constructs are available as yeast or *E. coli* transformants. With some restrictions, this collection can also be used for *trans*-complementation studies.

J. Selection of Haploid Segregants (Magic Marker Selection)

The 'Magic Marker' selection system (SGA reporter) developed by Boone and co-workers (Tong *et al.*, 2001) facilitates the conversion of heterozygous diploid deletion strains into haploid mutants (see Chapter 16, this volume). This method enables mass segregations as it selects for haploid *MAT***a** segregants after sporulation. Magic Marker selection is the method of choice for mass segregations, however, for individual gene analysis tetrad analysis (see also Section I.B) still provides much more information as the segregation and the size of the segregant colonies provide additional information about possible secondary mutations.

For "Magic Marker" selection the *HIS3* prototrophy is expressed via the *MFA1*[21] promoter together with the recessive canavanine resistance (*can1*^{r}).[22] Consequently, only *MAT***a** cells are histidine prototrophs and can be selected on histidine minus media. Furthermore, the recessive *can1* mutation further strengthens the selection for haploids if the selection medium is also supplied with canavanine.

In each of the heterozygous diploid deletion strains the "Magic Marker" cassette (*CAN1L-LEU2*$^+$*-MFA1pr-HIS3-CAN1R*[23]) from plasmid pXP346 (Pan *et al.,* 2004) was integrated at the *CAN1* locus resulting in a disruption of this locus (can1*Δ*^{r}*::LEU2-MFA1pr-HIS3).* The resulting double heterozygous deletion strains are histidine auxotroph, leucine prototroph and canavanine sensitive. After sporulation haploid *MAT***a** segregants can easily be selected as they are canavanine resistant and prototrophic for histidine and leucine.

Individual strains or the whole collection are available from Open Biosystems with the following genotypes:

$$\frac{MAT\mathbf{a}}{MAT\alpha}\quad\frac{orfX::kanMX4}{ORFX}\quad\frac{ura3\text{-}\Delta0}{ura3\text{-}\Delta0}\quad\frac{leu2\text{-}\Delta0}{leu2\text{-}\Delta0}\quad\frac{his3\text{-}\Delta1}{his3\text{-}\Delta1}$$

$$\frac{lys2\text{-}\Delta0}{LYS2}\quad\frac{met15\text{-}\Delta0}{MET15}\quad\frac{can1\Delta::LEU2\text{-}MFA1pr\text{-}HIS3}{CAN1}$$

♦♦♦♦♦♦ IV. STRAIN STABILITY AND SERVICES

Each deletion mutant underlies an unavoidable selection pressure and compensatory genetic changes may occur. Those mutations with a strong phenotype particularly accumulate secondary mutations. For haploid or homozygous diploid deletion strains the selection pressure for compensatory genetic exchanges (secondary mutations) is certainly much stronger than for the heterozygous deletion strains. As a result, the genetic quality of such strains deteriorates with the number of generations and amplification diminishes the quality of the collection. Therefore, EUROSCARF keeps the number of generations needed for the amplification of the

[21]In *S. cerevisiae* the mating type is strongly controlled by silencing and the *MFA1* promoter, which encodes the pheromone is only expressed in haploid *MAT***a** cells.

[22]Canavanine is an analogue of arginine. The *can1*^{r} mutation makes cells resistant against canavanine as the *CAN1* permease is mutated and therefore the uptake of arginine as well as canavanine is prevented.

[23]The *LEU2-MFA1pr-HIS3* construct was inserted into the *CAN1* gene so that the construct is flanked by the left part of the *CAN1* gene (= *CAN1L*) and the right part of the *CAN1* gene (= *CAN1R*).

deletion collection as low as possible. Especially, for the genome-wide analysis of yeast genetic networks the accumulation of secondary mutations needs to be considered. In heterozygous diploid deletion strains the mutation is covered by the wild-type allele so that the accumulation of compensatory mutations as well as strong mutant phenotypes are largely absent. For the large-scale generation of *MAT***a** haploid deletion mutants from heterozygous diploids the "Magic Marker" system (see Section III.J is very suitable), however limits the functional analysis to the *MAT***a** mating type. For the identification of possible secondary mutations tetrad analysis is still the most efficient method as secondary mutations result in a 3:1 segregation in tetratype asci (see Section I.B).

In many cases the genetic background of the yeast strain strongly influences the phenotype observed, and hence a strain collection can never be comprehensive for all scientific questions. Therefore, EUROSCARF and other collections also provide routine services such as transferring bar-coded deletions to other strains, tetrad analysis and haplotoxicity tests. Such services can be obtained upon request. Furthermore, the collections are continuously extended and provide a number of useful tools for yeast functional analysis.

♦♦♦♦♦♦ V. ACCESS TO THE STRAIN AND PLASMID COLLECTIONS

Strains and plasmids described in here are accessible from the strain collections mentioned above (see Section II.A–II.D). For non-commercial fundamental research most if not all strains and plasmids are free from patent restrictions. If strains and plasmids are used for commercial research in some cases patent restrictions have to be considered and license fees may be requested from the respective patent holders.

Tables 6 and 7 summarize the biological materials described in here and also refers to the respective mutant and plasmid collections which provide the respective strains and plasmids.

Distributing collections: EUROSCARF: web.uni-frankfurt.de/fb15/mikro/euroscarf), ATCC: www.atcc.org, Open Biosystems: www.openbiosystems.com and Invitrogen: www.invitrogen.com.

♦♦♦♦♦♦ VI. CONCLUSIONS AND IMPACT OF THE DELETION COLLECTIONS ON EUKARYOTIC RESEARCH

For the isolation of mutants, two basic principles can be followed. First, a large number of yeast mutants can be screened individually

Table 6. Genetically modified *S. cerevisiae* strains for gene/protein functional analysis

Strains	Strain background	Described in chapter	No. of strains available[a]	Source/reference	Collection
Deletion mutants	CEN.PK2	Sections I.E and II.A	800	BMBF + EUROFAN I	EUROSCARF
	W.303	Sections I.E and II.A	300	EUROFAN I	EUROSCARF
	FY	Sections I.E and II.A	1000	EUROFAN I	EUROSCARF
Bar-coded deletions	BY (FY derivative)	Sections I.E, II.A–D, III.A	6100	EUROFAN II	EUROSCARF, ATCC, Open Biosystems, Invitrogen
tet-regulated genes	YUG37 (FY derivative)	Section III.D	100	EUROFAN I	EUROSCARF
	R1158 (BY derivative)	Section III.D	600	Mnaimneh *et al.* (2004)	EUROSCARF
Degron strains	???	Section III.E	130	Kanemaki *et al.* (2003)	EUROSCARF
TAP-tagged ORFs	W303	Section III.F	5000	Gavin *et al.* (2002)	EUROSCARF
TAP-tagged ORFs	BY	Section III.F	5000	Ghaemmaghami *et al.* (2003)	Open Biosystems
Strains with HA_inserted ORFs	BY	Section III.G	3600	Ross-Macdonald *et al.* (1999) and Kumar *et al.* (2002)	Open Biosystems
GFP strains	BY	Section III.H	4000	Huh *et al.* (2003)	Invitrogen
Magic marker strains	BY	Section III.J	5000	Pan *et al.* (2004)	Open Biosystems
Strains with multiple use ORFs	BY	Section III.I	5500	Unpublished, see text	Open Biosystems

[a] These deletions are available either as *MAT***a** haploids, *MAT*α haploids, homozygous or heterozygous diploids.

Table 7. Plasmids for gene/protein functional analysis

Plasmid description	Genes/plasmid name	Described in chapter	Reference	Collection
Dominant markers for gene deletion	*KanMX, natMX, nphMX, LYS5, dsdA*	Section III.B	See text for citations	EUROSCARF
Heterologous markers for gene deletion	*CaURA3, CaLYS5*	Section III.B	See text for citations	EUROSCARF
loxP deletion markers	*kanMX, LYS5, CaLYS5, dsdA, KlURA3, KlLEU2*	Section III.B	See text for citations	EUROSCARF
Cre recombinase expression	pSH47, pSH62, pSH63, pSH65	Section III.B	Güldener *et al.* (2002)	EUROSCARF
Cognate clones		Section III.C	EUROFAN I	EUROSCARF
tet-regulated expression	TET-SET1, 2 and 3	Section III.D	See text for citations	EUROSCARF
Zero background reporter plasmids	*MEL1*	Section III.D	Melcher *et al.* (2000)	EUROSCARF
N- and C-terminal TAP-tagging	pBS1479, pBS1539, pBS1761	Section III.F	Puig *et al.* (2001)	EUROSCARF
GAL10 regulated N-terminal epitopes	2 × ProtA, 3 × *c-myc*, 8 × His	Section III.G	Lafontaine and Tollervey (1996)	EUROSCARF
N- and C-terminal epitope tagging tool box	Several heterologous promoters, several epitopes	Section III.G	Janke *et al.* (2004)	EUROSCARF
Internal and N-terminal epitope tagging	6 × HA, 9 × *c-myc*, TEV-GST-6 × His, ProtA, TEV-ProtA, TEV-ProtA-7 × His	Section III.G	Gauss *et al.* (2005)	EUROSCARF
GFP plasmids for N-terminal tagging	RedStar*, RedStar2, yeGFP	Section III.H	Janke *et al.* (2004) and Gauss *et al.* (2005)	EUROSCARF
Plasmids with multiple use ORFs	BG1805	Section III.I	Unpublished, see text	Open Biosystems

for a specific phenotype. An example of this screening approach is the replica-plating method (Lederberg and Lederberg, 1952). Second, growth conditions can be used that favor the growth of, and thereby enrich for, a particular set of mutants. The use of 5-FOA to select for uracil auxotrophic mutants is an example of this (Boeke *et al.*, 1984). In general, mutant isolation using the screening approach is more labor-intensive than the selective approach. However, sometimes selective systems are not sufficiently sensitive and many mutants are missed, especially those with reduced growth rates. In the era of whole genome analysis, these two approaches are also applicable for the analysis of deletion mutants. Screening analyses (mutant scanning) test each deletion mutant individually for a large number of possible phenotypes (Entian *et al.*, 1999), whereas selective growth conditions can be used to diminish mutants with a particular phenotype from a population. Selective growth conditions can be either applied to a transposon-mutagenized mutant population (Smith *et al.*, 1995, 1996) or even more effectively to the entire population of bar-coded deletion mutants using chip hybridization for the detection of diminished mutants (Giaever *et al.*, 2002).

After the *S. cerevisiae* genome had been unraveled and bar-coded deletion became available for nearly all yeast genes several analytical tools have been developed for the genome-wide analysis of gene functions. These include transcriptome, proteome and metabolome analysis. Such analytical tools generate a huge amount of data and many of these data are preliminary and need further experimental confirmation. Therefore, one of the major challenges for the future is the handling of the vast amounts of data. Today, it is not sufficient simply to generate large information-rich databases: establishing the quality of that data is of paramount importance. Although genome-wide screenings are important and useful for gene functional analysis, they only provide the basis for the gene/protein-specific functional analysis. The yeast deletion mutant collections are of major importance for the genome-wide and the specific gene-functional analysis as well. They make such valuable biological materials easily accessible to the scientific community. Especially, for the genome-wide analysis it is of major importance that the quality of the deletion collection is preserved and possible genetic instabilities (see above) are kept to a minimum.

Today, after several years of experience with the collection of yeast deletion mutants it is without doubt true that the labor- and cost-intensive generation of the genome-wide deletions marked a breakthrough in the functional analysis of the yeast genome. Many key methods for the functional analysis of eukaryotic cells have been developed with yeast and *S. cerevisiae* has as a result become an important eukaryotic model organism. Therefore, the collection of deletion mutants represents an important resource for the functional analysis of higher eukaryotes and eukaryotic human pathogens.

Since the deletion collections have been established more than a million of deletion strains[24] have been distributed. This shows that the deletion collections also have a strong socio-economic impact because of their relevance for the investigation of the molecular basis of a large number of human and plant diseases[25] and for the development of new therapeutic strategies to scope with eukaryotic diseases and infections.

Acknowledgments

The authors would like to thank the curator of EUROSCARF, Dr. M Rose, for valuable discussions and suggestions. EUROSCARF was funded by the German Minister for Education and Research (BMBF) and the European Union (EUROFAN I and II projects). It also received support from the Yeast Industrial Platform (YIP) and the Fonds der Chemischen Industrie, Germany.

References

Barnett, J. A. and Entian, K.-D. (2005). A history of research on yeasts 9: regulation of sugar metabolism. *Yeast* **22**, 835–894.

Baudin, A., Ozier-Kalogeropoulos, O., Denouel, A., Lacroute, F. and Cullin, C. (1993). A simple and efficient method for direct gene deletion in *Saccharomyces cerevisiae*. *Nucleic Acids Res.* **21**, 3329–3330.

Belli, G., Gari, E., Aldea, M. and Herrero, E. (1998a). Functional analysis of yeast essential genes using a promoter-substitution cassette and the tetracycline-regulatable dual expression system. *Yeast* **14**, 1127–1138.

Belli, G., Gari, E., Piedrafita, L., Aldea, M. and Herrero, E. (1998b). An activator/repressor dual system allows tight tetracycline-regulated gene expression in budding yeast. [erratum appears in *Nucleic Acids Res* 1998 Apr 1;26(7):following 1855]. *Nucleic Acids Res.* **26**, 942–947.

Boeke, J. D., LaCroute, F. and Fink, G. R. (1984). A positive selection for mutants lacking orotidine-5′-phosphate decarboxylase activity in yeast: 5-fluoro-orotic acid resistance. *Mol. Gen. Genet.* **197**, 345–346.

Brachmann, C. B., Davies, A., Cost, G. J., Caputo, E., Li, J., Hieter, P. and Boeke, J. D. (1998). Designer deletion strains derived from *Saccharomyces cerevisiae* S288C: a useful set of strains and plasmids for PCR-mediated gene disruption and other applications. *Yeast* **14**, 115–132.

Carle, G. and Olson, M. (1984). Separation of chromosomal DNA molecules from yeast by orthogonal-field-alternation gel electrophoresis. *Nucleic Acids Res.* **12**, 5647–5664.

Cliften, P., Sudarsanam, P., Desikan, A., Fulton, L., Fulton, B., Majors, J., Waterston, R., Cohen, B. A. and Johnston, M. (2003). Finding functional features in *Saccharomyces* genomes by phylogenetic footprinting. *Science* **301**, 71–76.

[24]This is a roughly estimated number extrapolated from the number of EUROSCARF requests (M. Rose, personal communication).

[25]This counts for those yeast genes (proteins) which share functional homologies to genes (proteins) which are implied in human and plant diseases and infections.

Dohmen, R. J., Wu, P. and Varshavsky, A. (1994). Heat-inducible degron: a method for constructing temperature-sensitive mutants. *Science* **263**, 1273–1276.

Entian, K. D., Schuster, T., Hegemann, J. H., Becher, D., Feldmann, H., Güldener, U., Gotz, R., Hansen, M., Hollenberg, C. P., Jansen, G., Kramer, W., Klein, S., Kötter, P., Kricke, J., Launhardt, H., Mannhaupt, G., Maierl, A., Meyer, P., Mewes, W., Munder, T., Niedenthal, R. K., Ramezani Rad, M., Röhmer, A., Romer, A. and Hinnen, A. (1999). Functional analysis of 150 deletion mutants in *Saccharomyces cerevisiae* by a systematic approach. *Mol. Gen. Genet.* **262**, 683–702.

Gari, E., Piedrafita, L., Aldea, M. and Herrero, E. (1997). A set of vectors with a tetracycline-regulatable promoter system for modulated gene expression in *Saccharomyces cerevisiae*. *Yeast* **13**, 837–848.

Gauss, R., Trautwein, M., Sommer, T. and Spang, A. (2005). New modules for the repeated internal and N-terminal epitope tagging of genes in *Saccharomyces cerevisiae*. *Yeast* **22**, 1–12.

Gavin, A. C., Aloy, P., Grandi, P., Krause, R., Boesche, M., Marzioch, M., Rau, C., Jensen, L. J., Bastuck, S., Dumpelfeld, B., Edelmann, A., Heurtier, M. A., Hoffman, V., Hoefert, C., Klein, K., Hudak, M., Michon, A. M., Schelder, M., Schirle, M., Remor, M., Rudi, T., Hooper, S., Bauer, A., Bouwmeester, T., Casari, G., Drewes, G., Neubauer, G., Rick, J. M., Kuster, B., Bork, P., Russell, R. B. and Superti-Furga, G. (2006). Proteome survey reveals modularity of the yeast cell machinery. *Nature* **440**, 631–636.

Gavin, A. C., Bosche, M., Krause, R., Grandi, P., Marzioch, M., Bauer, A., Schultz, J., Rick, J. M., Michon, A. M., Cruciat, C. M., Remor, M., Hofert, C., Schelder, M., Brajenovic, M., Ruffner, H., Merino, A., Klein, K., Hudak, M., Dickson, D., Rudi, T., Gnau, V., Bauch, A., Bastuck, S., Huhse, B., Leutwein, C., Heurtier, M. A., Copley, R. R., Edelmann, A., Querfurth, E., Rybin, V., Drewes, G., Raida, M., Bouwmeester, T., Bork, P., Seraphin, B., Kuster, B., Neubauer, G. and Superti-Furga, G. (2002). Functional organization of the yeast proteome by systematic analysis of protein complexes. *Nature* **415**, 141–147.

Ghaemmaghami, S., Huh, W.-K., Bower, K., Howson, R. W., Belle, A., Dephoure, N., O'Shea, E. K. and Weissman, J. S. (2003). Global analysis of protein expression in yeast. *Nature* **425**, 737–741.

Giaever, G., Chu, A. M., Ni, L., Connelly, C., Riles, L., Veronneau, S., Dow, S., Lucau-Danila, A., Anderson, K., Andre, B., Arkin, A. P., Astromoff, A., El-Bakkoury, M., Bangham, R., Benito, R., Brachat, S., Campanaro, S., Curtiss, M., Davis, K., Deutschbauer, A., Entian, K. D., Flaherty, P., Foury, F., Garfinkel, D. J., Gerstein, M., Gotte, D., Güldener, U., Hegemann, J. H., Hempel, S., Herman, Z., Jaramillo, D. F., Kelly, D. E., Kelly, S. L., Kötter, P., LaBonte, D., Lamb, D. C., Lan, N., Liang, H., Liao, H., Liu, L., Luo, C., Lussier, M., Mao, R., Menard, P., Ooi, S. L., Revuelta, J. L., Roberts, C. J., Rose, M., Ross-Macdonald, P., Scherens, B., Schimmack, G., Shafer, B., Shoemaker, D. D., Sookhai-Mahadeo, S., Storms, R. K., Strathern, J. N., Valle, G., Voet, M., Volckaert, G., Wang, C. Y., Ward, T. R., Wilhelmy, J., Winzeler, E. A., Yang, Y., Yen, G., Youngman, E., Yu, K., Bussey, H., Boeke, J. D., Snyder, M., Philippsen, P., Davis, R. W. and Johnston, M. (2002). Functional profiling of the *Saccharomyces cerevisiae* genome. *Nature* **418**, 387–391.

Giaever, G., Shoemaker, D. D., Jones, T. W., Liang, H., Winzeler, E. A., Astromoff, A. and Davis, R. W. (1999). Genomic profiling of drug sensitivities via induced haploinsufficiency. *Nat. Genet.* **21**, 278–283.

Goffeau, A., Barrell, B. G., Bussey, H., Davis, R. W., Dujon, B., Feldmann, H., Galibert, F., Hoheisel, J. D., Jacq, C., Johnston, M., Louis, E. J., Mewes, H. W., Murakami, Y., Philippsen, P., Tettelin, H. and Oliver, S. G. (1996). Life with 6000 genes. *Science* **274**, 546–567.

Goldstein, A. L. and McCusker, J. H. (1999). Three new dominant drug resistance cassettes for gene disruption in *Saccharomyces cerevisiae*. *Yeast* **15**, 1541–1553.

Goldstein, A. L., Pan, X. and McCusker, J. H. (1999). Heterologous URA3MX cassettes for gene replacement in *Saccharomyces cerevisiae*. [erratum appears in *Yeast* 1999 Sep 15;15(12):1297]. *Yeast* **15**, 507–511.

Gossen, M. and Bujard, H. (1992). Tight control of gene expression in mammalian cells by tetracycline-responsive promoters. *Proc. Natl. Acad. Sci. USA* **89**, 5547–5551.

Güldener, U., Heck, S., Fielder, T., Beinhauer, J. and Hegemann, J. H. (1996). A new efficient gene disruption cassette for repeated use in budding yeast. *Nucleic Acids Res.* **24**, 2519–2524.

Güldener, U., Heinisch, J., Koehler, G. J., Voss, D. and Hegemann, J. H. (2002). A second set of *loxP* marker cassettes for Cre-mediated multiple gene knockouts in budding yeast. *Nucleic Acids Res.* **30**, e23.

Hampsey, M. (1997). A review of phenotypes in *Saccharomyces cerevisiae*. *Yeast* **13**, 1099–1133.

Huh, W. K., Falvo, J. V., Gerke, L. C., Carrol, A. S., Howson, R. W., Weissman, J. S. and O'Shea, E. K. (2003). Global analysis of protein localization in budding yeast. *Nature* **425**, 686–691.

Ito-Harashima, S. and McCusker, J. H. (2004). Positive and negative selection LYS5MX gene replacement cassettes for use in *Saccharomyces cerevisiae*. *Yeast* **21**, 53–61.

Janke, C., Magiera, M. M., Rathfelder, N., Taxis, C., Reber, S., Maekawa, H., Moreno-Borchart, A., Doenges, G., Schwob, E., Schiebel, E. and Knop, M. (2004). A versatile toolbox for PCR-based tagging of yeast genes: new fluorescent proteins, more markers and promoter substitution cassettes. *Yeast* **21**, 947–962.

Kanemaki, M., Sanchez-Diaz, A., Gambus, A. and Labib, K. (2003). Functional proteomic identification of DNA replication proteins by induced proteolysis *in vivo*. *Nature* **423**, 720–724.

Kowalczuk, M., Mackiewicz, P., Gierlik, A., Dudek, M. R. and Cebrat, S. (1999). Total number of coding open reading frames in the yeast genome. *Yeast* **15**, 1031–1034.

Kumar, A., Agarwal, S., Heyman, J. A., Matson, S., Heidtman, M., Piccirillo, S., Umansky, L., Drawid, A., Jansen, R., Liu, Y., Cheung, K.-H., Miller, P., Gerstein, M., Roeder, G. S. and Snyder, M. (2002). Subcellular localization of the yeast proteome. *Genes Dev.* **16**, 707–719.

Labib, K., Tercero, J. A. and Diffley, J. F. (2000). Uninterrupted MCM2-7 function required for DNA replication fork progression. *Science* **288**, 1643–1647.

Lafontaine, D. and Tollervey, D. (1996). One-step PCR mediated strategy for the construction of conditionally expressed and epitope tagged yeast proteins. *Nucleic Acids Res.* **24**, 3469–3471.

Lazo, P. S., Ochoa, A. G. and Gascon, S. (1977). Alpha-galactosidase from *Saccharomyces carlsbergensis*. Cellular localization, and purification of the external enzyme. *Eur. J. Biochem.* **77**, 375–382.

Lederberg, J. and Lederberg, E. M. (1952). Replica plating and indirect selection of bacterial mutants. *J. Bacteriol.* **63**, 399–406.

Lindegren, C. C. and Lindegren, G. (1943). A new method for hybridizing yeast. *Proc. Natl. Acad. Sci. USA* **29**, 306–308.

Link, A. J. and Olson, M. V. (1991). Physical map of the *Saccharomyces cerevisiae* genome at 110-kilobase resolution. *Genetics* **127**, 681–698.

Mackiewicz, P., Kowalczuk, M., Mackiewicz, D., Nowicka, A., Dudkiewicz, M., Laszkiewicz, A., Dudek, M. R. and Cebrat, S. (2002). How many protein-coding genes are there in the *Saccharomyces cerevisiae* genome?. *Yeast* **19**, 619–629.

Melcher, K. (2000). A modular set of prokaryotic and eukaryotic expression vectors. [erratum appears in *Anal Biochem* 2000 Jul 1;282(2):266]. *Anal. Biochem.* **277**, 109–120.

Melcher, K., Sharma, B., Ding, W. V. and Nolden, M. (2000). Zero background yeast reporter plasmids. *Gene* **247**, 53–61.

Mnaimneh, S., Davierwala, A. P., Haynes, J., Moffat, J., Peng, W. T., Zhang, W., Yang, X., Pootoolal, J., Chua, G., Lopez, A., Trochesset, M., Morse, D., Krogan, N. J., Hiley, S. L., Li, Z., Morris, Q., Grigull, J., Mitsakakis, N., Roberts, C. J., Greenblatt, J. F., Boone, C., Kaiser, C. A., Andrews, B. J. and Hughes, T. R. (2004). Exploration of essential gene functions via titratable promoter alleles. *Cell* **118**, 31–44.

Mortimer, R. K. and Schild, D. (1980). Genetic map of *Saccharomyces cerevisiae*. *Microbiol. Rev.* **44**, 519–571.

Mortimer, R. K. and Schild, D. (1985). Genetic map of *Saccharomyces cerevisiae*, edition 9. *Microbiol. Rev.* **49**, 181–213.

Mortimer, R. K., Schild, D., Contopoulou, C. R. and Kans, J. A. (1989). Genetic map of *Saccharomyces cerevisiae*, edition 10. *Yeast* **5**, 321–403.

Oliver, S. (1996). A network approach to the systematic analysis of yeast gene function. *Trends Genet.* **12**, 241–242.

Olson, M. V., Dutchik, J. E., Graham, M. Y., Brodeur, G. M., Helms, C., Frank, M., MacCollin, M., Scheinman, R. and Frank, T. (1986). Random-clone strategy for genomic restriction mapping in yeast. *Proc. Natl. Acad. Sci. USA* **83**, 7826–7830.

Ostergaard, S., Olsson, L., Johnston, M. and Nielsen, J. (2000). Increasing galactose consumption by *Saccharomyces cerevisiae* through metabolic engineering of the *GAL* gene regulatory network. *Nat. Biotechnol.* **18**, 1283–1286.

Pan, X., Yuan, D. S., Xiang, D., Wang, X., Sookhai-Mahadeo, S., Bader, J. S., Hieter, P., Spencer, F. and Boeke, J. D. (2004). A robust toolkit for functional profiling of the yeast genome. *Mol. Cell* **16**, 487–496.

Patterson, G. H. and Lippincott-Schwartz, J. (2002). A photoactivatable GFP for selective photolabeling of proteins and cells. *Science* **297**, 1873–1877.

Puig, O., Caspary, F., Rigaut, G., Rutz, B., Bouveret, E., Bragado-Nilsson, E., Wilm, M. and Seraphin, B. (2001). The tandem affinity purification (TAP) method: a general procedure of protein complex purification. *Methods (Duluth)* **24**, 218–229.

Rieger, K. J., El-Alama, M., Stein, G., Bradshaw, C., Slonimski, P. P. and Maundrell, K. (1999). Chemotyping of yeast mutants using robotics. *Yeast* **15**, 973–986.

Rieger, K. J., Kaniak, A., Coppee, J. Y., Aljinovic, G., Baudin-Baillieu, A., Orlowska, G., Gromadka, R., Groudinsky, O., Di Rago, J. P. and Slonimski, P. P. (1997). Large-scale phenotypic analysis – the pilot project on yeast chromosome III. *Yeast* **13**, 1547–1562.

Rigaut, G., Shevchenko, A., Rutz, B., Wilm, M., Mann, M. and Seraphin, B. (1999). A generic protein purification method for protein complex characterization and proteome exploration. *Nat. Biotechnol.* **17**, 1030–1032.

Riles, L., Dutchik, J. E., Baktha, A., McCauley, B. K., Thayer, E. C., Leckie, M. P., Braden, V. V., Depke, J. E. and Olson, M. V. (1993). Physical maps of the six smallest chromosomes of *Saccharomyces cerevisiae* at a resolution of 2.6 kilobase pairs. *Genetics* **134**, 81–150.

Ross-Macdonald, P., Coelho, P. S., Roemer, T., Agarwal, S., Kumar, A., Jansen, R., Cheung, K. H., Sheehan, A., Symoniatis, D., Umansky, L., Heidtman, M., Nelson, F. K., Iwasaki, H., Hager, K., Gerstein, M., Miller, P., Roeder, G. S. and Snyder, M. (1999). Large-scale analysis of the yeast genome by transposon tagging and gene disruption. *Nature* **402**, 413–418.

Rothstein, R. (1991). Targeting, disruption, replacement, and allele rescue: integrative DNA transformation in yeast. *Methods Enzymol.* **194**, 281–301.

Rothstein, R. J. (1983). One-step gene disruption in yeast. *Methods Enzymol.* **101**, 202–211.

Saldanha, A. J., Brauer, M. J. and Botstein, D. (2004). Nutritional homeostasis in batch and steady-state culture of yeast. *Mol. Biol. Cell* **15**, 4089–4104.

Schneider, J. C. and Guarente, L. (1991). Vectors for expression of cloned genes in yeast: regulation, overproduction, and underproduction. *Methods Enzymol.* **194**, 373–388.

Shoemaker, D. D., Lashkari, D. A., Morris, D., Mittmann, M. and Davis, R. W. (1996). Quantitative phenotypic analysis of yeast deletion mutants using a highly parallel molecular bar-coding strategy. *Nat. Genet.* **14**, 450–456.

Singh, A. and Manney, T. R. (1974). Genetic analysis of mutations affecting growth of *Saccharomyces cerevisiae* at low temperature. *Genetics* **77**, 651–659.

Smith, V., Botstein, D. and Brown, P. O. (1995). Genetic footprinting: a genomic strategy for determining a gene's function given its sequence. *Proc. Natl. Acad. Sci. USA* **92**, 6479–6483.

Smith, V., Chou, K. N., Lashkari, D., Botstein, D. and Brown, P. O. (1996). Functional analysis of the genes of yeast chromosome V by genetic footprinting. *Science* **274**, 2069–2074.

Stückrath, I., Lange, H. C., Kötter, P., van Gulik, W. M., Entian, K. D. and Heijnen, J. J. (2002). Characterization of null mutants of the glyoxylate cycle and gluconeogenic enzymes in *S. cerevisiae* through metabolic network modeling verified by chemostat cultivation. *Biotechnol. Bioeng.* **77**, 61–72.

Tong, A. H. Y., Evangelista, M., Parsons, A. B., Xu, H., Bader, G. D., Page, N., Robinson, M., Raghibizadeh, S., Hogue, C. W. V., Bussey, H., Andrews, B., Tyers, M. and Boone, C. (2001). Systematic genetic analysis with ordered arrays of yeast deletion mutants. *Science* **294**, 2364–2368.

van Dijken, J. P., Bauer, J., Brambilla, L., Duboc, P., Francois, J. M., Gancedo, C., Giuseppin, M. L., Heijnen, J. J., Hoare, M., Lange, H. C., Madden, E. A., Niederberger, P., Nielsen, J., Parrou, J. L., Petit, T., Porro, D., Reuss, M., van Riel, N., Rizzi, M., Steensma, H. Y., Verrips, C. T., Vindelov, J. and Pronk, J. T. (2000). An interlaboratory comparison of physiological and genetic properties of four *Saccharomyces cerevisiae* strains. *Enzyme Microb. Technol.* **26**, 706–714.

Vorachek-Warren, M. K. and McCusker, J. H. (2004). DsdA (d-serine deaminase): a new heterologous MX cassette for gene disruption and selection in *Saccharomyces cerevisiae*. *Yeast* **21**, 163–171.
Wach, A. (1996). PCR-synthesis of marker cassettes with long flanking homology regions for gene disruptions in *S. cerevisiae*. *Yeast* **12**, 259–265.
Wach, A., Brachat, A., Pohlmann, R. and Philippsen, P. (1994). New heterologous modules for classical or PCR-based gene disruptions in *Saccharomyces cerevisiae*. *Yeast* **10**, 1793–1808.
Wieczorke, R., Krampe, S., Weierstall, T., Freidel, K., Hollenberg, C. P. and Boles, E. (1999). Concurrent knock-out of at least 20 transporter genes is required to block uptake of hexoses in *Saccharomyces cerevisiae*. *FEBS Lett.* **464**, 123–128.
Winston, F., Dollard, C. and Ricupero-Hovasse, S. L. (1995). Construction of a set of convenient *Saccharomyces cerevisiae* strains that are isogenic to S288C. *Yeast* **11**, 53–55.
Winzeler, E. A., Shoemaker, D. D., Astromoff, A., Liang, H., Anderson, K., Andre, B., Bangham, R., Benito, R., Boeke, J. D., Bussey, H., Chu, A. M., Connelly, C., Davis, K., Dietrich, F., Dow, S. W., El Bakkoury, M., Foury, F., Friend, S. H., Gentalen, E., Giaever, G., Hegemann, J. H., Jones, T., Laub, M., Liao, H., Davis, R. W. *et al.* (1999). Functional characterization of the *S. cerevisiae* genome by gene deletion and parallel analysis. *Science* **285**, 901–906.
Wood, V. K. M. R., Ivens, A., Rajandream, M.-A. and Barrell, B. (2001). A re-annotation of the *Saccharomyces cerevisiae* genome. *Comp. Funct. Genom.* **2**, 143–154.
Zimmermann, F. K. and Eaton, N. R. (1974). Genetics of induction and catabolite repression of maltase synthesis in *Saccharomyces cerevisiae*. *Mol. Gen. Genet.* **134**, 261–272.

26 Yeast Gene Analysis: The Remaining Challenges

Michael JR Stark[1] **and Ian Stansfield**[2]
[1] *Division of Gene Regulation and Expression, College of Life Sciences, University of Dundee, Dundee, UK;* [2] *University of Aberdeen, School of Medical Sciences, Institute of Medical Sciences, Foresterhill, Aberdeen, UK*

♦♦

CONTENTS

Introduction
The new technologies: past achievements and future prospects for yeast gene analysis
Genome-wide versus focused studies: an enduring need for both
Towards a predictive model of the yeast cell
Concluding remarks

List of abbreviations

BLAST	Basic Local Alignment Search Tool
DASH	*Dam1/Duo1*, *Ask1*, *Spc34/Spc19* and *Hsk1* complex
FRET	Fluorescence resonance energy transfer
FRAP	Fluorescence recovery after photobleaching
FT-ICR	Fourier transform-ion cyclotron resonance
GFP	Green fluorescent protein
MALDI	Matrix-assisted laser desorption ionisation
RFP	Red fluorescent protein
TAP	Tandem affinity purification
TOF	Time of flight

♦♦♦♦♦♦ I. INTRODUCTION

Developing a thorough understanding of how the yeast cell functions and reproduces itself was always going to be a huge undertaking. Ten years on since the completion of the *Saccharomyces cerevisiae* genome sequence, the true scale of the task can perhaps now be better appreciated, and it remains a formidable one. A key objective has

METHODS IN MICROBIOLOGY, VOLUME 36
0580-9517 DOI:10.1016/S0580-9517(06)36026-6

been to assign a function to every yeast gene. As various strategies have been applied to this goal the experience accumulated has guided successive experimental approaches, and functions are constantly, if gradually, still being assigned to ORFs with unknown roles. Nevertheless, the law of diminishing returns will almost certainly mean that for some such ORFs, specific, time-consuming and individually designed approaches will be needed to elucidate their function.

A major challenge for yeast researchers over the next decade is therefore clear; to assign functions to genes of currently unknown function, to identify molecular functions for those gene products only loosely assigned to a cellular process and to correct the mis-assigned functions that must surely exist at some level in the data set. There are many instances where ORF annotation currently relies on just a single piece of information, and so incorrect assignment of function is to be expected in a significant proportion of these. However, even apparently well-understood cases can also hold surprises. For example, the Elp protein complex (Elongator) was originally proposed on the basis of a variety of evidence to be involved in transcriptional elongation. However, it now seems that Elongator is primarily involved in tRNA modification and the transcriptional defects in Elongator-deficient strains are downstream effects of the tRNA modification deficiency (Esberg *et al.*, 2006). Robust annotation will be helped both by the growing body of high-throughput data and by detailed, focused studies, undoubtedly leading to many reassigned functions. That we have a partial answer to the question, 'what do all the yeast genes do?' is a testimony to the efforts of the high-throughput 'surveyors', as well as the efforts of those concentrating on 'sub-systems', working outwards from a small set of gene products. However, many genes still remain to have a function assigned, in some cases because their true roles are masked by redundancy, in others perhaps because the phenotypes generated are revealed properly only under a limited set of specific conditions or because the knockout phenotype is just too subtle. How marginal a contribution to fitness a gene can make while ensuring its retention in the genome is not at all clear, and yet the answer to that question defines the required detection threshold for subtle effects during phenotypic screening. Finally, since like all organisms yeast is actively evolving, it is possible that some genes in the genome no longer confer a selective advantage, and will in due course be lost. Considerable experimental ingenuity will thus be required to complete the functional assignment task.

♦♦♦♦♦♦ II. THE NEW TECHNOLOGIES: PAST ACHIEVEMENTS AND FUTURE PROSPECTS FOR YEAST GENE ANALYSIS

The first edition of this book was published at the dawn of the yeast post-genomic era and made several predictions about how yeast gene analysis was likely to develop. While in some cases the details of the new technologies that emerged were not quite as envisaged, many of these predictions have turned out to be surprisingly prescient. For example, the use of DNA microarrays has emerged as a major technology just as anticipated, and has proved a powerful technique not just for transcript profiling, but as a means of working with populations of bar-coded yeast deletion strains (see Chapter 17). The microarray technology and the informatics approaches required for analysing the data are now very well-developed (Chapter 9), and continue to yield a wealth of information on the transcriptional regulation of the yeast genome. The development of high-density protein microarrays was also correctly predicted, although these have so far utilised recombinant polypeptides rather than the capture of proteins using an array of specific antibodies as originally anticipated. The importance of mass spectrometry as an emerging technology that would have a major impact on proteomics, glycomics and metabolomics was also realised, although its application to the high-throughput characterisation of affinity isolated protein complexes was not foreseen. Finally, it was realised that the sequencing of additional yeast genomes would play an important role in yeast gene analysis, although the power of comparative sequence data for confirming ORFs (especially the small ones) and for identifying regulatory elements has been an unexpected spin-off (Chapter 22). In what follows, we will consider how successful these new technologies have been, and how they might continue to develop over the next decade.

A. Proteomics

Of all the post-genomic technologies, transcriptomics has arguably played one of the central roles in expanding our understanding of cell function. However, without quantitative knowledge of how that transcriptome is translated into the proteome, essential pieces of the jigsaw are missing. Reports in the literature offer mixed views of the degree of correlation between mRNA and protein levels. While good correlations have been reported for some highly expressed proteins, in other investigations significant uncoupling was apparent between the mRNA and protein abundances, indicative of post-transcriptional control of gene expression (Ideker *et al.*, 2001; Griffin *et al.*, 2002). Moreover, experience tells us that such dissonance could be particularly marked following sudden shifts in environmental condition, such as heat shock, when it is well-established that only a subset of mRNAs are translated. Thus, extrapolating from transcriptome analyses to protein levels is risky. Qualitatively, we have no clear idea of whether a significant number of yeast mRNAs are alternatively spliced: the best example to date (*HAC1* mRNA) is clearly a special case (Sidrauski and Walter, 1997). Equally unclear is the extent to which alternative translation initiation and termination codons are used to change the nature of the translated product, as exemplified by the Mod5 and Pde2 proteins, respectively (Slusher *et al.*, 1991; Namy *et al.*, 2002). Finally, we do not yet have a comprehensive view of which proteins experience post-translational modifications, nor in most instances how such modifications affect cell physiology. Given on the one hand the extent of uncertainty in these areas, coupled on the other with the wealth of knowledge to be gained from a comprehensive understanding of the proteome, there is clearly a compelling case for quantitative proteomics to be developed into a robust technology capable of dealing with the significant complexity of the yeast proteome.

While recent advances in mass spectrometry technology have markedly improved our ability to analyse the proteome, the challenge still remains of analysing the approximately 4800 proteins expressed in growing yeast cells (Newman *et al.*, 2006), whose abundances vary over five orders of magnitude and whose modifications are multiple and varied. The different forms of mass spectrometer, from matrix-assisted laser desorption ionisation time-of-flight (MALDI-TOF) to quadrupole time-of-flight (Q-Q-ToF) and the recently developed Fourier transform ion cyclotron resonance (FT-ICR) and Orbitrap machines, all offer different advantages: some are highly accurate, some offer high-throughput capability, others are optimised for quantitative analysis or identification of protein modifications. At present, no one machine is capable of performing all mass spectrometry-based approaches well (Domon and Aebersold, 2006). Comparative and quantitative proteomics has been rendered more accurate by stable isotope labelling methods such as stable isotope labelling by amino acids in cell culture (SILAC; Mann, 2006) and the use of tandem mass tags (reviewed in Domon and Aebersold, 2006), while absolute quantitation is also now a possibility (see Beynon *et al.*, 2005). It is likely that these types of approach will be applied much more widely in the future, helped by further advances in mass spectrometry technology. The ability to perform quantitative analysis by mass spectrometry should at least partly overcome the limitations of two-dimensional gel analysis, which is biased towards the detection of a subset of the proteome. Perhaps the best way of assessing the current challenge is to look at some recent attempts to carry out complete yeast proteome analysis; using advanced mass spectrometers, one recent study succeeded in identifying 2000 of the roughly 4800 expressed yeast proteins (de Godoy *et al.*, 2006), while another identified 3109 proteins (Wei *et al.*, 2005). Although this represents a significant achievement it clearly shows the detection 'gap' still to be plugged in proteomic analysis, even before the issue of variation in protein composition and modification is addressed. However, new developments in instrumentation and software for data analysis hold promise for a much improved coverage of the proteome in the near future (de Godoy *et al.*, 2006). Nevertheless, the wide dynamic range of protein levels in the cell may preclude achieving near-complete coverage of the proteome for some time yet.

Regarding high-throughput analysis of protein modification, the recent advances in mass spectrometry have now made quantitative investigations of protein modification on a proteome-wide scale a realistic prospect: a recent study combining the SILAC method with approaches for detecting and sequencing phosphopeptides identified 139 phosphorylation sites that were differentially regulated at least two-fold in response to mating pheromone treatment (Gruhler *et al.*, 2005). In principle such approaches can be applied to any posttranslational modification, and so in the future we can anticipate much-improved data on a wider range of modifications. In particular, the ability to perform at least relative quantitation of posttranslational modifications holds great promise for developing a much more comprehensive picture of how the proteome is regulated at this level. At present, a major problem with high-throughput analysis of post-translational modifications such as phosphorylation is the labour-intensive nature of the data analysis that is required. New approaches aimed at automating these aspects are likely to have a significant impact on future studies in yeast (Beausoleil *et al.*, 2006).

Looking to the future, the development of protein microarrays (see Chapter 14) is an exciting advance where we can hope to see significant progress in the next decade. While the ability to generate such arrays is a major feat involving the individual expression and purification of a large fraction of the yeast proteome, their great potential is currently limited by the lack of complete coverage. Amongst other things, this reflects the difficulty in expressing many proteins individually in soluble recombinant form: in a recent study of the yeast DASH complex it was necessary to co-express all 10 different components to achieve this (Miranda *et al.*, 2005). New developments in protein chip fabrication, for example, by the capture of nascent polypeptides following *in vitro* translation of *in vitro* transcribed or synthetic mRNAs are likely to help address the problem of proteome coverage and lead to an improved second generation of protein chips (Tao and Zhu, 2006). Furthermore, by combining the developing knowledge of protein complexes with protein chip technology it may be possible to generate microarrays carrying individual protein complexes for a variety of proteome-wide studies.

B. The yeast Interactome

Our understanding of protein complex formation surely represents another hugely challenging frontier to explore in spite of the progress already made. It is likely that the next 10 years will see increasing attempts to understand how protein association is controlled both over time and by environmental factors. In a complementary approach to genome-wide two-hybrid analysis of protein–protein interaction (Chapter 7), tandem affinity purification

(TAP)-tagging (outlined in Chapter 18) has already been applied on a genome-wide scale and has the potential to reveal rapidly the complexities of the yeast interactome (Gavin *et al.*, 2002; Ho *et al.*, 2002). The most recent studies have revealed the existence of literally hundreds of hitherto unrecognised protein complexes (Gavin *et al.*, 2006; Krogan *et al.*, 2006). However, TAP-tagging is a sensitive biochemical tool, and which precise complexes are detected using TAP-tagging is, of course, responsive to variations of salt and detergent concentration during purification, just as for any standard protein isolation. Proteins that associate weakly or transiently with a given complex might not be detected under certain TAP-tag conditions, and yet may play an important role in the biology of a particular assembly. Conversely, TAP-tag experimental conditions must ideally eliminate non-specific interactions. Techniques will need to be developed to distinguish more clearly between weak and non-specific classes of interaction.

Protein complex membership can be sophisticated in its arrangement. For example, it has been established for a while that the MAP kinase pathway transcription factor Ste12p is a part of different MAP kinase complexes and participates in two distinct signalling events, leading either to mating or to pseudohyphal growth. In this respect, Ste12p appeared to be somewhat different from the average run-of-the mill protein (reviewed in Madhani and Fink, 1998; Chou *et al.*, 2006). However, perhaps one of the most fundamental discoveries to emerge from the TAP-tag studies is the degree to which many proteins appear to be integral components of multiple and distinct complexes (Gavin *et al.*, 2006; Krogan *et al.*, 2006); this property of Ste12p may prove to be the rule rather than the exception. Furthermore, the ability of individual proteins to participate in different protein complexes at different times, and under different environmental conditions, means the interactome is a varied and fluid entity that will require new tools not only just to investigate it but also to represent it in a searchable form. Most current databases are two dimensional (**a** interacts with **b**) with some three-dimensional information (**a** interacts with **b** in the nucleus). The paucity of information on the fluid nature of complex membership has hitherto not required the development of database tools to annotate systematically detailed protein association behaviour that might vary over time in response to altered environmental conditions. A related requirement for database tools to interrogate such networks of information using Basic Local Alignment Search Tool (BLAST) type methodology has already been recognised (Kelley *et al.*, 2004), and further developments in these areas are to be expected.

C. Fluorescent Tagging of Proteins

The growing focus on the three-dimensional distribution of proteins within the yeast cell is now revealing a great deal of information and genome-wide tagging of proteins with the Green Fluorescent Protein (GFP) that has already provided a vast resource of information (Huh *et al.*, 2003). Tagging with fluorescent proteins has also been crucial in establishing protein co-localisation, an indicator of a potentially common function that in many cases is distinct from the issue of protein complex formation. Co-localisation of proteins can be a key indicator of their participation in a common biochemical process, including sequential interactions such as during ribosome assembly in the nucleolus, or mRNA degradation in the *P*-bodies (Sheth and Parker, 2003; Sheth and Parker, 2006). The combination of the synthetic lethal screen data (Tong *et al.*, 2001; Chapter 16, this volume) and the protein co-localisation studies could be a rich area for exploration in the future.

Temporal variation in protein complex composition and the idea that the cytoplasm is not homogeneous, but perhaps 'zoned', are also concepts ripe for exploration with an expanding suite of fluorescent tools for use in tagging proteins, including variants of GFP and other proteins such as RFP. With these tools in place, FRET can be employed to explore the *in vivo* association of proteins in complexes (Chapter 12) and has already been applied to the study of complex systems in yeast (Damelin and Silver, 2000). Fluorescence recovery after photobleaching (FRAP), which allows a pulse-chase type monitoring of protein traffic in and out of complexes or specific regions of the cell, is another under-exploited technique that is likely to see increasingly wide usage in yeast studies. For example, FRAP has been employed to show movement of eIF2B (a translation initiation factor), in and out of eIF2B 'organising centres' (Campbell *et al.*, 2005). As more of the complexes identified through tandem affinity tagging are investigated individually and in detail, it is likely that these types of fluorescent tagging techniques will see much more widespread applications, particularly since a given protein may be dynamically a member of multiple complexes (Gavin *et al.*, 2006; Krogan *et al.*, 2006). Protein marking by fluorescent tags can also be employed to follow other, non-protein macromolecules including whole chromosomes (see Chapter 10) and specific mRNAs. For instance, the movement within the cell of mRNA molecules marked with U1A binding RNA hairpins can be tracked using a U1A (Mud1p)-GFP fusion protein (Brodsky and Silver, 2000; Sheth and Parker, 2006), while an analogous approach has been used to monitor the movement of *ASH1* mRNA (Bertrand *et al.*, 1998). There is now an opportunity to apply such tools on a wider scale to investigate biological complex formation and dissociation involving not just proteins but RNA or DNA.

Factors currently limiting exploration of the complex membership and protein association include the sensitivity of the fluorescent tags, particularly when dealing with the low-abundance proteins, and their tendency to become bleached during monitoring. It is certain that such challenges will be addressed in the near future by a combination of improved optical technologies, better image processing, and improved fluorescent tags. New tools are emerging and will undoubtedly continue to be developed to address problems with the existing technology. For example, proteins tagged with GFP may exhibit artefactual behaviour such as altered stability or biological activity, and tagging is limited to the attachment of the tag to either end of the protein of interest. One recent development that allows the fluorescent tagging of proteins using a very short genetic tag is the tetracysteine biarsenical system, in which a 12-amino acid tag attached to the protein of interest reacts with a membrane-permeable fluorescent dye (FlAsH or ReAsH) with picomolar affinity (Griffin *et al.*, 1998). While such tags have not to date been widely used in yeast, there is some evidence already that they can offer a solution for problem cases where GFP tagging causes loss of function of the parent polypeptide (e.g. β-tubulin tagging; Andresen *et al.*, 2004), and might also enable proteins to be tagged at additional sites that are exposed on their surface.

The ability to add fluorescent tags such as GFP to proteins combined with increasingly sophisticated optics and image processing, opens many other doors. It is possible to monitor the expression of proteins in single cells, giving information on expression levels and the noise of expression levels. Measurement of gene expression noise in turn reveals the degree of precision with which individual genes are controlled; high precision/low noise control can indicate the existence of feedback control of expression (Newman *et al.*, 2006). Furthermore, the ability to monitor fluorescence in single cells allows monitoring of oscillatory levels of proteins such as cyclins on a per cell basis (Bean *et al.*, 2006). For certain classes of proteins, monitoring single cells using fluorescently tagged proteins will give important insights into the properties of gene products, and it seems likely that these approaches will see increasing application in the coming years. The ability to generate spotted cell microarrays of living yeast cells for phenotypic screening (Narayanaswamy *et al.*, 2006) coupled with the use of fluorescently tagged proteins is another area that could be exploited in the future.

D. Genome Sequencing and High-Throughput Screens

What of genome sequencing in this post-genomic era? It seems inevitable that sequencing will continue, adding at a steady rate to the stock of completed genomes. Completion of other fungal genomes will be of particular interest to yeast researchers and within the yeasts, sequencing efforts recently completed have focused on the *Saccharomyces sensu stricto* species (Cliften *et al.*, 2003; Kellis *et al.*, 2003), helping to confirm or eliminate 'doubtful' ORFs. However, the output from such studies has proved of much wider value, representing a rich store of information about selection for functional motifs within the genome, such as promoter elements (Cliften *et al.*, 2003), nucleosome positioning elements (Ioshikhes *et al.*, 2006), origins of DNA replication (Nieduszynski *et al.*, 2006) and small RNA genes (McCutcheon and Eddy, 2003), all of which are difficult to detect with reference to only a single genome. The *sensu stricto* genome resource is likely to continue to be of immense value to other researchers in the coming years. The evolutionary aspects of the *Saccharomyces* genus, and of the species *S. cerevisiae* itself will almost certainly benefit from the further sequencing of multiple strains of the species, helping unpick the tangled family history of this model organism. A major 'resequencing' effort is already underway (see http://www.sanger.ac.uk/Teams/Team71/durbin/sgrp/) and by the time this book is published we will have access to the genome sequence of many of the yeast community's favourite strains including SK1, W303 and Y55, a valuable new resource.

Recent years have seen an explosive acceleration in the implementation of high-throughput screens, including yeast two hybrid (Uetz *et al.*, 2000), titratable promoter allele (Mnaimneh *et al.*, 2004), synthetic lethal (Tong *et al.*, 2001) and gene overexpression (Sopko *et al.*, 2006). Novel ways have been found of utilising the bar-coded gene deletion set, including chemogenomics (see Chapter 17). In this approach, chemical insult is employed to identify unfit genotypes from the heterozygous deletion population, allowing identification of the targets of small molecule inhibitors (Giaever *et al.*, 2004). The development of newer and better versions of the 'TAG' arrays for monitoring the bar-coded strains (Pierce *et al.*, 2006) will add to the power of such approaches, enabling the generation of more robust and comprehensive datasets. New insight will undoubtedly come from a combinatorial use of many of the above resources, although clearly such multiplexing will often require the application of robotics to manage the sheer numbers of strains generated. Nevertheless, the reward for such effort is the potential for such combined use of reagents to tease out novel phenotypes for at least a proportion of the remaining genes for which no assigned function currently exists.

It is likely that ever more sophisticated high-throughput screens will be designed in the coming years, adding more to the enormous body of data already garnered. For many researchers engaged in focused studies on specific aspects of yeast, the challenge is to ensure that they profit from this output. New information management systems are undoubtedly required to take advantage of the every new lead and new piece of evidence. Really simple syndication (RSS) feeds linked to ORF locus pages in SGD (http://www.yeastgenome.org/) and MIPS (http://mips.gsf.de/genre/proj/yeast/) to which a researcher could subscribe might be useful in this regard, allowing researchers to be alerted automatically once any new information on a given gene or gene product become available.

♦♦♦♦♦♦ III. GENOME-WIDE VERSUS FOCUSED STUDIES: AN ENDURING NEED FOR BOTH

Without doubt, one of the most exciting and significant developments in yeast research over the past decade has been the development of the means to perform genome-wide and proteome-wide studies, and the contribution of such studies to our understanding of yeast cannot be underestimated. As discussed above, such studies are clearly set to continue and to evolve into more and more robust and sophisticated approaches. So, where does this leave the researcher working in a focused manner on a small subset of genes? Clearly, such investigations are now immeasurably more effective given the wealth of knowledge available, whether it is a question of identifying the genes present on the insert of a yeast gene library clone or designing better experiments based on a significant body of existing knowledge. However, despite an almost exponential increase in genome-wide studies over the past decade, the rate at which functions have been assigned to unassigned ORFs has not significantly increased, and there are still many ORFs whose annotation relies on scant information (Hughes *et al.*, 2004). While this may well change as the variety and robustness of genome-wide approaches increases, it emphasises the continued need for focused studies that can take a multifaceted approach to produce detailed information about a particular subset of genes and their products in a functional context. The power of the high-throughput, genome-wide approaches is their ability to generate a global, comprehensive picture, but in doing so they inevitably cannot go into the details. For example, knowing the entire set of pairwise synthetic lethal interactions between yeast gene knockouts will be a highly valuable resource, but its true value will be realised only when we know *why* particular combinations are synthetic lethal, and the answer to this type of question requires further, more focused work. The value of the genome-wide approaches is that they enable new connections to be made, providing a fantastic framework for investigating the details of specific aspects of yeast cell biology. There is clearly still plenty of room for small-scale investigations to work out how subsets of genes and their products identified in high-throughput approaches function together in specific processes. Furthermore, as we move towards a 'systems biology' approach, the focused studies will be required to fill in the detailed parameters needed to build, test and refine models for each of the relevant pathways or processes.

With regard to the high-throughput approaches, one problem has been the issue of coverage, already mentioned above in the context of protein chips. Within the generally available knockout collections, for example, there are a significant number of ORFs that are still not represented for one reason or another, and perhaps as many as 8% of knockout strains have in fact been found to retain a wild-type copy of the deleted locus (Scherens and Goffeau, 2004). Other problems have included the question of barcode identification using microarrays, with some bar-coded strains giving a poor hybridisation signal. Thus, there is still room to improve the resources available for the high-throughput approaches, and new developments such as better-designed microarrays (Pierce *et al.*, 2006) will do a lot to help.

For the more focused studies, there are now a wide variety of approaches both traditional and novel that can be brought to bear on a specific biological question, many of which are described in this volume. Over the next decade, we can expect to see even more new approaches developed. The advantage now for such studies is that they can usually be approached from the standpoint of a comprehensive 'parts list' derived from protein interaction studies and genetic interaction data, with knowledge of how expression of the relevant genes may be regulated under different conditions, and where the gene products are localised. With such a starting point, better experiments can be designed to address the deeper questions about the system of study.

For both types of approach an exciting development is the ability to look at events occurring in single cells. Whether it is the dynamic behaviour of chromosomes (Chapter 10) or the localisation or abundance of proteins detected through fluorescence tagging, this provides an important extra dimension over the traditional approaches that have necessarily dealt with populations of cells, as required in all approaches that involve making cell extracts, for example, biochemical assays or protein purification. Perhaps one major technical challenge is to roll out 'single cell technology' to a wider range of applications, adding precision to the data and enabling different parameters to be correlated in individual yeast cells. Such a capability would undoubtedly enable a deeper knowledge to be gained of many cellular processes.

♦♦♦♦♦♦ IV. TOWARDS A PREDICTIVE MODEL OF THE YEAST CELL

Recent times have seen an explosion of interest in systems biology, with yeast researchers in the vanguard of this arguably new area. The increasing volume and quality of information on yeast gene function is certain to make yeast an attractive system with which to begin to make representative and predictive mathematical models of biological systems. Such modelling can and has taken place at the network level, seeking to understand the actions and interactions of whole gene networks and their products. Transcript profile data, particularly those datasets that contain samples collected over a time-series, enable sophisticated network analysis to be carried out, identifying the different temporal stages of a response to an environmental cue, and seeking to identify feed-forward and feedback control mechanism that regulate gene expression (Lee *et al.*, 2002; Bar-Joseph *et al.*, 2003; Luscombe *et al.*, 2004). In other cases modellers are starting small, seeking to develop a model of a sub-system, perhaps one particularly well understood at the biochemical level, and working outwards. As a model system, yeast is well positioned to make a substantial contribution to systems biology approaches, given the developed state of functional genomics in this organism.

At the level of individual pathways, overtures into systems biology modelling have produced some very encouraging results. Recent attempts to model the osmotic stress response pathway in yeast have produced ordinary differential equation-based (ODE) models that predict osmotic stress transcriptional responses with an encouraging degree of accuracy (Klipp *et al.*, 2005). In this case, the researchers needed to 'knit' together modelled transcriptional responses, signal transduction responses, and the metabolic behaviour of the yeast glycolysis pathway, itself the subject of a number of modelling attempts (e.g. Teusink *et al.*, 2000). Another example is the generation of a quantitative model for mitotic exit, based on the now detailed understanding of this process (Queralt *et al.*, 2006). It is likely that time will see the development of a number of such models of 'sub-systems', with the ultimate aim of making them link and cross-talk. Some modelling activities are likely to be impeded by a dearth of hard biochemical information on rates, substrate affinities and allosteric control mechanisms. It might be predicted that high-throughput biochemical characterisation techniques will need to be developed to acquire a basic biochemical understanding of the 'nuts and bolts' of yeast biochemistry before large scale overtures in modelling are begun. However, the degree of success of some modelling already published makes clear that a good deal of data is already out there, and special literature search algorithms have been developed to facilitate its extraction (Hakenberg *et al.*, 2004).

Ultimately, the goal of systems biology is to use mathematical models to make novel and perhaps surprising discoveries about the way a system is predicted to function. This then generates verifiable experimental hypotheses, whose testing allows model refinement, and yet more hypothesis testing, closing the circle of mathematical discovery of system behaviour. Good examples of this iterative process include those that have helped identify components of circadian rhythm control circuits in *Arabidopsis* (Locke *et al.*, 2005). Robust models that can accurately represent true system function will be capable of prompting such discoveries, with the model used to explore both environmental factors and system input 'space', while monitoring system outputs.

♦♦♦♦♦♦ V. CONCLUDING REMARKS

The first international yeast meeting at Carbondale in 1961 acknowledged the importance of the yeast research community that had at that time already become established. Since then, the yeast community has moved from classical genetics through the molecular age and finally into the post-genomic era. Along the way the number of researchers using yeast as their model organism has expanded far beyond anyone's expectation, and the tools that have been developed have brought yeast to the point where it is arguably the most powerful eukaryotic model system available for studying basic cellular processes. Studies in yeast have made major contributions to our understanding of a wide variety of fundamental processes, including the cell division cycle and protein targeting to name just two. The post-genomic era has enabled both genome-wide and focused studies in yeast to advance at an amazing pace, and yet our knowledge of yeast is still far from complete. There is still much to learn and the wide variety of techniques that can be used now, many of which are described in this volume, add to the power of yeast as an experimental system. We are working now in exciting times, in which basic technical issues have largely been overcome, and in which only our ingenuity, and our ability to design experiments that ask searching questions, limits progress. Yeast research over the next decade promises to yield many more fundamental insights into the workings of the eukaryotic cell.

References

Andresen, M., Schmitz-Salue, R. and Jakobs, S. (2004). Short tetracysteine tags to beta-tubulin demonstrate the significance of small labels for live cell imaging. *Mol. Biol. Cell* **15**, 5616–5622.

Bar-Joseph, Z., Gerber, G. K., Lee, T. I., Rinaldi, N. J., Yoo, J. Y., Robert, F., Gordon, D. B., Fraenkel, E., Jaakkola, T. S., Young, R. A. and Gifford, D. K. (2003). Computational discovery of gene modules and regulatory networks. *Nat. Biotechnol.* **21**, 1337–1342.

Bean, J. M., Siggia, E. D. and Cross, F. R. (2006). Coherence and timing of cell cycle start examined at single-cell resolution. *Mol. Cell* **21**, 3–14.

Beausoleil, S. A., Villen, J., Gerber, S. A., Rush, J. and Gygi, S. P. (2006). A probability-based approach for high-throughput protein phosphorylation analysis and site localization. *Nat. Biotechnol.* **24**, 1285–1292.

Bertrand, E., Chartrand, P., Schaefer, M., Shenoy, S. M., Singer, R. H. and Long, R. M. (1998). Localization of *ASH1* mRNA particles in living yeast. *Mol. Cell* **2**, 437–445.

Beynon, R. J., Doherty, M. K., Pratt, J. M. and Gaskell, S. J. (2005). Multiplexed absolute quantification in proteomics using artificial QCAT proteins of concatenated signature peptides. *Nat. Methods* **2**, 587–589.

Brodsky, A. S. and Silver, P. A. (2000). Pre-mRNA processing factors are required for nuclear export. *RNA* **6**, 1737–1749.

Campbell, S. G., Hoyle, N. P. and Ashe, M. P. (2005). Dynamic cycling of eIF2 through a large eIF2B-containing cytoplasmic body: implications for translation control. *J. Cell Biol.* **170**, 925–934.

Chou, S., Lane, S. and Liu, H. (2006). Regulation of mating and filamentation genes by two distinct Ste12 complexes in *Saccharomyces cerevisiae*. *Mol. Cell Biol.* **26**, 4794–4805.

Cliften, P., Sudarsanam, P., Desikan, A., Fulton, L., Fulton, B., Majors, J., Waterston, R., Cohen, B. A. and Johnston, M. (2003). Finding functional features in *Saccharomyces genomes* by phylogenetic footprinting. *Science* **301**, 71–76.

Damelin, M. and Silver, P. A. (2000). Mapping interactions between nuclear transport factors in living cells reveals pathways through the nuclear pore complex. *Mol. Cell* **5**, 133–140.

de Godoy, L. M., Olsen, J. V., de Souza, G. A., Li, G., Mortensen, P. and Mann, M. (2006). Status of complete proteome analysis by mass spectrometry: SILAC labeled yeast as a model system. *Genome. Biol.* **7**, R50.

Domon, B. and Aebersold, R. (2006). Mass spectrometry and protein analysis. *Science* **312**, 212–217.

Esberg, A., Huang, B., Johansson, M. J. and Bystrom, A. S. (2006). Elevated levels of two tRNA species bypass the requirement for elongator complex in transcription and exocytosis. *Mol. Cell* **24**, 139–148.

Gavin, A. C., Aloy, P., Grandi, P., Krause, R., Boesche, M., Marzioch, M., Rau, C., Jensen, L. J., Bastuck, S., Dumpelfeld, B., Edelmann, A., Heurtier, M. A., Hoffman, V., Hoefert, C., Klein, K., Hudak, M., Michon, A. M., Schelder, M., Schirle, M., Remor, M., Rudi, T., Hooper, S., Bauer, A., Bouwmeester, T., Casari, G., Drewes, G., Neubauer, G., Rick, J. M., Kuster, B., Bork, P., Russell, R. B. and Superti-Furga, G. (2006). Proteome survey reveals modularity of the yeast cell machinery. *Nature* **440**, 631–636.

Gavin, A. C., Bosche, M., Krause, R., Grandi, P., Marzioch, M., Bauer, A., Schultz, J., Rick, J. M., Michon, A. M., Cruciat, C. M. *et al.* (2002). Functional organization of the yeast proteome by systematic analysis of protein complexes. *Nature* **415**, 141–147.

Giaever, G., Flaherty, P., Kumm, J., Proctor, M., Nislow, C., Jaramillo, D. F., Chu, A. M., Jordan, M. I., Arkin, A. P. and Davis, R. W. (2004). Chemogenomic profiling: identifying the functional interactions of small molecules in yeast. *Proc. Natl. Acad. Sci. USA* **101**, 793–798.

Griffin, B. A., Adams, S. R. and Tsien, R. Y. (1998). Specific covalent labeling of recombinant protein molecules inside live cells. *Science* **281**, 269–272.

Griffin, T. J., Gygi, S. P., Ideker, T., Rist, B., Eng, J., Hood, L. and Aebersold, R. (2002). Complementary profiling of gene expression at the transcriptome and proteome levels in *Saccharomyces cerevisiae*. *Mol. Cell Proteomics* **1**, 323–333.

Gruhler, A., Olsen, J. V., Mohammed, S., Mortensen, P., Faergeman, N. J., Mann, M. and Jensen, O. N. (2005). Quantitative phosphoproteomics applied to the yeast pheromone signaling pathway. *Mol. Cell Proteomics* **4**, 310–327.

Hakenberg, J., Schmeier, S., Kowald, A., Klipp, E. and Leser, U. (2004). Finding kinetic parameters using text mining. *Omics* **8**, 131–152.

Ho, Y., Gruhler, A., Heilbut, A., Bader, G. D., Moore, L., Adams, S. L., Millar, A., Taylor, P., Bennett, K., Boutilier, K., Yang, L., Wolting, C., Donaldson, I., Schandorff, S., Shewnarane, J., Vo, M., Taggart, J., Goudreault, M., Muskat, B., Alfarano, C., Dewar, D., Lin, Z., Michalickova, K., Willems, A. R., Sassi, H., Nielsen, P. A., Rasmussen, K. J., Andersen, J. R., Johansen, L. E., Hansen, L. H., Jespersen, H., Podtelejnikov, A., Nielsen, E., Crawford, J., Poulsen, V., Sorensen, B. D., Matthiesen, J., Hendrickson, R. C., Gleeson, F., Pawson, T., Moran, M. F., Durocher, D., Mann, M., Hogue, C. W., Figeys, D. and Tyers, M. (2002). Systematic identification of

protein complexes in *Saccharomyces cerevisiae* by mass spectrometry. *Nature* **415**, 180–183.
Hughes, T. R., Robinson, M. D., Mitsakakis, N. and Johnston, M. (2004). The promise of functional genomics: completing the encyclopedia of a cell. *Curr. Opin. Microbiol.* **7**, 546–554.
Huh, W. K., Falvo, J. V., Gerke, L. C., Carroll, A. S., Howson, R. W., Weissman, J. S. and O'Shea, E. K. (2003). Global analysis of protein localization in budding yeast. *Nature* **425**, 686–691.
Ideker, T., Thorsson, V., Ranish, J. A., Christmas, R., Buhler, J., Eng, J. K., Bumgarner, R., Goodlett, D. R., Aebersold, R. and Hood, L. (2001). Integrated genomic and proteomic analyses of a systematically perturbed metabolic network. *Science* **292**, 929–934.
Ioshikhes, I. P., Albert, I., Zanton, S. J. and Pugh, B. F. (2006). Nucleosome positions predicted through comparative genomics. *Nat. Genet.* **38**, 1210–1215.
Kelley, B. P., Yuan, B., Lewitter, F., Sharan, R., Stockwell, B. R. and Ideker, T. (2004). PathBLAST: a tool for alignment of protein interaction networks. *Nucleic Acids Res.* **32** W83–W88.
Kellis, M., Patterson, N., Endrizzi, M., Birren, B. and Lander, E. S. (2003). Sequencing and comparison of yeast species to identify genes and regulatory elements. *Nature* **423**, 241–254.
Klipp, E., Nordlander, B., Kruger, R., Gennemark, P. and Hohmann, S. (2005). Integrative model of the response of yeast to osmotic shock. *Nat. Biotechnol.* **23**, 975–982.
Krogan, N. J., Cagney, G., Yu, H., Zhong, G., Guo, X., Ignatchenko, A., Li, J., Pu, S., Datta, N., Tikuisis, A. P., Punna, T., Peregrin-Alvarez, J. M., Shales, M., Zhang, X., Davey, M., Robinson, M. D., Paccanaro, A., Bray, J. E., Sheung, A., Beattie, B., Richards, D. P., Canadien, V., Lalev, A., Mena, F., Wong, P., Starostine, A., Canete, M. M., Vlasblom, J., Wu, S., Orsi, C., Collins, S. R., Chandran, S., Haw, R., Rilstone, J. J., Gandi, K., Thompson, N. J., Musso, G., St Onge, P., Ghanny, S., Lam, M. H., Butland, G., Altaf-Ul, A. M., Kanaya, S., Shilatifard, A., O'Shea, E., Weissman, J. S., Ingles, C. J., Hughes, T. R., Parkinson, J., Gerstein, M., Wodak, S. J., Emili, A. and Greenblatt, J. F. (2006). Global landscape of protein complexes in the yeast *Saccharomyces cerevisiae*. *Nature* **440**, 637–643.
Lee, T. I., Rinaldi, N. J., Robert, F., Odom, D. T., Bar-Joseph, Z., Gerber, G. K., Hannett, N. M., Harbison, C. T., Thompson, C. M., Simon, I., Zeitlinger, J., Jennings, E. G., Murray, H. L., Gordon, D. B., Ren, B., Wyrick, J. J., Tagne, J. B., Volkert, T. L., Fraenkel, E., Gifford, D. K. and Young, R. A. (2002). Transcriptional regulatory networks in *Saccharomyces cerevisiae*. *Science* **298**, 799–804.
Locke, J. C., Southern, M. M., Kozma-Bognar, L., Hibberd, V., Brown, P. E., Turner, M. S. and Millar, A. J. (2005). Extension of a genetic network model by iterative experimentation and mathematical analysis. *Mol. Syst. Biol.* **1**, 0013.
Luscombe, N. M., Babu, M. M., Yu, H., Snyder, M., Teichmann, S. A. and Gerstein, M. (2004). Genomic analysis of regulatory network dynamics reveals large topological changes. *Nature* **431**, 308–312.
Madhani, H. D. and Fink, G. R. (1998). The riddle of MAP kinase signaling specificity. *Trends Genet* **14**, 151–155.
Mann, M. (2006). Functional and quantitative proteomics using SILAC. *Nat. Rev. Mol. Cell Biol.* **7**, 952–959.

McCutcheon, J. P. and Eddy, S. R. (2003). Computational identification of non-coding RNAs in *Saccharomyces cerevisiae* by comparative genomics. *Nucleic Acids Res.* **31**, 4119–4128.

Miranda, J. J., De Wulf, P., Sorger, P. K. and Harrison, S. C. (2005). The yeast DASH complex forms closed rings on microtubules. *Nat. Struct. Mol. Biol.* **12**, 138–143.

Mnaimneh, S., Davierwala, A. P., Haynes, J., Moffat, J., Peng, W. T., Zhang, W., Yang, X., Pootoolal, J., Chua, G., Lopez, A., Trochesset, M., Morse, D., Krogan, N. J., Hiley, S. L., Li, Z., Morris, Q., Grigull, J., Mitsakakis, N., Roberts, C. J., Greenblatt, J. F., Boone, C., Kaiser, C. A., Andrews, B. J. and Hughes, T. R. (2004). Exploration of essential gene functions via titratable promoter alleles. *Cell* **118**, 31–44.

Namy, O., Duchateau-Nguyen, G. and Rousset, J. P. (2002). Translational readthrough of the *PDE2* stop codon modulates cAMP levels in *Saccharomyces cerevisiae*. *Mol. Microbiol.* **43**, 641–652.

Narayanaswamy, R., Niu, W., Scouras, A. D., Hart, G. T., Davies, J., Ellington, A. D., Iyer, V. R. and Marcotte, E. M. (2006). Systematic profiling of cellular phenotypes with spotted cell microarrays reveals mating-pheromone response genes. *Genome. Biol.* **7**, R6.

Newman, J. R., Ghaemmaghami, S., Ihmels, J., Breslow, D. K., Noble, M., DeRisi, J. L. and Weissman, J. S. (2006). Single-cell proteomic analysis of *S. cerevisiae* reveals the architecture of biological noise. *Nature* **441**, 840–846.

Nieduszynski, C. A., Knox, Y. and Donaldson, A. D. (2006). Genome-wide identification of replication origins in yeast by comparative genomics. *Genes Dev.* **20**, 1874–1879.

Pierce, S. E., Fung, E. L., Jaramillo, D. F., Chu, A. M., Davis, R. W., Nislow, C. and Giaever, G. (2006). A unique and universal molecular barcode array. *Nat. Methods* **3**, 601–603.

Queralt, E., Lehane, C., Novak, B. and Uhlmann, F. (2006). Downregulation of PP2A(Cdc55) phosphatase by separase initiates mitotic exit in budding yeast. *Cell* **125**, 719–732.

Scherens, B. and Goffeau, A. (2004). The uses of genome-wide yeast mutant collections. *Genome Biol.* **5**, 229.

Sheth, U. and Parker, R. (2003). Decapping and decay of messenger RNA occur in cytoplasmic processing bodies. *Science* **300**, 805–808.

Sheth, U. and Parker, R. (2006). Targeting of aberrant mRNAs to cytoplasmic processing bodies. *Cell* **125**, 1095–1109.

Sidrauski, C. and Walter, P. (1997). The transmembrane kinase Ire1p is a site-specific endonuclease that initiates mRNA splicing in the unfolded protein response. *Cell* **90**, 1031–1099.

Slusher, L. B., Gillman, E. C., Martin, N. C. and Hopper, A. K. (1991). mRNA leader length and initiation codon context determine alternative AUG selection for the yeast gene *MOD5*. *Proc. Natl. Acad. Sci. USA* **88**, 9789–9793.

Sopko, R., Huang, D., Preston, N., Chua, G., Papp, B., Kafadar, K., Snyder, M., Oliver, S. G., Cyert, M., Hughes, T. R., Boone, C. and Andrews, B. (2006). Mapping pathways and phenotypes by systematic gene overexpression. *Mol. Cell* **21**, 319–330.

Tao, S. C. and Zhu, H. (2006). Protein chip fabrication by capture of nascent polypeptides. *Nat. Biotechnol.* **24**, 1253–1254.

Teusink, B., Passarge, J., Reijenga, C. A., Esgalhado, E., van der Weijden, C. C., Schepper, M., Walsh, M. C., Bakker, B. M., van Dam, K., Westerhoff,

H. V. and Snoep, J. L. (2000). Can yeast glycolysis be understood in terms of *in vitro* kinetics of the constituent enzymes? Testing biochemistry. *Eur. J. Biochem.* **267**, 5313–5329.

Tong, A. H., Evangelista, M., Parsons, A. B., Xu, H., Bader, G. D., Page, N., Robinson, M., Raghibizadeh, S., Hogue, C. W., Bussey, H., Andrews, B., Tyers, M. and Boone, C. (2001). Systematic genetic analysis with ordered arrays of yeast deletion mutants. *Science* **294**, 2364–2368.

Uetz, P., Giot, L., Cagney, G., Mansfield, T. A., Judson, R. S., Knight, J. R., Lockshon, D., Narayan, V., Srinivasan, M., Pochart, P., Qureshi-Emili, A., Li, Y., Godwin, B., Conover, D., Kalbfleisch, T., Vijayadamodar, G., Yang, M., Johnston, M., Fields, S. and Rothberg, J. M. (2000). A comprehensive analysis of protein-protein interactions in *Saccharomyces cerevisiae*. *Nature* **403**, 623–627.

Wei, J., Sun, J., Yu, W., Jones, A., Oeller, P., Keller, M., Woodnutt, G. and Short, J. M. (2005). Global proteome discovery using an online three-dimensional LC-MS/MS. *J. Proteome. Res.* **4**, 801–808.

Index

Page numbers in italics refer to figures and tables

α-Aggulutinin, 288
α-Aminoadipate reductase, *61*
α-Aminoadipic acid (α-AA), 89, 103, 121, 648
α-Factor, 338, 348
α-Factor block-release, 98
α-Galactosidase reporter gene, 158, 651
α-1,6-Mannosyl transferase, 569
α-tubulin, 344
3-Amino triazole (3-AT), 151
6-Aminophenanthridine, 510, 512
Acetate, 583
Actin cytoskeleton, 260, 267, 512
Actin, 345
Actinomycin D, 423
ADE1, 592
  use as a reporter gene, 170, 171
ADE2, 158
  use as a reporter gene, 170, 171
Adenylate cyclase, 351
Adrenomedullin receptor, 587
Affinity chromatography, 9
Affinity isolated protein complexes, 669
Affinity purification, 599, 655
Affinity tags (hexahistidine-GST), 655, see also *Epitope tags*
Affymetrix GeneChip® system, 190, 191, 198
Affymetrix report files, 200
Agarose gels (denaturant), 420
Agarose pad, 226, *227*
Aging, yeast cell, 336, see also *Life span*
Alcian blue, 293
Alcohol dehydrogenase-1 (*ADH1* or *ADC1*), 580
ALD6, 561
Aldehyde dehydrogenase, 561
Algorithmic classifiers, 616
AlignAce, 560
Allantoate, 498
Allele-specific suppressors, 99
Alverine citrate, 399
American type culture collection (ATCC), 27, 582, 641
Amino acid biosynthetic enzymes, 545
Aminoglycoside, 26, 646
Amphimixis, 5
Amphiphilic α-helices, 459
Amyloid, 492, 496
  fibres, 517
Analysis of variance, 195, 207, 527
Anaphase, 223, 232
Anaphase progression, 236
Anaphase promotion complex (APC), 223
Aneuploidy, 641
Annotation errors (gene function), 573
Annotation tools, gene ontology (GO), 211
ANOVA (analysis of variance between groups), 195, 207, 527
ANP1, 569
Antagonist peptides, 120
Antibiotic resistance markers, phenotypic recovery, 72
Antibodies, 337
  affinity purification, 243
  commercially available, 243
  conformation specific, 248
  determination of specificity, 249
  glycoprotein detection, 289
  horse radish peroxide conjugated, 252
  lack of specificity, 253
  monoclonal, 242
  polyclonal, 242, 263
  production of, 242
  radiolabelled, 252
  secondary, 252
Antibody microarrays, 306
  sandwich assay, 307
Antimycin, 455
Antisense transcripts, 7
Apoptosis, 455
Arabidopsis, 678
Arginine analogue, 37, see also *Canavanine*
Arginine permease, 37, 375, see also *CAN1*
Array screen, 142
Array-based screens, 139
Ascomycete species, 555
Ascospores, 24, 334
Ascus, 4, 24, 34
ASH1 mRNA, 673
Ashbya gossypii, 4, 555
Aspartate transcarbamylase, 501
Aspartyl protease, 27
a-specific genes, 373
a-specific promoter, 372
Atomic force microscopy (AFM), 518
ATP synthase, 449, 452
ATP synthesis, *446*
Autofluorescence, 75, 176, 225
Automixis, 5
Autonomously replicating sequence (ARS), 339
AVID prediction framework, 616

β-1,3-Glucan, 282
β-1,3-Glucanase (Quantazyme), 34
  use in releasing cell wall proteins, 284
β-1,6-Glucan, 282
β-1,6-Glucanase, use in releasing cell wall proteins, 284
β-Galactosidase (*lacZ*), 105, 109, 500
  activity, measurement with ONPG substrate, *174*
  assays, colony and whole-cell based 173
  filter lift assay, 160
  use as a reporter gene, 173, 105
β-Isopropyl malate dehydrogenase, *61*
β-oxidation, 580
β-Tubulin, 674
Bacteriophage lambda, 149

Bacteriophage MS2 coat protein, 436
Barcode microarrays, 409
Bar-coded gene deletion strains, 388, 401, 409, *640*, 641, 642, 643, 646, 660, 669
Barth syndrome, 588
Bayes' theorem, 608
Bayesian networks, 614, 615, 619
Bayesian statistics, 607
BCIP, western blotting detection, *251*
Benomyl, *644*
BESTFIT, sequence alignment, 557
Beta oxidation, 580
Bifunctional cross-linkers, 475
Bik1p, 234
Bim1p, 234
Biochemical systems theory (BST), 528
Bioinformatics, 589
  online resources, 561, *562*
Biolistic gun, 451
Biomolecular interaction network database (BIND), 599
Biotin, 337
  Biotin staining, *396*
  Biotin tagging, 117
BLAST, 555
  alignment, 557
  searches, 55
Blasticidin, 581
Blebbistatin, 399
Bleomycin resistance cassette, *60*
ble^{R}-phleomycin resistance marker, *648*
Blue fluorescent protein (BFP), 271
Blue native gels (BN-PAGE), 475
Bovine alkaline phosphatase (BAP), 426
Box C/D small nucleolar RNAs, 416
Bromodeoxyuridine (BrdU), 445, 464
Buchnera, 12
Bud formation, 357, 360
Bud scars, 337
Bud tip, 269

Ca^{2+}-dependent cellular signalling, 477
Caenorhabditis elegans, 149
Calcitonin receptor-like receptor, 587
Calf intestinal phosphatase (CIP), 426
Calmodulin binding peptide, *436*, 652
Calmodulin Sepharose, *436*
CaLYS5MX cassettes, 648
cAMP phosphodiesterase, 351
CAN1 gene, 37, 375
Canavanine, 37, 371, 375
Candida albicans, non-canonical codon usage, 181
Candida glabrata, 553
Capillary electrophoresis, 192
Capillary zone electrophoresis, 8
Carboplatin, 403
Cardiolipin, 455
Carnitine acylcarnitine translocase (CACT), 591
Caspase, 352, 455
Cassettes for PCR targeting, 70
CBF3 complex, 234
CDC2, 588
CDC6, 338
cDNA expression libraries, 588
cDNA expression, 579
Cell cycle, 13, 222, 342, 552
Cell division cycle, 343
  START, 338
Cell microarrays, 408, 674
Cell polarity, 357, 360, 370
Cell size, 338
  Cell size mutants, 338
Cell surface display of heterologous proteins, 295
Cell wall, 11, 245
Cell wall breakage, use of FastPrep instrument, 285
Cell wall integrity, 294, 393
Cell wall isolation, 285
Cell wall permeability, 293
Cell wall phenotypes
  β-1,3-glucanase resistance, *294*
  Calcofluor white hypersensitivity, 294, *295*
  Congo Red hypersensitivity, 294
Cell wall protein, 282
  alkali extraction, *286*
  detection using antibodies, 289
  detection, *288*
  Endo-H digestion to remove N-glycans, 293
  glycosylation, 282, 292
  identification by mass spectrometry, 290
  in silico prediction, 289
  in situ proteolytic digestion for mass spectrometry, *291*
  polysaccharide complexes, 282–4, 287, 289
    by lectin blotting, *288*
  problems in identification by mass spectrometry, 292
  release and identification, 284, 287
  release by HF-pyridine cleavage, *286*
  release using Quantazyme, 287
  size exclusion chromatography, 287
  strategies for tagging, 288
Cell wall stress, 283
Centrifugal elutriation, 98
Centromere, conditional, 229
Centromere-linked markers, 4
Centromeres, 223, 633
Centromeric (*CEN*) sequence, 339, 581
Centromeric plasmids, 356, 648
Centrosomes, 399
Chaperones, 448, 452, 494
Chaperonin, 453
Chemical crosslinking, 458–9, 475
Chemical genomics (Chemogenomics), 389, 675
Chemostat cultures, 195, 528, 534
Chithin deacetylases, 335
Chithin synthase, 335
Chitin, 282, 335
Chitinase, 285, 335
Chitosan, 335
Chloramphenicol acetyltransferase (CAT), 166
  activation by cell extracts, 183
  fluorescent substrates, 173
  use as a reporter, 173
Cholesterol, 586
Cholesterol biosynthesis, 591
Chromatin immunoprecipitation (ChIP), *305*, 540, 599
Chromophore, 270
Chromosome bi-orientation, 223, 234

Chromosome dots, 237
Chromosome instability, 590
Chromosome mis-segregation, 236
Chromosome movement, 235
Chromosome non-dysjunction, 237
Chromosome segregation, 222, 224
Chromosomes, linkage map, 630
Cincreasin, 399
Cisplatin, 402
Clark electrode, 471
clonNAT, *376*, 377, see also *Nourseothricin-resistance marker* (natMX)
ClustalW, 557
Clusters of orthologous groups (KOGs), 606
Codon usage, 585
Co-immunoprecipitation, 245, 256, 458, 620
Co-inheritance of genes, 601, *602*
Colchicine, 399
Cold sensitivity, Trp auxotrophs, 635
Cold-sensitive (Cs^-) alleles, 80, 87, 344
Colony sectoring, 334, 345
  assay, 352, 353, 370
Colourless native gels (CN-PAGE), 475
Comparative genomic hybridisation, 2, 3
Comparative genomics, 3, 601
Competitive growth assay, 6, *390*
Competitive inhibitor, 151
Complementation cloning, 370
Comprehensive yeast genome database (CYGD), 606
Concanavalin A, 226
Conditional mutants, 79, 80, 87, 331, 342, 348, 461
  analysis, 98
Conditional overexpression, 385
Confocal microscope, *179*
Congo Red, 492, 517
Conservation of gene neighbors, *602*
Constitutive promoters, 579
Contingency-gene loss, 12
Continuous culture, 637
Control of metabolic fluxes, *533*, 537
Cosmids, 641
Counter-selectable marker, 67
Counterselection, 334, 337, 342, 344, 648
cre recombinase, 56, 67, 338, 634, 648
cre/loxP system, 67, *68*, 147, 648, 654
Creator cloning system 313
Cre-recombination, 654
Crick strand, of yeast chromosome, 633
Cristae, mitochondrial, 446, 448, 454, 464
cRNA, biotinylated, 193
Cross-complementation, 30
Crossing yeast strains, 28
Cryptic unstable transcript (CUT), 418
CTA1 promoter, 580
C-terminal tagging of proteins, PCR-mediated, 58, *63*
C-terminally tagged proteins, antibody detection, 244
Ctf19 complex, 234
CUP1 promoter (copper-inducible), 64, 80, 85, 86, 508
CUP1 protein, 651
Cy3-labelled streptavidin, 319
Cyan Fluorescent Protein (CFP), *177*, 221, 269, 270
CYC1 promoter, 347
Cyclin-dependent kinase, 339, 343
Cyclins, 674
Cycloheximide, 121, 431
CYH2, 121
  CYH2 counter-selection, 122
Cystinosin, 589, 591
Cystinosis, 589
Cytochalasins, 399
Cytochrome oxidase, 455, 468
Cytoduction, 503
Cytoplasmic petite, 450
Cytoskeleton, 345
Cytosol, 11

2-D gel protein electrophoresis, 245
2-D IEF gel
  fluorescence difference, *540*
2-Dimensional protein electrophoresis, *305*
DAL5 promoter, 501
DAPI (4′,6-diamidino-2-phenylindole), 236, 267, 445, 456
DASH complex, 234, 671
Data integration for predicting gene function, 614
Data mining, 605
Database of interacting proteins (DIP), 599
dChip, 199, 200
DEAH box protein, 621
Debaryomyces hansenii, 555
Decision trees, 616
Degron, 80, 85, 651
  tagging, *85*, 86, 651
Deletion cassettes, 634
Deletion collection, 5–7, 30, 335, 369, 378, 388, *411*
  pool construction, *391*
Desmosterol, 586
Desmosterol reductase, 586, 591
Dialign, 558
Diauxic shift, 458
Differential centrifugation, 505
Differential equations, ordinary (ODEs), 13
Digitonin, 465
Dihydrofolate reductase (DHFR), 393, *398*, 473
Dikaryon, 503
Diploid synthetic lethality analyzed by microarray (dSLAM), 409
Disulphide bonds, 296
Dityrosine, 334
Diuron, 455
DNA repair, 370, 402
DNA replication origin, 675
DNA staining, 267
DNA-binding proteins, 117, 551
DNA-damage-response, 403
DNA-damaging agents, 403
DNA-microarrays, see *Microarrays*
DNA–protein interactions, 118
  methylation-dependent, 118
Dominance, genetic, 340
Dominant lethal, 345
Dominant markers, 56, 581, 634
Dominant suppressor, 351
Dosage lethality, 353, 384
  screens, 331
Dosage suppression, 99, 361, 384
Double interaction screen, 118
Doubling times, 638

Index

Doxycyclin-regulated expression, 83, 333, 461, 580, 649
Drosophila, 106
Drug target identification, 392
Drug-induced haploinsufficiency, 389, *391*
dsdA gene, 647
dsdAMX marker, 646
D-serine deaminase, 634, 647
DsRed (*Discosoma* red fluorescent protein), 180, 271, 445, 463
Dyclonine, 399

EBFP (Enhanced Blue Fluorescent Protein), *177*
ECFP (Enhanced Cyan Fluorescent Protein), 230
Ectopic integration, 96
EGFP (Enhanced Green Fluorescent Protein), *177*, 230
eIF2B, 673
Electron transport, 469
Electrophoretic transfer, 245, *250*
Elp protein complex, 668
Elutriation centrifugation, 336
Endo-H, 292, 322
Endometabolome, 11
Endoplasmic reticulum, 125, 282, 465, 587
Endosymbiotic bacteria, 12
Environmental sensing, 531
Enzyme-conjugated second antibody, 244
Epifluorescence microscopy, 270
Epigenetic determinants, 492
Episomal vectors, 339, 340, 356
Epistatic mutations, 631
Epitope tags, 63, 146, 242, 243, 264, 348, 583, 586
  2 × ProtA, 654
  3 × *c-myc*, 654
  8 × His, 654
  haemagluttinin (HA) tag, 64, 81, 244, 436, 583, 654
  hexa-histidine (His_6), 244, 518, 583
  myc tag, 64, 85, 244, 583
  Protein A tag, 583
  TAP tag, 103, 107, 417, 433, 652, 653, 671
Epitope-tagged strains and plasmids, 654
Epothilones, 399
eRF1 (Sup45p), 495
eRF3 (Sup35p), 495
ERS1, 589
Essential genes, 6, 7
Estrogen response element, 590
Ethanol, 450, 580
Ether killing in random spore analysis, 36
EUROFAN, 5, 630, 639
EUROSCARF, 70, 83, 86, 333, 378, 581, 582, 630, 639, 657
EXG1, use as a reporter gene, 172
Exoglucanases, 172
Exometabolome, 11, 539
Exons, 578
Exosome, 418
Exponential growth, 509
Expressing human genes in yeast, 578
  choice of expression vector, 580
  choice of growth conditions, 582
  choice of strain, 581
  promoter choice, 579
Expression library, 307, 309
Expression of recombinant proteins, 655
Expression plasmids, 578
Expression profiles as diagnostic tools, 198
Extragenic suppressors, 99, 331, 341, 338, 349
Extramitochondrial proteins, 456
EYFP, 230

5-Fluorootic acid (5-FOA) counterselection, 89, 94, 96, *97*, 103, 119, 121, 334, 648
5-Fluorouracil (5-FU), 398
F_1F_0-ATP synthase, 447
FAD, 581
False discovery rate (FDR), microarray analysis, 207, 209
Fanconi anemia, 590
Farnesylation, 123
FASTA, 557
Fatty acid oxidation, 586
Fed-batch, 637
Fenpropimorph, 399
Fermentable carbon sources, 462
Fermentation, 3, 450, 458
Fermentative growth, 450
Ferritin, 337
Fe–S proteins, 447
Filamentous ascomycetes, 3
FinalNet, *617*
Firefly luciferase, 173, 175, 500, 592
Fission yeast, 223
Fitness defect score, 391
Fitness profiling, 401
FLAsH (fluorescein arsenical helix binder), 177, 178, 180
FLAsH tag, 674
Flocculation, 295, 635
Flow cytometry, 172
Fluorescein di-β-D-galactopyranoside, 175
Fluorescence 2-D difference gel electrophoresis (2D-DIGE), *540*
Fluorescence activated cell sorting (FACS), 337
  use in cell isolation, 337
Fluorescence microscopy, 75, *305*, 457
Fluorescence recovery after photobleaching (FRAP), 673
Fluorescence resonance energy transfer (FRET), 179, 221, 230, 269, 270, 673
  comparison of FRET partners for CFP, *272*
  constructing double tagged strains, 273
  fluorescence calculations, 277
  image acquisition, 275
  image analysis, 277
  preparation of cells for imaging, 274
  preparing the microscope slide, 274
  strategy for determining if two proteins interact, 278
  use of flexible linker in protein tagging, 272
  vectors for use in protein tagging, *272*
Fluorescent proteins, 673
  BFP, 271
  CFP (Cyan Fluorescent Protein) *177*, 221, 269, 270
  DsRed (Discosoma red fluorescent protein), 180, 271, 445, 463
  EBFP (Enhanced Blue Fluorescent Protein), *177*

ECFP (Enhanced Cyan Fluorescent Protein), 230
EGFP (Enhanced Green Fluorescent Protein), *177*, 230
EYFP, 230
GFP, 6, 63, 58, 121, 143, 221, 229, 230, 231, 445, 457, 491, 655, 673
mOrange, 271
Red fluorescent protein (RFP), 221, 230, 673
RedStar2, 654
T-Sapphire, 271
Topaz DsRed, *177*
Venus, 271
YFP (Yellow fluorescent protein), 177, 221, 269, 270
Fluorescent reporters, 176, 177
Fluorescent tagging of proteins, 58, 64, 231, 655, 672
Fluorescent tags, 673
Fluorophore, 270
Alexa™ dyes, 265, 317
Cy3 dye, 265, 317
Cy5, 317
FITC, 266
rhodamine, 266, 317
Texas Red, 266
Flux control
genetic model, 531
Flux control coefficient, 529
FMN, 581
Förster distance, 270
French press, *434*
FRET, 269–280
Förster distance, 270
$FRET_R$, 271, 277
FT-ICR mass spectrometry, 670
Functional analysis by co-responses in yeast (FANCY), 527, 531, *540*
Functional annotations, 381
FunAssociate, 381
FunSpec, 381
Functional complementation, 339, 340, 582, 588
Functional genomics data
assessing error, 604
use of reference sets, 604
Functional genomics, *600*
Functional linkages, 616
Functional redundancy, 339

G1 cyclins, 339
G418 resistance gene, 646
G418, 26, 371, 377, *390*, 581, 646
GAL promoter (galactose-inducible), 64, 80–3, 225, 309, 461, 508, 580, 584, 649
problems of overexpression, 82
regulatory region, 347
GAL1, 7
GAL1L promoter, 82, 347
GAL1S promoter, 82
GAL4, 552
GAL4 promoter, 553
Gal4p, 104
Gal80p, 104
Galactokinase, 122, 580
Galactose, 580
metabolism, 552
Galactose-1-phosphate, 122
Gap repair of plasmids *in vivo*, 89, 92, *93*
GAP, sequence alignment algorithm, 557
Gateway® cloning, 147, 149
Gateway® vector system (Invitrogen), 108, 118, 311, 313
GC content (genome), 585
GCN4, 166, 167
Gene conversion, 24, 41, 96, 374
Gene deletion, 57, *58*, 633
Gene fusions, 603, *611*
C-terminal, 56
N-terminal, 56
Gene nomenclature, 632
Gene ontology, 211, 605
database, 214
molecular function, 381
Gene regulation, 4
Gene regulatory elements, 551, 552
Gene targeting through flanking homology, 90
Gene truncations, 65, *66*
General linear model, 207
General repository for interaction databases (GRID), 599
Generation times, cell growth, 630
Genetic background, 26, 57
Genetic crosses, spore inviability, 27
Genetic foot-printing, 6
Genetic interactions, 334
Genetic linkage, 25, 26, 630
Genetic map, 630, 631
Genetic networks, 333
Genetic screens, 331–62
Geneticin, see *G418*
Genolevures, 332
Genome sequencing, 674
Genomic resources for *Saccharomyces* and related yeasts, *556*
GEO database, 601
GFP, see *Green Fluorescent Protein*
Gln3p, 497
Glucoamylase, use as a reporter gene, 172
Glutathione-S-transferase-His_6 protein tagging, 7, 309, see also *Epitope tags*
Glyceraldehyde-3-phosphate dehydrogenase (GDP or *TDH3*), 579
Glycerol, 122, 580
gradients, 428
Glycolysis, 13, 458, 678
Glycolytic enzymes, 580
Glycolytic pathway, 542
Glycomics, 669
Glycoproteins, 322, 586
Glycosylation, protein, 282-283, 292-293, 304, 320, 586
N-linked glycosylation, 282-283, 586
O-linked glycosylation, , 282-283, 586
GO annotations, 621
Golgi apparatus, 125
GPI anchor attachment, prediction, 290
GPI anchor attachment signal, 288, 289
GPI-anchored proteins, 282, 321
G-protein, 122, 591
G-protein coupled receptors (GPCRs), 591
Gradient centrifugation, 465
Graph partitioning, 616
Graph theory, 10

Green fluorescent protein (GFP), 58, 121, 143, 221, 445, 457, 491, 655, 673
GFP-based microscopy, 222
GFP-based microscopy, image acquisition, 226
GFP-tagged strains and plasmids, 655
GFP tagging, 6, 63
GFP tagging of kinetochore and spindle proteins, 229
GFP-tubulin fusion, 231
GFP variants, 230
use as a reporter of gene expression, 176
Growth media
complete, 24
defined, 24
dropout, 24
for SGA, 377
GNA, *33*
presporulation, *33*
sporulation, *33*
synthetic complete, 24
synthetic dextrose (SD), 24
VB sporulation, *33*
YPAD, 47
YPD (YEPD), 24, 156
Growth rate reporters, 169
GST-tagged proteins, high-throughput expression and purification, *310*
Guanidine hydrochloride (GdnHCl), 491, 507, 510
Guilt by association, gene function assignment, 598, 612

H/ACA snoRNAs, 416
HA epitope tag, 64, 81, 244, 436, 583, 654
HAC1, 669
Haploid deletion mutants, 643
Haploinsufficiency, 5, 6, 11, 333, *391*, 644
Haploinsufficiency profiling (HIP), 388, 389, *391*
caveats, 400
data analysis, 391
procedure, *395*
inhibitor dose optimization for screening, *394*
use in drug discovery, 398
validation of results, 404
Haploproficiency, 6
Haplo-selfing, 5
Haplotoxicity screening, 644
Heat shock, 669
Helix–loop–helix protein, 118
Heme biosynthesis, *446*
Hemiascomycetes, 2, 553, 555
Hemizygote, 6
Hep1 (mtHsp70 escort protein 1), *469*
Heterologous auxotrophic markers, 72
Heterologous prototrophic markers, 647
Heterothallism, 4, 24
Heterotypic interactions, 345
Heterozygous deletants, 6, 675
Heterozygous mutants, 644
Het-s, 492
Hexose transporter family, 648
Hierarchical clustering, 618
in microarray analysis, 211
High copy suppressors, 99, 361, 384
High flux control (HFC) genes, 6
High-throughput screens, 674
HIP and HOP assays combined, 404
his5^{+} (*Schizosaccharomyces pombe*), 372, 647
Histone acetyltransferase Gcn5, 114
HO gene, 4, 24, 409, 503
HO promoter, mother cell specificity, 338
Hoechst 33342, DNA stain, 236
Homologous recombination, 26, 56, 92, 108, *146*, 229, 409, 588, 631
cloning by, 147
Homothallism, 5
Homotypic interaction, 345
Homozygous diploid deletion mutants, 335, 643
Homozygous profiling (HOP), 401
caveats, 405
validation of results, 404
'Hook', for use with non-protein baits in the yeast two-hybrid system, 115-116
HOP profiles, 402
hphMX marker, 646, 647
Hsp104 chaperone, 511
Hsp40, 513
Human estrogen receptor alpha, 590
Hybridisation-array analysis, 2, 7, *540*
Hydrophobic residues, 345
Hygromycin B, 293, 591, 634
Hygromycin B phosphotransferase, *60*, 647
Hypergeometric error model, 612
Hypomorphs, 333
Hypoxia, 457

IgG binding domains, 652
IgG Sepharose, 435
Imidazoleglycerol phosphate hydrolase (*HIS3* gene product), *62*
Immune autofluorescence, 258
Immunoblotting, 244, *251*
Immunodetection, 583
Immunofluorescence, 179, 241, 463
Immunofluorescence microscopy, 256, 257, 586
antifade mounting solutions, 266
cell fixation, 257
cell wall digestion, 262
choice of microscope, 267
demonstration of specific staining, 264
DNA co-staining, 267
double labelling, 265
epitope masking problems, 263
formaldehyde fixation, 258, *259*
mounting cells on slides, 262
mounting solution, 266
poly-L-lysine, *259*, 262
primary antibodies, 263
protein localization, 267
SDS treatment of cells, *260*
secondary antibodies, 265
solvent fixation, 258, *259*, *261*
Immunofluorescence staining, effects of yeast growth conditions, 257
Immunogold electron microscopy, 457

Immunoprecipitation, 241, 253, 428, 462, 475
of radiolabelled cells, *255*
pre-clearing lysates, 254
preparation of cell lysates, 254
protein A agarose use in, 253, *255*, 256
protein G agarose use in 253, 256
In organello labelling, 472
In vitro mutagenesis, 342, 344
Independent assortment (of genetic markers), 25
Inducible promoters, 119, 580
Integral membrane protein, 123, 448
Integrative transformation, 341
Interaction networks, 144
Interactome, 9, 607, 619, 671, 672
Intragenic complementation, 99
Intergenic regions, 418, 556, 564
Intragenic suppressors, 349
Intracellular transport, 590
Intra-tetrad mating, 5
Introns, 573, 578
Invasive growth, 295
Ipl1p kinase, 234, 235
Ire1p, 125
Iron regulatory protein 1 (IRP1), 439
Iron responsive element (IRE), 438, 439
IRE binding protein, 438
Iron-sulfur clusters, 447
Isoelectric focusing (IEF), 245
Isotope tags for quantitative proteomics analysis (iTRAQ), 9, 535

K. lactis LAC4 gene, 181
Kanamycin resistance cassette (*kanMX*), 26, *59*,72, 371, *390*, 634, 646
Kanamycin, 371, 634
Kar1, 503
Kar3p, 234, 235
Karyogamy, 503
Kastellpaolitines, 510, 512
Kernel canonical correlation analysis, 616
Kex2 cleavage site, 290
Kinesin, 399
Kinetic constants, 13
Kinetic metabolic models, 544
Kinetochores, 222–5, 269, 360
attachment to the spindle microtubules, 233
capture by microtubules, 233, 234
Klebsiella pneumoniae, 647
KlURA3, 654
Kluyveromyces lactis, 4, 555
Kluyveromyces waltii, 4
K-means clustering, microarray analysis, 211
Krebs cycle, *446*, 458
Kyoto encyclopedia of genes and genomes (KEGG), 605

lac repressor, 348
*lac*O/GFP-*lacI* system, 228, 234
Lactose, 580, 584
lacZ reporter, see *β-galactosidase*
Lagging chromosomes, anaphase, 236
Lariat RNA, 621
Latrunculin A, 510, 512
Learning association rules, 616
Lectin blotting, detection of mannosylated proteins, 287
Lectins, 334
LexA two-hybrid system, 143
LexA, 104
LIANG, gene function annotation algorithm, *617*
Life span (yeast), 336, 337, 445
Linkage, 38
Lipid biosynthetic pathways, 455
Liquid chromatography, 8
Live cell microscopy, 224-237, 269
Immobilizing cells for imaging, 225
Logarithmic growth, 338
Logical model, 14
Longevity, yeast cell, see *Lifespan*
loxP recombination sites, 56, *59*, 147, 634, 635, 648, see also *Cre/loxP system*
loxP-kanMX-loxP, *59*, 635, 648
Luciferase, use as a reporter, 173, 175, 500, 592
Lyp1Δ selectable marker, 375
Lys2[+] counterselection, 89
lys5, 647
LYS5MX cassette, *61*, 648
Lysine biosynthesis, 536
Lysine biosynthetic pathway, 542
Lysine permease, 375
Lysosomal cystine transporter, 589
Lyticase, 34, 519

2-(4′-Methylaminophenyl) benzothiazole (BTA-1), 505
2-μm plasmid, 339, 581
Machine-learning methods, 198
MAGIC, gene function prediction algorithm, *617*
Magic marker selection system, 655
Magnetic beads, 337
use for cell isolation, 337
MALDI-TOF mass spectrometry, 652, 670
Maltose utilization, 638
Mannose, 293, 586
Manual pinning tools, 377
MAP kinase, 672
MAPK pathway, 592
Mass spectrometry, 8, 9, 107, 304, 332, 599, *611*, 652, 669, 670
MAT locus, 5, 24
Mating, 27-32, *152*, 672
Mating efficiency, *155*
Mating figures, *31*
Mating pheromone, 27, 98, 671
Mating type, 5, 24, *29*,
determination, *29*
loci, 348
specific promoter, 372
switch, 503
switching, 4, 24
tester strains, 27
Mating-based split-ubiquitin system (mbSUS), 124
Matrix processing peptidase (MPP), 453, 459
maxdView, microarray cluster analysis, 211
Meiosis, 4, 335
Meiotic recombination, 384

Index

MEL1, 651
Membrane potential, *469*
Membrane proteins, 122, 591
Membrane transporters, 393
Membrane yeast two-hybrid system (MYTHS), 125
Mendelian genetics, 25
MET25 promoter, 64, 82
MET3 promoter, 64, 82, 83, 580
Metabolic control analysis (MCA), 6, 13, 527, 528, 530
  connectivity theorem, 530
  distributed control, 529
  modular theory, 532, 539
Metabolic control, 527, 528, *541*
  frameworks, 528
  hierarchical regulation, 544
  integrative studies, 539, *540*
  multiple enzyme regulation, 536
Metabolic co-responses studies, 531
Metabolic engineering, 528, 545
Metabolic flux analysis (MFA), 527, 529, 635
  physiological control, 534
Metabolic network, 13, 528, 531, 532, 544
  E.coli, 12
Metabolic profiling, 144
Metabolic regulation coefficient, 542
  robustness, 542
Metabolome, 10, 528, 531, 660
Metabolomics, 10, 528, 537, 669
  co-responses of metabolites, 539
Metaphase, 232
Methionine, 580
Methotrexate, 392, 393, *398*, 473
Methylation-protein, 320
MIAME (microarray standards), 215, 216
Michaelis–Menten enzyme kinetics, 538, 542
Microarray scanning, 193
Microarray-based transcript analysis, 189
  clustering, 211
  computational analysis, 198
  data mining, 206
  dealing with low signal intensity, 209, 210
  experimental design, 194
  false negative data, 196
  false positive data, 196, 209
  limitations, 190, 191
  pooling and replicates, 197
  probe labelling and hybridisation, 193
  random error, 195
  replicates, 195, 196
  RNA extraction, 191, *192*
  statistical significance, 207
  systematic error, 194, 205
Microarrays, 3, 303, 332, 336, 389, 407, 419, 429, *540*, 599, 669, 674
  data analysis pipeline, 189, 190, 198
  expression analysis and normalization, 202, *203*
  quality control, 199, *201*
  use in identifying non-coding RNA genes, 417
Micro-dismembrator, *192*
Micromanipulator, 30, 34, 35, 338
Microtubules, 221, 257
  stabilizers, 399
Mig1 repressor, 553
Minimal gene set, 12
Minimal metabolic network, 12
Minimum information requested in the annotation of biochemical models (MIRIAM), 544
MIPS, 675
Misincorporation of dNTPs, 91
mit⁻ mutants, 451
Mitochondria, 11, 445
  biogenesis, 472
  discriminating nuclear and mitochondrial mutations, 455
  fusion competence, 463
  inner membrane folding, 449
  integrity after isolation, 465
  intermembrane space, 473
  isolation, 464, *466*
  nuclear-encoded proteins, 451
  sub-fractionation, 465, *470*
  targeting fluorescent proteins to different sub-compartments, 462
  transformation, 451
  ultrastructure, 446, 463
  visualization, 462
Mitochondrial biogenesis, 450, 458
Mitochondrial carnitine-acylcarnitine translocase, 586
Mitochondrial cytochromes, 468
Mitochondrial DNA, *446*, 447
  transfer to nuclear genome, 453
Mitochondrial FAD transporter, 581
Mitochondrial flavoproteins, 581, 582
Mitochondrial function, *446*
  advantages of studying in yeast, 450
Mitochondrial genes
  identification by *in silico* methods, 456
Mitochondrial genome, 448, 456, 464
  loss, 450, 460
  mutation, 450
Mitochondrial heteroplasmy, 456
Mitochondrial Hsp70 (mtHsp70), *469*
Mitochondrial membranes, 455
Mitochondrial morphology, 448, *449*, 454, 463
Mitochondrial mutants
  common phenotypes, 451
  drug sensitivity, 455
Mitochondrial nucleoids, 464
Mitochondrial oxidative metabolism, 32
Mitochondrial protease, 454
Mitochondrial protein complexes, 475
Mitochondrial protein import, 448, 462, 472, *474*
Mitochondrial proteins
  chemical crosslinking, *476*
  essential, 453
Mitochondrial proteome, 458
Mitochondrial rRNA, 415
Mitochondrial targeting sequence (MTS), 459
Mitochondrial transporters, 447, 581
Mitochondrial uncoupling, 471
Mitoplasts, 465, *469*
MITOPRED, 459
Mitosis, 222, 399
Mitotic exit (model), 678
Mitotic recombination, 5, 375
Mitotic spindle checkpoint, 399
MluI-box binding protein, 552
Mod5p, 670
Modelling, 13

Molecular-bar codes, 5, 388
Monastrol, 399
Monoclonal antibodies, 306
Monosodium glutamic acid (MSG), 377
mOrange, 271
MORF (moveable ORF) collection, 311
Motif-finding algorithms, 557, 559
Motif-finding programs, 559
  AlignAce, 560
  Consensus, 560
  Footprinter, 560
  Gibbs sampler, 560
  MEME, 560
  Phylocon, 560
Mounting cells for live cell imaging, 225
Mps1p, 399
mRNA degradation, 673
mRNA expression, 601
mRNA processing, 621
mRNA stability, 579
MRP, 415
MS2 coat protein, 116
MS2 recognition RNA motif, 436
mtDNA, 448, 450, 456, 460, 464
Multicopy suppression, 352
Multicopy suppressor, 341, 349, 350
  screens, 81
Multiple integration, 41
Multiple sequence alignments (MSAs), 558
  evaluation, 571
Multiplexed sequencing, 440
Multi-spectral imaging, 230
Mutagenic PCR, 90
  optimisation, 92
Mutant alleles
  DNA sequencing, 94, 97, *98*
  generating a marked copy, 97, *98*
  integration at their genomic locus, 95
  selection, 93
  verification of integrated copies, 96
  verification, 94, 95
Mutant isolation, 334
myc epitope tag, 64, 85, 244, 583
Myxothiazol, 455

3′-Non-coding region, 578
5′-Non-coding region, 578
'N-degron' fusions, 80
^{15}N label, 9
N-acetylgalactosamine, 586
NADH-dehydrogenases, 447
NADP, 563
Nail polish, cytotoxicity to yeast, 226, *227*
Native gel electrophoresis, 458, 475
natMX, 26, *60*, 72, 371, 646
Naturing polyacrylamide gel, 424
NCBI BLAST searches, 555
Ndt80p, 553
Needleman–Wunsch algorithm 557
Nested PCR, 426
Network analysis, 677
Network models, 616
Neurospora crassa, 4
N-glycanase, 292
Nitro blue tetrazolium, western blot detection, *251*
Nitrogen catabolic gene expression, 497
Nitrogen starvation, 4
N-linked carbohydrate, 282
NMR spectrometry, 10
NNPSL, 459
Nocodazole, 234
Non-coding RNA gene search software, 419
Non-coding RNA genes, 191, 415
  identification, 416
Non-coding sequences, 568
Non-essential genes, 63
Non-fermentable carbon sources, 351, 450, 462
Non-homologous endo-joining (NHEJ) pathway, 409
Non-Mendelian inheritance, 494, 502
Non-parental ditype tetrads, 37
Nonsense suppression, quantitative assays, 500
Nonsense suppressors, 349
Northern blot analysis, 8
Northern blot hybridization, 417, 419, 420, 421
Nourseothricin, 26, *60*, 634, 647
Nourseothricin N-acetyltransferase, 646
Nourseothricin-resistance marker (*natMX*), 371
N-terminal tagging of proteins, 64, 244, 654
  6 × HA, 654
  9 × *c-myc*, 654
  ProtA, 654
  TEV-GST-6 × HIS, 654
  TEV-ProtA, 654
  TEV-ProtA-7 × HIS, 654
  yEGFP, 654
Nuclear division, 222
Nuclear localisation signal (NLS), 228
Nuclear localization, 142
Nuclear periphery, visualization during live cell imaging, 231
Nuclear polyadenylation complex (TRAMP), 418
Nuclear pore complex, 304, 354
Nuclease protection, 424
Nucleic acid programmable protein array (NAPPA), 316
Nucleolar U3 RNA, 432
Nucleosome positioning elements, 675
Nucleotide modifications, 415
Nucleus, 11, 465

O-glycosylation, 282
Oleic acid, 580
Oligo(dT), 425
Oligomycin, 455
Oligonucleotide arrays, 3, 407, 417
Oligonucleotide probes, 190
One-and-a-half hybrid, 118
One-hybrid, 114
  assay, 108
ONPG (*o*-nitrophenyl-β-D-galactopyranoside), 109, 173
Open biosystems, 657
Operons, 602
Orbitrap mass spectrometer, 670
Ordinary differential equations (ODE), 13
  ODE-based models, 678
ORF annotation, 668
ORFeome, 146, 149

Index

Origin of replication, 149, 339
Ornithine transcarbamylase (OTC), 453
Orotic acid decarboxylase, 344
Orthologous genes, 552, 553, 556
Orthologs, 602, 603
Osmotic shock, 465
Osmotic stress response pathway, 678
Overexpression array, 385
Overexpression phenotypes, 7
Oxidative metabolism, 590
Oxidative phosphorylation, *446*, 447, 468, 471
Oxidative stress, 337, 447
Oxygen radicals, 337

[PIN^+] prion phenotype, 501
[PIN^+] prion, 497, 498
[PSI^+] prion, 494, 495, 592
  strong and weak variants, 496, 500
[RNQ^+] prion, 494
[$URE3^+$] prion, 494, 497
  phenotype, 501
Papillary growth, 640
Paralogs, 564
Parental ditype tetrads, 37, 341
PAS8 gene, 582
patMX, 646
P-bodies, 673
PCR amplification, 72
PCR mutagenesis, *91*, 344
PCR primers
  design, 70
  quality, 71
PCR verification
  of gene knockouts, 38, *39*, 42, 58
  of tagged genes, *63*, *65*, *68*, *69*
PCR-mediated deletion, 624
PCR-mediated mutagenesis, 87
PCR-mediated promoter substitution, 64, 650
PCR-mediated targeting, 55, 624
Pde2p, 670
Peptide antisera, 242
Peptide mass fingerprinting, 292
Permissive temperature, 340
Peroxisomal catalase, 580
Peroxisomes, 11, 465, 580
  biogenesis, 588
petite (*pet*) mutants, 451, 519
petite phenotype, 456
PEX genes, 588
Phenanthridine, 512
Pheromone signaling, 592
PHO5 promoter, 166
Phosphate buffered saline (PBS), *251*
Phosphinothricine-N-acetyltransferase, *60*, 647
Phosphinotricin, 634
  bialaphos, 647
Phosphoglycerate kinase-1 (*PGK1*), 579
Phosphopantetheinyl transferase, *61*
Phosphopeptides, 671
Phosphoribosylamino-imidazole, 225
Phosphoribosylanthranilate isomerase (*TRP1* gene product), *62*
Phosphorylation-dependent interactions, 120
Phosphorylome, 320
Phospho-specific antibodies, 320
Phosphotyrosine, 113
Photobleaching, 224, 275, *276*, 318
Photo-toxicity, 224
Phylogenetic footprinting, 551, 552
Phylogenetic profiles, 602, 603, *609*, 620
Phylogenetic relationships within the *Saccharomyces* and related yeasts, 555
Phylogenetic trees, 2
Phylogenetics, 2
Phylogeny, 2, 3
Physical-map distance, 26, 631
Pileup (Wisconsin package), 558
Pinning, 408
  Manual tools, 377
  Robotic tools, 378
pir gene, 149
Pir proteins, 283
PIT, *617*
Plasma membrane, 335, 351, 393, 465
Plasmid integration, 40
Plasmid shuffling, 81, 94, 342, 344, 384
Plasmogamy, 503
Plex, *617*
PNGase F, 322
Podospora anserina, 492
Poly(A) tail, 418, 425, 427
Polyacrylamide denaturing gels, 418
Polyadenylated transcripts, 418
Polyclonal antibodies, 306
Polyribosomes, 429
Polysome analysis, 350, 429
Pop-in, pop out, 96
Positional cloning, 341
Post-transcriptional control of gene expression, 166, 427, 669
Post-translational control of gene expression, 535
Post-translational modifications, 112, 142, 253, 304, 307, 320, 670, 671
Potential-dependent vital dyes, 462
  DASPMI, 462
  MitoTrans, 462
Prediction of gene function, 597
Predictome, *617*
Pre-mRNA processing, 356
Pre-mRNA splicing, 438, 621
Pre-ribosomes, 432
Pre-rRNAs, 420
Primer extension, 422, 423
Principal component analysis (PCA), 194, 204, *205*
Prion elimination, 512
Prion protein aggregates, 493, 503
  isolation, 505
  separation using agarose gels, 506
  visualization in cells, 509
Prion protein oligomers, 507
Prion protein polymerisation *in vitro*, 517
Prion proteins, protease resistance, 503
Prion-associated phenotypes, 499
Prion-eliminating compounds, 512
Prions, *493*
  criteria for establishing prion nature, 519
  de novo formation, 498
  elimination, 510
  elimination by ethyl methyl sulphonate (EMS), 510

elimination by ultraviolet light (UV), 510
estimation of propagon number, 513, 515, *516*
forming domain, 491
genetics, 502
GFP fusion technology, 508
in vitro polymerisation studies, 518
self-propagation mechanisms, 492
transmission by cytoduction, 503, *504*
Probabilistic assignment of function, 616
Programmed cell death, 455
Prolinks, *617*
Promoter elements, 675
Promoter shut-off, 80
Promoter substitution, 64, 650
Promoters, for heterologous expression, 578
Propagon, 491, 493, 511
PROPHECY project, 408
Propidium iodide, 267
Prospore, 335
Protease protection assay, 465
Proteasome, 167, 339
Protein A, 63, 652
fusion proteins, high-throughput expression and purification, 312
Protein acetylation, 587
Protein aggregates, 491, 587
Protein chips, see *Protein microarrays*
Protein complexes, 672
Protein concentration in cell lysates using TCA/acetone, *246*
Protein conformational diseases, 592
Protein depletion, using promoter shut-off, 81
Protein destabilization, 347
Protein expression, problems with aggregation, 309
Protein extraction for Western blot analysis, *76*
Protein folds, 597
Protein glycosylation, 282-283, 292-293, 304, 320, 586
Protein half-life, 348
Protein interaction, 9
databases, *110*, 599
mapping, 598
maps, 105
networks, 9
Protein kinase A, 320, 339, 351
Protein localization, 269, 601
Protein methylation, 587
Protein microarray slide
nickel coated, 315
streptavidin coated, 315
Protein microarrays, 7, 146, 303–24, 332, 669, 671
advantages and limitations, 307
applications, 319
detection methods, 316
DNA-protein interaction, 319
expressing the proteins, 309
fabrication, 314–316, 671
future directions, 323
history, 305
printing, 315, *316*
slide surface chemistry, 314
studying protein–lipid interactions, 319
studying protein–protein interactions, 319
technical aspects, 314
use for studying posttranslational modification, 320
use in assaying enzyme activities, 322
use in identifying enzyme substrates, 307
use in kinase assays, *321*
use in screening antibodies, 306
use in small molecule screens, 319, 323
use in studying glycosylated proteins, 321
Protein mislocalization, 64
Protein phosphorylation, 304, 309, 539, 587
Protein processing, 587
Protein tagging, 243
with glutathione-S-transferase, 253
Proteinase K, 465, 503
Protein-binding motifs, 559
Protein–drug interactions, 115
Protein–protein interactions, 9, 99, 106, 139, 142, 306, 475, 599, 613
Protein–RNA interactions, 115, 120
Proteolytic degradation, 584, 651
Proteome, 11, 303, 528, 531, 660, 669
Proteome–transcriptome correlations, 535
Proteomics, 8, 528, 669
aims and objectives, *305*
Proton gradient, 447
Prototrophic markers, 634, 647
PrP prion protein, 491
Prp20p, 233
PSORT, 459
Pseudohyphal growth, 672
pTEF1/ZEO, 581
pTEF1Bsd, 581
Puf proteins, 427
Pulse-chase, 433
labelling of proteins, 253
Pulsed-field gel electrophoresis, 630
pYES vector, 581
Pyruvate kinase, 343

QCat polypeptide, 9
QCat, 9
QRNA, 419
QRT-PCR, *540*
Quadrupole time-of-flight (Q-TOF) mass spectrometer, 291, 670
Quantazyme, *287*
Quantitative phenomics, 409
Quantitative phenotypes, 6
Quantitative proteomics, 670
Quantum dots, 318

[rho°] petite, 504
Radioactive labelling *in vivo*, 462
with [^{35}S] methionine, *255*
Raffinose, 86, 122
Ran GTPase, 233
Random mutagenesis, 344
Random oligonucleotide hybridization, 417
Random primed probes, 420
Random spore analysis, 36, 382, *383*
Rapamycin, 319
Rapid amplification of cDNA ends (RACE), 425
Rapid transformation, 47
RAS genes, 351
RAS/cAMP pathway, 123, 343, 350, 351
Ras-recruitment systems (RRS/rRRS), 123

Rate-limiting step, 529
Reactive oxygen species (ROS), 455
Real-time PCR, 437
ReAsH tag, 178, 674
ReAsH, 178, 180
Reb1p, 553
Recall–precision curve, *609*, *613*
Receiver operating characteristic (ROC) curve, 608, 609
Receptor activity modifying protein (RAMP), 587
Recessivity, genetic, 340
Recombination frequency, 26, 631, 641
Recombination-based cloning system, 67, *68*, 147, 149, 648, 654
Red fluorescent protein (RFP), 221, 230
  RedStar2, 654
Redundancy of gene function, 668
Regulatable promoter, 7, 81-86, 347, 461, 584
  copper-inducible, 85-86, 461
  doxycyclin, 83-85, 461
  galactose, 64, 80–3, 225, 309, 347, 461, 508, 580, 584, 649
  methionine, 81-85, 461
Regulation analysis theory, 541
RegulonDB, 602
Relative growth rate, 644
Release factor 1, eukaryote (eRF1), 495
Release factor 3, eukaryote (eRF3), 495
Renilla (sea pansy) luciferase, 175
Replica-plating, 37, 96, 123, 335, 344, 517, 640, 660
Replication forks, stalled, 403
Replication origins, 231, 340
Replication protein p, 149
Replicator, 377
  sterilization, 377
  see also *Pinning*
Reporter assays
  influence of clonal variation, 183
  influence of reporter half-life, 184
  linearity, 181–3
Reporter gene constructs, 167, *169*
  generation by PCR, *171*
Reporter gene, polymerase III responsive, 119
Reporter genes, 165–84
  advantages, 166
  studying post-transcriptional regulation, 167
Repressed TransActivator (RTA), 119
Respiration, 450, 458
Respiratory carbon sources, 452
Respiratory chain, 448, 458, 469
Respiratory deficiency, 452, 454, 462
Respiratory deficient phenotype, 581
Respiratory-competent, 462
Restrictive temperature, 340
Retrotransposons, 3
Reverse transcriptase, 422, 423, 579
Reverse transcription, 422
Reverse two-hybrid system, *GAL1* reporter, 122
Reversed phase HPLC, 335
Reverse-hybrid system, 119
RFP, 221, 230, 673
rho$^-$ (cytoplasmic petite) yeast, 450, 454
rho factor, 450
rho$^\circ$ yeast, 450
Ribonucleoprotein (RNP) complexes, 415, 427
  isolation, 433
Ribonucleoside vanadyl complexes, *434*
Ribosomal internally transcribed spacer (ITS), 2
Ribosomal protein L29, 121, 345
Ribosomal RNA, 350, 433
  maturation, 428, 433
  processing, 350
Ribosomal subunits, 427, *429*
Ribosome biogenesis, 348, 349, 350, 598, 623
Ribosome density mapping, 429
Ribosomes, 429, 579
Rickettsia prowazekii, 459
RLM-RACE, 426
RL-PCR, 426
RNA binding protein, 116, 144
RNA extraction, quality control, 192
RNA genes, 415
RNA helicase, 621
RNA interference, *651*
RNA polymerase holoenzyme, 428
RNA polymerase II, 428
RNA polymerase III promoter, 439
Rna1p, 234
RNase A, *424*
RNase H, *424*
  RNase H digestion, 424
RNase inhibitors, *434*
RNase P, 415
RNase protection, 423, *424*
RNase T1, *424*
Rnq1p, 492
Robotic manipulation, 144
Robotic pin tools, see *Pinning*
Robotics, 147, 332, 675
Rolling circle amplification (RCA), 317
RPN3 promoter, *558*
Rpn4 binding site, *558*
rRNA rRNA
rRNA, *446*
  biogenesis, 621
  processing, 398, 621

60S ribosomal subunit, 350
6S RNA, 428
7SK human RNA, 428
S. boulardii, 3
S. castellii, 3
S. pombe, 223, 224, 231
S288c, 4
Saccharomyces 'sensu lato', 3, 555
Saccharomyces 'sensu stricto', 3, 555
Saccharomyces bayanus, *554*
Saccharomyces castellii, 555
Saccharomyces genome database (SGD), 25, 561, 675, xiv
Saccharomyces kluyveri, 555
Saccharomyces kudriavzevil, *554*
Saccharomyces mikatae, *554*
Saccharomyces paradoxus, *554*
Scale-free networks, 10
Scanning transmission electron microscopy (STEM), 518

Schizosaccharomyces pombe, 223-224, 228, 232, 234, 338, 588, xiv
Screening for interactions between extracellular proteins (SCINEX-P), 125
SDS-PAGE, 244, 505
Sea coral fluorescent proteins, 180
Second-site mutations, 640
Secretion, 370
Secretory pathway, 282
Secretory proteins, 282-284, 586
Sectored colony, 357
Sedimentation analysis, 428
Selectable marker, 26
  recycling, 67
Selection of haploid segregants (Magic marker selection), 655
Self-organising maps, microarray analysis, 211
Semi-denaturing detergent agarose gel electrophoresis (SDD-AGE), 506, *507*
Sensu lato species, 3, 555
Sensu stricto species, 3, 555
Serial analysis of gene expression (SAGE), 7, 8, 191, 418, *540*
  Long-SAGE, 418
Serine deaminase, *60*
SGA (synthetic gene array) 333, 600
  analysis, 370
  applications, 384
  array design, *380*
  confirmation of results, 382
  controls, 379
  mapping (SGAM), 384
  procedures, *376*
  scoring putative interactions, 379, *381*
  technology, 408
SGA reporters, construction, 373
SGD, 25, 561, 675, xiv
Shmoos, *31*, 339
Short flanking homology, 56
  yeast gene deletion strategy, 90, 646
Short flanking recombination sequences, 634
Shuttle vectors, 339, 344
Sialic acid, 586
Signal peptidase, prediction of cleavage sites, 292
Signal peptide, 282, 288
Signal recognition particle, 415
Signal transduction, 9, 13, 112, 532, 678
  pathway, 349, 544, 591
  regulatory networks, 531
Silicon cell, 544
Single cell studies, 674, 677
Single chromosome tag, 236
Single-nucleotide polymorphisms, 3
Sis1p, 499
Sister chromatids, 223
Site-directed integration, 580
Site-directed mutagenesis, 66, 67, *69*
Site-specific recombination, 108, 149
Slow-growth phenotype, 401
Small nuclear RNAs (snRNAs), 415, 416
Small ORFs, xiii
Small ribosomal subunit, 433
Small RNA genes, 675
Small-molecule inhibitors, 144, 675
Smith–Waterman algorithm 557
snoRNAs, 415
SOS-recruitment system (SRS), 123
Southern blot analysis, 456
Spc42-GFP, 234
Spheroplasts, 45, *247*, 262, *467*, 519
  preparation, 464
Sphingolipid biosynthesis, 13
Spindle, 222-224, 399
  assembly, 231
  checkpoint, 222
  elongation, 231, 232, 236
  microtubules, 222, 224
Spindle pole body (SPB), 221–4, 258, 601
  duplication, 223
Splice variants, 304, 588
Spliceosomal RNAs, 416
Spliceosome, 433, 438
Splicing by overlap extension (SOE-PCR), 168, *170*, 171
Splicing, 433, 438, 578
Split-hybrid, 122
Split-ubiquitin system, 108, 123, 124,143
Spore coat, 34, 335
Spore viability, 637
Spores, 4, 25, 334
Sporulation, 24, 32, 58, 333, 335, 377, 410, 553
  frequency, 635
Ssb1p, 513
SSD1, 26
SST1 (*BAR1*) gene, 27
Stable isotope labeling by amino acids in cell culture (SILAC), 670, 671
Stanford microarray database, 601
Stationary phase, 509, 640
Ste12p, 672
Stochasticity in gene expression, *540*
Stoichiometric model, 12–4
Strain construction, 25
Strain stability, 656
Streptavidin, 117, 438
Streptavidin–phycoerythrin, 193, 337, *396*
Streptomyces hygroscopicus, 647
Streptomyces noursei, 646
Streptomyces viridochromogenes, 647
STRING, algorithm for gene function prediction, *617*
Structural genomics, 144
Structure activity relationships (SAR), 399
Stu2p, 234
Sub-cellular fractionation, 11, 457
Sub-telomeric regions, 3
Succinate dehydrogenase, 468
Sucrose density gradients, 336
Sucrose gradient, 338, 350, 429, *430*, 432
  centrifugation, 245, 507
Sucrose utilization, 638
Summation theorem, 529
  for the regulation of flux, 542
Sup35p, see *Release factor, eukaryote (eRF3)*
Support vector machines, microarray classification, 198
Suppressees, 349, 352
Suppressor mutations, 349, 351
Suppressor tRNA, 500
Suppressors, 349

Index

Surface enhanced laser desorption/ionization (SELDI), 318
Surface plasmon resonance (SPR), 315, 318
SWISS-PROT, 621
Synchronous cultures, 98
Synthetic dosage lethality, 361
  screens, 360, 409
Synthetic gene interactions, 12
Synthetic genetic array, see *SGA*
Synthetic lethality, 10, 64, 331, 333, 348, 353, 355, 356, 359, 369, *600*
  analyzed by microarray (SLAM), 409
  screens, 144, 345, 369, 370
Synthetic phenotypes, 58
Systems biology, 527, 544, 676–8
Systems biology mark-up language (SBML), 527, 544

2,2,7-trimethylguanosine cap, 433
Tag arrays, *396*, 675
Tandem affinity purification (TAP) fusion libraries, 385
Tandem affinity purification (TAP) tagging, 103, 107, 417, 433, 652, 653, 671
Tandem mass tags, 670
TAP-purified protein complexes, 653
TAP-tagged strains and plasmids, 652
TargetP, mitochondrial targeting signal prediction, 459
TATA box, 573
Taxol, 399
Taxotere, 399
TBLASTN, 563
T-Coffee, multiple sequence alignment, 558
Telomerase, 415
Telomere, 231
Temperature-sensitive (Ts^-) mutants, 30, 80, 87, 271, 331, 334, 340, 342, 461, 633, 651
Temperature-sensitive *pet* mutants, 452
Terminal deoxytransferase, 425
Tet repressor, 122
Tethered catalysis, 114
tetO array/*tetR*-GFP system, 228, 233, 234
tetO promoter, 7, 348, 580, 581, 649
tetO-CYC1 promoter, 649
Tetracycline operator sequence, 228
Tetracycline-inducible repressor tetR, 649
Tetracycline-regulated promoter, 7, 83, 348, 371, 384, 580, 581, 649
Tetracysteine biarsenical system, 177, 178, 674
Tetrad analysis, 58, 81, 341, 384, 631
Tetrad dissection, 58
Tetrads, 4, 32
  Non-parental ditype, 37
  Parental ditype, 37, 341
  Tetratype, 38, 350, 631
TEV protease (Tobacco etch virus), *435*, 652
TFIIIC factor, 119
The Yeast Genome S98 Array (YG-S98), 190
Thialysine, 371, 375, 377
Thioflavin-S, 505
Thioflavin-T, 517
Three-hybrid system, 108, 428, 438, 439
Thymidylate synthase, 398
Tiling arrays, 3, 8, 191, 417, 419, xiii
Tim Hughes collection (yTHC), 348, 384, 650
Time-lapse microscopy of yeast cells, *179*, *180*, 224, 226, *227*
Time-lapse microscopy, *179*, *180*
Titratable promoter allele, 675
Tobacco acid pyrophosphatase (TAP), 426
Tobramycin binding aptamers, 438
Topaz DsRed, fluorescent protein, *177*
Topogenesis mediating complex of outer membrane beta-barrel proteins (TOB), *446*
TOR pathway, 339, 352
Trans-complementation, 648
Transcript profiling, 669, see also *Microarray-based transcript analysis*
Transcription, 578
  factors, 4, 104, 118, 551, 553, 599
  regulation, 351, 590
Transcriptional activation domain, 117
Transcriptional activator, 104
Transcriptional elongation, 668
Transcriptome, 8, 11, 190, 457, 531, 660
Transcriptomics, 2, 7, see also *Microarray-based transcript analysis*
Transformation, yeast, see *Yeast transformation*
Transformation of yeast by prion proteins, 519
Transformation-competent frozen yeast cells, 72
Translation initiation, 579, 669–70
Translation regulatory elements, 578
Translation termination, 494, 670
Translational control of gene expression, 534, 535
Translational elongation factor 1α, 646
Translocase complex of the outer membrane (TOM), *446*, 472
Translocase of inner membrane (TIM), 472
Transmissible spongiform encephalopathies (TSE), 491
Transmission electron microscopy (TEM), 518
Transposon-based mutagenesis, 6
Transposon-tagging, 6
Trehalose pathway, 322
Treponema pallidum, 149
Trimethyl guanosine cap, 416
Triple mutant genetic interactions, 384
TRIPLES (Transposon-insertion phenotypes, localization, and expression in *Saccharomyces*), 607
TriZol reagent, *192*
tRNA, 415
  modification, 668
trp1 strains, cold-sensitivity, 87
Trypsin, 9, 465
Tryptic peptides, 9
T-Sapphire, fluorescent protein, 271
tTA (tetracycline-dependent transactivator protein), 347, 649
Tubulin, 234, 344, 644, 674
Tunicamycin, 293, *391*
Two-dimensional gel analysis, 670
Two-dimensional gel electrophoresis, 8
Two-hybrid system, see *Yeast two-hybrid*
Tyrosine kinase, 113
Tyrosine phosphorylation, 113

[URE3], 592
U1A (Mud1p)-GFP fusion, 673
U1A pre-mRNA, 436
U1A, 673
U1A protein, human (hU1A), 436
U2/U5.U6 spliceosomal complex, 621
U3 RNA, 436
U3 snoRNP, 436
Ubiquitin, 123, 348
  conjugation, 86
Ubiquitin-conjugating enzyme, 651
Ubiquitin-specific proteases, 124
Ubiquitylation, 223, 320, 651
UBR1, 86, 651
Ume6p, 553
Unfolded protein response (UPR), 103, 125, 587
Univector cloning, 147
Untranslated transcripts, 8
Upstream activator sequence (UAS), *554*, 651
Upstream regions of genes, 4
Upstream regulatory sequences, 332
Upstream repressing sequences (URS), 651
URA3 gene (*Candida albicans*), 647
URA3MX cassette, *61*, 648
Uracil biosynthesis, 344
Ure2p, 492, 592
Ureidosuccinic acid (USA), 498

Vacuolar endopeptidase Pep4p, 585
Vacuolar hydrolases, 585
Vacuole, 11
Vanadate, 293
Vegetative incompatibility, 492
Vinca alkaloids, 399
Volcano plot, 207
VP16 activator domain, 649

Watson strand, of yeast chromosome, 633
Weighted sum method, 615
Western blot staining, Ponceau S, 248, *250*
Western blot, antibody stripping, *251*
Western blotting, 241, 244, 462, 505
  alkaline phosphatase detection, *251*
  blocking agents, 248, 253
  chemiluminescent detection, *251*
  nitrocellulose, use in, 245
  PVDF membranes in, 245
  semi-dry transfer, 248
  troubleshooting, 253
Wheat germ agglutin-FITC, 337
Whole chromosome analysis, 221
  choice of microscopy equipment, 224
  GFP tagging strategies, 224
Whole-genome arrays, 144
Whole-genome duplication, 4, 555, 556
Wigglesworthia, 12

X-Gal (5-bromo-4-chloro-3-indolyl-β-D-galactopyranoside), 109, 173

Yarrowia lipolitica, 555
Ydj1p, 513
Yeast as immobilized biocatalysts, 295
Yeast cell cycle, 223
Yeast cell lysates
  denaturing conditions, *246*
  native conditions, *247*, *249*
  preparation, 245
Yeast cell lysis using glass beads, *246*
Yeast centromeric plasmid (YCp), 89
Yeast centromeric vectors, 648
Yeast colony PCR, *41*, *75*
Yeast deletion consortium, 388
Yeast genetic analysis, 28, 37, 630
Yeast genetic stock center (YGSC), 641
Yeast genome 2.0 array, 191
Yeast GFP fusion localization database, 606
Yeast integrative plasmid (YIp), 40, 96
Yeast interactome, 672
Yeast mating, *31*
Yeast mating type, 27
Yeast plasmids, 30
  YCp50, 339
  YCplac22, 30
  YEp13, 339
Yeast prions, 491
Yeast protein localization database (YPL.db), 607
Yeast selectable marker cassettes, *59-62*, *647*
Yeast strains
  BY4741, 30
  BY4742, 30
  BY4743, 635
  CEN.PK2, 635
  DC14, 27
  DC17, 27
  FY1679, 635
  RC629, 27
  S288c, 26
  Sigma1278B, 26
  SK1, 26, 675
  W303, 26, 635, 675
  Y55, 675
Yeast strains for SGA, *374*
Yeast transformation, 40, 45–54, *145*, 641
  by prion proteins, 519
  carrier DNA, 46
  efficiency, 48, 637
  frozen competent cells, 50, *53*
  high efficiency, 47, *49*
  microtitre plate transformation, 50, *145*
  mix, *48*
  rapid, *48*
Yeast one-hybrid, 103, 117, 118
Yeast three-hybrid (Y3H), 114–6
Yeast two-hybrid (Y2H), 9, 103-105, 139, *141*, 459, 589
  acetylation-dependent interactions, 114
  activating baits, 118
  commercial sources of reagents, *108*
  controls, 108
  dual-bait, 119
  false negatives, 107, 112, 157
  false positives, 106, 110, 142
  genome-wide screening, 144

LexA DNA binding domain, 105
merits of different platforms, 106
phosphorylation-dependent interactions, 113, 114
reverse, 119, 121
RNA polymerase III dependent, 119
small molecule–protein interactions, 117
split, 119
verification of bait and pray expression, 108
verification of positive clones, 110, 157–*9*
VP16 activation domain, 105

Yeast two-hybrid screening, commercial services, *111*
Yeast, use in bioremediation, 295
Yellow Fluorescent Protein (YFP), *177*, 221, 269, 270

Z-buffer (β-galactosidase assay), *174*
Zellweger syndrome, 589
Zeocin, 581
Zero background reporter plasmids, 650
Zonal centrifugation, 339
Zygotes, isolation by micromanipulation, 30, 31
Zymolyase, 34, 247, *259*, *261*, 262

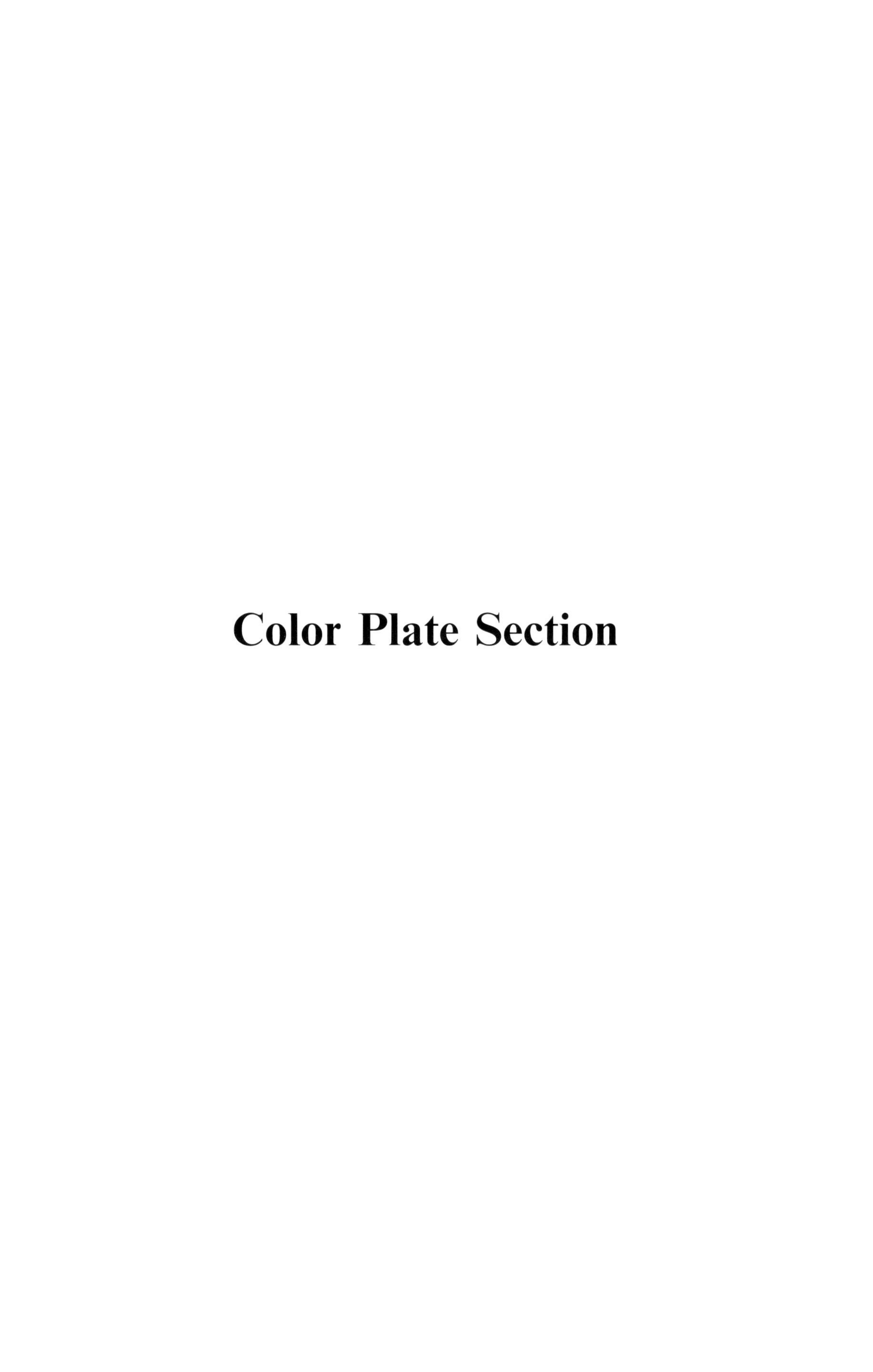

Color Plate Section

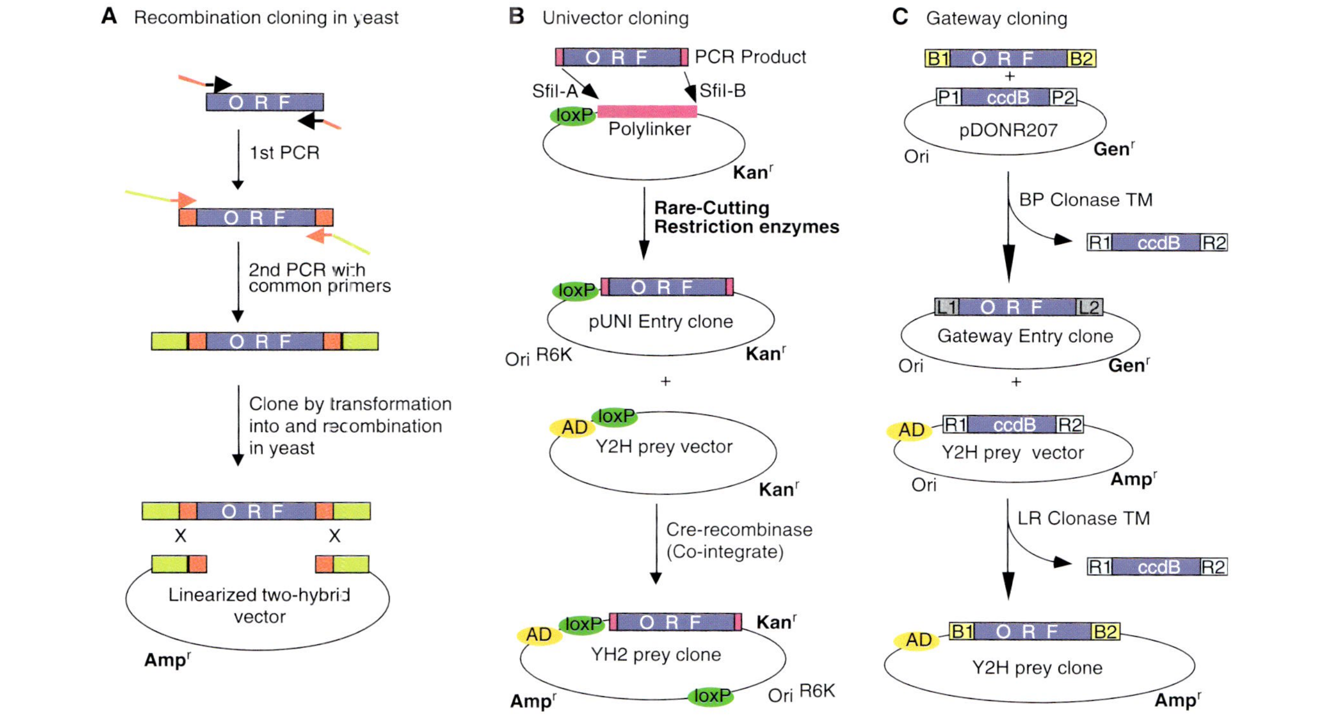

Plate 1. Cloning strategies for creating baits and preys. (A) *Homologous recombination*. ORFs are amplified (first PCR) with specific primers that generate a product with common 5′ and 3′ 20-nucleotide tails. A second PCR generate a product with common 5′ and 3′ 70-nucleotide tails. The common 70-nucleotide ends allow cloning into linearized two-hybrid expression vectors by co-transformation into yeast. The endogenous yeast recombination machinery performs the recombination reaction and results in a circular plasmid. (B) *Univector plasmid fusion system*. ORFs are amplified with specific primers that generate a product with common 5′ and 3′ Rare-Cutting restriction sites. The PCR product is cloned into a pUNI entry vector by DNA ligation. Cre–*loxP*-mediated site-specific recombination fuses the pUNI entry clone and yeast two-hybrid expression plasmids (bait/prey) at the *loxP* site. As a result, the gene of interest is placed under the control of the yeast two-hybrid expression vector promoter. (C) *Gateway cloning*. The ORFs are amplified with specific primers that generate a product with common 5′ and 3′ recombination sites (*attB*). The entry clones are made by recombining the ORFs of interest with the flanking *att*B sites into the *att*P sites of a suitable Gateway entry vector (such as pDONR201 or pDONR207) mediated by the Gateway BP Clonase (Invitrogen) Enzyme Mix. Subsequently, the fragment in the entry clone can be transferred to any yeast two-hybrid expression vector that contains the *att*R sites by mixing both plasmids and by using the Gateway LR Clonase Enzyme Mix. (See also page 148 of this volume).

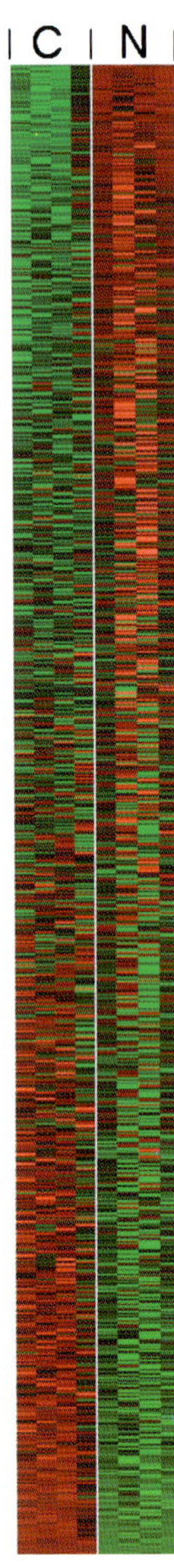

Plate 2. Ranking the sample data by statistical significance (*p*-value). For visualization purposes the data has been normalized by *z*-transformation (set the mean expression to zero and standard deviation to 1 for each gene in the dataset) and separated in terms of N- vs. C-limitation into up- (above) and down-regulated genes (below). These are colour coded in Red and Green respectively. (See also page 208 of this volume).

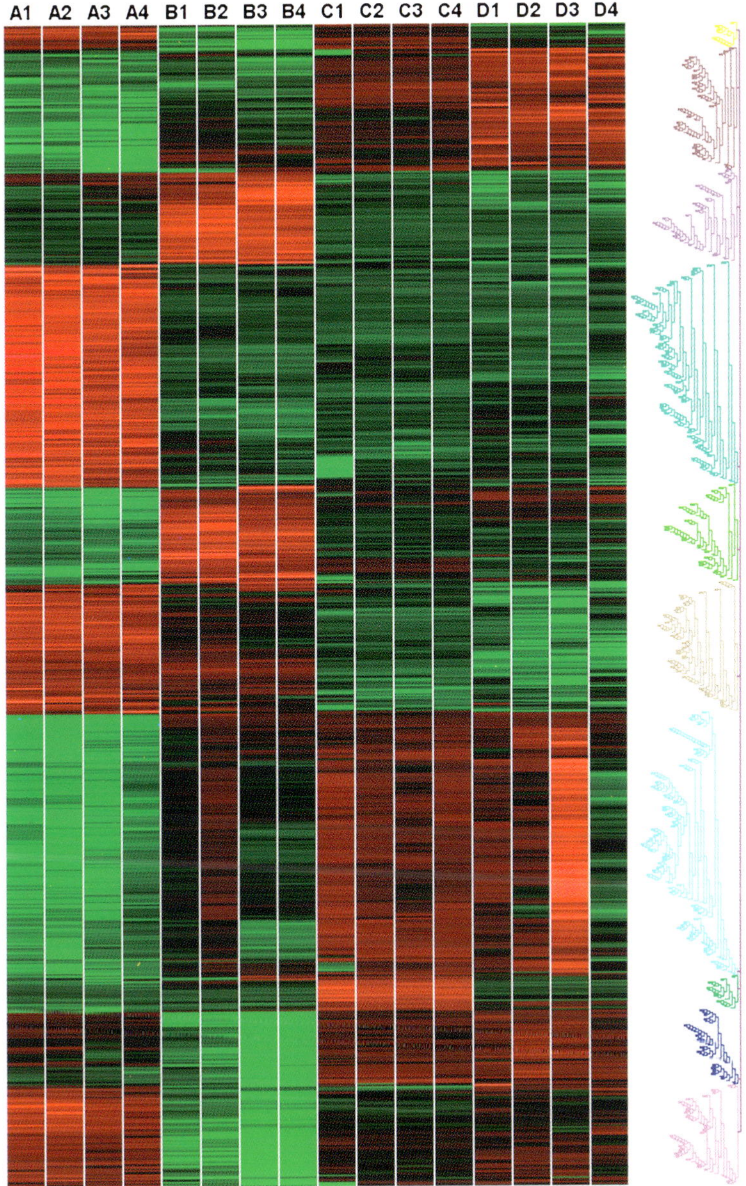

Plate 3. Clustering analysis of data from an experiment similar to that shown in Figure 14A. Following a one-way ANOVA analysis, the data were filtered down to 800 genes by statistical significance (*p*-value), fold change (fold change of 2 or more in any 2-way comparison of conditions) and expression level (mean expression level above 100). The $\log_2$ data values were normalized by *z*-transformation (set the mean expression to zero and standard deviation to 1 for each gene in the dataset) prior to clustering by the k-means method (performed in maxdView: http://bioinf.man.ac.uk/microarray/maxd/ using the XCluster plug-in http://genetics.stanford.edu/~sherlock/cluster.html). (See also page 213 of this volume).

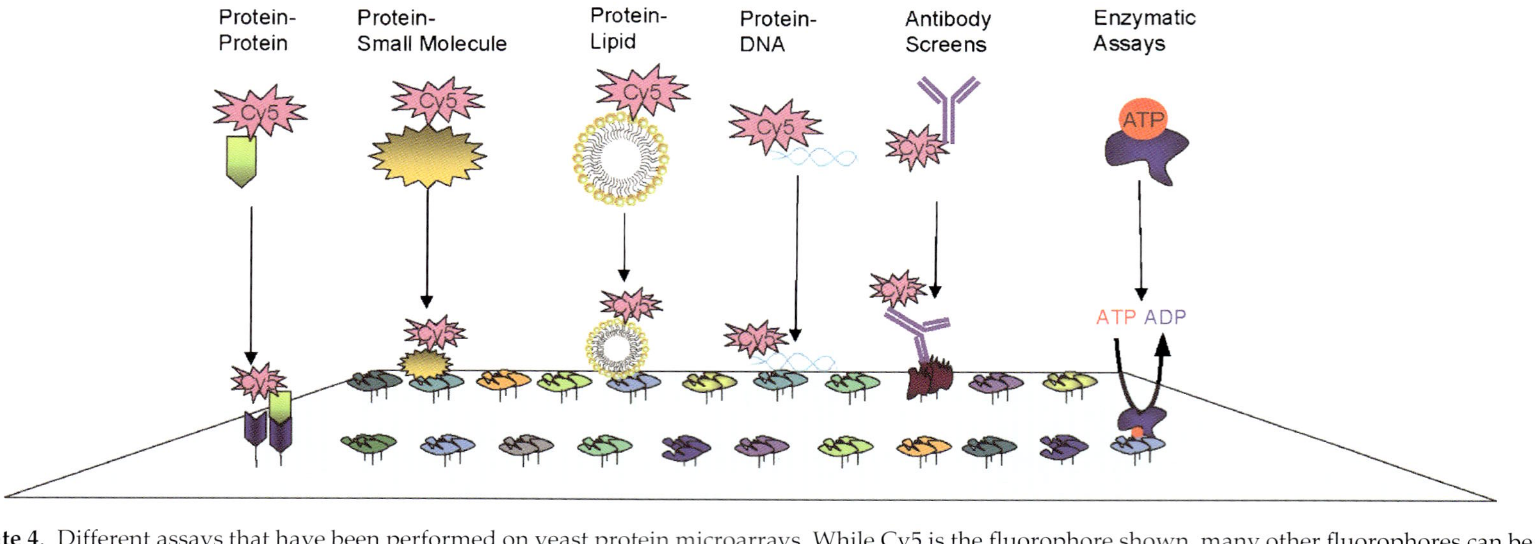

Plate 4. Different assays that have been performed on yeast protein microarrays. While Cy5 is the fluorophore shown, many other fluorophores can be used for detection. (See also page 308 of this volume).

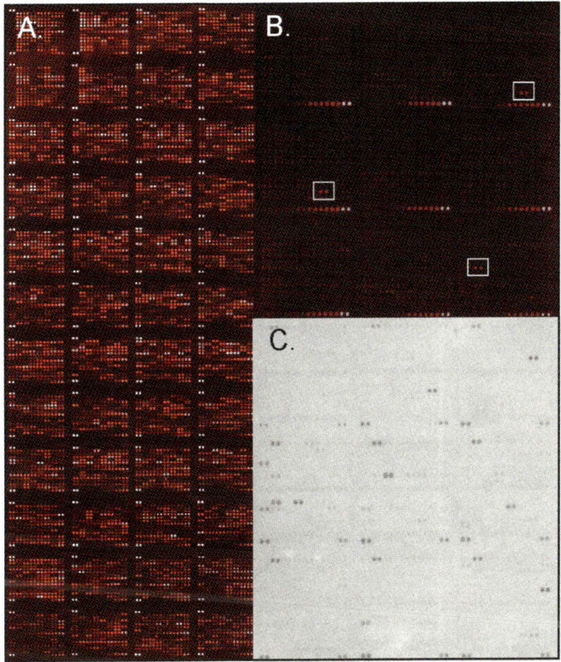

Plate 5. Yeast protein microarrays in use. (A) A yeast proteome microarray probed with anti-GST antibodies followed by Cy5-labeled anti-rabbit antibodies. All 48 blocks are shown. (B) Biotinylated Cpr7p probed against the yeast proteome array, followed by Cy5-labeled streptavidin. A dilution of Cy5 labeled controls can also be seen in each of the 9 blocks shown. (C) A kinase assay with [γ-^{33}P]ATP and active kinase on a yeast proteome microarray. Dark spots represent radiolabeled phosphorylated substrates. Kinases that autophosphorylate are printed in each of the nine blocks shown and serve as reference points on the slide. (See also page 317 of this volume).

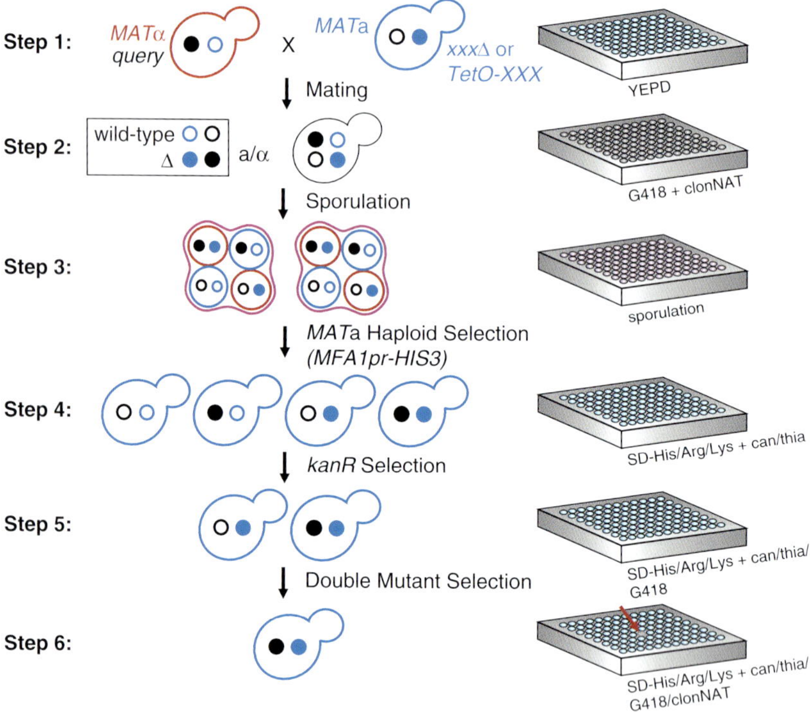

Plate 6. SGA methodology. Step 1, a *MATα* strain carrying a query mutation (*bni1Δ*) linked to a dominant selectable marker, such as the nourseothricin-resistance marker *natMX* that confers resistance to the antibiotics nourseothricin (clonNAT), and the *MFA1pr-HIS3*, *can1Δ* and *lyp1Δ* reporters is crossed to an ordered array of *MAT***a** viable deletion mutants (*xxxΔ*), each carrying a gene deletion mutation linked to a kanamycin-resistance marker *kanMX* that confers resistance to the antibiotic geneticin (G418). To score genetic interactions amongst essential genes, the query strain can be crossed to an array of conditional yeast mutants. For example, an array in which each mutant carries a different essential gene placed under the control of the conditional Tetracycline-regulated promoter (*TetO-XXX*); however, when screening the conditional array the selection conditions at each step differ from those outlined here as described previously (Mnaimneh *et al.*, 2004; Davierwala *et al.*, 2005). Step 2, growth of resultant zygotes is selected for on medium containing nourseothricin and geneticin. Step 3, the heterozygous diploids are transferred to medium with reduced levels of carbon and nitrogen to induce sporulation and the formation of haploid meiotic spore progeny. Step 4, spores are transferred to synthetic medium lacking histidine, which allows for selective germination of *MAT***a** meiotic progeny because only these cells express the *MFA1pr-HIS3* reporter, and containing canavanine and thialysine, which allows for selective germination of meiotic progeny that carries the *can1Δ* and *lyp1Δ* markers. Step 5, the *MAT***a** meiotic progeny are then transferred to medium that contains G418, which selects for growth of meiotic progeny that carries the gene deletion mutation (*xxxΔ::kanR*). Finally, the *MAT***a** meiotic progeny are transferred to medium that contains both clonNAT and G418, which then selects for growth of double mutant (*bni1Δ::natR xxxΔ::kanR*). (See also page 371 of this volume).

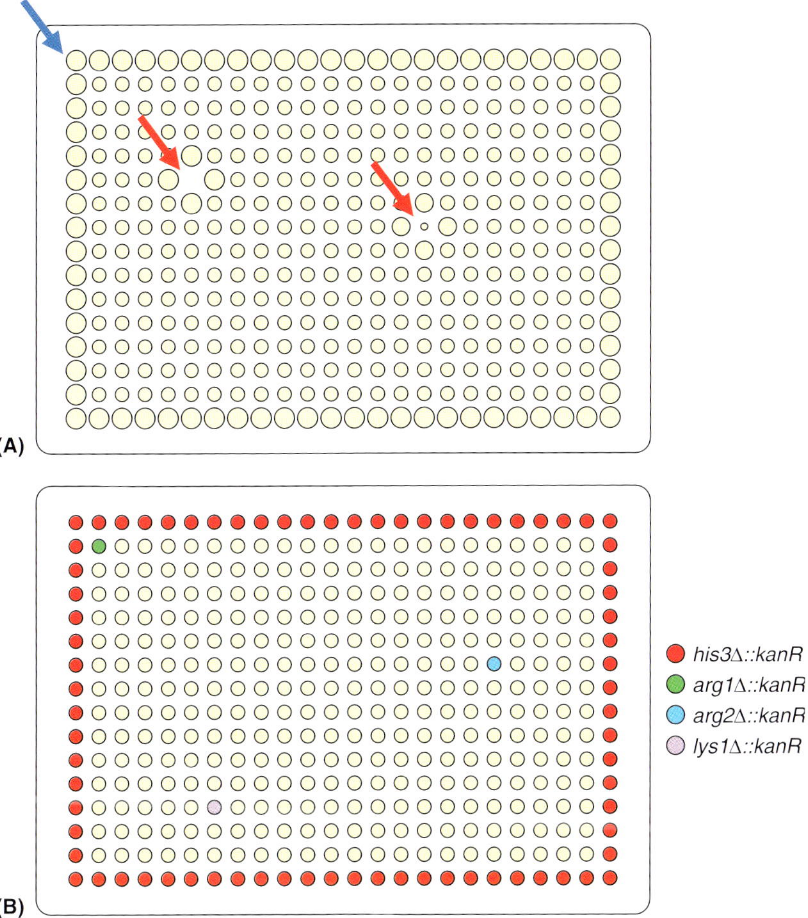

Plate 7. Array Design. Each spot represents a yeast colony growing in a 384-density array. (A) Yeast colonies surrounding an empty spot or a slow-growing strain (red arrows), and those positioned along the edges of the array (blue arrow), have access to more nutrients in the medium and therefore, tend to be larger than the ones positioned in a dense area away from the edges. (B) An ideal array layout for SGA analysis should facilitate accurate output readout and include the following: (i) removal of slow-growing strains from the regular array to a special array containing only mutants with a slow growth rate; (ii) a border around the edges of the plate, i.e. the outermost layer of colonies on four edges of the plate, using a neutral strain carrying all the markers required in the experimental procedure, for example the *MAT***a** *his3Δ::kanR* deletion strain (red colonies); (iii) filled in gaps or empty spots to make the array more robust for examining subtle differences in fitness amongst the deletion mutants; (iv) a number of auxotrophic mutants to be used as a unique plate identification system, for example, the *MAT***a** *ura4Δ::kanR* deletion strain (green colony), the *MAT***a** *trp1Δ::kanR* deletion strain (blue colony), and the *MAT***a** *lys1Δ::kanR* deletion strain (purple colony), are unable to grow on medium lacking uracil, tryptophan, and lysine, respectively. (See also page 380 of this volume).

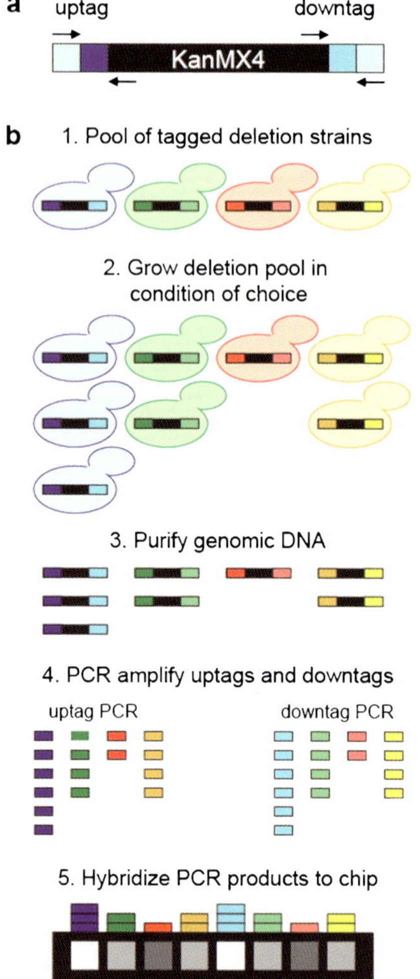

Plate 8. Description of the competitive growth assay. (A) The deletion cassette module is shown. The uptag and dntags (shown in darker blue hues) are flanked by universal primers common to every strain in the collection (shown by arrows). The KanMX4 marker is required for selection of transformants with G418, and 45bp of flanking genomic homology (light blue) to ensure proper insertion by homologous recombination. (B) Cartoon outlining the experimental protocol. (1) Strains are first pooled at approximately equal abundance. (2) The pool is grown competitively under a selection of choice; strains deleted for genes required for growth under a particular condition will grow more slowly and become under-represented in the pool. (3) The pool is sampled over time and genomic DNA isolated. (4) Genomic DNA is PCR amplified using the universal primers in two PCR reactions, one for the uptag and one for the dntag. (5) PCR products are then hybridized to the array carrying the tag complements. Array intensity is related to the amount of each strain present. Those strains that are under-represented are likely deleted for genes important for survival under the condition selected. Modified from Scherens and Goffeau (2004). (See also page 390 of this volume).

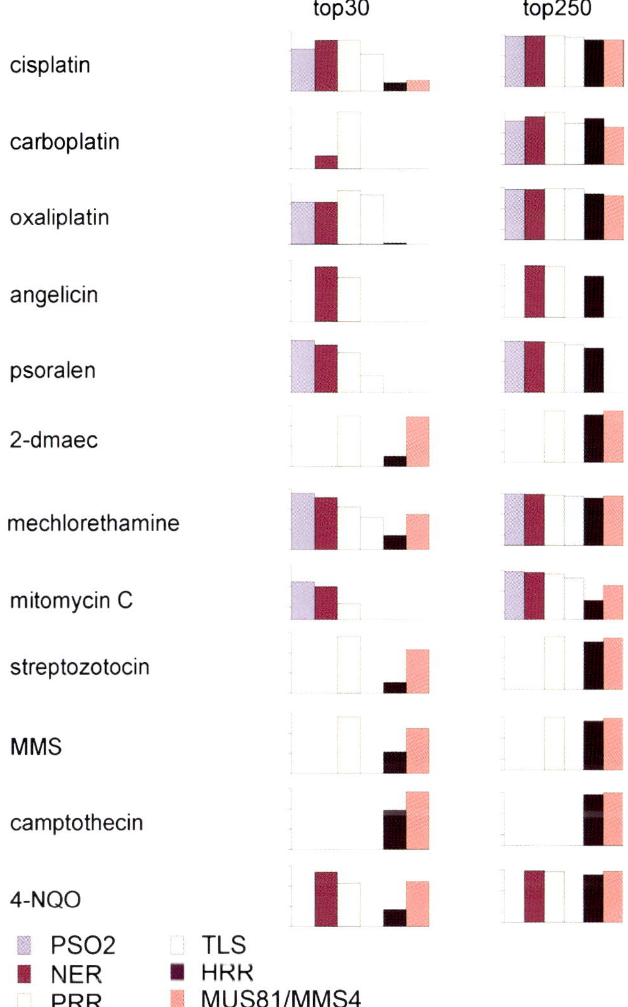

Plate 9. Relative importance of the DNA damage response modules for resistance to DNA-damaging agents. Each bar graph represents strains that were found to be among the top 30 (or 250) most sensitive strains in that compound and are known to be members of a well-characterized DNA-damage-response pathway. The bars represent the median rank for genes in each of the gene groups listed in the visual key. The gene groups were defined in the following way: x-linking genes (*PSO2*); NER (*RAD2*, *RAD4*, *RAD10*, *RAD1*, and *RAD14*); PRR (*RAD6*, *RAD18*, and *RAD5*); error-prone TLS (*REV1*, *REV3*, and *REV7*); HRR (*RAD57*, *RAD55*, *RAD51*, *RAD52*, *RAD54*, and *RAD59*); stalled replication-fork repair (*MUS81* and *MMS4*). Those compounds that form interstrand cross-links are labeled with an asterisk. Taken from Lee *et al.* (2005). (See also page 403 of this volume).

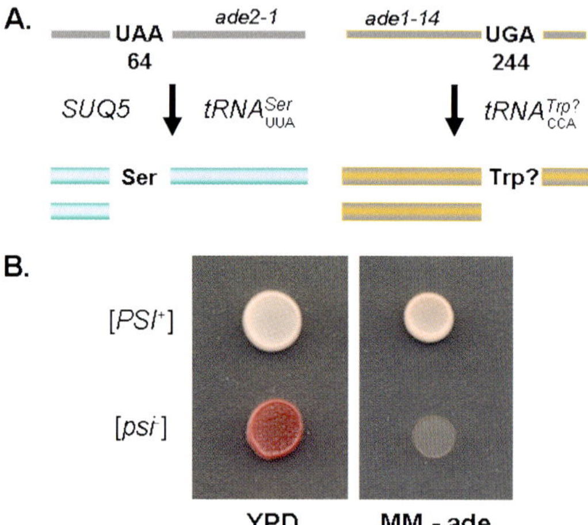

Plate 10. A simple colony colour assay for the presence of the [*PSI*$^+$] prion in *Saccharomyces cerevisiae*. Either of two different suppressible alleles can be used, the *ade2-1* allele and the *ade1-14* allele. In both the cases, when the mutation is expressed, i.e. in a [*psi*$^-$] strain, the cells form red colonies that signal an adenine prototrophic phenotype. In [*PSI*$^+$] cells, suppression of either allele leads to white colonies that are prototrophic and can grow without the provision of exogenous adenine. Note that strains carrying the *ade2-1* allele must also carry the weak ochre suppressor $tRNA^{Ser}$ encoded by the *SUQ5* (*SUP16*) gene (Cox, 1965), whereas the *ade1-14* allele can be suppressed directly by [*PSI*$^+$] in the absence of a suppressor tRNA. The identity of the amino acid inserted when the UGA codon in the *ade1-14* allele is suppressed is unknown but is likely to be tryptophan (encoded by the UGG codon). (See also page 495 of this volume).

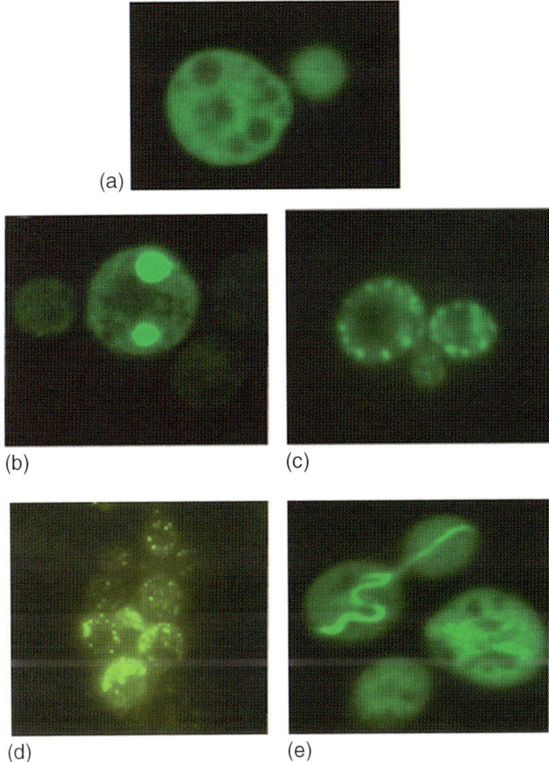

Plate 11. The use of green fluorescent protein (GFP) fusions to visualise Sup35p-based aggregates in [PSI^+] strains. All cells shown contain a plasmid expressing an identical Sup35pPrD+M-GFP fusion protein whose synthesis was induced using the copper-inducible *CUP1* promoter (Patino *et al.*, 1996); (a) a [psi^-] strain; (b)–(d) different [PSI^+] variants; (e) a [PIN^+][psi^-] strain undergoing *de novo* conversion to [PSI^+] as a consequence of the over expression of the Sup35pPrD+M-GFP fusion protein (see Zhou *et al.*, 2001). (See also page 509 of this volume).

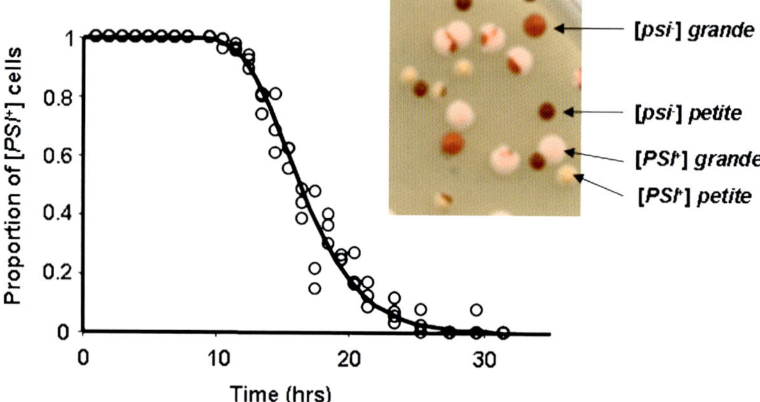

Plate 12. The elimination of the [PSI^+] prion from cells grown in the presence of 3 mM guanidine hydrochloride (GdnHCl) over a 30-h period. The experiment was carried out as described in the Protocol 1 and the percent [PSI^+] with time was plotted as shown. The data are combined from three independent experiments. The inset shows the types of colonies that one observes in such an experiment noting in particular that (a) GdnHCl induces a high frequency of mitochondrial *petites*, and (b) the *petite* mutation does cause a change in colouration when compared to *grande* strains with functional mitochondria. Colonies sectored red and white are counted as [PSI^+]. Further details can be found in the text. (See also page 511 of this volume).

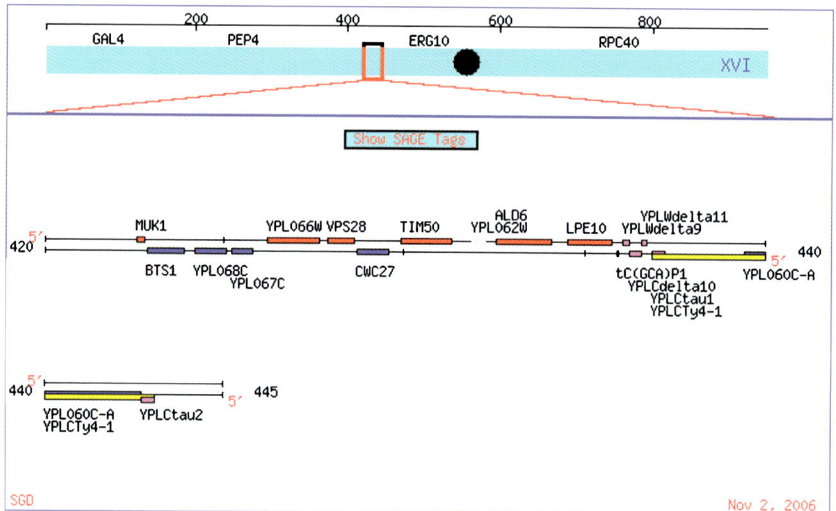

Plate 13. An SGD "Chromosomal Features Map" of the *Saccharomyces* genome near the *ALD6* gene. This expanded view is obtained by selecting the ORF map on the gene (*ALD6* in this case) summary page. The map provides a quick overview of genes or other chromosomal features adjacent to the gene of interest. (See also page 568 of this volume).